The American Society of
Mechanical Engineers

A17.1
HANDBOOK

SAFETY CODE FOR ELEVATORS AND ESCALATORS

2000 EDITION

Edward A. Donoghue

CONTENTS

Part 1
General

Part 2
Electric Elevators

iii

Part 3
Hydraulic Elevators

Part 4
Elevators With Other Types of Driving Machines

Part 5
Special Application Elevators

Part 6
Escalators and Moving Walks

Part 7
Dumbwaiters and Material Lifts

Part 8
General Requirements

Part 9
Reference Codes, Standards, and Specifications

Appendices

FOREWORD

The A17.1 Safety Code for Elevators and Escalators is written by a committee comprised of technically qualified persons who demonstrate a concern and competence in the subject within the A17 Committee's scope and a willingness to participate in the work of the committee. The A17 Standards Committee is restricted to a maximum of 35 members of which no more than one-third can be from the same interest category. In addition, there are over 250 members serving on the Regulatory Advisory Council, National Interest Review Group, Technical Committees, Administrative Committees, and Ad Hoc Committees.

ASME recognizes that the Code must be written in a form that is suitable for enforcement by state, municipal, and other jurisdictional authorities; and as such, the text is concise, without examples or explanations. It is also recognized that this Code cannot cover every situation nor can it cover new technology before it is developed and field experience is gained. For these reasons, ASME determined that a handbook would be useful to augment the Code by providing a commentary on the Code requirements.

This Handbook contains the rationale for Code requirements; explanations, examples, and illustrations of the implementations of the requirements; plus excerpts from other nationally recognized standards which are referenced by the Code. This information is intended to provide the users of the ASME A17.1 Code with a better understanding of, and appreciation for, the requirements contained in the Code. The net result should be increased safety for the manufacturers, installers, maintainers, consultants and users of the equipment covered by the ASME A17.1 Code.

The commentary in this Handbook was compiled from A17 Committee minutes, correspondence, and interpretations, as well as conversations with past and present committee members.

The original intent for many of the requirements in this Code is obscure in the Committee's records. The author, therefore, has tried to convey, through text and illustrations, the end result of the Code requirements as applied to equipment installed today. It should not be construed that the examples and illustrations are the only means for complying with the Code requirements, nor that all the illustrations necessarily represent requirements that are in the Code. Some illustrations represent general industry or specific company practices. With information of this type it is hoped that the reader will have a better comprehension of, and appreciation for, the requirements in the ASME A17.1 Code.

The 2000 Edition of the A17.1 Code was the result of a joint effort between the ASME A17 Elevator and Escalator Committee and the CSA B44 Technical Committee to harmonize the requirements in the ASME A17.1 and CSA B44 Codes. Because of that effort, the ASME A17.1 Code was reorganized and renumbered. Work on the 2000 Edition of the Handbook had to be delayed until that effort was completed. Your author was also required to review all Handbook commentary to follow the new format. The results should provide the reader with a greater insight into the Code requirements. While some of the Canadian requirements in ASME A17.1 are addressed, the commentary regarding the Canadian requirements is not as complete as the commentary on the requirements applicable in the United States.

The commentary contained in this Handbook is the opinion of the author. The commentary does not necessarily reflect the official position of ASME or the ASME A17 Standards Committee for Elevators and Escalators. When an official interpretation is required, the user should write to the Secretary of the ASME A17 Standards Committee in accordance with the instructions given in the Preface to the Code. Comments and suggestions for this and future editions of the Handbook should be addressed to: Secretary, A17 Standards Committee, The American Society of Mechanical Engineers, Three Park Avenue, New York, New York 10016-2300 or infocentralasme.org.

ASME Elevator and Escalator Courses. ASME Professional Development is a leader in delivering top quality elevator and escalator education. Courses range from an introduction to elevators and escalators, inspection techniques, equipment modernization code requirements, maintenance evaluation, and an in depth review of ASME A17.1 using this Handbook as the course text. The courses meet the needs for those with little or no elevator and escalator experience as well as those who have an extensive background in the industry. To obtain the latest catalog of course material, contact: ASME Professional Development, Three Park Avenue, New York, NY 10016-2300, Phone 800-THE-ASME, or 212-591-7604.

METRIC

This edition of the Handbook emphasizes metric units. Where ASME A17.1 Code recognizes either metric or imperial units, both are reflected in the Handbook commentary. The equivalent imperial unit is shown in parentheses.

ACKNOWLEDGEMENTS

The author gratefully acknowledges the time, effort, and dedication of the many people and organizations that assisted and contributed in the preparation of this 2000 Edition of the Handbook A17.1 Safety Code for Elevators and Escalators

The following deserve special recognition, for their participation on the A17.1 Handbook Review Group for this Edition:

Robert S. Caporale, Editor, Elevator World, Inc., Mobile, AL

Richard Gregory, Consultant, Vertex Corp., Chicago, IL

Andy Juhasz, Manager Codes & Standards, KONE, Inc., Moline, IL

Norman B. Martin, Chief Elevator Inspector, State of Ohio, North Canton, OH

Zack McCain, Jr., P.E., Consultant, McCain Engineering Associates, Inc., Norman, OK

David McColl, P. Eng., Manager Codes and Standards, Otis Canada, Inc., Mississauga, ON

I also want to acknowledge and thank the following for their assistance and contributions in this Edition of the Handbook:

Ralph Droste, Consultant, Avon, CT — Sections 2.25 and 2.26

Nick Marchitto, Otis Elevator Co., Farmington, CT — Section 2.26 (ANSI/NFPA 70 - 2002 Revisions)

John O'Donoghue, Massachusetts Firefighting Academy, Stow, MA — Sections 2.27.3 through 2.27.8

Albert Marchant, Alimak Elevator Company, Bridgeport, CT — Section 4.1

Thomas Barkand, US DOL MSHA, Pittsburgh, PA and Albert Saxer, Consultant, Avon, CT — Section 5.9

Hank E. Peelle III, The Peelle Company, Hauppauge, NY — Part 7

Albert Saxer, Consultant, Avon, CT — Section 8.7

My special thanks and appreciation are again extended to my partner, friend, and wife, Janet for her patience and understanding during the countless hours that it took to prepare this Handbook. She deserves as much credit as I do, for her assistance in the preparation of the manuscript was invaluable. My appreciation for her contributions cannot be expressed in words.

CROSS-REFERENCE TABLE
ASME A17.1-2000 vs. ASME A17.1-1996 Including A17.1a-1997 Through A17.1d-2000

A17.1-2000	A17.1-1996 With Addenda	A17.1-2000	A17.1-1996 With Addenda	A17.1-2000	A17.1-1996 With Addenda
1.1	1	2.5.1.3	108.1c	2.11.6	110.6
1.1.1	1.1	2.5.1.4	108.1d	2.11.7	110.7
1.1.2	1.2	2.5.1.5	108.1e	2.11.7.1	110.7a
1.1.3	1.3	2.5.1.6	–	2.11.7.2	110.7b
1.1.4	1.4	2.5.1.7	108.1f	2.11.8	110.8
1.2	2	2.6	109, 109.1	2.11.9	110.9
1.3	3	2.7	101	2.11.9.1	110.9a
2.1	100	2.7.1	101.1	2.11.9.2	110.9b
2.1.1	100.1	2.7.1.1	101.1a	2.11.10	110.10
2.1.1.1	100.1a	2.7.1.2	101.1b	2.11.10.1	110.10a
2.1.1.2	100.1b	2.7.2.1	101.2	2.11.10.2	110.10b
2.1.1.3	100.1c	2.7.2.2	–	2.11.10.3	110.10d
2.1.1.4	100.1d	2.7.3	101.3	2.11.11	110.11
2.1.1.5	100.1e	2.7.3.1	101.3a	2.11.11.1	110.11a
2.1.2	100.2	2.7.3.2	101.3b	2.11.11.2	110.11b
2.1.2.1	100.2a	2.7.3.3	101.3c	2.11.11.3	110.11c
2.1.2.2	100.2b	2.7.3.4	101.3d	2.11.11.4	110.11d
2.1.2.3	100.2c	2.7.3.5	101.3e	2.11.11.5	110.11e
2.1.3	100.3	2.7.4	101.4	2.11.11.6	110.11f
2.1.3.1	100.3a	2.7.5	101.5	2.11.11.7	110.11g
2.1.3.2	100.3b	2.7.5.1	101.5a	2.11.11.8	110.11h
2.1.3.3	100.3c	2.7.5.2	101.5b	2.11.11.9	110.11i
2.1.3.4	100.3d	2.7.6	101.6	2.11.11.10	110.11j
2.1.3.5	100.3e	2.7.7	101.7	2.11.12	110.12
2.1.3.6	100.3f	2.7.8	101.8	2.11.12.1	110.12a
2.1.4	100.4	2.8	102	2.11.12.2	110.12b
2.1.5	100.5	2.8.1	102.1	2.11.12.3	110.12c
2.1.6	100.6	2.8.2	102.2	2.11.12.4	110.12d
2.2	106	2.8.3	102.3	2.11.12.5	110.12e
2.2.1	106.1a	2.8.4	102.4	2.11.12.6	110.12f
2.2.2	106.1b	2.9	105	2.11.12.7	110.12g
2.2.3	106.1c	2.9.1	105.1	2.11.12.8	110.12h
2.2.4	106.1d	2.9.2	105.2	2.11.13	110.13
2.2.5	106.1e	2.9.2.1	105.2a	2.11.13.1	110.13a
2.2.6	106.1f	2.9.2.2	105.2b	2.11.13.2	110.13b
2.2.7	106.1g	2.9.3	105.3	2.11.13.3	110.13c
2.3	103	2.9.3.1	105.3a	2.11.13.4	110.13d
2.3.1	103.1	2.9.3.2	105.3b	2.11.13.5	110.13e
2.3.2	103.2	2.9.3.3	105.3c	2.11.14	110.14
2.3.3	103.3	2.9.3.4	105.3d	2.11.15	110.15
2.3.4	103.4	2.9.4	105.4	2.11.15.1	110.15a
2.4	107	2.9.5	105.5	2.11.15.2	110.15b
2.4.1	107.1a	2.9.6	–	2.11.15.3	110.15c
2.4.2	107.1b	2.10	104	2.11.16	110.16
2.4.3	107.1c	2.10.1	104.1	2.11.17	110.17
2.4.4	107.1d	2.10.2	–	2.11.18	110.18
2.4.5	107.1d	2.11	110	2.11.19	110.19
2.4.6	107.1e	2.11.1.1	110.1	2.12	111
2.4.7	107.1f	2.11.1.2	110.1	2.12.1	111.1, 111.1a, 111.1b
2.4.8	107.1g	2.11.1.3	–	2.12.2	111.2
2.4.9	107.1h	2.11.1.4	–	2.12.2.1	111.2a
2.4.10	107.1i	2.11.2	110.2	2.12.2.2	111.2b
2.4.11	107.1j	2.11.2.1	110.2a	2.12.2.3	111.2c
2.4.12	107.1k	2.11.2.2	110.2b	2.12.2.4	111.2d
2.5	108	2.11.2.3	110.2c	2.12.2.5	111.2e
2.5.1	108.1	2.11.3	110.3	2.12.2.6	111.2f
2.5.1.1	108.1a	2.11.4	110.4	2.12.3	111.3
2.5.1.2	108.1b	2.11.5	110.5	2.12.3.1	111.3a

A17.1-2000	A17.1-1996 With Addenda	A17.1-2000	A17.1-1996 With Addenda	A17.1-2000	A17.1-1996 With Addenda
2.12.3.2	111.3b	2.14.4.4	204.4d	2.16.4	207.4
2.12.3.3	111.3c	2.14.4.5	204.4e	2.16.5	207.5
2.12.3.4	111.3d	2.14.4.6	204.4f	2.16.5.1	207.5a
2.12.3.5	111.3e	2.14.4.7	204.4h	2.16.5.2	207.5b
2.12.4	111.4	2.14.4.8	204.4i	2.16.6	207.6
2.12.4.1	111.4a	2.14.4.9	204.4j	2.16.7	207.7
2.12.4.2	111.4b	2.14.4.10	204.4k	2.16.8	207.8
2.12.4.3	111.4c	2.14.4.11	204.4m	2.16.9	207.9
2.12.5	111.5	2.14.5	204.5	2.17	205
2.12.6	111.6	2.14.5.1	204.5a	2.17.1	205.1
2.12.6.1	111.6a	2.14.5.2	204.5b	2.17.2	205.2
2.12.6.2	111.6b	2.14.5.3	204.5c	2.17.3	205.3
2.12.7	111.7	2.14.5.4	204.5d	2.17.4	205.4
2.12.7.1	111.7a	2.14.5.5	204.5e	2.17.5	205.5
2.12.7.2	111.7b	2.14.5.6	204.5g	2.17.6	–
2.12.7.3	111.7c	2.14.5.7	204.5h	2.17.7	205.7
2.13	112	2.14.5.8	204.5i	2.17.8	205.8
2.13.1	112.1	2.14.6	204.6	2.17.8.1	205.8a
2.13.2	112.2	2.14.6.1	204.6a	2.17.8.2	205.8b
2.13.2.1	112.2a	2.14.6.2	204.6b	2.17.9	205.9
2.13.2.2	112.2b	2.14.6.3	204.6c	2.17.9.1	205.9a
2.13.3	112.3	2.14.7	204.7	2.17.9.2	205.9b
2.13.3.1	112.3a	2.14.7.1	204.7a	2.17.9.3	205.9c
2.13.3.2	112.3b	2.14.7.2	204.7b	2.17.9.4	205.9d
2.13.3.3	112.3c	2.14.7.3	204.7c	2.17.10	205.10
2.13.3.4	112.3d	2.14.7.4	204.7d	2.17.11	205.11
2.13.4.1	112.4	2.15	203	2.17.12	205.12
2.13.4.2.1	112.4	2.15.1	203.1	2.17.13	205.13
2.13.4.2.2	112.4	2.15.2	203.2	2.17.14	205.14
2.13.4.2.3	112.4	2.15.3	203.3	2.17.15	205.15
2.13.4.2.4	–	2.15.4	203.4	2.17.16	205.16
2.13.5	112.5	2.15.5	203.5	2.17.17	205.17
2.13.6	112.6	2.15.6	203.6	2.18	206
2.13.6.1	112.6a	2.15.6.1	203.6a	2.18.1	206.1
2.13.6.2	112.6b	2.15.6.2	203.6b	2.18.2	206.2
2.14	204	2.15.6.3	203.6c	2.18.2.1	206.2a
2.14.1	204.1	2.15.6.4	203.6d	2.18.2.2	206.2b
2.14.1.1	204.1a	2.15.7	203.7	2.18.3	206.3
2.14.1.2	204.1b	2.15.7.1	203.7a	2.18.4	206.4
2.14.1.3	204.1c	2.15.7.2	203.7b	2.18.4.1	206.4a
2.14.1.4	204.1d	2.15.7.3	203.7c	2.18.4.2	206.4b
2.14.1.5	204.1e	2.15.8	203.8	2.18.4.3	206.4c
2.14.1.6	204.1f	2.15.9	203.9	2.18.5	206.5
2.14.1.7	204.1g	2.15.10	203.10	2.18.5.1	206.5a
2.14.1.8	204.1h	2.15.11	203.11	2.18.5.2	206.5c
2.14.1.9	204.1i	2.15.12	203.12	2.18.5.3	206.5e
2.14.1.10	204.1j	2.15.13	203.13	2.18.6	206.6
2.14.2	204.2	2.15.14	203.14	2.18.7	206.7
2.14.2.1	204.2a	2.15.15	203.15	2.18.8	206.8
2.14.2.2	204.2b	2.15.16	203.16	2.18.9	206.9
2.14.2.3	204.2c	2.15.17	–	2.19	–
2.14.2.4	204.2d	2.16	207	2.20	212
2.14.2.5	204.2e	2.16.1.1	207.1	2.20.1	212.1
2.14.2.6	–	2.16.1.2	207.1a	2.20.2	212.2
2.14.3	204.3	2.16.1.3	207.1b	2.20.2.1	212.2a
2.14.3.1	204.3a	2.16.2	207.2	2.20.2.2	212.2b
2.14.3.2	204.3b	2.16.2.1	207.2a	2.20.3	212.3
2.14.3.3	204.3c	2.16.2.2	207.2b	2.20.4	212.4
2.14.4	204.4	2.16.3	207.3	2.20.5	212.5
2.14.4.1	204.4a	2.16.3.1	207.3a	2.20.6	212.6
2.14.4.2	204.4b	2.16.3.2	207.3b	2.20.7	212.7
2.14.4.3	204.4c	2.16.3.3	207.3c	2.20.8	212.8

A17.1-2000	A17.1-1996 With Addenda	A17.1-2000	A17.1-1996 With Addenda	A17.1-2000	A17.1-1996 With Addenda
2.20.9	212.9	2.23.7.1	200.7a	2.26.4	210.4
2.20.9.1	212.9a	2.23.7.2	200.7b	2.26.5	210.15
2.20.9.2	212.9b	2.23.8	200.8	2.26.6	210.6
2.20.9.3	212.9c	2.23.9	200.9	2.26.7	210.7
2.20.9.4	212.9d	2.23.9.1	200.9a	2.26.8	210.8
2.20.9.5	212.9e	2.23.9.2	200.9b	2.26.9	210.9
2.20.9.6	212.9f	2.23.9.3	–	2.26.10	210.10
2.20.9.7	212.9g	2.23.10	200.10	2.26.11	–
2.20.9.8	212.9h	2.23.10.1	200.10a	2.26.12	210.13
2.20.10	212.10	2.23.10.2	200.10b	2.27	211
2.21	202	2.23.10.3	200.10c	2.27.1	211.1
2.21.1	202.1	2.24	208	2.27.2	211.2
2.21.1.1	202.1a	2.24.1	208.1	2.27.3	211.3
2.21.1.2	202.1b	2.24.2	208.2	2.27.3.1	211.3a
2.21.1.3	202.1c	2.24.2.1	208.2a	2.27.3.2	211.3b
2.21.1.4	202.1d	2.24.2.2	208.2b	2.27.3.3	211.3c
2.21.2	202.2	2.24.2.3	208.2c	2.27.3.4	211.3d
2.21.2.1	202.2a	2.24.2.4	208.2d	2.27.3.5	211.3e
2.21.2.2	202.2b	2.24.3	208.3	2.27.4	211.4
2.21.2.3	202.2c	2.24.3.1	–	2.27.4.1	211.4a
2.21.2.4	202.2d	2.24.3.2	–	2.27.4.2	211.4b
2.21.2.5	202.2e	2.24.4	208.4	2.27.5	211.5
2.21.2.6	202.2f	2.24.5	208.5	2.27.6	211.6
2.21.3	202.3	2.24.6	208.6	2.27.7	211.7
2.21.4	202.4	2.24.7	208.7	2.27.8	211.8
2.22	201	2.24.8	–	2.28	214
2.22.1	201.1	2.24.8.1	–	2.28.1	214.1
2.22.1.1	201.1a	2.24.8.2	–	2.29.1	211.9
2.22.1.2	201.1b	2.24.8.3	208.8	2.29.2	100.7
2.22.2	201.2	2.24.8.4	–	3.1	300.1
2.22.3	201.3	2.24.8.5	–	3.1.1	300.1a
2.22.3.1	201.3a	2.24.8.6	–	3.1.2	300.1b
2.22.3.2	201.3b	2.24.9	208.9	3.2	300.7
2.22.3.3	201.3c	2.24.9.1	208.9a	3.2.1	300.7a
2.22.4	201.4	2.24.9.2	208.9b	3.3	300.4
2.22.4.1	201.4a	2.24.9.3	208.9c	3.4	300.8
2.22.4.2	201.4b	2.24.10	208.11	3.4.1	300.8a
2.22.4.3	201.4c	2.25	209	3.4.2	300.8b
2.22.4.4	201.4d	2.25.1	209.1	3.4.3	300.8c
2.22.4.5	201.4e	2.25.2	209.2	3.4.4	300.8d
2.22.4.6	201.4f	2.25.2.1	209.2a	3.4.5	300.8e
2.22.4.7	201.4g	2.25.2.2	209.2b	3.4.6	300.8f
2.22.4.8	201.4h	2.25.2.3	209.2c	3.4.7	300.8g
2.22.4.9	201.4i	2.25.3	209.3	3.4.8	300.8h
2.22.4.10	201.4j	2.25.3.1	209.3a	3.5	300.9
2.22.4.11	201.4k	2.25.3.2	209.3b	3.6	300.10
2.23	200	2.25.3.3	209.3c	3.7	300.2
2.23.1	200.1	2.25.3.4	209.3d	3.7.1	300.2a
2.23.2	200.2	2.25.3.5	209.3e	3.8	300.3
2.23.2.1	200.2a	2.25.4	209.4	3.9	300.6
2.23.2.2	200.2b	2.25.4.1	209.4a	3.10	300.5
2.23.3	200.3	2.25.4.2	209.4b	3.11	300.11
2.23.4	200.4	2.26	210	3.11.1	300.11a
2.23.4.1	200.4a	2.26.1	210.1	3.12	300.12
2.23.4.2	200.4b	2.26.1.1	210.1a	3.13	300.13
2.23.4.3	200.4c	2.26.1.2	210.1b	3.14	301.7
2.23.5	200.5	2.26.1.3	210.1c	3.15	301.6
2.23.5.1	200.5a	2.26.1.4	210.1d	3.15.1	301.6a
2.23.5.2	200.5b	2.26.1.5	210.14	3.15.2	301.6b
2.23.5.3	–	2.26.1.6	210.1e	3.15.3	301.6c
2.23.6	200.6	2.26.2	210.2	3.16	301.10
2.23.7	200.7	2.26.3	–	3.16.1	301.10

A17.1-2000	A17.1-1996 With Addenda	A17.1-2000	A17.1-1996 With Addenda	A17.1-2000	A17.1-1996 With Addenda
3.17	–	3.23	301.1	4.1.16	1607.1
3.17.1	301.8	3.23.1	301.1a	4.1.17	1608.1
3.17.2	301.9	3.23.2	301.1b	4.1.18	1604.2
3.18	302	3.24	304	4.2.1	1800.1
3.18.1	302.1	3.24.1	304.1	4.2.2	1800.3
3.18.1.1	302.1a	3.24.1.1	304.1a	4.2.2.1	1800.3a
3.18.1.2	302.1b	3.24.2	304.2	4.2.2.2	1800.3b
3.18.2	302.2	3.24.2.1	304.2a	4.2.2.3	1800.3c
3.18.2.1	302.2a	3.24.2.2	304.2b	4.2.2.4	1800.3d
3.18.2.2	302.2b	3.24.3	304.3	4.2.3	1800.4
3.18.2.3	302.2c	3.24.3.1	304.3a	4.2.4	1800.5
3.18.2.4	302.2d	3.24.3.2	304.3b	4.2.5	1800.2
3.18.2.5	302.2e	3.24.3.3	304.3c	4.2.6	1801.1
3.18.2.6	302.2f	3.24.4	304.4	4.2.7	1802.1
3.18.2.7	302.2g	3.24.5	307.2	4.2.8	1803.4
3.18.3	302.3	3.25	305	4.2.9	1803.3
3.18.3.1	302.3a	3.25.1	305.1	4.2.10	1803.7
3.18.3.2	302.3b	3.25.1.1	305.1a	4.2.11	1803.5
3.18.3.3	302.3c	3.25.1.2	305.1b	4.2.12	1803.6
3.18.3.4	302.3d	3.25.1.3	305.1c	4.2.13	1803.2
3.18.3.5	302.3e	3.25.1.4	305.1d	4.2.14	1803.1
3.18.3.6	302.3f	3.25.2	305.2	4.2.15	1804.1
3.18.3.7	302.3g	3.25.2.1	305.2a	4.2.16	1805
3.18.3.8	302.3h	3.25.2.2	305.2b	4.2.16.1	1805.1
3.18.3.9	302.3i	3.25.3	305.3	4.2.16.2	1805.2
3.18.4	302.4	3.26	306	4.2.16.3	1805.3
3.18.4.1	302.4a	3.26.1	306.1	4.2.17	1806
3.18.4.2	302.4b	3.26.2	306.2	4.2.17.1	1806.1
3.18.5	302.5	3.26.3	306.3	4.2.18	1807
3.19	303	3.26.3.1	306.3a	4.2.19	1808.1
3.19.1	303.1	3.26.3.2	306.3b	4.2.20	1804.2
3.19.1.1	303.1a	3.26.4	306.4	4.3.1	600
3.19.1.2	303.1b	3.26.5	306.5	4.3.2	600.3
3.19.1.3	303.1c	3.26.6	306.9	4.3.3	600.4
3.19.1.4	303.1d	3.26.7	306.13	4.3.3.1	600.4a
3.19.2	303.2	3.26.8	306.14	4.3.3.2	600.4b
3.19.2.1	303.2a	3.26.9	306.15	4.3.4	600.2
3.19.2.2	303.2b	3.26.10	–	4.3.5	608
3.19.2.3	303.2c	3.27	306.11	4.3.5.1	608.1
3.19.2.4	303.2d	3.28	308	4.3.5.2	608.2
3.19.3	303.3	3.28.1	308.1	4.3.6	600.5
3.19.3.1	303.3a	3.29	–	4.3.6.1	600.5a
3.19.3.2	303.3b	4.1.1	1600.1	4.3.6.2	600.5b
3.19.3.3	303.3c	4.1.2	1600.2	4.3.6.3	600.5c
3.19.4	303.4	4.1.3	1601	4.3.7	600.6
3.19.4.1	303.4a	4.1.4	1602	4.3.8	600.7
3.19.4.2	303.4b	4.1.5	1600.3	4.3.9	601.1
3.19.4.3	303.4c	4.1.6	1603.5	4.3.10	601.3
3.19.4.4	303.4d	4.1.7	1603.4	4.3.11	601.2
3.19.4.5	303.4e	4.1.8	1603.7	4.3.12	601.4
3.19.4.6	303.4f	4.1.9	1603.6	4.3.13	601.5
3.19.4.7	–	4.1.9.1	1603.6a	4.3.14	603.1
3.19.5	303.5	4.1.10	1603.3	4.3.14.1	603.1a
3.19.5.1	303.5a	4.1.11	1603.2	4.3.14.2	603.1b
3.19.5.2	303.5b	4.1.12	1603.1	4.3.15	602.1
3.19.6	303.6	4.1.13	1604.1	4.3.16	610
3.19.7	303.7	4.1.14	1605	4.3.16.1	610.1
3.20	307.1	4.1.14.1	1605.1	4.3.16.2	610.2
3.21	301.5	4.1.14.2	1605.2	4.3.16.3	610.3
3.22	–	4.1.15	1606	4.3.16.4	610.4
3.22.1	301.3	4.1.15.1	1606.1	4.3.16.5	610.5
3.22.2	301.4	4.1.15.2	1606.2	4.3.17	606

A17.1-2000	A17.1-1996 With Addenda	A17.1-2000	A17.1-1996 With Addenda	A17.1-2000	A17.1-1996 With Addenda
4.3.17.1	606.1	5.1.14.2	1710.2	5.2.1.11	2500.11
4.3.18	605	5.1.14.3	1710.3	5.2.1.12	2500.12
4.3.18.1	605.1	5.1.14.3.1	1710.3a	5.2.1.13	2500.13
4.3.18.2	605.2	5.1.14.3.2	1710.3b	5.2.1.14	2501.5
4.3.18.3	605.3	5.1.15	1710.4	5.2.1.15	2501.4
4.3.19	607	5.1.15.1	1710.4a	5.2.1.15.1	2501.4a
4.3.19.1	607.1	5.1.15.2	1710.4b	5.2.1.15.2	2501.4b
4.3.19.2	607.2	5.1.15.3	–	5.2.1.16	2501.8
4.3.20	609	5.1.16	1715	5.2.1.16.1	2501.8a
4.3.20.1	609.1	5.1.16.1	1715.1	5.2.1.16.2	2501.8b
4.3.21	611	5.1.17	1707	5.2.1.16.3	2501.8c
4.3.22	604	5.1.17.1	1707.1	5.2.1.16.4	2501.8d
5.1	Part XVII	5.1.17.2	1707.2	5.2.1.16.5	2501.8e
5.1.1	1700.1	5.1.17.3	1707.3	5.2.1.17	2501.6
5.1.1.1	1700.1	5.1.17.4	1707.4	5.2.1.17.1	2501.6a
5.1.1.2	1705	5.1.17.4.1	1707.4(a)	5.2.1.18	2501.7
5.1.2	–	5.1.17.4.2	1707.4(b)	5.2.1.19	–
5.1.2.1	1700.1a	5.1.17.4.3	1707.4(c)	5.2.1.20	2501.13
5.1.2.2	1700.1b	5.1.17.4.4	1707.4(d)	5.2.1.20.1	–
5.1.2.2.1	1700.1b(1)	5.1.17.4.5	1707.4(e)	5.2.1.20.2	–
5.1.2.2.2	1700.1b(2)	5.1.18	1706	5.2.1.21	2501.3
5.1.2.2.3	1700.1b(3)	5.1.18.1	1706.1	5.2.1.21.1	2501.3a
5.1.2.2.4	1700.1b(4)	5.1.18.2	1706.2	5.2.1.22	2501.2
5.1.3	1701	5.1.18.3	1706.3	5.2.1.22.1	2501.2a
5.1.3.1	1701.1	5.1.18.4	–	5.2.1.23	2501.1
5.1.3.2	–	5.1.19	1712	5.2.1.23.1	2501.1a
5.1.4	1701.2	5.1.19.1	1712.1	5.2.1.23.2	2501.1b
5.1.5	1702	5.1.19.2	1712.2	5.2.1.24	2501.9
5.1.5.1	1702.1	5.1.20	1713	5.2.1.24.1	–
5.1.5.2	1702.2	5.1.20.1	–	5.2.1.24.2	–
5.1.6	–	5.1.20.2	–	5.2.1.24.3	–
5.1.7	1717	5.1.20.3	–	5.2.1.25	2501.10
5.1.7.1	1717.1	5.1.20.4	–	5.2.1.26	2501.11
5.1.7.2	1717.2	5.1.20.5	1713.2	5.2.1.27	2501.12
5.1.8	1703	5.1.20.6	1713.3	5.2.1.28	2501.15
5.1.8.1	1703.1	5.1.20.6.1	1713.3(a)	5.2.1.29	–
5.1.8.2	1703.3	5.1.20.6.2	1713.3(b)	5.2.1.30	2501.14
5.1.9	–	5.1.21	1714	5.2.2	2502
5.1.10	1704	5.1.21.1	1714.1	5.2.2.1	2502.1
5.1.10.1	1704.1	5.1.22	1716	5.2.2.2	2502.1a
5.1.10.2	–	5.1.22.1	1716.1	5.2.2.3	2502.2
5.1.10.3	–	5.1.22.2	1716.2	5.2.2.4	2502.2d
5.1.11	–	5.1.22.3	1716.3	5.2.2.5	2502.2c
5.1.11.1	1708.3	5.1.22.4	1716.4	5.2.2.6	2502.2e
5.1.11.1.1	1708.3a	5.1.22.5	1716.5	5.2.2.7	2502.3
5.1.11.1.2	1708.3b	5.1.22.6	1716.6	5.2.2.8	2502.4
5.1.11.1.3	1708.3c	5.2	Part XXV	5.2.2.9	2502.2b
5.1.11.2	1708.4	5.2.1	2500	5.2.2.10	2502.2a
5.1.11.3	–	5.2.1.1	2500.1	5.2.2.11	2502.5
5.1.12	1708	5.2.1.2	2500.7	5.2.2.12	2502.6
5.1.12.1	1708.1	5.2.1.3	2500.4	5.2.2.13	2502.7
5.1.12.2	1708.2	5.2.1.4	2500.8	5.2.2.14	2502.8
5.1.12.2.1	1708.2(a)	5.2.1.4.1	2500.8a	5.2.2.15	–
5.1.12.2.2	1708.2(b)	5.2.1.4.2	2500.8b	5.3	Part V
5.1.12.2.3	1708.2(c)	5.2.1.4.3	2500.8c	5.3.1	–
5.1.12.2.4	1708.2(d)	5.2.1.4.4	2500.8d	5.3.1.1	500.1
5.1.12.2.5	1708.2(e)	5.2.1.5	2500.9	5.3.1.1.1	500.1a
5.1.13	1711	5.2.1.6	2500.10	5.3.1.1.2	500.1b
5.1.13.1	1711.1	5.2.1.7	2500.2	5.3.1.1.3	500.1c
5.1.13.2	1711.2	5.2.1.8	2500.3	5.3.1.1.4	500.1d
5.1.14	1710	5.2.1.9	2500.6	5.3.1.2	500.2
5.1.14.1	1710.1	5.2.1.10	2500.5	5.3.1.2.1	500.2a

A17.1-2000	A17.1-1996 With Addenda	A17.1-2000	A17.1-1996 With Addenda	A17.1-2000	A17.1-1996 With Addenda
5.3.1.2.2	500.2b	5.3.1.17	507	5.4.7.3	513.14c
5.3.1.3	500.3	5.3.1.17.1	507.1	5.4.7.4	513.14d
5.3.1.4	500.6	5.3.1.17.2	507.2	5.4.7.5	513.14e
5.3.1.4.1	500.6a	5.3.1.18	508	5.4.7.6	513.14f
5.3.1.4.2	500.6b	5.3.1.18.1	508.1	5.4.7.6.1	–
5.3.1.5	500.5	5.3.1.18.2	508.2	5.4.7.6.2	–
5.3.1.6	500.7	5.3.1.18.3	508.3	5.4.7.7	513.14g
5.3.1.6.1	500.7a	5.3.1.18.4	508.4	5.4.7.8	513.14h
5.3.1.6.2	500.7b	5.3.1.18.5	508.5	5.4.7.9	513.14i
5.3.1.7	500.4	5.3.1.18.6	508.6	5.4.8	513.18
5.3.1.7.1	500.4a	5.3.1.18.7	508.7	5.4.8.1	513.18a
5.3.1.7.2	500.4b	5.3.1.18.8	508.8	5.4.8.2	513.18b
5.3.1.7.3	500.4c	5.3.1.19	509	5.4.8.3	513.18c
5.3.1.7.4	500.4d	5.3.1.20	511	5.4.8.4	513.18d
5.3.1.7.5	500.4e	5.3.1.20.1	511.1	5.4.8.5	513.18e
5.3.1.7.6	500.4f	5.3.1.20.2	511.2	5.4.8.6	513.18f
5.3.1.7.7	500.4g	5.3.2	514	5.4.8.7	513.18g
5.3.1.8	–	5.3.2.1	514.1	5.4.8.8	513.18h
5.3.1.8.1	501.2	5.3.2.2	514.2	5.4.8.9	513.18i
5.3.1.8.1(a)	501.2(a)	5.3.2.2.1	514.2a	5.4.8.10	513.19
5.3.1.8.1(b)	501.2(b)	5.3.2.2.2	514.2b	5.4.9	513.13
5.3.1.8.1(c)	501.2(c)	5.3.2.3	514.3	5.4.9.1	513.13a
5.3.1.8.1(d)	501.2(d)	5.3.2.4	514.4	5.4.9.2	513.13b
5.3.1.8.1(e)	501.3	5.4	513	5.4.9.3	513.13c
5.3.1.8.2	501.2	5.4.1	513.1	5.4.10	513.3e
5.3.1.8.3	501.5	5.4.2	513.2	5.4.10.1	513.3a
5.3.1.9	501	5.4.2.1	513.2a	5.4.10.2	513.3b
5.3.1.9.1	501.1	5.4.2.2	513.2b	5.4.10.3	513.3c
5.3.1.9.2	–	5.4.2.3	513.2c	5.4.10.4	513.3d
5.3.1.10	510	5.4.2.4	513.2d	5.4.11	513.12
5.3.1.10.1	510.1	5.4.2.5	513.2e	5.4.11.1	513.12a
5.3.1.10.2	510.2	5.4.3	513.4	5.4.11.2	513.12b
5.3.1.10.3	510.3	5.4.3.1	513.4a	5.4.11.3	513.12c
5.3.1.11	503	5.4.3.2	513.4b	5.4.12	513.21
5.3.1.11.1	503.1	5.4.3.3	513.4c	5.1.13	513.15
5.3.1.11.2	503.2	5.4.3.4	513.4d	5.4.13.1	513.15a
5.3.1.11.3	503.3	5.4.4	–	5.4.13.2	513.15b
5.3.1.11.4	503.4	5.4.4.1	513.8	5.4.13.3	513.15c
5.3.1.11.5	503.5	5.4.4.1.1	513.8a	5.4.13.4	513.15d
5.3.1.11.6	503.6	5.4.4.1.2	513.8b	5.4.13.5	513.15e
5.3.1.11.7	503.7	5.4.4.1.3	513.8c	5.4.13.6	513.15f
5.3.1.12	512	5.4.4.2	513.9	5.4.13.7	513.15g
5.3.1.12.1	512.1	5.4.4.2.1	513.9a	5.4.13.8	513.15h
5.3.1.12.2	512.2	5.4.4.2.2	513.9b	5.4.13.9	513.15i
5.3.1.12.3	512.3	5.4.4.2.3	513.9c	5.4.13.10	513.22
5.3.1.12.4	512.4	5.4.4.2.4	513.9d	5.4.14	513.16
5.3.1.12.5	512.5	5.4.4.3	513.10	5.4.14.1	513.16a
5.3.1.12.6	512.6	5.4.4.3.1	513.10a	5.4.14.2	513.16b
5.3.1.12.7	512.7	5.4.4.3.2	513.10b	5.4.14.3	513.16c
5.3.1.13	502	5.4.4.3.3	513.40c	5.4.15	513.17
5.3.1.13.1	502.1	5.4.5	513.7	5.4.15.1	513.17a
5.3.1.13.2	502.2	5.4.5.1	513.7a	5.4.15.2	513.17b
5.3.1.14	505.1	5.4.5.2	513.7b	5.4.15.3	513.17c
5.3.1.14.1	505.1a	5.4.5.3	513.7c	5.4.15.3.1	–
5.3.1.14.2	505.1b	5.4.5.4	513.7d	5.4.15.3.2	–
5.3.1.14.3	505.1c	5.4.6	513.5	5.4.15.4	513.17d
5.3.1.15	504.1	5.4.6.1	513.5a	5.4.15.5	513.20
5.3.1.16	506	5.4.6.2	513.5b	5.4.15.5.1	513.20a
5.3.1.16.1	506.1	5.4.6.3	513.6	5.4.15.5.2	513.20b
5.3.1.16.2	506.2	5.4.7	513.14	5.4.16	513.11
5.3.1.16.3	506.3	5.4.7.1	513.14a	5.4.16.1	513.11a
5.3.1.16.4	506.4	5.4.7.2	513.14b	5.4.16.2	513.11b

A17.1-2000	A17.1-1996 With Addenda	A17.1-2000	A17.1-1996 With Addenda	A17.1-2000	A17.1-1996 With Addenda
5.5	Part IV	5.5.2.17	402.7	5.6.2.14	2302.2a
5.5.1	–	5.5.2.17.1	402.7	5.6.2.15	2302.5
5.5.1.1	400.1	5.5.2.17.2	402.7a	5.6.2.16	2302.6
5.5.1.2	400.7	5.5.2.17.3	402.7a	5.6.2.17	2302.7
5.5.1.3	400.4	5.5.2.18	402.8	5.6.2.17.1	2302.7a
5.5.1.4	400.8	5.6	Part XXIII	5.7	Part XV
5.5.1.5	400.9	5.6.1	–	5.7.1	1500
5.5.1.6	400.10	5.6.1.1	2300.1	5.7.1.1	1500.1
5.5.1.7	400.2	5.6.1.2	2300.7	5.7.1.2	1500.3e
5.5.1.8	400.3	5.6.1.3	2300.4	5.7.2	1500.3a
5.5.1.8.1	–	5.6.1.4	2300.8	5.7.3	1500.5
5.5.1.8.2	–	5.6.1.5	2300.9	5.7.3.1	1500.5a
5.5.1.8.3	–	5.6.1.6	2300.10	5.7.3.2	1500.5b
5.5.1.9	400.6	5.6.1.7	2300.2	5.7.4	–
5.5.1.10	400.5	5.6.1.8	2300.3	5.7.4.1	1500.3c
5.5.1.11	400.11	5.6.1.9	2300	5.7.4.2	1500.3d
5.5.1.11.1	400.11a	5.6.1.10	2300.5	5.7.5	1500.2
5.5.1.11.2	400.11b	5.6.1.11	2300.11	5.7.6	1500.3b
5.5.1.11.3	400.11c	5.6.1.11.1	2300.11a	5.7.7	1500.4
5.5.1.11.4	400.11d	5.6.1.11.2	2300.11b	5.7.7.1	1500.4a
5.5.1.12	400.12	5.6.1.11.3	2300.11c	5.7.7.2	1500.4b
5.5.1.13	400.13	5.6.1.11.4	2300.11d	5.7.7.3	1500.4c
5.5.1.14	401.5	5.6.1.11.5	2300.11e	5.7.7.4	1500.4d
5.5.1.14.1	401.5a	5.6.1.12	2300.12	5.7.8	1500.6
5.5.1.14.2	401.5b	5.6.1.13	2300.13	5.7.8.1	1500.6a
5.5.1.14.3	401.5c	5.6.1.14	2301.5	5.7.8.2	1500.6b
5.5.1.15	401.4	5.6.1.15	2301.4	5.7.8.3	1500.6c
5.5.1.15.1	401.4a	5.6.1.15.1	2301.4a	5.7.8.4	1500.6d
5.5.1.15.2	401.4b	5.6.1.15.2	2301.4b	5.7.8.5	1500.6e
5.5.1.16	401.8	5.6.1.16	2301.8	5.7.8.6	1500.6f
5.5.1.17	401.6	5.6.1.17	2301.6	5.7.9	1501.2
5.5.1.18	401.7	5.6.1.18	2301.7	5.7.10	1502.5
5.5.1.19	401.13	5.6.1.19	2301.13	5.7.10.1	1502.5a
5.5.1.20	401.3	5.6.1.20	2301.3	5.7.10.2	1502.5b
5.5.1.21	401.2	5.6.1.21	2301.2	5.7.10.3	1502.5c
5.5.1.22	401.1	5.6.1.22	2301.1	5.7.10.4	1502.5d
5.5.1.23	401.9	5.6.1.23	2301.9	5.7.10.5	1502.6
5.5.1.24	401.10	5.6.1.24	2301.10	5.7.11	1502.4
5.5.1.25	401.11	5.6.1.25	2301.11	5.7.11.1	1502.4a
5.5.1.25.1	401.11a	5.6.1.25.1	2301.11a	5.7.11.2	1502.4b
5.5.1.25.2	401.11b	5.6.1.25.2	2301.11b	5.7.11.3	1502.4c
5.5.1.25.3	401.11c	5.6.1.25.3	2301.11c	5.7.11.4	1502.4d
5.5.1.26	401.12	5.6.1.25.4	2301.11d	5.7.12	–
5.5.1.27	401.15	5.6.1.25.5	2301.11e	5.7.12.1	1502.8
5.5.1.28	401.14	5.6.1.26	2301.12	5.7.12.2	1502.9
5.5.2	402	5.6.1.27	2301.14	5.7.13	1502.7
5.5.2.1	402.1	5.6.2	2302	5.7.13.1	1502.7a
5.5.2.2	402.1b	5.6.2.1	2302.1	5.7.13.2	1502.7b
5.5.2.3	402.1c	5.6.2.2	2302.1b	5.7.13.3	1502.7c
5.5.2.4	402.1a	5.6.2.3	2302.1c	5.7.13.4	1502.7d
5.5.2.5	402.1d	5.6.2.4	2302.1a	5.7.13.5	1502.7e
5.5.2.6	402.2e	5.6.2.5	2302.1d	5.7.14	1502.11
5.5.2.7	402.2d	5.6.2.6	2302.2e	5.7.14.1	1502.11a
5.5.2.8	402.2g	5.6.2.7	2302.2d	5.7.14.2	1502.11b
5.5.2.9	402.2f	5.6.2.8	2302.2g	5.7.14.3	1502.11c
5.5.2.10	402.3	5.6.2.9	2302.2f	5.7.14.4	1302.11d
5.5.2.11	402.4	5.6.2.9.1	–	5.7.14.5	1502.11e
5.5.2.12	402.2c	5.6.2.9.2	–	5.7.14.6	1502.11f
5.5.2.13	402.2b	5.6.2.10	2302.3	5.7.14.7	1502.11g
5.5.2.14	402.2a	5.6.2.11	2302.4	5.7.14.8	1502.11h
5.5.2.15	402.5	5.6.2.12	2302.2c	5.7.15	1502.3
5.5.2.16	402.6	5.6.2.13	2302.2b	5.7.15.1	1502.3a

A17.1-2000	A17.1-1996 With Addenda	A17.1-2000	A17.1-1996 With Addenda	A17.1-2000	A17.1-1996 With Addenda
5.7.15.2	1502.3b	5.10.1.7.2	1900.1d	5.10.2.5	1903.1d
5.7.15.3	1502.3c	5.10.1.8	1900.4	5.10.2.6	1903.2
5.7.16	1502.2	5.10.1.9	1900.6	5.10.2.7	1903.3
5.7.17	1502.1	5.10.1.9.1	1900.6a	5.10.2.8	1903.4
5.7.17.1	1502.1a	5.10.1.9.2	1900.1e	5.10.2.9	1903.8
5.7.17.2	1502.1b	5.10.1.9.3	1900.6b	5.10.2.10	1903.5
5.7.17.3	1502.1c	5.10.1.9.4	1900.6c	5.10.2.11	1903.6
5.7.18	1502.10	5.10.1.9.5	1900.6d	5.10.2.12	1903.7
5.7.18.1	1502.10a	5.10.1.9.6	1900.6e	6.1	Part VIII
5.7.18.2	1502.10b	5.10.1.9.7	1900.6f	6.1.1	800
5.7.18.3	1502.10c	5.10.1.9.8	1900.6g	6.1.1.1	800.1
5.7.18.4	1502.10d	5.10.1.10	1902.5	6.1.2	801
5.7.18.5	1502.10e	5.10.1.10.1	1902.5a	6.1.2.1	801.1
5.7.18.6	1502.10f	5.10.1.10.2	1902.5b	6.1.3	802
5.7.18.7	1502.10g	5.10.1.10.3	1902.5c	6.1.3.1	802.1
5.7.18.8	1502.10h	5.10.1.10.4	1902.5d	6.1.3.2	802.2
5.7.18.9	1502.10i	5.10.1.10.5	1902.5e	6.1.3.3	802.3
5.7.18	1501	5.10.1.10.6	1902.5f	6.1.3.3.1	802.3a
5.7.20	1501.3	5.10.1.10.7	1902.6	6.1.3.3.2	802.3b
5.7.20.1	1501.3a	5.10.1.10.8	1902.7	6.1.3.3.3	802.3c
5.7.20.2	1501.3b	5.10.1.11	1902.4	6.1.3.3.4	802.3d
5.7.21	1502.12	5.10.1.12	1902.10	6.1.3.3.5	802.3e
5.7.22	–	5.10.1.12.1	1902.10a	6.1.3.3.6	802.3f
5.7.23	–	5.10.1.12.2	1902.10b	6.1.3.3.7	802.3k
5.8	Part XXII	5.10.1.12.3	1902.10c	6.1.3.3.8	802.3j
5.8.1	–	5.10.1.13	1902.8a	6.1.3.3.9	802.3g
5.8.1.1	2200.2	5.10.1.14	1902.8b	6.1.3.3.10	802.3h
5.8.1.2	2200.3	5.10.1.15	–	6.1.3.3.11	802.3i
5.8.1.3	2200.4	5.10.1.16	1902.12	6.1.3.4	802.4
5.8.1.4	2200.5	5.10.1.16.1	1902.12a	6.1.3.4.1	802.4a
5.8.1.5	2201.2	5.10.1.16.2	1902.12b	6.1.3.4.2	802.4b
5.8.1.6	2201.3	5.10.1.16.3	1902.12c	6.1.3.4.3	802.4c
5.8.1.7	2201.4	5.10.1.16.4	1902.12d	6.1.3.4.4	–
5.8.1.8	2201.5	5.10.1.16.5	1902.12e	6.1.3.4.5	–
5.8.1.9	2201.6	5.10.1.16.6	1902.12f	6.1.3.4.6	–
5.8.1.10	2201.7	5.10.1.16.7	1902.12g	6.1.3.5	802.5
5.8.1.11	2201.8	5.10.1.16.8	1902.12h	6.1.3.5.1	802.5a
5.8.2	2202	5.10.1.17	1902.3	6.1.3.5.2	802.5b
5.8.2.1	2202.3	5.10.1.18	1902.2	6.1.3.5.3	802.5c
5.8.2.2	2202.2	5.10.1.19	1902.1	6.1.3.5.4	–
5.8.2.3	2202.4	5.10.1.20	1902.11	6.1.3.5.5	802.5d
5.8.2.4	2202.5	5.10.1.20.1	1902.11a	6.1.3.5.6	–
5.8.3	2203	5.10.1.20.2	1902.11b	6.1.3.5.8	802.5e
5.8.3.1	2203.2	5.10.1.20.3	1902.11c	6.1.3.5.9	–
5.8.3.2	2203.3	5.10.1.20.4	1902.11d	6.1.3.6	802.6
5.8.3.3	2203.4	5.10.1.20.5	1902.11e	6.1.3.6.1	802.6a
5.9	–	5.10.1.20.6	1902.11f	6.1.3.6.2	802.6b
5.10	Part XIX	5.10.1.20.7	1902.11g	6.1.3.6.3	802.6c
5.10.1	1900	5.10.1.21	1901	6.1.3.6.4	802.6d
5.10.1.1	1900.1	5.10.1.21.1	1901.1	6.1.3.6.5	802.6e
5.10.1.1.1	1900.1a	5.10.1.21.2	1901.2	6.1.3.6.6	–
5.10.1.1.2	1900.1b	5.10.1.21.3	1901.3	6.1.3.7	802.7
5.10.1.2	1900.3a	5.10.1.22	1900.1f	6.1.3.8	802.8
5.10.1.3	1900.5	5.10.1.23	1902.9	6.1.3.9	802.9
5.10.1.3.1	1900.5a	5.10.1.23.1	1902.9a	6.1.3.9.1	802.9a
5.10.1.3.2	1900.5b	5.10.1.23.2	1902.9b	6.1.3.9.2	802.9b
5.10.1.3.3	1900.5c	5.10.1.23.3	1902.9c	6.1.3.9.3	802.9c
5.10.1.4	1900.3c	5.10.2	1903	6.1.3.9.4	802.9d
5.10.1.5	1900.2	5.10.2.1	1903.1	6.1.3.10	802.10
5.10.1.6	1900.3b	5.10.2.2	1903.1a	6.1.3.11	802.11
5.10.1.7	1900.1c	5.10.2.3	1903.1c	6.1.3.12	802.12
5.10.1.7.1	1900.1c	5.10.2.4	1903.1b	6.1.3.13	–

A17.1-2000	A17.1-1996 With Addenda	A17.1-2000	A17.1-1996 With Addenda	A17.1-2000	A17.1-1996 With Addenda
6.1.3.14	–	6.2.3.2	902.2	6.2.6.3.3	905.3c
6.1.4	803	6.2.3.3	902.3	6.2.6.3.4	905.3d
6.1.5	804	6.2.3.3.1	902.3a	6.2.6.3.5	905.3e
6.1.5.1	804.1	6.2.3.3.2	902.3b	6.2.6.3.6	905.3f
6.1.5.2	804.2	6.2.3.3.3	902.3c	6.2.6.3.7	905.3g
6.1.5.3	804.3	6.2.3.3.4	902.3d	6.2.6.3.8	905.3h
6.1.5.3.1	804.3a	6.2.3.3.5	902.3e	6.2.6.3.9	905.3i
6.1.5.3.2	804.3b	6.2.3.3.6	902.3f	6.2.6.3.10	905.3j
6.1.5.3.3	–	6.2.3.3.7	902.3g	6.2.6.3.11	905.3k
6.1.6	805	6.2.3.3.8	902.3h	6.2.6.3.12	–
6.1.6.1	805.1	6.2.3.4	902.4	6.2.6.4	905.4
6.1.6.1.1	–	6.2.3.4.1	902.4a	6.2.6.5	905.5
6.1.6.2.1	805.2	6.2.3.4.2	902.4b	6.2.6.6	905.6
6.1.6.2.2	–	6.2.3.4.3	902.4c	6.2.6.7	905.7
6.1.6.3	805.3	6.2.3.4.4	–	6.2.6.8	905.8
6.1.6.3.1	805.3a	6.2.3.4.5	–	6.2.6.8.1	905.8a
6.1.6.3.2	805.3b	6.2.3.5	902.5	6.2.6.8.2	905.8b
6.1.6.3.3	805.3c	6.2.3.5.1	902.5a	6.2.6.9	905.9
6.1.6.3.4	805.3d	6.2.3.5.2	902.5b	6.2.6.10	–
6.1.6.3.5	805.3e	6.2.3.5.3	902.5c	6.2.6.11	905.10
6.1.6.3.6	805.3f	6.2.3.5.4	902.5d	6.2.6.12	905.11
6.1.6.3.7	805.3g	6.2.3.5.5	902.5e	6.2.6.13	905.12
6.1.6.3.8	805.3h	6.2.3.5.6	–	6.2.6.14	–
6.1.6.3.9	805.3i	6.2.3.6	902.6	6.2.7	906
6.1.6.3.10	805.3j	6.2.3.6.1	902.6a	6.2.7.1	906.1
6.1.6.3.11	805.3k	6.2.3.6.2	902.6b	6.2.7.1.1	906.1a
6.1.6.3.12	805.3m	6.2.3.7	902.7	6.2.7.1.2	906.1b
6.1.6.3.13	805.3n	6.2.3.8	902.8	6.2.7.2	906.2
6.1.6.3.14	805.3p	6.2.3.8.1	902.8a	6.2.7.3	906.3
6.1.6.3.15	–	6.2.3.8.2	902.8b	6.2.7.4	906.4
6.1.6.4	805.4	6.2.3.8.3	902.8c	6.2.8	907
6.1.6.5	805.5	6.2.3.8.4	902.8d	6.2.8.1	907.1
6.1.6.6	805.6	6.2.3.8.5	–	6.2.8.2	907.2
6.1.6.7	805.7	6.2.3.9	902.9	6.2.8.3	907.3
6.1.6.8	805.8	6.2.3.9.1	902.9a	Part 7	Parts VII and XIV
6.1.6.9	805.9	6.2.3.10	902.10	7.1	700
6.1.6.9.1	805.9a	6.2.3.10.1	902.10a	7.1.1	700.1
6.1.6.9.2	805.9b	6.2.3.10.2	902.10b	7.1.2	700.7
6.1.6.10	805.10	6.2.3.10.3	902.10c	7.1.3	700.4
6.1.6.11	–	6.2.3.10.4	902.11	7.1.4	700.8
6.1.6.12	805.11	6.2.3.11	902.12	7.1.5	700.9
6.1.6.13	805.12	6.2.3.12	902.13	7.1.6	700.10
6.1.6.14	805.13	6.2.3.13	902.14	7.1.7	700.2
6.1.6.15	–	6.2.3.14	902.15	7.1.8	700.3
6.1.7	806	6.2.3.15	902.16	7.1.9	700.6
6.1.7.1	806.1	6.2.3.16	–	7.1.10	700.5
6.1.7.1.1	806.1a	6.2.3.17	–	7.1.11	700.11
6.1.7.1.2	806.1b	6.2.4	903	7.1.11.1	700.11a
6.1.7.2	806.2	6.2.5	904	7.1.11.2	700.11b
6.1.7.3	806.3	6.2.5.1	904.1	7.1.11.3	700.11c
6.1.7.4	806.4	6.2.5.2	904.2	7.1.11.4	700.11d
6.1.8	807	6.2.5.3	904.3	7.1.11.5	700.11e
6.1.8.1	807.1	6.2.5.3.1	904.3a	7.1.11.6	700.11f
6.1.8.2	807.2	6.2.5.3.2	904.3b	7.1.11.7	700.11g
6.1.8.3	807.3	6.2.6	905	7.1.11.8	700.11h
6.2	Part IX	6.2.6.1	905.1	7.1.11.9	700.11i
6.2.1	900	6.2.6.1.1	–	7.1.11.10	700.11j
6.2.1.1	900.1	6.2.6.2.1	905.2	7.1.11.11	700.11k
6.2.2	901	6.2.6.2.2	–	7.1.11.12	700.11m
6.2.2.1	901.1	6.2.6.3	905.3	7.1.11.13	700.11n
6.2.3	902	6.2.6.3.1	905.3a	7.1.11.14	–
6.2.3.1	902.1	6.2.6.3.2	905.3b	7.1.12	700.12

A17.1-2000	A17.1-1996 With Addenda	A17.1-2000	A17.1-1996 With Addenda	A17.1-2000	A17.1-1996 With Addenda
7.1.12.1	700.12a	7.3.6.3	–	7.7.2	1403.1(b)
7.1.12.2	700.12b	7.3.7	702.1e	7.7.3	1403.1(c)
7.1.12.3	700.12c	7.3.8	–	7.7.4	–
7.1.12.4	700.12d	7.3.8.1	702.1c	7.8	1402
7.1.13	700.13	7.3.8.2	702.1d	7.8.1	1402.1
7.1.14	–	7.3.9	702.1a	7.8.2	1402.2
7.2	701	7.3.10	702.3	7.8.3	1402.3
7.2.1	701.5	7.3.11	702.4	7.8.4	1402.4
7.2.1.1	701.5a	7.3.11.1	702.4a	7.9	1400
7.2.1.2	701.5b	7.3.11.2	702.4b	7.9.1	1400.1
7.2.1.3	701.5c	7.3.11.3	702.4c	7.9.2	1400.2
7.2.2	701.4	7.3.11.4	702.4d	7.10	1401
7.2.3	701.8	7.3.11.5	702.4e	7.11	1400.3
7.2.3.1	701.8a	7.3.11.6	702.4f	Part 8	–
7.2.3.2	701.8b	7.3.11.7	702.4g	8.1	–
7.2.3.3	701.8c	7.3.11.8	702.4h	8.1.1	–
7.2.3.4	701.8d	7.3.11.9	702.4i	8.1.2	–
7.2.4	701.6	7.3.12	702.6	8.1.3	–
7.2.4.1	701.6a	7.4	1406	8.1.4	–
7.2.4.2	701.6b	7.4.1	–	8.1.5	–
7.2.4.3	–	7.4.2	–	8.2	Part XIII
7.2.4.4	701.6c	7.4.3	1406.1, 1406.1a	8.2.1	1300.1
7.2.4.5	701.6d	7.4.4	1406.1g	8.2.2	1301
7.2.4.6	701.6g	7.4.5	1406.1d	8.2.2.1	1301.1
7.2.4.7	701.6e	7.4.6	1406.1h	8.2.2.1.1	1301.1a
7.2.4.8	701.6f	7.4.7	1406.1i	8.2.2.2	1301.2
7.2.5	701.7	7.4.8	1406.1j	8.2.2.3	1301.3
7.2.6	701.13	7.4.9	1406.1b	8.2.2.4	1301.4
7.2.6.1	701.13a	7.4.10	1406.1c	8.2.2.5	1301.5
7.2.6.2	701.13b	7.4.11	1406.1f	8.2.2.6	1301.6
7.2.6.3	701.13c	7.4.12	1406.1e	8.2.2.7	1301.7
7.2.6.4	701.13d	7.4.13	1406.1k	8.2.3	1308
7.2.6.5	701.13e	7.4.14	1406.1l	8.2.3.1	1308.1
7.2.6.6	701.13f	7.4.15	1406.1m	8.2.3.2	1308.2
7.2.6.7	–	7.4.16	–	8.2.4	1304
7.2.6.8	701.13g	7.5	1406.2	8.2.5	1305
7.2.7	701.3	7.5.1	1406.2e	8.2.6	1306
7.2.8	701.2	7.5.2	1406.2d	8.2.7	1307
7.2.8.1	701.2a	7.5.3	1406.2h	8.2.8	1302
7.2.8.2	701.2b	7.5.4	1406.2f	8.2.8.1	1302.1
7.2.9	701.1	7.5.5	1406.2g	8.2.8.2	1302.2
7.2.9.1	701.1a	7.5.6	1406.2m	8.2.8.3	1302.3
7.2.10	701.9	7.5.7	1406.2c	8.2.8.4	1302.4
7.2.10.1	701.9a	7.5.8	1406.2b	8.2.8.5	1302.5
7.2.10.2	701.9b	7.5.9	1406.2a	8.2.8.5.1	1302.5a
7.2.10.3	701.9c	7.5.10	1406.2i	8.2.8.5.2	1302.5b
7.2.10.4	701.9d	7.5.11	1406.2j	8.2.9	1303
7.2.11	701.10	7.5.12	1406.2k	8.2.9.1	1303.1
7.2.12	701.11	7.5.13	1406.2o	8.2.9.1.1	1303.1a
7.2.13	701.14	7.5.14	1406.2n	8.2.9.1.2	1303.1b
7.2.14	–	7.6	–	8.2.9.1.3	1303.1c
7.3	702	7.6.1	1406.3	8.2.9.1.4	1303.1d
7.3.1	702.1g	7.6.2	1406.3(a)	8.2.10	1309
7.3.2	702.1f	7.6.3	1406.3(b)	8.2.11	1310
7.3.3	702.1j	7.6.4	1406.3(c)	8.2.12	1405
7.3.4	–	7.6.5	1406.3(g)	8.3	Part XI
7.3.4.1	702.1h	7.6.6	1406.3(d)	8.3.1	–
7.3.4.2	702.1i	7.6.7	1406.3(e)	8.3.1.1	–
7.3.5	702.2	7.6.8	1406.3(f)	8.3.1.2	1100.2
7.3.6	702.5	7.6.9	1406.3(h)	8.3.1.3	–
7.3.6.1	702.5a	7.7	1403	8.3.1.4	–
7.3.6.2	702.5b	7.7.1	1403.1(a)	8.3.1.5	–

A17.1-2000	A17.1-1996 With Addenda	A17.1-2000	A17.1-1996 With Addenda	A17.1-2000	A17.1-1996 With Addenda
8.3.2	1100	8.4.4	2406	8.6.1.6.4	1206.2d
8.3.2.1	1100.1	8.4.4.1	2406.1	8.6.1.6.5	1206.1h
8.3.2.2	–	8.4.5	2405	8.6.1.6.6	–
8.3.2.3	1100.3	8.4.5.1	2405.1	8.6.1.6.7	–
8.3.2.4	1100.4	8.4.5.2	2405.2	8.6.2	1200.4
8.3.2.4.1	1100.4a	8.4.6	2407	8.6.2.1	1200.4a
8.3.2.4.2	1100.4b	8.4.6.1	2407.1	8.6.2.2	1200.4b, 1200.4c
8.3.2.4.3	1100.4c	8.4.7	2404	8.6.2.3	–
8.3.2.4.4	1100.4d	8.4.7.1	2404.1	8.6.2.4	–
8.3.2.5	1100.5	8.4.7.2	2404.2	8.6.2.5	1200.4d
8.3.2.5.1	1100.5a	8.4.8	2403	8.6.3	1200.5
8.3.2.5.2	1100.5b	8.4.8.1	2403.1	8.6.3.1	–
8.3.2.5.3	1100.5c	8.4.8.2	2403.2	8.6.3.2	1200.5a
8.3.2.5.4	1100.5d	8.4.8.3	2403.3	8.6.3.3	1200.5a
8.3.2.5.5	1100.5e	8.4.8.4	2403.4	8.6.3.4	1200.5a
8.3.2.6	1100.5f	8.4.8.5	2403.5	8.6.3.5	1200.4e
8.3.3	1101	8.4.8.6	2403.6	8.6.3.6	–
8.3.3.1	1101.1	8.4.8.7	2403.7	8.6.3.7	–
8.3.3.2	1101.2	8.4.8.8	2403.8	8.6.3.7.1	1200.5c
8.3.3.3	1101.3	8.4.8.9	2403.9	8.6.3.7.2	1200.4f
8.3.3.3.1	1101.3a	8.4.9	2408	8.6.3.8	1200.5d
8.3.3.3.2	1101.3b	8.4.9.1	2408.1	8.6.3.9	–
8.3.3.3.3	1101.3c	8.4.10	2409	8.6.3.10	–
8.3.3.3.4	1101.3d	8.4.10.1	2409.1	8.6.3.11	–
8.3.3.4	1101.4	8.4.10.1.1	2409.1a	8.6.4	–
8.3.4	1102	8.4.10.1.2	2409.1b	8.6.4.1	1206.1b
8.3.4.1	1102.1, 1102.2	8.4.10.1.3	2409.1c	8.6.4.2	1206.1c
8.3.5	1107	8.4.10.1.4	2409.1d	8.6.4.3	1206.1d
8.3.5.1	1107.1	8.4.11	2410	8.6.4.4	1206.1e
8.3.5.2	1107.2	8.4.11.1	2410.1	8.6.4.5	1206.1g
8.3.5.3	1107.3	8.4.11.2	2410.6	8.6.4.6	–
8.3.5.3.1	1107.3a	8.4.11.3	2410.5	8.6.4.7	1206.2a
8.3.5.3.2	1107.3b	8.4.11.4	2410.4	8.6.4.8	1206.2b
8.3.5.3.3	1107.3c	8.4.11.5	2410.2	8.6.4.9	1206.2c
8.3.5.3.4	1107.3d	8.4.11.6	2410.7	8.6.4.10	1206.3
8.3.5.3.5	1107.3e	8.4.11.7	2410.3	8.6.4.10.1	1206.3a
8.3.6	–	8.4.12	2411	8.6.4.10.2	1206.3b
8.3.7	1104	8.4.12.1	2411.1	8.6.4.10.3	1206.3c
8.3.7.1	1104.1	8.4.12.1.1	2411.1a	8.6.4.11	1206.8
8.3.7.2	1104.2	8.4.12.1.2	2411.1b	8.6.4.12	–
8.3.7.3	1104.3	8.4.12.2	2411.2	8.6.4.13	–
8.3.7.4	1104.4	8.4.12.2.1	2411.2a	8.6.4.14	–
8.3.7.5	1104.5	8.4.12.2.2	2411.2b	8.6.4.15	–
8.3.7.6	1104.6	8.4.13	–	8.6.4.16	–
8.3.8	1106	8.4.13.1	–	8.6.4.17	–
8.3.9	–	8.4.13.2	–	8.6.5	1206.5a
8.3.10	1103	8.5	–	8.6.5.1	1206.5b(1), 1206.5b(3)
8.3.10.1	1103.1	8.5.1	–	8.6.5.2	1206.5b(2)
8.3.10.2	1103.2	8.5.2	–	8.6.5.3	–
8.3.10.3	1103.3	8.5.3	–	8.6.5.4	–
8.3.11	1105	8.5.4	–	8.6.5.5	1206.5b(4), 1206.5b(5)
8.3.11.1	1105.1	8.6	1206	8.6.5.6	1206.5b(6)
8.4	Part XXIV	8.6.1	–	8.6.5.7	–
8.4.1	2402	8.6.1.1	–	8.6.5.8	–
8.4.1.1	2402.1	8.6.1.2	1206.1	8.6.5.9	–
8.4.2	2401	8.6.1.3	–	8.6.5.10	–
8.4.2.1	2401.1	8.6.1.4	–	8.6.5.11	–
8.4.2.2	2401.2	8.6.1.5	1206.1i	8.6.6	–
8.4.2.3	2401.3	8.6.1.6	–	8.6.6.1	–
8.4.3	2400	8.6.1.6.1	1206.4	8.6.6.2	–
8.4.3.1	2400.1	8.6.1.6.2	1206.1a	8.6.6.3	–
8.4.3.2	2400.2	8.6.1.6.3	1206.1f	8.6.7	–

A17.1-2000	A17.1-1996 With Addenda	A17.1-2000	A17.1-1996 With Addenda	A17.1-2000	A17.1-1996 With Addenda
8.6.7.1	–	8.7.2.7.6	1201.2f	8.7.3.9	1203.1c
8.6.7.2	–	8.7.2.7.7	1201.2g	8.7.3.10	1203.1j
8.6.7.3	1206.10	8.7.2.8	1201.3	8.7.3.11	1203.1k
8.6.7.4	–	8.7.2.9	1201.5	8.7.3.12	1203.1m
8.6.7.5	–	8.7.2.10	1201.10	8.7.3.13	1203.2e
8.6.7.6	–	8.7.2.10.1	1201.10a	8.7.3.14	1203.2d
8.6.7.7	–	8.7.2.10.2	1201.10b	8.7.3.15	1203.2f
8.6.7.8	–	8.7.2.10.3	1201.10c	8.7.3.16	1203.2g
8.6.7.9	–	8.7.2.10.4	1201.10d	8.7.3.17	1203.2h
8.6.7.10	–	8.7.2.10.5	1201.10e	8.7.3.18	1203.2i
8.6.8	1206.6	8.7.2.11	1201.11	8.7.3.19	1203.2j
8.6.8.1	–	8.7.2.11.1	1201.11a	8.7.3.20	1203.2k
8.6.8.2	–	8.7.2.11.2	1201.11b	8.7.3.21	1203.2m
8.6.8.3	1206.6c	8.7.2.11.3	1201.11c	8.7.3.22	1203.4
8.6.8.4	–	8.7.2.11.4	1201.11d	8.7.3.22.1	1203.4a
8.6.8.5	1206.6b	8.7.2.11.5	–	8.7.3.22.2	1203.4b
8.6.8.6	–	8.7.2.12	1201.12	8.7.3.22.3	1203.4c
8.6.8.7	–	8.7.2.13	–	8.7.3.23	1203.3
8.6.8.8	–	8.7.2.14	1202.5	8.7.3.23.1	1203.3a
8.6.8.9	–	8.7.2.15	1202.4	8.7.3.23.2	1203.3b
8.6.8.10	–	8.7.2.15.1	1202.4a	8.7.3.23.3	1203.3c
8.6.8.11	–	8.7.2.15.2	1202.4b	8.7.3.23.4	1203.3d
8.6.8.12	–	8.7.2.16	1202.8	8.7.3.23.5	1203.3e
8.6.8.13	–	8.7.2.16.1	1202.8a	8.7.3.23.6	1203.3f
8.6.8.14	1206.6a	8.7.2.16.2	1202.8b	8.7.3.24	1203.5
8.6.9	1206.5	8.7.2.16.3	1202.8c	8.7.3.25	1203.9
8.6.9.1	–	8.7.2.16.4	1202.8d	8.7.3.25.1	1203.9a
8.6.9.2	–	8.7.2.17	1202.10	8.7.3.25.2	1203.9c
8.6.10	–	8.7.2.17.1	1202.10a	8.7.3.26	1203.2c
8.6.10.1	1206.7	8.7.2.17.2	1202.10b	8.7.3.27	1203.2b
8.6.10.2	–	8.7.2.17.3	1202.10c	8.7.3.28	1203.2a
8.6.10.3	1206.9	8.7.2.18	1202.6	8.7.3.29	1203.6
8.6.10.4	–	8.7.2.19	1202.7	8.7.3.30	1203.7
8.6.10.5	–	8.7.2.20	–	8.7.3.31	1203.8
8.7	Part XII	8.7.2.21	1202.14	8.7.3.31.1	1203.8a
8.7.1	1200	8.7.2.21.1	1202.14a	8.7.3.31.2	1203.8b
8.7.1.1	1200.1	8.7.2.21.2	1202.14c	8.7.3.31.3	1203.8c
8.7.1.2	1200.2	8.7.2.21.3	1202.14d	8.7.3.31.4	1203.8d
8.7.1.3	1200.3	8.7.2.22	1202.3	8.7.3.31.5	1203.8e
8.7.1.4	1200.4b	8.7.2.23	1202.2	8.7.3.31.6	1203.8f
8.7.1.5	1200.4c	8.7.2.24	1202.1	8.7.3.31.7	1203.8g
8.7.1.6	–	8.7.2.25	1202.9	8.7.3.31.8	1203.8h
8.7.1.7	–	8.7.2.25.1	1202.9a	8.7.3.31.9	–
8.7.1.8	1200.6	8.7.2.25.2	1202.9b	8.7.4	–
8.7.2	1201	8.7.2.26	1202.11	8.7.4.1	1211
8.7.2.1	1201.1	8.7.2.27	1202.12	8.7.4.2	1213
8.7.2.1.1	1201.1a	8.7.2.27.1	1202.12a	8.7.4.3	1204, 1219
8.7.2.1.2	1201.1b	8.7.2.27.2	1202.12b	8.7.4.3.1	1219.1
8.7.2.1.3	1201.1c	8.7.2.27.3	1202.12c	8.7.4.3.2	1219.2
8.7.2.1.4	1201.1d	8.7.2.27.4	1202.12d	8.7.4.3.3	1219.3
8.7.2.1.5	1201.1e	8.7.2.27.5	1202.12e	8.7.4.3.4	1219.4
8.7.2.2	1201.6	8.7.2.27.6	1202.12f	8.7.4.3.5	1219.5
8.7.2.3	1201.4	8.7.2.28	1202.13	8.7.4.3.6	1219.6
8.7.2.4	1201.7	8.7.3	1203	8.7.4.3.7	1219.7
8.7.2.5	1201.8	8.7.3.1	1203.1a	8.7.4.3.8	1219.8
8.7.2.6	1201.9	8.7.3.2	1203.1f	8.7.4.3.9	1219.9
8.7.2.7	1201.2	8.7.3.3	1203.1d	8.7.4.3.10	1219.10
8.7.2.7.1	1201.2a	8.7.3.4	1203.1g	8.7.5	–
8.7.2.7.2	1201.2b	8.7.3.5	1203.1h	8.7.5.1	1212
8.7.2.7.3	1201.2c	8.7.3.6	1203.1i	8.7.5.2	–
8.7.2.7.4	1201.2d	8.7.3.7	1203.1b	8.7.5.3	–
8.7.2.7.5	1201.2e	8.7.3.8	1203.1e	8.7.5.4	–

A17.1-2000	A17.1-1996 With Addenda	A17.1-2000	A17.1-1996 With Addenda	A17.1-2000	A17.1-1996 With Addenda
8.11.5	1010	8.11.5.6	1010.6	8.11.5.12	1010.13
8.11.5.1	1010.1	8.11.5.7	1010.7	8.11.5.13	1011
8.11.5.2	1010.2	8.11.5.8	1010.8	8.11.5.13.1	1011.2
8.11.5.3	1010.3	8.11.5.9	1010.9	8.11.5.13.2	1011.3, 1011.4
8.11.5.4	1010.4	8.11.5.10	1010.11	8.11.5.13.3	1011.5
8.11.5.5	1010.5	8.11.5.11	1010.12	8.11.5.13.4	1011.6

CROSS-REFERENCE TABLE
ASME A17.1-1996 Including A17.1a-1997 Through A17.1d–2000 vs. ASME A17.1-2000

A17.1-1996 With Addenda	A17.1-2000	A17.1-1996 With Addenda	A17.1-2000	A17.1-1996 With Addenda	A17.1-2000
1	1.1	103.4	2.3.4	110.7a	2.11.7.1
1.1	1.1.1	104	2.10	110.7b	2.11.7.2
1.2	1.1.2	104.1	2.10.1	110.8	2.11.8
1.3	1.1.3	105	2.9	110.9	2.11.9
1.4	1.1.4	105.1	2.9.1	110.9a	2.11.9.1
2	1.2	105.2	2.9.2	110.9b	2.11.9.2
3	1.3	105.2a	2.9.2.1	110.10	2.11.10
4	Part 9	105.2b	2.9.2.2	110.10a	2.11.10.1
100	2.1	105.3	2.9.3	110.10b	2.11.10.2
100.1	2.1.1	105.3a	2.9.3.1	110.10c	–
100.1a	2.1.1.1	105.3b	2.9.3.2	110.10d	2.11.10.3
100.1b	2.1.1.2	105.3c	2.9.3.3	110.11	2.11.11
100.1c	2.1.1.3	105.3d	2.9.3.4	110.11a	2.11.11.1
100.1d	2.1.1.4	105.4	2.9.4	110.11b	2.11.11.2
100.1e	2.1.1.5	105.5	2.9.5	110.11c	2.11.11.3
100.2	2.1.2	106	2.2	110.11d	2.11.11.4
100.2a	2.1.2.1	106.1	–	110.11e	2.11.11.5
100.2b	2.1.2.2	106.1a	2.2.1	110.11f	2.11.11.6
100.2c	2.1.2.3	106.1b	2.2.2	110.11g	2.11.11.7
100.3	2.1.3	106.1c	2.2.3	110.11h	2.11.11.8
100.3a	2.1.3.1	106.1d	2.2.4	110.11i	2.11.11.9
100.3b	2.1.3.2	106.1e	2.2.5	110.11j	2.11.11.10
100.3c	2.1.3.3	106.1f	2.2.6	110.12	2.11.12
100.3d	2.1.3.4	106.1g	2.2.7	110.12a	2.11.12.1
100.3e	2.1.3.5	107	2.4	110.12b	2.11.12.2
100.3f	2.1.3.6	107.1	–	110.12c	2.11.12.3
100.4	2.1.4	107.1a	2.4.1	110.12d	2.11.12.4
100.5	2.1.5	107.1b	2.4.2	110.12e	2.11.12.5
100.6	2.1.6	107.1c	2.4.3	110.12f	2.11.12.6
100.7	2.29.2	107.1d	2.4.4, 2.4.5	110.12g	2.11.12.7
101	2.7	107.1e	2.4.6	110.12h	2.11.12.8
101.1	2.7.1	107.1f	2.4.7	110.13	2.11.13
101.1a	2.7.1.1	107.1g	2.4.8	110.13a	2.11.13.1
101.1b	2.7.1.2	107.1h	2.4.9	110.13b	2.11.13.2
101.2	2.7.2.1	107.1i	2.4.10	110.13c	2.11.13.3
101.3	2.7.3	107.1j	2.4.11	110.13d	2.11.13.4
101.3a	2.7.3.1	107.1k	2.4.12	110.13e	2.11.13.5
101.3b	2.7.3.2	108	2.5	110.14	2.11.14
101.3c	2.7.3.3	108.1	2.5.1	110.15	2.11.15
101.3d	2.7.3.4	108.1a	2.5.1.1	110.15a	2.11.15.1
101.3e	2.7.3.5	108.1b	2.5.1.2	110.15b	2.11.15.2
101.4	2.7.4	108.1c	2.5.1.3	110.15c	2.11.15.3
101.5	2.7.5	108.1d	2.5.1.4	110.16	2.11.16
101.5a	2.7.5.1	108.1e	2.5.1.5	110.17	2.11.17
101.5b	2.7.5.2	108.1f	2.5.1.7	110.18	2.11.18
101.5c	–	109	2.6	110.19	2.11.19
101.6	2.7.6	109.1	2.6	111	2.12
101.7	2.7.7	110	2.11	111.1	2.12.1
101.8	2.7.8	110.1	2.11.1.1, 2.11.1.2	111.1a	2.12.1
102	2.8	110.2	2.11.2	111.1b	2.12.1
102.1	2.8.1	110.2a	2.11.2.1	111.2	2.12.2
102.2	2.8.2	110.2b	2.11.2.2	111.2a	2.12.2.1
102.3	2.8.3	110.2c	2.11.2.3	111.2b	2.12.2.2
102.4	2.8.4	110.3	2.11.3	111.2c	2.12.2.3
103	2.3	110.4	2.11.4	111.2d	2.12.2.4
103.1	2.3.1	110.5	2.11.5	111.2e	2.12.2.5
103.2	2.3.2	110.6	2.11.6	111.2f	2.12.2.6
103.3	2.3.3	110.7	2.11.7	111.3	2.12.3

A17.1-1996 With Addenda	A17.1-2000	A17.1-1996 With Addenda	A17.1-2000	A17.1-1996 With Addenda	A17.1-2000
111.3a	2.12.3.1	201.1b	2.22.1.2	204.1d	2.14.1.4
111.3b	2.12.3.2	201.2	2.22.2	204.1e	2.14.1.5
111.3c	2.12.3.3	201.3	2.22.3	204.1f	2.14.1.6
111.3d	2.12.3.4	201.3a	2.22.3.1	204.1g	2.14.1.7
111.3e	2.12.3.5	201.3b	2.22.3.2	204.1h	2.14.1.8
111.4	2.12.4	201.3c	2.22.3.3	204.1i	2.14.1.9
111.4a	2.12.4.1	201.4	2.22.4	204.1j	2.14.1.10
111.4b	2.12.4.2	201.4a	2.22.4.1	204.2	2.14.2
111.4c	2.12.4.3	201.4b	2.22.4.2	204.2a	2.14.2.1
111.5	2.12.5	201.4c	2.22.4.3	204.2b	2.14.2.2
111.6	2.12.6	201.4d	2.22.4.4	204.2c	2.14.2.3
111.6a	2.12.6.1	201.4e	2.22.4.5	204.2d	2.14.2.4
111.6b	2.12.6.2	201.4f	2.22.4.6	204.2e	2.14.2.5
111.7	2.12.7	201.4g	2.22.4.7	204.3	2.14.3
111.7a	2.12.7.1	201.4h	2.22.4.8	204.3a	2.14.3.1
111.7b	2.12.7.2	201.4i	2.22.4.9	204.3b	2.14.3.2
111.7c	2.12.7.3	201.4j	2.22.4.10	204.3c	2.14.3.3
112	2.13	201.4k	2.22.4.11	204.4	2.14.4
112.1	2.13.1	202	2.21	204.4a	2.14.4.1
112.2	2.13.2	202.1	2.21.1	204.4b	2.14.4.2
112.2a	2.13.2.1	202.1a	2.21.1.1	204.4c	2.14.4.3
112.2b	2.13.2.2	202.1b	2.21.1.2	204.4d	2.14.4.4
112.3	2.13.3	202.1c	2.21.1.3	204.4e	2.14.4.5
112.3a	2.13.3.1	202.1d	2.21.1.4	204.4f	2.14.4.6
112.3b	2.13.3.2	202.2	2.21.2	204.4h	2.14.4.7
112.3c	2.13.3.3	202.2a	2.21.2.1	204.4i	2.14.4.8
112.3d	2.13.3.4	202.2b	2.21.2.2	204.4j	2.14.4.9
112.4	2.13.4.1, 2.13.4.2.1, 2.13.4.2.2, 2.13.4.2.3	202.2c	2.21.2.3	204.4k	2.14.4.10
		202.2d	2.21.2.4	204.4m	2.14.4.11
112.5	2.13.5	202.2e	2.21.2.5	204.5	2.14.5
112.6	2.13.6	202.2f	2.21.2.6	204.5a	2.14.5.1
112.6a	2.13.6.1	202.3	2.21.3	204.5b	2.14.5.2
112.6b	2.13.6.2	202.4	2.21.4	204.5c	2.14.5.3
112.6c	–	203	2.15	204.5d	2.14.5.4
200	2.23	203.1	2.15.1	204.5e	2.14.5.5
200.1	2.23.1	203.2	2.15.2	204.5f	–
200.2	2.23.2	203.3	2.15.3	204.5g	2.14.5.6
200.2a	2.23.2.1	203.4	2.15.4	204.5h	2.14.5.7
200.2b	2.23.2.2	203.5	2.15.5	204.5i	2.14.5.8
200.3	2.23.3	203.6	2.15.6	204.6	2.14.6
200.4	2.23.4	203.6a	2.15.6.1	204.6a	2.14.6.1
200.4a	2.23.4.1	203.6b	2.15.6.2	204.6b	2.14.6.2
200.4b	2.23.4.2	203.6c	2.15.6.3	204.6c	2.14.6.3
200.4c	2.23.4.3	203.6d	2.15.6.4	204.7	2.14.7
200.5	2.23.5	203.7	2.15.7	204.7a	2.14.7.1
200.5a	2.23.5.1	203.7a	2.15.7.1	204.7b	2.14.7.2
200.5b	2.23.5.2	203.7b	2.15.7.2	204.7c	2.14.7.3
200.6	2.23.6	203.7c	2.15.7.3	204.7d	2.14.7.4
200.7	2.23.7	203.8	2.15.8	205	2.17
200.7a	2.23.7.1	203.9	2.15.9	205.1	2.17.1
200.7b	2.23.7.2	203.10	2.15.10	205.2	2.17.2
200.8	2.23.8	203.11	2.15.11	205.3	2.17.3
200.9	2.23.9	203.12	2.15.12	205.4	2.17.4
200.9a	2.23.9.1	203.13	2.15.13	205.5	2.17.5
200.9b	2.23.9.2	203.14	2.15.14	205.6	–
200.10	2.23.10	203.15	2.15.15	205.7	2.17.7
200.10a	2.23.10.1	203.16	2.15.16	205.8	2.17.8
200.10b	2.23.10.2	204	2.14	205.8a	2.17.8.1
200.10c	2.23.10.3	204.1	2.14.1	205.8b	2.17.8.2
201	2.22	204.1a	2.14.1.1	205.9	2.17.9
201.1	2.22.1	204.1b	2.14.1.2	205.9a	2.17.9.1
201.1a	2.22.1.1	204.1c	2.14.1.3	205.9b	2.17.9.2

A17.1-1996 With Addenda	A17.1-2000	A17.1-1996 With Addenda	A17.1-2000	A17.1-1996 With Addenda	A17.1-2000
205.9c	2.17.9.3	208.9	2.24.9	212.2a	2.20.2.1
205.9d	2.17.9.4	208.9a	2.24.9.1	212.2b	2.20.2.2
205.10	2.17.10	208.9b	2.24.9.2	212.3	2.20.3
205.11	2.17.11	208.9c	2.24.9.3	212.4	2.20.4
205.12	2.17.12	208.9d	–	212.5	2.20.5
205.13	2.17.13	208.10	–	212.6	2.20.6
205.14	2.17.14	208.11	2.24.10	212.7	2.20.7
205.15	2.17.15	209	2.25	212.8	2.20.8
205.16	2.17.16	209.1	2.25.1	212.9	2.20.9
205.17	2.17.17	209.2	2.25.2	212.9a	2.20.9.1
206	2.18	209.2a	2.25.2.1	212.9b	2.20.9.2
206.1	2.18.1	209.2b	2.25.2.2	212.9c	2.20.9.3
206.2	2.18.2	209.2c	2.25.2.3	212.9d	2.20.9.4
206.2a	2.18.2.1	209.3	2.25.3	212.9e	2.20.9.5
206.2b	2.18.2.2	209.3a	2.25.3.1	212.9f	2.20.9.6
206.3	2.18.3	209.3b	2.25.3.2	212.9g	2.20.9.7
206.4	2.18.4	209.3c	2.25.3.3	212.9h	2.20.9.8
206.4a	2.18.4.1	209.3d	2.25.3.4	212.10	2.20.10
206.4b	2.18.4.2	209.3e	2.25.3.5	213	8.8
206.4c	2.18.4.3	209.4	2.25.4	213.1	8.8.1
206.5	2.18.5	209.4a	2.25.4.1	213.2	8.8.2
206.5a	2.18.5.1	209.4b	2.25.4.2	213.3	8.8.3
206.5b	–	210	2.26	214	2.28
206.5c	2.18.5.2	210.1	2.26.1	214.1	2.28.1
206.5d	–	210.1a	2.26.1.1	215	8.9
206.5e	2.18.5.3	210.1b	2.26.1.2	215.1	8.9.1, 8.9.2,
206.6	2.18.6	210.1c	2.26.1.3		8.9.3
206.7	2.18.7	210.1d	2.26.1.4	Part III	Part 3
206.8	2.18.8	210.1e	2.26.1.6	300	–
206.9	2.18.9	210.2	2.26.2	300.1	3.1
207	2.16	210.4	2.26.4	300.1a	3.1.1
207.1	2.16.1.1	210.5	–	300.1b	3.1.2
207.1a	2.16.1.2	210.6	2.26.6	300.2	3.7
207.1b	2.16.1.3	210.7	2.26.7	300.2a	3.7.1
207.2	2.16.2	210.8	2.26.8	300.3	3.8
207.2a	2.16.2.1	210.9	2.26.9	300.4	3.3
207.2b	2.16.2.2	210.10	2.26.10	300.5	3.10
207.3	2.16.3	210.11	–	300.6	3.9
207.3a	2.16.3.1	210.12	–	300.7	3.2
207.3b	2.16.3.2	210.13	2.26.12	300.7a	3.2.1
207.3c	2.16.3.3	210.14	2.26.1.5	300.8	3.4
207.4	2.16.4	210.15	2.26.5	300.8a	3.4.1
207.5	2.16.5	211	2.27	300.8b	3.4.2
207.5a	2.16.5.1	211.1	2.27.1	300.8c	3.4.3
207.5b	2.16.5.2	211.2	2.27.2	300.8d	3.4.4
207.6	2.16.6	211.3	2.27.3	300.8e	3.4.5
207.7	2.16.7	211.3a	2.27.3.1	300.8f	3.4.6
207.8	2.16.8	211.3b	2.27.3.2	300.8g	3.4.7
207.9	2.16.9	211.3c	2.27.3.3	300.8h	3.4.8
208	2.24	211.3d	2.27.3.4	300.9	3.5
208.1	2.24.1	211.3e	2.27.3.5	300.10	3.6
208.2	2.24.2	211.4	2.27.4	300.11	3.11
208.2a	2.24.2.1	211.4a	2.27.4.1	300.11a	3.11.1
208.2b	2.24.2.2	211.4b	2.27.4.2	300.12	3.12
208.2c	2.24.2.3	211.5	2.27.5	300.13	3.13
208.2d	2.24.2.4	211.6	2.27.6	301	–
208.3	2.24.3	211.7	2.27.7	301.1	3.23
208.4	2.24.4	211.8	2.27.8	301.1a	3.23.1
208.5	2.24.5	211.9	2.29.1	301.1b	3.23.2
208.6	2.24.6	212	2.20	301.3	3.22.1
208.7	2.24.7	212.1	2.20.1	301.4	3.22.2
208.8	2.24.8.3	212.2	2.20.2	301.5	3.21

A17.1-1996 With Addenda	A17.1-2000	A17.1-1996 With Addenda	A17.1-2000	A17.1-1996 With Addenda	A17.1-2000
301.6	3.15	304.1	3.24.1	400.12	5.5.1.12
301.6a	3.15.1	304.1a	3.24.1.1	400.13	5.5.1.13
301.6b	3.15.2	304.2	3.24.2	401	–
301.6c	3.15.3	304.2a	3.24.2.1	401.1	5.5.1.22
301.7	3.14	304.2b	3.24.2.2	401.2	5.5.1.21
301.8	3.17.1	304.3	3.24.3	401.3	5.5.1.20
301.9	3.17.2	304.3a	3.24.3.1	401.4	5.5.1.15
301.10	3.16	304.3b	3.24.3.2	401.4a	5.5.1.15.1
302	3.18	304.3c	3.24.3.3	401.4b	5.5.1.15.2
302.1	3.18.1	304.4	3.24.4	401.5	5.5.1.14
302.1a	3.18.1.1	305	3.25	401.5a	5.5.1.14.1
302.1b	3.18.1.2	305.1	3.25.1	401.5b	5.5.1.14.2
302.2	3.18.2	305.1a	3.25.1.1	401.5c	5.5.1.14.3
302.2a	3.18.2.1	305.1b	3.25.1.2	401.6	5.5.1.17
302.2b	3.18.2.2	305.1c	3.25.1.3	401.7	5.5.1.18
302.2c	3.18.2.3	305.1d	3.25.1.4	401.8	5.5.1.16
302.2d	3.18.2.4	305.2	3.25.2	401.9	5.5.1.23
302.2e	3.18.2.5	305.2a	3.25.2.1	401.10	5.5.1.24
302.2f	3.18.2.6	305.2b	3.25.2.2	401.11	5.5.1.25
302.2g	3.18.2.7	305.3	3.25.3	401.11a	5.5.1.25.1
302.3	3.18.3	306	3.26	401.11b	5.5.1.25.2
302.3a	3.18.3.1	306.1	3.26.1	401.11c	5.5.1.25.3
302.3b	3.18.3.2	306.2	3.26.2	401.11d	5.5.1.25.4
302.3c	3.18.3.3	306.3	3.26.3	401.12	5.5.1.26
302.3d	3.18.3.4	306.3a	3.26.3.1	401.13	5.5.1.19
302.3e	3.18.3.5	306.3b	3.26.3.2	401.14	5.5.1.28
302.3f	3.18.3.6	306.4	3.26.4	401.15	5.5.1.27
302.3g	3.18.3.7	306.5	3.26.5	401.16	8.9
302.3h	3.18.3.8	306.6	3.26	402	5.5.2
302.3l	3.18.3.9	306.8	3.26	402.1	5.5.2.1
302.4	3.18.4	306.9	3.26.6	402.1a	5.5.2.4
302.4a	3.18.4.1	306.10	–	402.1b	5.5.2.2
302.4b	3.18.4.2	306.11	3.27	402.1c	5.5.2.3
302.5	3.18.5	306.12	–	402.1d	5.5.2.5
303	3.19	306.13	3.26.7	402.2	–
303.1	3.19.1	306.14	3.26.8	402.2a	5.5.2.14
303.1a	3.19.1.1	306.15	3.26.9	402.2b	5.5.2.13
303.1b	3.19.1.2	307	–	402.2c	5.5.2.12
303.1c	3.19.1.3	307.1	3.20	402.2d	5.5.2.7
303.1d	3.19.1.4	307.2	3.24.5	402.2e	5.5.2.6
303.2	3.19.2	308	3.28	402.2f	5.5.2.9
303.2a	3.19.2.1	308.1	3.28.1	402.2g	5.5.2.8
303.2b	3.19.2.2	309	8.9	402.3	5.5.2.10
303.2c	3.19.2.3	309.1	8.9.1, 8.9.2, 8.9.3	402.4	5.5.2.11
303.2d	3.19.2.4			402.5	5.5.2.15
303.3	3.19.3	Part IV	5.5	402.6	5.5.2.16
303.3a	3.19.3.1	400	–	402.7	5.5.2.17.1
303.3b	3.19.3.2	400.1	5.5.1.1	402.7a	5.5.2.17.2, 5.5.2.17.3
303.3c	3.19.3.3	400.2	5.5.1.7	402.8	5.5.2.18
303.4	3.19.4	400.3	5.5.1.8	402.9	8.9
303.4a	3.19.4.1	400.4	5.5.1.3	Part V	5.3
303.4b	3.19.4.2	400.5	5.5.1.10	500	–
303.4c	3.19.4.3	400.6	5.5.1.9	500.1	5.3.1.1
303.4d	3.19.4.4	400.7	5.5.1.2	500.2	5.3.1.2
303.4e	3.19.4.5	400.8	5.5.1.4	500.2a	5.3.1.2.1
303.4f	3.19.4.6	400.9	5.5.1.5	500.2b	5.3.1.2.2
303.5	3.19.5	400.10	5.5.1.6	500.3	5.3.1.3
303.5a	3.19.5.1	400.11	5.5.1.11	500.4	5.3.1.7
303.5b	3.19.5.2	400.11a	5.5.11.1	500.4a	5.3.1.7.1
303.6	3.19.6	400.11b	5.5.1.11.2	500.4b	5.3.1.7.2
303.7	3.19.7	400.11c	5.5.1.11.3	500.4c	5.3.1.7.3
304	3.24	400.11d	5.5.1.11.4	500.4d	5.3.1.7.4

A17.1-1996 With Addenda	A17.1-2000	A17.1-1996 With Addenda	A17.1-2000	A17.1-1996 With Addenda	A17.1-2000
500.4e	5.3.1.7.5	508.1	5.3.1.18.1	513.9c	5.4.4.2.3
500.4f	5.3.1.7.6	508.2	5.3.1.18.2	513.9d	5.4.4.2.4
500.4g	5.3.1.7.7	508.3	5.3.1.18.3	513.10	5.4.4.3
500.5	5.3.1.5	508.4	5.3.1.18.4	513.10a	5.4.4.3.1
500.6	5.3.1.4	508.5	5.3.1.18.5	513.10b	5.4.4.3.2
500.6a	5.3.1.4.1	508.6	5.3.1.18.6	513.11	5.4.16
500.6b	5.3.1.4.2	508.7	5.3.1.18.7	513.11a	5.4.16.1
500.7	5.3.1.6	508.8	5.3.1.18.8	513.11b	5.4.16.2
500.7a	5.3.1.6.1	508	5.3.1.19	513.11c	8.9
500.7b	5.3.1.6.2	509.1	–	513.12	5.4.11
501	5.3.1.9	510	5.3.1.10	513.12a	5.4.11.1
501.1	5.3.1.9.1	510.1	5.3.1.10.1	513.12b	5.4.11.2
501.2	5.3.1.8.1	510.2	5.3.1.10.2	513.12c	5.4.11.3
501.2a	5.3.1.8.1(a)	510.3	5.3.1.10.3	513.13	5.4.9
501.2b	5.3.1.8.1(b)	511	5.3.1.20	513.13a	5.4.9.1
501.2c	5.3.1.8.1(c)	511.1	5.3.1.20.1	513.13b	5.4.9.2
501.2d	5.3.1.8.1(d)	511.2	5.3.1.20.2	513.13c	5.4.9.3
501.3	5.3.1.8.1(e)	511.3	8.9	513.14	5.4.7
501.4	5.3.1.8.2	512	5.3.1.12	513.14a	5.4.7.1
501.4a	5.3.1.8.2(a)	512.1	5.3.1.12.1	513.14b	5.4.7.2
501.4b	5.3.1.8.2(b)	512.2	5.3.1.12.2	513.14c	5.4.7.3
501.4c	5.3.1.8.2(c)	512.3	5.3.1.12.3	513.14d	5.4.7.4
501.5	5.3.1.8.3	512.4	5.3.1.12.4	513.14e	5.4.7.5
502	5.3.1.13	512.5	5.3.1.12.5	513.14f	5.4.7.6
502.1	5.3.1.13.1	512.6	5.3.1.12.6	513.14g	5.4.7.7
502.2	5.3.1.13.2	512.7	5.3.1.12.7	513.14h	5.4.7.8
502.2a	5.3.1.13.2(a)	512.8	5.3.1.12.8	513.14i	5.4.7.9
502.2b	5.3.1.13.2(b)	512.9	5.3.1.12.9	513.15	5.4.13
502.2c	5.3.1.13.2(c)	513	5.4	513.15a	5.4.13.1
503	5.3.1.11	513.1	5.4.1	513.15b	5.4.13.2
503.1	5.3.1.11.1	513.2	5.4.2	513.15c	5.4.13.3
503.2	5.3.1.11.2	513.2a	5.4.2.1	513.15d	5.4.13.4
503.3	5.3.1.11.3	513.2b	5.4.2.2	513.15e	5.4.13.5
503.4	5.3.1.11.4	513.2c	5.4.2.3	513.15f	5.4.13.6
503.5	5.3.1.11.5	513.2d	5.4.2.4	513.15g	5.4.13.7
503.6	5.3.1.11.6	513.2e	5.4.2.5	513.15h	5.4.13.8
503.7	5.3.1.11.7	513.3	5.4.10	513.15i	5.4.13.9
504	–	513.3a	5.4.10.1	513.16	5.4.14
504.1	5.3.1.15	513.3b	5.4.10.2	513.16a	5.4.14.1
505	–	513.3c	5.4.10.3	513.16b	5.4.14.2
505.1	5.3.1.14	513.3d	5.4.10.4	513.16c	5.4.14.3
506	5.3.1.16	513.4	5.4.3	513.17	5.4.15
506.1	5.3.1.16.1	513.4a	5.4.3.1	513.17a	5.4.15.1
506.1a	5.3.1.16.1(a)	513.4b	5.4.3.2	513.17b	5.4.15.2
506.1b	5.3.1.16.1(b)	513.4c	5.4.3.3	513.17c	5.4.15.3
506.1c	5.3.1.16.1(c)	513.4d	5.4.3.4	513.17d	5.4.15.4
506.2	5.3.1.16.2	513.5	5.4.6	513.18	5.4.8
506.2a	5.3.1.16.2(a)	513.5a	5.4.6.1	513.18a	5.4.8.1
506.2b	5.3.1.16.2(b)	513.5b	5.4.6.2	513.18b	5.4.8.2
506.2c	5.3.1.16.2(c)	513.6	5.4.6.3	513.18c	5.4.8.3
506.2d	5.3.1.16.2(d)	513.7	5.4.5	513.18d	5.4.8.4
506.2e	5.3.1.16.2(e)	513.7a	5.4.5.1	513.18e	5.4.8.5
506.2f	5.3.1.16.2(f)	513.7b	5.4.5.2	513.18f	5.4.8.6
506.2g	5.3.1.16.2(g)	513.7c	5.4.5.3	513.18g	5.4.8.7
506.2h	5.3.1.16.2(h)	513.7d	5.4.5.4	513.18h	5.4.8.8
506.2i	5.3.1.16.2(i)	513.8	5.4.4.1	513.18i	5.4.8.9
506.3	5.3.1.16.3	513.8a	5.4.4.1.1	513.19	5.4.10
506.4	5.3.1.16.4	513.8b	5.4.4.1.2	513.20	5.4.15.5
507	5.3.1.17	513.8c	5.4.4.1.3	513.20a	5.4.15.5.1
507.1	5.3.1.17.1	513.9	5.4.4.2	513.20b	5.4.15.5.2
507.2	5.3.1.17.2	513.9a	5.4.4.2.1	513.21	5.4.12
508	5.3.1.18	513.9b	5.4.4.2.2	513.22	5.4.13.10

A17.1-1996 With Addenda	A17.1-2000	A17.1-1996 With Addenda	A17.1-2000	A17.1-1996 With Addenda	A17.1-2000
802.5	6.1.3.5	806.1a	6.1.7.1.1	904.3	6.2.5.3
802.5a	6.1.3.5.1	806.1b	6.1.7.1.2	904.3a	6.2.5.3.1
802.5b	6.1.3.5.2	806.2	6.1.7.2	904.3b	6.2.5.3.2
802.5c	6.1.3.5.3	806.3	6.1.7.3	905	6.2.6
802.5d	6.1.3.5.5	806.4	6.1.7.4	905.1	6.2.6.1
802.5e	6.1.3.5.8	807	6.1.8	905.2	6.2.6.2.1
802.6	6.1.3.6	807.1	6.1.8.1	905.3	6.2.6.3
802.6a	6.1.3.6.1	807.2	6.1.8.2	905.3a	6.2.6.3.1
802.6b	6.1.3.6.2	807.3	6.1.8.3	905.3b	6.2.6.3.2
802.6c	6.1.3.6.3	808	8.9	905.3c	6.2.6.3.3
802.6d	6.1.3.6.4	Part IX	6.2	905.3d	6.2.6.3.4
802.6e	6.1.3.6.5	900	6.2.1	905.3e	6.2.6.3.5
802.7	6.1.3.7	900.1	6.2.1.1	905.3f	6.2.6.3.6
802.8	6.1.3.8	901	6.2.2	905.3g	6.2.6.3.7
802.9	6.1.3.9	901.1	6.2.2.1	905.3h	6.2.6.3.8
802.9a	6.1.3.9.1	902	6.2.3	905.3i	6.2.6.3.9
802.9b	6.1.3.9.2	902.1	6.2.3.1	905.3j	6.2.6.3.10
802.9c	6.1.3.9.3	902.2	6.2.3.2	905.3k	6.2.6.3.11
802.9d	6.1.3.9.4	902.3	6.2.3.3	905.4	6.2.6.4
802.10	6.1.3.10	902.3a	6.2.3.3.1	905.5	6.2.6.5
802.11	6.1.3.11	902.3b	6.2.3.3.2	905.6	6.2.6.6
802.12	6.1.3.12	902.3c	6.2.3.3.3	905.7	6.2.6.7
803	6.1.4	902.3d	6.2.3.3.4	905.8	6.2.6.8
803.1	6.1.4.1	902.3e	6.2.3.3.5	905.8a	6.2.6.8.1
804	6.1.5	902.3f	6.2.3.3.6	905.8b	6.2.6.8.2
804.1	6.1.5.1	902.3g	6.2.3.3.7	905.9	6.2.6.9
804.2	6.1.5.2	902.3h	6.2.3.3.8	905.10	6.2.6.11
804.3	6.1.5.3	902.4	6.2.3.4	905.11	6.2.6.12
804.3a	6.1.5.3.1	902.4a	6.2.3.4.1	905.12	6.2.6.13
804.3b	6.1.5.3.2	902.4b	6.2.3.4.2	906	6.2.7
805	6.1.6	902.4c	6.2.3.4.3	906.1	6.2.7.1
805.1	6.1.6.1	902.5	6.2.3.5	906.1a	6.2.7.1.1
805.2	6.1.6.2.1	902.5a	6.2.3.5.1	906.1b	6.2.7.1.2
805.3	6.1.6.3	902.5b	6.2.3.5.2	906.2	6.2.7.2
805.3a	6.1.6.3.1	902.5c	6.2.3.5.3	906.3	6.2.7.3
805.3b	6.1.6.3.2	902.5d	6.2.3.5.4	906.4	6.2.7.4
805.3c	6.1.6.3.3	902.5e	6.2.3.5.5	907	6.2.8
805.3d	6.1.6.3.4	902.6	6.2.3.6	907.1	6.2.8.1
805.3e	6.1.6.3.5	902.6a	6.2.3.6.1	907.2	6.2.8.2
805.3f	6.1.6.3.6	902.6b	6.2.3.6.2	907.3	6.2.8.3
805.3g	6.1.6.3.7	902.7	6.2.3.7	908	8.9
805.3h	6.1.6.3.8	902.8	6.2.3.8	Part X	8.10, 8.11
805.3i	6.1.6.3.9	902.8a	6.2.3.8.1	1000	8.10.1, 8.11.1
805.3j	6.1.6.3.10	902.8b	6.2.3.8.2	1000.1	8.10.1.1(c), 8.11.1.1
805.3k	6.1.6.3.11	902.8c	6.2.3.8.4	1000.1a	8.11.1.1.1, 8.11.1.1.2
805.3m	6.1.6.3.12	902.8d	6.2.3.8.5	1000.1b	8.11.1.1.1, 8.11.1.1.2
805.3n	6.1.6.3.13	902.9	6.2.3.9	1000.1c	8.10.1.1
805.3p	6.1.6.3.14	902.9a	6.2.3.9.1	1000.2	8.10.1.2, 8.11.1.2
805.4	6.1.6.4	902.10	6.2.3.10	1000.3	8.11.1.4
805.5	6.1.6.5	902.10a	6.2.3.10.1	1001	8.11.2
805.6	6.1.6.6	902.10b	6.2.3.10.2	1001.1	8.11.2.1
805.7	6.1.6.7	902.10c	6.2.3.10.3	1001.2	–
805.8	6.1.6.8	902.11	6.2.3.10.4	1001.2a	8.11.2.1.1
805.9	6.1.6.9	902.12	6.2.3.11	1001.2b	8.11.2.1.2
805.9a	6.1.6.9.1	902.13	6.2.3.12	1001.2c	8.11.2.1.3
805.9b	6.1.6.9.2	902.14	6.2.3.13	1001.2d	8.11.2.1.4
805.10	6.1.6.10	902.15	6.2.3.14	1001.2e	8.11.2.1.5
805.11	6.1.6.12	902.16	6.2.3.15	1001.2f	8.11.2.1.6
805.12	6.1.6.13	903	6.2.4	1002.1	8.11.2.2
805.13	6.1.6.14	904	6.2.5	1002.2a	8.11.2.2.1
806	6.1.7	904.1	6.2.5.1	1002.2b	8.11.2.2.2
806.1	6.1.7.1	904.2	6.2.5.2	1002.2c	8.11.2.2.3

A17.1-1996 With Addenda	A17.1-2000	A17.1-1996 With Addenda	A17.1-2000	A17.1-1996 With Addenda	A17.1-2000
1002.2d	8.11.2.2.4	1006.2h	8.10.3.2.3(r)	1011	8.10.5.10, 8.11.5.13
1002.2e	8.11.2.2.5	1006.3	8.10.3.3.2	1011.1	–
1002.2f	8.11.2.2.6	1006.4	8.10.3.3.2(o)	1011.2	8.11.5.13.1
1002.2g	8.11.2.2.7	1007	8.11.4	1011.3	8.11.5.13.2
1002.2h	8.11.2.2.8	1007.1	–	1011.4	8.11.5.13.2
1002.2i	8.11.2.2.9	1007.2	8.11.4.1	1011.5	8.11.5.13.3
1002.3	8.11.2.3	1008	8.11.4.2	1011.6	8.11.5.13.4
1002.3a	8.11.2.3.1	1008.1	8.11.4.2	1011.7	8.10.5.10
1002.3b	8.11.2.3.2	1008.2	–	1011.8	8.10.5.10
1002.3c	8.11.2.3.3	1008.2a	8.11.4.2.1	Part XI	8.3
1002.3d	8.11.2.3.4	1008.2b	8.11.4.2.2	1100	8.3.2
1002.3e	8.11.2.3.5	1008.2c	8.11.4.2.3	1100.1	8.3.2.1
1002.3f	8.11.2.3.6	1008.2d	8.11.4.2.4	1100.2	8.3.1.2
1002.3g	8.11.2.3.7	1008.2e	8.11.4.2.5	1100.2a	8.3.1.2.2
1002.3h	8.11.2.3.8	1008.2f	8.11.4.2.6	1100.2b	8.3.1.3.3
1002.3i	8.11.2.3.9	1008.2g	8.11.4.2.7	1100.2c	8.3.1.4.1
1003	8.10.2	1008.2h	8.11.4.2.8	1100.2d	8.3.1.3.4
1003.1	8.10.2.1	1008.2i	8.11.4.2.9	1100.3	8.3.2.3
1003.2	8.10.2.2	1008.2j	8.11.4.2.10	1100.4	8.3.2.4
1003.2a	8.10.2.2.2(cc)(1)	1008.2k	8.11.4.2.11	1100.4a	8.3.2.4.1
1003.2b	8.10.2.2.2(cc)(2)	1008.2l	8.11.4.2.12	1100.4b	8.3.2.4.2
1003.2c	8.10.2.2.2(cc)(3)	1008.2m	8.11.4.2.13	1100.4c	8.3.2.4.3
1003.2d	8.10.2.2.2(cc)(4)	1008.2n	–	1100.4d	8.3.2.4.4
1003.2e	8.10.2.2.5(c)	1008.2o	–	1100.5	8.3.2.5
1003.2f	8.10.2.2.2(z)	1008.2p	8.11.4.2.14	1100.5a	8.3.2.5.1
1003.2g	8.10.2.2.3(w)(3)	1008.2q	8.11.4.2.18	1100.5b	8.3.2.5.2
1003.2h	8.10.2.2.2(m)	1008.2r	8.11.4.2.13	1100.5c	8.3.2.5.3
1003.2i	8.10.2.2.2(m)(3)	1008.2s	8.11.4.2.19	1100.5d	8.3.2.5.4
1003.2j	8.10.2.2.2(t)	1008.2t	8.11.4.2.20	1100.5e	8.3.2.5.5
1003.2k	8.10.2.2.2(u)	1009	8.10.4	1100.5f	8.3.2.6
1003.3	8.10.2.3.2	1009.1	8.10.4.1	1101	8.3.3
1004	8.11.3	1009.2	8.10.4.1.1, 8.10.4.1.3	1101.1	8.3.3.1
1004.1	8.11.3.1	1009.2a	8.10.4.1.1(a)	1101.2	8.3.3.2
1004.2	–	1009.2b	8.10.4.1.1(b)	1101.3	8.3.3.3
1004.2a	8.11.3.1.1	1009.2c	8.10.4.1.1(c)	1101.3a	8.3.3.3.1
1004.2b	8.11.3.1.2	1009.2d	8.10.4.1.1(e)	1101.3b	8.3.3.3.2
1004.2c	8.11.3.1.3	1009.2e	8.10.4.1.1(g)	1101.3c	8.3.3.3.3
1004.2d	–	1009.2f	8.10.4.1.1(h)	1101.3d	8.3.3.3.4
1004.2e	8.11.3.1.5	1009.2g	8.10.4.1.1(n)	1101.4	8.3.3.4
1005	8.11.3.2	1009.2h	8.10.4.1.2(q)	1102	8.3.4
1005.1	–	1009.2i	8.10.4.1.1(q)	1102.1	8.3.4.1
1005.2	–	1009.2j	8.10.4.1.2(a)	1102.2	8.3.4.1
1005.2a	8.11.3.2.1	1009.2k	8.10.4.1.2(b)	1103	8.3.10
1005.2b	8.11.3.2.2	1009.2l	8.10.4.1.2(c)	1103.1	8.3.10.1
1005.2c	8.11.3.2.3	1009.2m	8.10.4.1.2(e)	1103.2	8.3.10.2
1005.2d	8.11.3.2.4	1009.2n	8.10.4.1.2(k)	1103.3	8.3.10.3
1005.2e	8.11.3.2.5	1009.2o	8.10.4.1.2(l)	1104	8.3.7
1005.2f	–	1009.2p	8.10.4.1.2(o)	1104.1	8.3.7.1
1005.3	8.11.3.3	1009.3	8.10.4.2	1104.2	8.3.7.2
1005.3a	8.11.3.3.1	1010	8.10.5, 8.11.5	1104.3	8.3.7.3
1005.3b	8.11.3.3.2	1010.1	8.10.5.1, 8.11.5.1	1104.4	8.3.7.4
1005.4	8.11.3.4	1010.2	8.10.5.2, 8.11.5.2	1104.5	8.3.7.5
1006	8.10.3	1010.3	8.10.5.3, 8.11.5.3	1104.6	8.3.7.6
1006.1	8.10.3.1	1010.4	8.10.5.4, 8.11.5.4	1105	8.3.11
1006.2	8.10.3.2	1010.5	8.10.5.5, 8.11.5.5	1105.1	8.3.11.1
1006.2a	8.10.3.2.2(s)	1010.6	8.10.5.6, 8.11.5.6	1106	8.3.8
1006.2b	8.10.3.2.2(m)	1010.7	8.10.5.7, 8.11.5.7	1107	8.3.5
1006.2c	8.10.3.2.3(d)	1010.8	8.10.5.8, 8.11.5.8	1107.1	8.3.5.1
1006.2d	8.10.3.2.5(b)	1010.9	8.10.5.9, 8.11.5.9	1107.2	8.3.5.2
1006.2e	8.10.3.2.5(c)	1010.11	8.10.5.11, 8.11.5.10	1107.3	8.3.5.3
1006.2f	8.10.3.2.2(q)	1010.12	8.10.5.12, 8.11.5.11	1107.3a	8.3.5.3.1
1006.2g	8.10.3.2.3(cc)	1010.13	8.10.5.13, 8.11.5.12	1107.3b	8.3.5.3.2

A17.1-1996 With Addenda	A17.1-2000	A17.1-1996 With Addenda	A17.1-2000	A17.1-1996 With Addenda	A17.1-2000
1107.3c	8.3.5.3.3	1202.4b	8.7.2.15.2	1203.3e	8.7.3.23.5
1107.3d	8.3.5.3.4	1202.5	8.7.2.14	1203.3f	8.7.3.23.6
1107.3e	8.3.5.3.5	1202.6	8.7.2.18	1203.4	8.7.3.22
Part XII	8.7	1202.7	8.7.2.19	1203.4a	8.7.3.22.1
1200	8.7.1	1202.8	8.7.2.16	1203.4b	8.7.3.22.2
1200.1	8.7.1.1	1202.8a	8.7.2.16.1	1203.4c	8.7.3.22.3
1200.2	8.7.1.2	1202.8b	8.7.2.16.2	1203.5	8.7.3.24
1200.3	8.7.1.3	1202.8c	8.7.2.16.3	1203.6	8.7.3.29
1200.4	8.6.2	1202.8d	8.7.2.16.4	1203.7	8.7.3.30
1200.4a	8.6.2.1	1202.9	8.7.2.25	1203.8	8.7.3.31
1200.4b	8.6.2.2	1202.9a	8.7.2.25.1	1203.8a	8.7.3.31.1
1200.4c	8.6.2.2	1202.9b	8.7.2.25.2	1203.8b	8.7.3.31.2
1200.4d	8.6.2.5	1202.10	8.7.2.17	1203.8c	8.7.3.31.3
1200.4e	8.6.3.5	1202.10a	8.7.2.17.1	1203.8d	8.7.3.31.4
1200.4f	8.6.3.7	1202.10b	8.7.2.17.2	1203.8e	8.7.3.31.5
1200.5	8.6.3	1202.10c	8.7.2.17.3	1203.8f	8.7.3.31.6
1200.5a	8.6.3.2, 8.6.3.3, 8.6.3.4	1202.11	8.7.2.26	1203.8g	8.7.3.31.7
		1202.12	8.7.2.27	1203.8h	8.7.3.31.8
1200.5b	–	1202.12a	8.7.2.27.1	1203.9	8.7.3.25
1200.5c	8.6.3.7	1202.12b	8.7.2.27.2	1203.9a	8.7.3.25.1
1200.5d	8.6.3.8	1202.12c	8.7.2.27.3	1203.9b	–
1200.6	8.7.1.8	1202.12d	8.7.2.27.4	1203.9c	8.7.3.25.2
1201	8.7.2	1202.12e	8.7.2.27.5	1204	8.7.4.3
1201.1	8.7.2.1	1202.12f	8.7.2.27.6	1204.1	–
1201.1a	8.7.2.1.1	1202.13	8.7.2.28	1205	–
1201.1b	8.7.2.1.2	1202.14	8.7.2.21	1205.1	8.7.7.2
1201.1c	8.7.2.1.3	1202.14a	8.7.2.21.1	1206	8.6
1201.1d	8.7.2.1.4	1202.14b	–	1206.1	8.6.1.2
1201.1e	8.7.2.1.5	1202.14c	8.7.2.21.2	1206.1a	8.6.1.6.2
1201.2	8.7.2.7	1202.14d	8.7.2.21.3	1206.1b	8.6.4.1
1201.2a	8.7.2.7.1	1203	8.7.3	1206.1c	8.6.4.2
1201.2b	8.7.2.7.2	1203.1	–	1206.1d	8.6.4.3
1201.2c	8.7.2.7.3	1203.1a	8.7.3.1	1206.1e	8.6.4.4
1201.2d	8.7.2.7.4	1203.1b	8.7.3.7	1206.1f	8.6.1.6.3
1201.2e	8.7.2.7.5	1203.1c	8.7.3.9	1206.1g	8.6.4.5
1201.2f	8.7.2.7.6	1203.1d	8.7.3.3	1206.1h	8.6.1.6.5
1201.2g	8.7.2.7.7	1203.1e	8.7.3.8	1206.1i	8.6.1.5
1201.3	8.7.2.8	1203.1f	8.7.3.2	1206.2	–
1201.4	8.7.2.3	1203.1g	8.7.3.4	1206.2a	8.6.4.7
1201.5	8.7.2.9	1203.1h	8.7.3.5	1206.2b	8.6.4.8
1201.6	8.7.2.2	1203.1i	8.7.3.6	1206.2c	8.6.4.9
1201.7	8.7.2.4	1203.1j	8.7.3.10	1206.2d	8.6.1.6.4
1201.8	8.7.2.5	1203.1k	8.7.3.11	1206.2e	–
1201.9	8.7.2.6	1203.1m	8.7.3.12	1206.3	8.6.4.10
1201.10	8.7.2.10	1203.2	–	1206.3a	8.6.4.10.1
1201.10a	8.7.2.10.1	1203.2a	8.7.3.28	1206.3b	8.6.4.10.2
1201.10b	8.7.2.10.2	1203.2b	8.7.3.27	1206.3c	8.6.4.10.3
1201.10c	8.7.2.10.3	1203.2c	8.7.3.26	1206.4	8.6.1.6.1
1201.10d	8.7.2.10.4	1203.2d	8.7.3.14	1206.5	8.6.5, 8.6.9
1201.10e	8.7.2.10.5	1203.2e	8.7.3.13	1206.5a	8.6.5
1201.11	8.7.2.11	1203.2f	8.7.3.15	1206.5b	8.6.5.1, 8.6.5.2, 8.6.5.5, 8.6.5.6
1201.11a	8.7.2.11.1	1203.2g	8.7.3.16		
1201.11b	8.7.2.11.2	1203.2h	8.7.3.17	1206.6	8.6.8
1201.11c	8.7.2.11.3	1203.2i	8.7.3.18	1206.6a	8.6.8.14
1201.11d	8.7.2.11.4	1203.2j	8.7.3.19	1206.6b	8.6.8.5
1201.12	8.7.2.12	1203.2k	8.7.3.20	1206.6c	8.6.8.3
1202	–	1203.2m	8.7.3.21	1206.7	8.6.10.1
1202.1	8.7.2.24	1203.3	8.7.3.23	1206.8	8.6.4.11
1202.2	8.7.2.23	1203.3a	8.7.3.23.1	1206.9	8.6.10.3
1202.3	8.7.2.22	1203.3b	8.7.3.23.2	1206.10	8.6.7.3
1202.4	8.7.2.15	1203.3c	8.7.3.23.3	1207	8.7.6.1
1202.4a	8.7.2.15.1	1203.3d	8.7.3.23.4	1207.1	8.7.6.1.1

A17.1-1996 With Addenda	A17.1-2000	A17.1-1996 With Addenda	A17.1-2000	A17.1-1996 With Addenda	A17.1-2000
1207.2	8.7.6.1.2	Part XIII	8.2	1406.1j	7.4.8
1207.3	8.7.6.1.3	1300	–	1406.1k	7.4.13
1207.4	8.7.6.1.4	1300.1	8.2.1	1406.1l	7.4.14
1207.5	8.7.6.1.5	1301	8.2.2	1406.1m	7.4.15
1207.6	8.7.6.1.6	1301.1	8.2.2.1	1406.2	7.5
1207.7	8.7.6.1.7	1301.1a	8.2.2.1.1	1406.2a	7.5.9
1207.8	8.7.6.1.8	1301.2	8.2.2.2	1406.2b	7.5.8
1207.9	8.7.6.1.9	1301.3	8.2.2.3	1406.2c	7.5.7
1207.10	8.7.6.1.10	1301.4	8.2.2.4	1406.2d	7.5.2
1207.11	8.7.6.1.11	1301.5	8.2.2.5	1406.2e	7.5.1
1207.12	8.7.6.1.12	1301.6	8.2.2.6	1406.2f	7.5.4
1207.13	8.7.6.1.13	1301.7	8.2.2.7	1406.2g	7.5.5
1207.14	8.7.6.1.14	1302	8.2.8	1406.2h	7.5.3
1207.15	8.7.6.1.15	1302.1	8.2.8.1	1406.2i	7.5.10
1208	8.7.6.2	1302.2	8.2.8.2	1406.2j	7.5.11
1208.1	8.7.6.2.1	1302.3	8.2.8.3	1406.2k	7.5.12
1208.2	8.7.6.2.2	1302.4	8.2.8.4	1406.2l	–
1208.3	8.7.6.2.3	1302.5	8.2.8.5	1406.2m	7.5.6
1208.4	8.7.6.2.4	1302.5a	8.2.8.5.1	1406.2n	7.5.14
1208.5	8.7.6.2.5	1302.5b	8.2.8.5.2	1406.2o	7.5.13
1208.6	8.7.6.2.6	1303	8.2.9	1406.3	7.6
1208.7	8.7.6.2.7	1303.1	8.2.9.1	Part XV	5.7
1208.8	8.7.6.2.8	1303.1a	8.2.9.1.1	1500	5.7.1
1208.9	8.7.6.2.9	1303.1b	8.2.9.1.2	1500.1	5.7.1.1
1208.10	8.7.6.2.10	1303.1c	8.2.9.1.3	1500.2	5.7.5
1208.11	8.7.6.2.11	1303.1d	8.2.9.1.4	1500.3	–
1208.12	8.7.6.2.12	1304	8.2.4	1500.3a	5.7.2
1208.13	8.7.6.2.13	1305	8.2.5	1500.3b	5.7.6
1208.14	8.7.6.2.14	1306	8.2.6	1500.3c	5.7.4.1
1209	8.7.7.3	1307	8.2.7	1500.3d	5.7.4.2
1210	8.7.5.7	1308	8.2.3	1500.3e	5.7.1.2
1211	8.7.4.1	1308.1	8.2.3.1	1500.4	5.7.7
1212	8.7.5.1	1308.2	8.2.3.2	1500.4a	5.7.7.1
1213	8.7.4.2	1309	8.2.10	1500.4b	5.7.7.2
1214	–	1310	8.2.11	1500.4c	5.7.7.3
1215	–	Part XIV	Part 7	1500.4d	5.7.7.4
1216	8.7.5.8	1400	7.9	1500.5	5.7.3
1217	8.7.5.6	1400.1	7.9.1	1500.5a	5.7.3.1
1218	8.7.7.1	1400.2	7.9.2	1500.5b	5.7.3.2
1218.1	8.7.7.1.1	1400.3	7.10	1500.6	5.7.8
1218.2	8.7.7.1.1	1401	7.12	1500.6a	5.7.8.1
1218.2	8.7.7.1.2	1402	7.8	1500.6b	5.7.8.2
1219	8.7.4.3	1402.1	7.8.1	1500.6c	5.7.8.3
1219.1	8.7.4.3.1	1402.2	7.8.2	1500.6d	5.7.8.4
1219.2	8.7.4.3.2	1402.3	7.8.3	1500.6e	5.7.8.5
1219.3	8.7.4.3.3	1402.4	7.8.4	1500.6f	5.7.8.6
1219.4	8.7.4.3.4	1403	7.7	1501	5.7.19
1219.5	8.7.4.3.5	1403.1	7.7.1, 7.7.2, 7.7.3, 7.7.4	1501.1	–
1219.6	8.7.4.3.6			1501.2	5.7.9
1219.7	8.7.4.3.7	1404	–	1501.3	5.7.20
1219.8	8.7.4.3.8	1405	8.2.12	1501.3a	5.7.20.1
1219.9	8.7.4.3.9	1406	7.4	1501.3b	5.7.20.2
1219.10	8.7.4.3.10	1406.1	7.4.3	1502	–
1220	8.7.5.5	1406.1a	7.4.3	1502.1	5.7.17
1220.1	8.7.5.5.5	1406.1b	7.4.9	1502.1a	5.7.17.1
1220.2	8.7.5.5.6	1406.1c	7.4.10	1502.1b	5.7.17.2
1220.3	8.7.5.5.2	1406.1d	7.4.5	1502.1c	5.7.17.3
1220.4	8.7.5.5.1	1406.1e	7.4.12	1502.2	5.7.16
1220.5	8.7.5.5.8	1406.1f	7.4.11	1502.3	5.7.15
1220.6	8.7.5.5.3	1406.1g	7.4.4	1502.3a	5.7.15.1
1220.7	8.7.5.5.2	1406.1h	7.4.6	1502.3b	5.7.15.2
1220.8	8.7.5.5.4	1406.1i	7.4.7	1502.3c	5.7.15.3

A17.1-1996 With Addenda	A17.1-2000	A17.1-1996 With Addenda	A17.1-2000	A17.1-1996 With Addenda	A17.1-2000
1502.4	5.7.11	1606.1	4.1.15.1	1711.1	5.1.13.1
1502.4a	5.7.11.1	1606.2	4.1.15.2	1711.2	5.1.13.2
1502.4b	5.7.11.2	1607	–	1712	5.1.19
1502.4c	5.7.11.3	1607.1	4.1.16	1712.1	5.1.19
1502.4d	5.7.11.4	1608	–	1712.2	5.1.19.1
1502.5	5.7.10	1608.1	4.1.17	1713	5.1.20
1502.5a	5.7.10.1	1609	8.9	1713.1	–
1502.5b	5.7.10.2	1609.1	8.9	1713.2	5.1.20.5
1502.5c	5.7.10.3	Part XVII	5.1	1713.3	5.1.20.6
1502.5d	5.7.10.4	1700	–	1714	5.1.21
1502.6	5.7.10.5	1700.1	5.1.1	1714.1	5.1.21.1
1502.7	5.7.13	1700.1a	5.1.2.1	1714.2	–
1502.7a	5.7.13.1	1700.1b	5.1.2.2	1715	5.1.16
1502.7b	5.7.13.2	1700.1c	–	1715.1	5.1.16.1
1502.7c	5.7.13.3	1700.2	–	1716	5.1.22
1502.7d	5.7.13.4	1701	5.1.3	1716.1	5.1.22.1
1502.7e	5.7.13.5	1701.1	5.1.3.1	1716.2	5.1.22.2
1502.8	5.7.12.1	1701.2	5.1.4	1716.3	5.1.22.3
1502.9	5.7.12.2	1702	5.1.5	1716.4	5.1.22.4
1502.10	5.7.18	1702.1	5.1.5.1	1716.5	5.1.22.5
1502.10a	5.7.18.1	1702.2	5.1.5.2	1716.6	5.1.22.6
1502.10b	5.7.18.2	1703	5.1.8	1717	5.1.7
1502.10c	5.7.18.3	1703.1	5.1.8.1	1717.1	5.1.7.1
1502.10d	5.7.18.4	1703.2	–	1717.2	5.1.7.2
1502.10e	5.7.18.5	1703.3	5.1.8.2	1800	–
1502.10f	5.7.18.6	1703.4	–	1800.1	4.2.1
1502.10g	5.7.18.7	1703.5	–	1800.2	4.2.5
1502.10h	5.7.18.8	1704	5.1.10	1800.3	4.2.2
1502.10i	5.7.18.9	1704.1	5.1.10.1	1800.3a	4.2.2.1
1502.11	5.7.14	1705	5.1.1.2	1800.3b	4.2.2.2
1502.11a	5.7.4.1	1706	5.1.18	1800.3c	4.2.2.3
1502.11b	5.7.4.2	1706.1	5.1.18.1	1800.3d	4.2.2.4
1502.11c	5.7.4.3	1706.2	5.1.18.2	1800.4	4.2.3
1502.11d	5.7.4.4	1706.3	5.1.18.3	1800.5	4.2.4
1502.11e	5.7.4.5	1707	5.1.17	1801	–
1502.11f	5.7.4.6	1707.1	5.1.17.1	1801.1	4.2.6
1502.11g	5.7.4.7	1707.2	5.1.17.2	1802	–
1502.11h	5.7.4.8	1707.3	5.1.17.3	1802.1	4.2.7
1502.12	5.7.21	1707.4	5.1.17.4	1803	–
1502.13	8.9	1708	5.1.12	1803.1	4.2.14
1600	–	1708.1	5.1.12.1	1803.2	4.2.13
1600.1	4.1.1	1708.2	5.1.12.2	1803.3	4.2.9
1600.2	4.1.2	1708.3	5.1.11.1	1803.4	4.2.8
1600.3	4.1.5	1708.3a	5.1.11.1.1	1803.5	4.2.11
1601	4.1.3	1708.3b	5.1.11.1.2	1803.6	4.2.12
1602	4.1.4	1708.3c	5.1.11.1.3	1803.7	4.2.10
1603	–	1708.4	5.1.11.2	1804	–
1603.1	4.1.12	1708.5	5.1.11.3	1804.1	4.2.15
1603.2	4.1.11	1709	–	1804.2	4.2.20
1603.3	4.1.10	1709.1	–	1805	4.2.16
1603.4	4.1.7	1709.2	–	1805.1	4.2.16.1
1603.5	4.1.6	1709.3	–	1805.2	4.2.16.2
1603.6	4.1.9	1710	5.1.14	1805.3	4.2.16.3
1603.6a	4.1.9.1	1710.1	5.1.14.1	1806	4.2.17
1603.7	4.1.8	1710.2	5.1.14.2	1806.1	4.2.17.1
1604	–	1710.3	5.1.14.3	1807	4.2.18
1604.1	4.1.13	1710.3a	5.1.14.3.1	1807.1	–
1604.2	4.1.18	1710.3b	5.1.14.3.2	1808	–
1605	4.1.14	1710.4	5.1.15	1808.1	4.2.19
1605.1	4.1.14.1	1710.4a	5.1.15.1	1809	8.9
1605.2	4.1.14.2	1710.4b	5.1.15.2	1809.1	–
1606	4.1.15	1711	5.1.13	Part XIX	5.10

A17.1-1996 With Addenda	A17.1-2000	A17.1-1996 With Addenda	A17.1-2000	A17.1-1996 With Addenda	A17.1-2000
1900	5.10.1	1902.11g	5.10.1.20.7	2300.11	5.6.1.11
1900.1	5.10.1.1	1902.12	5.10.1.16	2300.11a	5.6.1.11.1
1900.1a	5.10.1.1.1	1902.12a	5.10.1.16.1	2300.11b	5.6.1.11.2
1900.1b	5.10.1.1.2	1902.12b	5.10.1.16.2	2300.11c	5.6.1.11.3
1900.1c	5.10.1.7.1	1902.12c	5.10.1.16.3	2300.11d	5.6.1.11.4
1900.1d	5.10.1.7.2	1902.12d	5.10.1.16.4	2300.11e	5.6.1.11.5
1900.1e	5.10.1.9.2	1902.12e	5.10.1.16.5	2300.12	5.6.1.12
1900.1f	5.10.1.22	1902.12f	5.10.1.16.6	2300.13	5.6.1.13
1900.2	5.10.1.5	1902.12g	5.10.1.16.7	2301	–
1900.3	–	1902.12h	5.10.1.16.8	2301.1	5.6.1.22
1900.3a	5.10.1.2	1903	5.10.2	2301.2	5.6.1.21
1900.3b	5.10.1.6	1903.1	5.10.2.1	2301.3	5.6.1.20
1900.3c	5.10.1.4	1903.1a	5.10.2.2	2301.4	5.6.1.15
1900.4	5.10.1.8	1903.1b	5.10.2.3	2301.4a	5.6.1.15.1
1900.5	5.10.1.3	1903.1c	5.10.2.4	2301.4b	5.6.1.15.2
1900.5a	5.10.1.3.1	1903.1d	5.10.2.5	2301.5	5.6.1.14
1900.5b	5.10.1.3.2	1903.2	5.10.2.6	2301.6	5.6.1.17
1900.5c	5.10.1.3.3	1903.3	5.10.2.7	2301.7	5.6.1.18
1900.6	5.10.1.9	1903.4	5.10.2.8	2301.8	5.6.1.16
1900.6a	5.10.1.9.1	1903.5	5.10.2.10	2301.9	5.6.1.23
1900.6b	5.10.1.9.2	1903.6	5.10.2.11	2301.10	5.6.1.24
1900.6c	5.10.1.9.3	1903.7	5.10.2.12	2301.11	5.6.1.25
1900.6d	5.10.1.9.4	1903.8	5.10.2.9	2301.11a	5.6.1.25.1
1900.6e	5.10.1.9.5	Part XXII	5.8	2301.11b	5.6.1.25.2
1900.6f	5.10.1.9.6	2200	5.8.1	2301.11c	5.6.1.25.3
1900.6g	5.10.1.9.7	2200.1	5.8.1	2301.11d	5.6.1.25.4
1901	5.10.1.21	2200.2	5.8.1.1	2301.11e	5.6.1.25.5
1901.1	5.10.1.21.1	2200.3	5.8.1.2	2301.12	5.6.1.26
1901.2	5.10.1.21.2	2200.4	5.8.1.3	2301.13	5.6.1.19
1901.3	5.10.1.21.3	2200.5	5.8.1.4	2301.14	5.6.1.27
1902	–	2201	–	2301.15	8.9
1902.1	5.10.1.19	2201.1	5.8.1	2302	5.6.2
1902.2	5.10.1.18	2201.2	5.8.1.5	2302.1	5.6.2.1
1902.3	5.10.1.17	2201.3	5.8.1.6	2302.1a	5.6.2.4
1902.4	5.10.1.11	2201.4	5.8.1.7	2302.1b	5.6.2.2
1902.5	5.10.1.10	2201.5	5.8.1.8	2302.1c	5.6.2.3
1902.5a	5.10.1.10.1	2201.6	5.8.1.9	2302.1d	5.6.2.5
1902.5b	5.10.1.10.2	2201.7	5.8.1.10	2302.2	–
1902.5c	5.10.1.10.3	2201.8	5.8.1.11	2302.2a	5.6.2.14
1902.5d	5.10.1.10.4	2202	5.8.2	2302.2b	5.6.2.13
1902.5e	5.10.1.10.5	2202.1	5.8.2	2302.2c	5.6.2.12
1902.5f	5.10.1.10.6	2202.2	5.8.2.2	2302.2d	5.6.2.7
1902.6	5.10.1.10.7	2202.3	5.8.2.1	2302.2e	5.6.2.6
1902.7	5.10.1.10.8	2202.4	5.8.2.3	2302.2f	5.6.2.9
1902.8	–	2202.5	5.8.2.4	2302.2g	5.6.2.8
1902.8a	5.10.1.13	2203	5.8.3	2302.3	5.6.2.10
1902.8b	5.10.1.14	2203.1	5.8.3	2302.4	5.6.2.11
1902.9	5.10.1.23	2203.2	5.8.3.1	2302.5	5.6.2.15
1902.9a	5.10.1.23.1	2203.3	5.8.3.2	2302.6	5.6.2.16
1902.9b	5.10.1.23.2	2203.4	5.8.3.3	2302.7	5.6.2.17
1902.9c	5.10.1.23.3	Part XXIII	5.6	2302.7a	5.6.2.17.1
1902.10	5.10.1.12	2300	5.6.1	2302.8	8.9
1902.10a	5.10.1.12.1	2300.1	5.6.1.1	Part XXIV	8.4
1902.10b	5.10.1.12.2	2300.2	5.6.1.7	2400	8.4.3
1902.10c	5.10.1.12.3	2300.3	5.6.1.8	2400.1	8.4.3.1
1902.11	5.10.1.20	2300.4	5.6.1.3	2400.2	8.4.3.2
1902.11a	5.10.1.20.1	2300.5	5.6.1.10	2401	8.4.2
1902.11b	5.10.1.20.2	2300.6	5.6.1.9	2401.1	8.4.2.1
1902.11c	5.10.1.20.3	2300.7	5.6.1.2	2401.2	8.4.2.2
1902.11d	5.10.1.20.4	2300.8	5.6.1.4	2401.3	8.4.2.3
1902.11e	5.10.1.20.5	2300.9	5.6.1.5	2402	8.4.1
1902.11f	5.10.1.20.6	2300.10	5.6.1.6	2402.1	8.4.1.1

A17.1-1996 With Addenda	A17.1-2000
2403	8.4.8
2403.1	8.4.8.1
2403.2	8.4.8.2
2403.3	8.4.8.3
2403.4	8.4.8.4
2403.5	8.4.8.5
2403.6	8.4.8.6
2403.7	8.4.8.7
2403.8	8.4.8.8
2403.9	8.4.8.9
2404	8.4.7
2404.1	8.4.7.1
2404.2	8.4.7.2
2405	8.4.5
2405.1	8.4.5.1
2405.2	8.4.5.2
2406	8.4.4
2406.1	8.4.4.1
2407	8.4.6
2407.1	8.4.6.1
2408	8.4.9
2408.1	8.4.9.1
2409	8.4.10
2409.1	8.4.10.1
2409.1a	8.4.10.1.1
2409.1b	8.4.10.1.2
2409.1c	8.4.10.1.3
2409.1d	8.4.10.1.4
2410	8.4.11
2410.1	8.4.11.1
2410.2	8.4.11.5
2410.3	8.4.11.7
2410.4	8.4.11.4
2410.5	8.4.11.3
2410.6	8.4.11.2

A17.1-1996 With Addenda	A17.1-2000
2410.7	8.4.11.6
2411	8.4.12
2411.1	8.4.12.1
2411.1a	8.4.12.1.1
2411.1b	8.4.12.1.2
2411.2	8.4.12.2
2411.2a	8.4.12.2.1
2411.2b	8.4.12.2.2
Part XXV	5.2
2500	5.2.1
2500.1	5.2.1.1
2500.2	5.2.1.7
2500.3	5.2.1.8
2500.4	5.2.1.3
2500.5	5.2.1.10
2500.6	5.2.1.9
2500.7	5.2.1.2
2500.8	5.2.1.4
2500.8a	5.2.1.4.1
2500.8b	5.2.1.4.2
2500.8c	5.2.1.4.3
2500.8d	5.2.1.4.4
2500.9	5.2.1.5
2500.10	5.2.1.6
2500.11	5.2.1.11
2500.12	5.2.1.12
2500.13	5.2.1.13
2501	–
2501.1	5.2.1.23
2501.1a	5.2.1.23.1
2501.1b	5.2.1.23.2
2501.2	5.2.1.22
2501.2a	5.2.1.22.1
2501.3	5.2.1.21
2501.3a	5.2.1.21.1
2501.4	5.2.1.15

A17.1-1996 With Addenda	A17.1-2000
2501.4a	5.2.1.15.1
2501.4b	5.2.1.15.2
2501.5	5.2.1.14
2501.6	5.2.1.17
2501.6a	5.2.1.17.1
2501.7	5.2.1.18
2501.8	5.2.1.16
2501.8a	5.2.1.16.1
2501.8b	5.2.1.16.2
2501.8c	5.2.1.16.3
2501.8d	5.2.1.16.4
2501.8e	5.2.1.16.5
2501.9	5.2.1.24
2501.10	5.2.1.25
2501.11	5.2.1.26
2501.12	5.2.1.27
2501.13	5.2.1.20
2501.14	5.2.1.30
2501.15	5.2.1.28
2501.16	8.9
2502	5.2.2
2502.1	5.2.2.1
2502.1a	5.2.2.2
2502.2	5.2.2.3
2502.2a	5.2.2.10
2502.2b	5.2.2.9
2502.2c	5.2.2.5
2502.2d	5.2.2.4
2502.2e	5.2.2.6
2502.3	5.2.2.7
2502.4	5.2.2.8
2502.5	5.2.2.11
2502.6	5.2.2.12
2502.7	5.2.2.13
2502.8	5.2.2.14
2502.9	8.9

SUMMARY OF CODE CHANGES

In this edition of the Handbook there are two summaries of code changes due to the editorial reorganization of ASME A17.1–2000. The first includes the revisions made to ASME A17.1–1996 in Addenda A17.1a–1997 through A17.1d–2000. The second includes those revisions that appear for the first time in ASME A17.1–2000.

ASME A17.1a–1997 THROUGH ASME A17.1d–2000

In this summary of Code changes, the reasons for the revisions that were published in ASME A17.1a–1997 Addenda are identified by [97a]; ASME A17.1b–1998 Addenda are identified by [98b]; ASME A17.1c–1999 Addenda are identified by [99b]; and ASME A17.1d–2000 Addenda are identified by [00d]. The revisions that appeared for the first time in ASME A17.1–2000 are addressed in a separate summary of code changes.

The approved technical revisions can be found in the applicable published Addenda. The Reason reflects the position of the ASME A17 Main Committee, at the time of balloting, for revising the requirement.

The TR (technical revision) number in parentheses immediately following each Reason is an administrative number used by A17 Committee.

[99c] Section 1.1
REASON: See ASME A18.1. (TR# not assigned)

[99c] Section 1.2(x)
REASON: See ASME A18.1. (TR# not assigned)

[00d] Section 3 — deflector device
REASON: Define new term. (TR 96-10)

[97a] Section 3 — door
REASON: The definitions are being clarified to cover every application of a door. The requirements for use are in the rules. (TR 90-48)

[97a] Section 3 — door center opening
REASON: The requirement for the panels to be interconnected was deleted since it is not required by Rule 110.11g(2). Additionally, the requirement for the panels to operate simultaneously was deleted from the definition since it is covered in Rule 110.11g. (TR 90-48)

[97a] Section 3 — door, biparting
The requirement for the sections to be interconnected is in Rules 111.3b and 111.4c. (TR 90-48)

[97a] Section 3 — door or gate electric contact
REASON: The application of electric contacts is now basically found on cars and the term *hoistway* is misleading. (TR 88-38)

[98b] Section 3 — elevator, limited-use/limited application
REASON: To avoid a conflict in scope with other standards (such as CSA B355). (TR 96-50)

[97a] Section 3 — gate
REASON: The definitions are being clarified to cover every application of a gate. (TR 90-48)

[97a] Section 3 — hoistway-unit-system
REASON: Delete. The term *hoistway-unit-system* has been deleted for simplicity. (TR 88-38)

[99c] Section 3 — inclined stairway chairlift
REASON: Delete. See ASME A18.1. (TR# not assigned)

[98b] Section 3 — material lift
REASON: To provide definitions. (TR 92-55)

[97a] Section 3 — unlocking device, hoistway door
REASON: To define term used in Code. (TR 88-38)

[97a] Rules 102.2(c)(3) and (c)(5)
REASON: This proposal is similar to the one approved as TR 92-08 and subsequently withdrawn at the last minute to coordinate with Part XXIV. This proposal would coordinate the ASME A17.1 requirements with those found in NFPA 13 and 72. (The revision to NFPA 13 requires sprinklers to be installed 2 ft or less above the pit floor). There is no need to disconnect power when a sprinkler in the pit is discharged under the condition cited in the proposal. (TR 94-96)

[97a] Rule 110.1(e)
REASON: The electric contact requirements have been relocated. (TR 88-38)

[97a] Rules 110.7a(4) and (7)
REASON: To allow use of suitable materials other than wire glass. (TR 92-20)

[97a] Rule 110.7a(8)
REASON: To provide a steel thickness tolerance. To rephrase the term *tamper-proof*. To correct the opening hole size allowance. (TR 92-20)

[97a] **Rule 110.11g(2)**
REASON: The term *hoistway-unit-system* has been deleted. (TR 88-38)

[97a] **Rule 111.1**
REASON: This prohibits the condition where there is a locked hoistway door with the car at the landing, e.g., the use of straight-arm closers. (TR 88-38)

[97a] **Rule 111.2b**
REASON: Same as ASME A17.1–1996, Rule 111.7. (TR 88-38)

[97a] **Rule 111.2c**
REASON: Same as ASME A17.1–1996, Rule 111.3a except for references. (TR 88-38)

[97a] **Rule 111.2d**
REASON: Same requirements as ASME A17.1–1996, Rule 111.3b, except:
  (a) change the reference in (3) to Rule 111.2c;
  (b) omit the reference in (4)(a) to Rule 110.11g as horizontal slides may not be the only multisection-type door;
  (c) omit a hoistway-unit system from (4)(c)(2). (TR 88-38)

[97a] **Rule 111.2e**
REASON: Same requirements as ASME A17.1–1996, Rule 111.3c. (TR 88-38)

[97a] **Rule 111.2g**
REASON: Same requirements as ASME A17.1–1996, Rule 111.3d. (TR 88-38)

[97a] **Rule 111.3b**
REASON: Same requirements as ASME A17.1–1996, Rule 111.7 for vertically sliding doors. (TR 88-38)

[97a] **Rule 111.3d(4)(a)**
REASON: Reference in Rule 111.3d(4)(a) to Rule 110.11g removed as lock and contacts cannot be used with horizontal slide doors. (TR 88-38)

[97a] **Rule 111.5**
REASON: Same requirements as ASME A17.1–1996, Rule 111.12. (TR 88-38)

[97a] **Rule 111.8**
REASON: Requirements for parking devices deleted. (TR 88-38)

[97a] **Figure 111.5**
REASON: Same figure as ASME A17.1–1996, Fig. 111.12. (TR 88-38)

[97a] **Rule 111.6a**
REASON: Unlocking devices are necessary for the evacuation of passengers from stalled cars and to stop emergency personnel from using unauthorized methods for gaining access to the hoistway. (TR 88-38)

[97a] **Rule 111.7a**
REASON: Hoistway access switch permissive but not as a substitute for unlocking devices, as it does not permit access to the hoistway in an emergency. (TR 88-38)

[97a] **Rule 111.7c**
REASON: A requirement has been added to assure that the car cannot be run with more than one door open. (TR 88-38)

[97a] **Rule 111.7c(3)**
REASON: The requirements have been made the same for automatic elevators and continuous pressure operation elevators and are to require visual contact with the elevator at the time of switch operation. (TR 88-38)

[97a] **Rule 111.7c(8)**
REASON: To prevent misuse of the access switch to run the elevator and to assure safe operation. (TR 88-38)

[98b] **Rule 204.1h**
REASON: To correct the intent of the rule which is that each separate piece of glass, whether tested to ANSI Z97.1 or CFR, has to have the markings specified in Rule 204.1h(3)(c). The current Rule incorrectly suggests that only the glass tested to ANSI Z97.1 has to have markings on each separate piece of glass. (TR 96-70)

[97a] **Rule 201.4j**
REASON: The ISO Technical Report, TR 11071-1 included a recommendation that all standards should take into account the increased loads on buffers which result from tie-down compensation. (TR 92-67)

[97a] **Rule 204.1j(2)(g)**
REASON: The term *car-door* is proposed for deletion from Rules 210.2s and 204.1j(2)(g) to eliminate confusion between a car-door or a gate contact and a side emergency exit contact. A car-door or gate contact typically allows a car to run at leveling speeds, whereby an electrical protective device requires an immediate stop upon activation. (TR 93-86)

[97a] **Rule 204.2e**
REASONS:
  (a) The Consumer Product Safety Committee (CPSC) has declared that glass, other than wire glass in fire doors, in hazardous locations must meet 16 CFR 1201. TR 95-90 will be opened to replace reference to ANSI Z97.1 with 16 CFR Part 1201 in all A17 requirements for glass in hazardous locations.
  (b) To apply to manually operated car doors as well as power-operated doors.
  (c) Minimize the possibility of tampering with a vision panel. (TR 92-20)

[97a] **Rule 204.4a**
REASON: To allow only solid panel doors for passenger

elevators while continuing to allow both solid panel doors and gates for freight elevators. (TR 90-48)

[97a] Rule 204.4b
REASON: Electric contacts are more typically used on car doors, so this Rule has been relocated to Section 204. (TR 88-38)

[97a] Rule 204.4c
REASON: All doors for car entrances should meet the same material requirements. (TR 90-48)

[97a] Rule 204.4d
REASON: As collapsible gates are horizontally sliding, they are included but their specifics are transferred to Rule 204.6c. (TR 90-48)

[97a] Rule 204.4g
REASON: Relocate all collapsible specifics to Rule 204.6c. No other code Parts that allow the use of collapsible gates reference Part 2 Rules and, therefore, the only application is on freight elevators. (TR 90-48)

[97a] Rule 204.4h(3)
REASON: Use the specific Rule references rather than the full Rule. (TR 90-48)

[97a] Rule 204.5b
REASON: We see no reason to exclude this requirement for a new elevator in an existing building. (TR 90-48)

[97a] Rule 204.5c
REASON: To eliminate the use of gates on passenger elevators, Rules 204.5, 204.5b, 204.5c, 204.5d, and 204.5h. (TR 90-48)

[97a] Rule 204.5g
REASON: To provide specific clearances for passenger elevator car doors not now in A17.1 and similar to the clearances for hoistway doors in response to Inquiry 90-53. (TR 90-48)

[97a] Rules 204.5g(5) and (6)
REASON: Requirement identical to Rule 110.11e(4). (TR 90-48)

[97a] Rule 204.6
REASON: To provide the same type rule heading for freight elevators as for passenger elevators. (TR 90-48)

[97a] Rule 204.6b(2)
REASON: To coordinate with revisions to freight elevators authorized to carry passengers. (TR 90-48)

[97a] Rule 204.6b(3)(c)
REASON: Reopening device may not be on the gate. (TR 90-48)

[97a] Rule 204.6b(4)
REASON: Car entrance panels that are not now in A17.1. (TR 90-48)

[97a] Rules 204.6c(5) and (6)
REASON: To locate collapsible gate information under the freight elevator door and gate rules. Relocated from Rules 204.4d and 204.4g, respectively, without text change. (TR 90-48)

[97a] Rule 205.14
REASON: Including the safety activation force and the governor pull-through force on the nameplates provides the necessary information to ensure equipment compatibility and gives the data necessary to conduct the 5 year governor test required by Rule 1002.3b(2). The revision of Rule 206.6(b) provides clarification that the rope tension is the governor pull-through tension as defined in A17.1, Section 3. (TR 92-66 and 93-31)

[97a] Rule 206.6b
REASON: Including the safety activation force and the governor pull-through force on the nameplates provides the necessary information to ensure equipment compatibility and gives the data necessary to conduct the 5 year governor test required by Rule 1002.3b(2). The revision of Rule 206.6(b) provides clarification that the rope tension is the governor pull-through tension as defined in A17.1, Section 3. (TR 92-66 and 93-31)

[97a] Rule 206.9
REASON: Including the safety activation force and the governor pull-through force on the nameplates provides the necessary information to ensure equipment compatibility and gives the data necessary to conduct the 5 year governor test required by Rule 1002.3b(2). The revision of Rule 206.6(b) provides clarification that the rope tension is the governor pull-through tension as defined in A17.1, Section 3. (TR 92-66 and 93-31)

[97a] Rule 210.2s
REASON: The term *car-door* is proposed for deletion from Rules 210.2s and 204.1j(2)(g) to eliminate confusion between a car-door or a gate contact and a side emergency exit contact. A car-door or gate contact typically allows a car to run at leveling speeds, whereby an electrical protective device requires an immediate stop upon activation. (TR 93-86)

[97a] Rule 211.3a
REASONS:
   (*a*) To recognize that a Code has finally been written, at the request of the ASME A17 Committee, which contains requirements for fire alarm systems in all building types.
   (*b*) To recognize the expertise for automatically determining when conditions, due to a fire, require elevator recall are within the jurisdiction of the NFPA 72 National Fire Alarm Code Committee.
   (*c*) To recognize that devices other than smoke detectors may be more appropriate under certain conditions.
   (*d*) To recognize that the requirements for resetting of

the fire alarm is not within the Scope of ASME A17.1. (TR 94-142)

[97a] Rule 211.3b
REASON: This proposal is similar to the one approved as TR 92-08 and subsequently withdrawn at the last minute to coordinate with Part XXIV. This proposal would coordinate the ASME A17.1 requirements with those found in NFPA 13 and 72. (The revision to NFPA 13 requires sprinklers to be installed 2 ft or less above the pit floor). There is no need to disconnect power when a sprinkler in the pit is discharged under the condition cited in the proposal. (TR 94-96)

[97a] Rule 211.3b
REASONS:

(a) To recognize that a Code has finally been written, at the request of the ASME A17 Committee, which contains requirements for fire alarm systems in all building types.

(b) To recognize the expertise for automatically determining when conditions, due to a fire, require elevator recall are within the jurisdiction of the NFPA 72 National Fire Alarm Code Committee.

(c) To recognize that devices other than smoke detectors may be more appropriate under certain conditions.

(d) To recognize that the requirements for resetting of the fire alarm is not within the Scope of ASME A17.1. (TR 94-142)

[98b] Rule 211.3b(6)
REASON: This Rule is already approved in the joint A17/B44 ballot (no negatives were received during first letter ballot from either A17 Main Committee or B44 Technical Committee members), therefore, the Committee requests that the A17 Main Committee allow this Rule 211.3b(6) be placed in the 1998 Addenda with TR 94-117, revised Fig. 211.7b due to a commitment made to NFPA 72 that if a third signal was made available, we would use it and also have already approved the signage to use it in 1996. NFPA 72-1996 has given us the third signal, therefore, the Emergency Operations Committee feels that they have an obligation to proceed with the use of the signal as quickly as possible. (TR# not assigned)

[97a] Rule 211.4b
REASONS:

(a) To recognize that a Code has finally been written, at the request of the ASME A17 Committee, which contains requirements for fire alarm systems in all building types.

(b) To recognize the expertise for automatically determining when conditions, due to a fire, require elevator recall are within the jurisdiction of the NFPA 72 National Fire Alarm Code Committee.

(c) To recognize that devices other than smoke detectors may be more appropriate under certain conditions.

(d) To recognize that the requirements for resetting of the fire alarm is not within the Scope of ASME A17.1. (TR 94-142)

[98b] Figure 211.7(b)
REASON: To coordinate with revisions made by TR 94-117. (TR 94-30)

[97a] Rule 212.9c(5)
REASON: Thread specifications were added to the Rule to control the tolerance of mating manufactured parts to ensure that the strength of the threaded connection will meet the design requirements. (TR 94-133)

[97a] Rule 212.9h
REASON: To provide performance language for the installation of antirotation devices. (TR 95-96)

[97a] Rule 500.1
REASON: To correct an improper reference to Rule 100.1b. (TR 95-43)

[97a] Rule 513.16
REASON: To be consistent with other parts of the Code. (TR 94-105)

[97a] Rule 701.5c
REASONS:

(a) Rule 204.7c is not relevant to dumbwaiters, and Rule 207.7d provides sufficient protection.

(b) Rule 204.7e no longer exists. (TR 94-99)

[00d] Rule 802.3e
REASON: See Reason for Rule 802.3k.

[00d] Rule 802.3f
REASON: See Reason for Rule 802.3k.

[00d] Rule 802.3j and Fig. 802.3j
REASONS:

(a) Deflector systems are an uncomplicated method of deflecting the feet of riders away from the step/skirt panel interface and can be secured in various manners depending on the design.

(b) To prevent the use of common flat or phillips head screws the same terminology as in A17.1, Rule 802.3i is used. (TR 96-10)

[00d] Rule 803.3j(3)(a)
REASON: This dimension is consistent with the committee's expressed previous consensus and the majority, but not all, of what has been safely applied in other parts of the world. The dimension also conforms with what has been used in California without problems. It would be logical to consider this experience. The last two sentences have been added with similar terminology as A17.1, Rule 802.3f(3) for exactly the same reason of preventing footwear, etc., from being caught. (TR 96-10)

[00d] **Rule 802.3j(3)(b)**
REASON: Based on historical experience with use of these types of devices, this should provide sufficient clearance and angle to prevent footwear from catching between the step nose and horizontal protrusion. The proposed clearance is also the value stipulated in EN 115 clause 5.1.5.6 on deflector device application. Clause 5.1.5.6 was added a number of years ago specifically for deflectors, thus it would be logical to accept the experience obtained in other parts of the world. The 25 mm dimension is also the historical value deemed necessary in A17.1, Rule 802.3f(1) for may years. The 10 deg angle at this clearance is sufficient based on International history of installations at a 25 mm height over many years. (TR 96-10)

[00d] **Rule 803.3j(3)(c)**
REASON: Based on historical experience, this should provide sufficient clearance and angle to take care of the change in elevation at the comb and allow passage of footwear. (TR 96-10)

[00d] **Rule 803.3j(3)(d)**
REASON: This is a reasonable value that could withstand the force exerted by an adult's foot. Test plate area is specified; similar to rules for the step fatigue test, combplate, etc., a reasonable area is given. (TR 96-10)

[00d] **Rule 802.3j(4)(a)**
REASON: To minimize restriction of usable step area. (TR 96-10)

[00d] **Rule 802.3j(4)(b)**
REASON: The requirement that the flexible part be capable of bending to above the 10 deg line covers the concern of easy withdrawal of a foot. (TR 96-10)

[00d] **Rule 802.3j(4)(c)**
REASON: Although this would only apply to a very small area of contact at the step nose, general rules are added to prevent damage. For example, to avoid the possibility of a brush providing a flat stepping surface by extending over two or more step cleats. Consideration was: center to center distance of step cleats in A17.1, Rule 802.5e. (TR 96-10)

[00d] **Rule 802.3j(4)(d)**
REASON: Any step contact with continuous flexible elements could potentially cause damage to the elements. (TR 96-10)

[00d] **Rule 802.3k**
REASON: National Elevator Industry, Inc. (NEII) contracted with Arthur D. Little (ADL) to develop an Escalator Step/Skirt Performance Index. These recommended revisions are the result of this comprehensive study. The Index, valued from zero to one, represents the relative potential for entrapment of objects in the step-to-skirt gap. A lower Index represents a lower potential.

This proposal is based on the results of the ADL study and is summarized in the following table:

Code	Step/Skirt Performance Index	Lubrication Allowed	Loaded Gap Required
< A17.1d–2000 and A17.3	< 0.15 Skirt deflector not required > 0.15 to 0.4 Skirt deflector required > 0.4 Not in compliance	Yes	Indirectly
A17.1d–2000	< 0.15 Skirt deflector not required > 0.15 to 0.4 Skirt deflector required > 0.4 Not in compliance	No	Yes
A17.1–2000	< 0.15 Skirt deflector not required > 0.15 to 0.25 Skirt deflector required > 0.25 Not in compliance	No	Yes

Based on the ADL study, the Index value recommendations were established for existing/current escalators, and future escalators. The Index values were established from the following criteria:

(a) the nominal estimated Index value of the current ASME A17 Code;

(b) the desired Index value for a low entrapment potential for hands;

(c) the desired Index value for a low entrapment potential for leg calf.

Current A17 Code Index

The Step/Skirt Index value is based on estimates of the two primary escalator parameters in the current and prior ASME A17.1 Code. The loaded gap parameter (the value of the step-to-skirt gap under a spreading force of 25 lbf) and skirt coefficient of friction parameters were estimated due to ambiguity or nonexistent ASME A17.1 Code requirements. The current ASME A17.1 Code specification of 0.19 in. max. step-to-skirt clearance does not address additional parameters of step stiffness and step dead band movement identified in the ADL study. These contribute to the loaded gap index parameter, and so were estimated. This additional gap value is nominally 0.05 in. resulting in a gap of 0.24 in. An additional gap increase due to the ASME A17.1 Code specified skirt stiffness of 0.06 in. at 150 lbf or 0.01 in. at 25 lbf increases the gap to 0.25 in. loaded gap.

The current ASME A17.1 Code specifies that the skirt be "made from a low coefficient of friction material or treated with a friction reducing material." A conservative coefficient of friction of 0.4 between the skirt and polycarbonate (the friction test sample) is estimated based upon ADL tests in both the lab and field.

The estimated loaded gap of 0.25 in. and a polycarbonate coefficient of friction of 0.4 results in an Index of 0.4 for escalators that comply with current ASME A17.1 Code requirements. However, it is also possible that an existing escalator could have an Index as high as 0.7 and still comply with current ASME A17.1 Code requirements.

ADL Study Index Values

The ADL study included a series of highly stressed tests to try and introduce entrapments of artificial shoes and body parts (referred to as Sawbones parts). These tests indicate that an Index of 0.15 is needed for low entrapment potential of all objects studied, including the leg calf that had the highest incidents of entrapment. One method of complying with the proposed Rules is to provide escalators with an Index of 0.15 or below. However, this low Index cannot realistically be achieved and maintained on all existing and new escalator designs, thus additional design options are needed.

Skirt deflector devices installed in compliance with the proposed ASME A17.1, Rule 802.3j, should be effective in the prevention of leg calf entrapments and the proposal makes them mandatory for escalators with an Index above 0.15. However, skirt deflectors may not be as effective in the prevention of other entrapments such as those involving hands and shoes. Therefore, an Index value up to 0.4 is proposed, for escalators installed under ASME A17.1d–2000 and earlier editions, as an alternative, when used in conjunction with skirt deflector devices.

An Index of 0.2 is clearly a valid threshold based on Sawbones hand entrapment tests (see ADL report Figure 5-2 and Table 5-1). However, other factors show this Index value to be conservative. First, the Index was derived from severe test conditions with test sample placement into the gap and maximum expected entrapping force applied. Second, for the leg calf, entrapments at these low index values were sometimes actually pinches that were classified as entrapments. Third, the object coefficient of friction with stainless steel for the test sample Sawbones hand and calf at 0.8 is significantly higher than real skin at 0.5. This makes the entrapment of Sawbones hand samples significantly more likely than real hands at the same loaded gap.

Estimation can be made of the effect of Sawbones sample coefficient of friction as compared to real skin coefficient of friction on the Index. For Sawbones hand tests, the ADL study showed a low entrapment potential below an object Index of 0.4, with object coefficient of friction of 0.8 and loaded gap of 0.16 in. The index curves show that an object Index of 0.4 and a low entrapment potential are maintained when the coefficient of friction is decreased from 0.8 (Sawbones skin) to 0.5 (real skin)

while the loaded gap is increased from 0.16 in. to 0.23 in. Thus, the loaded gap (0.23 in.) at which real hand entrapment potential diminishes is larger than the loaded gap (0.16 in.) associated with diminished Sawbones hand entrapment potential. The Index of an escalator will be measured with a polycarbonate test sample with an assumed coefficient of friction of 0.4. Using this test sample, a Sawbones hand Index of 0.4 with a loaded gap of 0.16 in. would be reduced to a measured Index of 0.2. In like manner, a real hand Index of 0.4 with a loaded gap of 0.23 in. would be reduced to a measured Index of 0.35. Therefore, an escalator with a loaded gap of 0.23 and an Index of 0.35 has a low real hand entrapment potential. A real hand would have the same low entrapment potential at an Index of 0.35 as the test Sawbones hand at an Index of 0.2.

Similar analysis can be made for the leg calf. Low entrapment potential exists for the Sawbones calf below object Index of 0.2, with object coefficient of friction of 0.8 and loaded gap of 0.06 in. Equivalently low entrapment potential exists for the real calf at object Index of 0.2 with object coefficient of friction decreased to 0.5 and loaded gap increased to 0.13 in. The equivalent measured Index is 0.1 for Sawbones and 0.16 for real calves. Entrapment potential is low for real calves at an Index of 0.16 or below.

The above assumptions provides the basis and support for the index recommendations:

(a) The current ASME A17.1 Code reflects an Index of 0.4 nominal.

(b) The ADL study indicates low entrapment potential with an Index of 0.2 for Sawbones hand that translates to an Index of 0.35 for a real hand.

(c) The ADL study indicates low entrapment potential with an Index of 0.1 for Sawbones calf that translates to an Index of 0.16 for a real calf.

(d) These are conservative numbers due to the severity of testing and the conservative classification of calf entrapments.

Because the nominal ASME A17.1 Code Index of 0.4 is reasonably close to the low hand entrapment potential, a maximum Index of 0.4 is proposed for escalators installed under ASME A17.1d–2000 and earlier editions. A lower entrapment potential is desired for the future. Therefore, a maximum Index of 0.25 is proposed beginning with ASME A17.1–2000. This should allow sufficient time for manufacturers to design for and achieve the desired Index. These Index thresholds, when used in conjunction with the required skirt deflector devices, will significantly reduce the entrapment potential on existing and new escalators. An Index of 0.15 or below will allow an escalator to be installed without skirt deflector devices.

Even though the Index provides a comprehensive measure of entrapment potential, a requirement for a loaded gap in the new ASME A17.1 Code requirement is desirable. The loaded gap parameter provides an additional margin of control of escalators that rely heavily upon low coefficient of friction that can be difficult to maintain in the field. In addition, the loaded gap parameter provides a sound means of monitoring step band wear and need for correction. A loaded gap of 5 mm is proposed for escalators installed under ASME A17.1d-2000 and ASME A17.1–2000.

The current ASME A17.1 Code parameters of step-to-skirt gap and skirt treatment are redundant and replaced by the proposed new ASME A17.1 Code requirements. If these requirements were maintained, they would be misleading, as compliance with the old requirements would not assure compliance with the new loaded gap and step/skirt performance index requirements.

[97a] **Rule 805.1u**
REASON: Experience backed up by test data indicate that the present level of forces required are too low and create false stops. Further test data show that the new figures would prevent casual contact with the comb from tripping the comb-step impact device. (TR 95-14)

[00d] **Rule 807.2**
REASON: Brushes and other devices, which could turn into hazardous blocks of ice, are addressed. (TR 96-10)

[97a] **Rule 902.3h(3)**
REASON: To discourage and/or prevent the removal of some or all of the barricade fasteners and/or the barricade itself, which would create a dangerous falling condition. This proposal is identical to a proposed revision to Rule 802.3i for escalators. (TR 95-71)

[97a] **Rule 903.1**
REASON: A proportion of moving walk accidents are related to falls. A problem would occur if the speed was varied during operation without going through a stop and restart sequence. This in fact could be hazardous. Misgauging of the speed could result in a fall anywhere on the moving walk.

NOTE: Variable frequency acceleration to provide controlled starts or Wye-delta motor control is not prohibited by the rules. This TR is similar to TR 93-63 for escalators.

[97a] **Rule 905.3j**
REASON: For moving walks equipped with this type of device, it is possible to open the end where the handrail is exiting and insert body parts into the moving machinery. This proposal is to prevent that from occurring. This proposal is identical to a proposed revision to Rule 805.3m for escalators. (TR 95-72)

[97a] **Rule 1001.2(c)(29)**
REASON: Rope replacement criteria will be in A17.1. (TR 94-13)

[98b] **Rule 1002.3d**
REASON: To add testing to verify that the brake can hold the car with load in compliance with Rules 208.8 and 207.8. See also Inquiry 92-71. (TR 94-128)

[98b] **Rule 1003.1**
REASONS:
(a) To correct an oversight since the 1984 Edition of A17.1.
(b) For new installations, current wording omits certain testing requirements not specified in Rules 1001.2, 1002.2, and 1003.2 — added references will now require testing of: brakes, standby power operation, emergency terminal stopping and speed limiting devices, power opening of doors, leveling zone and leveling speed, and inner landing zone. (TR 94-52)

[97a] **Rule 1002.3f**
REASONS:
(a) The rated speed when emergency terminal stopping devices are required was changed.
(b) Emergency Terminal Speed Limiting Devices are covered under Item 5.3, but listed under Item 2.26.
(c) Emergency Terminal Stopping Devices can be in the hoistway instead of the machine room.

[97a] **Rule 1006.4**
REASON: To add testing requirements where replacements are made. (TR 93-19)

[98b] **Rule 1008.2q**
REASON: The handrail entry devices are subject to damage and miss-adjustment. Therefore, the annual test should be required and witnessed by the inspector. Revising A17.1 and A17.2.3 to be in concert with approved TR 85-35. (TR 95-31)

[97a] **Rule 1008.2(r)**
REASON: To require testing of combplate impact devices required by TR 87-71. These devices are subject to damage and maladjustment, therefore, their annual test should be witnessed by the inspector. (TR 95-34)

[00d] **Rule 1008.2s and Fig. 1008.2s(5)**
REASON: See Reason for Rule 802.3k.

[00d] **Rule 1008.2t**
REASON: See Reason for Rule 802.3k.

[99c] **Rule 1010.2**
REASON: See ASME A18.1. (TR# not assigned)

[98b] **Rule 1010.4**
REASON: Clarifies the intent that inspection is not to take place from the top-of-car unless required safety devices are provided. (TR 97-39)

[99c] **Rule 1010.10**
REASON: See ASME A18.1. (TR# not assigned)

[97a] **Rule 1200.4f**
REASON: To clarify that alterations to labeled components must be equivalent to the original. Labeling by the certifying agency is not required, however, the part manufacturer name and part number is. The component size makes labeling by the certifying agency impractical. The manufacturer's identification will permit identification of the part for cross checking with the certifying agencies published listing. This is a clarification to current code requirements. See Inquiries 89-15 (Book 13) and 93-12 (Book 18). (TR 93-14a)

[97a] **Rule 1202.10c**
REASON: To provide rules when an alteration involves a decrease in rated speed of an electric elevator, thus ensuring that the level of safety has not been diminished. (TR 93-102)

[97a] **Rule 1203.4c**
REASON: To provide rules when an alteration involves a decrease in rated speed of a hydraulic elevator, thus ensuring that the level of safety has not been diminished. (TR 93-102)

[00d] **Rule 1206.6c**
REASON: See Reason for Rule 802.3k.

[00d] **Rule 1207.5c**
REASON: The vertical dimensions of existing escalator skirt panels may not allow full compliance with the rules. See third paragraph of Section 2, which states "The specific requirements of this Code may be modified by the authority having jurisdiction based upon technical documentation or physical performance verification to allow alternative arrangements that will assure safety equivalent to that which would be provided by conformance to the corresponding requirements of this Code." (TR 96-10)

[99c] **Section 1214**
REASON: See ASME A18.1. (TR# not assigned)

[99c] **Section 1215**
REASON: See ASME A18.1. (TR# not assigned)

[98b] **Part XIV — Title**
REASON: To provide appropriate titles for heading section. (TR 92-55)

[98b] **Section 1406**
REASONS:
Rule 1406.1a(1) — Rule 100.7 serves no safety purpose on material lifts.
Rule 1406.1c(1) — Type SF-2 wire is not required, Firemans' service is not applicable.
Rule 1406.1c(2) — Provides sufficient protection for material carrying equipment upon sprinkler activation.
Rule 1406.1h — Means are provided for refuge space and in the overhead when appropriate conditions are present. As material lifts do not carry passengers, less restrictive clearances and runbys are permitted.
Rule 1406.1h(3)(d) — Hoistway construction and machinery location in the hoistway may not provide adequate refuge space when the car is at the upper landing.
Rule 1406.1h(4) — Tighter tolerances are permitted on these relatively smaller cars and the passengers are not carried. The use of a single rail for both the car and counterweight requires tighter clearances.
Rule 1406.1j — References are made to the appropriate material lift sections.
Rule 1406.1k(1) — Emergency access doors to blind hoistways are not required for material lifts that do not carry people. Opening size is determined by material lift usage.
Rule 1406.1k(2) — The requirements for passenger elevators are not applicable.
Rule 1406.1k(3) — This Rule was modified to suit door styles chosen with proviso that doors be closed when not being used for loading or unloading.
Rule 1406.1k(4) — Rule 110.4 does not apply because the lifts cannot be operated from within the car and passengers are not permitted.
Rule 1406.1k(5) — Since there is no means to operate the material lift from the car, a risk of locking someone in the car who is in the process of loading and unloading should not be allowed.
Rule 1406.1k(6) — Hoistway door vision panels serve no safety purpose.
Rule 1406.1k(7) — The inclusion of glass on a device designated to carry materials adds an unnecessary risk that the materials may shift and cause breakage of the glass.
Rule 1406.1k(8) — Landing sill guards are not required on non-passenger elevators.
Rule 1406.1k(9) — There are instances where control over the emerging carts is desirable.
Rule 1406.1k(10) — Bridging devices may be desirable to help with the transfer of carts from the landing to the car.
Rule 1406.1k(11) — This Rule provides clearance for personnel who might be in the area of emerging carts.
Rule 1406.1k(12) — This Rule prevents undesirable motion of the carts after they have been unloaded.
Rule 1406.1k(13) — Pull straps inside the car are not necessary since the doors should not be closed from inside the car.
Rule 1406.1k(14) — Combination horizontally sliding and swinging panels are not allowed.
Rule 1406.1l(1) — The requirements for passenger elevators are not applicable.
Rule 1406.1l(2) — Doors should be closed before the lift starts to travel.
Rule 1406.1l(3) — Parking devices are not needed for doors that are unlocked when the lift is at the landing.

Rule 1406.1l(4) — The requirements for passenger elevators are not applicable.

Rule 1406.1m(1) — No door or car operating pushbuttons will be in the car.

Rule 1406.1m(2) — No door or car opening pushbuttons will be in the car. The momentary button called for in Rule 112.3c(2) is placed at the landing.

Rule 1406.1m(3) — There is no safety reason for sequence operation. There will be no door or car operating pushbuttons in the car.

Rule 1406.1m(4) — No passengers will be in the car.

Rule 1406.1m(5) — There is no safety reason for sequence operation. No passengers will be in the car.

Rule 1406.2a — It is common practice for small elevators such as residence elevators, personnel elevators, and dumbwaiters to have the car and counterweight run in a common rail.

Rule 1406.2a(1) — Allows for rails with smaller section modulus and moment of inertia consistent with cars and counterweights of smaller size, weight, and rated capacity.

Rule 1406.2a(2) — Eliminates the requirement for approval of the authority having jurisdiction for different design rail joints.

Rule 1406.2a(3) — Does not dictate the size of bolts required for fastening the guide rail to the rail brackets, however, Rule 200.10a, which defines the strength requirements, remains applicable.

Rule 1406.2a(4) — Adds the requirement that the allowable deflection, in addition to compliance with that specified in Section 200, cannot allow the safety to disengage from the rail.

Rule 1406.2b — Retardation force on a non-passenger carrying elevator can meet less stringent requirements. Same as in Rules 1400.2(a), (b), and (c). Incorporates hydraulic bumper requirements.

Rule 1406.2d(1) — It is not the intent to dictate the design of the equipment, only to ensure that it is designed to meet its intended purpose.

Rule 1406.2d(2) — Exceptions to specific platform design parameters and specific loading classifications.

Rule 1406.2d(3) — A material lift does not carry passengers. Protection of the platform against fire is irrelevant.

Rule 1406.2d(4) — 3 in. plus leveling truck or landing zone provides adequate shear protection.

Rule 1406.2d(5) — An overlap of 3 in. is required, which eliminates the need for the bevel.

Rule 1406.2d(6) — The inclusion of Rule 203.10, which limits the maximum allowable stresses, provides sufficient safety factors and design criteria for a non-passenger carrying material lift.

Rule 1406.2e(1) — To prohibit the use of a fork-lift truck when placing a load on the car.

Rule 1406.2e(1)(a) — This Rule contains the essence of the requirement, the additional wordage contained in Rule 204.1b relates to passenger type elevators.

Rule 1406.2e(1)(b) — The criteria for design of the enclosure walls is more closely related to the actual load being carried than to a generalized design criteria suitable for a passenger-type application.

Rule 1406.2e(1)(c) — Material lifts may have multiple compartments, fixed and/or removable shelves.

Rule 1406.2e(1)(d) — Emergency exits are not required in non-passenger carrying material lifts.

Rule 1406.2e(1)(e) — The inclusion of glass on a device designed to carry materials adds an unnecessary risk that the materials may shift and cause breakage of the glass.

Rule 1406.2e(1)(f) — This Rule contains the essence of the requirement, the additional wordage contained in Rule 204.1(i) relates to passenger-type elevators.

Rule 1406.2e(1)(g) — Emergency exits are not required in non-passenger carrying material lifts.

Rule 1406.2e(1)(h) — Specific requirements for passenger elevators.

Rule 1406.2e(1)(i) — Grille or perforated construction may be used for the full height and car top (if provided). Enclosures may be less than 6 ft.

Rule 1406.2e(1)(j) — If ventilation grilles are provided, their location is irrelevant in a non-passenger carrying material lift.

Rule 1406.2e(2)(a) — This Rule ensures that the lift will not run unless the car door or gate is in the closed position.

Rule 1406.2e(2)(b) — As there are no passengers, there is no safety hazard to personnel if a car door or gate is not provided. The purpose of a car door or gate is to assist in retaining the materials carried within the car. This Rule provides minimum design criteria for a car door or gate.

Rule 1406.2e(2)(c) — As there are no passengers, there is no safety hazard to personnel if a car door or gate is not provided. The purpose of a car door or gate is to assist in retaining the materials carried within the car. This Rule provides minimum design criteria for a car door or gate.

Rule 1406.2e(2)(d) — Since this is a non-passenger carrying device, the location of the gate handle in relation to the car opening device is irrelevant. Permits placement of the gate handle most suitable for the design and operation of the lift.

Rule 1406.2e(2)(e) — Paragraphs (b), (c), and (e) above incorporate the required design criteria.

Rule 1406.2e(2)(f) — As there are no occupants on a material lift, there is no hazard if the counterweight is located within the car enclosure. Guides or counterweight boxes prevent the weights from swinging freely and causing incidental damage plus provisions for suitable restraint in the event of suspension member failure is included.

Rule 1406.2e(2)(g) — References proper Rule.

Rule 1406.2e(2)(h) — Specific requirements for passenger elevators.

Rule 1406.2e(2)(i) — Specific requirements for freight elevators. Paragraphs (b), (c), and (d) above incorporate the required design criteria.

Rule 1406.2e(3) — Provides for a minimum level of illumination within the car for loading and plus minimum design criteria for the light.

Rule 1406.2f(1) — The objective is to stop the car should a failure occur. Since there are no passengers on the car, the retardation force is of little importance. Type B safeties are permitted.

Rule 1406.2f(2) — Same as Rule 1400.2(j). Passengers are not carried.

Rule 1406.2f(3) — The objective is to stop the car should a failure occur. Since there are no passengers on the car, the retardation force is of little importance. Type B safeties are permitted. When the speed is slow, broken rope safeties are considered sufficient, see Rule 1400.2(k).

Rule 1406.2f(4) — Provision to allow for rails of different design.

Rule 1406.2g(1) — Same as Rule 1400.2(m). Passengers are not carried.

Rule 1406.2g(2) — To conform with previous changes in rated speed from 150 ft/min to 200 ft/min.

Rule 1406.2g(3) — Incorporate provisions for governor ropes smaller than $3/8$ in. when the suspension means is less than $3/8$ in.

Rule 1406.2h(1) — This is not a passenger elevator.

Rule 1406.2h(2) — Similar to rated load requirements for freight elevators. Provisions are included for the weight of hand truck. These material lifts are designed for hand truck use but not intended for fork lift use.

Rule 1406.2h(3) — Riders are not permitted on material lifts.

Rule 1406.2i — Clarifies that drum machines are permissive but with restrictions.

Rule 1406.2i(1) — A material only handling device can meet a lower standard for pitch diameter.

Rule 1406.2i(2) — Permits the use of chain drives and provides for general design criteria.

Rule 1406.2i(3) — Rule 208.9 applies in its entirety when an indirect drive machine is provided. A17.1–1993 does not list a Rule 208.9(i).

Rule 1406.2j — A final terminal stopping device within the confines of the hoistway in addition to the normal stopping means provides an adequate level of redundancy and sufficient overtravel protection on a device designed solely to carry materials.

Rule 1406.2k(1) — Passengers are not permitted on the car.

Rule 1406.2k(2) — One-piece loads greater than the rated load are not permitted.

Rule 1406.2k(4) — Passengers are not permitted on the car.

Rule 1406.2k(5) — Similar intent but modified for material lift application.

Rule 1406.2k(6) — Machine room normal terminal stopping devices are not allowed in Rule 1406.2i(b).

Rule 1406.2k(8) — See Rule 1406.2f(3).

Rule 1406.2k(9) — Appropriate Rule reference.

Rule 1406.2k(10) — Passengers are not carried.

Rule 1406.2k(11) — While a switch is required, an approved switch is not required because passengers are not carried on the car.

Rule 1406.2k(12) — Emergency exits are not permitted. Passengers are not carried.

Rule 1406.2k(13) — When the device is retained on the landing by the hoistway door or retained in the car by the car door or gate, an electric contact is not required.

Rule 1406.2k(14) — See Rule 1406.2k(5).

Rule 1406.2k(15) — Single-phase AC motors may be used.

Rule 1406.2k(16) — Floating platforms are not permitted.

Rule 1406.2k(17) — Material lifts are not designed for use by the disabled.

Rule 1406.2k(18) — To prevent material lifts from being operated from the car.

Rule 1406.2l — Not relevant to a material lift.

Rule 1406.2m(1) — Permits the use of chains and wire ropes other than elevator wire rope.

Rule 1406.2m(2) — Includes requirements for chain data similar to that required for wire ropes.

Rule 1406.2m(4) — To include an appropriate safety factor for chains. Note that the safety factor for chains in Table 701.13d is 1.25 times the safety factor for wire ropes.

Rule 1406.2m(5) — Two ropes provide sufficient redundancy on a non-passenger carrying material lift. Allows single-bar-type equalizers on all drives when only two ropes are provided.

Rule 1406.2m(6) — Adds suspension fastening requirements for chains.

Rule 1406.3(a) — Rule 301.2 duplicates Rule 214 which is referenced by Rule 1406.2o. Rule 1406.3(h)(1) contains the essential elements for layout data.

Rule 1406.3(a)(1) — Provides appropriate reference to rules in the material lift section. A total of two ropes is equivalent to that required for electric material lifts. See Rules 1406.2i and 1406.2m.

Rule 1406.3(e)(1) — Provides equivalency to Rule 1406.2j.

Rule 1406.3(g)(1) — Provides proper referencing to appropriate material lift sections.

Rule 1406.3(h)(1) — Rule 308.1 duplicates Rule 214 which is referenced by Rule 1406.2o. The specific hydraulic requirements have been referenced. (TR 92-11)

[97a] Rule 1709.3
REASON: To conform to wording in Part I. (TR 93-95)

[97a] Section 1714
REASON: To conform to Part II. (TR 95-05)

[97a] Rule 1714.2
REASON: These elevators do not provide firefighters

access into buildings nor to floors of buildings as they do not have entrance into buildings or are not near a building or share a common area with escalators or moving walks without separation by fire-resistive construction. (TR 95-05)

[99c] **Part XX**
REASON: Delete. See ASME A18.1. (TR# not assigned)

[97a] **Rule 2000.1b(1)**
REASON: This provides running clearance requirements for stationary as well as telescoping panels where none currently exist. This TR resulted from the Wheelchair Lift Committee's review of Inquiry 92-75. (TR 93-04)

[97a] **Rule 2000.1b(7)**
REASON: To clarify the existing rules as to when and where additional guarding is required. (TR 93-103)

[98b] **Rule 2000.10c(3)**
REASON: To provide consistency and to clarify the intended operation of equipment covered by Parts XX and XXI. (TR 94-40)

[97a] **Rule 2000.11**
REASON: It is the intent of Part XX that the user be in control of the equipment at all times. This requirement is especially needed where there may be a problem that has caused the user to summon help. There currently exists the requirement for constant pressure controls and emergency stop switches on all lifts to ensure user control.

Under current rules for summoning help, due to a separate alarm control, the users control is left to any other individual operating a hall station which could cause a hazard. No additional safety is added by requiring a second switch for this function. This code change eliminates that potential hazard by allowing only one control, which must be labeled as both a stop and alarm. (TR 95-19)

[99c] **Part XXI**
REASON: Delete. See ASME A18.1. (TR# not assigned)

[98b] **Rule 2001.10c(3)**
REASON: To provide consistency and to clarify the intended operation of equipment covered by Parts XX and XXI. (TR 94-40)

[98b] **Rule 2002.10c(3)**
REASON: To provide consistency and to clarify the intended operation of equipment covered by Parts XX and XXI. (TR 94-40)

[98b] **Rule 2100.10c(3)**
REASON: To provide consistency and to clarify the intended operation of equipment covered by Parts XX and XXI. (TR 94-40)

[98b] **Rule 2101.10c(3)**
REASON: To provide consistency and to clarify the intended operation of equipment covered by Parts XX and XXI. (TR 94-40)

[98b] **Rule 2102.10c(3)**
REASON: To provide consistency and to clarify the intended operation of equipment covered by Parts XX and XXI. (TR 94-40)

[97a] **Appendix C**
REASON: Delete, no longer necessary. (TR 88-38)

[97a] **Appendix C**
REASON: Control System diagram to be added to A17.1 Appendix, to help define *control system* definition in A17.1–1996. (TR 90-49)

ASME A17.1–2000
SUMMARY OF CHANGES

This summary of Code changes includes the reasons for the noteworthy revisions in Parts 2, 3, and 6 and Sections 5.2, 8.3, 8.7, 8.8, and 8.9, which were published in ASME A17.1–2000. Due to the shear amount of revisions, the reasons for all the changes are not included in this summary. The revisions that were made in Addenda, ASME A17.1a–1997 through ASME A17.1d–2000, are addressed in a separate summary of Code changes.

The approved technical revisions can be found in the Code. The reason reflects the position of the ASME A17 Main Committee at the time of balloting.

TR (technical revision) numbers were not assigned to the revisions that were made during the balloting of ASME A17.1–2000, thus none are referenced in this summary. The number in parenthesis, following the ASME A17.1–2000 requirement, is the Rule number used during Committee balloting. The Rule number used during balloting may not correspond to a Rule number in ASME A17.1d–2000 or earlier editions of the Code.

Reference to "Clause" is to requirements in CAN/ CSA-B44-94.

2.1.1.1 [Rule 100.1a(4)]
Relocated to Rule 110.14a. To recognize an elevator requirement in the National Building Code of Canada.

2.1.1.1.2 [Rule 100.1a(2)]
There is no rationale for this partition to be unperforated when you recognize there has to be openings for equipment to pass through.

2.1.1.1.3 [Rule 100.1a(3)]
Clarify the requirement and recognize that the building codes establish the requirements.

2.1.1.2.2 [Rule 100.1b(2)]
Reference changes made due to reorganization.

2.1.1.2.2(a) [Rule 100.1b(2)(a)]
The enclosure adjacent to the car guide rails should be guarded similar to the counterweight runway for safety. Editorial reorganization of (a), (b), and (c).

2.1.1.2.2(d) [Rule 100.1b(2)(d)]
To clarify the intent that the standard applicable to the particular jurisdiction should be applied. Canadian glass standard added.

2.1.1.3 (Rule 100.1c)
Deleted (1) and (2) as unnecessary due to clarification of Rules 100.1a and 100.1b. The intent is to provide protection where the people may come into contact with elevator equipment in areas accessible to nonelevator personnel.

2.1.2.3 [Rule 100.2c]
Relocated from Rule 109.1 and Clause 2.7.1.3(c), for clarity.

2.1.2.3(a) [Rule 100.2c(1)]
To agree with B44 revisions and Section 109.

2.1.3.1.1 [Rule 100.3a(1)]
Construction requirement addressed in Rule 100.3d.

2.1.3.1.2(b) [Rule 100.3a(2)(b)]
Clarification of intent. Permits elimination of access doors, provided you can service and test the governor by other means.

2.1.3.1.2(b)(1) [Rule 100.3a(2)(a)]
Permits elimination of access doors, provided you can service and test the governor.

2.1.3.3 (Rule 100.3c)
Rounded metric conversion, which gives approximately equivalent load pressure.

2.1.3.5.1 [Rule 100.3e(1)]
To make use of the new Rule 104.2 for standard railings.

2.1.3.5.2 [Rule 100.3e(2)]
To add requirements for safe access. To make use of the new Rule for standard railings (Rule 104.2).

2.1.3.6 (Rule 100.3f)
These are OSHA requirements.

2.1.4 (Rule 100.4)
Recognition that means other than ventilations are recognized by the building codes for control of hot gases. This is now in performance language.

2.1.5 (Rule 100.5)
Requirements for windows are under the jurisdiction of the NBCC.

2.1.6 (Rule 100.6)
Unnecessary and nonspecific wording deleted.

2.1.6.2(b) [Rule 100.6(b)(2)]
Redundant; the subject is covered in the opening sentence.

2.2.2.4 [Rule 106.1b(4)]
Rewritten in performance language to prohibit the use of a trap that can dry out.

2.2.2.5 [Rule 106.1b(3)]
For protection of firefighters. Rewritten in performance language.

2.2.2.6 [Rule 106.1b(5)]
Eliminating potential tripping hazard for personnel working in the pit.

2.2.3.1 [Rule 106.1c(1)]
Delete "of more than 2 ft." Covered in Rule 106.1c(2).

2.2.3.2 [Rule 106.1c(2)]
To correct unenforceable language.

2.2.4.2 [Rule 106.1d(2)]
Increase dimension for safety of the personnel using the ladder. To provide a safe means for personnel to exit the pit.

2.2.4.2 [Rule 106.1d(2)] — Second paragraph
To provide requirements for safe access to a pit.

2.2.4.4 [Rule 106.1d(4)]
Editorial clarification. Rules 106.1d(4)(b) and (d) added to provide for safety of persons entering the pit.

2.2.4.4(e) [Rule 106.1d(4)(e)]
Editorial clarification. Covered by Rule 106.1d(4).

2.2.5.1 [Rule 106.1e(1)]
To agree with B44 requirement.

2.2.8 (Rule 106.1h)
An elevated work platform without a 42 in. guard rail is a recognized safety hazard. Due to the space constraints, a platform with a proper guard rail is not feasible. This Rule requires means to be installed or permanently stored, which does not prohibit a platform. An 18 in. guard rail is unsafe.

2.3.2.1 [Rule 103.2(a)]
Perforations allow visual inspections behind the guard and still meet the criteria of unintentional; "or" is necessary.

2.3.2.1(a) [Rule 103.2(a)(1)]
Enhances safety.

2.3.2.1(b) [Rule 103.2(a)(2)]
Note is superfluous, directs to next Rule.

2.3.3 (Rule 103.3)
If mechanical compensation or counterweight safeties are used, this Rule is not applicable.

2.3.3.3 [Rule 103.3(c)]
Should specify an enclosed stop switch meeting the requirements of Rules 210.2(e)(1), (2), and (3) only, as Rule 210.2(e)(4) (audible device to sound) does not apply.

2.4.1.6 [Rule 107.1a(6)]
Incorrect terminology. To clarify that the intent is that the standard applicable to the particular jurisdiction should be applied. To require marking of the spaces where clearance of 600 mm is not provided throughout the entire cross-sectional area of the pit

2.4.6.2(a) [Rule 107.1e(1)]
Clarification and correct reference.

2.4.6.2(b) [Rule 107.1(e)(2)]
These requirements are being added as part of the new requirements (TR 85-23) for ascending car overspeed and unintended car motion protection.

2.4.6.2(c) [Rule 107.1(e)(3)]
To advise the reader that refuge space may impact on car top clearance. Same intent, use B44 for clarity.

2.4.6.2(d) [Rule 107.1(e)(4)]
To provide Rule related to clearances, when compensating rope tie-down is installed.

2.4.7(b) [Rule 107.1f(2)]
To advise the reader that refuge space may impact car top clearance.

2.4.8(b) [Rule 107.1g(2)(d)]
To advise the reader that refuge space may impact car top clearance.

2.4.12.1 [Rule 107.1k(1)]
To require marking of the spaces where refuge clearance is not provided throughout the entire cross-sectional area of the pit or the top of the car enclosure. Increase refuge space for greater safety. Has been moved. See Rule 204.1e(1)(f).

2.4.12.2 [Rule 107.1k(2)]
To clarify that the intent is that the standard applicable to the particular jurisdiction should be applied.

2.5.1.4 (Rule 108.1d)
Dimensions change to coordinate with the accessibility standards.

2.5.1.5.2 [Rule 108.1e(2)]
Already addressed under Part 2 of A17.1.

2.5.1.5.3 [Rule 108.1e(3)]
Adopt B44 Clause.

2.5.1.6 (Rule 108.1f)
To protect personnel in pit from an area of possible entrapment.

2.6.1 (Rule 109.1)
By separating space underneath car and counterweight, you allow more flexibility in design. Rule 109.1(c) deleted. See Rule 100.2c.

2.7.1 (Rule 101.1)
Relocated from Rule 101.1a(3).

2.7.1.1 (Rule 101.1a)
Editorial clarification and to be consistent with Rule 100.1b.

2.7.1.1.2 [Rule 101.1a(2)]
Openings do not have a fire rating, but things that close openings do. Subject germane to the building code.

2.7.1.2 (Rule 101.1b)
Where perforated materials are used, a perforation size is required.

2.7.2.2 (Rule 101.2b)
Performance language; to provide minimum clearances for servicing the equipment.

2.7.3.1 (Rule 101.3a)
Less subjective.

2.7.3.2.1 [Rule 101.3b(1)]
Horizontal door is not defined, replaced with swinging door, which is defined. To comply with definitions.

2.7.3.2.2 [Rule 101.3b(2)]
Correct editorial error. To provide protection for personnel accessing the machine rooms.

2.7.3.3.2 [Rule 101.3c(1)]
Use more stringent A17 requirement. Eliminate undefined language.

2.7.3.3.3 [Rule 101.3c(2)]
Use more stringent A17 requirement. Eliminate undefined language.

2.7.3.3.5 [Rule 101.3c(5)]
To provide adequate space for a person to stand when they open the door. This is a building code issue in the NBCC.

2.7.3.4.1(a) [Rule 101.3d(1)(a)]
Canadian building code requires every door to be 2 030 mm high.

2.7.3.4.1(d) [Rule 101.3d(1)(d)]
Editorially corrected to direct to Section 5, Security.

2.7.3.4.3(c) [Rule 101.3d(3)(c)]
Editorially corrected to direct to Section 5, Security.

2.7.3.5 (Rule 101.3e)
Editorial correction. Equivalent protection is provided with the switch in either location.

2.7.5.1 (Rule 101.5a)
200 lx at floor levels improves level of safety.

2.7.5.2 (Rule 101.5b)
Assist enforcement of requirements. Performance language.

2.7.6 (Rule 101.6)
Delete redundant language.

2.7.7.2 [Rule 101.7(b)]
Editorial correction. Rounded metric conversion that gives approximate equivalent load pressure.

2.7.8 (Rule 101.8)
Editorial correction.

2.8 (Section 102)
Editorial simplification of title.

2.8.1.1 (Rule 102.1) — First paragraph
To clarify the intent that the standard applicable to the particular jurisdiction should be applied. Editorial.

2.8.1.3 (Rule 102.1) — Third paragraph
To clarify the intent that the standard applicable to the particular jurisdiction should be applied. Not previously addressed.

2.8.2.1.1 [Rule 102.2(a)(1)]
B44 requirement, not originally in A17.1. 15 psi is more realistic.

2.8.2.1.2 [Rule 102.2(a)(2)]
To address a need to get heating pipes into machine rooms in cold environments.

2.8.2.2 [Rule 102.2(b)]
To leave the required clearances intact.

2.8.2.3.1 [Rule 102.2(c)(1)]
To address a need to get sprinkler pipes into machine rooms in cold environments.

2.8.2.3.2 [Rule 102.2(c)(2)]
Editorially restated in positive language. Subject within the NBCC.

2.8.2.4 [Rule 102.2(c)(4)(d)]
To address drainage from penthouse or machine room roofs.

2.8.3 (Rule 102.3)
Surface temperature is addressed by test standard utilized for listed/certified electric heaters.

2.9.2.1(b) [Rule 105.2a(2)]
Revision incorporates Note into Rule. Further rewritten for clarity.

2.9.3.3.5 and 2.9.3.3.6 (Rule 105.3c) — Second and third paragraphs
To provide design specificity. Rule 203.13 makes reference to this so requirements must be included here.

2.9.3.4 (Rule 105.3d)
ASTM E8 is the reference standard for elongation.

2.9.4.1 (Rule 105.4)
To coordinate with revisions to Rule 105.6.

2.9.4.1(b) [Rule 105.4(b)]
Add Canadian Standard requirements. Revised for clarification and consistency.

2.9.6 (Rule 105.6)
To provide criteria for the safe design of the respective components/structures subject to forces developed during the retardation phase of the emergency braking and all other loading acting simultaneously, if applicable, with factors of safety consistent with the types of equipment/structures involved.
See also Rules 200.5c, 202.2c(3), 203.10(b), 208.3d, 216.4, and the Note to Clause 3.2.

2.10.1(d) [Rule 104.1(d)]
Adopted from Clause 2.3.8.3.1(a) to provide personnel safety.

2.10.2 (Rule 104.2)
Incorporates OSHA (US) regulations. Puts requirement for standard railing in one Rule which can be referred to from many other Rules in A17/B44 [e.g., Rules 100.3e(1), 100.3e(2), 100.3f, 101.3b(2)]. "Referred to in" is being deleted as it is superfluous.

2.11.1.1 (Rule 110.1a)
Height of openings for all elevators should be 6 ft 8 in.

2.11.1.2(a) [Rule 110.1b(1)]
Use of standard height doors.

2.11.1.2(e) [Rule 110.1b(5)]
Improve safety by using an electromechanical device that will prevent motion of car if door is not closed and locked.

2.11.1.2(g) [Rule 110.1b(7)]
To address a condition that is not currently addressed by the Code.

2.11.1.2(h) [Rule 110.1b(8)]
Editorially corrected to direct to Section 5, Security.

2.11.1.2(i) [Rule 110.1b(9)]
To guard against a fall hazard.

2.11.1.3 (Rule 110.1c)
To provide an alternative where emergency access doors are not feasible.

2.11.1.4 (Rule 110.1d)
Procedures for cleaning will be addressed in the maintenance Section. To add the required key for this access door to the security key list. Rule 210.2(ff) recognizes that a locking device is necessary when access openings for cleaning car and hoistway enclosures are required in accordance with Rule 110.1d. This revision is not a new requirement, it is a clarification.

2.11.2.1 (Rule 110.2a)
To modify carrying of passengers on freight elevator requirements with additional passenger elevator attributes to justify carrying passengers without the need to request permission from an AHJ. Change of swing and slide to swinging and sliding for editorial consistency.

2.11.2.2 (Rule 110.2b)
To modify carrying of passengers on freight elevator requirements with additional passenger elevator attributes to justify carrying passengers without the need to request permission from an AHJ.

2.11.2.3 (Rule 110.2c)
To clarify terminology.

2.11.3.2 [Rule 110.3(b)]
Phase I will close and open the door in event of a fire.

2.11.3.2(d) [Rule 110.3(b)(4)]
The door is permitted to remain open only when fire alarm initiating devices are provided per Rule 211.3b.

2.11.4.2(a) [Rule 110.4(b)(1)]
Clarification of intent.

2.11.4.2(b) [Rule 110.4(b)(2)]
Clarification.

2.11.6.1 [Rule 110.6(a)]
To eliminate duplication between this and Rule 111.1 and reorganize editorially.

2.11.6.3 [Rule 110.6(c)]
Beyond the scope of the Code.

2.11.7.1.4 [Rule 110.7a(4)]
To clarify that the intent is that the standard applicable to the particular jurisdiction should be applied. Equivalent Canadian Standards added.

2.11.7.1.5 [Rule 110.7a(5)]
So as not to penalize existing designs referencing B44.

2.11.7.1.7 [Rule 110.7a(7)]
To provide a steel thickness tolerance. To rephrase the term *tamper proof*. To correct the opening hole size allowance.

2.11.7.2.1 [Rule 110.7b(1)]
This glass location is classified as a hazardous location by the model building codes.

2.11.9.1 (Rule 110.9a)
Parking devices no longer addressed in Section 111 and access switches are not subject to this Rule.

2.11.9.2 (Rule 110.9b)
To simplify the title; the text itself differentiates between "power-operated," the defined term and opened/closed "by power" that are generic terms.

2.11.10.1.2 [Rule 110.10a(2)]
This requirement should apply to all elevators for the safety of passengers.

2.11.10.2 (Rule 110.10b)
50 lx is too dark. 100 lx is appropriate lighting value. Improves level of safety.

2.11.11.1(b) [Rule 110.11a(2)]
This is beyond the scope of the Elevator Code.

2.11.11.2 (Rule 110.11b)
Requirements should be applicable to all doors, not only power operated.

2.11.11.3.1 [Rule 110.11c(1)]
Use B44 Clause 2.11.10.2. Same intent as Rule 110.11c(1), (2), and (3).

2.11.11.3.2 [Rule 110.11c(2)]
This provision is applicable only on listed/certified frames. To coordinate with similar requirement for panels in Rule 110.11e(6).

2.11.11.5.1 [Rule 110.11e(1)]
Use B44 requirement of 13 mm to not restrict Canadian designs without a good reason.

2.11.11.5.5 [Rule 110.11e(5)]
Adopt B44 requirement to increase safety.

2.11.11.5.6 [Rule 110.11e(6)]
This provision is applicable only to listed/certified panels.

2.11.11.5.8 [Rule 110.11e(8)]
A17 clearance reduced in all cases. B44 clearance reduced for tall narrow doors.

2.11.11.7.4 [Rule 110.11g(4)]
Enhance safety, not covered in A17.

2.11.11.8 (Rule 110.11h) — Third paragraph
The retaining means is a backup in case the normal means fails and the backup should not be subject to wear.

2.11.12.4.3(a) [Rule 110.12d(3)(a)]
To provide compression capability and penetration protection.

2.11.12.4.3(c) [Rule 110.12d(3)(c)]
To provide compression capability between the meeting of the panels.

2.11.12.4.6 [Rule 110.12d(6)]
Use B44 as modified. More stringent requirement.

2.11.12.8 (Rule 110.12h)
Simplify the wording. Conformance to Rule 110.1a.

2.11.13.1(b) [Rule 110.13a(2)]
Beyond the scope of the Elevator Code.

2.11.13.2.2 [Rule 110.13b(2)]
This provision is applicable only on listed/certified panels. Revised for consistency with Rules 110.11c(3) and 110.11e(6).

2.11.13.3.4 [Rule 110.13c(4)]
This provision is applicable only on listed/certified panels.

2.11.13.3.5 [Rule 110.13c(5)]
Rewritten for clarity and to add more stringent requirement from B44.

2.11.13.3.6 [Rule 110.13c(6)]
Appropriate subject for certifying laboratory to address.

2.11.14.1 [Rule 110.14(a)]
Relocated from Rule 100.1a(4). The requirements in Rules 110.14 through 110.18 apply in the US but not in Canada due to the building code requirements in Canada and US.

2.11.15.1 (Rule 110.15a)
To clarify labeling requirements.

2.11.17 (Rule 110.17)
Fixed side panels are considered to be the same as transoms.

2.11.19.1 [Rule 110.19(a)]
To clarify that the intent is that the standard applicable to the particular jurisdiction should be applied. Add equivalent B44 standard, delete out of print standard.

2.11.19.4 [Rule 110.19(d)]
Editorial. To be consistent with Rule 110.16.

2.12.1 (Rule 111.1) A17/B44
Rationale: Clarify intent of the Rule. This sentence and Clause 2.12.1 have been relocated to Rule 110.6 to eliminate duplication. Clarification.

2.12.2.3(c) [Rule 111.2c(3)]
Editorial coordination with Rule 204.4b(2)(3).

2.12.2.4.1 [Rule 111.2d(1)(a)]
Cross-reference added. To resist tampering.

2.12.2.4.1 [Rule 111.2d(1)(b)]
To ensure the reliability of the contact, by adding a force requirement.

2.12.2.4.3 [Rule 111.2d(3)]
Revised to include requirements from Rule 111.2d(7) of A17.1a–1997 and B44 Clause 2.12.2.1.4. Metric correction. Revised to incorporate requirements from Rule 111.2d(7) as rationalized.

2.12.2.4.4(c)(2) [Rule 111.2d(4)(c)(2)]
Editorial rewrite for clarification. Revised for clarity. Add reference to Rule 204.4b.

2.12.2.4.6 [Rule 111.2d(6)]
Force level is compatible with the force created by normal human effort without the use of tools. To prevent disengagement of the locking member and quantify forces.

2.12.3.1 (Rule 111.3a)
Clarification of intent.

2.12.3.3(c) [Rule 111.3c(3)]
Editorial coordination with Rule 204.4b(2)(3).

2.12.3.4.2 [Rule 111.3d(2)]
Editorial clarification. Cross-reference added for convenience.

2.12.3.4.5 [Rule 111.3d(5)]
Force level is compatible with the force created by normal human effort without the use of tools. To prevent disengagement of the locking member and quantify forces.

2.12.4.1(b) (Rule 111.4a) — Second paragraph
Editorial clarification. Appendices deleted. Add Canadian grandfather clause.

2.12.4.2 (Rule 111.4b)
The certifying organization is the more appropriate party to make this determination. Not covered in A17, add B44 car-door interlocks.

2.12.4.3 (Rule 111.4c)
To clarify the requirement; "only one identification" is not clear.

2.12.4.3(a) [Rule 111.4c(1)]
To coordinate with changes made to Rule 110.15a.

2.12.5.3 [Rule 111.5(c)]
Clarification. The intent is that the doors be openable, but not necessarily unlocked. The term *unlocking zone* is defined in Section 3.

2.12.6.1 (Rule 111.6a)
Recognizes that this is in most applications a safe way of getting into the hoistway. In some occupancies, it may not be appropriate.

2.12.6.2.3 [Rule 111.6b(3)]
This new Rule was adopted from B44 Clause 2.12.9.2.3.

2.12.6.2.4 [Rule 111.6b(4)]
Editorially corrected to direct to Section 5, Security.

2.12.6.2.5 [Rule 111.6b(5)]
To include height restriction for locked panel.

2.12.7.1 (Rule 111.7a)
Unsafe to allow access on every floor due to wiring mishaps. The intent is that the switch be used only by elevator personnel.

2.12.7.2.2 [Rule 111.7b(2)]
Editorially corrected to direct to Section 5, Security.

2.12.7.2.3 [Rule 111.7b(3)]
To address reliability of switch action, same as an electrical protective device requirement.

2.12.7.3.3 [Rule 111.7c(3)]
Editorially corrected to direct to Section 5, Security.

2.12.7.3.5 [Rule 111.7c(5)]
Modification defines hoistway access operation as a lower priority than top-of-car and in-car inspection operations.

2.12.7.3.8 [Rule 111.7c(8)]
To coordinate with changes made to Rule 210.

2.13.3.3.2 [Rule 112.3c(2)]
Reference added for button markings.

2.13.4.2.1 [Rule 112.4b(1)(a)]
Instantaneous kinetic energy is introduced. Its numerical value is derived from sinusoidal velocity dictation where the maximum value is equal to 1.57 squared times the average, where the average results in a kinetic energy equal to the historical value of 7 ft-lbf. Therefore, the instantaneous value is equal to $7.0 (1.57)^2 = 17.25$ ft-lbf $\times 1.356 = 23.4$ J, rounded to 23 J. The same rationale is used to quantify the maximum instantaneous value of the reduced KE.

2.13.5.1 [Rule 112.5(a)]
Clarification of intent. To allow less than full reopening.

2.13.5.4 [Rule 112.5(d)]
Editorial clarification of intent.

2.13.6.1.2 [Rule 112.6a(2)]
This is now covered in revised Rule 207.4 but is included here for clarification.

2.14.1.2.1 (Rule 204.1b)
Take into account new requirements for overspeed protection.

2.14.1.4.1 [Rule 204.1d(1)]
Clarification for multi-compartment elevators carrying freight.

2.14.1.5.1 [Rule 204.1e(1)]
To coordinate with revisions to 204.1e(2).

2.14.1.5.1(b) [Rule 204.1e(1)(b)]
Clear passageway is the safety objective and it is defined by other dimensions required by the Rule.

2.14.1.5.1(b)(2) [Rule 204.1e(1)(b)(2)]
Clarification. Not previously covered in A17.

2.14.1.5.1(c) [Rule 204.1e(1)(c)]
Clarification of existing requirements for elevators with two compartments.

2.14.1.5.1(e) [Rule 204.1e(1)(e)]
Requirements for hydraulic elevators belong in that Section.

2.14.1.5.1(f) [Rule 204.1e(1)(f)]
Editorial relocation of existing requirement. Addition from TR 91-12. Revised to clarify "all elevators." To prevent operation with emergency exit open.

2.14.1.5.2 [Rule 204.1e(2)]
Adopt B44 Clause 3.6.1.5.2, as modified, as it gives more options for the elevator designer and for passenger evacuation. Top emergency exit is now permitted due to provisions for railings on top of the car.

2.14.1.7.1 [Rule 204.1g(1)]
For protection of personnel on top of cars. Clarification of intent.

2.14.1.8.1 [Rule 204.1h(1)]
To provide for the protection of the passengers in case glass is broken.

2.14.1.8.1(a) [Rule 204.1h(1)(a)]
Wired glass is not allowed to be used in hazardous locations, according to CPSC requirements.

2.14.1.8.2 [Rule 204.1h(2)(b)]
This paragraph is rewritten in performance language.

2.14.1.8.2(d) [Rule 204.1h(2)(d)]
Put requirements in performance terms.

2.14.1.8.3 [Rule 204.1h(3)]
To allow Canadian jurisdictions to use type 3C film reinforced mirror glass.

2.14.1.9.1(d) [Rule 204.1i(1)(d)]
Reference to Rule 207.1b(3) has been deleted.

2.14.1.9.1(e) [Rule 204.1i(1)(e)]
Relocated requirement from Rule 110.10c and clarified.

2.14.1.9.1(f) [Rule 204.1i(1)(f)]
To require secure fastenings for the safety of passengers.

2.14.1.10.2(f) [Rule 204.1j(2)(f)]
Editorially corrected to direct to Section 5, Security.

2.14.2 [Rule 204.2a(1)(c)]
To incorporate both A17 and B44 requirements for car enclosure.

2.14.2.1 (Rule 204.2a)
Based upon an agreement of all members of the committee to change to a flame spread rating of 0 to 75, the committee has taken the Canadian requirements and the American requirements and harmonized them. As a result, the only change that effects the jurisdictions enforcing the NBCC is that the maximum flame spread rating has been changed for elevators not classified as firefighters elevators from 0 to 150 to 0 to 75 and a smoke development rating has been added. This eliminated the need for separate requirements for jurisdictions enforcing the NBCC and the United States.

2.14.2.1.2 [Rule 204.2a(1)(d)]
Added due to requirements in NBCC.

2.14.2.1.5 [Rule 204.2a(2)]
Reference both floor covering requirements.

2.14.2.2(f) [Rule 204.2b(6)]
Deleting this statement throughout Code, and to provide coordination with Rule 204.2f.

2.14.2.6 (Rule 204.2f)
To provide access for cleaning glass. Rule 1206.9 contains procedures for cleaning glass.

2.14.2.6(b) [Rule 204.2f(2)]
Editorially corrected to direct to Section 5, Security.

2.14.3.1 (Rule 204.3a) — Second paragraph
Revised to provide a closer tolerance.

2.14.4.2.1 [Rule 204.4b(1)]
Side emergency exits are covered in Rule 204.1j. Alternative to required specified clearances in Rule 108.1e. Clarification of intent. Not covered in A17, use B44.

2.14.4.2.2 [Rule 204.4b(1)(a)]
Not covered in A17, use B44.

2.14.4.2.3(a) [Rule 204.4b(2)(a)]
Not covered in A17, use B44.

2.14.4.5.1(d) [Rule 204.4e(1)(d)]
To allow a larger clearance on freight elevators not accessible to the general public.

2.14.4.7.3 [Rule 204.4h(3)]
To reduce pinch and shearing hazards.

2.14.4.9 (Rule 204.4j)
To reduce the probability of failure of suspension means.

2.14.4.11(a) [Rule 204.4m(1)]
To require clearances similar to a hoistway door in a car door interlock situation.

2.14.4.11(c) [Rule 204.4m(3)]
To require clearances similar to a hoistway door in a car door interlock situation. Prior clearance allows an excessive gap not needed in modern elevators. It is outdated.

2.14.5.1 (Rule 204.5a)
Reference to AHJ deleted in accordance with ASME and CSA guidelines.

2.14.5.6.1 [Rule 204.5g(1)]
To be consistent with door panel molding requirements.

2.14.5.8.2(a) [Rule 204.5i(2)(a)]
To maintain continuity with Rule 204.1h.

2.14.5.8.2(c) [Rule 204.5i(2)(c)]
Items deleted not a safety issue.

2.14.5.8.2(f) [Rule 204.5i(2)(f)]
Structural requirements will dictate the thickness of the glass. To be consistent with Rule 110.7b(7) — clarification.

2.14.6.2.3(a) [Rule 204.6b(3)(a)]
Minimum height requirement is in Rule 110.1a.

2.14.6.3.1 [Rule 204.6c(1)]
To be consistent with Rule 204.5b.

2.14.6.3.5 [Rule 204.6c(5)]
Relocated from Rule 204.4d in TR 90-48. Editorial.

2.14.6.3.6 [Rule 204.6c(6)]
Relocated from Rule 204.4g in TR 90-48.

2.14.7.1.3(a) [Rule 204.7a(3)(a)]
To allow for activation delay.

2.14.7.1.3(b) [Rule 204.7a(3)(b)]
Add performance requirements for battery lighting.

2.14.7.1.3(b)(3) [Rule 204.7(a)(3)(b)(3)]
To conform to NEC/CEC requirements.

2.14.7.2.1(b) [Rule 204.7b(1)(b)]
Editorially corrected to direct to Section 5, Security.

2.14.7.4(a) [Rule 204.7d(1)]
To allow drop ceilings to be considered adequate guarding. Clarification of intent.

2.15.2 (Rule 203.2)
To provide protection against possible collision between car and other equipment in the hoistway.

2.15.5.5 [Rule 203.5(e)]
Equivalent Canadian Standard added.

2.15.6.2.1(a) [Rule 203.6b(1)(a)]
Equivalent Canadian Standard added.

2.15.6.2.1(c) [Rule 203.6b(3)]
Reference appropriate standard for elongation.

2.15.8 (Rule 203.8)
Acceptance criteria rewritten in performance terms measurable by the same standardized tests as required for evaluating materials for car enclosures, Rule 204.2a.

2.15.8(b)(1) [Rule 203.8(b)(1)]
To correct terminology.

2.15.9.2 [Rule 203.9(b)]
Since the car door is locked (see door restrictor requirements, Rule 111.5) when the elevator is out of the unlocking zone, a long toe guard is not necessary to prevent people from falling under the car on hydraulic elevators. We do not have the hazard, which can be created if the car is stopped out of floor level with the door open because of uncontrolled low speed movement detection.

Table 2.15.10.1 (Table 203.10)
AISC formula is more appropriate.

2.15.10.2 [Rule 203.10(b)]
To provide criteria for the safe design of the respective components/structures subject to forces developed during the retardation phase of the emergency braking and all other loading acting simultaneously, if applicable, with factors of safety consistent with the types of equipment/structures involved. See also Rules 105.6, 200.5c, 202.2c, 208.3d, 216.4, and the Note to Clause 3.2 in B44.

2.15.17 (Rule 203.17)
See rationale for Rule 202.4.

2.16.2.2.3 [Rule 207.2b(3)]
Revised definitions of Class C1, C2, and C3 for clarity. Also, see Section 3 for proposed definitions of *load, dynamic*; *load, static*; and *load, impact*.

2.16.3.2.2(e) [Rule 207.3b(2)(d)]
Lubrication instructions are important as it may affect stopping performance of safety.

2.16.4 (Rule 207.4)
To modify requirements for carrying of passengers on new freight elevators.

2.16.4.4 [Rule 207.4(d)]
To require the use of interlocks and eliminate mechanical lock and contacts. Editorial reorganization. Editorial rewrite. To require that car doors be used, which meet passenger-type construction.

2.16.4.7 [Rule 207.4(g)]
To restrain passengers from exiting the hoistway at any point outside the unlocking zone.

2.16.4.8 [Rule 207.4(h)]
To ensure adequate safety factor for passenger elevator use.

2.16.4.9(b) [Rule 207.4i(1)]
All door types should be equipped with noncontact reopening devices.

2.16.4.9(c) [Rule 207.4i(3)]
To require sequence opening and closing of power hoistway doors for passenger protection.

2.16.4.9(d) [Rule 207.4i(4)]
To require a smooth interior door face.

2.16.4.9(e) [Rule 207.4i(5)]
Do not consider the visual warning necessary with momentary pressure operation.

2.16.5.1.1 [Rule 207.5a(1)]
Editorial, as signs in both kg and lb should not be mandatory. Can use signs with kg, lb, or both.

2.16.8 (Rule 207.8)
For clarification that this requirement is for the down direction only.

2.16.8(c) [Rule 207.8(c)]
Editorial coordination.

2.16.8(h) [Rule 207.8(h)]
See rationale for Rule 208.8.

2.16.8(i) [Rule 207.8(i)]
Inadvertently omitted from both Codes. Rule 211.2a refers to this requirement.

Table 2.17.3 [Table 205.3]
"Metric Units" and "Imperial Units" have been added for clarity.

2.17.4 (Rule 205.4)
See rationale for Rule 208.8.

2.17.6 (Rule 205.6)
Deleted to avoid conflict with ascending car and overspeed protection requirements in Section 216.

2.17.7.1 [Rule 205.7(a)]
See rationale for Rule 208.8.

2.17.7.2 [Rule 205.7(b)]
The requirements from Rule 206.4 are now in Rule 205.7(c); the reference to Rule 206.4 is deleted.

2.17.7.3 and 2.17.7.4 [Rules 205.7(c) and 205.7(d)]
Requirements have been relocated from Rules 206.4a and 206.4c.

2.17.8.2.7 [Rule 205.8b(7)]
Clarification.

2.17.8.2.8 [Rule 205.8b(8)]
Clarification.

2.17.9.2 (Rule 205.9b)
The Committee is attempting to delete Notes where appropriate.

2.17.12.1 [Rule 205.12(a)]
The factor of safety definition includes the ultimate strength. ASTM E8 is the appropriate standard for the elongation test.

2.17.12.4 [Rule 205.12(d)]
Because of the limited availability of phosphor bronze rope, the Wire Rope Technical Board has recommended galvanized steel rope as an alternative. However, a United States source of phosphor bronze rope has been identified. The revised wording allows for the usage of either type.

2.17.14(d) [Rule 205.14(d)]
Requirement was added per TRs 92-66 and 93-31, which were approved for A17.1a–1997. TRs also included revisions to Rules 206.6(b) and 206.9(c).

2.17.14(e) [Rule 205.14(e)]
Added for traceability.

2.17.16 (Rule 205.16)
Cross-reference added for clarity.

2.18.1.1 [Rule 206.1(a)]
See rational for Rule 208.8.

Table 2.18.2.1 (Table 206.2a)
Governor overspeed switches are now always required. See proposed Rule 206.4a.

2.18.3.3 [Rule 206.3(c)]
Adopt B44 Clause 3.8.3.4 not previously covered in A17.1.

2.18.4 (Rule 206.4)
Reorganization of Rule 206.4 to cover speed governors only and Section 205 to cover car safeties. Inherent risk of overspeed exists regardless of speed or type of control (e.g., the motor is always connected to the line but can also have a failure of contactor to the main line.)

2.18.4.1.2 [Rule 206.4a(2)]
Clarification.

2.18.4.3 [Rule 206.4b(6)]
Should try to electrically shut system down before you activate safety. To correct an incorrect reference.

2.18.5 (Rule 206.5c)
This requirement will be in the new Maintenance section.

2.18.5.1 (Rule 206.5a)
Delete, material not necessary. Covered in Rule 1202.7.

2.18.5.3(h) [Rule 206.5d(8)]
For editorial consistency.

2.18.6.1 [Rule 206.6(a)]
Clarification of intent.

2.18.6.2 [Rule 206.6(b)]
Revised per TRs 92-66 and 93-31, which were approved for A17.1a–1997. TRs also included revisions to Rules 205.14d and 206.9(c).

2.18.7 (Rule 206.7)
This switch will no longer be an electrical protective device but it still needs to be a switch to allow car to go to next level before car is shut down, so as not to trap passengers.

2.18.8.1 [Rule 206.8(a)]
ASTM E8 is the appropriate elongation standard.

2.18.9(c) [Rule 206.9(c)]
Added per TRs 92-66 and 93-31, which were approved for A17.1a–1997.

2.18.9(d) [Rule 206.9(d)]
For traceability.

2.18.9(e) [Rule 206.9(e)]
To ensure that pull-through force is not adversely affected by lubrication. Rule 1206.1c states that governor rope shall not be lubricated.

2.19.1 (Rule 216.1)
See rationale for Rule 208.8.

2.19.1.1 (Rule 216.1a)
Redundant wording deleted.

2.19.3.2 (Rule 216.3b)
More than one emergency brake may be appropriate to satisfy various modes of failure.

2.19.4 (Rule 216.4)
To provide criteria for the safe design of the respective components/structures subject to forces developed during the retardation phase of the emergency braking and all other loading acting simultaneously, if applicable,

with factors of safety consistent with the types of equipment/structures involved. See also Rules 105.6, 200.5c, 202.2c(3), 203.10(b), 208.3d, and the Note to Clause 3.2 in B44.

2.20.2.2(i) [Rule 212.2b(9)]
Editorial consistency.

2.20.2.2(j) [Rule 212.2b(10)]
Important to know what type of lubricant to use and not use.

2.20.5 (Rule 212.5)
To remove the requirement for fatigue testing and to clarify the approval requirements for suspension rope equalizers of types other than the individual compressive spring type. If the operation of the equalizers is determined to be proper when inspected as required by Rule 1001.2c(28) of A17.1b–1995, load fluctuations will not be large enough to cause stress reversals and, therefore, induce fatigue failures. See also new definition for *engineering test*.

2.20.6 (Rule 212.6)
Deleting administrative requirements.

2.20.8 (Rule 212.8)
Covered by Rule 1200.4d.

2.20.9.1(a) [Rule 212.9a(1)]
Deleting administrative requirements. To coordinate with revisions to Part XI. See also new definition for *engineering test*.

2.20.9.3.4 [Rule 212.9c(4)]
ASTM Standards E8 is the elongation test standard.

2.20.9.5.1 [Rule 212.9e(1)]
See definition for *engineering test*.

2.20.9.7.4 [Rule 212.9g(3)]
EPA lists trichloroethane as a hazardous air pollutant.

2.20.9.8 (Rule 212.9h)
To provide performance language for the installation of antirotation devices.

2.20.10.1 [Rule 212.10(b)]
See new definition for *engineering test*.

2.20.10.9 [Rule 212.10(i)]
Editorial consistency.

2.21.1.2 (Rule 202.1b)
Additional clarification that weights should not be dislodged in the event of buffering or safety application.

2.21.1.3 (Rule 202.1c)
To provide protection against possible collision between counterweight and other hoistway equipment.

2.21.2.3.3 [Rule 202.2c(3)]
To provide criteria for the safe design of the respective components/structures subject to forces developed during the retardation phase of the emergency braking and all other loading acting simultaneously, if applicable, with factors of safety consistent with the types of equipment/structures involved. See also Rules 105.6, 200.5c, 203.10(b), 208.3d, 216.4, and the Note to Clause 3.2 in B44.

2.21.4 (Rule 202.4)
The Rule is rewritten and expanded to cover the whole compensation system. There is no safety justification for requirements that the means must be fastened to the frame, as long as they meet the strength requirements. In second paragraph, reference to a specific design deleted.

2.22.3.2.3 [Rule 201.3b(3)]
Reference revised to coordinate with revised Rule 109.1. To make it clear that requirement is in addition to (1) and (2) of the same Rule.

2.22.4.1 (Rule 201.4a)
Note revised for editorial consistency. The Committee felt this last paragraph of Rule 201.4a is helpful information to the reader and should remain; however, it is not a requirement.

Table 2.22.4.1 (Table 201.4a)
"SI Units" and "Imperial Units" have been added for clarity.

2.22.4.3 (Rule 201.4c)
Clarification of requirement, adding standard for conducting elongation test and editorial revision/reorganization.

2.22.4.5(c) [Rule 201.4e(3)]
Rule 210.2(v) contains the deleted requirements. Can be used for electric and hydraulic elevators.

2.22.4.7 (Rule 201.4g)
The deceleration is dependent upon the stroke of the buffer. The Rule ensures that the longest stroke for a given buffer type is tested. The Rule allows buffers previously qualified to be used without retesting.

2.22.4.7.3 (Rule 201.4g)
If the buffer is listed/labeled with a reference to a specific A17.1 or B44 editions, there is no need for any statement by the organization installing the buffer.

2.22.4.10 (Rule 201.4j)
To aid the user.

2.22.4.11(g) [Rule 201.4k(7)]
For traceability and for consistency with marking requirements throughout Code.

2.22.4.11(h) [Rule 201.4k(8)]
For verification and certification.

2.23.1 (Rule 201.1a)
Editorial reorganization.

2.23.2.1(a) [Rule 200.2a(1)]
Clarification of requirement. Adding standard for conducting elongation test.

2.23.3 (Rule 200.3)
Restates Rule in purely performance language exclusive of local administrative requirements.

2.23.4.1(c) [Rule 200.4a(3)]
Example added for those using metric dimensions.

2.23.4.3.1 [Rule 200.4c(1)]
Revise to take into account the two smaller rails (6.25, 5.75).

Table 2.23.4.3.1 [Table 200.4c(1)]
"SI Units" and "Imperial Units" have been added for clarity. Adopt the two smaller rails.

Table 2.23.4.3.3 [Table 200.4c(2)]
Good engineering practice would not have limits fall on top of one another.

2.23.5.3 (Rule 200.5c)
To provide criteria for the safe design of the respective components/structures subject to forces developed during the retardation phase of the emergency braking and all other loading acting simultaneously, if applicable, with factors of safety consistent with the types of equipment/structures involved. See also Rules 105.6, 202.2c(3), 203.10(b), 208.3d, 216.4, and the Note to Clause 3.2 in B44.

2.23.6 (Rule 200.6)
Reference for Rule on lubrication added.

Table 2.23.7.2.1 [Table 200.7b]
"SI Units" and "Imperial Units" have been added for clarity. Two smaller rails were added.

2.23.7.2.2 [Rule 200.7b(2)]
States Rule in purely performance language exclusive of local administrative requirements.

2.23.9.1.2 [Rule 200.9a(3)]
For clarity, the Note in the second paragraph has been incorporated into Rule as Rule 200.9a(3). The Note is similar to Note after B44 Clause 3.2.9.2.

2.23.9.2 (Rule 200.9b)
Rewritten in performance language.

2.23.9.2.1(b) [Rule 200.9b(2)]
Clips not covered in A17.1. Use B44.

2.23.9.3 (Rule 200.9c)
B44 requirements in Clause 3.2.9.2.2 rewritten in performance language.

Table 2.23.10.2 (Table 200.10b)
"SI Units" and "Imperial Units" have been added for clarity. Two smaller rails were added.

2.24.1 (Rule 208.1)
Editorial clarification.

2.24.2.3.1 [Rule 208.2c(1)]
To clarify requirements.

2.24.2.4 (Rule 208.2d)
To clarify requirement.

2.24.3 (Rule 208.3)
Delete as this is covered in the definition.

2.24.3(a) [Rule 208.3(a)]
Clarification of requirement. ASTM Standard E8 contains the elongation test.

2.24.3.1.1 [Rule 208.3c(1)]
To be consistent with sound engineering practice.

NOTE: the B44 definition of endurance limit (as modified), is being adopted in addition to Clause 3.10.3.2.

2.24.3.1.2 [Rule 208.3c(2)]
The number of cycles removed from definition and relocated into this Rule.

2.24.3.2 (Rule 208.3d)
To provide criteria for the safe design of the respective components/structures subject to forces developed during the retardation phase of the emergency braking and all other loading acting simultaneously, if applicable, with factors of safety consistent with the types of equipment/structures involved. See also Rules 105.6, 200.5c, 202.2c, 203.10(b), 216.4, and the Note to Clause 3.2 in B44.

2.24.8 (Rule 208.8)
Machine brake system requirements have been completely revised concurrently with the introduction of ascending car/uncontrolled low speed protection and are harmonized with the comparable requirements in the B44 Code.

To provide protection against ascending car overspeed to prevent collision of the elevator with the building overhead. The proposed A17.1 requirements harmonize with B44 (see Rule 216.1)

To prevent movement of the elevator away from the landing when the car and hoistway doors are open. The proposed A17.1 requirements harmonize with B44 (see Rule 216.2).

To ensure the safe functioning of the detection means, redundancy is required so that no single component fault will lead to an unsafe condition.

Once the detection means has indicated an overspeed or unintended car movement away from the landing with the doors open, a manual reset of the detection means is required.

The reason for specifying that power be removed from the drive motor of the motor-generator set is because of the concern for protection against the failure of the "suicide circuit."

An important consideration in formulating this requirement is the fact that it will prevent regeneration back to the AC line, and in so doing, may extend the braking distance.

2.24.8.1 (Rule 208.8a)
See new definition for *braking system*. Rule 210.8 was an incorrect reference as it deals with release and application of the DM brake, not the braking system.

2.24.8.2.1 and 2.24.8.2.2 (Rules 208.8b and 208.8c)
ISO report recommended defining braking system and adding requirements for brake capacity.

2.24.8.4 (Rule 208.8d)
Performance requirements to allow for the use of manual brake release tools.

2.24.8.5 (Rule 208.8e)
Vital for maintaining brakes in safe operating condition.

2.24.8.6 (Rule 208.8f)
To ensure brakes continued safe operation.

2.24.9 (Rule 208.9)
To coordinate with editorial changes to Section 3, Definitions.

2.25.2.3 (Rule 209.2c)
Changes made are to harmonize and are editorial, intent remains the same.

2.25.2.3.3 [Rule 209.2c(3)]
Editorial clarification.

2.26 (Rule 210.12)
Delete Rule 210.12, it is outdated. The Code clearly says that the car must not move with open doors except as specified in Rule 204.4b(2).

2.26.1.4.1 [Rule 210.1d(1)]
General requirements for inspection operation and transfer switch have been grouped under the heading "General Requirements." A17.1–1996, Rules 210.1d(2)(c), 210.1d(3)(a), and 210.1d (4)(a) consolidated in one general Rule and clarified. Transfer switch labeling requirements added. Transfer switch functional requirements are defined for the "INSPECTION" position.

2.26.1.4.1(b)(4)(b) [Rule 210.1d(1)(b)(4)(b)]
Clarified that "positively opened contacts" are only required when transferring from "NORMAL" to "INSPECTION" and not from "INSPECTION" to "NORMAL."

2.26.1.4.1(b)(4)(c) [Rule 210.1d(1)(b)(4)(c)]
Require that automatic power door opening and closing shall be disabled anytime that the inspection transfer switch is in the "INSPECTION" position.

2.26.1.4.1(b)(5) [Rule 210.1d(1)(b)(5)]
Transfer switch functional requirements are defined for the "NORMAL" position.

2.26.1.4.1(c) [Rules 210.1d(1)(d)(1) and (2)]
General requirements for inspection operating devices have been added, including labeling requirements. "Operation" separated from "operating devices."

2.26.1.4.1(d)(3) [Rule 210.1d(1)(c)]
Reference to bypass switch operation. To clarify that doors may be held closed electrically.

2.26.1.4.2 [Rule 210.1d(2)]
Requirements for top-of-car inspection operation are grouped together in this requirement.

2.26.1.4.2(a) [Rule 210.1d(2)(a)]
Requirements added for top-of-car switches' permanent location.

2.26.1.4.2(c) [Rule 210.1d(2)(c)]
To include a B44 requirement not previously covered in A17.1. Personnel safety has been adequately addressed by requiring the stop switch to be readily accessible from the hoistway entrance. Requirements added for an "ENABLE" device to protect against accidental activation of the inspection operating devices or the failure of the inspecting operating device (i.e., failure to release of a single button).

2.26.1.4.2(e) [Rule 210.1d(2)(e)]
The inspection operating device is permitted to be portable as long as the "ENABLE" and stop switch are included within the portable unit.

2.26.1.4.3 [Rule 210.1d(3)]
Requirements for in-car inspection are grouped in this requirement.

2.26.1.4.3(d) [Rule 210.1d(3)(d)]
A separate switch or switch position is now required to enable hoistway access switches.

2.26.1.5 (Rule 210.1e)
Revisions based on B44 proposal, which includes renumbering as shown to include as part of the inspection requirements. Editorial change made, less restrictive requirement.

2.26.1.5.2 [Rule 210.1e(2)]
Editorial revisions made based on letter ballot comments.

2.26.1.5.5 [Rule 210.1e(5)]
Need to add this to refer to conditions for specific types of "inspection" operation.

2.26.1.5.6 [Rule 210.1e(6)]
Need to add this to refer to conditions for specific types of "inspection" operation.

2.26.1.5.7 [Rule 210.1e(7)]
This requirement is added to avoid the need for future interpretations. Not currently in A17.1, therefore, used B44 language.

2.26.1.5.8 [Rule 210.1e(8)]
Rules 210.1e(1) through (7) contain requirements for when one switch only is activated. Rule 210.1e(8) contains requirements when both switches are activated.

2.26.1.5.10(c) [Rule 210.1e(10)(c)]
Harmonized proposal provides further clarification of requirements.

2.26.1.5.10(e) [Rule 210.1e(10)(e)]
Harmonized proposal provides further clarification of requirements.

2.26.1.5.10(e)(4) [Rule 210.2e(10)(e)(4)]
To address concerns regarding safety of persons that are in the car, if they do not answer the call and the safety of moving the car even if there is no person inside.

2.26.2 (Rule 210.2)
New wording to establish general requirements for all electrical protective devices under Rule 210.2. All electrical protective devices will have the same response when "activated," i.e., cause power to be removed from the elevator driving machine motor and brake. In addition, cross-reference of Rules, which refer back to Rule 210.2 requirements. A number of Sections have editorial changes for clarity. Deleted "cause power to be removed from the elevator driving machine motor and brake," since this requirement is now covered in the opening paragraph. In some cases, clarified what occurs to activate the electrical protective device, and referenced the Rule that requires the electrical protective device.

2.26.2.4 [Rule 2.10.2(d)]
Electric power should be removed from the motor armature not necessarily from the field.

2.26.2.5 [Rule 210.2(e)]
Editorial clarification.

2.26.2.6 [Rule 210.2(f)]
Broken rope, tape, or chain switch is the same electrical protective device, regardless of the function of the rope, tape, or chain; to activate the normal terminal stopping device or the emergency terminal speed limiting device or anything else. Furthermore, there is no need to repeat performance requirements specified in other Rules [Rule 209.2c(2) or 209.4a(8)(b)].

2.26.2.9 [Rule 210.2(i)]
Rules 206.4a and 206.4c no longer apply.

2.26.2.15 [Rule 210.2(o)]
This contact is not required on all elevators anymore.

2.26.2.26 [Rule 210.2(z)]
Harmonized to cover the same requirements as currently listed in B44.

2.26.2.27 [Rule 210.2(aa)]
Rule 103.3(c) requires a stop switch conforming to Rules 210.2(e)(1), (2), and (3) much the same as Rule 101.3e requires a stop switch in overhead machinery spaces. Generally, Rule 210.2 is viewed as a total shopping list of all of the electrical protective devices required in the Code. As such, this switch required by Rule 103.3(c) should be included in Rule 210.2.

2.26.2.28 [Rule 210.2(bb)]
Harmonized to cover the same requirements as currently listed in B44.

2.26.2.31 [Rule 210.2(dd)]
New requirements added in Rule 204.2f(5), which are electrical protective devices.

2.26.2.32 [Rule 210.2(ff)]
Recognizes that a locking device is necessary when access openings for cleaning of car and hoistway enclosures is required in accordance with Rule 110.1d.

2.26.3 (Rule 210.3)
To clarify that springs may be used but spring failure shall not cause the monitoring contact(s) to indicate an open critical circuit contact if the critical circuit contact is not open.

2.26.4 (Rule 210.4)
There is a difference between A17.1 and B44 with regard to electrical equipment and wiring; therefore, exceptions [Rules 210.4(a), (b), and (c)] are listed until this requirement can be moved to Section 38 of the Canadian Electrical Code.

2.26.4.2 [Rule 210.4(b)]
References both Canadian and US requirements. Rule 210.4(b) adopted B44 language.

2.26.4.3 [Rule 210.4(c)]
Adopted B44 language in Clause 3.12.4.3. Consolidates the requirements for "electrical protective devices" under one Rule.

2.26.4.4 [Rule 210.4(d)]

There have been a number of reports of car movement initiated by two-way radios and other forms of RF transmission. This new Rule is proposed to address this issue. The Rule directly references EN 12016, a European Standard being developed specifically for elevators, escalators and other forms of passenger conveyors. The Rule specifies control equipment requirements, which must be complied with when tested to the interference levels given within the draft EN standard. Warning sign requirements are given for the case where doors or other equipment must not be removed in order to meet the immunity requirements.

2.26.8 (Rule 210.8)

Editorially harmonized, intent remains the same.

2.26.9.3 [Rule 210.9(c)]

To introduce the concept of a software failure and the effects on elevator safety. Software may fail in numerous ways and at any time without warning. Software failures describe software output that does not do what is intended (specified).

Redundancy may be used to meet the requirements of Rule 210.9c. If redundant devices such as relays, contactors, or solid-state devices are used, the implementation of such hardware devices must ensure the integrity of the redundancy, and monitoring (checking) is also required.

Software may be used to implement the redundancy if the software is not the only means available to disconnect the driving motor and brake.

The use of software is increasingly becoming a part of elevator control systems. As with hardware, failures in software can occur. The software may be involved directly or indirectly with the elevator safety. Such a failure must result in a "nondangerous" state of elevator operation as listed. (Also, see definition for "software system failure"). To introduce the concept of single failures and affects on elevator safety when on hoistway access, inspection bypass and leveling operation.

2.26.9.4 [Rule 210.9(d)]

New requirement not previously in A17, to check redundancy. Where software is used, removal of power from the elevator driving machine motor and brake cannot be solely dependent on software. Additional clarification: "checked prior to each start of the elevator from a landing" means checking for failures anytime since the last start from a landing.

2.26.11 (Rule 210.11)

To harmonize with requirements in B44 Clause 3.12.1.6 and acknowledge jurisdictions enforcing ANSI A117.1 and ADAAG, in response to comments received on both the 1st and 2nd letter ballots. Former Rule 210.11 is deleted, because setting the voluntary provided load weighing device as any percentage of the rated load would in no case diminish the safety level assured by the enforcement of all mandatory Code Rules.

2.26.13 (Rule 210.13)

To coordinate with requirements of CABO A117.1-1997 edition. Note symbols if not shown remain the same as in the first ballot.

2.27 (Section 211)

Note added. To recognize that requirements may be found in additional Codes.

2.27.1.1.2 [Rule 211.1(a)(2)]

For elevators with travel under 18 m, you can communicate from outside the hoistway to within the car without the need of a system providing two-way conversation. In addition, the Emergency Operations Committee is of the opinion that even where the communication system is provide, it is not used in low-rise installations.

2.27.1.2 [Rule 211.1(b)]

Editorial clarification; reference to authorized personnel is defined in Section 3.

2.27.2 (Rule 211.2)

Emergency Power Systems is used since the title recognizes terminology used in the NFPA 70 and the Canadian Electrical Code. To establish performance criteria.

2.27.2.1 [Rule 211.2(a)]

Facilitate the availability of power for the elevator system. Speeds up the evacuation process, minimizes entrapments, and facilitates firefighting operations.

2.27.2.3 [Rule 211.2(c)]

To indicate that the emergency or standby power is in effect.

2.27.2.4.1 [Rule 211.2(d)(1)]

Providing requirements to automatically recall elevators, when emergency or standby power is provided, to facilitate passenger evacuation. To provide firefighters with control over the allocation of emergency or standby power. In addition, B44 requirements have been included as they are more suitable for general understanding or interpretation.

2.27.2.5 [Rule 211.2(e)]

This is equivalent to former Rules 211.2(a) and (b).

2.27.3 (Rule 211.3)

Editorial clarification of the type of operation being described within this requirement. This exception is taken to recognize that in Canada the scoping requirements for providing Emergency Operations is within the National Building Code of Canada.

2.27.3.1 (Rule 211.3a)

To prevent the switch from being located in unknown

or unrelated locations for fire department operations. Labeling is to indicate to the end user (firefighter) the identity and functions of the switch. The "BYPASS" position was developed as a result of early, less reliable smoke detection equipment, to give building owners and managers a means of operating elevators in the event of nuisance alarms caused by elevator recall smoke detectors. However, system-type smoke detectors that are monitored by the fire alarm system have replaced single station detection devices. System-type detectors have evolved to become much more reliable, and are much less prone to nuisance alarms. System-type smoke detectors permit bypassing of a single malfunctioning fire alarm initiating device. This will only prevent this device from initiating a false alarm signal, and removal will produce a trouble signal at the fire alarm control panel. Trouble signals will not initiate elevator recall and the cause of the signal must be corrected by the building owner/manager. Removal of this device will not prevent operation of any other initiating device used on the system, including those used for elevator recall. In contrast, the "BYPASS" position eliminates all detectors from initiating elevator recall. A17.1 recognizes that a standard for fire alarm systems used in elevator recall has been developed.

This Committee also wishes to delegate responsibility of bypassing fire alarm equipment to the appropriate technical committee. "Be readily accessible" was added to use performance terminology. See new definition.

The light advises that Phase I has been activated. It will provide an indication that a secondary switch or fire alarm initiating device is still activated within the building, when attempting to restore normal elevator operation.

2.27.3.1.6(a) [Rule 211.3a(1)]
Editorial clarification. Requirement remains unchanged.

2.27.3.1.6(b) [Rule 211.3a(2)]
Editorial clarification, intent is still similar.

2.27.3.1.6(c) [Rule 211.3a(3)]
The emergency stop switch and the in-car stop switch are both electrical protective devices and, thus there is no reason to treat them differently.

2.27.3.1.6(d)(3) [Rule 211.3a(4)(c)]
Changed to editorially conform with Rules 211.3(a)(4)(a) and (b).

2.27.3.1.6(f) [Rule 211.3a(6)]
Editorial clarification, intent is still similar.

2.27.3.1.6(g) [Rule 211.3a(7)]
Editorial clarification, intent is still similar.

2.27.3.1.6(h) [Rule 211.3a(8)]
Visual signal is used to coordinate with Fig. 211.7(b) and indicates that the car has been recalled and is available for Phase II. After the doors are closed, the audible signal serves no useful purpose after 5 s. To clarify if the car is already at the designated level.

Figure 2.27.3.1(h) (Fig. 211.3a)
Note added for clarification.

2.27.3.1.6(i) [Rule 211.3a(9)]
To clarify the operation of the "DOOR OPEN" button. The second sentence clarifies that the "DOOR OPEN" button will not be operable on an up-running elevator when the car stops and reverses direction.

2.27.3.1.6(j) [Rule 211.3a(10)]
This is to ensure that there is no accidental recall to the designated level, by switch failure or operator error.

2.27.3.1.6(k) [Rule 211.3a(11)]
This is to ensure that there is no accidental restoration of normal operation by operator error. A key switch left in the "RESET" position shall not restore the elevator to automatic. In addition, to coordinate changes to the "FIRE RECALL" switch. Also, covers former Rule 211.3b(4).

2.27.3.1.6(l) [Rule 211.3a(12)]
Firefighting is considered to be of a higher priority than the convenience of building personnel to park elevators, such as for building security or energy conservation, etc. In addition, at times there may not be anyone available to return the elevators to service for fire personnel.

2.27.3.1.6(o) [Rule 211.3a(13)]
To coordinate with the deletion of Rule 210.11. To permit carrying the maximum load during Firefighters' Emergency Operation.

2.27.3.2 (Rule 211.3b)
Editorially reformatted. Rules 211.3b(1) through (3).

2.27.3.2.1 [Rule 211.3b(2)]
To recognize that only Manual Emergency recall is required by the NBCC. Note that within this Rule the terms "smoke detector" and "elevator lobby" have been utilized because of requirements in the NBCC.

2.27.3.2.3(a) [Rule 211.3b(3)(a)]
Editorial clarification, the intent is still the same.

2.27.3.2.3(b) [Rule 211.3b(3)(b)]
Editorial clarification, the intent is still the same.

2.27.3.2.3(c) [Rule 211.3b(3)(c)]
Editorial clarification.

2.27.3.2.4 [Rule 211.3b(4)(b)]
Editorial clarification. Position indicators at the designated level are required to be operational but not to be operational at any other level. To eliminate any confusion that a key switch is required at an alternate level.

2.27.3.2.5 [Rule 211.3b(5)]
Editorial clarification.

2.27.3.2.6 [Rule 211.3b(6)]
To provide crucial information to firefighters on potential problems with the elevator system and to clarify that only those elevators in imminent danger be notified.

2.27.3.3 (Rule 211.3c)
Labeling is to indicate to the end user (firefighter) the identity and functions of the switch. Editorial clarification, the intent remains the same.

2.27.3.3.1 [Rule 211.3c(1)]
Editorial clarification, the intent remains the same. See also definition of *emergency personnel*.

2.27.3.3.1(b) [Rule 211.3c(1)(b)]
Operating hall position indicators may convey a message the elevator can be used. Only those places where elevator location is required by firefighters, should hall position indicators remain operative. Car position indicators are useful for firefighters. Additional change is an editorial clarification of intent.

2.27.3.3.1(d) [Rule 211.3c(1)(d)]
Editorial clarification of intent.

2.27.3.3.1(e) [Rule 211.3c(1)(e)]
Editorial clarification of intent.

2.27.3.3.1(f) [Rule 211.3c(1)(f)]
Sentence has been moved to Rule 211.3c(1)(c).

2.27.3.3.1(g) [Rule 211.3c(1)(g)]
Clarification of intent.

2.27.3.3.1(h) [Rule 211.3c(1)(h)]
Editorial. The "CALL CANCEL" button is equivalent to the "HOLD." The intent remains the same. To coordinate with Rule 211.3c(7).

2.27.3.3.1(i) [Rule 211.3c(1)(i)]
Clarification of intent.

2.27.3.3.1(k) [Rule 211.3c(1)(k)]
Lobby parking feature. Firefighting is considered to be of a higher priority than the convenience of building personnel to park elevators, such as for building security or energy conservation, etc. In addition, at times there may not be anyone available to return the elevators to service for fire personnel. Additional deletion made because it can be any means.

2.27.3.3.2 [Rule 211.3c(2)]
Editorial clarification, intent remains the same.

2.27.3.3.3 [Rule 211.3c(3)]
To recognize that you may have gone onto Phase II from an alternate level. Erroneously removed since Committee believed it was covered in Rule 211.3c(4).

2.27.3.3.3(b) [Rule 211.3c(3)(b)]
To coordinate with new Rule 211.3c(4). Rule 211.3c(3)(c) is now covered in Rule 211.3c(4).

2.27.3.3.4 [Rule 211.3c(4)]
Provides firefighters a simplified means to safely reverse the elevator when moving towards the fire floor. It also provides consistent operation to firefighters, any time fire operation switch is turned "OFF" car will return to the designated level and eliminate potential of leaving car stranded after a fire emergency. Additional changes are further clarification of intent.

2.27.3.3.5 [Rule 211.3c(5)]
Additional editorial clarification.

2.27.3.3.6 [Rule 211.3c(6)]
During Phase II Operation, it is probable that fire or water will cause shorts or grounds to occur in the hall mounted devices or associated wiring. Since these hall devices are not crucial to operate under Phase II, they should not be allowed to interrupt Phase II Firefighters' Operation. Some of these devices may be required by Rule 211.3c. The requirement recognizes that the required devices may become inoperable, however, Phase II must not be disabled.

2.27.3.4 (Rule 211.3d)
Editorial clarification.

2.27.3.5 (Rule 211.3e)
Editorial clarification, the intent remains the same.

2.27.4 (Rule 211.4)
Editorial clarification, the intent remains the same. Changes within this requirement are made to correlate with changes made in Rule 211.3. This exception is taken to recognize that in Canada the scoping requirements provided for Emergency Operations is within the National Building Code.

2.27.4.1 (Rule 211.4a)
Editorial clarification, the intent remains the same. Changes within this requirement are made to correlate with changes made in Rule 211.3. The original intent of "BYPASS" position (possibly faulty smoke detector) no longer exists.

2.27.4.2 (Rule 211.4b)
Changes within this requirement are made to correlate with changes made in Rule 211.3(b). To provide crucial information to firefighters on potential problems with the elevator system and to clarify that only those elevators in imminent danger be notified.

2.27.5 (Rule 211.5)
Changes within this requirement are made to correlate with changes made in Rule 211.3.

2.27.5.2(b) [Rule 211.5(b)(2)]
Compromise between the A17 and B44 Codes in order to harmonize language.

2.27.6 (Rule 211.6)
Changes within this requirement are made to correlate with changes made in Rule 211.3. See also Rule 211.3a(3). To recognize that taking away elevator from mechanic may be hazardous.

2.27.7 (Rule 211.7)
Changes within this requirement are made to correlate with changes made in Rule 211.3.

2.27.7.1 [Rule 211.7(a)]
The sign is needed for the safety of the firefighters due to the changes in fire service for elevators in the recent years.

2.27.7.4 [Rule 211.7(d)]
To recognize a NBCC requirement.

2.27.8 (Rule 211.8)
Editorial clarification. The intent remains the same. To provide a reference to a recognized product standards for those jurisdictions that wish to use lock boxes and to ensure adequate product quality for design, construction, and security of both lock boxes and keys.

2.28.1 (Rule 214.1)
See rationale for Rule 208.8.

2.28.1(i) [Rule 113.1(i)]
Moved from Rule 100.7 and B44 Clause 3.12.15.7.

2.29 (Section 113)
New Section is derived from B44 Clause 2.3.9 and A17.1 Rules 113.1 and 100.7 with this addition; Rules 211.9 and 208.10 have been deleted.

2.29.1 (Rule 211.9)
See Section 113 for emergency identification of elevators.

3.4.1.6 [Rule 300.8(a)(6)]
To require marking of the spaces where refuge clearance is not provided throughout the entire cross-sectional area of the pit or the top of the car enclosure. The marking is the color for danger.

3.12.1 [Rule 300.12(a)]
Changes made to be consistent with changes made in Parts I and II. Deleted "elevator parking device" because Part III refers to Section 111 and refers to Section 204 in Rule 301.6.

3.14 (Rule 301.6)
Since cars can move with loss of oil, Rule 204.1j can only be permitted when a car has a safety.

3.16.1 (Rule 301.9)
Similar, intent remains.

3.18 (Section 302)
Similar, intent remains.

3.18.1.1 (Rule 302.1a)
Editorial clarification. Can be misconstrued to be other than a mechanical connection.

3.18.1.2.1 [Rule 302.1b(1)]
Use of B44 wording, since it is in clearer performance language. Editorial clarification, intent remains the same.

3.18.1.2.8 [Rule 302.1b(8)]
Use B44, better performance language.

3.18.2.7.1 [Rule 302.2g(1)]
If the maximum free length is exceeded, returning the car to the lowest landing and rendering the elevator inoperative will maintain the elevator in a safe condition. Additionally, the paragraph order was reformatted for clarification. Editorial.

3.18.2.7.2 [Rule 302.2g(2)]
Clarifies the intent of the Rule.

3.18.3.4 (Rule 302.3d)
Same intent.

3.18.3.8 (Rule 302.3h)
This will allow several methods of protection with the same level of safety. In addition, the Rule is stated in performance language.

3.18.3.8.3 [Rule 302.3h(3)]
It was not the intent that the monitored cathodic protection be approved by the authority having jurisdiction.

3.18.5 (Rule 302.5)
Editorial clarification, intent remains the same.

3.19.3.3.1(a) [Rule 303.3c(1)(a)]
Reference is corrected.

3.19.3.3.1(f) [Rule 303.3c(1)(f)]
Change due to new terminology.

3.19.4.6.2 [Rule 303.4f(2)]
Editorial coordination of labeling requirements.

3.19.4.7 (Rule 303.4g)
To cover the design, installation, and testing of overspeed values when provided.

3.19.4.7.3(b) [Rule 303.4g(3)(b)(2)]
To be consistent for dual jack systems.

3.19.5 (Rule 303.5)
Editorial clarification, intent remains the same.

3.19.6.1 [Rule 303.6(a)]
Harmonized wording, clarification of intent of requirements.

3.19.6.2 [Rule 303.6(b)]
To permit field welding to be done by certified field welders.

3.24 (Rule 304.4)
Delete to Rule 304.4. To harmonize with B44. In addition, this application is not being applied to new construction.

3.24.1.1 (Rule 304.1a)
Editorial clarification, same intent.

3.24.5 (Rule 304.4)
To add requirements to address welding of hydraulic machine components as within other Sections.

3.25 (Section 305)
The editorial changes proposed to the title of this device are to provide further clarification to avoid possible confusion with the emergency terminal speed limiting device. On the hydraulic elevator this device, which may be either mechanical or electrical, should slow the car down but not permanently remove power from the operating devices (i.e., motor and valve). These changes only change the name of the device and not the current Code requirements.

3.25.2.2.4 [Rule 305.2b(4)]
Because Rule 306.4 in A17.1–1993 edition has been deleted, these requirements in Rules 305.2b(4)(a) and 306.3a(4) are no longer applicable and should have been deleted at the time. In addition, editorial correction has been made to Rule 305.2b(1) (removal of the word "should").

3.25.2.2.5 [Rule 305.2b(5)]
The proposed changes are to bring Section 305 in sync with Rules 209.2, 209.4a(5), and 209.4a(9).

3.26.1 (Rule 306.1)
By referencing all of Section 210 the following requirements became mandatory:
(a) Inspection operation (Rule 210.1d)
(b) Inspection operation with open door circuits (Rule 210.1e)
(c) Contactors and relays in critical operating circuits (Rule 210.3)
(d) Electrical equipment and wiring (Rule 210.4)
(e) Monitoring of door circuits (Rule 210.5)

(f) Making electrical protective devices inoperative (Rule 210.7)
(g) Redundancy and checking of redundancy [Rules 210.9(c) and (d)]
Anticreep and leveling operation are covered by Rule 306.3.
Phase reversal and failure protection requirement are covered by Rule 306.5.
Machine brake requirements are not applicable.
Requirements in Rules 210.9(a) and (b) are covered in Rule 306.6(a) and (b). Requirements in Rules 210.9(e), (f), and (g) do not apply to hydraulic elevators.

3.26.2 (Rule 306.2)
See rationale for Rule 210.1d.

3.26.3 (Rule 306.3)
Editorial change due to reformatting of Section 210.

3.26.4 (Rule 306.4)
Editorial clarification, intent remains the same. See definition of *hydraulic driving machine*.

3.26.4.1 [Rule 306.4(a)]
Editorial clarification, intent remains the same. Previously Rule implied that devices not listed in Rule 306.4 do not have to meet Rule 210.2. This was not the intent. The following devices in Rule 210.2 could be found on a hydraulic lift, but are not listed under Rule 306.4: broken rope, tape, or chain switch; buffer switch on Type C safeties; gas spring return oil buffer; pit access door switch and others. To avoid adding more items, revision was made to general reference of Rule 210.2.

3.26.4.2 [Rule 306.4(b)]
Current technology may increase deceleration rates.

3.26.4.2(d) [Rule 306.4(b)(4)]
These are now electrical protective devices.

3.26.4.2(f) [Rule 306.4(b)(6)]
Editorial clarification, intent remains the same.

3.26.9 (Rule 306.10)
Editorially reorganized.

3.26.10 (Rule 306.11)
To describe the normal function of auxiliary power lowering operation, which is covered in B44.

3.26.10.3 (Rule 306.11c)
To allow for the fact that since the door buttons remain operative, there is a possibility that unexpected repetitive door cycling and long periods of holding a car at an intermediate stop could drain the battery. To cover requirements formerly in Rule 306.4a(8).

3.27 (Rule 306.7)
If an elevator is already parked out of service, the passengers already out of it and the firemen can't use it. There is no point in doing anything.

The firefighter community has advised that in their opinion passengers should not be restricted from exiting an elevator even if the elevator is not capable of reaching the recall level. It is safer to exit, even at a potential fire floor since the fire is most likely in its early stages; if passengers are not allowed to exit, they may be overcome by smoke.

The low oil indicator means the car cannot reach all floors, but it may still be able to reach the fire recall level; we should make the attempt.

A car with low oil or only emergency lowering will not be useful to the firemen, after Phase I recall and may sink away from the floor, so once the passengers are out at the recall level, we should shut the doors.

Once a fireman is in the car on Phase II, we should give him as much control of the cars as possible.

Provide firefighters with an indicator (flashing, Fig. 211.3a) when an elevator should not be used.

Requirements are covered in Rule 306.1 by reference to Section 210.

5.2 (Part XXV)

Limits on size, capacity, speed, and rise will regulate how this type of elevator is used. It is anticipated that other regulations (i.e., the ADA, CABO/ANSI A117.1, Model Building Codes, etc.) may further dictate where this equipment will be acceptable for accessibility.

5.2.1.4.2(d) [Rule 2500.8b(4)]
For consistency.

5.2.1.4.4(a) [Rule 2500.8d(1)]
Editorial clarification.

5.2.1.11 [Rule 2500.11(c)]
Rule 110.10d has been deleted from Part I.

5.2.1.12(b) [Rule 2500.12(b)]
To coordinate with renumbering to Section 111.

5.2.1.13 (Rule 2500.13)
To provide requirements for power closing of doors by continuous pressure means. While the Committee does not anticipate this to be used, there is no safety reason to prohibit.

5.2.1.14 (Rule 2501.5)
Rule 204.4g no longer exists. The references to Rules 204.4h and 204.4j updated.

5.2.1.15.2 (Rule 2501.4b)
This allows the use of toe guards of less than 48 in. which may be necessary because of shallow pits.

5.2.1.16.5 (Rule 2501.8e)
Relocated from Rule 2501.9a. The allowance for the maximum rated speed of 0.15 m/s (30 ft/min) in Rule 2501.8d is less than the 0.25 m/s (50 ft/min) in Rule 208.1 per reference Rule 208.1(b) — the rated speed of the elevator shall not exceed 0.25 m/s (50 ft/min). The allowance for the maximum travel of 7.6 m (25 ft) in Rule 2501.8e is less than the 12.2 m (40 ft) in Rule 208.1 per reference Rule 208.1(c) — the travel of the elevator car shall not exceed 12.2 m (40 ft).

5.2.1.18(c) [Rule 2501.7(c)]
For consistency with Rule 206.5.

5.2.1.20 (Rule 2501.13)
Expanded to cover the modifications proposed in Rule 2501.13a.

5.2.1.20.1 (Rule 2501.13a)
This Rule clarifies the types of ropes allowed for LU/LA elevators and that ropes previously installed may not be reused.

The substantiation is based on aircraft cable fully tested to a Military Specification and successfully utilized in aircraft. Aircraft cable is currently allowed for Part 5 elevators in the A17.1 Code. These elevators run at 40 ft/min, which is faster than the 30 ft/min allowed for LU/LA. They also are allowed up to 50 ft of rise, which is twice as high as the 25 ft of rise allowed for LU/LA. Residence elevators are currently allowed up to 340 Kg (750 lb) capacity. A LU/LA elevator is allowed up to 630 Kg (1,400 lb) and based on the limited use/limited application, lower speed, lower rise, tight cable specification versus no specification, and information package provided should allow this cable.

5.2.1.20.1(a) [Rule 2501.13a(1)]
This is the same wording as used in Part II and is used to be consistent.

5.2.1.20.1(b) [Rule 2501.13a(2)]
Only Mil Spec cable is allowed in accord with the Wire Rope Technical Board advisement that Most aircraft cable does not meet MIL-W-83420. See also reference for wording on MIL-W-83420 Section 3.2.4. See also reference for wording on MIL-W-83420 Section 3.5.2.

5.2.1.20.2(a) [Rule 2501.13b(1)]
Per the reference and the LU/LA speed of 30 ft/min this would be a factor of safety of 7.5.

5.2.1.20.2(b) [Rule 2501.13b(2)]
The number used is the same factor of safety of 7.5 as obtained for (1). Since this applies to LU/LA, only a specific number for 30 ft/min can be given.

5.2.1.24 (Rule 2501.9a)
Some of the conditions in Section 208 allow winding drum machine use beyond that generally stipulated for LU/LA. Although somewhat repetitive, the LU/LA conditions have been restated here so that winding drum machines are no longer allowed to exceed the general stipulation via a reference.

5.2.1.24.1 (Rule 2501.9a)
In addition the condition that multiple layers of wrap are not intended has been clearly stated.

5.2.1.24.2(a) [Rule 2501.9b(1)]
This is a condition that is the same as in Part II.

5.2.1.24.2(b) [Rule 2501.9b(2)]
The Elevator Wire Rope Technical Board stated that 6 × 19 WSC (7 × 19) aircraft cable is made as a cross-laid strand design and, therefore, should not be subjected to crushing pressure in application. Thus, no V-type or undercut grooves are proposed; only finished "U" grooves.

5.2.1.24.3(a) [Rule 2501.9c(1)]
This is a condition that is the same as that allowed by Rule 506.2b for Residence Elevators.

5.2.1.24.3(b) [Rule 2501.9c(2)]
This is a condition, which is not mentioned in Rule 506.2b, but the Committee felt that compensating ropes, although not likely to be widely used, should also be clearly covered.

5.2.1.24.3(c) [Rule 2501.9c(3)]
This is a condition that is the same as that allowed by Rule 506.2b for Residence Elevators.

5.2.1.27 (Rule 2501.12)
Deleted reference no longer necessary as Rule 211.1(a)(2) has been changed to not apply to travel less than 18 m.

5.2.1.29 (Rule 2501.17)
New requirements for Layout Data were added to A17.1b–1992 per TR 86-23 to consolidate all necessary layout data in one Section or Rule for each type of equipment. See Sections 214, 308, 611, 1608, and 1808, and Rules 401.15, 402.8, 701.14, and 702.6 of the current Code for copies of the requirements. At the time TR 86-23 was developed, Part XXV did not exist. Reordering of Rules for consistency with corresponding requirements in other Parts of Code.

5.2.2.7 (Rule 2502.3)
Traveling Sheaves attached to the upper end of the plunger are covered in Rules 302.1b(5), (6), and (8) so both the title should be corrected and the references changed to reflect Rules 2501.9 and 2501.13.

5.2.2.15 (Rule 2502.10)
New requirements for Layout Data were added to A17.1b–1992 per TR 86-23 to consolidate all necessary layout data in one Section or Rule for each type of equipment. See Sections 214, 308, 611, 1608, and 1808, and Rules 401.15, 402.8, 701.14, and 702.6 of the current Code for copies of the requirements. At the time TR 86-23 was developed, Part XXV did not exist.

6.1.3.1 (Rule 802.1)
Both Codes are almost identical in technical content, but the language in B44 is easier to read. The revision in A17.1 for nonlinear escalators is retained.

6.1.3.2.1 [Rule 802.2(a)]
Both Codes are identical except for the A17.1 reference to "the next whole inch." This should be dropped since the Code will be in metric dimensions.

6.1.3.2.2 [Rule 802.2(b)]
Both Codes are the same except for the 241/254 mm difference for the dimension from the HR centerline to the edge of the step. A17.1 revised this dimension from 254 mm to 241 mm when it was discovered that the 254 dimension was in error with other references in the escalator Code. The correct dimension is 241 mm (subsequently rounded to 240 mm) and should be retained. The B44 terminology for the reduction in clearance was adopted as it is in performance language.

6.1.3.3.1(a) [Rule 802.3a(1)]
Both Codes have the same technical content but A17.1 is more explicit in the manner it covers the intent. Metric dimensions have replaced the old imperial units

6.1.3.3.1(b)(2) [Rule 802.3a(2)(b)]
Rounded for constancy.

6.1.3.3.2 (Rule 802.3b)
Both Code requirements are similar. Converted to metric dimensions. Force is the correct term for N/m.

6.1.3.3.3 (Rule 802.3c)
This requirement is the same in both Codes, except for reference standards. All referenced standards have been included in the binational Rule. Editorial consistency.

6.1.3.3.4 (Rule 802.3d)
Both Codes have the same intent, but the maximum angle of the interior deck profile in A17.1 should be retained. It provides a limit that would assist in preventing a youngster from slipping while attempting to walk on the interior decking.

6.1.3.3.4(b) [Rule 802.3d(2)]
The wording has been revised for clarity and is now similar to that used in Rule 802.4e.

6.1.3.3.4(c) [Rule 802.3d(3)]
Editorially revised.

6.1.3.3.5 (Rule 802.3e)
See Summary of Code changes ASME A17.1a–1997 through ASME A17.1d–2000 for Rule 802.3e.

6.1.3.3.7 (Rule 802.3k)
See Summary of Code changes ASME A17.1a–1997 through ASME A17.1d–2000 for Rule 802.3k.

6.1.3.3.8 (Rule 802.3j)
See Summary of Code changes ASME A17.1a–1997 through ASME A17.1d–2000 for Rule 802.3j.

6.1.3.3.9 (Rule 802.3g)
A17.1 language is retained since it covers a larger variety of conditions. Metric conversions have been added.

6.1.3.3.9(c) [Rule 802.3g(3)]
Provides more room to extract your body if you get caught.

6.1.3.3.9(e) [Rule 802.3g(5)]
1 in. provides less opportunity for injury to a person contacting edge of guard.

6.1.3.3.10 (Rule 802.3h)
Reworded to address critical dimensions. Knobs are typically 2 in. to 4 in. in diameter.

6.1.3.3.11 (Rule 802.3i)
Both Codes are similar. The language of A17.1 is retained with the new Code revision changing the absolute phrase "to prevent walking" to a realistic "to restrict access."

6.1.3.4.1 (Rule 802.4a)
The requirements of both Codes are combined and revised to clarify intent. A17.1 has language covering curved escalators is adopted as it is necessary since this type has been installed in North America. B44 has language that requires a minimum handrail driving force. This is necessary to ensure that the handrail continues to operate under a passenger load.

6.1.3.4.4 (Rule 802.4d)
There was no comparable Rule in A17.1. Although field splicing of escalator handrails is rare, it is done on occasion and needs to be addressed.

6.1.3.4.5 (Rule 802.4e)
The prime reason for the moving handrail and the height of the handrail proposed in Rule 802.4e is to provide support for passengers riding the escalator. The moving handrail was never intended to provide protection as a barricade. This was an unaddressed requirement that was deemed important to define and included as a safety requirement for decreasing max. from 1 070 mm to 1 000 mm: provides access to a greater number of users of the escalator; and floor opening protection has been addressed in new Rule 802.6f.

6.1.3.5.1(a) (Rule 802.5a)
Reference to the Canadian National Standard has been added.

6.1.3.5.1(c) [Rule 802.5a(3)]
Rewritten in performance language.

6.1.3.5.4 (Rule 802.5d)
This requirement will reduce the risk of entrapment between adjacent steps on the horizontal.

6.1.3.5.6 (Rule 802.5f)
There is no equivalent requirement in A17.1. This original B44 Clause should be added, as amended, to provide an improved method of highlighting the step area that boarding passengers should step onto. ADA does not address this as they are expecting the harmonized Code to address. The $1\frac{1}{2}$ in. was a result of suggestions from CABO/ANSI A117; a MARTA study found that yellow is the best color, particularly for site impaired. CABO/ANSI A117 also felt yellow is appropriate.

6.1.3.5.7 (Rule 802.5g)
Reference corrected to coordinate with changes to Part XI.

6.1.3.5.8 (Rule 802.5h)
There have been a number of instances where the trailing wheels have become separated from the step. Although the front of the step is still attached to the step chain, the other end can pivot down into the escalator interior. If the wheels are located inside the step, then when the step is pivoted down it would be stopped by the step wheel track located below. However, if the wheels are located outside the width of the step, the wheel support track is located beyond the edge of the step and there is nothing to prevent the step from pivoting into the escalator interior. If this happens, passengers standing on the step could fall into the escalator interior and be seriously hurt. This requirement is intended to prevent that.

6.1.3.6.1(a) [Rule 802.6a(1)]
Provides a clearer distinction between the combs and the combplate.

6.1.3.6.1(d) [Rule 802.6a(4)]
Ensures that the combs and combplate do not drag on the steps when a passenger is standing on them.

6.1.3.6.4 (Rule 802.6d)
The requirement provides a minimum guideline for the safety zones at both landings. It was not the Committee's intent to be specific on any given traffic but rather to caution to take into account any traffic in the safety zone.

6.1.3.6.5 (Rule 802.6e)
The A17.1 language is retained in this Rule, with the addition of the reference to Rule 802.5b, which requires 400 mm per step. The B44 wording does not fit the A17 format. It combines the definition of flat steps along with the requirement for the number of flat steps. Additionally, the 4 mm step height tolerance can be misinterpreted as extending the distance being considered as flat steps.

6.1.3.6.6 (Rule 802.6f)
The Clauses in B44 involve building structures and the protection of floor openings. These are design features that are controlled by building codes and are under the control of the architect and building owners. The elevator industry does not have the expertise to evaluate the needs of floor opening protection and should not undertake the design responsibility of providing such protection. The A17 Committee has submitted a proposal to the International Code Council for inclusion in the International Building Code on behalf of the A17 Committee so that requirements for floor protection will be included in the building codes. The 915 mm requirement is now covered in new Rule 802.4e. The prime reason for the moving handrail and the height of the handrail proposed in Rule 802.4e is to provide support for passengers riding the escalator. The moving handrail was never intended to provide protection as a barricade.

6.1.3.7 (Rule 802.7)
The text was identical in both A17.1 and B44. The revision is being proposed to clarify that the purpose of the requirement is to prevent any part of the running gear or passengers on the running gear from falling through the truss to the space below. This is as a result of Inquiry 95-46. See also proposed definition of *running gear*.

6.1.3.9.1 (Rule 802.9a)
The Rule for structural load is the same in both Codes but the A17.1 terminology covers a broader spectrum of conditions by including a definition of terms that are compatible to both curved and linear escalators. Metric dimensions have been added.

6.1.3.9.2 (Rule 802.9b)
The Rule for machinery load is the same in both Codes but the A17.1 terminology covers a broader spectrum of conditions by including a definition of terms that are compatible to both curved and linear escalators as well as escalators with multiple drive units in the step band. Metric dimensions have been added.

6.1.3.9.3 (Rule 802.9c)
The Rule for brake load is the same in both Codes but the A17.1 terminology covers a broader spectrum of conditions by including a definition of terms that are compatible to both curved and linear escalators as well as escalators with multiple drive units in the step band.

6.1.3.9.4 (Rule 802.9d)
The Rule is the same in both Codes. The A17.1 format is retained with metric dimensions replacing the old imperial units.

6.1.3.10 (Rule 802.10)
The basic requirements for safety factors are the same in both Codes. References are included for both the AISC specifications and the CSA standards as they apply to the trusses and supporting structures.

6.1.3.12 (Rule 802.12)
The A17.1 format is retained with metric dimensions replacing the old imperial units.

6.1.3.13 (Rule 802.13)
Welding requirements are included in Section 213 of the A17 Code. This represents an important construction area that impacts the safety of the total design and should be part of the new binational Code. The part of the B44 requirement that gives permission to regulatory authorities to approve other standards, has been deleted. The Code can only state requirements and has no control over the regulatory authorities.

6.1.4.1 (Rule 803.1)
The lower maximum speed is being proposed as a required speed. Passengers become accustomed to a set speed when boarding escalators. Making this speed a constant throughout the industry would provide a consistency that would help to reduce falls when boarding.

6.1.5.3.1(c) [Rule 804.3a(3)]
New Fig. D6 is identical to B44 Fig. 15, except the note has been added.

6.1.5.3.1(d) [Rule 804.3a(4)]
The brake nameplate has been modified to include the minimum distance from the skirt-switch to the comb-plate, and the minimum stopping distance.

6.1.5.3.2 (Rule 804.3b)
Metric dimensions have been added.

6.1.5.3.3 (Rule 804.3c)
The cross-references are necessary due to the new type test requirements of Rule 1105.1 and the new general requirements for tests and certification in Section 1100. Section 1100 is applicable to all elevators/escalator components/equipment that require certification or engineering tests, including escalator brakes and steps.

6.1.6.1 (Rule 805.1a)
Clarification that this requirement is not in conflict with Section 805.

6.1.6.1.1 (Rule 805.1b)
Relocated from Rule 805.2(c), and modified to address stopping as well as starting.

6.1.6.2 (Rule 805.2)
To include the possibility of a single or multiple starting switch.

6.1.6.2.1(a)(2) [Rule 805.2(a)(2)]
Key operated, added for clarification.

6.1.6.2.1(c) [Rule 805.2(c)]

An emergency stop button should be accessible from the starting switch at the same time by the same person. Note that this Rule will limit the installation of start switches to a location, which is in the vicinity of the stop switches. The intent of Clause 8.6.1.3 is covered in this revised Rule and in new Section 5, Security. The A17.1 format is retained since it is more explicit in defining the switch and its operation.

6.1.6.2.2 [Rule 805.2(b)]

To provide requirements that correspond to those of inspection operation on the elevator and to encompass comments from members of the Electrical Committee. Accidents can occur through inadvertent starting/running of escalators by mechanics when attempting to position escalators in order to check/adjust/repair components as it becomes necessary from time to time. The intent is to prevent such accidents by providing a switch for use during maintenance and repairs, which will prevent the escalator from running unless the key switch (part switch) is manually held in the "ON" position.

6.1.6.3.1 (Rule 805.3a)

A17.1 is retained as it is more definitive in the location of the stop buttons and provides superior visual markings. The phrase "and automatic stopping" deleted as unnecessary because it is covered in Rule 805.1b.

6.1.6.3.2 (Rule 805.3b)

A17.1 is retained with modifications. It has the same requirements as B44, plus it addresses variable frequency drives.

6.1.6.3.2(c) [Rule 805.3b(3)]

Editorial clarification.

6.1.6.3.4 (Rule 805.3d)

Both Codes are similar, except for the manual reset and the permissible delay of the machine brake requirements in A17.1, which is included in the proposal. Metric dimensions have been added.

6.1.6.3.5 (Rule 805.3e)

By dropping the reference to Rule 210.2(e) in this rewrite, and due to a reorganization of Rule 210.2 which moved a requirement of this switch to Rule 210.4c, a very important safety requirement was lost. That requirement is now being proposed as Rule 805.3e(5).

6.1.6.3.6 (Rule 805.3f)

Both Codes have the same requirements. The A17.1 format is retained. The "accidentally caught" in B44 is changed to "caught." Whether an object is accidentally caught or deliberately caught is irrelevant.

6.1.6.3.7 (Rule 805.3g)

The A17.1 Rule is modified to cover other exiting restrictors other than rolling shutters to provide for the

safety of passengers "EXIT" replaced by "EGRESS" for consistency throughout definitions, Parts VIII and IX.

6.1.6.3.12(b) [Rule 805.3l(2)]

Revised for clarity.

6.1.6.3.13(b) [Rule 805.3m(2)]

Language revised for consistency to Rule 805.3m(1) and to clarify intent.

6.1.6.3.15 (Rule 805.3o)

In order to include inspection control (Rule 805.2b) in the list of electrical protective devices. The reason this Rule was not combined with Rule 805.3e was because this switch is not required to comply with all of the requirements of Rule 805.3e.

6.1.6.8 (Rule 805.8)

The term "smoke detector" is retained, rather than "fire alarm initiating device" because the smoke detector was the specific element of fire protection that was considered at the time the Rule was developed. Editorial Clarification.

6.1.6.9 (Rule 805.9)

Both Codes are identical in this Rule, but the A17 sign has a no wheeled vehicle symbol on it, which has been added to the binational sign (see proposed revision to Fig. 805.2) as a reminder to passengers not to take carts on escalators.

6.1.6.9.2 (Rule 805.9b)

There is no equivalent Clause in B44. This Rule is necessary to prevent additional signs from confusing the passenger and distracting attention away from the Code sign.

6.1.6.10 (Rule 805.10)

This Rule has been coordinated with revisions to electrical requirements in Part II.

6.1.6.10.2 [Rule 805.10(b)]

Clarification. Specific wording to coordinate with electrical requirements in Part II will be addressed after harmonization. Revision to opening sentence of Rule 805.10b clarifies the requirements and makes these proposed Rules inappropriate and unnecessary.

6.1.6.11 [Rule 805.10(d)]

To require protection against single point failures. Editorial.

6.1.6.12 (Rule 805.11)

Stand-alone Rule for electrically powered safety devices for clarification.

6.1.6.13 (Rule 805.12)

Revised to reflect new definition of *electrically-assisted braking* included in Section 3. Dynamic braking is not a defined term. Electrically assisted is not used in Parts VIII or IX and is not permitted. See Rule 804.3a.

6.1.6.15 (Rule 805.15)
This Rule specifies the requirements for single failure protection, i.e., redundancy, in the design of specific control and operating circuits [see Rule 805.10(a)], and requires the checking of that redundancy [see Rule 805.10(b)]. This may require the use of two relays instead of one in these circuits, and the checking of the operation of these relays.

When electromechanical relays and contactors are being checked for operation, specifically drop out operation, by utilizing contacts on the relays or contactors themselves, care must be taken that the contact used for monitoring the state of the contact used in the critical circuit is a true and correct representation of the critical circuit contact.

This new Rule addresses this concern. The required operation of the monitoring contact is specified, and it is clarified that springs in the design of the relay, contactor, or contact design may be used, but that spring failure shall not cause the monitoring contact to indicate an open critical circuit contact if the critical circuit contact is not open.

6.1.7.1.1 (Rule 806.1a)
To rewrite in more definable terms and to coordinate with B44 language.

6.1.7.3 (Rule 806.3)
The requirements of both Codes are incorporated in the proposal since they all relate to important issues of safety.

6.1.7.3.3 [Rule 806.3(c)]
The key security requirements are now covered in the proposed Section 5.

6.1.7.3.4 [Rule 806.3(d)]
The requirements of both Codes are incorporated in the proposal since they all relate to important issues of safety.

6.1.7.4 (Rule 806.4)
To coordinate with changes to Parts I and II. See comments for Rule 210.4(a).

6.2 (Part IX)
Moving walks consistent with Part VIII, except as noted.

6.2.1.1 (Rule 900.1)
Consistent with Rule 800.1.

6.2.2.1 (Rule 901.1)
Consistent with Rule 801.1.

6.2.3.1 (Rule 902.1)
The word "exit" replaced by "egress ends" for consistency with terminology in Rule 902.8.

6.2.3.2.1 [Rule 902.2(a)]
For consistency with Rule 802.2(a).

6.2.3.2.2 [Rule 902.2(b)]
For consistency with Rule 802.4(e). See rationale for Rule 802.4(e) for additional information.

6.2.3.3.2 (Rule 902.3b)
For consistency with Rule 802.3c.

6.2.3.3.3 (Rule 902.3c)
For consistency with Rule 802.3c.

6.2.3.3.4(b) [Rule 902.3d(2)]
The wording has been revised for clarity and to coordinate with Rule 802.3d.

6.2.3.3.5 (Rule 902.3e)
Round metric number preferred. Clarification.

6.2.3.3.6(a) [Rule 902.3f(1)]
Round metric number preferred.

6.2.3.3.7 (Rule 902.3g)
For consistency with Rule 802.3g.

6.2.3.3.7(f) [Rule 902.3g(6)]
Consistent with Rule 802.3g.

6.2.3.3.8 [Rule 902.3h(1)]
Consistent with Rule 802.3i.

6.2.3.4.1 (Rule 902.4a)
Consistent with Rule 802.4a.

6.2.3.4.2 (Rule 902.4b)
Consistent with Rule 802.4b.

6.2.3.5 (Rule 902.5)
Consistent with Rule 802.5e.

6.2.3.6.2 (Rule 902.6b)
Committee agreed not to round the metric dimensions in this case so that molds would not have to be recast. No valid safety reasons to round. The correct conversion from $\frac{3}{16}$ in. is 0.188 as is consistent with Rules 905.3a(2) and 805.3a(2).

6.2.3.8.1(d) [Rule 902.8a(4)]
Consistent with Rule 802.6a(4).

6.2.3.8.4 (Rule 902.8d)
To coordinate with Rule 802.6d.

6.2.3.8.5 (Rule 902.8e)
To coordinate with Rule 802.6e.

6.2.3.9.1(a) [Rule 902.9a(1)]
For consistency with definitions.

6.2.3.9.1(d) [Rule 902.9a(4)]
Consistent with Rule 805.3k.

6.2.3.13 (Rule 902.14)
B29.1M is an ASME Standard.

6.2.5.3.1(c) [Rule 904.3a(3)]
Consistent with Rule 804.3a(3).

6.2.5.3.1(d)(4) [Rule 904.3a(4)(d)]
Consistent with Rule 804.3a(4)(d).

6.2.5.3.2 (Rule 904.3b)
To coordinate with Rule 804.3b.

6.2.6.1.2 (Rule 905.1b)
Relocated from Rule 905.2(c). Consistent with Rule 805.1b.

6.2.6.2 (Rule 905.2)
To coordinate with Rule 805.2.

6.2.6.2.1(3) [Rule 905.2(a)(3)]
Relocated from Rule 905.2(b).

6.2.6.2.2 (Rule 905.2b)
For consistency with Rule 805.2b.

6.2.6.3.1 (Rule 905.3a)
For consistency with Rule 805.3a.

6.2.6.3.1(c) [Rule 905.3a(3)]
Rounding 4.8 mm to 5 mm is not appropriate as it will require a change in design for no known safety reason. The correct conversion for $^{3}/_{16}$ in. is 0.188 as is consistent with the conversions in Rules 902.6b and 805.3a(2).

6.2.6.3.2(d) [Rule 905.3b(4)]
To coordinate with Rule 805.3b(3).

6.2.6.3.5 (Rule 905.3e)
To coordinate with Rule 805.3e.

6.2.6.3.6 (Rule 905.3f)
The Rule is being modified to coordinate with Part VIII and to cover other exiting restrictors other than rolling shutters to provide for the safety of passengers. The word "exit" replaced by "egress" for consistency throughout definitions, Parts VIII and IX.

6.2.6.3.9 (Rule 905.3i)
Consistent with Rule 805.3k.

6.2.6.3.10 (Rule 905.3j)
For consistency with Rule 805.3l.

6.2.6.3.12 (Rule 905.3o)
To coordinate with Rule 805.3o.

6.2.6.7 (Rule 905.7)
Consistent with Rule 805.8.

6.2.6.8.1 (Rule 905.8a)
Consistent with Rule 805.9a.

6.2.6.8.2 (Rule 905.8b)
Consistent with Rule 805.9b.

6.2.6.9.2 [Rule 905.9(b)]
Consistent with Rule 805.10(b).

6.2.6.9.3(a)(1) [Rule 905.9(c)(1)(b)]
Consistent with Rule 805.10(b).

6.2.6.10 (Rule 905.10)
Stand-alone Rule for electrically powered safety devices for clarification.

6.2.6.12 (Rule 905.11)
Consistent with Rule 805.12.

6.2.6.14 (Rule 905.13)
Consistent with Rule 805.14.

6.2.7.1.1 [Rule 906.1(a)]
Consistent with Rule 806.

6.2.7.3 (Rule 906.3)
Editorially revised to be consistent with Rule 806.3.

8.3 (Section 1100)
The intent of new Section 1100 is to apply procedural and administrative Rules, currently applicable only to buffers in A17.1, to certification of other elevator/escalator components. This has been adopted from B44 with additional revisions made due to new definitions of type test, engineering test, and certifier and for editorial clarification of intent.

8.3.1.3.1(d) [Rule 1100.3a(4)]
Editorial consistency.

8.3.1.3.1(i) [Rule 1100.3a(9)]
The certificate should state what edition of the Code the testing was done to.

8.3.2.1.1 (Rule 1101.1a)
Current A17.1 requirement covered in new Scope. Incorporates testing requirements previously contained in Rule 201.4g.

8.3.2.1.2 (Rule 1101.1b)
Former Rule 1100.2a now covered in new Rule 1100.2b(b).

8.3.2.2 (Rule 1101.2)
Former Rule 1100.2a(2) now covered in new Rule 1101.1a. Incorporates requirements from Rule 201.4g.

8.3.2.2(b) [Rule 1101.2(b)]
Former Rule 1100.2b now covered in new Rule 1100.3c. Former Rule 1100.2c now covered in new Rule 1100.4a. Second paragraph now covered in new Rule 1100.4b. Former Rule 1100.2d now covered in new Rule 1100.3d.

8.3.2.5.1(b) [Rule 1101.5a(2)]
Editorial clarification.

8.3.2.5.2(b) [Rule 1101.5b(2)]
Editorial clarification of cross-referenced Rule.

8.3.2.5.3 (Rule 1101.5c)
Editorial clarification of second paragraph using correct terminology (see Rule 1101.5e).

8.3.2.6 (Rule 1101.6a)
Incorporates requirements from Rule 201.4g.

8.3.3.1 (Rule 1102.1)
Covered by scope and definitions. Editorial.

8.3.3.2 (Rule 1102.2)
Clarification.

8.3.3.4.1 [Rule 1102.4(a)]
Wheelchair lift not covered within A17/B44 Code.

8.3.3.4.4 [Rule 1102.4(d)]
Clarification of intent.

8.3.3.4.8 [Rule 1102.4(h)]
Editorial clarification.

8.3.3.4.11 [Rule 1102.4(k)]
Testing requirement added due to new Rule 111.2d(1)(b).

8.3.4.2 (Rule 1103.1)
Editorial.

8.3.4.1.2 (Rule 1103.1)
Exception added for NBCC, to second paragraph.

8.3.5 (Section 1107)
Similar intent, revisions due to changes made to Rule 1100.2.

8.3.7 (Section 1106)
Editorial clarification.

8.3.7.4 (Rule 1106.4)
Editorial.

8.3.9 (Rule 1108.1)
To cover the design, installation, and testing of overspeed valves when provided.

8.7 (Part XII)
Compliance with A17.3 deleted because A17.3 is a stand-alone document.

8.7.1.1 (Rule 1200.1)
Clarification.

8.7.1.2 (Rule 1200.2)
Deleted material is not necessary with the Rules as revised.

8.7.1.5 (Rule 1200.5)
This assures that the level of safety is not diminished.

8.7.1.6 (Rule 1200.6)
To recognize the need for temporary wiring during alterations.

8.7.1.8 (Rule 1200.7)
Deleted material is addressed in Rule 2000.3.

8.7.2.1.1 (Rule 1201.1a)
Deleted because Rule 100.1 now references the Building Code.

8.7.2.7.1 (Rule 1201.2a)
Add equivalent Canadian standard.

8.7.2.3 (Rule 1201.4)
Presently not covered, use B44.

8.7.2.4 (Rule 1201.7)
Editorial clarification.

8.7.2.5 (Rule 1201.8)
Editorial clarification.

8.7.2.9 (Rule 1201.5)
Coordination with Rule 1200.5.

8.7.2.10 (Rule 1201.10)
Editorial.

8.7.2.10.1(b) [Rule 1201.10a(2)]
Editorial clarification.

8.7.2.10.1(e) [Rule 1201.10a(5)]
To recognize new requirement for access openings for cleaning hoistway enclosures.

8.7.2.10.2(c) [Rule 1201.10b(3)]
Editorial clarification.

8.7.2.10.2(f) [Rule 1201.10b(6)]
Requirements for door retainers added.

8.7.2.10.5 (Rule 1201.10e)
To recognize that a new rated door panel used in existing frames, etc., is not a rated entrance; however, it meets the requirements of this Code. Editorial change to clarify any one or more of the three conditions is acceptable.

8.7.2.11.1 (Rule 1201.11a)
Relocated from Rule 1100.4b due to reorganization of Section 111.

8.7.2.11.3 (Rule 1201.11c)
To address requirements no longer applicable to new installations. Relocated from A17.1–1996, Rule 111.8b. To incorporate new Code requirements for inspection operation.

8.7.2.11.5 (Rule 1201.11e)
To address an oversight in current requirements.

8.7.2.12 (Rule 1201.12)
Editorial clarification.

8.7.2.13 (Rule 1201.13)
Used similar wording in old Rule 1200.5d to cover alteration requirements for door reopening devices.

8.7.2.14.3 [Rule 1202.5(c)]
Recognize requirement is the jurisdiction of the NBCC in this area.

8.7.2.14.4 [Rule 1202.5(d)]
Recognize requirement is the jurisdiction of the NBCC in this area.

8.7.2.14.5 [Rule 1202.5(e)]
Rule 1200.5b moved here and renumbered.

8.7.2.16.1(d) [Rule 1202.8a(4)]
Same intent.

8.7.2.16.1(h) [Rule 1202.8a(8)]
The elevator should be provided with car overspeed protection and car unintended movement protection.

8.7.2.16.3 (Rule 1202.8c)
Editorial clarification.

8.7.2.16.4(b) [Rule 1202.8d(5)]
References were revised to coordinate with new Code requirement.

8.7.2.16.4(e) [Rule 1202.8d(8)]
Exception for drum machine deleted.

8.7.2.16.4(f) [Rule 1202.8d(14)]
The elevator should be provided with car overspeed protection, car unintended movement protection, monitoring of door contacts and door bypass switches. Due to the extent of work to increase capacity, the addition of Section 216 is consistent with other requirements for this type of alteration.

8.7.2.17.1(b) [Rule 1202.10a(2)]
Editorial.

8.7.2.17.1(c)(2) [Rule 1202.10a(3)(b)]
Presently not covered, use B44.

8.7.2.17.2(b)(9) [Rule 1202.10b(2)(i)]
To reflect changes made to Rule 210.4.

8.7.2.17.2(b)(11) [Rule 1202.10b(2)(k)]
Editorial clarification. Requirements of car overspeed protection, car unintended movement protection, added. Requirements of access clearance added. Due to the extent of work to increase capacity, the addition of Section 216 is consistent with other requirements for this type of alteration.

8.7.2.17.3 (Rule 1202.10c)
To provide new requirements when an alteration involves a decrease in rated speed of an electric elevator.

8.7.2.17.3 [Rule 1202.10c(5)]
Update to show correct reference.

8.7.2.18.3 [Rule 1202.6(c)]
Presently not covered.

8.7.2.19 (Rule 1202.7)
Requirements for replacement of ropes as part of repair will be covered in the new Part XX. These requirements are more specific to alteration.

8.7.2.20 (Rule 1202.15)
Not presently covered.

8.7.2.21.1 (Rule 1202.14a)
To eliminate a term, which may cause confusion.

8.7.2.21.1 (Rule 1202.14a)
To eliminate a term, which may cause confusion.

8.7.2.22.2 [Rule 1202.3(b)]
Editorial.

8.7.2.24 (Rule 1202.1)
Editorial.

8.7.2.25.1(a) [Rule 1202.9a(1)]
The elevator should be provided with car overspeed protection, car unintended movement protection, monitoring of door contacts, and door bypass switches. Due to the extent of work to increase capacity, the addition of Section 216 is consistent with other requirements for this type of alteration.

8.7.2.25.2 (Rule 1202.9b)
Requirements regarding accessibility, guarding, traction, and identification added.

8.7.2.27.1 (Rule 1202.12a)
Editorial.

8.7.2.27.2 (Rule 1202.12b)
Editorial.

8.7.2.27.3 (Rule 1202.12c)
Editorial.

8.7.2.27.3(c) [Rule 1202.12c(3)]
Update to show correct reference.

8.7.2.27.4(a) [Rule 1202.12d(1)]
To emphasize alteration. Requirements of door bypass and door monitoring added.

8.7.2.27.4(b) [Rule 1202.12d(2)]
To emphasize alteration.

8.7.2.27.5 (Rule 1202.12e)
To suit new definition.

8.7.2.27.5(b) [Rule 1202.12e(2)]
Update to show correct references.

8.7.2.27.5(e) [Rule 1202.12e(5)]
Requirements of car overspeed protection, car unintended movement protection added. Requirements of access clearance added. Due to the extent of work to increase capacity the addition of Section 216 is consistent with other requirements for this type of alteration.

8.7.2.27.6 (Rule 1202.12f)
To suit new definition.

8.7.2.27.6(a) [Rule 1202.12f(1)]
To be specific about changes that constitutes alteration.

8.7.2.28 (Rule 1202.13)
To suite new wording of the referenced Rule. To cover requirements for the addition of an elevator to a group.

8.7.3.4 (Rule 1203.1g)
Editorial clarification.

8.7.3.5 (Rule 1203.1h)
Editorial clarification.

8.7.3.7 (Rule 1203.1b)
Add Canadian references.

8.7.3.9 (Rule 1203.1e)
Coordination with Rule 1200.5.

8.7.3.10 (Rule 1203.1j)
Editorial clarification.

8.7.3.12 (Rule 1203.1m)
Editorial clarification. Addressed in new cross-reference (see Rule 1203.1j).

8.7.3.19 (Rule 1203.2j)
Editorial clarification. To be consistent with Rule 207.4.

8.7.3.22.1 (Rule 1203.4a)
Editorial clarification.

8.7.3.22.1(d) [Rule 1203.4a(4)]
Presently not covered, use B44.

8.7.3.22.2(d) [Rule 1203.4b(4)]
Deflection is covered in Rule 301.6.

8.7.3.22.2(h) [Rule 1203.4b(8)]
Rule 306.8 has been deleted.

8.7.3.22.3 (Rule 1203.4c)
Editorial clarification. To provide requirements when

an alteration involves a decrease in rated speed of a hydraulic elevator, thus ensuring that the level of safety has not been diminished.

8.7.3.23.1 (Rule 1203.3a)
Coordination with new definition.

8.7.3.23.2 (Rule 1203.3b)
Coordination with new definition.

8.7.3.23.3 (Rule 1203.3c)
Coordination with new definition.

8.7.3.23.4 (Rule 1203.3d)
Exception addressed items not practical for existing equipment.

8.7.3.23.5 (Rule 1203.3e)
To conform to new definition.

8.7.3.23.6 (Rule 1203.3f)
To conform to new definition.

8.7.3.24 (Rule 1203.5)
To emphasize alteration.

8.7.3.29 (Rule 1203.6)
Editorial clarification.

8.7.3.31.5(a) [Rule 1203.8e(1)]
To conform to new definition.

8.7.3.31.6 (Rule 1203.8f)
To suit new definition.

8.7.3.31.7 (Rule 1203.8g)
To conform to new definition.

8.7.3.31.8(c) [Rule 1203.8h(3)]
To coordinate with new terminology.

8.7.3.31.9 (Rule 1203.8i)
To address the new Code requirements for auxiliary power lowering operation.

8.7.4.3 (Section 1204)
Alterations to hand elevators appears twice in Part XII. The entire Section 1219 has been redesignated as Section 1204 as an editorial correction.

8.7.4.3.8 (Rule 1204.8)
To emphasize alteration.

8.7.5.5.1 (Rule 1220.4)
Editorial. Items relating to replacement have been reviewed and relocated.

8.7.5.5.2 (Rule 1220.3)
Editorial. Items relating to replacement have been reviewed and relocated.

8.7.5.5.3 (Rule 1220.6)
Editorial. Items relating to replacement have been reviewed and relocated.

8.7.5.5.4 (Rule 1220.8)
Editorial. Items relating to replacement have been reviewed and relocated.

8.7.5.5.7 (Rule 1220.7)
Editorial. Items relating to replacement have been reviewed and relocated.

8.7.5.5.8 (Rule 1220.5)
Editorial clarification.

8.7.6.1.1 (Rule 1207.1)
Escalator harmonization has relocated Rule 805.2c. This is properly addressed in Rule 1207.2.

8.7.6.1.2 (Rule 1207.2)
Editorial reorganization. It is impractical to require two flat steps on a relocated escalator.

8.7.6.1.7 (Rule 1207.7)
Editorial clarification.

8.7.6.1.9 (Rule 1207.9)
Accept B44, clarification of intent. It is not practical to require two flat steps during relocation.

8.7.6.1.11 (Rule 1207.11)
Editorial clarification.

8.7.6.1.12(c) (Rule 1207.12c)
Last paragraph is redundant; covered by Rules 1207.12a, 1207.12b, and 1207.12c.

8.7.6.2.2 (Rule 1208.2)
To be consistent with Rule 1207.2.

8.7.6.2.4 (Rule 1208.4)
Editorial clarification.

8.7.6.2.7(b) [Rule 1208.7(b)]
To permit the existing width to be maintained, but not be reduced to below current requirements.

8.7.6.2.9 (Rule 1208.9)
For consistency with Rule 1207.9.

8.7.6.2.11 (Rule 1208.11)
Editorial clarification.

8.7.7.1.1 (Section 1218)
General requirements are covered in Rule 1200.

8.7.7.3.3 (Section 1209)
B44 requirement not covered in A17.1.

8.7.7.3.4 (Section 1209)
B44 requirement not covered in A17.1.

8.8.1(a) [Rule 213.1(b)]
Allows for deviations in Canada.

8.8.2 (Rule 213.2)
Modified to include Canadian requirements.

8.9 (Rule 215.1)
Part X requires that inspection and tests be performed to confirm the equipment complies with the Code under which it was installed and/or altered. This information should be readily available at the site for inspection and maintenance personnel.

PART 1
GENERAL

SECTION 1.1
SCOPE

The ASME A17.1 Safety Code for Elevators and Escalators is the accepted guide for the design, construction, installation, operation, inspection, testing, maintenance, alteration, and repair of elevators, dumbwaiters, escalators, moving walks, material lifts, and dumbwaiters with automatic transfer devices. It is the basis in total or in part for the elevator codes used throughout the United States and Canada.

The Safety Code for Elevators and Escalators is only a guide unless adopted as law or regulation by the authority having jurisdiction.

Local jurisdictions in their adopting legislation may sometimes revise and/or include requirements in addition to those found in this Code. It is therefore advisable to check with the local jurisdiction before applying these requirements in any area.

Requirement 1.1.2 outlines examples of equipment not covered by the ASME A17.1 Code. Requirement 1.1.3 specifies those Parts and requirements of the Code, which apply only to new installations, as well as those Parts and requirements, which apply to both new and existing installations.

SECTION 1.2
PURPOSE AND EXCEPTIONS

Present Code requirements provide a framework for standards of safety for current products whose technologies have become state-of-the-art and commonplace. The A17 Committee has demonstrated its responsiveness to prepare new requirements throughout its long history, which spans over 80 years.

In a progressive elevator industry, new designs emerge as a result of new technology. As a result, new products will embody concepts or the application of new technology not specifically covered in the A17.1 Code. Since Code requirements cannot anticipate all future development and innovation in the elevator art, they are written to reflect the state-of-the-art technologies following their introduction.

Accordingly, new products must, of necessity, be introduced prior to formal requirements being adopted into the A17.1 Code.

The A17 Code Committee has continued its long-standing policy of revising the Code to keep it abreast of developments in the elevator art. References to this precedent can be traced through all the editions over the course of 80 years. This very spirit is evident in the Foreword to the Code.

The A17 Committee has responded to new developments throughout its history as evidenced by the inclusion of requirements to recognize technological advances such as increased car speeds, solid state electronic devices, observation elevators, installation of counterweights in separate hoistways, material lifts and dumbwaiters with automatic transfer devices, special purpose personnel elevators, inclined elevators, elevators used for construction, limited-use/limited-application elevators, and shipboard elevators, just to name a few.

There is no question that the local authorities having jurisdiction have acted wisely in recognizing technical advances and have granted waivers until the Code could respond with new requirements to cover the changing elevator art.

With the emerging new materials and processes in the mechanical, structural, electronic, and optics areas, in addition to the analytical capabilities now possible, the need for flexibility to introduce new products resulting from maturing technical developments is heightened.

While the A17 Committee can neither grant specific waivers of the requirements nor anticipate revisions to suit the state of a changing technology in the future, Section 1.2 and the paragraphs covering new technology in the Preface will serve to encourage the proper use of waivers by authorities having jurisdiction.

It is the goal of responsible manufacturers to continue to bring innovative products, which are safe for public use, into the marketplace. In the exercise of this responsibility, they must provide safeguards to ensure safe operation and, accordingly, must be prepared to document that the new product is equivalent or superior in quality, strength, stability, fire resistance, effectiveness, durability, and safety to that intended by the present Code.

Some of the A17.1 Code requirements differ depending on whether the Code is being enforced in the United States or Canada. The user should not assume that these differing requirements are equivalent. While in many instances they may be equivalent, that was not evaluated by the ASME A17 Committee. Where there are different requirements, it is only recognition that ASME A17.1 does not apply as the referenced national

1

standards (e.g., building code, etc.) has jurisdiction. Authorities Having Jurisdiction should keep this in mind when an application for a variance is based on compliance with the requirements from another country.

SECTION 1.3
DEFINITIONS

This Section defines the terms used in the Code. The user of the Code should become familiar with all the terms contained in Section 1.3 of the Code. Where terms are not defined, they have their ordinarily accepted meanings or such as the context may imply.

Words used in the present tense include the future. Words in the masculine gender include the feminine and neuter. Words in the feminine and neuter gender include the masculine. The singular number includes the plural and the plural number includes the singular.

Many interpretation requests can be avoided if the time is taken to become familiar with the definitions in this Section.

PERFORMANCE TERMINOLOGY

Industry standards for dimensional, performance, application, electrical, and evaluation of building transportation equipment are published in NEII-1-2000, *Building Transportation Standards and Guidelines*. Performance terms are defined and measurement standards specified.

NEII-1-2000 incorporates the Code requirements in ASME A17.1, NFPA 70, etc. It is an invaluable reference for architects, engineers, consultants, building owners and managers, elevator manufacturers, contractors, and suppliers. Procurement information for NEII-1-2000 can be found in Part 9 of this Handbook.

PART 2
ELECTRIC ELEVATORS

SCOPE

Part 2 applies to traction and winding drum electric elevators. Other types of elevators, such as hydraulic, private residence, screw column, sidewalk, hand, inclined, rack and pinion, shipboard, rooftop, limited-use/limited-application and special purpose personnel elevators, elevators used for construction, dumbwaiters, and material lifts reference many of the requirements in Part 2 in their respective Sections of the Code.

Electric elevators installed at an angle of 70 deg or less from the horizontal must comply with Section 5.1 or if installed in a private residence, Section 5.4.

SECTION 2.1
CONSTRUCTION OF HOISTWAYS AND HOISTWAY ENCLOSURES

2.1.1 Hoistway Enclosures

2.1.1.1 Fire-Resistive Construction. Fire-resistive construction controls the spread of fire, and retards the penetration of hot gases and smoke from one floor or area to another.

The building code is an ordinance, which sets forth requirements for building design and construction. Where such an ordinance has not been enacted, one of the following model codes must be complied with:

(a) National Building Code (NBC)
(b) Standard Building Code (SBC)
(c) Uniform Building Code (UBC)
(d) National Building Code of Canada (NBCC)

Most of the building codes in the United States are based on one of the model building codes. The NBC is developed and published by the Building Officials and Code Administrators International (BOCA) and is widely adopted in the Northeast and Midwest. Prior to the 1984 Edition, it was known as the Basic Building Code. The 1984 Edition was titled the Basic/National Building Code. These changes in titles reflect an agreement between BOCA and the American Insurance Association, which from 1905 through 1976 published the National Building Code. The SBC, formerly Southern Standard Building Code, is developed and published by the Southern Building Code Congress International (SBCCI), and is widely adopted in the southern states. The UBC is developed and published by the International Conference of Building Officials (ICBO) and is widely adopted west of the Mississippi River. In the

spring of 2000, the three U.S. model building codes joined and published the International Building Code (IBC), which was promulgated by the International Code Council (ICC). It is anticipated that the IBC will replace the other model codes over the next few years. The model building code used in Canada is the NBCC, which is promulgated by National Research Council of Canada.

Building codes normally regulate the properties of materials and the methods of construction as they pertain to the hazards presented by various occupancies. They are based on the findings of previous experiences such as fires, earthquakes, and structural collapses. Requirements found in building codes that have an impact on fire protection include enclosures of vertical openings such as stair shafts, elevator hoistways (shafts), pipe chases, exit requirements, flame spread requirements for interior finishes, and sprinkler requirements.

The evaluation of the risk to a building with respect to fire resistance has been a goal of building codes, fire codes, and insurance underwriters for years. For this reason, building codes have classified construction types and controlled the size of buildings based on the potential fire hazard. Each building code identifies various construction types, and although each code may classify the various types of construction differently, the classification system is essentially the same for all. Originally, there were only two classifications, "fireproof" and "nonfireproof." These terms were misleading, especially since "fireproof" conveyed a false sense of security. Thus, the term "fire-resistive" was coined to provide a more realistic assessment of the resistance of buildings to the effects of a fire. The use of this term also allows for the identification of relative fire resistance, resulting in five basic construction classifications: fire-resistive; noncombustible or limited-combustible; heavy timber; ordinary; and wood frame.

With the advent of modern construction systems that mix traditional materials such as wood and steel with new materials such as plastics, this descriptive classification system also became misleading. Thus, most of the building codes adopted a numerical or alphabetical type of identification for construction classifications. This system is more general and does not relate to a construction technique but rather to fire protection performance.

In building construction, the interior walls and partitions act as barriers to the spread of fire. Proper construction of these barriers provides an effective means to protect against the spread of fire. Firewalls are customarily self-supporting and are designed to maintain structural integrity even in case of complete collapse of the structure on either side of the firewall. A fire partition normally possesses somewhat less fire resistance than a firewall and does not extend from the basement through a roof, as does a firewall. Usually, a fire partition is used to subdivide a floor or an area and is erected to extend from one floor to the underside of the floor above. Fire partitions are constructed of noncombustible, limited-combustible, or protected combustible materials and are attached to, and supported by, structural members having a fire-resistance rating at least equal to that of the partition. Elevator hoistway enclosure walls are normally fire partitions with a fire-resistance rating of 2 h. In some types of construction, other fire-resistance ratings are specified by the building code.

Current practice is to use an hourly rating designation. Elevator hoistway fire-resistive entrances have been commonly rated at $1\frac{1}{2}$ h, but may be rated at 3 h, 1 h, or $\frac{3}{4}$ h. In the past, alphabetical designations were used. Class A openings are in walls separating buildings or dividing a building into fire areas. Class B openings are in enclosures of vertical communication through buildings (stairs, hoistways, etc.). There are also C, D, E, and F class openings. Usually both the hourly rating and a letter designation are used. The letter designation normally applied to fire-resistance-rated elevator entrances is "B." The "B" may not have much value, but since many codes still refer to the letter classification system, the letter has usually been retained by most manufacturers. Opening protection (entrances) in fire-resistive enclosures are generally required to have less resistance to fire than is provided by the wall. Thus the requirement for $1\frac{1}{2}$ h, fire-resistance rating for an entrance in 2 h fire-resistance-rated construction. This standard was established years ago because easily ignited combustibles are not normally found on the opposite side of the entrance.

Elevator hoistway entrances are an important fire control means, which should remain closed and structurally sound on any floor involved in a fire. A labeled entrance (see 2.11.15) ensures that the hoistway entrance doors meet the fire safety requirements specified by the Code. A detailed discussion of the testing of fire-resistive entrances is contained in the commentary on requirement 2.11.14.

Drywall construction is a wall system that has replaced the use of masonry wall construction. The early elevator entrance installations in drywall were similar to masonry installations, but in order to assure the integrity of the entrance to drywall interface, fire door assemblies were subjected to the $1\frac{1}{2}$ h test. Successful tests showed that the drywall entrance provided the building occupants with the same measure of safety provided by entrances in masonry construction. Floor-to-floor or floor-to-beam struts are now a typical industry standard with drywall applications as shown in Diagram 2.1.1.1(a). Brackets are placed within the entrance frame to accept the "J" strut that is furnished by the drywall contractor. The "J" strut, including its proper fastening to the entrance frame as well as the drywall, is of vital importance in assuring a proper entrance to drywall interface. There usually is no difference in the door panels. The drywall fire test was usually run in a standard gypsum wall. This established a minimum jamb depth of 133 mm ($5\frac{1}{4}$ in.) as shown in Diagram 2.1.1.1(b). Diagram 2.1.1.1(b) provides a cutaway view of the drywall interface. Entrance frames installed in drywall must be fire-tested and labeled for said construction if a fire-resistive entrance is required.

Hall call fixture boxes and hall position fixture boxes, etc., occur in every elevator shaft wall. These boxes often penetrate the hoistway wall and will invalidate the fire rating unless proper protection is provided. Diagrams 2.1.1.1(c) and 2.1.1.1(d) show two examples of fixture boxes in a 2 h fire-resistive wall. Both designs have successfully passed the 2 h fire endurance test specified in ASTM E 119. For additional information, contact the United States Gypsum Company (see Part 9 of the A17.1 Handbook).

The requirements for partitions between a fire-resistive hoistway and machine room having fire-resistive enclosures are illustrated in Diagram 2.1.1.1(e).

While the A17.1 Code has no requirements for elevator lobbies, the model building codes do.

2.1.1.2 Non-Fire-Resistive Construction. Where the building code (see commentary on 2.1.1.1) permits non-fire-resistive construction, this requirement applies. Diagram 2.1.1.2(a) illustrates the application of the requirements contained in 2.1.1.2.2. The hoistway enclosure may consist of some walls, which must be fire-rated, and others, which do not. Laminated glass is required by 2.1.1.2.2(d) because it will not shatter if broken, not for fire protection reasons. Organic-coated glass, wired or tempered glass are not acceptable alternatives to laminated glass. See Handbook commentary (8.6.10.3) on cleaning of observation elevator glass. Those requirements apply to both new and existing observation elevators.

2.1.1.3 Partially Enclosed Hoistways. Observation and nonobservation elevators may not require fully enclosed hoistways. If a portion of the hoistway enclosure is required to be fire resistive, it must conform to the requirements of 2.1.1.1. Other portions of the enclosure would have to conform to the requirements of 2.1.1.2. See Diagram 2.1.1.3.

Hoistways do not need to be fully enclosed, but protection must be provided adjacent to areas permitting the passage of people. Many different forms of protection have been provided over the years. Additionally 2.5.1.5 specifies a maximum clearance that must be

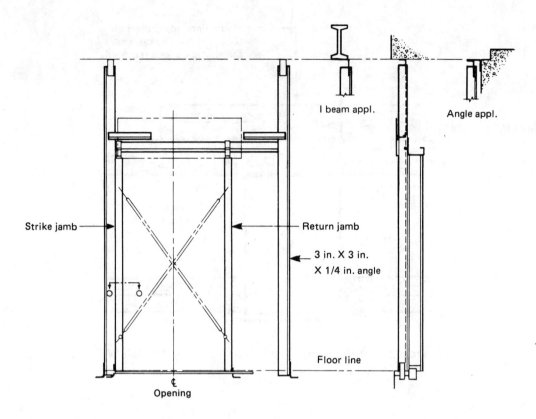

Diagram 2.1.1.1(a)   Typical Drywall Struts

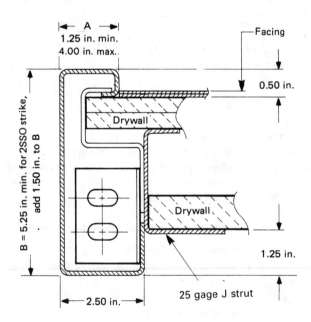

Diagram 2.1.1.1(b)   Typical Door Jamb Detail

maintained between the car sill and the adjacent enclosure or facia plate. This requirement prohibits a design where there is no adjacent hoistway wall or facia opposite a car entrance, unless a car door interlock conforming to 2.14.4.2.3 is provided and the car door meets the requirement of 2.5.1.5.3. If these requirements are

adhered to the exposure at the car door is similar to that found at a hoistway door. See commentary on 2.5.1.5.

The requirement protects the public from coming into contact with elevator equipment.

2.1.1.4 Multiple Hoistways. The maximum number of elevators permitted in a single hoistway is controlled by the building code (see 1.3, Definitions) in order to limit the potential of a fire disabling all elevator service in a structure.

Most building codes require that where there are three or fewer elevators in a building, they can be in the same hoistway enclosure. Where there are four elevators, they must be in at least two separate hoistway enclosures. Where there are more than four elevators, not more than four can be in the same hoistway enclosure.

Diagram 2.1.1.4 illustrates the multiple hoistway requirements in the model building codes.

2.1.1.5 Strength of Enclosure. If the wall is load bearing (masonry, brick, concrete block, etc.), the hoistway entrance operating mechanisms and locking devices can be supported by the wall. When the wall is not load bearing (gypsum drywall, gypsum block, etc.), the hoistway entrance operating mechanism and locking devices must be supported by other building structure (structural steel, floor, etc.). See Diagrams 2.1.1.5(a) and 2.1.1.5(b).

Elevator pressure measurements were made by U.S.G. Research in three typical Chicago high-rise structures.

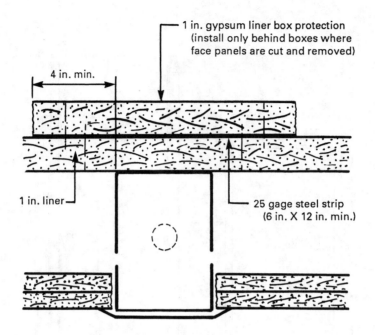

Diagram 2.1.1.1(c)   Typical Fixture Box in 2 h Fire-Resistive Wall

The buildings involved were: United States Gypsum Building (18 stories); the First National Bank Building (65 stories); and the John Hancock Building (100 stories). These buildings encompass elevators having velocities ranging from 3.5 m/s or 700 ft/min to 9 m/s or 1,800 ft/min.

The maximum elevator hoistway pressure measured was 5.15 lbf/ft^2 in a single operating hoistway at the John Hancock Building. In a multiple three-car hoistway, a maximum pressure of 1.35 lbf/ft^2 was measured.

The following recommendations are derived from these tests. Loads were calculated according to the following formula:

(a) For shafts with one or two elevators:

(Imperial Units)

$$P_s = P_c + (2/3)(V_e/400)^{1.32}$$

(b) For shafts with three or more elevators:

(Imperial Units)

$$P_s = P_c + (2/3)(V_e/1000)^{1.32}$$

where

P_s = design pressure or negative pressure in shaft in lbf/ft^2

P_c = code design pressure or designed static pressure for surrounding partitions with elevators in lbf/ft^2

V_e = elevator velocity in ft/min.

Elevator Velocity, ft/min	1 or 2 Elevators Per Shaft, lbf/ft^2	3 Plus Elevators Per Shaft, lbf/ft^2
0 to 180	5.0	5.0
180 to 1,000	7.5	5.0
1,000 to 1,800	10.0	7.5
Over 1,800	15.0	7.5

This information is intended only as a guide and is taken from the United States Gypsum Co. Technical Data Sheet. Hoistway pressure is dependent on variable factors such as hoistway size, elevator speed, number of elevators per hoistway, outside wind velocity, temperature, etc. See Diagram 2.1.1.5(c).

2.1.2 Construction at Top and Bottom of the Hoistway

2.1.2.1 Construction at Top of Hoistway. The building code requirements can generally be summarized as follows:

(a) Where a hoistway extends into the top floor of a building, fire-resistive hoistway, or machinery-space enclosures, where required, shall be carried to the underside of the roof if the roof is of fire-resistive construction, and at least 900 mm or 3 ft above the top surface of the roof if the roof is of non-fire-resistive construction.

(b) Where a hoistway does not extend into the top floor of a building, the top of the hoistway shall be enclosed with fire-resistive construction having a fire-resistance rating of at least equal to that required for hoistway enclosure.

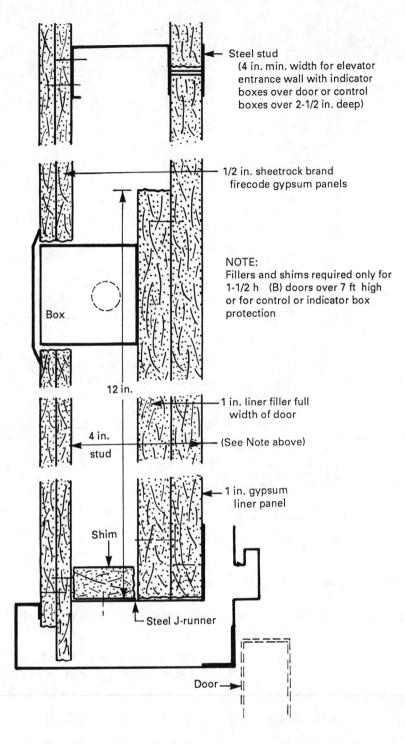

Diagram 2.1.1.1(d) Typical Fixture Box in 2 h Fire-Resistive Wall

The requirement to extend the hoistway 900 mm or 3 ft above a non-fire-resistive roof is necessary in order to protect the roof in case of a fire. The 900 mm or 3 ft of hoistway above the roof will act as a heat sink if hot gases from a fire enter the hoistway. If this is not done, the possibility exists of the roof igniting from the buildup of heat in the top of the hoistway. Fire-resistive construction at the roof over the hoistway does not meet this requirement.

The ceiling construction above a hoistway not extending to the top floor of a building must have the same fire-resistive rating as required for the hoistway

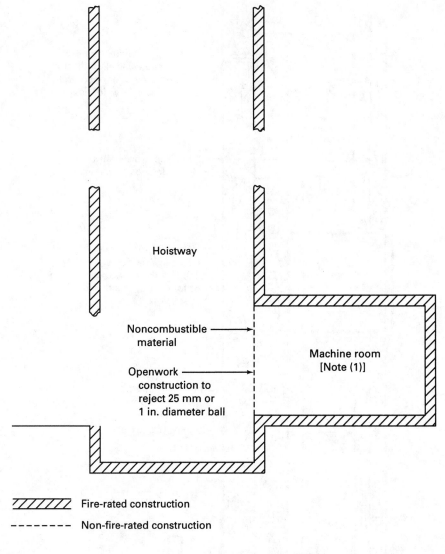

Fire-rated construction

Non-fire-rated construction

NOTE:
(1) Machine room may be located at other locations below top of hoistway.

Diagram 2.1.1.1(e)   Basement and Mid-Level Machine Room/Hoistway Separation

enclosure walls in 2.1.1.1. When a 2 h fire-resistive hoistway enclosure is provided, a 2 h fire-resistive ceiling is required over the hoistway. Similarly, a 1 h enclosure requires a 1 h ceiling. The local building code should be reviewed as it may contain additional requirements. In the design and construction of a building, this is one item that is often overlooked. Diagrams 2.1.2.1(a) and 2.1.2.1(b) illustrate these requirements.

2.1.2.2 Construction at Bottom of Hoistway. If ground water is evident at a building site, the pit design should incorporate a means of keeping the water from entering the pit. Also, see 2.2.2 and 8.6.4.7.

2.1.2.3 Strength of Pit Floor. The pit floor structure must be strong enough for both the compressive load and tension loads caused by compensating sheave tie

down or other device tie downs. The impact loads are based on 125% of rated speed since the governor-safety system will prevent higher impacts from occurring.

2.1.3 Floor Over Hoistways

2.1.3.1 General Requirements. A floor is not required at the top of the hoistway when the elevator machine is located below or at the side of the hoistway, and where elevator equipment can be serviced and inspected from outside the hoistway or from the top of the car. When an elevator machine is located over the hoistway, a floor is not required below secondary deflecting sheaves, if the elevator equipment can be serviced and inspected from the top of the car or from outside the hoistway. Access from the top of an adjacent car would not be

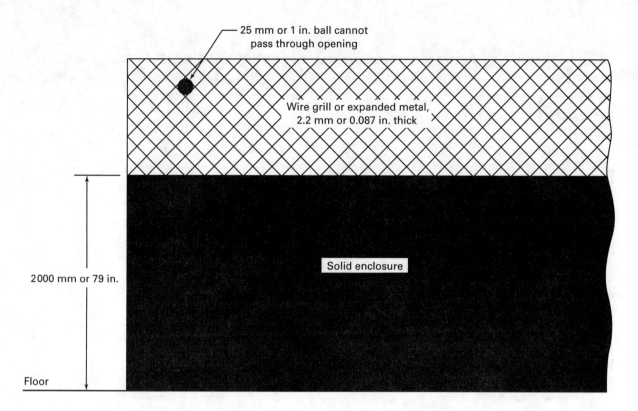

Diagram 2.1.1.2(a)   Non-Fire-Resistive Hoistway and Machine Room Construction

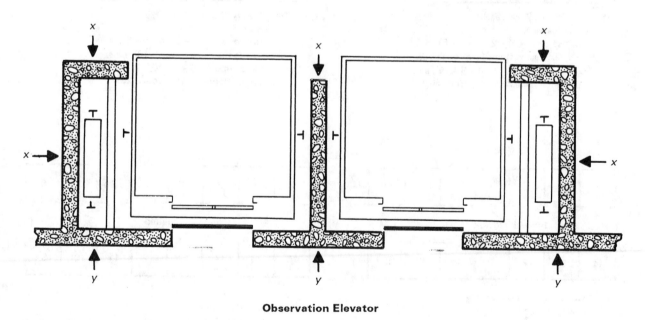

Observation Elevator

GENERAL NOTE:
x = Non-fire-rated
y = Fire rated

Diagram 2.1.1.3   Typical Observation Elevator Hoistway Arrangement

I. MULTIPLE ELEVATOR GROUPS [ASSUME 20 STORY BUILDING (20 FLOORS)]

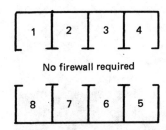

No firewall required

A1. Two Groups (4 cars each)
All cars serve all floors

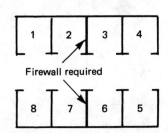

Firewall required

A2. Two Groups (4 cars each)
Low rise (1 – 2 – 3 – 4), all serve
same floors (1 through 10).
High rise (5 – 6 – 7 – 8), all serve
same floors (1, 10 through 20).

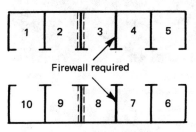

Firewall required

B. Two Groups (5 cars each)
All cars serve all floors, or
Low rise groups (1 – 2 – 3 – 4 – 5), all
cars serve same floors (1 through 10), plus
High rise group (6 – 7 – 8 – 9 – 19), all
cars serve same floors (1, 10 through 20).

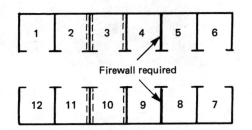

Firewall required

C. Two Groups (6 cars each)
All cars serve all floors, or
Low rise group (1 – 2 – 3 – 4 – 5 – 6), all
serve same floors (1 through 10), plus
High rise group (7 – 8 – 9 – 10 – 11 – 12), all
serve same floors (1, 10 through 20).

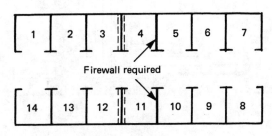

Firewall required

D. Two Groups (7 cars each)
All cars serve all floors, or
Low rise group (1 through 7), all
serve same floors (1 through 10), plus
High rise group (8 through 14), all
serve same floors (1, 10 through 20).

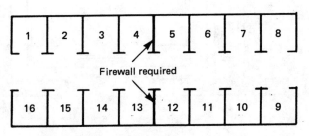

Firewall required

E. Two Groups (8 cars each)
All cars serve all floors, or
Low rise group (1 through 8), all
serve same floors (1 through 10), plus
High rise group (9 through 16), all
serve same floors (1, 10 through 20).

Diagram 2.1.1.4   Illustration of Multiple Hoistway Requirements in Building Codes

II. SINGLE GROUP

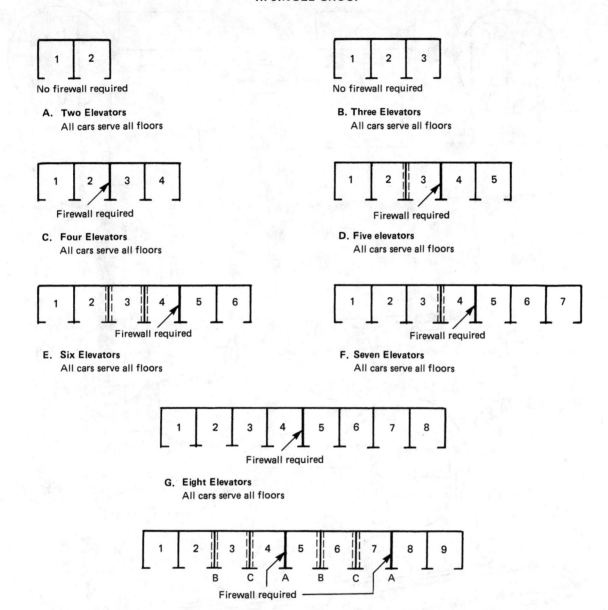

A17.1–2000 Handbook

A. **Two Elevators**
All cars serve all floors
No firewall required

B. **Three Elevators**
All cars serve all floors
No firewall required

C. **Four Elevators**
All cars serve all floors
Firewall required

D. **Five elevators**
All cars serve all floors
Firewall required

E. **Six Elevators**
All cars serve all floors
Firewall required

F. **Seven Elevators**
All cars serve all floors
Firewall required

G. **Eight Elevators**
All cars serve all floors
Firewall required

H. **Nine Elevators**
All cars serve all floors
[Note (1)]
Firewall required

GENERAL NOTES:
(a) Not more than four elevators are located in one hoistway.
(b) Grouped elevators (Part I) serve same floors (same portion of building).
(c) Alternate firewall locations are shown dotted.
(d) It is recommended that not less than two elevators be located in one hoistway. While 2.1.1.4 does not require this, isolating one car into a single hoistway should be avoided for evacuation reasons. With two elevators in a single hoistway, side emergency exits on each car can be utilized to evacuate passengers if one of the cars becomes stalled between floors.

NOTE:
(1) Alternate firewall locations B-B or C-C may be used in lieu of firewall location A-A in sketch "H," Part II.

Diagram 2.1.1.4   Illustration of Multiple Hoistway Requirements in Building Codes (Cont'd)

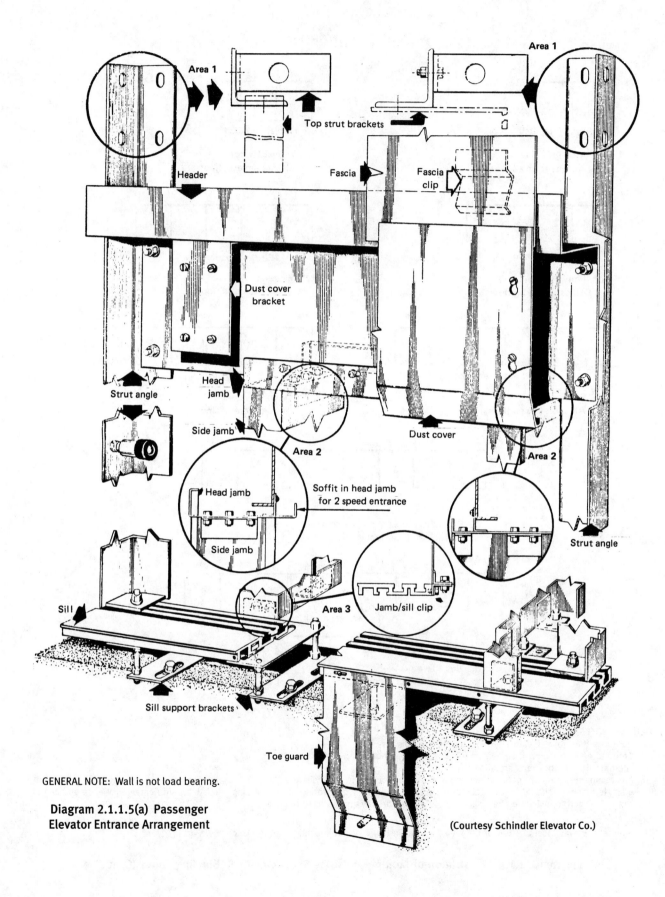

Area 1

Area 1

Top strut brackets

Header

Fascia

Fascia clip

Dust cover bracket

Strut angle

Head jamb

Side jamb

Area 2

Dust cover

Area 2

Head jamb

Soffit in head jamb for 2 speed entrance

Side jamb

Strut angle

Sill

Area 3

Jamb/sill clip

Sill support brackets

Toe guard

GENERAL NOTE: Wall is not load bearing.

Diagram 2.1.1.5(a) Passenger Elevator Entrance Arrangement

(Courtesy Schindler Elevator Co.)

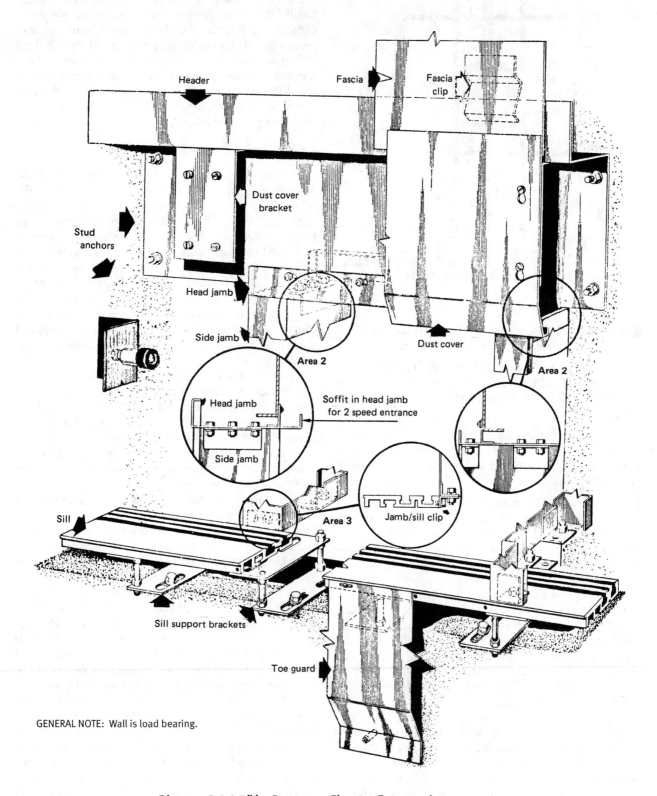

Header

Fascia

Fascia clip

Dust cover bracket

Stud anchors

Head jamb

Side jamb

Area 2

Dust cover

Area 2

Head jamb

Soffit in head jamb for 2 speed entrance

Side jamb

Sill

Jamb/sill clip

Area 3

Sill support brackets

Toe guard

GENERAL NOTE: Wall is load bearing.

Diagram 2.1.1.5(b)   Passenger Elevator Entrance Arrangement
(Courtesy Schindler Elevator Co.)

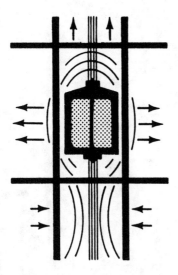

Diagram 2.1.1.5(c)   Piston Effect of Elevators Resulting in Oscillation of Hoistway Walls

considered to be from outside the hoistway and does not conform to the requirements.

2.1.3.2 Location of Floor. This requirement minimizes structural tripping hazards in the equipment area. It also positions the floor in order to facilitate the servicing of the equipment.

2.1.3.3 Strength of Floor. The structural requirements provide for a floor capable of supporting the normal loads that may be anticipated during maintenance and repair of elevator equipment.

2.1.3.5 Area to Be Covered by Floor. A machine room at the top with a hoistway having a cross-sectional area more than 10 m^2 (108 ft^2) does not require a floor covering the entire area over the hoistway. In this case, the floor need only extend 600 mm or 2 ft beyond the general contour of the machine, sheaves, or other equipment and to the entrance to the machinery space.

This requirement does not require a minimum clearance of 600 mm or 2 ft between the general contour of the machine, sheaves, governor, or other equipment to a wall. If more than 600 mm or 2 ft clearance is provided, the floor must extend a minimum of 600 mm or 2 ft; if less than 600 mm or 2 ft is provided to the wall, the floor must extend to the wall. There are minimum maintenance clearance requirements specified in 2.7.2.2. Minimum clearance requirements about electrical equipment are dictated by the National Electrical Code (see 2.26.4).

Where the floor does not cover the entire area of the hoistway, the open sides must be provided with a standard railing and toe board (see 2.10.2) to protect the elevator mechanic, inspector, or other authorized persons from falling down the hoistway and to protect the elevator car from objects being accidentally kicked off the machine room floor.

2.1.3.6 Difference in Floor Levels. Machine room floors ideally should be designed and constructed without a difference in floor levels. Where there is a difference of more than 400 mm or 16 in., a railing and stair or fixed ladder is required to reduce the hazard to an elevator mechanic or inspector from falling or tripping. Where the difference is less than 400 mm or 16 in., the area where the floor area differs should be marked to clearly show the potential hazard.

2.1.4 Control of Smoke and Hot Gases

The majority of deaths in fires are a result of smoke asphyxiation. In the MGM fire, for example, 70 of the 84 deaths occurred on the upper floors where smoke concentration was the greatest. This could be attributed to stack effect, which is a phenomenon that exists in high-rise buildings. Stack effect describes the movement of air inside and outside a building. During a fire, the presence of stack effect generally results in the movement of smoke and combustion products from lower levels to upper levels of the building. Stack effect may also have an adverse effect on elevator door closing, and it may be necessary to sense pressure differentials to ensure door closing.

An incident that illustrates how crucial this situation can be occurred during a time of high wind conditions. The winds blew out a revolving door on the first floor of a high-rise building. The ensuing stack effect was so severe that the elevator doors at the first floor of an eight-car group could not close, thus rendering the entire group of cars out of service. This could have been even more serious in the event of a fire.

All of the model building codes require venting of elevator hoistways. Most require venting of the hoistway for elevators serving three or more stories. Some local codes permit venting by means of floor grates into the machine room with mechanical or natural venting from the machine room to the outside. The model building codes prohibit this practice and require the hoistway to be vented directly to the outside of the building. They permit venting through the machine room only through enclosed ducts. Some building codes required cable slots and other openings between the machine room and hoistway to be sleeved from the machine room floor to a point not less than 305 mm (12 in.) below the hoistway vent to inhibit the passage of smoke into the machine room. See Diagram 2.1.4. When this requirement exists, caution should be taken, since overhead clearances are affected. Most building codes now state that holes in machine room floors are only permitted for the passage of ropes, cables, or other moving elevator equipment and shall be limited so as to provide no greater than 51 mm (2 in.) clearance on all sides.

Some building codes allow the vents to be closed with dampers that are automatically opened when a smoke detector in any elevator lobby is activated. These smoke

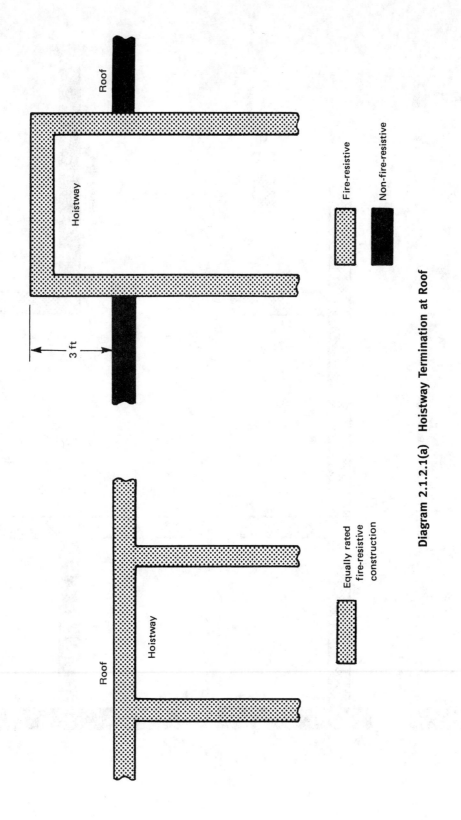

Diagram 2.1.2.1(a)   Hoistway Termination at Roof

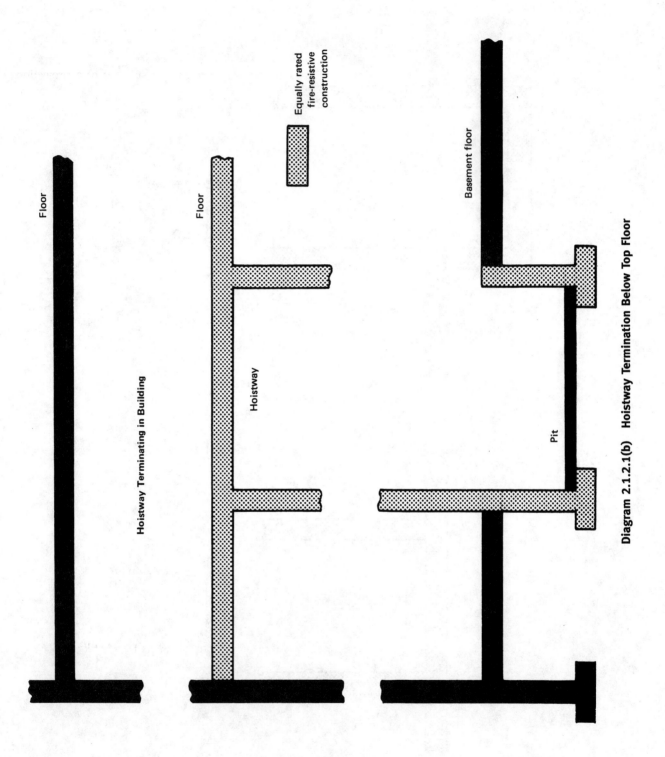

Equally rated
fire-resistive
construction

Floor

Floor

Hoistway

Basement floor

Pit

Hoistway Terminating in Building

Diagram 2.1.2.1(b)   Hoistway Termination Below Top Floor

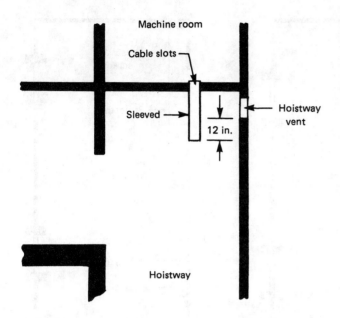

Diagram 2.1.4 Building Code Rope Slot Sleeving Requirements

detectors may be the same smoke detectors that initiate Phase I Emergency Recall Operation. The dampers should also be designed to open in case of power failure.

Some local codes prohibit venting and require pressurization of the hoistway to a specified minimum positive pressure above the elevator lobby. When a pressurized hoistway is provided, it should contain a vent that would open automatically in case of power failure or from a buildup of smoke in the hoistway.

The Americans with Disabilities Act Accessibility Guidelines (ADAAG) require "Areas of Rescue Assistance" in new buildings. ADAAG Section 4.3.11.1(7) recognizes an elevator lobby as an area of rescue assistance when the lobby and adjacent shaft are pressurized. The model building codes also have requirements for "areas of refuge," which may be required to be accessible to an elevator.

Due to numerous local regulations and the rapid developments in this area, it would be wise to check local and Federal ADA requirements.

2.1.5 Windows and Skylights

Windows and skylights are permitted in the machine room. Skylights are also permitted in the hoistway. Windows are prohibited in the hoistway, as a fireman entering a building through a window in a fire situation may not realize that he is in the elevator hoistway and that there is no floor on which to stand. A glass curtain wall is not considered a window and is permitted in the hoistway by 2.1.1.2.1(d).

The building code should be consulted for the detailed requirements on window and skylight glazing, framing and sash construction, and guarding. This requirement

does not apply in Canada as it is outside the Scope of the elevator code. See NBCC.

2.1.6 Projections, Recesses, and Setbacks in Hoistway Enclosures

The intent of these requirements are to prevent a person from standing on the top surfaces of a projection or laying tools or equipment on the top surfaces of a projection and not to eliminate a shear hazard. This requirement also applies to mullions for glass curtain walls. Bevels are not necessary on nonhorizontal projections such as diagonal bracing making an angle greater than 75 deg from the horizontal, except where they intersect with beams or columns. Projections and setbacks may be screened instead of beveled. The requirements for the screening material are the same as 5.10.1.1.1. See Diagram 2.1.6.2.

It is not a requirement, to guard the underside of projections on sides of hoistways not used for loading and unloading.

SECTION 2.2
PITS

2.2.2 Design and Construction of Pits

When the hoistway is enclosed with fire-resistive construction, the pit must also be enclosed with fire-resistive construction not less than that which is required for the hoistway. If the hoistway enclosure is rated for 2 h, the pit enclosure must be rated for at least 2 h. The fire-resistance rating of the pit access door must be as specified in the building code (see 1.3, Definitions). See Handbook commentary on 2.1.1.1 and 2.1.1.2.

Structural frames for buffers, compensating sheaves, and frames, which are on top of the floor, are permitted. The requirement refers to the floor only and not equipment mounted on it. A hydraulic jack casing extending above the floor is also permitted.

A sump pump or drain is required in every pit for elevators provided with firefighters' service (see 2.2.2.5), as water from sprinklers or other sources (i.e., water from fire suppression) should not be allowed to accumulate in a pit. If a sump pump is to be utilized, it must be permanently installed. This provision is intended to assist in maintaining elevator service during a fire emergency. The intent of 2.2.2.4 is to eliminate the possibility of sewer gases entering the hoistway. The sump pump or drain typically discharges into the open air outside the pit and hoistway. The Committee is aware that this requirement poses a need in many jurisdictions to provide an oil water separator or equivalent protection as required by the plumbing code. However, the need of firefighters to use elevators in a fire situation outweighs the cost associated with compliance to the plumbing code requirements.

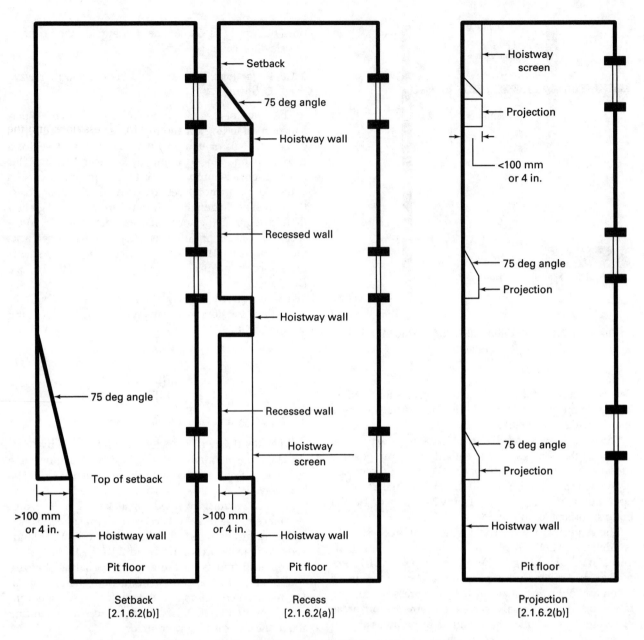

Diagram 2.1.6.2   Hoistway Setback, Projection, and Recess

Sumps in pit floors must be covered to reduce a tripping hazard to maintenance and inspection personnel. In Canada a sump pump is not permitted to be installed in pits, in order to avoid nonelevator personnel from having to enter the pit to service the pump.

2.2.3 Guards Between Adjacent Pits

A guard or railing is required only when adjacent pit floors are at different levels. The guard or railing is to protect the mechanic or inspector from falling due to the differences in elevations between adjacent pit floors.

2.2.4 Access to Pits

When the pit floor is more than 900 mm or 3 ft below the sill of the pit access door, a fixed ladder must be installed. If access to the pit in a multiple hoistway is through the lowest landing (hoistway) door and a ladder is required, then a ladder must be provided for each elevator in the multiple hoistway. One ladder is not acceptable.

The A17.1 requirements are based on the relevant OSHA regulations 29 CFR 1910.27(c)(4).

American National Standards Institute (ANSI) A14.3, for Ladders — Fixed-Safety Requirements, is the source standard for 29 CFR 1910.27.

See Diagram 2.2.4 for the recommended location of pit ladders.

2.2.4.2 See Diagram 2.2.4.2.

2.2.5 Illumination of Pits

Caution should be taken when installing lighting fixtures in the pit. The fixture and its lamp must be mounted in a location where they will not be struck by a car or counterweight when on their fully compressed buffers. A good suggestion is to mount lights at diagonal corners of the pit. The light switch must be accessible from the access door to allow for turning on the lights before accessing the pit.

The requirement for a duplex receptacle was removed from 2.2.1.5 as it is required by the National Electrical Code® and Canadian Electrical Code (see 2.26.4). A duplex receptacle is required as power is often necessary for droplights and power tools used for servicing elevators. The National Electrical Code® Section 620-85 requires that all receptacles in the pit be provided with ground-fault circuit interrupter protection, except a single receptacle supplying a permanently installed sump pump.

2.2.5.2 The guard on the light bulb shall be sufficient to prevent breakage and contact with voltage if it is accidentally contacted by a person working in the pit. Flexible guards would not meet this requirement.

2.2.6 Stop Switch in Pits

A pit stop switch must be provided for every elevator to allow the mechanic or inspector to control the movement of the elevator before entering the pit. Pit stop switches must be of the manually operated and enclosed type, have red operating handles or buttons, be conspicuously and permanently marked "STOP," shall indicate the stop and run positions, be positively opened mechanically, and their openings must not be solely dependent on springs. The switch must be located adjacent to the nearest point of access and where access to the pit is through the lowest landing hoistway door approximately 450 mm or 18 in. above the floor level of the access landing.

OSHA requires, under certain conditions, that equipment being worked on be locked-out and tagged-out of service. Locking a pit stop switch such as shown in Diagram 2.2.6 should not be used to facilitate locking the elevator out of service from the pit area. Only the main line disconnect switch should be used to de-energize the electrical power source prior to the system being locked-out and tagged-out. Detailed information on the proper lock-out/tag-out procedure can be found in the Elevator Industry Field Employees' Safety Handbook. Procurement information can be found in Part 9 of this Handbook. Detailed information on the application of OSHA confined space regulations applicable to elevator pits can be found at www.neii.org.

In a pit where the point of access is more than 1 700 mm or 67 in. above the pit floor, two pit stop switches which are wired in series must be provided: one approximately 450 mm or 18 in. above the access floor, and the second 1 200 mm or 47 in. above the pit floor. This allows personnel entering the pit to engage the upper stop switch, 450 mm or 18 in. above the access floor, then descend the pit ladder to the pit floor and engage the lower switch located 1 200 mm or 47 in. above the pit floor. If operation of the elevator is then necessary, elevator personnel can ascend the pit ladder and place the upper stop switch in the run position. The elevator can then be controlled from the lower stop switch. When leaving the pit, the upper stop switch is to be placed in the stop position before the lower stop switch is placed in the run position.

SECTION 2.3
LOCATION AND GUARDING OF COUNTERWEIGHTS

2.3.1 Location of Counterweights

Counterweights located in the hoistway of the elevator they serve facilitate the inspection and maintenance of the counterweight and its ropes. The term "mechanical compensation" as used in this requirement refers to physically connecting the bottom of the car to the bottom of the counterweight. When counterweights are located in remote (separate) hoistways, the requirements of 2.3.3 provide access for inspection and maintenance. Counterweights with mechanical compensation or with a counterweight safety must be located in the same hoistway with the elevator they serve.

2.3.2 Counterweight Guards

The counterweight guards protect a mechanic or inspector from accidentally stepping into the runway of the counterweight. When the counterweight cannot descend less than 2 130 mm or 84 in. above the pit floor, no guard is required, as an individual could not be struck by the counterweight while standing on the pit floor. When compensation is used, guards are not required when in compliance with 2.3.2.1(a), because the compensating ropes or chains alert the individual of the presence of the counterweight runway.

The height of the guards assures protection, but does not obscure the equipment, in order to facilitate inspection and maintenance. Perforations allow for visual inspection behind the guard yet provide protection against accidental intrusion into the counterweight runway. See Diagram 2.3.2.

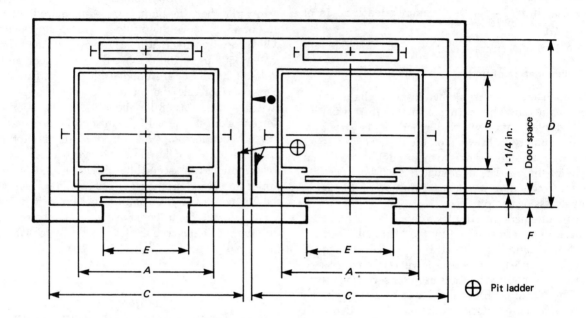

Center Opening Entrance

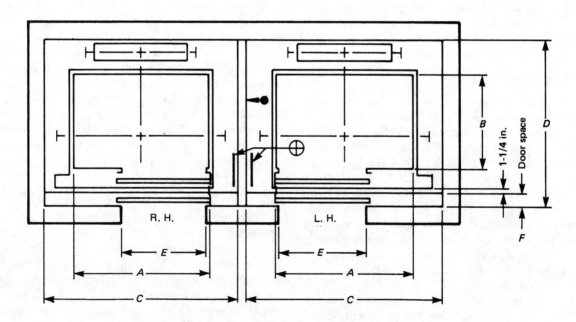

Diagram 2.2.4   Typical Pit Ladder Locations
(Courtesy Building Transportation Standards and
Guidelines, NEII-1 2000, © 2000, National
Elevator Industry, Inc., Teaneck, NJ)

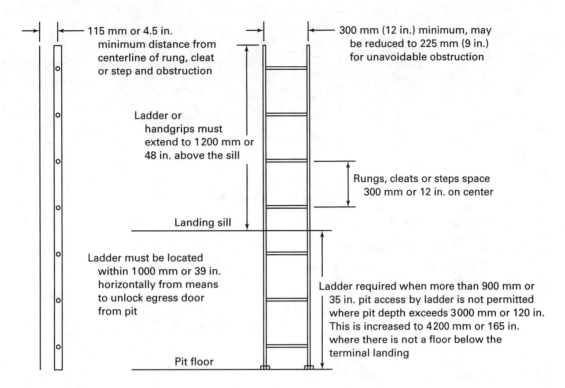

Diagram 2.2.4.2   Pit Ladder Requirements
(Courtesy Zack McCain, Jr., P.E.)

Diagram 2.2.6   Lockable Pit Stop Switch
(Courtesy Nylube Products Co.)

2.3.3 Remote Counterweight Hoistways

This requirement only applies to elevators, which do not have mechanical compensation or counterweight safeties. When a separate counterweight hoistway is provided, fire protection must be maintained, and safe, convenient access for maintenance and inspection personnel must be provided. Additionally, the requirements protect the public from moving counterweights. Also, see commentary on referenced requirements.

2.3.4 Counterweight Runway Enclosures

This requirement permits the screening of the counterweight runway from the remainder of the hoistway. This screen may run the full height of the hoistway. Provisions must be made to allow removal of the screen for inspection of the counterweight rails, brackets, ropes, etc.

SECTION 2.4
VERTICAL CLEARANCES AND RUNBYS FOR CARS AND COUNTERWEIGHTS

2.4.1 Bottom Car Clearances

Bottom car clearance is defined as the clear vertical distance from the pit floor to the lowest structural or mechanical part, equipment, or device installed beneath the car platform except guide shoes or rollers, safety device assemblies and platform aprons or guards when the car rests on its fully compressed buffer or solid bumper. The purpose of this requirement is to provide refuge space below the car even when resting on its fully compressed buffer or solid bumper. The 600 mm or 24 in. measurement specified in 2.4.1.1 is taken to the pit floor even though there may be a pit channel running across the pit floor. However, the 600 mm or 24 in. and

21

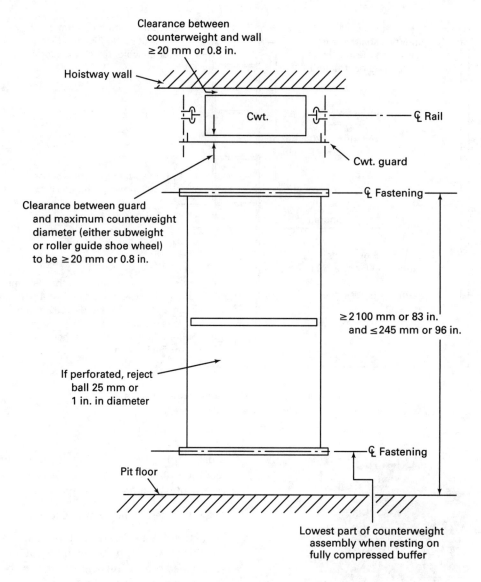

Clearance between
counterweight and wall
≥ 20 mm or 0.8 in.

Hoistway wall

Cwt.

℄ Rail

Cwt. guard

Clearance between guard
and maximum counterweight
diameter (either subweight
or roller guide shoe wheel)
to be ≥ 20 mm or 0.8 in.

℄ Fastening

≥ 2100 mm or 83 in.
and ≤ 245 mm or 96 in.

If perforated, reject
ball 25 mm or
1 in. in diameter

℄ Fastening

Pit floor

Lowest part of counterweight
assembly when resting on
fully compressed buffer

Diagram 2.3.2   Counterweight Pit Guards

1 070 mm or 42 in. minimum heights specified in 2.4.1.3 would be measured from the top of the channel. See Diagrams 2.4.1(a) and 2.4.1(b).

2.4.1.6 The marking and warning sign is required in the case where a car is on its fully compressed buffer and the vertical clearance is less than 600 mm or 24 in. If the only area where the 600 mm or 24 in. vertical clearance is not provided is the area below the platform guard and/or guiding members, then the marking and warning sign is not required. Elevator personnel need to be advised where there is a low clearance condition.

2.4.2 Minimum Bottom Runby for Counterweighted Elevators

2.4.3 Minimum Bottom Runby for Uncounterweighted Elevators

Requirements 2.4.2 and 2.4.3, Bottom Car Runby, is defined as the distance between the car buffer striker plate and the striking surface of the car buffer or solid bumper when the car floor is level with the bottom terminal landing. Bottom counterweight runby is defined as the distance between the counterweight buffer striker plate and the striking surface of the counterweight buffer or solid bumper when the car floor is

22

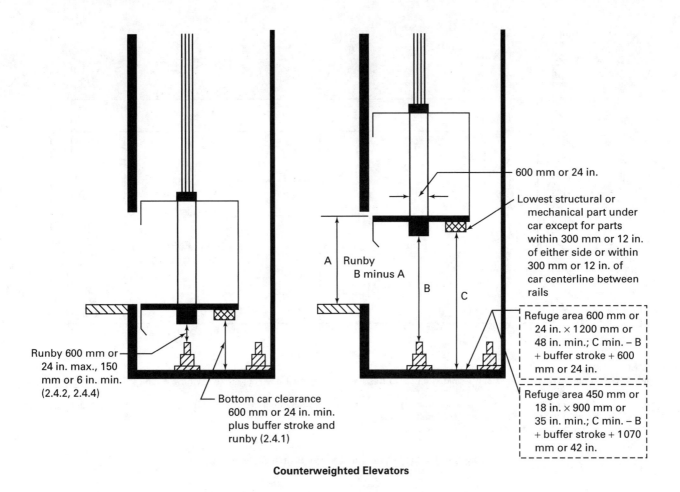

Counterweighted Elevators

Diagram 2.4.1(a)   Bottom Car Clearance and Runby
(Courtesy Zack McCain, Jr., P.E.)

level with the top terminal landing. Sufficient distance is allowed by this requirement for the elevator to stop normally without striking the buffers or solid bumpers.

Bottom runby is control dependent, because landing accuracy is control dependent. With rheostatic control, the speed is varied by changing the resistance and/or reactance in the armature and/or field circuit of the driving machine motor. Since the elevator is stopped on the brake with this type of control, the runby requirement increases with the rated speed. A modern control system such as thyristor voltage control can have a closer tolerance for landing accuracy. The requirements in Table 2.4.2.2 were empirically derived. Normal cable stretch should be taken into account during installation keeping in mind the runby.

The runby specified in 2.4.2 and 2.4.3 may be reduced below that which is specified after the installation is accepted by the authority having jurisdiction. The requirements for maintenance of runby are specified in 8.6.4.11. See also Handbook commentary on 8.6.4.11.

2.4.4 Maximum Bottom Runby

If additional runby was allowed, evacuation from a stalled elevator could be dangerous and/or impossible. This is a concern since a malfunctioning elevator has the tendency to runby a terminal landing. See Diagrams 2.4.1(a) and 2.4.4.

2.4.5 Counterweight Runby Data Plate

The data plate is a means of advising as to the designed runby. If the runby is not correct, the car could strike the overhead.

2.4.6 Top Car Clearances for Counterweighted Elevators

Top car clearance is defined as the shortest vertical distance between the top of the car crosshead, or between the top of the car when no crosshead is provided, and the nearest part of the overhead structure or any other obstruction when the car floor is level with the top terminal landing.

23

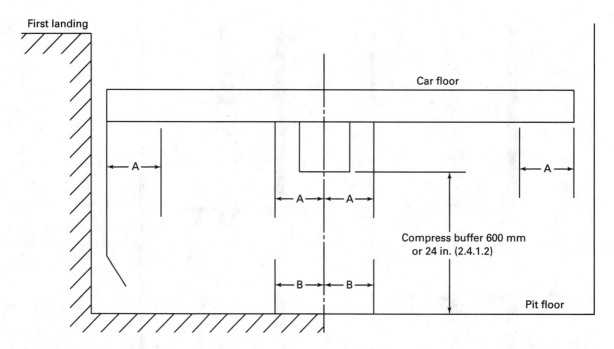

GENERAL NOTES:
(a) A = 300 mm or 12 in. equipment permitted on car in this area as long as it does not strike anything when car is on fully compressed buffer [2.4.1.2(a) and (b)].
(b) B = 300 mm or 12 in. equipment permitted on pit floor in this area as long as it is not struck by car, plunger follower guide, or other equipment when car is on fully compressed buffer [2.4.1.2(c)].
(c) See also minimum refuge space requirement (2.4.1.3).

Diagram 2.4.1(b) Bottom Car Clearance Does Not apply

The clearances differ depending on the type of counterweight buffers that are used, and whether or not compensating rope tie-down is provided (see 2.17.17). These clearances are shown in Diagram 2.4.6.

2.4.7 Top Car Clearance for Uncounterweighted Elevators

See Diagram 2.4.7.

2.4.8 Vertical Clearances With Underslung Car Frames

An underslung car frame is one to which the hoisting rope fastenings or hoisting rope sheaves are attached at or below the car platform. See Diagram 2.4.8(a). The intent of this requirement is to provide a minimum distance of 150 mm or 6 in. between the overhead car rope, dead-end hitch or car overhead sheave, and the portions of the structure vertically below them when the car reaches its maximum upward travel. See Diagram 2.4.8(b).

2.4.9 Top Counterweight Clearances

The top counterweight clearance is defined as the shortest vertical distance between any part of the counterweight structure and the nearest part of the overhead structure or any other obstruction when the car floor is level with the bottom terminal landing. See Diagram 2.4.9.

2.4.10 Overhead Clearances Where Overhead Beams Are Not Over Car Crosshead

2.4.10.2 The 600 mm or 24 in. horizontal clearance reduces a shearing hazard for elevator personnel riding on top of the car. Without this clearance, a person could be caught between the crosshead and an overhead beam. See Diagram 2.4.10.

2.4.12 Refuge Space on Top of Car Enclosure

2.4.12.1 The refuge area provides a clear, unobstructed space for mechanics and inspection personnel when on the top of the car. The top emergency exit may open into this area if there is also enough room for someone to stand adjacent to the opening. The refuge space does not have to be adjacent to the top emergency exit. The refuge space may be under the crosshead if there is 1 100 mm or 43 in. clearance between the top of the car and the underside of the crosshead. The term "maximum upward movement" is defined as the position of the car when the counterweight is on its fully compressed buffer plus the jump. See Diagram 2.4.12.

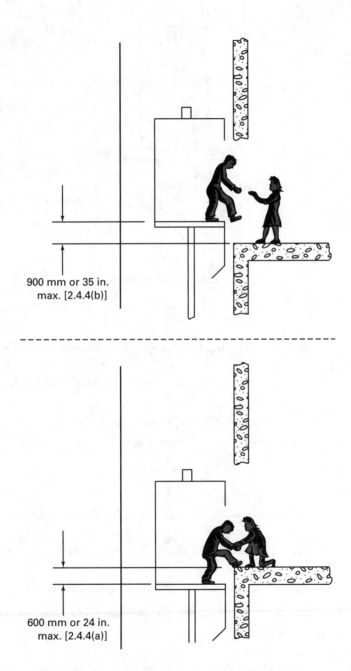

900 mm or 35 in. max. [2.4.4(b)]

600 mm or 24 in. max. [2.4.4(a)]

Diagram 2.4.4   Maximum Top and Bottom Runby

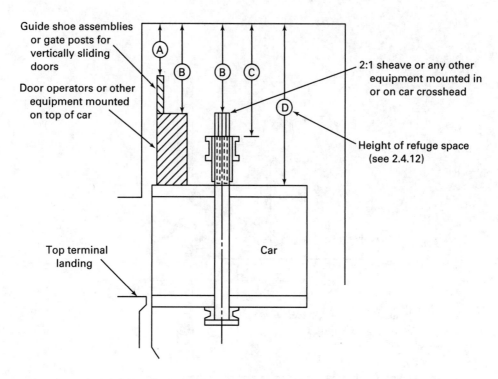

When the car floor is level with the top terminal landing, all of the following conditions must be met:

$$A > t$$
$$B \geq t + 6 \text{ in.}$$
$$C \geq t + 24 \text{ in.}$$

Where t = maximum possible travel of the car above the top terminal landing, as shown below:

Type of Cwt. Buffers	Comp. Rope Tie Down	Maximum Travel Above Top Landing, in., t
Oil buffers which are compressed with car at top landing	Yes	$S - S_c$
	No	$S - S_c + V_r^2 \, (3.423 \times 10^{-5})$
Reduced stroke oil buffers	Yes	$R + S$
	No	$R + 1.5S$
Other oil buffers	Yes	$R + S$
	No	$R + S + V_r^2 \, (3.423 \times 10^{-5})$
Spring Buffers	No	$R + S + V_g^2 \, (2.588 \times 10^{-5})$

where
R = bottom counterweight runby, in.
S = counterweight buffer stroke, in.
S_c = distance counterweight buffer is compressed when the car is at the top terminal landing, in. (see 2.22.4.8)
V_r = rated speed, ft/min
V_g = governor tripping speed, ft/min

Diagram 2.4.6 Top Car Clearances (Imperial Units)

26

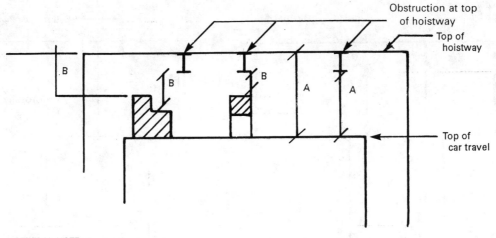

GENERAL NOTE:
A = 750 mm or 29.5 in.
B = 150 mm or 6 in.

Diagram 2.4.7   Top Car Clearances for Uncounterweighted Elevators
(Courtesy James Filippone)

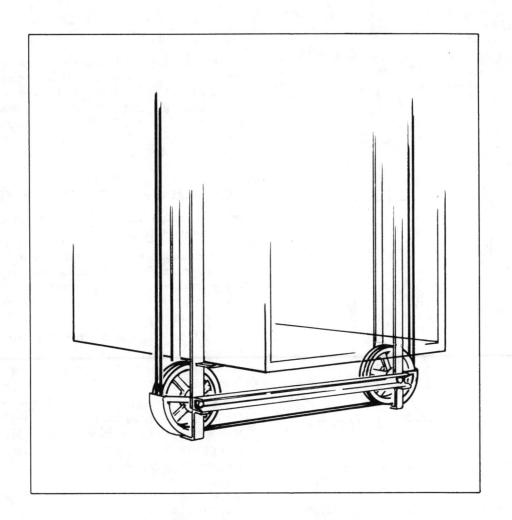

Diagram 2.4.8(a)   Underslung Car Frame
(Courtesy National Elevator Industry Educational Program)

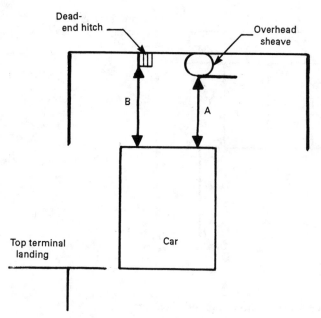

Diagram 2.4.8(b)   Vertical Clearances With Underslung Car Frames
(Courtesy James Filippone)

GENERAL NOTES:
(a) No counterweight: A ≥ 9, and B ≥ 9
(b) Counterweight: A ≥ t + 6, B ≥ t + 6, where t is defined in Diagram 2.4.6.

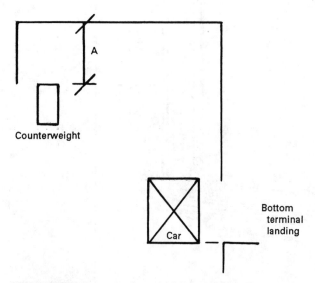

GENERAL NOTE: A ≥ t + 6, where t is defined in Diagram 2.4.6, except substitute car for counterweight and vice versa.

Diagram 2.4.9   Top Counterweight Clearance
(Courtesy James Filippone)

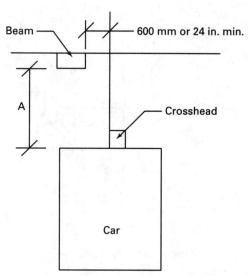

GENERAL NOTE: A ≥ clearance specified in 2.4.6 and 2.4.7.

Diagram 2.4.10   Overhead Clearance Where Overhead Beam Is Not Over Car Crosshead
(Courtesy James Filippone)

2.4.12.2 The markings and warning sign is required only when the 1 100 mm or 43 in. vertical clearance specified in 2.4.12.1 is not provided above the entire cross-sectional area of the car top. In this situation elevator personnel need to be advised where there is a low clearance.

SECTION 2.5
HORIZONTAL CAR AND COUNTERWEIGHT CLEARANCES

2.5.1 Clearances Between Cars, Counterweights, and Hoistway Enclosures

The clearances specified are measured from the car, excluding attachments such as vanes, conduit, etc., except as stated in 2.5.1.3. Requirement 2.5.1.3 includes attachments to the cars in measuring the clearance between cars in a multiple hoistway. See Diagram 2.5.1.

2.5.1.1 Between Car and Hoistway Enclosures. This requirement refers to the clearance between the car and hoistway enclosure. It does not refer to equipment that is in the hoistway.

2.5.1.5 Clearance Between Loading Side of Car Platforms and Hoistway. At the lower end of the hoistway the specified clearance must be maintained to the location of the car sill when the car is resting on its fully compressed buffer. At the upper end of the hoistway, the clearance must be maintained to the location of the car sill when it has reached its maximum upward travel.

Fascia plates are normally provided to maintain the required clearances, unless the hoistway wall line is located such that the clearance is provided.

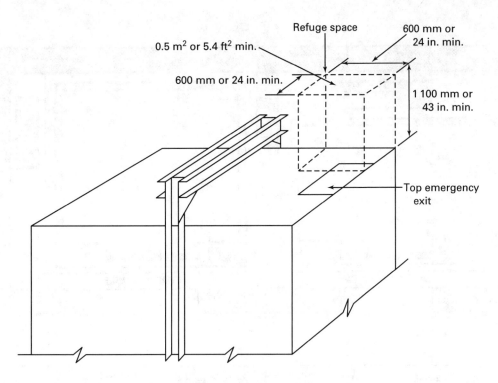

Refuge space

0.5 m² or 5.4 ft² min.

600 mm or 24 in. min.

600 mm or 24 in. min.

1 100 mm or 43 in. min.

Top emergency exit

Diagram 2.4.12   Top of Car Refuge Space

2.5.1.5.3 If the car door is equipped with a car door interlock (2.14.4.2.3) and meets the structural requirements in 2.5.1.5.3(b) the maximum clearances stated in 2.5.1.5.1 do not apply.

SECTION 2.6
PROTECTION OF SPACES BELOW HOISTWAYS

This requirement provides protection for people in spaces below an elevator hoistway. Where the hoistway does not extend to the lowest level of a building and the space under the hoistway is not permanently secured against access (see Diagram 2.6), the requirements of 2.6.1 and 2.6.2 apply.

2.6.1 Hoistways Not Extending to the Lowest Floor of the Building

The counterweight, in addition to the car, is required to be equipped with a safety device to stop and hold the counterweight in case of free fall or overspeed. This is intended to prevent building damage, which would endanger people below the hoistway, if the counterweight were to strike its buffer at excessive speed. These requirements are to be met when any of the spaces, on each succeeding floor directly below the hoistway, are not permanently secured against access.

See commentary on 2.7.6 when a machine room or control room is located below the hoistway.

SECTION 2.7
MACHINE ROOMS AND MACHINERY SPACES

2.7.1 Enclosure of Machine Rooms and Machinery Spaces

Machine room enclosures must conform to the building code. If the building code requires fire-resistive construction, follow the requirements in 2.7.1.1; otherwise, you can follow the requirements in 2.7.1.2. See commentary on 2.1.1.1 for important information on the building codes. The International Building Code (IBC), Section 3006.4 requires the machine room enclosure to have the same fire-resistance rating as the elevator hoistway.

2.7.1.1 Fire-Resistive Construction. Machine room enclosures must be fire resistive, as required by the building code. The fire-resistance rating is typically required to be equal to that required for the hoistway. Where a 2 h fire-resistive hoistway enclosure is required, the machine room and/or machinery space would be enclosed by a 2 h fire-resistive enclosure. Where a 1 h hoistway enclosure is required, a 1 h machine room and/or machinery space enclosure is required. Entrances to machine rooms and/or machinery spaces that are enclosed by fire-resistive construction typically would be rated at $1\frac{1}{2}$ h when a 2 h fire-resistive hoistway enclosure is provided. See commentary on 2.1.1.1.

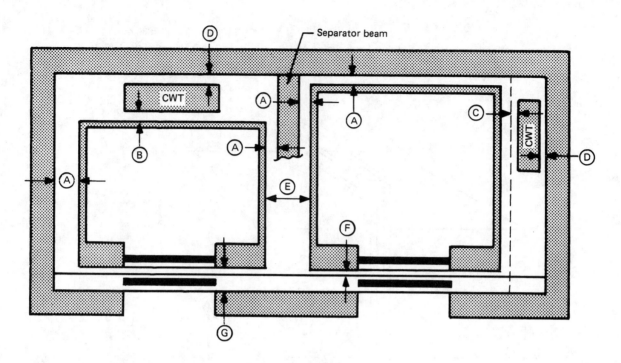

	Clearance	Requirement	Min., in./mm	Max., in./mm
A	Between car and hoistway enclosure (see also 2.14.1.3)	2.5.1.1	0.8/20	. . .
B	Between car and counterweight (see also 2.14.1.3)	2.5.1.2	1/25	. . .
C	Between counterweight and screen	2.5.1.2	0.8/20	. . .
D	Between counterweight and hoistway enclosure	2.5.1.2	0.8/20	. . .
E	Between cars (see also 2.14.1.3)	2.5.1.3	2/50	. . .
F	Between car sill and landing sill	2.5.1.4		
	Side-post construction		0.5/13	1.25/32
	Corner-post construction		0.8/32	1.25/32
G	Between car sill and hoistway enclosure or facia	2.5.1.5		
	Vertically sliding hoistway doors		. . .	7½/190
	Other doors		. . .	5/125

Diagram 2.5.1   Horizontal Car and Counterweight Clearances

2.7.1.2 Non-Fire-Resistive Construction. These requirements apply only if the building code does not require the enclosure to be of fire-resistive construction. See Diagram 2.1.1.2(a).

2.7.2 Equipment in Machine Rooms

Elevator machine room and/or machinery spaces should only contain elevator machinery and associated control equipment. Other equipment such as TV antenna controls, radio transmission, telephone equipment, etc., should not be in this room or space. Electronic and radio transmission equipment located in an elevator machine room has been found to cause interference with elevator equipment.

The intent of this requirement is to prevent unauthorized personnel from entering an area that is hazardous to those that are not trained in the safe maintenance or repair of elevator equipment. Unqualified persons may accidentally cause an elevator shutdown, trapping passengers in a stalled car. They also may be exposed to the moving machinery causing injury to themselves.

By not permitting any other equipment in the machine room, it reduces the potential of fire and sprinkler activation. Firefighters' require elevator service, especially in

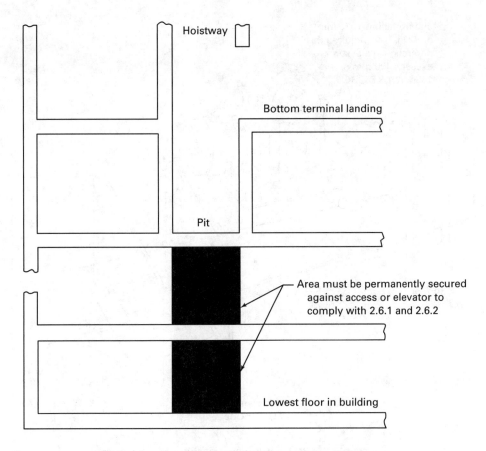

Diagram 2.6   Protection of Space Below Hoistway

high-rise buildings, thus the need to control the potential ignition sources in an elevator machine room. A fire in a basement machine room would expose the hoistway to the fire as there is no way to provide fire-resistive construction between the hoistway and machine room.

Elevator machine rooms are not permitted to be used as passageways to other areas inside and outside the building. As an example a roof scuttle used to check the condition of the roof is not allowed.

2.7.3 Access to Machine Rooms and Machinery Spaces

These requirements are intended to prohibit access to the machine room by unqualified persons, but provides safe and convenient access for elevator personnel (see 1.3, Definitions).

2.7.3.2 Access Across Roofs

2.7.3.2.1   The access requirement starts at the top floor in a building. Access over a flat roof is considered safe and convenient and no walkway is required, provided the roof has a parapet or guard rail at least 1 070 mm or 42 in. high. Access to and from the roof and the machine room must be by means of a stairway conforming to the requirements of 2.7.3.3. This will often prohibit the use of scuttles or other trap door-like openings to

the roof. The building code may also have specific requirements regarding roof access.

2.7.3.2.2   When access is over a sloping roof or a flat roof with no parapet or guard rail at least 1 070 mm or 42 in. high, a walkway must be provided to ensure safe and convenient access. Access to the walkway and from the walkway to the machine room must be by a stairway or ladder conforming to the requirements of 2.7.3.3.

2.7.3.3 Requirements for Means of Access. A vertical ladder is permitted as a means of access to machine rooms, machinery spaces, or different floor levels in a machine room if the difference in levels is 900 mm or 35 in. or less. However, if a machinery space contains only such equipment as overhead sheaves, deflecting sheaves, governors, or auxiliary equipment, 2.7.3.3.2 permits a vertical ladder to be provided for access to this area even though the difference in level exceeds 900 mm or 35 in. Diagrams 2.7.3.3(a) and 2.7.3.3(b) detail the access requirements in multilevel machine rooms. When a ladder is permitted for access from a roof, the roof must still be accessible by stairs. The exposures are different and the A17 Committee feels that a ladder is safe for one situation but not for the other. Ladders over

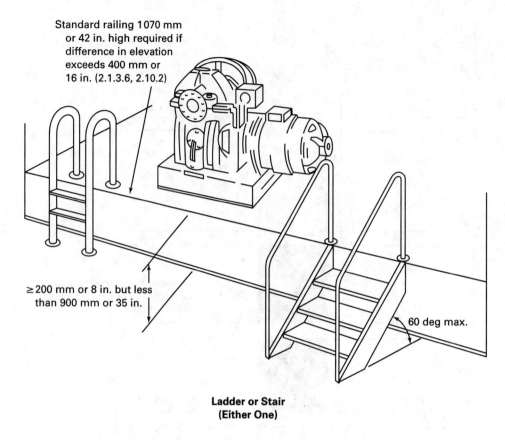

Standard railing 1070 mm
or 42 in. high required if
difference in elevation
exceeds 400 mm or
16 in. (2.1.3.6, 2.10.2)

≥ 200 mm or 8 in. but less
than 900 mm or 35 in.

60 deg max.

Ladder or Stair
(Either One)

Diagram 2.7.3.3(a)   Multilevel Machine Room Access
(Courtesy Charles Culp)

3.04 m (10 ft) high will have to be guarded to comply with OSHA requirements.

The requirement for a platform at the top of the stairs assures that there is a safe place to stand when entering and exiting and when opening and closing the machine room door. Diagrams 2.7.3.3(c) and 2.7.3.3(d) detail the access requirements to machine rooms. This requirement does not apply in Canada as it is within the Scope of NBCC.

2.7.3.4 Access Doors and Openings. In addition to the requirements found in this requirement, access doors in fire-resistive construction must have a fire-resistance rating as specified in the building code. These doors must be self-closing and self-locking to ensure access only by authorized personnel. The key must be kept in the building and available only to authorized personnel. See Handbook commentary on 8.1.1.

The Code does not limit the number of access doors. However, the doors must be for gaining access to the machine room. If the purpose of a door is to allow access, such as to a roof or any equipment located adjacent to the machine room, by passage through the machine room, then it is not permitted.

2.7.3.5 Stop Switch in Overhead Machinery Space in the Hoistway. The stop switch(es) provides a means for elevator personnel to control elevator movement. The exposure in the overhead is different from the exposure in the pit, thus the differences in the requirements.

2.7.4 Headroom in Machine Rooms and Overhead Machinery Spaces

The dimensions specified in this requirement are for clear headroom. Clear headroom measurements are taken from the floor to the bottom of the lowest obstruction below the ceiling (e.g., wiring gutters, conduit, beams). See Diagram 2.7.4.

2.7.5 Lighting, Temperature, and Humidity in Machine Rooms and Machinery Spaces

2.7.5.2 Temperature and Humidity. Machine rooms must be maintained within a prescribed temperature and humidity to assure that the elevator equipment does not overheat. The Building Transportation Standards and Guidelines, NEII-1-2000 recommends that the machine room temperature be maintained between 13°C (55°F) and 32°C (90°F) with a relative humidity not to

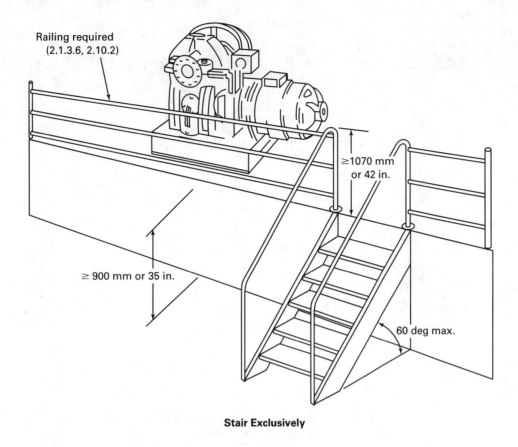

Stair Exclusively

Diagram 2.7.3.3(b)   Multilevel Machine Room Access
(Courtesy Charles Culp)

exceed 80%. The specific temperature and humidity range required by the elevator equipment manufacturer have to be permanently posted in the machine room. The Code does not specify the means used to control the temperature and humidity. It can be by natural means such as open vents or by mechanical means such as motorized vents, central air conditioning, or room air conditioners. Consideration should be given in colder climates to provide sufficient heat in the machine rooms to maintain a minimum 13°C (55°F) temperature. Control of machine room temperatures is critical when solid-state devices are utilized.

The building codes are beginning to recognize the need to control machine room temperatures during a fire to assure continued firefighters' service elevator operation. The International Building Code requires the machine room ventilation and air conditioning to be connected to standby power when the elevator is connected to standby power.

2.7.6 Location of Machine Rooms and Control Rooms

Drive and deflector sheaves may project into the hoistway from overhead machine rooms as well as adjacent (i.e., basement) machine rooms.

2.7.7 Machine and Control Rooms Underneath the Hoistway

A machine room located directly under the hoistway is not commonly found, but is permitted as long as the requirements of 2.7.7.1 through 2.7.7.5 are adhered to. If the counterweight runway does not extend into the machine room, or if there is occupiable space below the machine room, a counterweight safety is required. Also, see Handbook commentary on referenced requirements.

2.7.8 Remote Machine and Control Rooms

A remote machine and control room is defined as a room that does not share a common wall, floor, or ceiling with the hoistway. The requirement addresses the need for access to elevator equipment and communications between the car and machine room for inspection, maintenance, and repair.

SECTION 2.8
EQUIPMENT IN HOISTWAYS AND MACHINE ROOMS

2.8.1 Electrical Equipment and Wiring

See Handbook commentary on 2.26.4 for a discussion of the National Electrical Code (NEC®) also known as

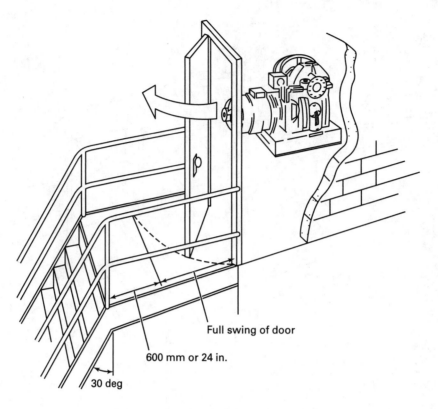

Diagram 2.7.3.3(c)   Machine Room Access
(Courtesy Charles Culp)

ANSI/NFPA 70 requirements that pertain to elevators. The requirements eliminate the need for persons other than elevator personnel from having access to the hoistway to install or maintain non-elevator wiring and piping. The hoistway is usually the most convenient location in a building to run wires and pipes from the basement to the roof; thus, a careful inspection should be made of all wiring and piping to ensure that no foreign wiring or piping has been installed. The only wiring permitted in the hoistway and machine room is elevator wiring. The main feeders for the elevator are not allowed (NEC®, Section 620-37) to be run in the hoistway unless by special permission, or unless the driving machine motor is located in the hoistway on either the car or counterweight. No wiring that is to service other equipment in the building is permitted to pass through these spaces. See Diagram 2.8.1.

NEC®, Section 620-71 requires only the "motor controller (power converter)" and not the "auxiliary control equipment" to be located in a separate machine room or enclosure. This clarifies the need to guard the motor controller as defined in 430-81 so as not to confuse them with motor control circuits as defined in 430-71. The disconnect switch is to be located within sight of the motor controller on nongenerator field control systems, and located within sight of the motor controller for the driving motor of the motor generator set on generator field control system. It is in the interest of safety that

a means to prevent starting be located near a driving machine and in view of it. It is appropriate that power circuits for the elevator driving motor be required to remain in the machine room. Requiring that the relay panel or logic panel (i.e., operation logic) be located in the machine room serves no safety purpose and penalizes modern advanced designs. It is common practice today to locate door operator control devices on car tops adjacent to the operating motors. There is no safety reason to require all devices and circuits covered under the "controller" definition to be located in the machine room.

The insulation on standard wire would readily melt in a fire causing an electrical short, which could allow an elevator to operate with the door(s) open. This condition could also place an elevator out of service for emergency personnel. To overcome this problem, NEC®, Section 620-11(a) requires wiring from the hoistway door interlock to the hoistway riser shall be flame retardant and suitable for temperatures not less than 200°C (392°F). The most commonly used wire insulation meeting this specification is Type SF. See NEC®, Section 620-11(a)/Table 402-3. This requirement does not apply to the factory wiring within the interlock.

Section 620-44 of NEC® (see also 2.26.4) requires that the traveling cable be enclosed at each end in a raceway such as rigid metal conduit, intermediate metal conduit,

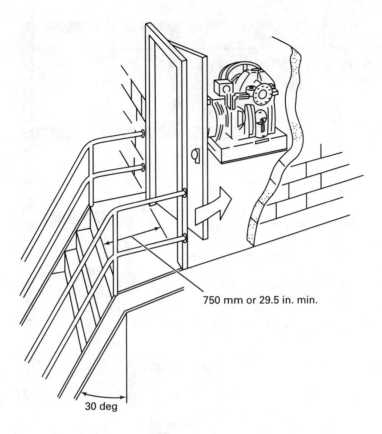

750 mm or 29.5 in. min.

30 deg

Diagram 2.7.3.3(d)   Machine Room Access
(Courtesy Charles Culp)

electrical metallic tubing, or wireways within a distance not exceeding 1.83 m (6 ft) after it is secured.

Traveling cables need not be encased in a raceway (see NEC®, Section 620-21) between the two fixed suspension points. One fixed suspension point will be somewhere on the car and the other will be in the hoistway or in the machine room at the top of the hoistway. If the fixed suspension on the car is at the car top, the traveling cable may be exposed on the side of the car. Suitable guards may be needed to protect the cables against damage. See NEC®, Sections 620-43 and 620-44. If the fixed suspension point is under the car, and the traveling cable is continued up the side of the car to the car top junction box, the continued length of cable if over 1.83 m (6 ft) from the point of support, is considered fixed wiring and has to be enclosed in a raceway. See NEC® Section 620-44.

The lightning down conductors are not allowed to be run in the hoistway. Lightning Protection Codes NFPA-780 and CAN/CSA-B72 addresses bonding of long vertical metal bodies in close proximity to the lightning protection system grounding down conductor in order to prevent side flash during a lightning strike. The results of this side flash (arcing) between the lightning protection system grounding down conductor and the vertical metal body may be personal injury, fire, or destruction

of equipment due to the extremely high voltage levels and electromagnetic forces. The use of bonding conductors will equalize the potential differences between the lightning protection system grounding down conductor and the vertical metal bodies and keep the "air gap" voltage between the two to a minimum to prevent an arc strike.

2.8.2 Pipes, Ducts, Tanks, and Sprinklers

If a pipe or duct conveying gases, vapors, or liquids passed through a hoistway, machine room, or machinery space and burst, the operation of the elevator could be affected in such a manner as to make it unsafe. Water could cause the malfunctioning of hoistway door interlocks allowing an elevator to leave a floor with the hoistway door open. A broken steam pipe could scald a passenger in the elevator. Additionally, the hoistway, machine room, and machine space are secured against access except to elevator personnel who are familiar with the precautions necessary when working around moving equipment and the dangers associated with tampering with the equipment. See Diagram 2.8.1 for examples of typical applications that comply and do not comply. Ducts for heating, cooling, ventilating, and venting of the machine rooms and machinery spaces

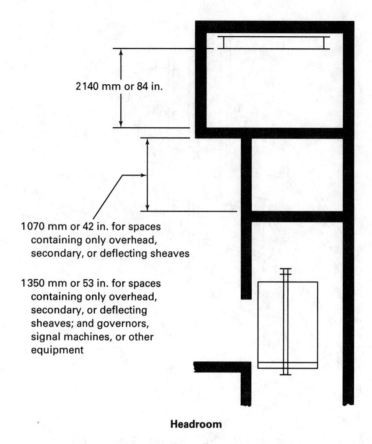

2140 mm or 84 in.

1070 mm or 42 in. for spaces
containing only overhead,
secondary, or deflecting sheaves

1350 mm or 53 in. for spaces
containing only overhead,
secondary, or deflecting
sheaves; and governors,
signal machines, or other
equipment

Headroom

Diagram 2.7.4   Headroom in Machine Rooms and Overhead Machinery Spaces

may be in the space, and also continue on to serve other portions of the buildings.

2.8.2.3 Sprinklers have been recognized by the A17.1 Code in elevator machine rooms and hoistways since the 1955 edition of the Code, although until the early 1980s they were not normally found in these areas. In recent years, building codes began to recognize sprinklers as one of the most effective means of controlling fires in buildings. The building codes at first did not require buildings to be fully sprinklered but encouraged their use by allowing "trade-offs," which reduced the cost of construction by relaxing code requirements in areas such as fire-resistance ratings, distances to exits, and area and height limitations, in exchange for full sprinkler protection. The theory behind this was that the cost of construction would be comparable for a fully sprinklered building vs. a nonsprinklered building and the levels of fire protection would be equal or greater in a sprinklered building. The payoff for a fully sprinklered building is a reduced premium for fire insurance throughout the life of the building, thus lowering the operating cost.

In the last few years, the building codes have begun to require fully sprinklered buildings for certain types of occupancies, such as high-rise office buildings and multiple family dwellings. This has been successful, and today it is common for new buildings especially high-rise buildings to be fully sprinklered.

In the early 1980s, the A17 Committee became aware of this trend and initiated a study of the hazards of water being discharged on elevator equipment. The possible effects of water on brakes, shorting an elevator safety circuit, motors, generators, or transformers, with the resulting hazards to people on the elevator, should not be too difficult to envision. Some of the reported hazardous conditions are cars operating with car and/or hoistway doors open, loss of braking, and loss of traction. The A17 Committee could not blindly prohibit sprinklers in elevator machine rooms and hoistways, for if a building code or fire code required sprinklers in all areas of the building, that code would be enforced, regardless of any contrary requirement that would appear in the elevator code.

The Committee was aware that something must be done to ensure that, when sprinklers are installed in areas with elevator equipment, the safety of the passengers on the car would be addressed. In early 1982, contact was made with the National Automatic Sprinkler and Fire Control Association, Inc., the trade association for the sprinkler industry. The potential hazards were discussed in detail, and the following response was sent to A17:

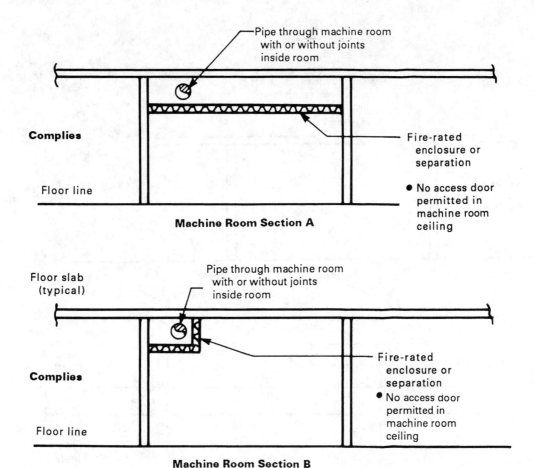

Diagram 2.8.1   Machine Room Pipe, etc., Separation

"Our Committee members and the insurance authorities are unanimous that sprinkler protection should be provided in all building areas, and that there is no record of adverse experience with sprinkler protection of electrical rooms or computer rooms or other rooms of this type, especially as compared to consequences of manual fire-fighting efforts with hose streams. As such, our Committee has recommended that the elevator machine rooms be protected in one of three ways.

(a) standard sprinkler protection in accordance with NFPA 13, with raintype rated electrical equipment specified for all equipment or otherwise shielded with noncombustible hoods;

(b) standard sprinkler protection in accordance with NFPA 13, but with a water flow switch on the branch line to the elevator machinery room, such that flow of sprinklers in the elevator machinery room causes automatic shutdown of elevators; or

(c) protection of the elevator machinery room with non-water automatic fire suppression systems such as carbon dioxide or Halon. In such

case, the special system should be provided with a connected reserve."

The A17 Committee reviewed this response and felt that the first recommendation was impractical. Even if complied with at the time of installation, shields and raintype covers would, over a period of time, be removed or become ineffective, thus permitting the equipment to become subject to water from a discharged sprinkler. The third recommendation also was discarded, as its effectiveness was questionable. Non-water automatic fire suppression systems such as carbon dioxide or Halon normally are effective in areas with only minimal air movement. It was agreed that air movement resulting from the stack effect in the hoistway would make these systems impractical and ineffective in controlling a fire. Recommendation two was acceptable, and was reworded in performance language, balloted, and approved.

If a fire developed in an elevator machine room or hoistway, the sequence of events would typically follow this scenario.

(a) Smoke in the machine room or hoistway during

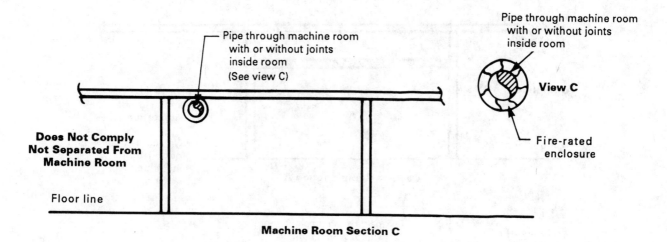

Pipe through machine room
with or without joints
inside room
(See view C)

Pipe through machine room
with or without joints
inside room

View C

**Does Not Comply
Not Separated From
Machine Room**

Floor line

Fire-rated
enclosure

Machine Room Section C

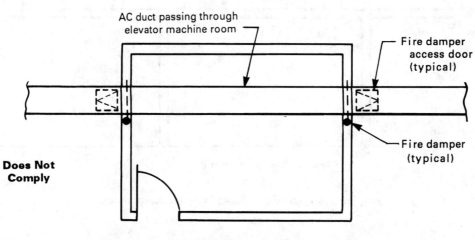

AC duct passing through
elevator machine room

Fire damper
access door
(typical)

**Does Not
Comply**

Fire damper
(typical)

Machine Room Section D

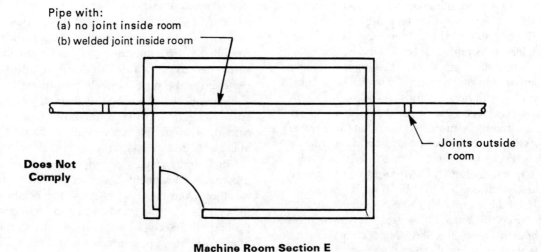

Pipe with:
(a) no joint inside room
(b) welded joint inside room

Joints outside
room

**Does Not
Comply**

Machine Room Section E

Diagram 2.8.1   Machine Room Pipe, etc., Separation (Cont'd)

the initial stage of the fire will activate the smoke detector required by 2.27.3.2, recalling all elevators to the designated level on Phase I Emergency Recall Operation.

(b) As the intensity of the fire builds, the sprinkler system would be activated and the power to the elevator driving machine would be interrupted. Power would be removed even if the elevator was operating on Phase I or Phase II firefighters' service.

It should also be pointed out that the Committee gave serious consideration to the possibility of a car stopping with passengers and even firefighters' between floors. All types of recommendations were made; such as not removing power until Phase I emergency recall was complete. All elevator controls (electromechanical and static) are limited to some maximum ambient operating temperature above which the performance of the devices is unpredictable, nonrepeatable, and unreliable. This may lead to a failure in the elevator control resulting in unsafe elevator operation. The Committee also recognized that in all probability the elevator system might be on fire. In Committee deliberations, it was concluded that if there was sufficient heat to activate the sprinkler, a fire was in progress and there was no way of assuring control of the elevator except by disconnecting the main line power supply.

Recently, many questions have been raised in the field as to what type of system could be installed that would meet the intent of this requirement. Following are three methods, the first being the most economical.

(a) First Method. Rate-of-rise/fixed-temperature heat detectors, in the elevator machine room and/or hoistway, would be arranged to automatically disconnect the main line power supply. These detectors would be placed near each sprinkler. The sprinkler rating would exceed the heat detector ratings. The detectors could be independent of the sprinkler system. The heat detector would cause a shunt trip circuit breaker to disconnect the main line power to the affected elevators prior to the application of water.

By using combination rate of rise and fixed temperature detectors, we can assume fast response from the rate of rise portion whenever we have a fast fire. However, in slower fires or in larger spaces it is more likely that the rate of rise portion will not respond and, eventually, the fixed temperature portion of the heat detector will actuate. With fixed temperature detectors, (including sprinklers that are heat detectors with water behind them) temperature rating is not the only important factor in determining its response time. It is also important to consider the mass of the element, which must be heated. It is very easy to show that two units with the same temperature rating, but different masses, will not respond at the same time. It can also be demonstrated that a higher temperature unit may respond faster than a lower temperature unit having greater mass. The factor being considered here is called "thermal lag."

Mechanical and electrical engineers measure thermal lag as a "time constant" at a given hot gas velocity or electrical current level. Higher time constants means longer response times with all other conditions being equal. In the fire protection community the latest term used is Response Time Index (RTI), which is related to a unit's time constant. The lower the RTI of a detector or sprinkler, the faster its response will be to a given fire. [See Charts 2.8.2.3(a) and 2.8.2.3(b).]

If color of the detector is the same as the color code marking for the detector it should be marked in a contrasting color with either a ring on the surface or numbers at least 9.38 mm ($^3/_8$ in.) high indicating the temperature rating.

This is not often a problem since most heat detectors have low RTI's compared to sprinkler heads. However, the growing popularity of "fast response sprinklers" means that it is possible to have a 73.8°C (165°F) sprinkler respond before a 57.2°C (135°F) heat detector.

Authorities having jurisdiction should require engineers and designers of such systems to demonstrate that the devices will operate in the correct order. This can be done by using calculation methods contained in National Fire Alarm Code®, NFPA 72. The calculations should be done for both slow and fast fires and should include appropriate factors of safety.

(b) Second Method. The sprinkler system in the elevator machine room and the hoistway would be a preaction system. A preaction sprinkler system employs automatic sprinklers attached to a piping system containing air or nitrogen that may or may not be under pressure, with a supplemental fire detection system installed in the same areas as the sprinklers. Actuation of a heat detector from a fire opens a valve that permits the air or nitrogen to escape and water to flow into the sprinkler piping system and to be discharged from any sprinkler head that may be open. The heat detector or flow valve in the sprinkler piping would cause a shunt trip circuit breaker to disconnect the main line power to the affected elevators at the time the flow valve opens. The preaction system should not be activated by a smoke detector, as this would shut the elevator down prematurely. Requirement 2.8.2.3.4 prohibits activation by a smoke detector, as this would shut the elevator down without any chance for Phase I recall as described above.

(c) Third Method. The sprinkler system in the machine room or hoistway would be a dry-pipe system. A dry-pipe switch would be installed in the sprinkler piping where it enters the elevator machine room or hoistway. The pipe would contain air or nitrogen under pressure from the sprinkler head to dry-pipe switch and water to the opposite side of the dry-pipe switch. The heat from a fire would open the sprinkler head allowing the air to escape and water to flow into the system. The dry-pipe switch would trip and cause a shunt trip circuit breaker to disconnect the main line power to the affected elevators.

Chart 2.8.2.3(a) NFPA 13-1999 Temperature Ratings, Classifications, and Color Codings for Sprinklers

Reprinted with permission from NFPA 13, *Installation of Sprinkler Systems*, Copyright© 1999, National Fire Protection Association, Quincy, MA 02269. This reprinted material is not the complete and official position of the National Fire Protection Association on the referenced subject which is represented only by the standard in its entirety.)

Maximum Ceiling Temperature		Temperature Rating		Temperature Classification	Color Code	Glass Bulb Colors
°F	°C	°F	°C			
100	38	135–170	57–77	Ordinary	Uncolored or black	Orange or red
150	66	175–225	79–107	Intermediate	White	Yellow or green
225	107	250–300	121–149	High	Blue	Blue
300	149	325–375	163–191	Extra high	Red	Purple
375	191	400–475	204–246	Very extra high	Green	Black
475	246	500–575	260–302	Ultra high	Orange	Black
625	329	650	343	Ultra high	Orange	Black

GENERAL NOTES:
(a) *Sprinkler Frame Arm Color Coding Exceptions*
  (1) *Exception No. 1:* A dot on the top of the deflector, the color of the coating material, or colored frame arms shall be permitted for color identification of corrosion-resistant sprinklers.
  (2) *Exception No. 2:* Color identification shall not be required for ornamental sprinklers such as factory-plated or factory-painted sprinklers or for recessed, flush, or concealed sprinklers.
  (3) *Exception No. 3:* The frame arms of bulb-type sprinklers shall not be required to be color coded.
(b) The temperature rating is usually stamped on the solder link for solder-style sprinklers. Other styles of sprinklers may have the temperature rating stamped on one of the releasing parts.
(c) *Color Code Exceptions*
  (1) For corrosion-resistant sprinklers, the color coding may be a dot on top of the deflector, the color of coating material, or colored frame arm.
  (2) Color identification is not required for ornamental sprinklers such as factory-plated or factory-painted sprinklers or flush or concealed sprinklers.

Chart 2.8.2.3(b) NFPA 72-1999 Temperature Classification for Heat-Sensing Detectors

Reprinted with permission from NFPA 72, *National Fire Alarm Code®*, Copyright© 1999, National Fire Protection Association, Quincy, MA 02269. This reprinted material is not the complete and official position of the National Fire Protection Association on the referenced subject which is represented only by the standard in its entirety.

Temperature Classification	Temperature Rating Range		Maximum Ceiling Temperateture		Color Code
	°F	°C	°F	°C	
Low [Note (1)]	100–134	39–57	20 below	11 below	Uncolored
Ordinary	135–174	58–79	100	38	Uncolored
Intermediate	175–249	80–121	150	66	White
High	250–324	122–162	225	107	Blue
Extra high	325–399	163–204	300	149	Red
Very extra high	400–499	205–259	375	191	Green
Ultra high	500–575	260–302	475	246	Orange

NOTE:
(1) Intended only for installation in controlled ambient areas. Units shall be marked to indicate maximum ambient installation temperature.

One final point: The A17.1 Code requires that the power removal be independent of the elevator control. System designs that require elevators to complete Phase I recall are permitted, provided the elevator control system is not required to send a signal that Phase I has been accomplished. Recently there have been designs where a heat detector in the machine room activates Phase I and starts a timer in the fire alarm system. After a predetermined time, the shunt trip is activated and water is allowed to flow from the sprinklers. The "predetermined time" is calculated based on the maximum time necessary to complete Phase I recall. Passengers may still be trapped if Phase I has not been completed; however, it significantly reduces that possibility. If the elevator controller is relied on to signal that Phase I is completed, it may never come and the sprinklers never activated. The elevator control may in all likelihood be the source of the fire and thus cannot be relied on to respond.

In 1994, the Sprinkler Standard NFPA 13 revised the requirement for sprinkler protection in elevator hoistways. Sprinklers installed at the bottom of the shaft must be positioned so that they do not interfere with the platform guard (see NFPA 13, 5-13.6, A-5-13.6.1, and A-5-13.6.2). Sprinklers may be omitted from the top of the noncombustible elevator hoistways of passenger elevators complying with ASME A17.1 (see NFPA 13, 5-13.6.3). The standards will also permit the elimination of sprinklers at the bottom of the shaft when the elevator shaft is enclosed with noncombustible materials and combustible hydraulic fluids are not present in the shaft (see NFPA 13, Exception to 5-13.6.1). Clearly this provision applies to hydraulic elevators. A potential problem is the interpretation of this provision on an electric traction installation with oil buffers.

See also Chart 2.8.2.3(c), which contains excerpts from the *National Fire Alarm Code®* Handbook on this issue.

NOTE: *National Fire Alarm Code®* Handbook is registered trademark of the National Fire Protection Association, Inc., Quincy, MA 02269.

2.8.4 Air Conditioning

This requirement controls the location and installation of air conditioning equipment in machine rooms and machinery spaces to ensure that they do not adversely effect elevator operation. Since air conditioner maintenance personnel may have to be in the machine room, the equipment should not be located over elevator equipment. Piping is to be routed in such a fashion that any leaks or condensation would not effect elevator operation. Drains connected directly to sewers could allow sewer gases to enter the machine room if the trap dried out. Adequate clearance must be provided to ensure the safety of air conditioning service personnel. Guards are to be provided on exposed moving parts to protect personnel from entanglement.

SECTION 2.9
MACHINERY AND SHEAVE BEAMS, SUPPORTS, AND FOUNDATIONS

Machine beams, where used, transfer their loads to the building structure. The Code does not limit the design of the machine and its structural support system, if they comply with the applicable requirements. Depending on the specific design configuration, the machine beams may be located directly beneath the machine, bedplate, blocking, or stacking assembly.

2.9.1 Beams and Supports Required

It is the intent that machines, machinery, and sheaves be so supported and maintained in place to prevent any part from becoming loose or displaced under the conditions imposed in service. If a structural slab is provided, a machine beam is not required. Additional details on machine beams and structural slabs can be found in the Building Transportation Standards and Guidelines, NEII-1-2000.

2.9.2 Loads on Machinery and Sheave Beams, Floors or Foundations, and Their Supports

2.9.2.2 Foundations, Beams, and Floors for Machinery and Sheaves Not Located Directly Over the Hoistway. The traction sheave is often located at a considerable distance from the center of gravity of the foundation mass in order to permit a particular roping arrangement. In such cases, the dead weight of the foundation might have to be heavier than two times the vertical components in order to have the capability of withstanding two times the vertical component and the overturning moment.

2.9.3 Securing of Machinery and Equipment to Beams, Foundations, or Floors

2.9.3.2 Beams or Foundations Supporting Machinery and Sheaves Not Located Directly Over the Hoistway. The bolt specification referenced, ASTM A 307, covers carbon steel threaded fasteners that are not quenched or tempered. The minimum tensile strength allowed by this standard is 414 MPa (60,000 psi); thus, the load of 82.7 MPa (12,000 psi) allowed by this requirement provides for a factor of safety of 5. The standard requires bolt heads to be marked with the manufacturer's identification. This standard requires the nuts to conform to ASTM A 563. Since there is no activity to accredit fastener manufacturers and there have been cases of spuriously marked bolts, the manufacturer's identification and source of bolt supply is worthy of careful attention.

2.9.4 Allowable Stresses for Machinery and Sheave Beams or Floors and Their Supports

The purpose of specifying allowable deflections and stresses for the equipment is to establish a safe level of design consistent with sound structural engineering practice for rigidity and strength.

2.9.5 Allowable Deflections of Machinery and Sheave Beams and Their Supports

The term "static load" means the actual weight of all equipment and applicable portions of the flooring supported by the beams as defined by 2.9.2.1(b) in addition to the sum of the tensions in all wire ropes supported by the beams with rated load in the car. There is no increase to these static loads to take care of impact, accelerations, etc.

Chart 2.8.2.3(c)
Excerpts From NFPA National Fire Alarm Code® Handbook

3-9.4 Elevator Shutdown

Subsection 3-9.4 is the result of additional sprinkler protection requirements of other codes for elevator machine rooms and hoistways.

The purpose of elevator shutdown prior to sprinkler operation is to avoid the hazards of a wet elevator braking system and electrical shock. If the elevator brakes are wet, there is the danger of the elevator rising uncontrollably to the top of the hoistway or to the bottom of the hoistway, depending on the load in the cab.

3-9.4.1 Where heat detectors are used to shut down elevator power prior to sprinkler operation, the detector shall have both a lower temperature rating and a higher sensitivity as compared to the sprinkler.

Paragraph 3-9.4.1 is extremely important. Often, it is not understood that a 135°F (57.2°C) heat detector may not respond prior to a 165°F (73°C) sprinkler head despite the obvious differences in temperature sensitivity. The response time is based on the RTI of both devices and must be known prior to design and installation of heat detectors for elevator shutdown. Because the RTI for heat detectors is not readily available, one must use a sensitive fixed temperature heat detector with a listed spacing equal to or greater than 40 ft (12.2 m) on center (or use a rate-of-rise-type heat detector) in the locations where heat detectors are required.

The 1998 edition of A17.1 no longer requires elevator shutdown for waterflow from less than 24 in. (610 mm) from the floor of the hoistway pit. Therefore, heat detectors are not required in those cases.

A-3-9.4.1 A lower response time index is intended to provide detector response prior to the sprinkler response, because a lower temperature rating alone might not provide earlier response. The listed spacing rating of the heat detector should be 25 ft (7.6 m) or greater.

3-9.4.2 If heat detectors are used to shut down elevator power prior to sprinkler operation, they shall be placed within 2 ft (610 mm) of each sprinkler head and be installed in accordance with the requirements of Chapter 2. Alternatively, engineering methods, such as specified in Appendix B, shall be permitted to be used to select and place heat detectors to ensure response prior to any sprinkler head operation under a variety of fire growth rate scenarios.

3-9.4.3 If pressure or waterflow switches are used to shut down elevator power immediately upon or prior to the discharge of water from sprinklers, the use of devices with time delay switches or time delay capability shall not be permitted.

The intent is to shut down elevator power as soon as water flows in the automatic sprinkler system that is protecting the elevator hoistway and machine room. These waterflow devices should be installed in the cross main or branch line serving the automatic sprinkler system in those areas. The requirement is for a device with no retard or time delay mechanism, not a device with its time delay or retard feature set to zero.

A-3-9.4.3 Care should be taken to ensure that elevator power cannot be interrupted due to water pressure surges in the sprinkler system. The intent of the code is to ensure that the switch and the system as a whole do not have the capability of introducing a time delay into the sequence. The use of a switch with a time delay mechanism set to zero does not meet the intent of the code, because it is possible to introduce a time delay after the system has been accepted. This might occur in response to unwanted alarms caused by surges or water

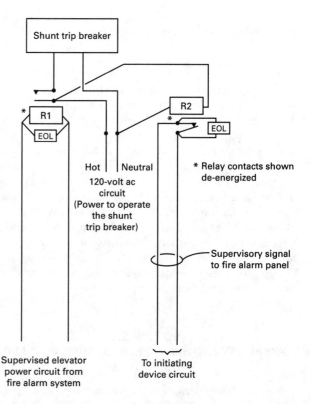

Figure A-3-9.4.4 Typical Method of Providing Elevator Power Shunt Trip Supervisory Signal

(continued)

Chart 2.8.2.3(c)
Excerpts From NFPA National Fire Alarm Code® Handbook (Cont'd)

movement, rather than addressing the underlying cause of the surges or water movement (often due to air in the piping). Permanently disabling the delay in accordance with the manufacturer's printed instructions should be considered acceptable. Systems that have software that can introduce a delay in the sequence should be programmed to require a security password to make such a change.

3-9.4.4 Control circuits to shut down elevator power shall be monitored for presence of operating voltage. Loss of voltage to the control circuit for the disconnecting means shall cause a supervisory signal to be indicated at the control unit and required remote annunciators.

There have been cases where the operating power for elevator shunt trip circuits has been de-energized. This is a dangerous condition because the elevator will not be shut down in the event of waterflow in the machine room or hoistway. The new requirement in 3-9.4.4 was added to the 1999 edition of the code to monitor the integrity of the operating power for the shunt trip control circuit. Monitoring the integrity of the control power is similar to monitoring the integrity of the power for an electric motor-driven fire pump.

A-3-9.4.4 Figure A-3-9.4.4 illustrates one method of monitoring elevator shunt trip control power for integrity.

The allowable deflection is intended to apply to machinery and sheave beams and any intermediate supports that comprise the structural interface between these structural elements and the building structure.

The basis for the derivation of the $\frac{1}{1666}$ figure cannot be found in the Committee records. The Committee is of the opinion that the intent was to have a stiff set of machine and sheave beams. A word of caution: $\frac{1}{1666}$ can be difficult to meet if the span of the structural support is long. The practical span limit that can be accommodated while still maintaining the $\frac{1}{1666}$ deflection is about 6 m or 20 ft between supports.

2.9.6 Allowable Stresses Due to Emergency Braking

To provide criteria for the safe design of the respective components/structures subject to forces developed during the retardation phase of the emergency braking and all other loading acting simultaneously, if applicable, with factors of safety consistent with the types of equipment/structures involved. As the A17.1–2000 provides for stopping an ascending car, additional forces may need to be considered depending on the type of device used. For example if a rope brake is used, the forces necessary to stop the descending counterweight would be transferred to the point that the rope brake is mounted.

SECTION 2.10
GUARDING OF EXPOSED EQUIPMENT

2.10.1 Guarding

The requirements protect mechanics and inspectors from contact with sheaves and ropes, which extend beyond the base of the machinery and overhead sheaves as well as moving parts of selectors and floor controllers or signal machines. Guarding of hand winding wheels, etc., may not be practical, and where they are not guarded, they must be clearly identifiable by yellow markings.

2.10.2 Standard Railing

Places requirement for standard railing in one place, which is then referred, by many other requirements in the Code. Incorporates United States OSHA and Ontario OHSA regulations. Requirement assures railing is structurally sound and of adequate height to guard against a fall hazard.

SECTION 2.11
PROTECTION OF HOISTWAY-LANDING OPENINGS

See Diagram 2.11, which illustrates hoistway components typically supplied on the entrance side of the hoistway wall.

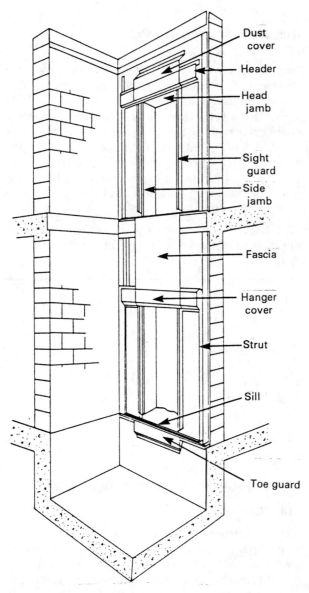

Typical Hoistway Elevation

Labels: Dust cover, Header, Head jamb, Sight guard, Side jamb, Fascia, Hanger cover, Strut, Sill, Toe guard

Diagram 2.11   Typical Hoistway Elevation

2.11.1.1 Hoistway Landing Entrances. The minimum height of the entrance is consistent with the dimensions for egress doorways required by the building codes.

2.11.1.2 Entrance Doors in Blind Hoistways. A single-blind hoistway is that portion of a single hoistway that passes floors or other landings at which no normal landing entrances are provided. The requirement for emergency doors is to provide access in order to evacuate passengers from a stalled elevator. The required distance between emergency doors was calculated to assure that an extension ladder placed on top of the disabled car could reach the floor at the access door. The sign [2.11.1.2(d)] warns building personnel that the door does

not lead to an office, slop sink, janitor's closet, etc. The requirement for a cylinder-type lock [2.11.1.2(f)], which will not be unlocked by any key that will open any other lock or device in the building, assures that building personnel will not be able to enter this door accidentally.

The electromechanical device [2.11.1.2(e)] will monitor the door to assure it is closed and locked and prevent elevator operation if it is not. This will check that the door has not been left in the open position. The barrier [2.11.1.2(i)] will guard the opening when the door is in the open position.

2.11.1.3 Telephone as Alternative to Emergency Doors. In some installations, there are no floors (2.11.1.2) between landings. This requirement applies to those applications. If intermediate floors as specified in 2.11.1.2 are available for emergency access doors, this requirement does not apply.

2.11.1.4 Access Openings for Cleaning of Car and Hoistway Enclosures. The Committee concluded it was safer to have an access panel in the car enclosure for cleaning hoistway glass than doing it from the top of the car. Maintenance requirement 8.6.10.3.2 requires a written procedure for cleaning the glass, for every installation. See Handbook commentary on 8.6.10.3.

2.11.2 Types of Entrances

2.11.2.1 For Passenger Elevators and Freight Elevators Authorized to Carry Passengers. See Diagrams 2.11.2.1(a)(1) through 2.11.2.1(d).

2.11.2.2 For Freight Elevators. See Diagrams 2.11.2.1(a)(1) through 2.11.2.1(d) and Diagrams 2.11.2.2(e)(1), 2.11.2.2(e)(2), and 2.11.2.2(f).

2.11.3 Closing of Hoistway Doors

The self-closing feature required for horizontally sliding or swing doors of automatic operation elevators provides an automatic mechanical means of closing the hoistway door opening if for some reason the elevator malfunctions and moves more than 450 mm or 18 in. away from the landing. The requirement for location of the door closer (2.11.3.3) is to assure that related panels are all closed if the normal relating means fails.

Consideration is given in 2.11.3.2 to the possible propagation of fire inherent to any open hoistway; thus the need for controlling the open door exposure. The intent of this requirement is to assure that hoistway doors are closed when there is a fire at a landing.

The writers of the A17.1 Code recognized that differences exist between vertical, and horizontal sliding doors, and swing doors. If a person was in the path of a closing horizontally sliding door, the door would strike individuals in such a manner that they would likely not be injured. As the individuals would be struck at the side of their bodies, they probably would retain their

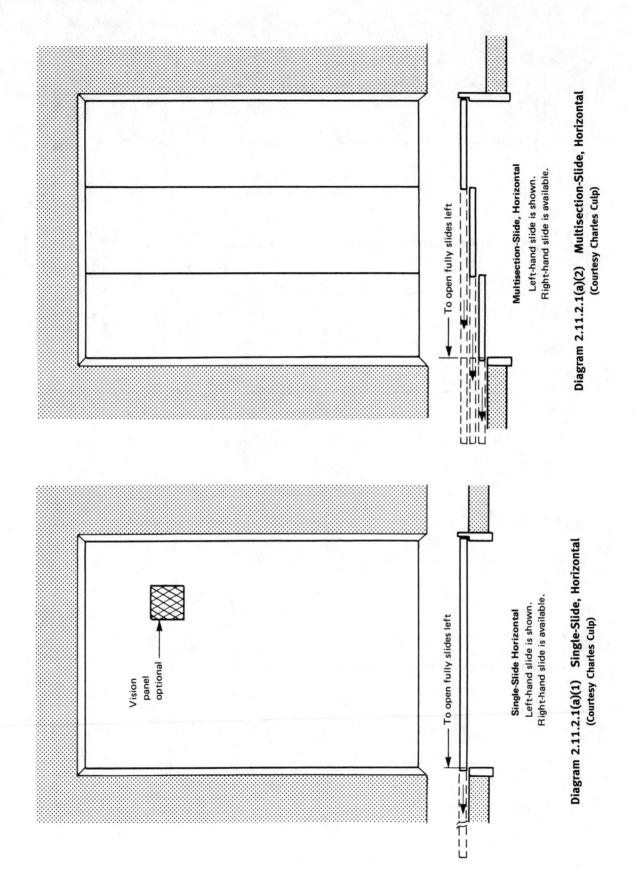

To open fully slides left

Multisection-Slide, Horizontal
Left-hand slide is shown.
Right-hand slide is available.

Diagram 2.11.2.1(a)(2) Multisection-Slide, Horizontal
(Courtesy Charles Culp)

Vision
panel
optional

To open fully slides left

Single-Slide Horizontal
Left-hand slide is shown.
Right-hand slide is available.

Diagram 2.11.2.1(a)(1) Single-Slide, Horizontal
(Courtesy Charles Culp)

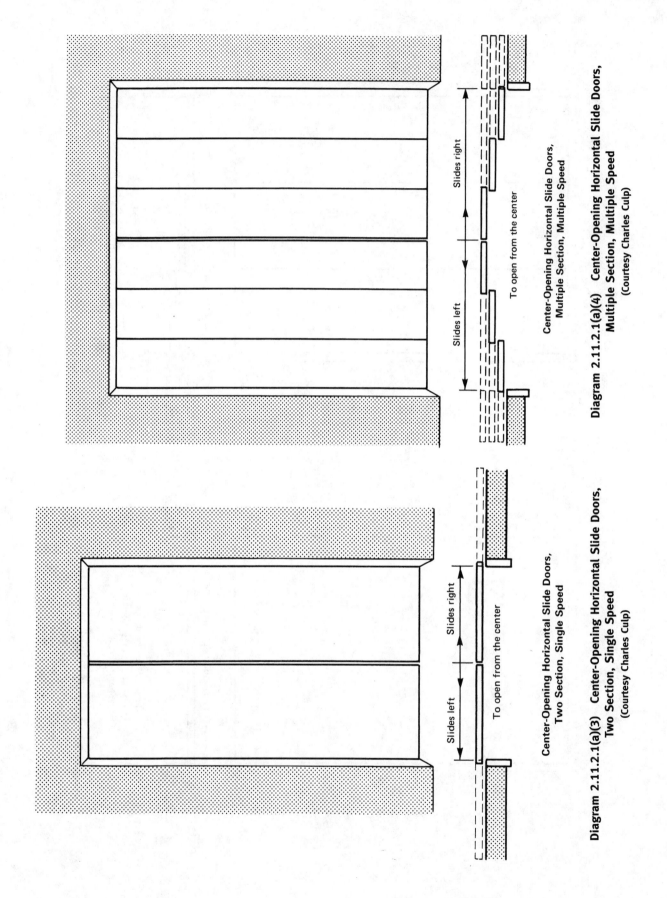

Slides right

Slides left

To open from the center

Center-Opening Horizontal Slide Doors,
Multiple Section, Multiple Speed

Diagram 2.11.2.1(a)(4)   Center-Opening Horizontal Slide Doors,
Multiple Section, Multiple Speed
(Courtesy Charles Culp)

Slides right

Slides left

To open from the center

Center-Opening Horizontal Slide Doors,
Two Section, Single Speed

Diagram 2.11.2.1(a)(3)   Center-Opening Horizontal Slide Doors,
Two Section, Single Speed
(Courtesy Charles Culp)

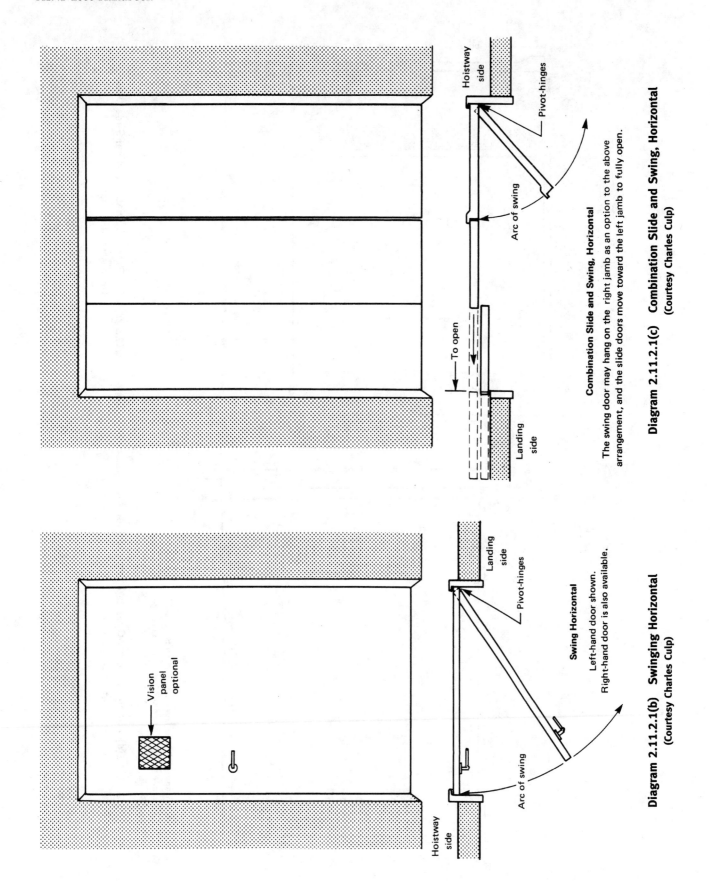

Combination Slide and Swing, Horizontal

The swing door may hang on the right jamb as an option to the above arrangement, and the slide doors move toward the left jamb to fully open.

Diagram 2.11.2.1(c) Combination Slide and Swing, Horizontal
(Courtesy Charles Culp)

Swing Horizontal

Left-hand door shown.
Right-hand door is also available.

Diagram 2.11.2.1(b) Swinging Horizontal
(Courtesy Charles Culp)

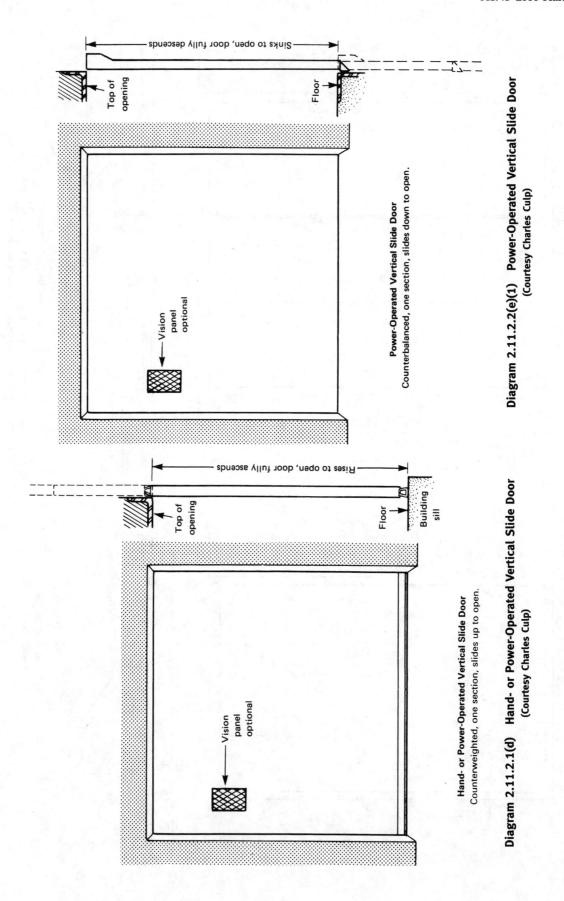

Power-Operated Vertical Slide Door
Counterbalanced, one section, slides down to open.

Diagram 2.11.2.2(e)(1)   Power-Operated Vertical Slide Door
(Courtesy Charles Culp)

Hand- or Power-Operated Vertical Slide Door
Counterweighted, one section, slides up to open.

Diagram 2.11.2.1(d)   Hand- or Power-Operated Vertical Slide Door
(Courtesy Charles Culp)

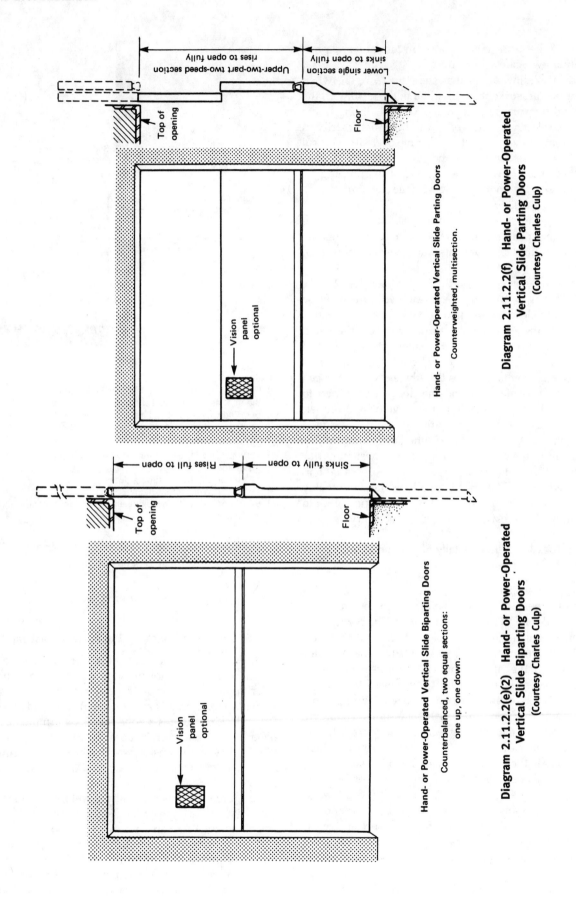

Hand- or Power-Operated Vertical Slide Parting Doors
Counterweighted, multisection.

Diagram 2.11.2.2(f) Hand- or Power-Operated Vertical Slide Parting Doors
(Courtesy Charles Culp)

Hand- or Power-Operated Vertical Slide Biparting Doors
Counterbalanced, two equal sections:
one up, one down.

Diagram 2.11.2.2(e)(2) Hand- or Power-Operated Vertical Slide Biparting Doors
(Courtesy Charles Culp)

balance. Self-closing swing doors are common and present no known safety hazard.

Vertical sliding doors that slide up to open, when closing would strike the individual in the head with the potential for very serious injury. Vertical sliding doors that slide down to open have the potential to trip an individual as they begin to close unexpectedly. Vertical biparting doors again have the same safety exposures. Both scenarios have the real potential of trapping an individual in the door opening with the elevator moving away from the landing.

At one time self-closing vertical sliding doors were utilized in North America. This was recognized as the cause of numerous accidents and they are now prohibited.

When the elevator is on firefighters' service (2.27.3), the following conditions are considered, which revise the need to close the doors of a parked car. Under this operation, the elevator is returned to the designated level or alternate level and the doors shall open and remain open (Phase I Emergency Recall Operation). This sequence of events assumes that the elevator has been returned to a safe floor and the doors are opened to allow passengers to exit from the elevator and are left open to assist arriving fire fighting personnel. Arriving fire fighting personnel want to immediately confirm that all elevators have been returned to the designated or alternate level, and that no passengers are caught in stalled elevators. The open door allows them to quickly assess this condition and gives them the opportunity to start immediate rescue operations if necessary.

2.11.4 Location of Horizontally Sliding or Swinging Hoistway Doors

The purpose of this requirement is to reduce the possibility of a person or object being caught between the hoistway door and the car door or gate. The permissible distance is greater for elevators, which can only be operated from within the car (2.11.4.1), since the operator has control of the car and would not operate it if a person or object were in this area. The permissible distance (2.11.4.2) is less for swinging doors than for sliding doors since a swinging door can close behind someone, forcing the person into the area between the two doors. This is not the case with sliding doors. The reason for these differences in the points of measurement is also based on reducing the possibility of entrapping a person between doors. With a swinging hoistway door, the largest entrapment area is between the hoistway door and the section of the car door furthest from it. With sliding car and hoistway doors, the leading sections are the ones, which could cause entrapment, as these sections are the ones nearest to each other.

Additionally, the specified distance eliminates the possibility of a person standing on the hoistway-landing

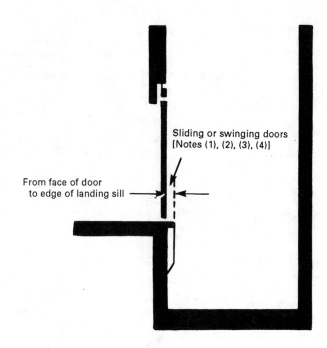

GENERAL NOTES:
(a) For multispeed doors, the distance is measured from the face of the door section nearest the car.
(b) See Code for exceptions related to new elevators installed in existing multiple hoistways or replacement doors.

NOTES:
(1) 100 mm or 4 in. where the elevator can be operated only from inside the car.
(2) 57 mm or $2^{1}/_{4}$ in. for sliding doors.
(3) 19 mm or $^{3}/_{4}$ in. for swinging doors.
(4) 57 mm or $2^{1}/_{4}$ in. for swinging emergency doors complying with 2.11.4.2(b).

Diagram 2.11.4   Door Face to Edge of Landing Sill
(Courtesy Zack McCain, Jr., P.E.)

sill when the hoistway door is closed. See Diagram 2.11.4.

2.11.6 Opening of Hoistway Doors From Hoistway Side

When a passenger elevator car is outside the unlocking zone, the hoistway door or car door must be so arranged that one or the other or both cannot be opened more than 100 mm or 4 in. from inside the car. The unlocking zone is a zone that is determined by the elevator manufacturer in accordance with 2.12.5. The 100 mm or 4 in. dimension was the maximum permitted opening under which a hoistway door and car door or gate could be considered in the closed position when this requirement was originally developed in the early 1980s. When the passenger elevator car is within the unlocking zone, 75 mm or 3 in. up to 450 mm or 18 in. or less above or below the landing, passengers may be able to open the door by hand from within the car. The passenger must be able to open the door from inside the car within 75 mm or 3 in. above or below the landing. See Diagram 2.12.5 (A17.1–2000, Fig. B1).

The requirements pertaining to locking doors out of service (2.11.6.2) applies to both passenger and freight elevators. "Locked out of service" refers to a physical locking of the doors. A keyed car and/or corridor switch does not constitute a locked-out-of-service means since it allows the doors to be opened manually. The hoistway door must not be locked out of service at the top or bottom terminal landing, as a malfunctioning elevator has a tendency to home-in on the terminal landing. Also, if these doors were locked out of service, evacuation of passengers would be severely hampered. The designated and alternate floor landing doors must not be locked out of service when Phase I is effective, as this is the location that elevators return to when operated under firefighters' service (2.27.3.1). If the designated and alternate floor doors were locked, passengers would not be able to exit the car and emergency personnel, firefighters, etc., would not be able to enter the car.

If the elevator is equipped with Phase II Service, no landing is permitted to be locked out of service when Phase II is in effect. Firefighters must be able to access a floor during a fire. A locked out of service entrance would not allow access.

2.11.7 Glass in Hoistway Doors

2.11.7.1 Vision Panels. A manually operated hoistway door is one that is opened and closed by hand. A self-closing hoistway door is one that is opened manually and that closes automatically when released. Vision panels or hall position indicators are only required for manually operated doors or doors equipped with self-closing devices such as swing doors. Neither a vision panel nor position indicator is required for doors that open and close under power even if such doors are equipped with self-closers. The vision panel construction requirements are applicable whenever hoistway door vision panels are provided, whether they are required or not.

Wire glass in a vision panel should never be replaced by, or covered by, any other material such as Plexiglas since this could adversely affect the fire-resistance rating of the door. Other transparent glazing materials are permitted [2.11.7.1.4(b)]; however, if the door is a labeled door, the glazing material must be approved by the certifying agency. See also Handbook commentary on 8.6.3.7. The size of each panel precludes someone from sticking his head through the opening if the glass is missing. The height of the panel from the floor places it at a location that facilitates its use.

Vision panel protective grills (2.11.7.1.7) were first required by New York City and have proven to be effective in reducing injuries and fatalities due to broken vision panels.

2.11.7.2 Glass Doors. The Code makes a clear distinction between vision panels and glass doors. The minimum area requirement for the glass door ensures that these requirements are not used to install oversized vision panels. Laminated glass is required to minimize the possibility of an opening into the hoistway if the glass is broken. The Consumer Product Safety Commission (CPSC) standard 16 CFR Part 120 would apply as the glazing is classified by the building codes and the CPSC as being in a location subject to human impact loads.

Protection is required on glass door edges as door panels tend to be struck by carts, etc. If the glass was not protected, the edges might become hazardous to people passing through the doorway.

In addition to the requirement for glass doors in 2.11.7.2, the requirements in 2.11.11 are also applicable. As an example, any area depressed or raised more than 3.125 mm or $\frac{1}{8}$ in. would require beveling in compliance with the requirements in 2.11.11.5.5.

2.11.10 Landing-Sill Guards, Landing-Sill Illumination, Hinged Landing Sills, and Tracks on Landings

2.11.10.1 Landing-Sill Guards. See Diagrams 2.11.10(a) and 2.11.10(b).

2.11.10.2 Illumination at Landing Sills. Just as automatic cameras do not focus on a long myriad of tightly spaced horizontal lines, our eyes are unable to perceive depth or contrast in closely spaced fine lines. The standard elevator door guide is generally 13 mm ($\frac{1}{2}$ in.) wide and anywhere from 13 mm ($\frac{1}{2}$ in.) to 25 mm (1 in.) deep. There are generally two of these guides at the threshold of an elevator, one in the car sill, and the other in the landing sill. In between these sills is the gap between the elevator car and the elevator hoistway. This running clearance is typically 25 mm (1 in.) to 32 mm ($1\frac{1}{4}$ in.). An elevator threshold resembles such a group of lines, which creates a competition for one's vision between the very visible gap between car and hoistway sills and the less visible door guides. The gap between car and hoistway wall is visible because it is sufficiently wide and dark. The door guides are camouflaged because they blend into the threshold pattern, which is generally a profusion of deceptive horizontal lines, thereby creating a lack of conspicuity. Dimly lighted areas render the inconspicuous guides more difficult to see, as do glares, shadows, or heavy reflection.

2.11.10.3 Hinged Hoistway Landing Sills. Hinged landing sills were common on old single or two speed A/C elevator that did not level to allow easy transport of handcarts from sill to sill. They are seldom if ever found on modern equipment. This requirement reduces the possibility that the elevator can be operated before the hinged landing sill is in the upright position.

2.11.11 Entrances, Horizontal Slide Type

2.11.11.1 Landing Sills

2.11.11.1(b) This requirement reduces a potential tripping hazard at the entrance-landing sill.

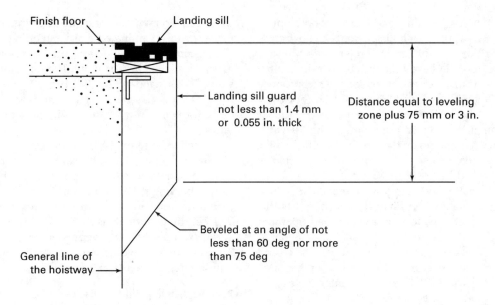

Finish floor   Landing sill

Landing sill guard
not less than 1.4 mm
or 0.055 in. thick

Distance equal to leveling
zone plus 75 mm or 3 in.

Beveled at an angle of not
less than 60 deg nor more
than 75 deg

General line of
the hoistway

GENERAL NOTE: Beveled bottom edge may be eliminated if landing sill guard extends to the top of door hanger pocket of the entrance immediately below.

Diagram 2.11.10(a)   Typical Landing-Sill Guard Where a Car-Leveling Device is Provided

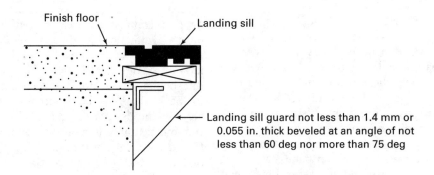

Finish floor   Landing sill

Landing sill guard not less than 1.4 mm or
0.055 in. thick beveled at an angle of not
less than 60 deg nor more than 75 deg

GENERAL NOTE: Beveled edge may be eliminated if landing sill guard extends to the top of door hanger pocket of the entrance immediately below.

Diagram 2.11.10(b)   Typical Landing-Sill Guard Where No Car-Leveling Device is Provided

2.11.11.3 Entrance Frames. The frame return leg must be parallel to the plane of the door panels but need not be in the same plane (flush) with the return wall of the hoistway.

Entrance frames are not designed to carry the weight of the wall above the frame. When inspecting entrance frames, be sure that a lintel is provided with masonry construction and that in other construction, the wall framing is designed to carry the weight of the wall. Entrance frames normally cannot use the wall alone for structural support. They should be secured to the building structure or to the track supports in addition to the hoistway wall. See Handbook commentary on 2.1.1.5. The wall anchor is provided to ensure fire integrity.

Passenger elevator doors installed in masonry walls would perform the same under fire exposure conditions without a frame, providing the doors overlapped the masonry opening the same distance as an assembly incorporating a frame. Masonry type walls are generally considered to consist of brick, concrete block, or reinforced concrete construction. All three of these materials have demonstrated their resistance to fire exposure.

Masonry and drywall constructions, which have veneers of marble or granite, have not been investigated by testing laboratories under fire conditions. Engineering judgment can be applied when comparing granite- or marble-faced drywalls with solid masonry walls. The passenger elevator doors should perform the same under fire conditions in a granite- or marble-faced wall, providing the doors overlapped the masonry wall opening, or the steel framed opening in a drywall construction, the same distance as an opening without the marble or granite facing. The thickness of the facing should not be considered in determining the correct overlap of door opening. This judgment can only be made by assuming that the application of the marble or granite facing does not adversely affect the structural integrity of the wall under fire conditions. In the case of drywall construction, the interface between the door and steel frame could be affected by the application of the facing material.

The stringent fire test of an entrance that subjects a test assembly to 982°C (1,800°F) necessitates the use of steel for doors and frames. Metals, such as aluminum or bronze, and other materials, such as marble, wood, etc., cannot withstand 982°C (1,800°F); therefore, a steel subframe is necessary in order to preserve the fire integrity. Diagram 2.11.11.3 illustrates the use of a steel subframe with stone facing. Note that in order to preserve the fire integrity, the door must overlap the steel subframe and not the marble facing.

2.11.11.5 Panels

2.11.11.5.3 A finger or hand caught between the landing edge of the door panels and a strike jamb pocket could cause serious injury. When there is no pocket, a finger or hand could be more easily removed, and not so likely to be wedged in place.

2.11.11.5.7 It is not intended that this requirement be field tested.

2.11.11.5.8 This requirement controls the amount the door can be forced open with the door in the closed and locked position. A simple field-test to determine compliance can be performed as follows:

> On the landing side, hook a spring scale around the leading edge of the hoistway door at the furthest point from the interlock. Let the door close and the car leave the floor. Exert a pull of 135 N or 30 lbf at the furthest point from the interlock, typically the bottom of the panel, in the opening direction and measure the door movement. For center parting doors exert the 135 N or 30 lbf pull on each panel measuring the door movement between the panels. Test only one panel at a time. Do not apply the force to both panels at the same time.

2.11.11.6 Bottom Guides. The 6 mm or $\frac{1}{4}$ in. engagement of the door guide in the sill is necessary to prevent disengagement of the guide from the sill. This would allow the door panel to swing into the hoistway, an extremely hazardous condition.

2.11.11.7 Multipanel Entrances. An interconnecting means complying with 2.11.11.7.1 or interlocks on each panel complying with 2.11.11.7.2 must be provided.

When you comply with 2.11.11.7.1, a door closer on the panel that does not have the interlock is required by 2.11.3.3. This will close the nondriven door if the interconnecting means fails. When an interconnecting means complying with 2.11.11.7.1 is not provided, interlocks are required by 2.11.11.7.2 on each driven panel. This is to assure that the elevator cannot run with an open hoistway door panel.

On multispeed doors, 2.11.11.7.3 requires a mechanical-interconnecting means between panels moving in the same direction, to assure the panels will not separate leaving a open hoistway door panel.

2.11.11.8 Hoistway Door Safety Retainers. It is not intended that this requirement be field-tested. See also 2.28.1.

2.11.11.9 Beams, Walls, Floors, and Supports. This requirement is necessary to assure that the doors are secure. It is not intended that this requirement be field-tested. However, they should be visually examined to assure they are in place and not subject to wear or stress during normal operation.

2.11.11.10 Hoistway Door to Sill Clearance. The horizontal distance from the leading edge of the hoistway doors or sight guard to the edge of the hoistway sill is limited, by these provisions, to reduce the open gap between the hoistway and car doors. The vertical distance between a sight guard and the landing sill is limited to reduce a pinching hazard. A sight guard is defined as a vertical member mounted on the hoistway side, leading edge of the hoistway door. It is used to reduce the opening between the leading edge, the hoistway door, and car door and inhibit placing objects in this space.

2.11.12 Entrances, Vertical Slide Type

See Diagram 2.11.12.

2.11.12.1 Landing Sills. See Diagram 2.11.12.1.

2.11.12.2 Entrance Frames. The walls may not be able to sustain the loads and forces applied to them by the entrance. This is the reason why the entrance frames normally are secured to the building structure in addition to the wall itself. The wall anchor is provided to ensure fire integrity. See Handbook commentary on 2.1.1.5. See Diagram 2.11.12.2 for typical drywall frame arrangement.

2.11.12.3 Rails. The walls themselves may be incapable of withstanding the loads and forces applied on them during the normal use of the entrance. These loads and forces are transmitted to the door guide rails and they in turn must transmit the load to the building structure.

2.11.12.4 Panels. See Diagram 2.11.12.4.

2.11.12.4.3 A rigid astragal is a horizontal molding attached to the meeting edge on a pair of panels for

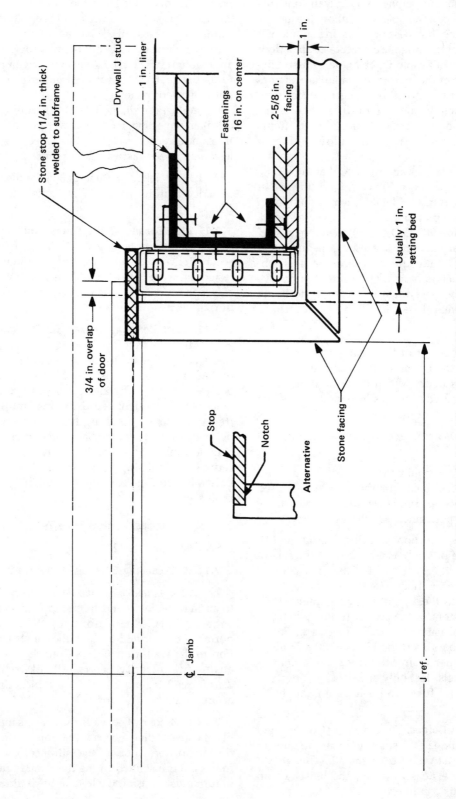

Stone stop (1/4 in. thick) welded to subframe

Drywall J stud

1 in. liner

Fastenings 16 in. on center

2-5/8 in. facing

1 in.

Usually 1 in. setting bed

3/4 in. overlap of door

Stop

Notch

Alternative

Stone facing

Jamb

J ref.

Diagram 2.11.11.3   Door Frame With Stone Facing

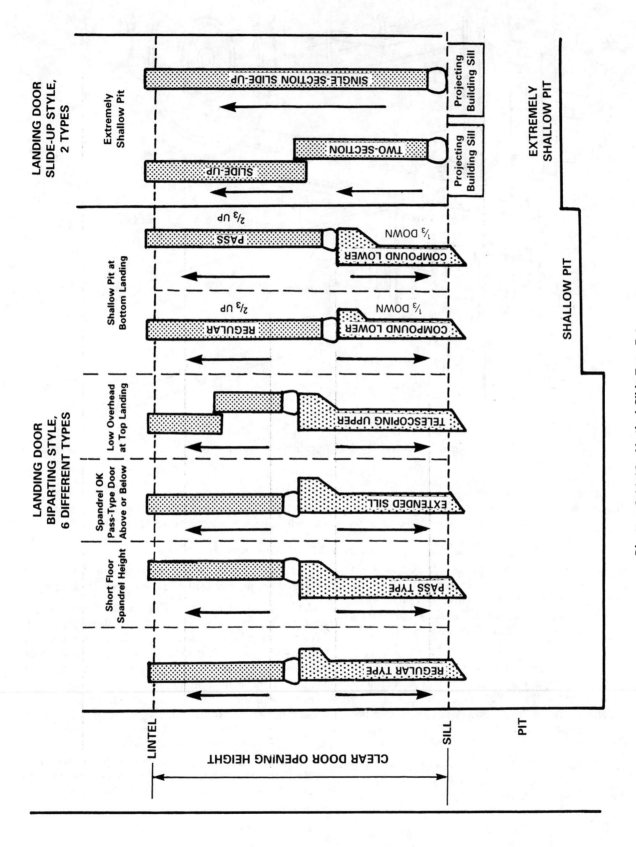

Diagram 2.11.12   Vertical Slide-Type Entrances
(Courtesy The Peelle Co.)

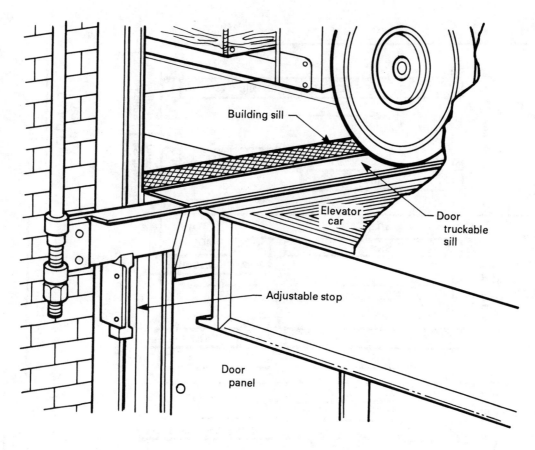

Diagram 2.11.12.1   Typical Trucking Sill Arrangement

protection against weather conditions and/or to retard the passage of smoke, flame, or gases during a fire. Rigid astragals are prohibited as they were found to be a shearing hazard. A fire-resistive, nonshearing, and noncrushing astragal is to be provided on the upper panel and provides the same protection that the rigid astragal does but in a safe manner. A nonshearing and noncrushing astragal reduces the effects of injury should someone's hand, etc., get caught between door panels during the closing operation of the door. See Diagram 2.11.12.4.3.

2.11.12.4.7   A pass-type entrance is a biparting door having the upper panel offset from the lower panel. It is used when the floor height is not sufficient for the upper panel at one landing to open without interference with the lower panel of the door at the landing immediately above. See Diagram 2.11.12.4.7.

2.11.12.4.8   This requirement controls the amount the door can be forced open with the door in the closed and locked position. A simple field test to determine compliance can be performed as follows:

> On the landing side a spring scale around the leading edge of the hoistway door at the furthest point from the interlock. Let the door close and the car leave the floor. Exert a pull of 135 N or 30

lbf at the furthest point from the interlock, typically the bottom of the panel, in the opening direction and measure the door movement.

2.11.13 Entrances, Swing Type

2.11.13.1 Landing Sills. This requirement reduces a potential tripping hazard at the entrance landing sills.

2.11.13.2 Entrance Frames. Entrance frames are not designed to carry the weight of the wall above the frame. When inspecting entrance frames, be sure a lintel is provided for masonry construction and that in all other construction the wall framing is designed to carry the weight of the wall. Entrance frames normally cannot use the wall alone for their structural support. They should be secured to the building structure or to track supports in addition to the hoistway walls. See Handbook commentary on 2.1.1.5. The wall anchoring is provided only to ensure fire integrity. When installed in other than masonry walls, the interface between the frame and wall construction should be approved for such use.

2.11.3.3 Panels

2.11.13.3.4   The use of brass, bronze, stainless steel, woven metal, or plastic laminate facings on UL classified passenger elevator doors is currently covered

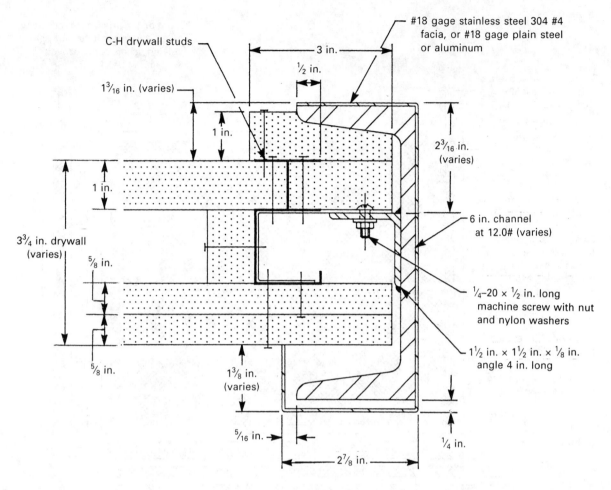

Diagram 2.11.12.2 Vertical Slide-Type Entrance Typical Section Through Frame Showing Connection to Drywall
(Courtesy The Peelle Co.)

in each individual manufacturer's follow-up service procedure and is based upon investigation by fire test. The facings are installed at the manufacturer's plant.

The particular type of mechanical fastening and/or adhesive fastening method investigated is specifically described. The fastening methods have been investigated per the test specified in 8.3.1.2 for their effect on the structural performance of the door (in the case of mechanically fastened systems) or flaming characteristics (in the case of adhesively fastened systems) while the assembly is under fire exposure conditions.

The addition of architectural facing materials, in the field, to labeled entrances already installed may not be authorized by the certifying organization and could affect the fire performance of the door assembly.

2.11.14 Fire Tests

Most elevator entrances are rated for $1\frac{1}{2}$ h and are tested in the furnace for $1\frac{1}{2}$ h. If an hourly rating of less or more is required, the test time is so adjusted. The corridor side of the door placed in a furnace enclosure

is exposed to a temperature rise that peaks at approximately 982°C (1,800°F) at the 90 min point. During this time, the test assembly must preserve its integrity by complying with limitations on flame spread and with no occurrence of through openings in the test assembly.

On conclusion of the fire test, the test assembly is moved away from the heat source and is subjected to a high-pressure stream of water for a varying length of time that is a function of the total square footage of the test specimen. This is known as the hose stream test. The hose stream test, with its resulting extreme temperature differential, tests for structural integrity simulating an explosion during a fire. As a result of the hose stream test, the space between the door panel and frame cannot exceed 73 mm ($2\frac{7}{8}$ in.).

As Underwriters Laboratories, Inc. (UL) is the predominant certifying organization in the United States, it will be referenced in the commentary. Other certifying organizations are approved in many areas and the information below should apply to them.

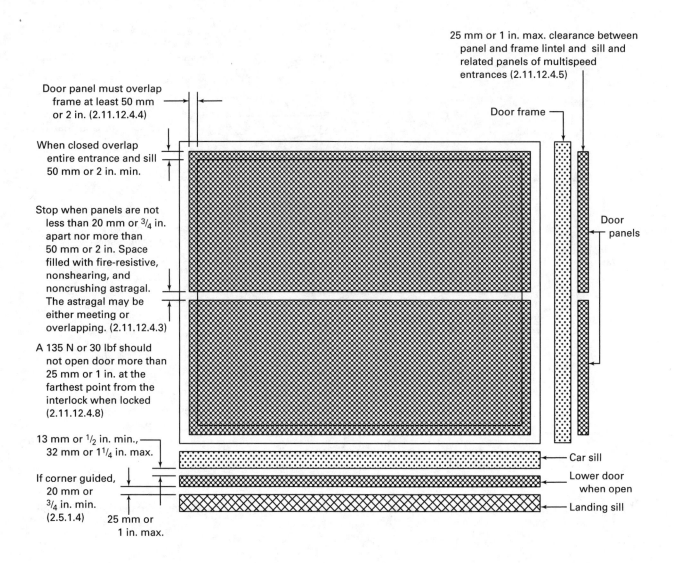

25 mm or 1 in. max. clearance between panel and frame lintel and sill and related panels of multispeed entrances (2.11.12.4.5)

Door frame

Door panel must overlap frame at least 50 mm or 2 in. (2.11.12.4.4)

Door panels

When closed overlap entire entrance and sill 50 mm or 2 in. min.

Stop when panels are not less than 20 mm or $^{3}/_{4}$ in. apart nor more than 50 mm or 2 in. Space filled with fire-resistive, nonshearing, and noncrushing astragal. The astragal may be either meeting or overlapping. (2.11.12.4.3)

A 135 N or 30 lbf should not open door more than 25 mm or 1 in. at the farthest point from the interlock when locked (2.11.12.4.8)

13 mm or $^{1}/_{2}$ in. min., 32 mm or $1^{1}/_{4}$ in. max.

Car sill

Lower door when open

If corner guided, 20 mm or $^{3}/_{4}$ in. min. (2.5.1.4)

Landing sill

25 mm or 1 in. max.

Diagram 2.11.12.4   Vertical Slide-Type Entrance Panels
(Courtesy Zack McCain, Jr., P.E.)

At the time elevator fire door assemblies were first investigated by UL, they were intended for installation in masonry walls only. UL did not establish listings for the frames because the doors were not attached or supported by the frame and the frame primarily served the purpose of finishing the masonry opening. At the time, it was judged that elevator doors installed in masonry walls would perform the same under fire exposure conditions with or without a frame, provided the doors overlapped the masonry opening the same distance as an assembly incorporating a frame. In later years, passenger and eventually freight elevator door assemblies were evaluated for installation in drywall construction. In the case of drywall construction, it was found that the interface detail of the wall and frame was an important factor in the overall fire performance of the assembly. Accordingly, the frame detail is an important factor in the performance of elevator fire door assemblies

installed in drywall construction. As a result, elevator door frames intended for installation in drywall construction are listed under the follow-up service program of UL.

If a manufacturer wanted to establish classification, fire tests were conducted on both the single-slide opening and center-opening entrance assemblies incorporating the largest size door panel the manufacturer intended to include under UL's follow-up service program. As the result of a successful investigation, a manufacturer would be in a position to provide elevator fire doors incorporating the classification mark (label) for use on single-slide entrances, multispeed side-slide entrances, center-opening entrances, and multispeed center-opening entrances, provided the door panel sizes do not exceed the maximum size of the panels tested.

At the time UL was requested to establish classification for passenger elevator fire doors installed in drywall

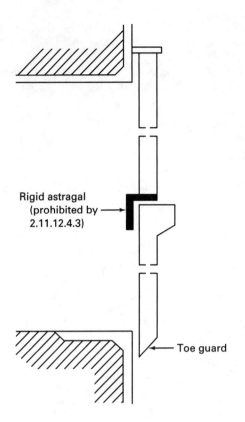

Rigid astragal
(prohibited by → 2.11.12.4.3)

Toe guard →

Diagram 2.11.12.4.3 Vertical Biparting Entrance With Rigid Astragal

construction, they reviewed the past fire performance data of single-slide and center-opening entrances previously tested in masonry walls. From that data, it was determined that, in general, the center-opening doors were more critical with respect to fire performance. Accordingly, passenger elevator fire door manufacturers who had established their classification as the result of tests conducted on single-slide and center-opening entrances in masonry construction could obtain coverage for their doors in a drywall construction based upon a successful fire test of a center-opening entrance. As a result of the performance of the center-opening assembly, UL was in a position to establish coverage for the single-slide and multispeed entrances in drywall construction. Manufacturers submitting their doors for the first time, or existing classified manufacturers submitting a new door design, would still need to fire test a single-slide opening and center-opening entrance assembly.

Based upon a review of past fire performance data of masonry and drywall elevator door entrances, it was determined that, in general, the drywall entrance assembly was more critical with respect to fire performance. As a result of the performance of a drywall entrance assembly, UL was in a position to establish coverage for the same assembly in masonry construction.

The UL building materials directory contains a current list of classified passenger and freight elevator fire door manufacturers, listed passenger and freight elevator fire door frame manufacturers, and listed elevator hardware manufacturers.

UL has been classifying sliding passenger elevator entrances and listing passenger elevator door hardware (see 1.3, Definitions) for approximately 30 years. The product categories of passenger elevator type fire doors (GSUX) and passenger elevator door hardware (GZKZ) were established in 1970. The listing of passenger elevator door frames (GVST) was formally established in 1979.

The classification marking (label) on passenger elevator doors covers the design and construction of the door or door panels only.

Passenger elevator door hardware consists of a header, track hangers, pendent bolts, floor sill with guides, sill support plates, sill brackets, retaining angles, and closer assemblies. Listing marks (labels) are applied to each header, track, hanger, aluminum sill, sill support plate, sill bracket, retaining angle, and closer assembly.

Elevator door frames are intended for use with sliding passenger elevator fire door designs for use in drywall shaft construction. In addition to the "Fire Door Frame" label, each frame bears a supplementary marking indicating "passenger elevator door frame for use with 'ABC' company's passenger elevator type fire doors to be installed in drywall shaft construction."

Between the date on which UL conducted the first successful test (1975) for passenger elevator door frames in a drywall and the date on which listing and UL follow-up service was promulgated (1979), manufacturers were providing copies of their UL test reports on drywall shaft construction to code authorities.

When a frame manufacturer wants to obtain the broadest listings of frame design options, based on a single fire test investigation, the certifying agency requires that he produce a frame sample for test having the minimum face and throat dimensions, minimum material gage, and the maximum height and width to be constructed under the follow-up service program.

If the test results are successful, door frames having a greater face and throat dimension, greater material thickness, and a lesser opening height and width providing the frame profile is similar, the frame construction identical, and the type and number of anchors the same as originally tested, are authorized for listing.

To date, UL has tested passenger elevator door frames for use in drywall construction with face dimensions ranging from 31.8 mm ($1\frac{1}{4}$ in.) to 51 mm (2 in.). Due to the importance of the wall and frame interface, frames with face dimensions less than those originally tested by a particular manufacturer would not be authorized for listing without further test.

Frames with face dimensions greater than 51 mm (2 in.) may be judged acceptable by engineering study.

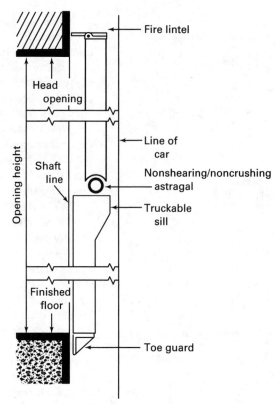

Vertical Section Pass Type Doors

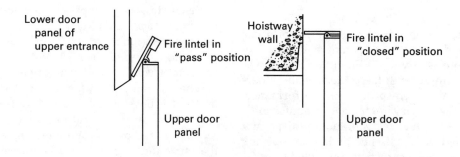

Operation of the Fire Lintel

GENERAL NOTE: The upper half of the pass-type entrance is offset into the hoistway so that when it opens it will pass the lower closed panel of the entrance on the floor immediately above.

Diagram 2.11.12.4.7   Typical Pass-Type Entrance
(Courtesy National Elevator Industry Educational Program)

However, UL would need to review detailed drawings illustrating not only the frame/wall interface, but also the method of wall construction. Based upon the present shaft wall designs, face dimensions greater than 51 mm (2 in.) may make proper wall installation very difficult.

UL also classifies gasketing materials for use on fire door and frame assemblies. The individual gasketing materials are currently investigated for use on particular frame or door types (i.e., hollow metal, wood composite,

passenger elevator, etc.) and for specific fire duration periods (i.e., $1\frac{1}{2}$ h, 2 h, 20 min, etc.).

The basic standard used to investigate gasketing materials applied to fire-resistive entrances is UL 10B or NFPA 252 (fire tests of door assemblies). The gasketing materials are investigated as applied to a door or frame type selected by the gasket material manufacturers and installed in accordance with the gasket manufacturer's instructions. The fire performance of the gasketing is observed during the test with particular attention to flaming characteristics and effect on the door or frame design. The assembly, with gasketing material applied, must comply with the conditions of acceptance specified in 8.3.4. The operation of the door assembly cannot be adversely affected by the installed gasketing material so as to restrict door operation.

Note that test standard UL 10B does not provide for evaluation of the door assembly or gasket material relative to the prevention of smoke or other products of combustion. Also, see the commentary on 2.1.1.1 and 2.11.19.

2.11.15 Marking

See Handbook commentary on 2.11.14.

2.11.15.1 Labeling of Tested Assemblies. This requirement provides a mechanism for identifying all entrance components of a labeled or listed entrance assembly.

2.11.15.3 Entrances Larger Than Tested Assemblies. The practice is to limit the assembly size bearing the labels to the maximum size actually fire tested. In some fire door categories (i.e., large freight doors, etc.), some certifying agencies have made available an oversize labeling program. In this case, an oversize door (i.e., one larger than the maximum size that can be fire tested) is provided with an "oversize label." The "oversize label" does not indicate that such doors are capable of furnishing standard fire protection, but only that they conform to the construction details of the doors successfully fire tested except for size.

In order to be eligible for an "oversize label" program, the door manufacturer must have tested the largest door the testing laboratories furnace would accept. They must then demonstrate that the oversize design and corresponding increase in hardware size and mounting technique is satisfactory for use through operational tests, structural calculations, drawings, or a combination of such data.

2.11.16 Factory Inspections

Recognized testing laboratories are required to have a follow-up inspection service to ensure that their label is being placed on assemblies that comply with the test reports.

2.11.17 Transoms and Fixed Side Panels

As UL is the predominant certifying organization, it will be referenced in the commentary. Other certifying organizations are approved in many areas and the information below will apply to them.

This requirement states in part that the opening closed by the transom and fixed side panels must not exceed in width or height the dimensions of the entrance above which it is installed.

This means if there is a 1.06 m (42 in.) wide by 2.13 m (84 in.) high entrance, the transom cannot exceed 1.06

m (42 in.) in width or 2.13 m (84 in.) in height. There are two types of transoms: flush and offset. The flush transom, as illustrated in Diagrams 2.11.17, sketches (a) and (e), requires a 69.9 mm ($2\frac{3}{4}$ in.) hoistway door panel. Offset transom panels, as shown in Diagrams 2.11.17, sketches (b), (c), and (d) can be of two types, offset pan transom and offset fire test transom. The offset fire test transom has a construction comparable to a door panel. The pan transom must be backed up with a fire-resistive wall.

It has been determined that the interface detail of the gypsum drywall design and the entrance frame are an important factor in the overall fire performance of the door assembly. Passenger elevator entrance frames and transoms are covered under UL's follow-up service program. A manufacturer can obtain UL listings for transom panels by having an engineering study conducted or by fire test. In the case of an engineering study, the manufacturer's request and drawings would be reviewed with regard to the following guidelines:

(a) The opening closed by the transom assembly cannot exceed the maximum width or height dimensions of the entrance above which it is installed.

(b) The transom panels must be constructed in a manner equivalent to the door construction eligible for labeling.

(c) The transom panel must be securely anchored.

(d) The passenger elevator door overlaps the transom panel by the same dimension as it overlaps the frame head when a transom panel is not installed.

Transom panel and door constructions incorporating a flush design requiring rabbeted edges or other unique design combinations would need to be investigated by fire test in order to determine their acceptability.

Requirement 2.11.15.1.2(c) requires transom panels to be labeled as passenger elevator frame transom panels by fire test or engineering study. When covered under the UL follow-up service program, the transom panel bears its own listing mark (label) reading "transom panel" along with any applicable supplementary markings.

2.11.18 Installation Instructions

This has been a condition for UL labeling for many years and is a means of assuring construction of the entrance assembly in accordance with the labeling procedure. Inspectors should check for conformance with the installation instructions.

2.11.19 Gasketing of Hoistway Entrances

The application of gasketing to hoistway entrances may affect the performance of the entrance. The reference to the tests in 2.11.19.1 is to evaluate the material's performance when subjected to a fire door test. Requirement 2.11.19.2 specifies a maximum temperature in accordance with a recognized national standard.

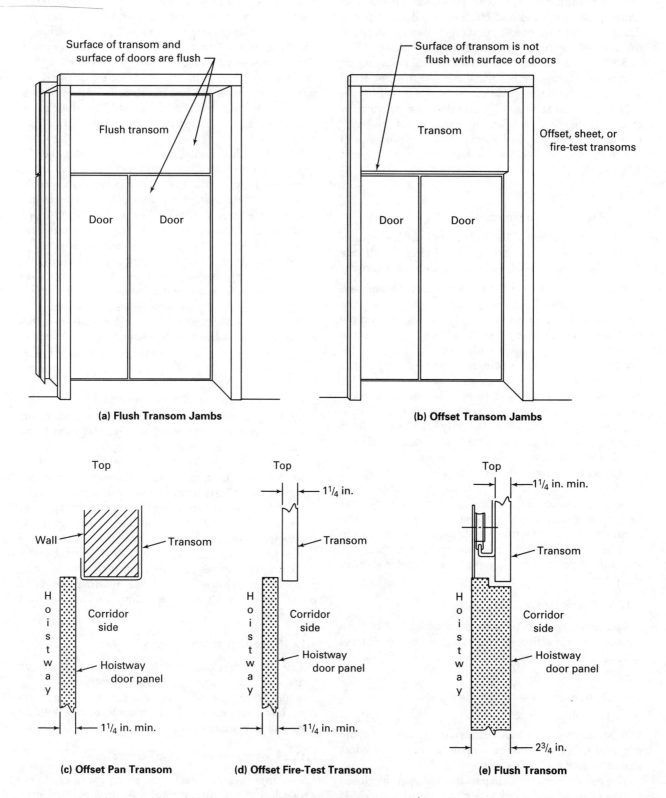

Diagram 2.11.17   Transoms

Note 1 to 2.11.19 is for information and suggests the Code requirements, which may be impacted when gasketing is applied to an elevator entrance.

Note 2 to 2.11.19 points out that these requirements do not attest to the performance of gasketing material in terms of air and/or smoke leakage performance. Gasketing quite often is specified for smoke and draft control purposes. Where gasketing has been installed for this purpose, it may not prove to be effective. Due to the limits placed on kinetic energy and door closing forces, a positive sealing may not be possible under field conditions.

SECTION 2.12
HOISTWAY-DOOR LOCKING DEVICES, CAR DOOR OR GATE ELECTRIC CONTACTS, AND HOISTWAY ACCESS SWITCHES

2.12.1 General

Requirement 2.12.1.1 prohibits the condition where there is a locked hoistway door with the car at the landing such as found with a straight arm closer. When this requirement was first introduced in the Code (ASME A17.1–1997), it eliminated the need for parking device requirements on new elevators. As parking devices may still be needed on existing elevators, the requirements were relocated to 8.7.2.11.3.

2.12.2.1 For Passenger Elevators. The hoistway door interlock is a device having two related and independent functions, which are:

(a) to prevent the operation of the driving machine by the normal operating device unless the hoistway door is locked in the closed position; and

(b) to prevent the opening of the hoistway door from the landing side, unless the car is within the landing zone and is either stopped or being stopped. See Diagram 2.12.2.1.

2.12.2.2 Closed Position of Hoistway Doors. When the hoistway doors are in the closed position [see Diagram 2.12.2.2, sketch (a)], the car may run normally. Operation of the driving machine, when the hoistway door is unlocked and not in the closed position is also permitted by a car leveling or trucking zone device (2.26.1.6), when a hoistway access switch is operated (2.12.7), or when a bypass switch is operated (2.26.1.5).

This requirement defines what is meant by the closed position of the hoistway doors. See Diagram 2.12.2.2, sketch (b).

2.12.2.3 Operation of the Driving Machine With a Hoistway Door Unlocked or Not in the Closed Position. This requirement allows controlled and restricted movement of the elevator when the hoistway door interlock is in the open position or the hoistway combination mechanical lock and electrical contact is in the open position.

2.12.2.4 General Design Requirements

2.12.2.4.1 Switches that depend solely on "snap action" to open their contacts do not meet the requirements.

2.12.2.4.5 This requirement prohibits the use of an interlock system that employs the mechanical locking portion of an interlock at each hoistway door and a single electric contact for a group of hoistway doors, which is actuated by a series of rods and/or cables attached to the mechanical portion of the interlock on each door.

2.12.2.4.6 The force level is compatible with the force created by normal human effort without the use of tools.

2.12.2.4.7 Misuse of mercury tube switches can be hazardous. They were permitted by the Code prior to the 1987 edition; it is believed that they have not been used for years.

2.12.2.5 Interlock Retiring Cam Device. A hoistway door interlock retiring cam device is a device that consists of a retractable cam with its actuating mechanism and that is entirely independent of the car door or hoistway door power operator. See Diagram 2.12.2.5.

2.12.3 Hoistway-Door Combination Mechanical Locks and Electric Contacts

2.12.3.1 Where Permitted. Hoistway door combination mechanical locks and electric contacts are permitted on manually opened, vertically sliding counterweighted doors or vertically sliding biparting counterbalanced doors on freight elevators only under the following conditions: if the pit depth is 1 525 mm or 60 in. or less, the lower door may have a mechanical lock and electrical contact. The upper hoistway door must be equipped with hoistway unit system hoistway door interlocks described in 2.12.2. However, if the elevator has a travel of 4 570 mm or 15 ft or less, the top landing door and any door whose sill is located not more than 1 225 mm or 48 in. below the sill of the top landing door may be equipped with combination mechanical locks and electrical contacts. See Diagrams 2.12.3.1(a) and 2.12.3.1(b).

A hoistway combination mechanical lock and electrical contact is a mechanical and electrical device with two related but entirely independent functions, which are:

(a) to prevent operation of the driving machine by the normal operating device unless the hoistway door is in the closed position; and

(b) to lock the hoistway door in the closed position and to prevent it from being opened from the landing side unless the car is within the landing zone. See Diagram 2.12.3.1(a).

As there is no positive mechanical connection between the electric contact and the door locking mechanism, this device only indicates that the door is closed but

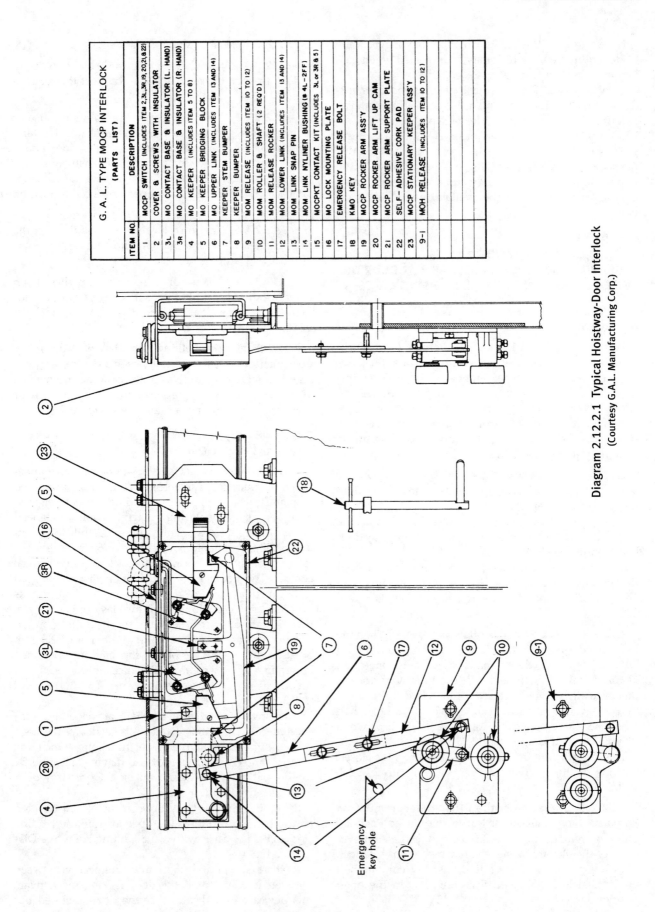

ITEM NO.	DESCRIPTION
1	MOCP SWITCH (INCLUDES ITEM 2,3L,3R,19,20,21,& 22)
2	COVER & SCREWS WITH INSULATOR
3L	MO CONTACT BASE & INSULATOR (L. HAND)
3R	MO CONTACT BASE & INSULATOR (R. HAND)
4	MO KEEPER (INCLUDES ITEM 5 TO 8)
5	MO KEEPER BRIDGING BLOCK
6	MO UPPER LINK (INCLUDES ITEM 13 AND 14)
7	KEEPER STEM BUMPER
8	KEEPER BUMPER
9	MOM RELEASE (INCLUDES ITEM 10 TO 12)
10	MOM ROLLER & SHAFT (2 REQ'D)
11	MOM RELEASE ROCKER
12	MOM LOWER LINK (INCLUDES ITEM 13 AND 14)
13	MOM LINK SNAP PIN
14	MOM LINK NYLINER BUSHING (# 4L-2FF)
15	MOCPKT CONTACT KIT (INCLUDES 3L or 3R & 5)
16	MO LOCK MOUNTING PLATE
17	EMERGENCY RELEASE BOLT
18	KMO KEY
19	MOCP ROCKER ARM ASS'Y
20	MOCP ROCKER ARM LIFT UP CAM
21	MOCP ROCKER ARM SUPPORT PLATE
22	SELF-ADHESIVE CORK PAD
23	MOCP STATIONARY KEEPER ASS'Y
9-1	MOH RELEASE (INCLUDES ITEM 10 TO 12)

G. A. L. TYPE MOCP INTERLOCK (PARTS LIST)

Emergency key hole

Diagram 2.12.2.1 Typical Hoistway-Door Interlock
(Courtesy G.A.L. Manufacturing Corp.)

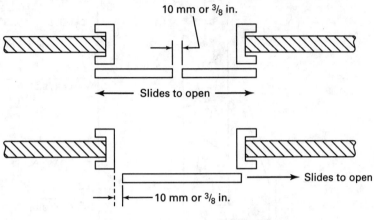

(a) Closed Position Horizontally Sliding Doors

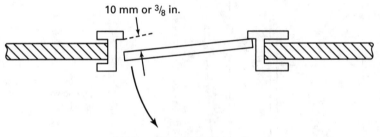

(b) Closed Position Swing Doors

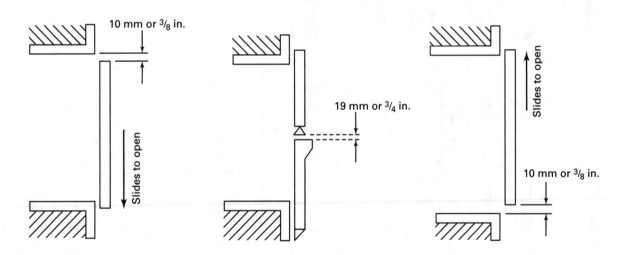

(c) Closed Position Vertically Sliding Doors

Diagram 2.12.2.2   Closed Position of Hoistway Doors

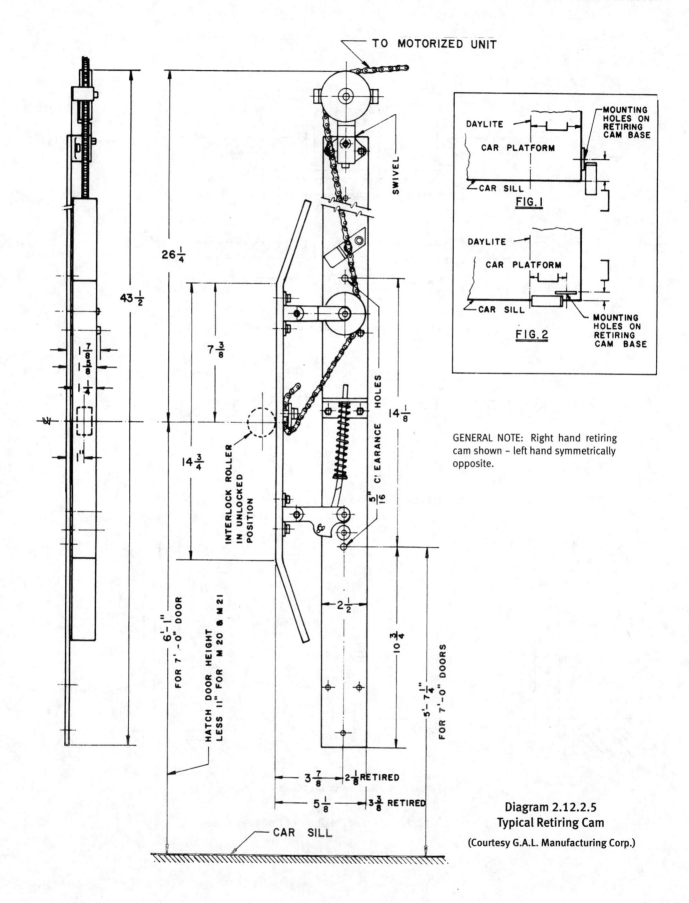

TO MOTORIZED UNIT

SWIVEL

26 1/4

43 1/2

7 3/8

7/8
5/8
1 1/4

1"

INTERLOCK ROLLER IN UNLOCKED POSITION

14 3/4

14 1/8

5/16 CLEARANCE HOLES

2 1/2

10 3/4

FOR 7'-0" DOOR

6'-1"

HATCH DOOR HEIGHT LESS 11" FOR M 20 & M 21

FOR 7'-0" DOORS

5'-7 1/4"

3 7/8   2 1/8 RETIRED

5 1/8   3 3/8 RETIRED

CAR SILL

DAYLITE

CAR PLATFORM

CAR SILL

MOUNTING HOLES ON RETIRING CAM BASE

FIG.1

DAYLITE

CAR PLATFORM

CAR SILL

MOUNTING HOLES ON RETIRING CAM BASE

FIG.2

GENERAL NOTE: Right hand retiring cam shown – left hand symmetrically opposite.

Diagram 2.12.2.5
Typical Retiring Cam

(Courtesy G.A.L. Manufacturing Corp.)

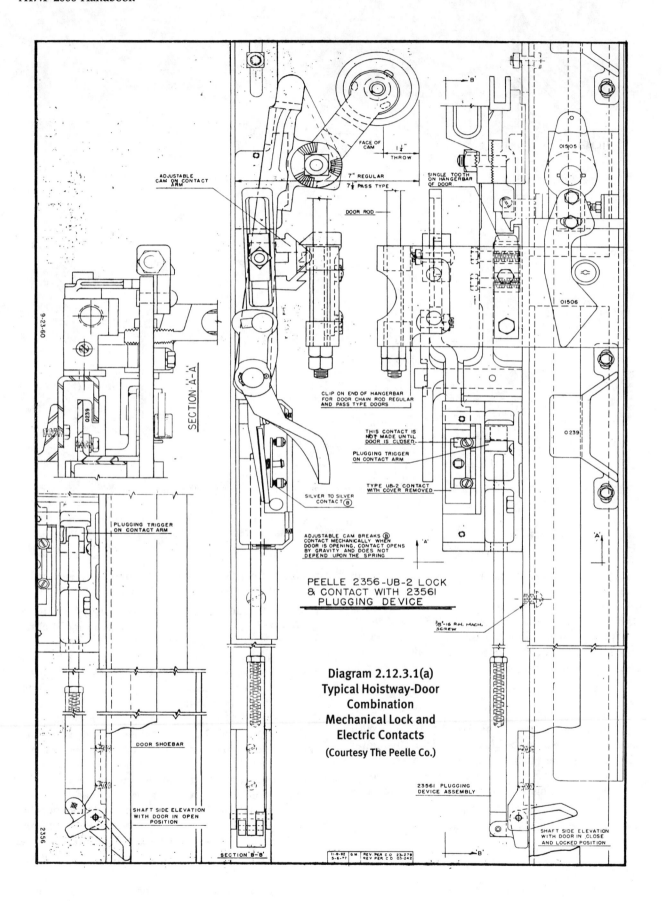

FACE OF CAM

1¼" THROW

7" REGULAR
7½" PASS TYPE

ADJUSTABLE CAM ON CONTACT ARM

DOOR ROD

SINGLE TOOTH ON HANGERBAR OF DOOR

0I505

0I506

9-23-60

SECTION 'A'-'A'

0239

PLUGGING TRIGGER ON CONTACT ARM

CLIP ON END OF HANGERBAR FOR DOOR CHAIN ROD REGULAR AND PASS TYPE DOORS

THIS CONTACT IS NOT MADE UNTIL DOOR IS CLOSED

PLUGGING TRIGGER ON CONTACT ARM

TYPE UB-2 CONTACT WITH COVER REMOVED

SILVER TO SILVER CONTACT Ⓑ

ADJUSTABLE CAM BREAKS Ⓑ CONTACT MECHANICALLY WHEN DOOR IS OPENING, CONTACT OPENS BY GRAVITY AND DOES NOT DEPEND UPON THE SPRING

0239

PEELLE 2356-UB-2 LOCK & CONTACT WITH 2356I PLUGGING DEVICE

⅜"-16 R.H. MACH. SCREW

DOOR SHOEBAR

SHAFT SIDE ELEVATION WITH DOOR IN OPEN POSITION

**Diagram 2.12.3.1(a)
Typical Hoistway-Door
Combination
Mechanical Lock and
Electric Contacts**
(Courtesy The Peelle Co.)

23561 PLUGGING DEVICE ASSEMBLY

SHAFT SIDE ELEVATION WITH DOOR IN CLOSE AND LOCKED POSITION

2356

SECTION 'B'-'B'

| 11-8-82 | GM | REV PER C.O 23-279 |
| 5-6-77 | | REV PER C.O 03-242 |

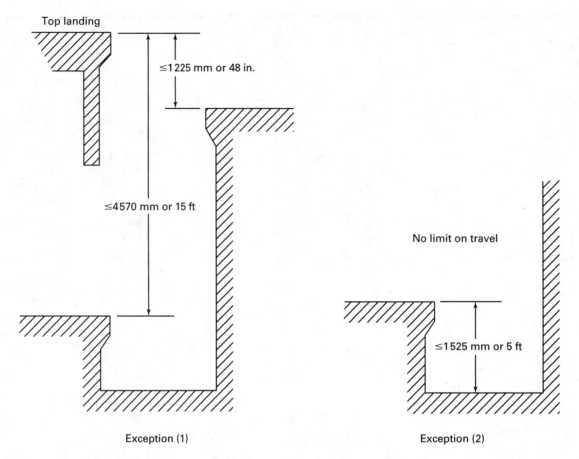

GENERAL NOTE: Hoistway-door interlocks are not required for the conditions shown above on freight elevators with manually operated vertically sliding doors.

Diagram 2.12.3.1(b)   Permitted Use of Hoistway-Door Combination Mechanical Lock and Electric Contacts

not necessarily locked when the car leaves the landing. Should the lock mechanism fail to operate, as intended, when released by a stationary or retiring cam device, the door can be opened from the landing side even though the car is not at the landing. If operated by a stationary cam device, it does not prevent opening the door from the landing side as the car passes the floor.

2.12.3.2   Closed Position of Hoistway Doors. See Handbook commentary on 2.12.2.2 and Diagram 2.12.2.2.

2.12.3.3   Operation of the Driving Machine With a Hoistway Door Not in the Closed Position. This requirement allows controlled and restricted movement of an elevator when the hoistway door interlock is in the open position or the hoistway combination mechanical lock and electrical contact is in the open position.

2.12.3.4   General Design Requirements

2.12.3.4.5   See Handbook commentary on 2.12.2.4.6.

2.12.3.4.6   See Handbook commentary on 2.12.2.4.7.

2.12.4   Listing/Certification Door Locking Devices and Door or Gate Electric Contacts

2.12.4.1   Type Tests. Type tests consist of an endurance test, current interruption test, test in moist atmospheres, test without lubricant, misalignment test, insulation test, and force and movement tests. See 8.3.3 for the test requirements for devices tested August 1, 1996, and later. Devices tested prior to August 1, 1996, are required to be tested to the requirements in A17.1a–1994, Section 1101 or in Canadian Jurisdiction CSA B44S1-1997, Clause 11.4. Devices tested to these requirements have an excellent safety record; thus, the Committee in developing 8.3.3 concluded there was no justifiable reason to require retesting to the new requirement. With computer aided design, new products could have been built such that they passed the former requirements but did not perform adequately in the field. The current 8.3.3 was prepared to address that concern. The label required by 2.12.4.3 requires a test date to assist in determining which test is appropriate.

2.12.4.3   Identification Marking. At the time of installation, it should be verified that the hoistway door interlock, hoistway combination mechanical lock and electric contact, and door or gate electric contacts not only have been tested as specified in 2.12.4.1, but have also been installed in accordance with all the requirements found in 2.12.

A testing laboratory can make a determination that an interlock is the same basic type as previously tested, based upon their technical evaluation, and testing would not be required for the modified interlock.

Approved devices must be labeled. The label among other requirements must identify the manufacturer, certifying agency, and date of test. The test date on the label will assist in identifying what test requirements are applicable. See also Handbook commentary on 8.6.3.7 replacement of listed devices and or their component parts.

2.12.5 Restricted Opening of Hoistway Doors or Car Doors

When a passenger elevator is outside the unlocking zone, it is unsafe for a passenger to try to exit through the elevator entrance unassisted. When a car is above the unlocking zone, the platform guard (apron) may not be long enough to close off the hoistway opening below the car. If this condition exists, a person exiting the elevator is exposed to the open hoistway. In fact, there have been many reports of fatalities due to this condition. A person inside the car should not be able to accomplish their own emergency evacuation through a hoistway door when the car is located outside of the unlocking zone. This requirement may be met by restricting the opening of the car door or the hoistway door. When the car is located in a blind portion of the hoistway away from any hoistway door and no hoistway door can be opened from inside the car, opening of the hoistway door is deemed restricted, and the requirement is met. See commentary on 2.11.6 and Diagram 2.12.5.

2.12.6 Hoistway Door Unlocking Devices

2.12.6.1 A hoistway door unlocking device is a device that permits the unlocking and opening of the hoistway door from the access landing irrespective of the position of the car. Hoistway door unlocking devices are required on all floors for emergency access unless restricted by the authority having jurisdiction. These requirements vary in many jurisdictions and it would be advisable to check the local ordinances. Unlocking devices are necessary for the evacuation of passengers from stalled cars and to stop emergency personnel from using unauthorized methods for gaining access to the hoistway.

It is recognized that there are some inherent dangers in the use of a hoistway door unlocking device; however, it is not possible to overcome them since access to an open hoistway is sometimes required. Hoistway door

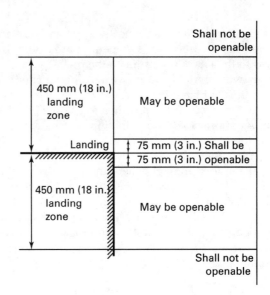

Diagram 2.12.5   Unlocking Zone

unlocking devices are provided for use by trained personnel for the purpose of maintenance, inspection, and emergency access. A trained person would only need to use the unlocking device when the car is not at the landing. Therefore, adding a warning device would not provide any additional safety since he already knows that the car may not be there (otherwise he would not be opening the doors through this means).

2.12.6.2 Location and Design. The device is to be designed to prevent unlocking the door with common tools (2.12.6.2.1). A screwdriver is a common tool and cannot be considered an acceptable means for opening hoistway doors. A lunar-shaped key is often used since it is not easily duplicated. When it is used, the escutcheon plate should be of hardened steel to prevent the tab from being distorted. If the tab is distorted, the escutcheon plate should be replaced.

Previous editions of the Code required the operating means for unlocking the device to be kept in a break-glass covered box, which would be accessible to the general public. This requirement was removed from the Code as it was found that vandals were misusing the key. The key should be in a location readily accessible to authorized persons but not where it is accessible to the general public.

The keyway or locked panel (see 2.12.6.2.3) is required to be at a height of 2 100 mm or 83 in. or less so that the inspector or mechanic can conveniently reach it while standing on the lobby floor (2.12.6.2.5). Keyways at greater heights might require the use of a stepladder, which could be dangerous if the person were to lose balance after the doors had opened.

2.12.7 Hoistway Access Switch

The hoistway access switch is defined as "a switch, located at a landing, the function of which is to permit operation of the car with the hoistway door at this landing and the car door or gate open, in order to permit access to the top of the car or to the pit."

Hoistway switches are required for all cars with speeds greater than 0.75 m/s or 150 ft/min at the lowest landing for access to the pit when a separate pit access door is not provided and at the top landing for access to the top-of-car. When car speeds are 0.75 m/s or 150 ft/min or less, hoistway access switches are required only at the top landings where the distance from the top-of-car to the access landing sill is greater that 900 mm or 35 in., with the car level at the landing below.

It was generally agreed that hoistway access switch operation provided the best control for gaining access to the car top for top-of-car inspection operation. See Diagram 2.12.7.

2.12.7.1 General

2.12.7.1.1 Speed is limited to inspection speed as car is moving with hoistway door in open position.

2.12.7.1.2 It is extremely difficult to position a car to allow safe access to the top-of-car when an access switch is not provided.

2.12.7.2 Location and Design. It is not the intent of this requirement to prohibit the use of a single key for the hoistway access switch and for other key switches in the car that perform functions as required by 2.12.7.3.3. Requirement 2.12.7.2.7 prohibits a lock, which can be operated by a key that is intended for other building uses. See also Handbook commentary on 8.1.

2.12.7.3 Operating Requirements. These requirements assure that the inspector or maintenance or repairperson that is operating the hoistway access switch has complete control over the car. However, it still assures that the hoistway door interlocks or electric contacts and mechanical locks at other landings are still operational. See Diagram 2.12.7.3.

2.12.7.3.6 See Diagram 2.12.7.3.6.

2.12.7.3.7 See Diagram 2.12.7.3.7.

SECTION 2.13
POWER OPERATION OF HOISTWAY DOORS AND CAR DOORS OR GATES

2.13.1 Types of Doors and Gates Permitted

This requirement reduces the potential of passengers being struck by closing doors or trapped between closing hoistway doors and the facing car door or gate.

2.13.2 Power-Opening

2.13.2.1 Power-Opening of Car Doors or Gates

2.13.2.1.1 The first requirements governing solid-state devices were published in ASME A17.1e–1975.

Strong arguments were made for and against allowing advance door opening before the elevator was level at a landing. Most committee members agreed that the point where the hoistway doors physically start to open was more relevant to safety than the point where door opening is initiated. Since high performance, fast-response elevator static control systems can cause the car to move very rapidly in a short interval of time, the initial application of power to the door operator was restricted to the following:

(a) The car must be leveling into a landing.

(b) The car must be within 300 mm or 12 in. of the landing.

(c) The car speed must not exceed 0.75 m/s or 150 ft/min).

The 300 mm or 12 in. dimension was a compromise between those who wanted no advance door opening and those who wanted no change to the Code requirement. This provides the necessary safety protections and, at the same time, does not severely hamper present-day performance standards of elevators.

2.13.2.1.2 The restriction on the distance that a collapsible car gate can be power-opened is to reduce the possibility of a hand or finger being pinched between the openings in the car gate.

2.13.2.2 Power-Opening of Hoistway Doors. This requirement does not prohibit power opening of the hoistway door at a landing if the car is stopped within the unlocking zone by other means such as a malfunctioning of a device or the operation of an emergency stop switch.

2.13.2.2.2 The phrase "when stopping under normal operating conditions" describes the condition required to be met only when the power opening is initiated automatically through control circuits.

2.13.3 Power-Closing

2.13.3.1 Power-Closing or Automatic Self-Closing of Car Doors or Gates Where Used With Manually Operated or Self-Closing Hoistway Doors

(a) This requirement is applicable only to passenger elevators with automatic or continuous-pressure operation when:

(1) the car door or gate is of the automatically released, self-closing type; and

(2) the hoistway door is:

(a) a door or gate that is opened and closed by hand; or

(b) a door or gate that is manually opened and, when released, closes automatically.

(b) This requirement is also applicable to freight elevators, which are permitted to carry passengers (see 2.16.4), when door or gate closing is:

(1) controlled only by a constant pressure switch; or

(2) by means of the car operating device.

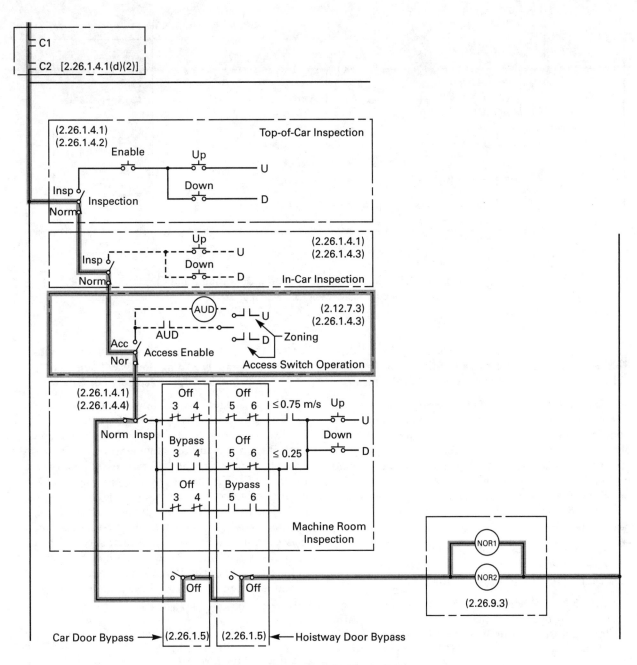

Diagram 2.12.7   Hoistway Access Switch Operation
(Courtesy Otis Elevator Co./Ralph Droste)

Release of the door-close switch or operating device must stop, or stop and reopen, the car door or gate.

There is no requirement for a reopening device (2.13.5) on an automatic car door opposite a manual hoistway door because the car door is not permitted to close until the hoistway door is closed.

2.13.3.2 Power-Closing of Hoistway Doors and Car Doors or Gates by Continuous-Pressure Means. Whenever continuous-pressure means for power closing of either or both the car door or hoistway gate is provided, this requirement applies. Release of the door or gate-closing switch must stop or stop and reopen the door or gate. A switch must be provided at each landing and control the hoistway and car door or gate at that landing only. For example, one cannot close the second floor hoistway door from the first or third floor, only from the second floor. When two openings are provided at a landing, a closing switch must be provided at each opening. The switch can control only that hoistway and car door or gate adjacent to the switch. A separate switch should be provided for each car door or gate and its corresponding hoistway door.

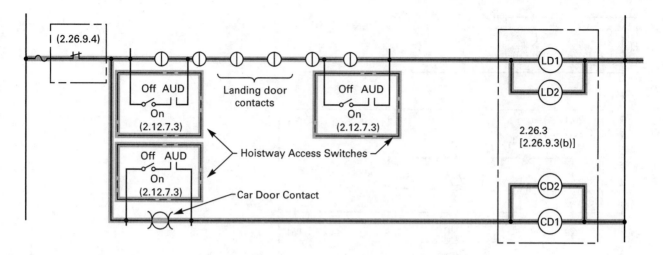

Diagram 2.12.7   Hoistway Access Switch Operation (Cont'd)
(Courtesy Otis Elevator Co./Ralph Droste)

2.13.3.3 Power-Closing of Horizontally Sliding Hoistway Doors and Horizontally Sliding Car Doors or Gates by Momentary Pressure or by Automatic Means

2.13.3.3.2 The momentary-pressure door-open switch allows a person to stop or stop and reopen the door if additional time is needed to enter or exit the elevator. The switch can only be of the momentary pressure type and cannot be a toggle switch, which would not conform to the requirements of 2.11.3. A safety edge or photoelectric door opener cannot be considered an appropriate substitute for the required door-open or door-stop button.

2.13.3.4 Power-Closing of Vertically Sliding Hoistway Doors and Vertically Sliding Car Doors or Gates by Momentary Pressure or by Automatic Means

2.13.3.4.1 When automatic means are used for power-closing of the door, the bell gives adequate warning for a person entering or exiting the elevator to get out of the way of the closing car door or gates. When the doors or gates are closed by a switch in the car, there is no need for the advanced warning, as the operator can visually observe the entrance area.

2.13.3.4.2 Sequence operation is required with automatic closing systems when the elevator is equipped with vertically sliding doors. When a biparting vertically sliding door faces a biparting vertically sliding car door or gate, there is no additional protection provided by sequence operation.

2.13.3.4.3 See Handbook commentary on 2.13.5.

2.13.3.4.4 A momentary-pressure switch within the car and at each landing gives a person the opportunity to retain control of the elevator and reverse the automatic closing operation of the door.

2.13.4 Closing Limitations for Power Door Operated Horizontally Sliding Hoistway Doors and Horizontally Sliding Car Doors or Gates

Since passenger elevator power door systems consist of moving masses, it is necessary to regulate the effects of the door system on passengers who may come into contact with moving or stationary doors. These effects relate to impact and/or direct forces and are expressed in terms of conventional physical parameters of kinetic energy and force, respectively. Kinetic energy, expressed in 2.13.4.2.1, quantifies the dynamic state of the door system by relating all the rigidly connected moving masses in the door system and their speeds. The direct force, expressed in 2.13.4.2.3 quantifies the static state of the stalled door condition. These two parameters, i.e., kinetic energy and force, are design requirements.

2.13.4.2 Closing Mechanism

2.13.4.2.1 The kinetic energy of completed door systems includes the applicable translational effects of the doors, linkages, hangers, vanes, interlocks, etc., as well as the rotational effects of the door operator motor and transmission.

Requirement 2.13.4 discusses two functions of door closing: kinetic energy (2.13.4.2.1) and force limitation (2.13.4.2.3) for power door operators on horizontally sliding hoistway doors and horizontally sliding car doors or gates. The kinetic energy of the hoistway and car doors and all parts originally connected thereto, computed for the average closing speed, within the Code zone distance (2.13.4.2.2) must not exceed 10 J or 7.37 ft-lbf where a reopening device for the power-operated car door or gate is used, and must not exceed 3.5 J or 2.5 ft-lbf where a reopening device is not used.

When hoistway pressurization is utilized in a building during a fire, it may be necessary to increase the forces

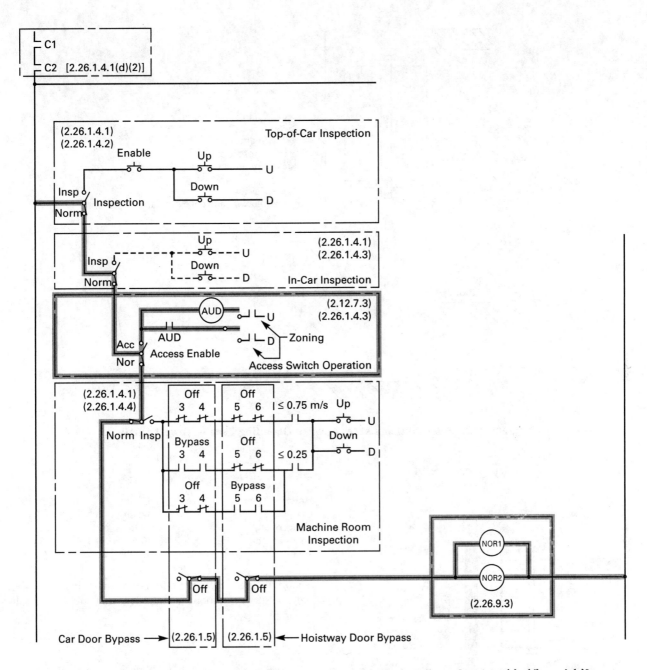

Diagram 2.12.7.3 Hoistway Access Switch Operation and Single Failures [2.26.9.3(c), (d), and (e)]
(Courtesy Otis Elevator Co./Ralph Droste)

only as much as required to overcome the forces caused by pressurization.

Kinetic energy is determined by using the standard formula:

(Imperial Units)

$$\text{Kinetic energy} = \tfrac{1}{2}mv^2$$

where m is the mass of the door system and v is the average door closing speed as specified in 2.13.4.2.2(a) or (b). The maximum average kinetic energy value of

7.37 ft-lbf was based on determining the kinetic energy developed by a typical door system with an average closing speed of 0.305 m/s (1.0 ft/s). The door system used in this analysis was 2.13 m (7 ft) high with a panel width of 1.07 m (42 in.) plus a small overlap at the jamb side yielding an area of approximately 2.32 m² (25 ft²). It is further assumed that the hoistway door weighs 34.2 kg/m² (7 lb/ft²), the car door weighs 23.6 kg/m² (5 lb/ ft²), and that there are two hangers weighing 11.3 kg (25 lb) each. The weight of the door system can then be computed as follows:

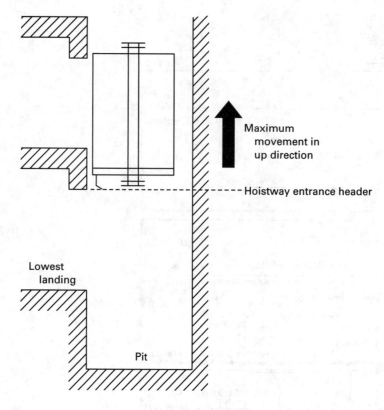

Diagram 2.12.7.3.6 Hoistway Access Switch Operating Zone — Lowest Landing

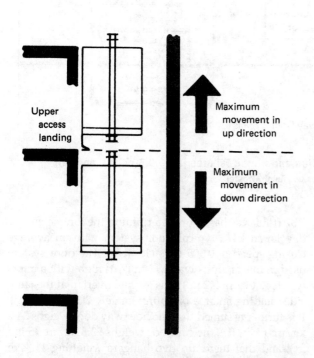

Diagram 2.12.7.3.7 Hoistway Access Switch Operating Zone — Upper Landing

Hoistway door	175 lb	(79.4 kg)
Car door	175 lb	(56.7 kg)
Hangers	50 lb	(22.7 kg)
Vanes and hardware	10 lb	(4.5 kg)
Total	360 lb	(163.3 kg)

Converting this weight to a mass and inserting the values into the kinetic energy formula yields the following:

(Imperial Units)

$$\text{Kinetic energy} = \frac{1}{2}\left(\frac{360}{32.2}\right)(1.0)^2 = 5.59 \text{ ft-lbf}$$

(SI Units)

$$\text{Kinetic energy} = \frac{1}{2}(163.3)(0.305)^2 = 7.6 \text{ J}$$

The rotating kinetic energy of the door operator (motor, pulley, etc.) was not included in this analysis and 25% seems to have been added to round the value up to 10 J (7.37 ft-lbf).

Kinetic energy is based on an average closing speed. Field measurement of kinetic energy by any mechanical device will normally result in an instantaneous reading. Instantaneous kinetic energy was introduced in ASME A17.1–2000, 2.13.4.2.1(b)(1). Its numerical value is derived from sinusoidal velocity dictation where the maximum value is equal to 1.57^2 times the average, where the average results in a kinetic energy equal to the historical value of 7 ft-lbf. Therefore, the instantaneous value is equal to $7.0 \ (1.57)^2 = 17.25$ ft-lbf × 1.356 = 23.4 J, rounded to 23 J. The same rationale is used to quantify the maximum instantaneous value of the reduced kinetic energy.

Attempting to improve floor-to-floor time and its influence on elevator capacity may lead to the temptation to decrease door time resulting in increase in kinetic energy, which may create an unsafe condition for elevator users. The reason is that every second used for the door closing usually means decrease in elevator capacity of 5% or more. This might be an incentive for compromising safety. Door operators that comply with requirements in present standards (based on average speed) could still create extremely high kinetic energy peaks as a result of poor adjustment.

It should be noted that calculated kinetic energy often gives considerably greater values than measured kinetic energy, because of the elasticity of door hangers and couplers. Therefore, reliance on calculated rather than physically analyzed values of kinetic energy often provides conservative results.

"High-quality" doors require as rigid as possible links between the door operator, car and landing doors, that means as little elasticity as possible, which in turn will cause the increase in the door force.

2.13.4.2.2 Door Travel in Code Zone Distance. The Code zone when taken in conjunction with 2.13.4.2.4 provides the means of field verification of compliance with the kinetic energy requirement in 2.13.4.2.1(b)(1).

2.13.4.2.3 Door Force. Any spring gauge readings taken when stopping an already moving door system will result in higher forces due to the dynamics of the system.

A reasonably accurate reading on the spring gauge is accomplished when the forward thrust of the door, i.e., door force, is balanced by the restoring force of the spring gauge. This is accomplished by pushing the spring gauge against the stopped door, removing the stop so that the door is held stationary by the spring gauge, and then slowly backing off on the spring gauge until the point the door just starts to move forward. At this point of impending motion, the door and spring forces are in equilibrium and the spring force can be read. Any motion will result in an incorrect reading.

The 133 N (30 lbf) closing force was arrived at based on a survey of the car doors on the New York City subway system. The committee determined that subway car doors had a closing force of 178 N (40 lbf). They concluded that this was too high and settled on a closing force of 133 N (30 lbf) for elevator door applications. The 133 N (30 lbf) force applies only to a door at rest between one-third and two-thirds of its travel, as

described in this requirement and Item 1.8 of the ANSI/ASME A17.2.1 and A17.2.2 Inspectors' Manual. There are no force requirements for the doors in other positions.

2.13.5 Reopening Device for Power-Operated Car Doors or Gates

A door with a reopening device may have its average kinetic energy reduced from 10 J or 7.37 ft-lbf to 3.5 J or 2½ ft-lbf when the reopening device is rendered inoperative.

The reopening device may be a mechanical shoe, photoelectrical device, electronic detector, etc. This Code is not a design handbook and, therefore, no one type of device is specified. The devices must be effective for substantially the full vertical opening of the door (2.13.5.1). The intent of this requirement is that any obstruction of a size normally related to body dimensions or those appliances relating to impaired mobility in the path of the closing car door(s) will cause the reopening device to function to stop and reopen the car door(s). The doors are not required to fully reopen only a distance sufficient to allow passenger transfer. Many photoelectrical devices will not meet this requirement. The reopening device is required to stop and reopen the car door or gate and the hoistway door in the event that the car door or gate is obstructed while closing. Full reopening of the door is not required.

There are many devices currently on the market designed to protect passengers and freight from being struck by elevator doors. These fall into a number of categories. Included in the electromechanical systems category would be all devices, which require physical contact between the "protective edge" and person or object being detected, i.e., they are mechanically activated though the trigger signal may be electrical. These range from protective edge in which a resilient or solid bumper, traveling ahead of the door, must be physically pushed in the opposite direction to that of the door motion causing the making/breaking of relay contacts and thus door stopping and reversal, to pressure sensitive membranes which give an electrical signal when the surface has pressure applied to it.

Light is a form of electromagnetic radiation whose characteristics vary with changes in wavelength and/or frequency. Only a small portion of electromagnetic radiation is visible to the human eye (visible light). Infrared light has a frequency too low to be sensitive to the naked eye. The main advantages of using infrared light are its greater penetrating ability with regard to smoke and dust, and the greater efficiency (hence range) of infrared transmitting and receiving diodes. Originally an electric eye comprised of a separate visible light transmitter and corresponding receiver, which would be linked by a direct beam of light. This light beam would span the elevator door opening and upon interruption cause a signal to be sent to the controller, thus reopening the door and holding it open until the beam obstruction was removed or the device was overridden by a timer.

This basic format has greatly changed over the years with new device configurations and different types of light being utilized. Without beam modulation two lenses must be used to finely focus light, otherwise one transmitter could feed both receivers. However, with finely focused beams alignment becomes critical. With modulation the transmitters can operate with a very wide angle of dispersion, as the upper receiver will only recognize a signal from the upper transmitter while the lower receiver/transmitter pair behave the same. Some systems have a feature that allows the beams to operate independently. If one beam malfunctions, it can be disabled and the elevator allowed to run with only one beam. Unfortunately, this encourages people to postpone the repair and allow the elevator to run without strict adherence to handicapped codes. This type does not meet the A17.1 Code requirements for door closing protection. The two beam arrangement properly placed only satisfies the requirement of the accessibility standards.

Another type of device is the retro reflective. This device includes separate pairs of transmitters and receivers arranged so they are mounted side by side. The beam is made by the transmitter sending light across the door opening where it is reflected back to the receiver by a reflective pad, usually mounted on the strike post. For reflective eyes to function properly any obstruction that is to be detected, cannot be allowed to reflect the beam back to the receiver or it will appear invisible. Hospital beds, stretchers, wheelchairs, etc., are all generally of a polished metal construction and thus highly reflective.

A third type of device is the single beam twin path. With these systems there is but a single transmitter and receiver. These are made to give the appearance of being two beams by passing the transmitted light through a periscope from which it emerges in the direction of the receiver.

One type of optical safety edge is a multiple electric eye system, where instead of two eyes there are twenty-four or more beams extending from approximately 25 mm (1 in.) above ground level to a height of approximately 1 575 mm (62 in.). This is achieved by having separate transmitter/receiver pairs and having each pair examined in sequence to check that the beam path is intact. It is only necessary for one beam to be blocked for a trigger signal to be sent to the controller to reopen the doors. A problem peculiar to this type of device, that the user should be aware of, is that there may be blind spots due to the approximate 50 mm (2 in.) to 63.6 mm (2½ in.) gap between the beams.

In multiple electric eyes with cross scanning in a system (instead of a simple set of parallel horizontal beams), a crisscross matrix is set up by having each receiver scan a multitude of transmitters. The device may have an approximate total of forty beams to hundreds of beams, but it may still suffer from blind spots and some early devices required absolute critical alignment. If the object to be detected is moving through the doorway, the blind spots should not be an issue if all the beams are functioning since any potential blind spots are also moving at right angles to the object.

Reflective infrared edges consist of a multiplicity of adjacent infrared transmitting and receiving diodes. These are arranged in such a way to allow the transmitted beams to be reflected off an object in the door path and such reflection to be detected by the receivers, thus causing a trigger signal to the controller to halt and reverse the door. In theory, this looks very effective but the user should be aware of some potential drawbacks. The effective range of detection depends greatly upon the obstruction's ability to reflect infrared light. Dark, dull surfaces tend to absorb such light and can sometimes appear virtually invisible to such a device. We are now seeing three-dimensional (3D) reflective infrared edges of two types. One has the infrared reflected back to the originating edge. The other has the infrared reflected across to the opposite edge. The 3D is more of an advance warning to the door to reopen, and augments, not replaces the multiple beams.

Recently noncontact infrared sensors are being used on power-operated vertical slide doors and gates. The principle is the same as described above; however, the sensor monitors the full width of the door or gate below its leading edge. The area of protection moves with the door or gate and thus is less prone to accidental damage from freight handlers. See Diagram 2.13.5.3.

According to the Firefighters' Service requirements, [2.27.3.1.6(e)] "Door reopening devices for power-operated doors which are sensitive to smoke or flame shall be rendered inoperative without delay." No optical device can claim to be insensitive to smoke/flame. With any optical protective edge currently on the market, there would be no protective door-reopening device operative under conditions of Firefighters' Service and hence the reduced kinetic energy requirement in the Code is required for door closing.

The electrostatic sensitive edge is a familiar device in the elevator industry and has appeared in many different forms since its origins in the 1950s. All such sensors rely on changes of induced high frequency signal caused by the proximity of a conductive object (i.e., the human body) to an antenna mounted on the door edge. Most early devices used the "balance bridge," a principle in which a pair of equal length antennas formed two arms of a whetstone bridge and was balanced by capacitors with the electronics. A 50 kHz oscillator provided the signal of about 50 V peak to peak, which was applied between the electronics and the car frame. Such circuits were of a simple design and relied upon DC coupling,

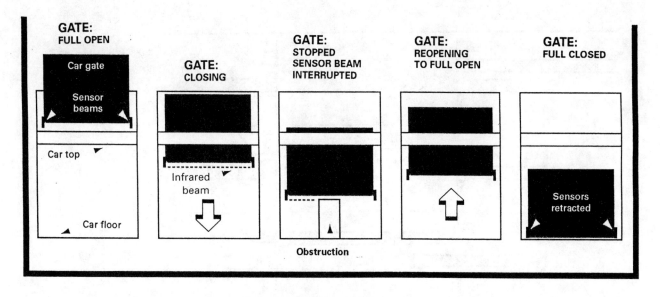

Diagram 2.13.5.3 Noncontact Door Reopening Device for Vertical Slide Door/Gate
(Courtesy The Peelle Co.)

which gave a poor temperature/humidity stability combined with a short practical range. More recently, the advent of low cost complex integrated circuits has led to AC coupled multiple antenna systems, which use far more stable techniques to achieve essentially drift-free operation. The designs extend the performance of the electrostatic detector from the early days of 25 mm (1 in.) range to stable ranges of the order 305 mm (12 in.) in free space. The concerns regarding electrostatic protective edges are the sensitivity to environment (wobbly doors, etc.) and the lack of sensitivity to nonconductive objects such as plastic containers. Recently, however, both of these problems have been mitigated by incorporating two infrared electric eyes into the edge at the appropriate levels to comply with accessibility codes.

The intent of this requirement is that any obstruction of a size normally related to body dimensions of those appliances relating to impaired mobility in the path of the closing car door(s) will cause the reopening device to function to stop and reopen the car door(s). A photo eye, which only protects a limited area, would not meet this requirement since an obstruction in an unprotected area would not cause the doors to function as required.

The effectiveness of a door protection device using a system of infrared beams is wholly dependent upon the placement and number of infrared emitters and receivers within a given door opening size.

If the resulting field of detection sensed the presence of an obstruction within the opening and functioned to reopen accordingly, then the reopening device would comply with this requirement.

Compliance of the infrared system with 2.13.5 must therefore be evaluated in terms of its effectiveness in satisfying the intent of the requirements for protection of passengers within the opening.

The reversing device is often selected based on the door mass and speed. If the kinetic energy on the system is 3.5 J or $2\frac{1}{2}$ ft-lbf or less, no reversing device is required. If the kinetic energy is more than 3.5 J or $2\frac{1}{2}$ ft-lbf, but 10 J or 7.37 ft-lbf or less, a reopening device is necessary. If a reopening device is provided, and the kinetic energy is reduced during the closing sequence to 3.5 J or $2\frac{1}{2}$ ft-lbf or less, the reopening device can be rendered inoperable.

The Americans with Disabilities Act Accessibility Guidelines (ADAAG) and ICC/ANSI A117.1–1992 require doors closed by automatic means to be provided with a door reopening device, which will function to stop and reopen a car door and adjacent hoistway door in case the car door is obstructed while closing. The reopening device is required to also be capable of sensing an object or person in the path of the closing door without requiring contact for activation at a nominal 127 mm (5 in.) and 3 277 mm (29 in.) above the floor. These requirements apply regardless of the kinetic energy on the closing door or gate.

2.13.6 Sequence Operation for Power-Operated Hoistway Doors With Car Doors or Gates

See Diagrams 2.13.6.2(a) and 2.13.6.2(b).

SECTION 2.14
CAR ENCLOSURES, CAR DOORS AND GATES, AND CAR ILLUMINATION

2.14.1 Passenger and Freight Enclosures, General

2.14.1.1 Enclosure Required. The car enclosure includes the walls as well as the car top. It would be unthinkable today to allow an elevator to operate without a car enclosure, but elevators were installed before

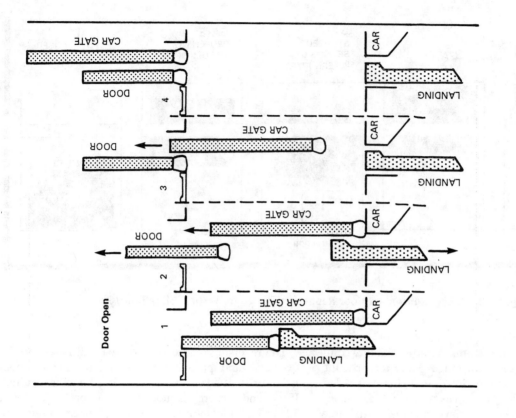

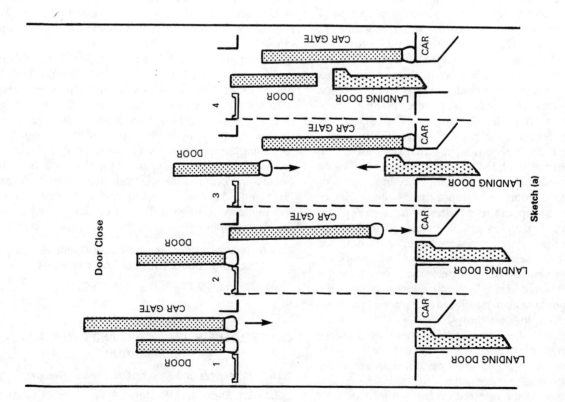

Diagram 2.13.6.2(a)   Sequence Operation
(Courtesy The Peelle Co.)

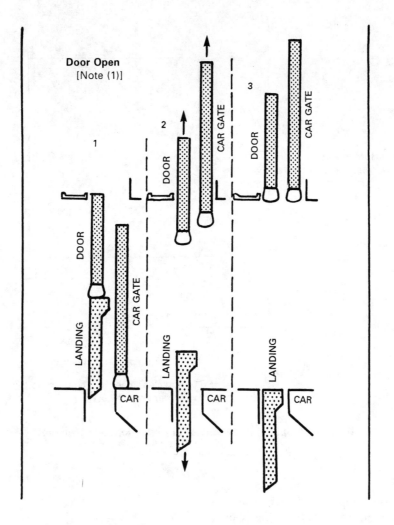

NOTE:
(1) Permitted only where sequence operation not required (2.13.6.1.1).

Diagram 2.13.6.2(b)   Simultaneous Operation
(Courtesy The Peelle Co.)

the first edition of this Code that did not have complete car enclosures. The car enclosure is intended to protect the passenger from falling objects and the possibility of extending a hand, etc., through an opening.

2.14.1.2 Securing of Enclosures. The car enclosure must be constructed in such a manner that it cannot be dismantled from within the car. Decorative wall panels that are removable from inside the car must be backed up by a car enclosure meeting the requirements of 2.14.1.2.2. Holes in the car enclosure for attachment of removable panels that can be removed from inside the car are acceptable as long as the holes conform to the requirements of 2.14.1.2.4. The requirements provide occupants of elevators the protection of not having to ride an elevator exposed to an open hoistway, counterweights, rail brackets, etc., when the decorative panels of a chassis-type car are removed for repair or refinishing. They

further prevent passengers from interfering with the travel of elevators by projecting objects through apertures where hanging decorative panels were attached and removed for repair or refinishing. Lighting deflectors and suspended ceiling should remain in place when the tests required by 8.10 and 8.11 are performed. If they remain in place during the test, it can be assumed that they conform to this requirement.

2.14.1.3 Strength and Deflection of Enclosure Walls. The strength requirements are based on limiting the allowable stress to the yield point of the material. See Diagram 2.14.1.3.

2.14.1.4 Number of Compartments in Passenger and Freight Elevator Cars. A multideck elevator is an elevator having two separate platforms and compartments, one above the other and supported within the same car

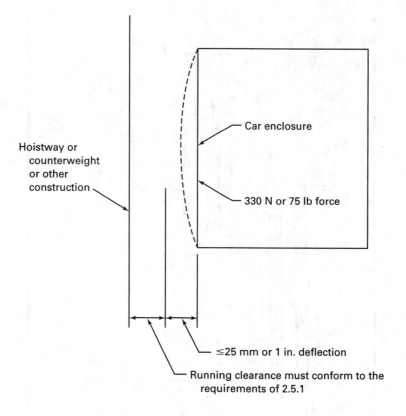

Diagram 2.14.1.3   Enclosure Wall Deflection Requirements

frame. A multideck elevator is often referred to as a double-deck elevator. Both upper and lower decks may be loaded and unloaded at the same time. Cars can be arranged to serve either odd or even floors or all floors, whichever operation is desired.

2.14.1.5 Top Emergency Exits. The top emergency exit is usually the preferred means of evacuating passengers from a stalled elevator when exit through the car door or gates and hoistway door is not possible. As such, the exit must be clear of all obstructions, i.e., fixed elevator equipment. See Diagram 2.14.1.5.1(b)(2). Most manufacturers locate the top emergency exit within the refuge space on top of the car enclosure (2.4.12). This is not required by the Code, but is the usual arrangement due to the limited space available. The exit cover latch must be openable from the top of the car only, unless required to comply with 8.4.4.1.1 seismic provisions. On many older installations, the exit cover was installed so that it opened from within the car only. A passenger in a stalled elevator should not be given the opportunity to exit the elevator without the assistance of trained elevator personnel. The exit cover should be latched in such a manner that it can be easily unlatched from the top of the car without the use of tools. Window sash latches and bolt latches are commonly used.

The exit cover is required to be attached by chain or hinges to the car top at all times. The purpose is to

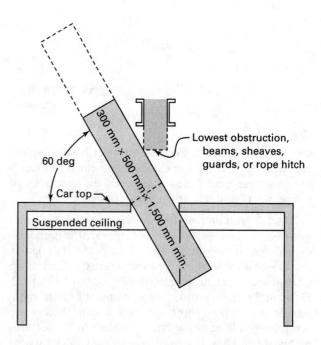

Parallelepiped Volume Orientations [2.14.1.5.1(b)(2)]

Diagram 2.14.1.5.1(b)(2)   Location of Top Emergency Exit

assure that the exit cover is not removed or cannot fall off the top of the car.

Starting with ASME A17.1–2000 a car top emergency exit electrical device (2.26.2.18) is required. The device will monitor the position of the top-emergency-exit and not allow operation of the elevator when the top-emergency-exit is in the open position, except as permitted by 8.4.4.1.2.

For additional information, refer to the Guide for Emergency Personnel, ASME A17.4.

2.14.1.5.2 Evacuation of passengers through the top-emergency-exit is not considered safe from elevators in partially enclosed hoistways. A top-emergency-exit is allowed but cannot be the means employed to evacuate passengers. A written evacuation plan is required by 8.6.10.4 for all elevators.

2.14.1.6 Car Enclosure Tops. This requirement assures that the car top will be able to carry the weight of maintenance and inspection personnel, or emergency personnel who might be on top of the car evacuating people from a stalled elevator. It also limits the deflection to assure no equipment damage due to ceiling deflection.

2.14.1.7 Railing and Equipment on Top of Cars. A standard railing (2.10.2) is required for protection of personnel working on top of completed cars whenever the specified horizontal clearance is exceeded. The vertical clearance above the top rail is 150 mm or 6 in. when the car has reached its maximum upward movement [2.4.6.2(c)].

2.14.1.8 Glass in Elevator Cars. Laminated glass when shattered will normally remain in place, thus assuring the safety of the passengers in the elevator. In an observation elevator, if shattered glass will not stay in place, a railing or framing is required that will guard the opening.

2.14.1.8.2 Requires all glass to meet the requirements for laminated glass in ANSI Z97.1-1984 or 16 CFR Part 1201, Sections 1201.1 and 1202.2 or CAN/CGSB-12.1. The Consumer Product Safety Commission Standard 16 CFR, Part 1201 is more stringent than ANSI Z97.1 is. See commentary on 2.11.7.2. Laminated glass is defined by 16 CFR as, "glazing material composed of two or more pieces of glass, each piece being either tempered glass, heat strengthened glass, annealed glass, or wired glass, bonded to an intervening layer or layers of resilient plastic material." The ANSI Z97.1 definition is similar.

Bonded glass, which is equivalent to laminated glass, is permitted by 2.14.1.8.2(c) and prohibits the use of film coatings, which are easily damaged rendering them ineffective (organic-coated glass, etc.). It provides performance language requirements to bond glass fragments and retain broken glass.

Glass is required to be tested in accordance with the applicable regulations. Testing requirement for laminated glass applies to both laminated and bonded glass. It recognizes alternative technology for glass products that are equivalent to laminated glass but do not meet the definition for laminated glass. Field marking will provide inspectors with a means of identifying glass that complies with these requirements.

The ANSI Z97.1 standard states in part:

> "After having successfully passed the appropriate tests in this standard, like products and materials produced in the same manner as specimens submitted per test shall be legibly and permanently marked in one corner with the words 'American National Standard Z97.1-1984' and shall be marked also with the manufacturer's distinctive mark or designation."

It is the intent that this marking be on each separate piece of glass. Certain specific types of glazing material have additional marking requirements.

All glass used in the elevator must be installed in such a manner that it does not become dislodged during the test specified in 8.10 and 8.11.

2.14.1.9 Equipment Inside Cars. An ashtray is one type of equipment or apparatus, which is prohibited by this requirement from being installed inside elevator cars. A fixed bench is permitted. The only equipment or apparatus permitted to be installed in the car, which is not used in conjunction with the use of the elevator, is that equipment which is listed such as: conveyor, tracks, lift hooks, and suspension support beams. Lighting, heating, ventilation, and air-conditioning equipment are also permitted.

Depending on the design and construction of the equipment installed in the car, it may be necessary to evaluate the total construction of the item and part of the car wall to which it is fastened, in accordance with the requirements of 2.14.2.1.1. Consultation with the testing laboratory will be necessary to determine the specified testing criteria.

Picture frames, graphic display boards, etc., are subject to the fire test requirements in 2.14.2.1. In their end use configuration, they must have a flame spread rating of 0 to 75 and smoke contribution of 0 to 450. The projection from the wall is limited to reduce the possibility of injury to the occupants of the car.

2.14.1.10 Side Emergency Exits. The safest means for evacuating passengers is through the hoistway entrance, if that is not possible then through the top emergency exit. The requirements for side emergency exits, when provided, apply to both passenger and freight elevators. Side emergency exits have not been required since ASME A17.1–1991. One of the conditions under which side emergency exits are permitted is that the distance between adjacent car platforms does not exceed 750 mm or 30 in. This dimension was selected with the intent of limiting the installation of side emergency exits where there is an excessive distance between elevator cars. A distance in excess of this is impractical to span for evacuation purposes. Most passengers would refuse to cross over to the rescue car on the evacuation bridge, and transporting an evacuation bridge the length necessary to accommodate a span greater than 750 mm or 30 in. in the rescue car would probably be impossible.

If the door opened outwardly and was accidentally opened when the car was moving, it could strike an obstruction in the hoistway. The door size requirements ensure that it is usable. The side emergency exit door must be openable from outside the car in order to provide access for emergency personnel into the stalled elevator. The door must be openable from within the car by a specially shaped removable key to permit evacuation personnel to open the door on the rescue car. The side emergency door must be provided with a car door electric contact to ensure that the elevator will not run when the side emergency exit door is opened.

For detailed information on the evacuation of passengers from a stalled elevator, refer to ASME A17.4, Guide for Emergency Personnel.

It should be noted that many jurisdictions prohibit side emergency exits.

2.14.2 Passenger-Car Enclosures

2.14.2.1 Material for Car Enclosures, Enclosure Linings, and Floor Coverings. The requirements for the materials used in the 1984 and earlier editions of the Code left a lot to the imagination. Those requirements were intended to protect a car from a fire to which, only the exterior of a car was exposed. Requirements were developed for the 1955 edition of A17.1, and were correct for the construction and conditions that were prevalent at that time. They covered conditions that included oiled rails for car guide shoes and leaking geared machines, the oil from which could go down the hoistway landing on the top of the car, which could lead to a fire in the hoistway.

Construction of elevator cars has changed over the years, resulting in a decreased fire exposure to the exterior of the car. With the introduction of new materials in the late 1970s, some argued that there is a fire exposure to the interior. Regardless, fire protection to the interior of passenger car enclosures was not addressed in the earlier editions of the Code. A study of this subject had begun in the early 1980s when it became apparent that the building codes were poised to act. The results of that study were revisions that first appeared in ASME A17.1a–1985.

There is no requirement for labeling of any of the material used in passenger car enclosures, in keeping with standard construction practice. If an enforcing authority, building owner, etc., questions the material used, all they need to do is request a copy of the test report from the supplier. The test report required by all of the test procedures discussed below will give sufficient information to evaluate the material supplied to assure that it complies with the requirements of the Code.

2.14.2.1.1 This requirement addresses the materials used for car construction. It requires all material in its end use configuration, except metal or glass, to be subjected to a Steiner tunnel test specified in ASTM E 84, NFPA 252, and UL 723 or CAN/ULC S102.2. In the test, a 508 mm (20 in.) by 7.6 m (25 ft) specimen is placed on a ledge in the top of the Steiner tunnel furnace in a face-down position, leaving an exposed area 0.45 m (17.5 in.) by 7.6 m (25 ft). A double-jet gas burner located 305 mm (1 ft) from the air intake of the tunnel is adjusted to provide approximately 5,000 BTU/min for a test period of 10 min. This pulls the gas flame downstream for approximately 1.4 m (4.5 ft) at the beginning of the test, leaving approximately 5.9 m (19.5 ft) of specimen for the flame to advance during the test. Flame travel is observed through sealed windows, and forms the basis for the flame spread rating. Furnace temperature and smoke density are also recorded, and these figures are the basis for calculating fuel contribution and smoke development. Ratings are based on an arbitrarily assigned 0 rating for asbestos-cement board and 100 rating for red oak flooring. It should be noted that there is not necessarily a relationship between flame spread, smoke development, and fuel contribution. The A17.1 Code specifies that acceptable passenger car enclosure materials have a flame spread rating of 0 to 75 and smoke development of 0 to 450. Fuel contribution is not specified and is normally not a requirement in any modern code. In general, material is classified by the building codes based on a flame spread rating as:

(a) 0 to 25 Class A or 1

(b) 26 to 75 Class B or 2

(c) 76 to 200 Class C or 3

You will find codes that mandate Class A or 1 material in certain occupancies. An example is the NBCC requirements for the designed firefighters' elevator (2.14.2.1.2).

It is the intent of 2.14.2.1 that metal and its painted or lacquered finish does not have to meet the requirements of 2.14.2.1.1 through 2.14.2.1.5. The application of a paint or lacquer finish to metal does not significantly add to the fire exposure of the car enclosure material.

A key requirement is that the material must be tested in its end use configuration. That statement requires car enclosure walls of sandwich construction to be tested as one complete sample. It is not permitted to take the individual component of the sandwich construction and test them individually and average all the results to get the flame spread and smoke development rating. When these products are combined and tested as one, the results will most likely be entirely different. The materials are also required to be tested on both the side exposed to the car interior and the side exposed to the hoistway.

In reviewing the requirement for testing of the enclosure material in its end use configuration with manufacturers who have tested walls, it has been shown that wall veneers, their finishes, and adhesives have a direct bearing on the results of the test. Testing has shown that minor changes of any one component, even when that component was acceptable when tested individually, have resulted in unacceptable flame spread and smoke development ratings for the car enclosure in their end use configuration.

2.14.2.1.3 In the past, napped, tufted, woven, looped (i.e., carpet), and similar material used on car enclosure walls was prohibited. This material is now permitted as long as it passes the vertical burn test as called for in 8.3.7 or in Canada the NBCC or National Fire Code of Canada. The vertical burn test (8.3.7) is based on the applicable portions of an FAA test. A specimen is placed in a draft-free cabinet 19 mm (0.75 in.) above a Bunsen burner with a flame length of 38 mm (1.5 in.) applied for 12 s, then removed. After the burner is removed, flame time, burn length and flaming, and time of the drippings is recorded. The material is acceptable if:

(a) average burn length does not exceed 203 mm (8 in.);

(b) average flame time after the removal of the burner does not exceed 15 s; and

(c) the drippings do not flame for more than 5 s.

The fabric or carpeting must meet the requirements of 2.14.2.1.3. The substrate must meet the requirements of 2.24.2.1.1.

Padded protective lining must be subjected to the Steiner tunnel test or vertical burn test and conform to the acceptable criteria for said test. Lining must also clear the floor by a minimum of 4 in. (102 mm), so they do not obstruct car vents.

2.14.2.1.5 The last subject covered is floor covering, underlayment, and its adhesive. This material is subjected to the critical radiant flux test using a radiant heat energy source, ASTM E 648 or NFPA 253. The test chamber consists of an air-gas fueled radiant heat energy panel inclined at 30 deg and directed at a horizontally mounted floor covering system specimen approximately 203 mm (8 in.) by 1 016 mm (40 in.). The floor-covering specimen should duplicate insofar as possible an actual field installation. Thermal energy is supplied by the radiant heat panel. A pilot light at one end provides the ignition source. Flame travel is then observed through a window in one side of the test chamber. The distance burned to flame-out is converted to watts per square centimeter from a flux profile graph and recorded as critical radiant flux w/cm^2. Unlike the other test addressed above, the higher the w/cm^2, the more resistant the floor covering is to the propagation of flame.

2.14.2.2 Openings Prohibited. This requirement prohibits openings in the sides of car enclosures to prevent a hazardous exposure to the passengers riding in the car. Panels that are designed to form an integral part of the enclosure and that are secured in place by means that are not accessible from inside the car are permitted.

2.14.2.3 Ventilation. Natural ventilation by means of vent openings is required on all passenger elevator cars. The openings are required to be equally divided between the top and bottom to facilitate natural airflow by convection. It is further stipulated that vent openings shall not be located in that portion of the side enclosure between a height of 300 mm or 12 in. and 1 825 mm or 72 in. above the car floor. It does not exclude the top of the car as a location for vent openings. The clearance between the car door panel and frame qualifies as a vent opening.

2.14.2.3.2 Observation elevators exposed to direct sunlight must be provided with forced (mechanical) ventilation. In observation elevators with glass walls exposed to direct sunlight, if a passenger was trapped in a stalled elevator and forced ventilation was not provided, the passenger could become overheated.

2.14.2.3.3 The cfm of the fan as installed must be equal to the volume inside the car. The auxiliary power to operate this ventilation for 1 h is required to provide time for rescue during a power outage.

2.14.2.5 Vision Panels. The restriction of the size of the vision panel is to reduce the hazard if the glass is broken. The term "total area" refers to the sum total of all vision panels. On older installations where larger vision panels were provided, there is a history of decapitations and amputations where panels were broken. Laminated glass panels and wire glass panels when broken will normally stay in one piece. On power-operated car doors, the glass must be located substantially flush with the inside surface of the door in order to reduce a pinching hazard when the door is opening or closing.

2.14.2.6 Access Panels. This new requirement was introduced in ASME A17.1–2000 and is intended to provide safe access for cleaning glass in the hoistway. Stringent requirements for their design and installation are specified to prevent misuse. A glass cleaning procedure is also required by 8.6.10.3.

2.14.3 Freight-Car Enclosure

2.14.3.1 Enclosure Material. Freight cars must be enclosed on all sides except the sides used for entrance and exit. The car enclosure must be so constructed that removable portions cannot be dismantled from within the car. Walls of freight cars must be solid to a height of 1 825 mm or 72 in. above the floor. Above that point, the enclosure may be perforated as long as it rejects a ball 25 mm or 1 in. in diameter. The top of the car enclosure may also be perforated as long as it also rejects

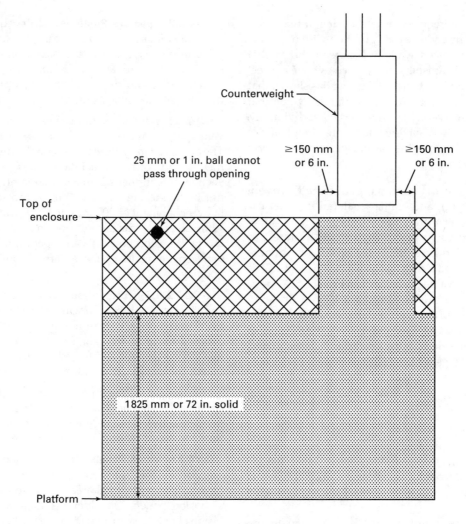

GENERAL NOTE: Does not apply to freight elevators permitted to carry passengers.

Diagram 2.14.3.1   Freight Car Enclosure Requirements

a ball of 25 mm or 1 in. in diameter. The car enclosure wall, in front of and to a point 150 mm or 6 in. on each side of the counterweight, must be solid. See Diagram 2.14.3.1.

If a freight elevator is permitted to carry passengers, perforations in the enclosure are not permitted.

2.14.3.2 Openings in Car Tops. Common freight car construction incorporates a top emergency exit opening that extends the full width of the car with a minimum depth of 400 mm or 16 in. Usually an emergency exit of this construction is located at the front of the car top enclosure.

2.14.4 Passenger and Freight Car Doors and Gates, General Requirements

2.14.4.1 Car-Door and Gate Electric Contacts. Car door or gate electric contacts must be so located that they are not readily accessible from inside the car. They must be

opened positively by a lever or other device attached to and operated by the door or gate. They must be maintained in the open position by the action of gravity or by a restrained compression spring, or by both, or by positive mechanical means.

2.14.4.2 Door and Gate Electric Contacts and Door Interlocks

2.14.4.2.2   A car-door interlock is the same as a hoistway door interlock. It is required to be provided on car doors when the clearances in 2.5.1.5 are exceeded. Another application is an observation elevator in a partially enclosed hoistway with front and rear entrances.

2.14.4.2.3   The car door or gate electric contact can be located within the power door operator enclosure. The contacts within the enclosure of the power door operator must be connected by a lever or other means

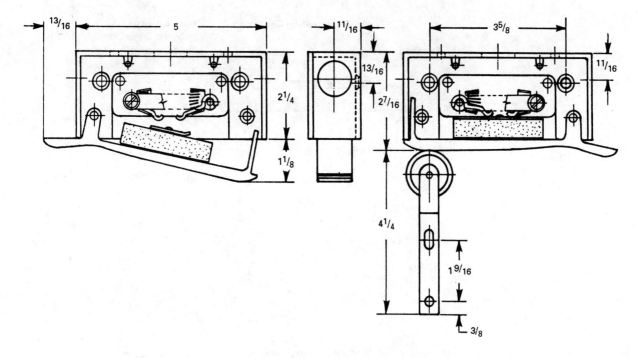

Diagram 2.14.4.2.3 Typical Car Door or Gate Electric Contact
(Courtesy G.A.L. Manufacturing Corp.)

to the door to be positively opened in order to conform to the requirements of this requirement.

Switches that depend solely on "snap action" to open their contacts do not meet the requirements of this requirement. See Diagram 2.14.4.2.3.

2.14.4.2.5 See commentary on 2.12.4.

2.14.4.3 Type and Material for Doors. See commentary on 2.11.1 for description and diagrams of horizontally and vertically sliding type doors. See also Handbook commentary on 2.14.2.1 and 2.14.3.1 for door material requirements.

2.14.4.4 Type of Gates. Horizontally sliding collapsible-type gates are permitted only on freight elevators designed for Type A loading (2.14.6.1).

2.14.4.5 Location. The distance between the hoistway doors and the car doors, or gates on automatic or continuous-pressure elevators, is restricted to eliminate the possibility of a person being caught between the car and hoistway doors or gates. See Diagrams 2.14.4.5(a) and 2.14.4.5(b).

On many older elevators installed before these requirements were incorporated into the Code, the bottom of the hoistway door is fitted on the hoistway side with a filler that is beveled so a person cannot stand on the hoistway sill when the door is closed. See ASME A17.3, Fig. A3 for door filler construction details.

2.14.4.7 Vertically Sliding Doors and Gates. Passenger elevators must have doors not gates (2.15.5.2).

2.14.4.8 Weights for Closing or Balancing Doors or Gates. If the weights are not guided or restrained from coming out of their runway, they could snag in the hoistway when the car was operated. If the weights were inside the car enclosure, there could be a hazard to the occupants of the elevator. The weights must be restrained if the suspension member fails to stop them from a free-fall within the hoistway.

2.14.4.10 Power-Operated and Power-Opened or Closed Doors or Gates. See Handbook commentary on 2.13 and the requirements within that Section.

2.14.4.11 Closed Position of Car Doors or Gates. See Diagram 2.14.4.11.

2.14.5 Passenger Car Doors and Gates

2.14.5.1 Number of Entrances Permitted. The more entrances, the greater the risk, as passengers do not know which door will open next. If they lean against a door, it might open, resulting in passenger injury.

2.14.5.2 Type Required. As a door is solid and a gate is perforated. A door offers less risk to the passengers.

2.14.5.3 Vertically Sliding Doors or Gates. When a slide-up-to-open door or gate faces other types of hoistway doors, even horizontally sliding doors, it is

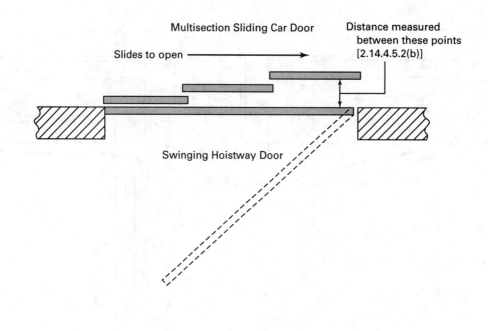

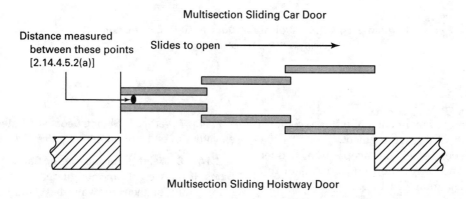

Diagram 2.14.4.5(a)   Measurement of Distance Between Car and Hoistway Doors

not necessary to have the car door or gate be power operated.

2.14.5.6 Door Panels

2.14.5.6.2 Indentations and moldings have a tendency to snag on clothing and could pinch a finger or hand during door operation.

2.14.5.7 Manual Opening of Car Doors or Gates. When a passenger elevator is within the unlocking zone (see Diagram 2.12.5) up to 450 mm or 18 in. above or below the landing, a person may be able to open the car door or gate and its related hoistway door from within the car. This requirement may be modified by a locked out of service entrance. See 2.11.6 for where a locked out of service entrance is permitted.

When the elevator is outside the unlocking zone, it is unsafe for a passenger to try to exit through the elevator entrance unassisted. When a car is above the unlocking zone, the platform guard (apron) may not be long enough to close off the hoistway opening below the car. If this condition exists, a person exiting the elevator is exposed to the open hoistway. In fact, there are many reports of fatalities due to this condition. Unlocking devices that rely on power to either lock or unlock the car door do not comply with 2.14.5.7, as they will eventually fail to comply with the Code when power is not available. The Code requirement, unlike auxiliary car lighting, emergency alarms, etc., does not permit the device to fail after a specified time limit.

2.14.5.8 Glass in Car Doors. This requirement makes a distinction between vision panels and glass doors. Laminated glass is required to maintain the integrity of the opening protection if the glass is broken. The minimum area requirement for the glass door ensures that these requirements are not used to install oversized vision panels. However, sufficient nonglass area may be needed for the mounting of interlocks, safety edges, etc. The Consumer Product Safety Commission (CPSC)

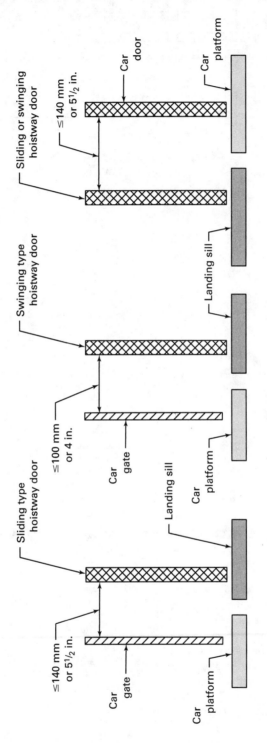

Diagram 2.14.4.5(b) Distance Between Hoistway Doors and Car Doors and/or Gates
(Courtesy Zack McCain, Jr., P.E.)

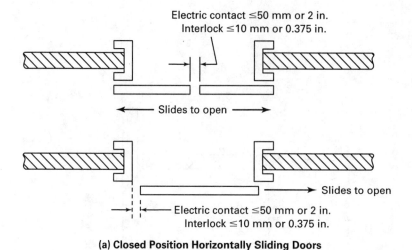

(a) Closed Position Horizontally Sliding Doors

(b) Closed Position Vertically Sliding Doors

Diagram 2.14.4.11   Closed Position of Car Doors or Gates

standard would apply, as the glazing would be classified by the building codes and the CPSC as being in a location subject to human impact loads.

Protection is required on glass door edges as door panels tend to be struck by carts, etc., and if glass was not protected, the edges might chip and become hazardous to people passing through the doorway. The thickness of the glass will be dictated by the structural requirements (2.14.4.6).

2.14.6 Special Requirements for Freight Elevator Car Doors and Gates

2.14.6.1 Type of Gates. This requirement specifies what type gate is permitted per class of freight loading. The requirements do not apply to doors.

2.14.6.2 Vertically Sliding Doors and Gates

2.14.6.2.3 If a reopening device is provided, and someone deliberately pushes his toe or foot against the reopening device and manages to bend it, he would have to extend the toe several inches beyond the car door before contacting the hoistway wall. See Diagram 2.14.6.2.3 for illustrations of typical car door/gate arrangements.

2.14.6.3 Collapsible-Type Gates. See Diagram 2.14.6.3 for illustrations of collapsible-type gates.

2.14.6.3.3 This assures that the gate in the extended position cannot be forced towards the hoistway and thus interfere with the running clearances of the car.

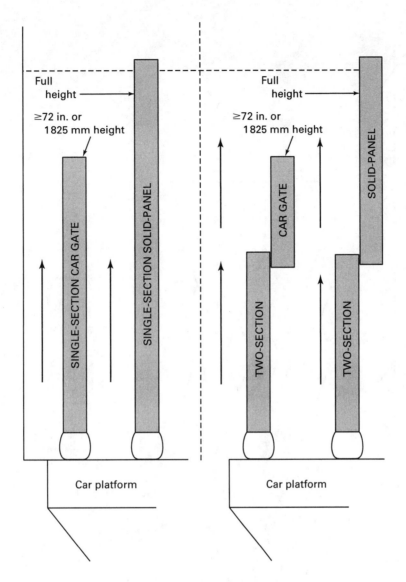

Diagram 2.14.6.2.3   Typical Vertically Sliding Car Door/Gate Arrangements
(Courtesy The Peelle Co.)

2.14.6.3.4 A restriction on the distance that a collapsible car gate can be power-opened (2.13.2.1.2) is specified to assure that a hand or finger will not be pinched in the opening of the car gate.

2.14.7 Illumination of Cars and Lighting Fixtures

2.14.7.1 Illumination and Outlets Required. Two lamps are required for both normal and emergency lighting to assure that if one burns out, the passengers in the car will not be placed in total darkness. Two fluorescent lamps powered from a single ballast do not conform to the intent of this requirement. The minimum illumination levels provide adequate lighting for safe ingress to and egress from the car. An auxiliary lighting power source assures that the passengers in an elevator will not be placed in total darkness if power to the normal lighting system fails.

2.14.7.1.3 The auxiliary lighting power source is required to be placed on each elevator. The building standby power generator is not a recognized alternative to this requirement as the auxiliary power source must be located on the car. The building standby power generator would normally not be activated unless the main building power was lost and would not respond to just the loss of the elevator lighting circuit. Moreover, the building standby power source would rely on the elevator traveling cable to maintain elevator lighting and would be useless if the traveling cable was the source of the lighting problem. Passengers in an elevator must be assured a source of reliable lighting.

2.14.7.1.4 Each elevator must also be provided with an electric light and convenience outlet fixture on the car top for the use of inspectors and maintenance

89

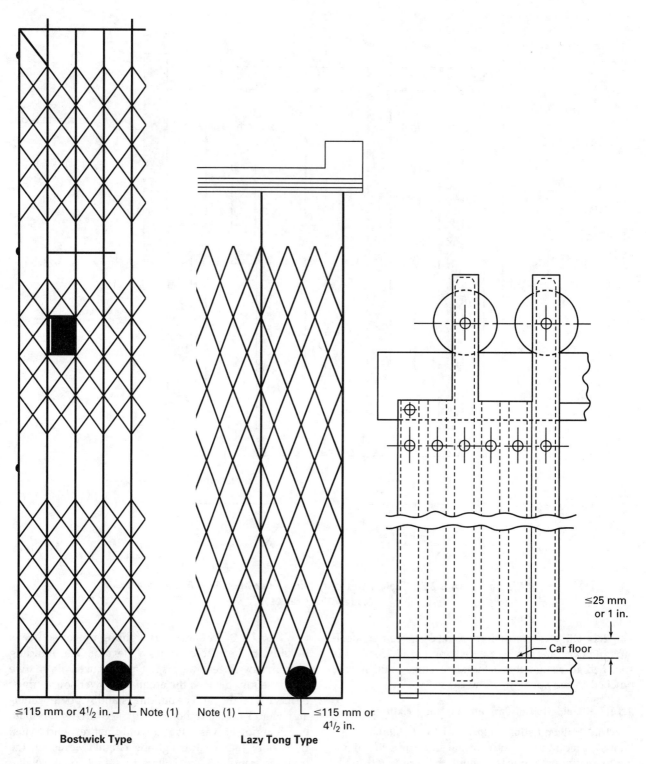

≤115 mm or 4¹/₂ in. ⌐ ∟ Note (1) Note (1) → ⌐ ∟ ≤115 mm or 4¹/₂ in.

≤25 mm or 1 in.

Car floor

Bostwick Type   **Lazy Tong Type**

NOTE:
(1) Every vertical member must be guided top and bottom.

Diagram 2.14.6.3   Collapsible-Type Gates

personnel. The NEC® requires that the light be guarded and the outlet be GFCI protected.

2.14.7.2 Light Control Switches

2.14.7.2.1 Interior car lighting may be controlled by a switch to facilitate maintenance of the lighting equipment. The switch must be either key operated or in a fixture with a locked cover to ensure that the lights can be turned off only by authorized personnel.

2.14.7.2.2 This requirement permits car lights to be automatically shut off. These requirements are intended to assure that before turning the lights off no one is in the car.

2.14.7.3 Car Lighting Devices. Glass used for lighting fixtures must be laminated glass meeting the requirements of 2.14.1.8. Glass shall be installed and guarded so as to provide adequate protection of passengers in case the glass panel breaks or is dislodged. The glass and the structure the glass is mounted in must withstand the required elevator test of 8.10 and 8.11 without damage.

If the light diffuser is other than metal or glass, it must be subjected to the test specified in 2.14.2.1.

2.14.7.4 Protection of Light Bulbs and Tubes. Light bulbs and tubes within the elevator car must be guarded to protect them from accidentally breaking. An exposed bulb or tube could be struck by an elevator passenger carrying an umbrella or similar object. If the bulb or tube was shattered, it could injure the elevator passenger. There are bulbs and tubes manufactured with an outside coating that resists breakage. This coating assures that the bulb or lamp, if broken, will not shatter.

The second part of this requirement addresses the structural soundness of the light fixture, its deflectors (egg crate, etc.), and its bulbs and tubes. There have been incidents of light bulbs or tubes, light fixtures, and their deflectors becoming dislodged during elevator car or counterweight safety application. This condition could cause serious injury to elevator passengers. The bulbs or tubes, fixtures, and deflectors should be in place during the periodic and acceptance safety and buffer tests. If they stay in place during the test, it can be assumed that the requirements have been satisfied.

SECTION 2.15
CAR FRAMES AND PLATFORMS

2.15.1 Car Frames Required

A car frame (sling) is the supporting frame to which the car platform, upper and lower sets of guide shoes, car safety, and hoisting ropes or hoisting rope sheaves, or the hydraulic elevator plunger or cylinder are attached. Two common designs of car frames that are used are the side post and the corner post. The side post construction has the guide rails located on the two

opposite sides and allows entrances on opposite sides of the car. When corner post construction is used, the guide rails are located at opposite corners of the platform. This type of construction allows for adjacent entrances.

Both side and corner post car frames can be either overslung or underslung. An overslung car frame is one to which the hoistway rope fastenings or hoisting rope sheaves are attached to the crosshead or top member of the car frame. An underslung car frame is one to which the hoistway rope fastenings or hoisting rope sheaves are attached at or below the car platform.

Another type of car frame that has been used is the sub-post, a car frame where all of the members are located below the car frame. See Diagrams 2.15.1(a) and 2.15.1(b).

2.15.2 Guiding Members

Elevator and counterweight guide shoes are either the sliding or roller type. Before the advent of the roller type, the swivel-type sliding shoe was used almost exclusively for passenger service elevators. In this type, the shoe is held in a bracket and arranged to turn so that it may adjust itself to bear evenly on the sides of the rail. In the direction against the face of the rail, the shoe is backed by a spring, held in the bracket, and adjustment is provided to obtain the desired clearance or float in this direction.

For freight elevators, the guide shoes are generally of the sliding type without provisions for swiveling or aligning themselves automatically to the sides of the rails; therefore, any misalignment due to inequalities of the car frame members must be corrected by means of shims. The shoes are provided with removable cast iron gibs, usually in one piece.

The roller guide shoe now supplants the swivel-type shoe on most passenger elevators. This guide allows the elevator car to ride smoothly even though the rails are not smooth and straight. The float is limited by the closeness of the safety jaws to each side of the rails. Each roller is generally mounted to a spring-loaded lever, which pivots about the roller guide stand.

Some of the principal advantages of the roller guide shoes are as follows.

(*a*) Oil and grease are eliminated from the hoistway, which cuts down the amount of maintenance, particularly in cleaning hoistway walls and pits, and also eliminates a serious fire hazard.

(*b*) Most of the knocking and scraping noises are eliminated.

(*c*) Riding quality is improved, especially on some high-speed elevators.

(*d*) Power consumption of the elevator motor is considerably decreased.

Each roller guide shoe assembly consists of three rubber-tired wheels resiliently mounted through springs or

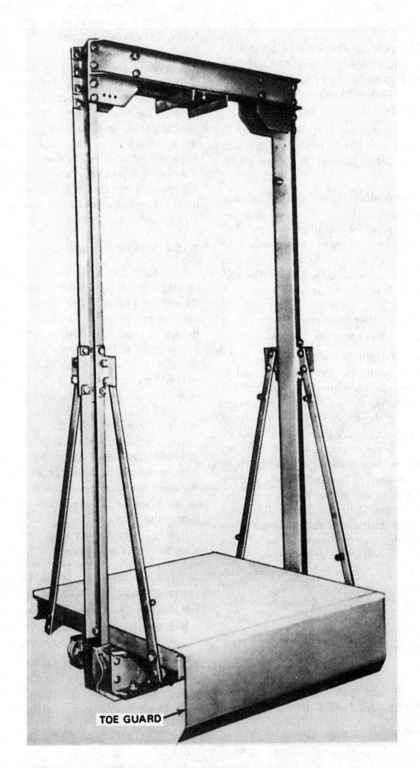

Diagram 2.15.1(a)   Side Post Car Frame
(Courtesy National Elevator Industry Educational Program)

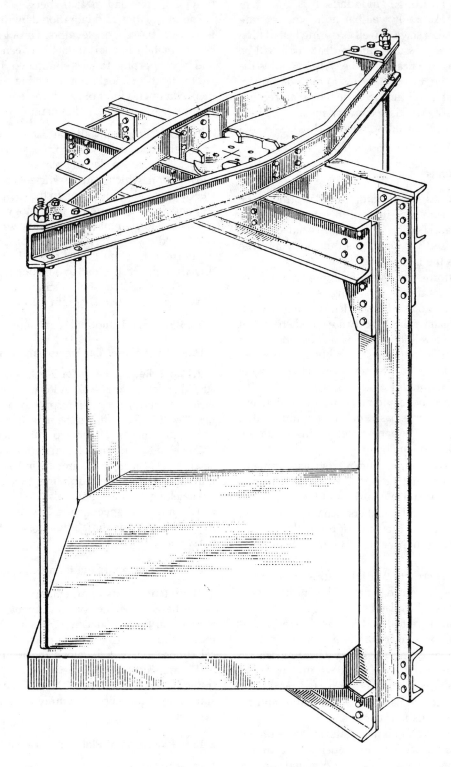

Diagram 2.15.1(b)   Corner Post Car Frame
(Courtesy National Elevator Industry Educational Program)

other means. The elevator, therefore, is riding on twelve wheels in contact with the guide rails. It is important that the cars be reasonably well balanced when setting a job up in the design stage as well as during installation so that guide shoe pressures due to dead load will be as close to zero as practical. ASME A17.1–2000 adds the requirement for the retention means to provide protection against possible collision between the car and equipment in the hoistway. This secondary means can be a part of the base of the guiding means. This has been a common practice of many suppliers.

2.15.5 Car Platforms

The Code defines the elevator car platform as the structure, which forms the floor of the car and directly supports the load. The basic function of an elevator platform is to directly support the duty load and the car that contains the load.

Passenger platforms are generally of three types: combination wood-steel design consisting of a steel frame on top of which is mounted a wooden flooring; the all-steel version, which might embody a welded structural frame to which a steel floor plate is welded forming a unitized construction; or the design might be comprised of several sections welded together wherein the stringers and floor plates are modules.

The Code defines the requirements for the design and construction of platforms based on the minimum rated load specified in 2.16. The general definitions of components in a platform assembly are:

(a) end channels the front and rear structural members of the platform that support the front and rear of the car, door threshold, toe guards, and the stringers;

(b) stringers structural members usually running in a front-to-rear direction. They are supported by the end channels and near their center by an intermediate support.

(c) floor plate a structural plate, either monolithic with the stringers, or welded to them. The finished flooring is laid on top of this plate.

A typical all-steel welded platform is shown in Diagram 2.15.5. The passenger platform rests on rubber blocks located at six points as shown. These blocks are supported by a sub-frame called the sound isolation support frame which, in turn, is supported by the car frame plank channels at its approximate center and by the side braces near its four corners.

As live load distributes on the platform, deflection of these rubber blocks occurs, thus causing measurable relative displacement between the platform and support frame. As the load increases in the car, the platform compresses the rubber. When a predetermined compression of the rubber is reached, load-weighing switches are activated and electrical circuits made or broken in order to render the car inoperative or bypass calls. More

sophisticated systems now on the market use strain gage technology for load measurement.

Sound isolation and vibration damping are important features of passenger elevators. Through the use of rubber, all metal-to-metal contact between the elevator car and its supporting frame is eliminated. The car is permitted to "float" on blocks of rubber that cushion car movement, dampen vibration, and prevent the transmission of sound. Also, see the commentary on 8.2.2.6.

2.15.5.4 Originally all platforms, whether passenger or freight, used a wood flooring supported by a steel frame. The modern elevator platform utilizes an all-steel construction, which is more efficient and lighter than its steel and wood predecessor. In the early 1970s, the Code responded to a changing technology and introduced provisions for laminated platforms, which consisted of steel-faced plywood. A laminated car platform is a self-supporting structure, which forms the floor of the car. It directly supports the load and is constructed of plywood with a bonded steel sheet facing on both the top and bottom surfaces. See Diagram 2.15.5.4.

2.15.5.5 See Handbook commentary on 2.15.5.4.

2.15.6 Materials for Car Frames and Platform Frames

2.15.6.2 Requirements for Steel. A detailed review of ASTM specifications A27, A36, A283, A307, A502, and A668 was performed with respect to mechanical properties. Chart 2.15.6.2 shows the tensile, yield, and elongation requirements for each specification.

The elongation range for materials cited in 2.15.6.2.1 is 21% to 24% in a 51 mm (2 in.) specimen. A 20% minimum elongation requirement would make sense.

The elongation for materials cited in 2.15.6.2.2 is 18% in a 51 mm (2 in.) specimen. A 20% minimum elongation requirement for materials for greater tensile does not make sense.

2.15.7 Car-Frame and Platform Connections

The intent of this requirement is to specify the type of mechanical fastener or attachment used in the platform-to-car frame connection. It is not the intent to limit the attachment to rivets, bolts, or welding. Any mechanical connection that develops the required strength to safely transmit the forces between the platform and car frame in accordance with 2.15.10 is acceptable. A clip, similar to a rail clip, is commonly used for this connection.

2.15.8 Protection of Platforms Against Fire

Most hoistway fires that develop today occur in the pit due to the accumulation of trash. At lower landings, the underside of the platform would be exposed to a pit fire and thus must be protected against such fires. The requirement is stated in performance language and is the same as required for car enclosures (2.14.2.1) with

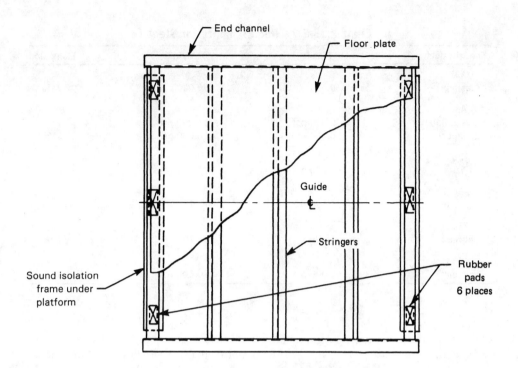

Diagram 2.15.5   Typical Passenger Elevator Platform
(Courtesy Otis Elevator Co.)

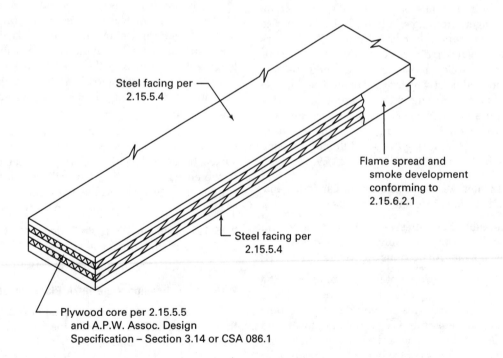

Diagram 2.15.5.4   Laminated Platform

95

Chart 2.15.6.2   Requirements for Steel

Specification	Tensile	Yield	Elongation
A27 (Castings)	60,000 psi	30,000 psi	24%
A36 (Plates & Bars)	58,000 psi	36,000 psi	28%
(Shapes)	58,000 psi	36,000 psi	21%
A283 (Plate)	60,000 psi	33,000 psi	23%
A307 (Bolts)	60,000 psi	N/A	18%
A502 (Rivets)	N/A	N/A	N/A
A668 (Forgings)	60,000 psi	30,000 psi	24%

one exception. Only the platform surface exposed to the hoistway must be tested.

2.15.9 Platform Guards (Aprons)

Platform guards (aprons) are required on all elevators. The 2000 edition of the Code requires the minimum length of the platform guard to be longer than required in previous edition of the Code. The 1 200 mm or 48 in. protect passengers from accessing the hoistway in case of unintended movement [2.19.2.2(b)]. Elevators equipped with leveling devices or truck zoning devices are allowed controlled movement when both the car door and gate and hoistway door are open. The purpose of the guard is to reduce the shearing action at the edge of the car sill and when the car is positioned above a floor sill to seal off the opening under the car from the landing thereby reducing the possibility of someone falling into the hoistway. See Diagram 2.15.9.

2.15.10 Maximum Allowable Stresses in Car-Frame and Platform Members and Connections

The permissible fiber stresses in the car frame and platform members and their connections are specified in 2.15.10. The allowable deflections are covered in 2.15.11.

2.15.10.1 The Code does not fix the amount that the platform edge may sink below the landing sill at the time the load is applied. This amount depends to some degree on the stiffness of the car frame, platform, and rails. Absolute rigidity can never be obtained; therefore, the allowable amount is controlled by the purpose for which the elevator is used.

Car frames for hydraulic elevators are constructed basically the same as those used for traction elevators, except that since the lifting is accomplished by a plunger fastened to the planks, the planks will usually be larger than the crosshead. Also, safeties are not required.

The uprights will be subjected to the same bending moments that were described for traction elevators, but in addition, they will have a compressive load in them due to the platform loads coming up through the side braces. Since this compressive load is quite appreciable, the slenderness ratio of the upright must be kept low enough to ensure that the member will not buckle. When long side braces are used, the slenderness ratio (L/r) is limited to 120; however, when short side braces are used for passenger application, an L/r of 160 is allowed since the compressive load from the braces acts over a shorter length of upright. L is the vertical distance from the lowest bolt hole in the crosshead to the uppermost bolt hole in the plank-to-upright connection, and r is the radius of gyration of the member.

2.15.10.2 The requirement provides criteria for the safe design of the respective components/structures subject to forces developed during the retardation phase of the emergency braking and all other loading acting simultaneously, if applicable, with factors of safety consistent with the types of equipment/structures involved.

2.15.11 Maximum Allowable Deflections of Car-Frame and Platform Members

See Handbook commentary on 2.15.10.

2.15.13 Suspension-Rope Hitch Plates or Shapes

The arrangement shown in Diagram 2.15.13 depicts that the vertical load acting on the hitch plate is distributed to all the lug plates, which, in turn, transmit their loads to the car frame through their fastenings to the webs of the crosshead channels. This design transmits the loads to the fasteners in shear, not in tension.

2.15.15 Platform Side Braces

The function of the side braces is to support the corners of the platform. One end of the brace is fastened

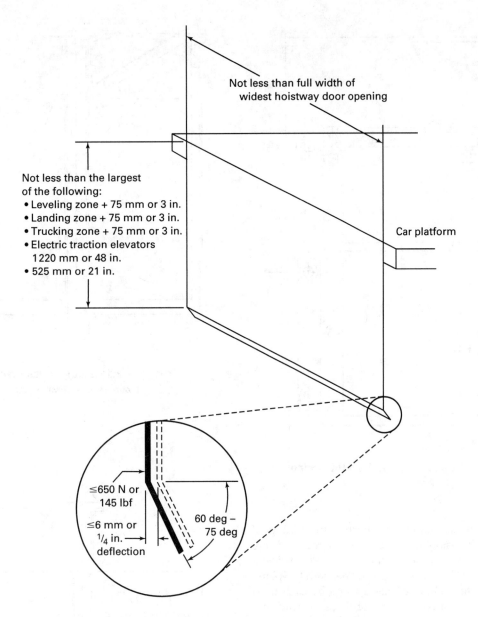

Not less than full width of
widest hoistway door opening

Not less than the largest
of the following:
• Leveling zone + 75 mm or 3 in.
• Landing zone + 75 mm or 3 in.
• Trucking zone + 75 mm or 3 in.
• Electric traction elevators
  1220 mm or 48 in.
• 525 mm or 21 in.

Car platform

≤650 N or
145 lbf

≤6 mm or
1/4 in.
deflection

60 deg –
75 deg

Diagram 2.15.9   Length of Platform Guard

to a bracket mounted on the underside of the platform and the other end to the car frame. Where short side braces are used, as on passenger elevators, the upper ends of the braces are fastened to the car frame upright.

The load in the side braces varies with the position of the live load in the car, the dead weight of the platform, enclosure, doors, and the angle it makes with the vertical. The effect of the side brace load basically takes three forms: it induces a direct axial load along the longitudinal axis of the upright; it produces bending, a corresponding deflection of the upright about its strong axis; and it causes the upright to twist due to the horizontal component of the brace load.

When side braces are designed so that the centerlines of the braces intersect the guide rail at the center of

the upper guide shoes, there will be no bending in the uprights caused by the force acting on the braces at this point. Since the brace load in this case is much higher than those associated with passenger loading, long side braces are usually used on freight elevators.

The lower end of each brace is fastened directly to the platform. At this point a bending of the upright is caused, for the reason that the thrust from the side brace through the platform is at a point above the lower guide shoes, which are mounted below the safety plank. However, this bending, acting in the strong axis of the upright and in close proximity to the lower guide shoes, is generally of small magnitude and in most cases not serious.

The basic design of the corner post truss embodies two structural members, usually bent channels, carrying

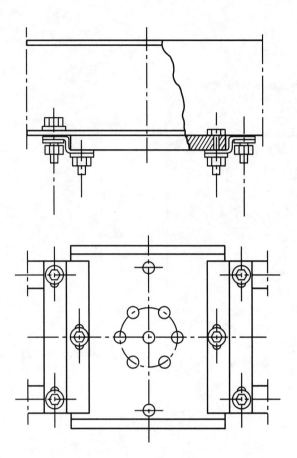

**Diagram 2.15.13 Typical Rope-Hitch Plate
Arrangement**

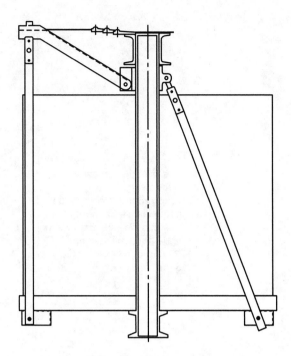

Diagram 2.15.15(a)   Truss Bracing
(Courtesy Otis Elevator Co.)

a tension rod at each end for supporting the corners of
the platform. Occasionally when the brace rod loads are
very high and the crosshead channels have insufficient
stiffness, an additional truss is placed beneath the plank
channels and the amount of load carried by each truss
depends on the relative stiffness of the crosshead and
planks.

For rectangular-shaped platforms, the differences in
length of the legs of the truss becomes so great that one
channel will carry practically all of the load, in which
case a different type of truss might be advisable. Since
corner post conditions vary so widely, no fixed set of
rules can be set up; therefore, each case must be treated
individually. It is important to stress the point again that
the use to which the elevator will be put is of extreme
importance when designing corner post structures. See
Diagrams 2.15.15(a) and 2.15.15(b).

SECTION 2.16
CAPACITY AND LOADING

Requirement 2.16.1 specifies the minimum rated load
for passenger elevators in terms of kilograms (pounds).
Requirement 2.16.3.2.1 requires that a capacity plate

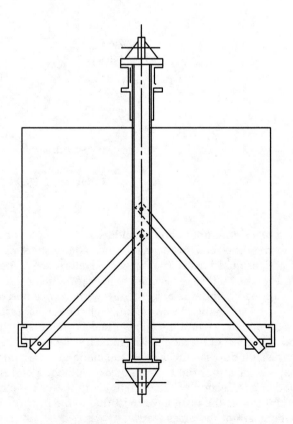

Diagram 2.15.15(b)   Short Side Braces
(Courtesy Otis Elevator Co.)

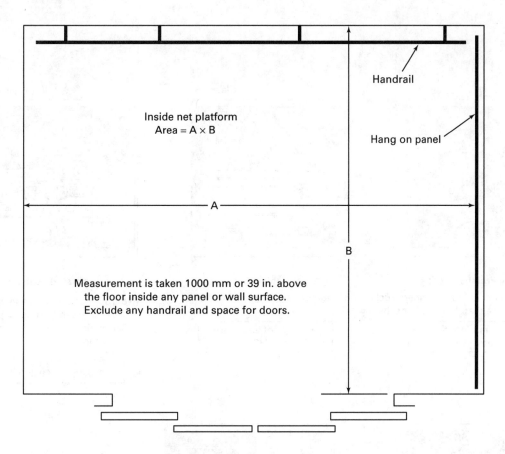

Diagram 2.16.1.1   Measurement of Passenger Elevator Inside Net Platform Area
(Courtesy Zack McCain, Jr., P.E.)

indicating the rated load in kilograms (pounds) be located inside the car.

When local ordinances require the elevator capacity to be also indicated in terms of persons, the number of persons should be calculated by dividing the rated load, if expressed in kilograms, by 72.5 or by 160 if expressed in pounds. The result (quotient) should be reduced to the next lowest whole number. For example, if the result is 14.97, the capacity in terms of persons should be 14.

2.16.1 Minimum Rated Load for Passenger Elevators

2.16.1.1 Minimum Load Permitted. These requirements and those in 8.2.1 establish a minimum rated load for the inside net platform area of passenger elevator cars. See Diagram 2.16.1.1. The Code does not prohibit a rated load greater than the minimum rated load required for a given net platform area. The minimum rated load for a given platform size is to mitigate the possibility of passengers overloading the elevator.

A car enclosure may be equipped with removable panels. When removed, the allowable increase in the net platform area of the car may not exceed 5%. If a 5% increase in the maximum net area were not allowed and the same car was rated at its maximum with the removable panels in place, the car would be in violation

of the Code if the panels were removed. The 5% increase is permitted since the actual weight of the removed panels would exceed any additional weight that may normally be placed on the car in the space made available when the panels are removed.

While the A17.1 Code does not have minimum car size requirements, the building codes, ADAAG, ICC/ANSI A117.1, and UFAS do have such requirements.

2.16.1.3 Carrying of Freight on Passenger Elevators. This application is typically referred to as a "service" elevator. The elevator must comply with all the requirements for a passenger elevator as well as be designed for applicable classes of freight loading.

2.16.2 Minimum Rated Load for Freight Elevators

2.16.2.2 Classes of Loading and Design Requirements. See Diagram 2.16.2.2.

2.16.3 Capacity and Data Plates

The capacity plate located in the car provides necessary information for the people using the elevator. The data plate located on the crosshead, on the other hand, provides vital information for maintenance and inspection personnel. The requirement specifies that the plates

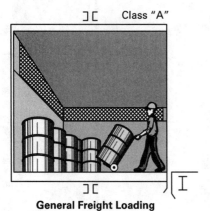

General Freight Loading

Where no item (including loaded truck) weighs more than $1/4$ rated capacity

Rating not less than
240 kg/m² (50 lb/ft²)

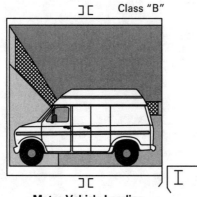

Motor Vehicle Loading

(Automobiles, trucks, buses)

Rating not less than
145 kg/m² (30 lb/ft²)

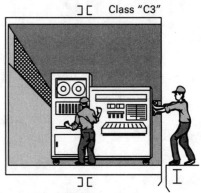

Concentrated Loading

(No truck used) but load increments are more than $1/4$ rated capacity. Carried load must not exceed rated capacity.

Rating not less than
240 kg/m² (50 lb/ft²)

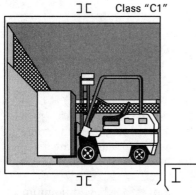

Industrial Truck Loading

Where truck is carried

Rating not less than
240 kg/m² (50 lb/ft²)

This loading applies where concentrated load including truck is more than $1/4$ rated capacity but carried load does not exceed rated capacity.

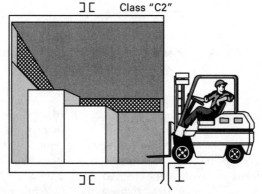

Industrial Truck Loading

Where truck is not carried, but is used for loading and unloading

Rating not less than
240 kg/m² (50 lb/ft²)

This loading applies where concentrated load including truck is more than $1/4$ rated capacity but carried load does not exceed rated capacity.
This loading also applies where increment loading is used but maximum load on car platform during loading or unloading does not exceed 150% of rated load.

Diagram 2.16.2.2   Freight Elevator Classes of Loading
(Courtesy Building Transportation Standards and Guidelines, NEII-1 2000, © 2000, National Elevator Industry, Inc., Teaneck, NJ)

be marked with letters that are stamped, etched, or cast on the surface. The metal photo process is not an acceptable alternative to this requirement, since it can be easily made unreadable.

2.16.3.2 Information Required on Plates

2.16.3.2.2 The data plate includes the year manufactured. This may not be the same as the Code edition (year) in effect at the time of installation. Typically, the edition of the Code the elevator is required to conform with is determined by the Code in effect (legally adopted) at the time the permit is issued for the installation. See also Handbook commentary on 8.9.

2.16.4 Carrying of Passengers on Freight Elevators

Prior to ASME A17.1–2000 permission had to be granted by the authority having jurisdiction to carry passengers on freight elevators. This provision has been removed and the Code permits carrying of passengers, provided the freight elevator complies with all the requirements in 2.16.4 including not being accessible to the general public.

2.16.4.4 Mechanical locks and electric contacts are not permitted.

2.16.4.5 Car gates are not permitted. The car door must be passenger-type construction.

2.16.4.7 Restrains passengers from exiting the car through the car/hoistway door outside the unlocking zone.

2.16.4.8 To require the same factors of safety that apply to passenger elevators.

2.16.4.9 Requires noncontact reopening devices, sequence operation of vertically sliding doors and a smooth interior finish to protect the passengers.

2.16.5 Signs Required in Freight Elevator Cars

The signs required for freight elevators are to advise the user of the limitations the elevator was designed and installed for and which personnel, if any, are permitted to ride on the elevator. The capacity displayed on the sign may be expressed in kilograms (kg) or pounds (lb) or both. The Code does not require freight capacity signs in passenger elevators designed to carry freight.

2.16.7 Carrying of One-Piece Loads Exceeding the Rated Load

"Capacity lifting one piece loads" is commonly referred to as a safe lift.

This requirement applies to a limited specific application, and when an elevator is used to lift a one-piece load (e.g., safe), which exceeds the rated load, these requirements must be adhered to. The elevator must have a locking device, which locks the car to the guide rails to take the load off the ropes during loading and unloading. The car must be designed to sustain the loads, but allowable stresses in the car frame, platform, ropes, etc., can be increased by 20%.

Safeties must stop and hold the one-piece load, with ropes intact; however, stopping distances do not have to be met. If there is occupied space below the hoistway, the safeties must be able to stop and hold the car independent of the ropes. If a safety was engaged near the bottom of the hoistway and there was insufficient sliding distance available before striking the buffer, the safety would absorb part of the kinetic energy of the descending mass and the buffer would absorb the balance. This condition could occur in an installation with a one-piece load in the car exceeding the rated load as well as on an installation operating without overloads; however, the maximum speed is limited to 0.75 m/s or 150 ft/min as covered in 2.26.1.3. Additional weight may also be added to the counterweight to increase traction.

2.16.7.10 Inspection operation takes precedence over "safe-lift" operation. Requirement 2.26.1.4.1-(b)(4)(b) states that during inspection operation the movement of the car shall be solely under the control of the inspection operating devices. That is, the operation of the car has to be exclusively (solely) under the control of the inspection device.

2.16.8 Additional Requirements for Passenger Overload

These requirements are design criteria. Compliance with these requirements is to field verified only to the extent specified in 8.10 and 8.11. See commentary on the referenced requirement.

2.16.9 Special Loading Means

The loads imposed by special loading devices or structures upon the car frame and platform must be included in the design.

SECTION 2.17
CAR AND COUNTERWEIGHT SAFETIES

NOTE: The author expresses his appreciation to George W. Gibson for his extensive contributions to Section 2.17.

The A17.1 Code does not specifically require that all safeties be designed to safely stop an overspeeding descending car or counterweight other than the condition where the suspension ropes are intact, that is, not free-fall.

In requiring conformance of Type B safeties to the stopping distances specified in 2.17.3, it is the intent that a Type B safety will, in a free-fall condition, develop a retardation sufficient to reduce the free-fall speed, resulting in an elongated safety slide or a combination safety and buffer stop within the allowable speed range of the buffer. While the intent is the natural result of a mathematical analysis of the stopping distance equations, it is not notably evident in the specific Code requirements. There are several considerations, which influence the formulation of an interpretation of this complex subject of safety stopping. The following provides a rational basis for that interpretation.

The basic function of the elevator mechanical safety system is to provide a safe stop for the passengers in the event of an overspeeding descending car, which cannot be electrically retarded and stopped. This basic function is consistent with the historical assurance of passenger safety evidenced in the demonstration of the first "safety hoister" at the Crystal Palace Exposition in 1853 where, upon the severing of the sole hoist rope, the car was safely stopped, thus signaling the beginning of safe passenger elevator transportation.

The sole embodiment of safety in such early "safety hoisters" to retard and stop a freely falling car was the broken rope wagon spring safety, which engaged the guide rails upon the parting of the single hemp hoist rope. These early "safety hoisters" predated later advances in the art, such as mechanical buffers in the pit, counterweighted elevators, steel wire hoist ropes, and traction machines.

Clearly, the historical basis for the stopping capability of car safeties must be viewed in relation to their basic function at any point in elevator history. The early safeties had to detect free-fall at its onset and apply immediately to retard and stop a freely falling car.

As elevator technology progressed and elevator safety codes were developed, higher degrees of passenger safety were embodied through mechanical and electrical safeguards, and the need for safeties to stop a freely falling car took on less importance throughout most of the U.S. Only a few states, notably Pennsylvania, Wisconsin, and California, and one of the federal government agencies (GSA), require free-fall safety tests. Industry practice was to conduct free-fall safety tests under the aegis of the former National Bureau of Standards, one of the early Co-Secretariats of the A17 Committee, in order to develop proprietary technical data on the stopping performance of safeties.

While the definition of a safety given in 1.3 states that free-fall is one of several conditions under which a safety is to stop and hold a car or counterweight, there are no design or test requirements in the A17.1 Code to validate such a reference to the free-fall stopping condition. The current definition has remained basically unchanged since it first appeared in the 1925 edition of the A17.1 Code. When viewed against the design requirements for safeties given in 2.17, the definition should have been assessed for consistency and revised accordingly. Therefore, the current definition should be viewed more for its descriptive than prescriptive value in setting forth the many conditions under which safeties may be required to operate.

The general requirements for the function of all types of safeties given by 2.17.3 do not specifically state that a car safety be designed to retard and stop a freely falling car whose suspension ropes failed.

Requirement 2.17.3 specifies that the safety have a capability of stopping and sustaining the entire car and rated load from governor tripping speed. Where the elevator carries passengers, 2.16.8, in conjunction with this requirement, assures a design capability for stopping and sustaining the entire car plus 125% rated load at the applicable speed for that load. The qualification of the car weight through the use of the word "entire" directly preceding it appears to have been included to emphasize that the safety had to have sufficient stopping capacity to develop a retardation in order to slow down and hold the car without mentioning the load state of the suspension ropes.

While certain requirements in 2.17 refer to safety application related to "...breaking or slackening of the suspension ropes," (2.17.7 and 2.18.1) and "...parting of the hoist ropes (free-fall)..." it is the intent of these requirements to activate the Type A safeties as quickly as possible in the event of a free-fall so as to avoid the buildup of the car speed during the time delays when the mechanical safety parts are being actuated but not delivering the full retarding force. This buildup of speed is more critical on slow-speed cars using Type A safeties, since the resulting speed of the car when the safety parts engage as a result of governor tripping would be a significantly higher percentage of the rated speed than would occur on higher speed cars using Type B safeties.

Nevertheless, while free-fall activation is required for Type A safeties, there are no requirements for the stopping performance in terms of retardation or stopping distance.

The absence of specific requirements for free-fall stopping capability for Type B safeties must be evaluated in relation to the range of stopping distances allowed by 2.17.3 and Table 2.17.3, which are derived from the formulas given in 8.2.6. These formulas for minimum and maximum safety stopping distances can be expressed in terms of the retardations of $1.0\,g$ and $0.35\,g$, respectively.

In order to determine whether the Code requires that a Type B safety can hold a freely falling car, it will be sufficient to analyze the singular case of a car which demonstrated a safety stopping distance equal to the maximum slide at the acceptance inspection, such as whose retardation $a = 0.35\,g = 11.25\ \text{ft/s}^2$ in a free-fall mode with rated load in the car.

Mathematical analysis will show that the capability of any Type B safety that complied with the acceptance inspections and tests in stopping a freely falling car is dependent upon several parameters, specifically the magnitude of the masses in the complete system, safety retarding force, vertical location of the car or counterweight in the hoistway when safety application occurs, available traction between the suspension ropes and driving machine sheave, and machine location.

The interrelationship of these parameters determines the dynamic performance of the safeties in controlling the descending car motion. If the analysis of a given system shows that the safety has insufficient retarding force to retard the freely falling car, then the descending car will, in fact, accelerate into the pit at an unsafe speed in excess of the maximum striking speed for which the buffer was designed.

However, if the analysis shows that there is sufficient retarding force to reduce the car speed to a value that is not greater than the speed for which the buffer is designed, then the system will undergo a combination safety and buffer stop, thus bringing the passengers to a safe stop, which is the basic intent.

Before analyzing the dynamics of the elevator system, a brief review of the pertinent A17.1 Code requirements will be made, since the acceptance test requirements differ from the actual failure mode in terms of the various switches, which are in effect normally but rendered inoperative during testing. During normal operation of the elevator system, which includes failure modes when overspeeding can occur, all electrical protection devices are operative. Those related to a safety stop include the governor overspeed switch (2.18.4.1) and safety mechanism switch (2.17.7.2), each of which removes power from the driving machine motor and brake. Accordingly, a certain amount of the stopping of the car at safety application derives from the dynamic braking of the machine imparted to the car by the traction between the drive sheave and ropes, and the major percentage of the stopping derives from the safety retarding force. However, during the acceptance testing of the safeties on the job site, these switches are temporarily adjusted to open as close as possible to the position at which the car safety mechanism is in the fully applied position. [See 8.10.2.2.2(bb)(4)(b).]

The A17.1 Code requires that the governor overspeed switch be inoperative during the overspeed test to ensure that the machine continues to be powered until the safety applies. The car safety mechanism switch is temporarily adjusted to open as close as possible to the position at which the safety is in the applied position to ensure that the brake does not apply any stopping to the car, but rather that the safety provides the entire retardation and stopping. This requirement ensures that the safety alone provides the required retardation.

The following analysis will show that because of the acceptance test requirements imposed by A17.1 8.10.2.2.2(bb)4, a Type B safety will not only stop an overspeeding car with the ropes intact but will also provide safe stopping for a freely falling car.

The analysis will be made for an elevator arranged with driving machine above, 1:1 roping, without rope or chain compensation, for the sake of simplicity; however, the general methodology is applicable with the appropriate modification to any elevator system.

(a) Overspeeding Car With Counterweight Attached (machine driving). Diagram 2.17(a) is representative of the field acceptance test required by 8.10.2.2.2(bb)(4) in which the driving machine keeps running during the stopping sequence.

The retardation impressed upon the car is found from the general method

(Imperial Units)

$$a = \Sigma F / \Sigma M$$

$$a = \left[\frac{W + \alpha\,(F - C - L - R_h)}{W + \alpha\,(C + L + R_h)} \right] g \qquad (1)$$

The following relationships are representative of the weights used in passenger elevators in the speed range using Type B safeties.

(Imperial Units)

$$\left. \begin{array}{c} C = L \\[1em] W = C + 0.45L = 1.45L \\[1em] R_h = 0.15L \end{array} \right\} \qquad (2)$$

Substituting Eq. (2) in Eq. (1) yields

(Imperial Units)

$$a = \left[\frac{1.45L + \alpha\,(F - 2.15L)}{1.45L + 2.15L\alpha} \right] g \qquad (3)$$

Equation (3) is evaluated for the maximum allowable stopping distance permitted by 2.17.3 and 8.2.6, in which case $a = 0.35g$. The available traction α for a double wrap traction gearless machine is 2. Therefore,

(Imperial Units)

$$F_{min.} = 2.43L \qquad (4)$$

where $F_{min.}$ is the minimum retarding force, which the safety can have.

The maximum retarding force will produce a stop equal to the minimum stopping distance allowed by 2.17.3 and is determined by letting $a = 1.0g$ and

(Imperial Units)

$$\alpha = 2$$

in Eq. (3). Therefore,

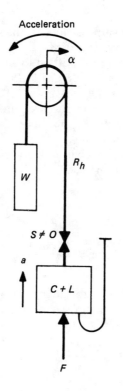

where

C = car weight, lb
L = rated load, lb
W = counterweight, lb
R_h = hoist rope weight, lb
F = safety retarding force, lb
a = retardation, ft/s²
g = gravity acceleration, 32.2 ft/s²
v = rated speed, ft/min
v_b = buffer striking speed, ft/min
v_g = governor tripping speed, ft/min
v_s = car speed at safety activation, ft/min
α = available traction

Diagram 2.17(a)   Overspeeding Car With Counterweight Attached (Machine Driving)

(Imperial Units)

$$F_{max.} = 4.3L \tag{5}$$

In that the safety stopping performance did not have to meet the test requirements of 8.10.2.2.2(bb)(4), the machine could be considered free-wheeling, similar to a frictionless pulley in which the tensions would be constant over the length of hoist rope passing over the machine. In this case, $\alpha = 1$ and Eq. (3) becomes

(Imperial Units)

$$a = \left[\frac{F - 0.7L}{3.6L}\right] g \tag{6}$$

For the maximum stopping distance, $a = 0.35g$ and Eq. (6) becomes

(Imperial Units)

$$F_{min.} = 1.96L \tag{7}$$

(b) Freely Falling Car

The retardation impressed upon the car is

(Imperial Units)

$$a = \left[\frac{F - C - L}{C + L}\right] g \tag{8}$$

Substituting Eq. (2) in Eq. (8) yields

(Imperial Units)

$$a = \left[\frac{F - 2L}{2L}\right] g \tag{9}$$

In order to develop a retardation

(Imperial Units)

$$F - 2L > 0 \tag{10}$$

Therefore,

(Imperial Units)

$$F > 2L \tag{11}$$

It is clear from Eqs. (7) and (11) that if the retarding force of the safety is too small, a free-falling car will never be retarded, but will be accelerated instead.

However, comparing Eq. (4) to Eq. (11) shows that by demonstrating conformance to the stopping distance during the overspeed test, in which the machine continues to run as required by 8.10.2.2.2(bb)(4), the safety will retard a free-falling car. This will be examined further.

(c) Overspeeding Car With Counterweight Attached (machine stopped)

Chart 2.17
Safety Retarding Forces

Case	Situation	Machine	$F_{min.}$ [Note(1)]	$F_{max.}$ [Note(2)]	Remarks
a	Overspeed	Driving	2.43L	4.3L	8.10.2.2.2(bb)(4)
b	Overspeed	Free Wheel	1.96L	4.3L	
c	Overspeed	Stopped	1.02L	4.3L	

NOTES:
(1) $F_{min.}$ = The safety retarding force that will produce a maximum stopping distance allowed by 2.17.3, i.e., $a = 0.35g$.
(2) $F_{min.}$ = The safety retarding force that will produce a maximum stopping distance allowed by 2.17.3, i.e., $a = 1.0g$.

This case is representative of the actual safety stop that would occur in an overspeeding system, in which case both the governor overspeed switch and safety mechanism switch would function to remove power from the driving machine motor and brake.

In setting up the system dynamics, the tractive effort at the drive sheave is just the opposite to case (a) above. Therefore, the system retardation is found to be

(Imperial Units)

$$a = \left[\frac{\alpha W - (C + L + R_h) + F}{\alpha W + (C + L + R_h)} \right] g \qquad (12)$$

Substituting the relationships from Eq. (2) yields

(Imperial Units)

$$a = \left[\frac{1.45L\,\alpha - 2.15L + F}{1.45L\,\alpha + 2.15L} \right] g \qquad (13)$$

The minimum safety retarding force that will produce the maximum allowable stopping distance is found as before by setting $a = 0.35g$ and $\alpha = 2$, from which Eq. (13) results in

(Imperial Units)

$$F_{min.} = 1.02L \qquad (14)$$

The maximum retarding force is found by substituting $a = 1.0g$ and $\alpha = 2$ in Eq. (13), from which

(Imperial Units)

$$F_{max.} = 4.3L \qquad (15)$$

(d) Summary of Safety Retarding Forces. The safety retarding forces from the above cases are summarized in Chart 2.17.

From Chart 2.17 it is seen that the minimum allowable retarding force must be not less than that shown for case (a) in order to

(1) meet the test requirements of 8.10.2.2.2(bb)(4); and

(2) retard a freely falling car.

Therefore, the following relationship must be met.

(Imperial Units)

$$2.43L \le F \le 4.3L \qquad (16)$$

(e) Evaluation of the Free-Fall Condition. Referring to Eq. (11), it is seen that if $F > 2L$, a retardation would be impressed upon a free-falling car. This retardation is given by Eq. (9) as:

(Imperial Units)

$$a = \left[\frac{F - 2L}{2L} \right] g$$

In order to pass the acceptance test requirements of 8.10.2.2.2(bb)(4), it was shown by Eq. (4) that the minimum retarding force must be

(Imperial Units)

$$F_{min.} = 2.43L$$

Substituting Eq. (4) in Eq. (9) yields

(Imperial Units)

$$a = 0.215g \qquad (17)$$

Diagram 2.17(b) is hypothesized in order to evaluate a free-fall condition.

The buffer striking speed is required by 2.22.4 to be 115% of the rated speed. It is assumed that the governor tripping speed is 25% greater than the rated car speed and that, in a free-fall condition, the car speed would continue to increase during the time it takes to activate the governor and safety system. It is assumed that the car speed at safety activation is 25% greater than governor speed.

Therefore,

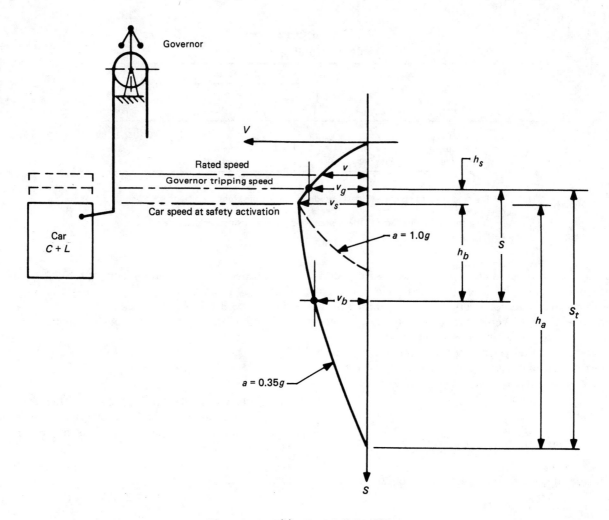

Diagram 2.17(b)   Free-Fall Conditions

(Imperial Units)

$$v_b = 1.15v \tag{18}$$

$$v_g = 1.25v \tag{19}$$

$$v_s = 1.25v_g = 1.5625v \tag{20}$$

The distance the car falls from governor tripping speed to the speed at which the safety activates and starts retarding the car is

(Imperial Units)

$$h_s = \frac{v_s^2 - v_g^2}{2g} \tag{21}$$

The distance the car falls from safety activation speed to buffer striking speed is

(Imperial Units)

$$h_b = \frac{v_s^2 - v_b^2}{2a} \tag{22}$$

The total distance the car falls from governor tripping speed to buffer striking speed is

(Imperial Units)

$$S = h_s + h_b \tag{23}$$

Substituting Eqs. (18) through (22) in (23) yields

(Imperial Units)

$$S = 0.44v^2 \left[\frac{1}{g} + \frac{1.273}{a} \right] \tag{24}$$

Substituting Eq. (17) in (24) yields

(Imperial Units)

$$S = \left[\frac{3.04v^2}{g} \right] \tag{25}$$

The necessary stopping distance to bring the car to a stop under the retardation of 0.215 g is as follows:

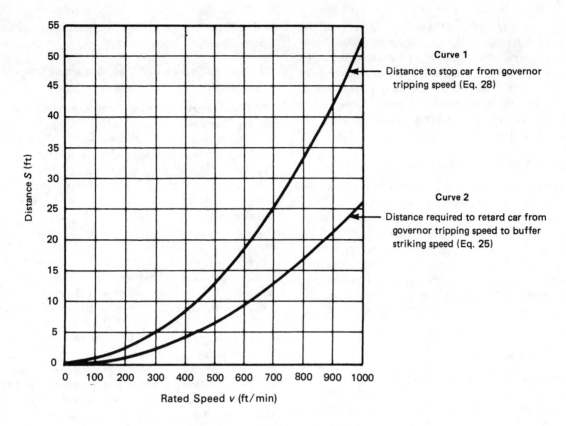

Diagram 2.17(c)   Stopping Distances

(Imperial Units)

$$S_t = h_s + h_a \tag{26}$$

where

(Imperial Units)

$$h_a = \frac{v_s^2}{2a} \tag{27}$$

Substituting Eqs. (21) and (27) in (26) yields

(Imperial Units)

$$S_t = \frac{6.12v^2}{g} \tag{28}$$

Equations (25) and (28) are graphed in Diagram 2.17(c).

(f) *Conclusion.* If a safety has been designed, installed, and successfully tested to meet the requirements of 8.10.2.2.2(bb)(4)(d), it will provide a retardation to slow down a freely falling car. The curves in Diagram 2.17(c) cover two conditions.

(1) Diagram 2.17(c) curve 1 gives the total distance from the point at which the governor trips to stop a freely falling car whose initial velocity was equal to the rated speed (Eq. 28).

(2) Diagram 2.17(c) curve 2 gives the total distance from the point at which the governor trips to retard a freely falling car down to the oil buffer striking speed (Eq. 25).

The analysis shows that in an extreme case, if the safety retarding force was only capable of stopping an overspeeding car, whose ropes were intact, within the maximum permissible stopping distance allowed by 2.17.3, it would produce sufficient retardation on a freely falling car to reduce its speed to the speed for which the pit buffer is designed, as required by 2.22.4.1, i.e., 115% rated speed.

As long as the striking speed of the car has been reduced to that which can be safely absorbed by the pit oil buffer, the car and passengers would be brought to a safe stop.

Since the high-speed elevators with speeds above 1000 ft/min typically use reduced stroke oil buffers arranged with an emergency terminal speed limiting device as allowed by 2.22.4.1, additional height would be needed to slow the car down to the rated speed of the reduced stroke buffer. From the foregoing discussion and analysis, it is evident that the maximum stopping distances allowed by the Code are indeed safe, not only for an overspeeding car but also for a freely falling car.

When a safety has retarding forces higher than the absolute minimum, which was the case described above, the distances required to stop a freely falling car are reduced accordingly.

The foregoing analysis was based upon a freely falling car carrying full rated load, which is a highly improbable failure scenario since elevators carry less than balanced load most of the time. The analysis given above can be easily modified to reflect a live load less than balanced load and will show that safeties designed to meet present A17.1 requirements will almost always stop a freely falling car before striking the buffer.

2.17.1 Where Required and Located

Prior to the adoption of this requirement, car safeties were occasionally attached to the car crosshead. There was a history of accidents resulting from the actuation of the car safety causing the car frame to part from the car crosshead. On existing installations where the car safety is attached to the car crosshead, particular attention should be paid during an inspection, as well as prior to and immediately following a test, to the structural condition of the car frame and its connections to the crosshead.

2.17.2 Duplex Safeties

Duplex safeties are pairs of safeties; one safety is located within or below the lower member of the car frame (safety plank) and the other safety normally is attached to the crosshead.

2.17.3 Function and Stopping Distance of Safeties

The stopping distance for Type B safeties provides a reasonable retarding force (minimum stopping distance) within a reasonable safe maximum distance of travel. The Code does not specify a minimum or maximum stopping distance for Type B safeties for an empty car load condition.

The reasonable retarding force for a Type C safety is provided by the use of one or more oil buffers interposed between the lower member of the car frame and the safety plank.

The Type A safety applies a high retarding force as soon as it is brought into action. They typically employ eccentrics and rollers without any flexible medium which would increase the stopping distance. Without such flexible medium, the energy absorbed in stopping the car or counterweight derives from the distortion of metal at the interface between the safety parts and guide rails. The exact amount of distortion cannot be predicted. Accordingly, the Code does not specify minimum and maximum allowable stopping distances. For this reason, Type A safeties are limited to elevators having a rated speed of not more than 0.75 m/s or 150 ft/min.

2.17.4 Counterweight Safeties

Counterweight safeties are necessary only when there is occupiable space below the hoistway. Counterweight safeties are also permitted to be used as emergency

brakes [2.19.3.2(a)(1)]. See Handbook commentary on 2.6 and 2.19.

2.17.5 Identification and Classification of Types of Safeties

2.17.5.1 Type A Safeties. Type A instantaneous safeties are designed to apply a high retarding force as soon as they are brought into action. It generally consists of:

(*a*) a roller normally located in a pocket but operating between a sloping surface and a guide rail; or

(*b*) an eccentric member pivoted on the car or counterweight structure and brought into contact with the guide rail surfaces.

Once the eccentric member or roller is in contact with the guide rail, this device is self-actuated by the operating forces being derived from the mass and motion of the car or counterweight. The governor rope acts only to bring the roller or eccentric member into contact with the guide rail. It is frequently designed to be applied by the inertia of the governor rigging. In addition to operation by means of a speed governor, the Code requires operation without appreciable delay in the event of free-fall, which may be accomplished by the inertia of the governor and governor rigging or by springs and mechanical linkage actuated by failure or slackening of the suspension ropes. Because it does not afford appreciable slide, the amount of energy it can absorb is distinctly limited; hence, the Code limits the maximum-rated speed from which it may be used to 0.75 m/s or 150 ft/min. This safety is normally released by raising the car or counterweight. See Diagram 2.17.5.1.

2.17.5.2 Type B Safeties. Type B wedge clamp safeties are safeties in which a wedge is driven between two pivoted members, the opposite ends of which form or carry the guide rail gripping surfaces. Travel of the wedge increases the pressure on the jaws. No elastic member is provided in the jaw assembly but one may be provided in the actuating mechanism. The wedges are normally operated by a rotation of right- and left-hand threaded screws working within a drum on which a wire rope attached to the governor rope is wound. These screws may either push or pull the wedges. The operating force is derived from the tension in the governor rope. Because of the inertia effects of the governor and governor rigging and elasticity of the governor rope, the tension on the safety drum rope varies not only from installation to installation, but also varies in the same installation with position of the elevator in the hoistway and with the speed in which the governor jaws apply. Uneven guide-rail thicknesses or bad guide-rail joints produce relatively large variations in the retardation of the car. The drum-type wedge clamp safety is released by means of a wrench from within the car. The wrench generally carries a beveled gear pinion or worm, which engages suitable teeth on the safety drum. Turning the

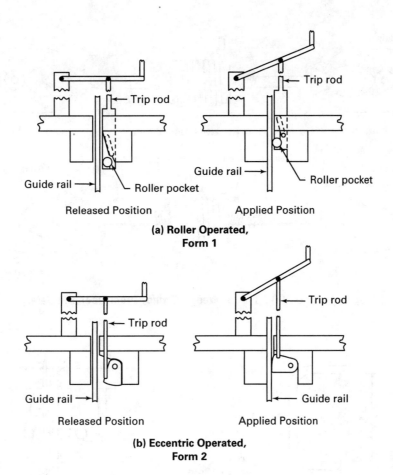

Released Position   Applied Position

**(a) Roller Operated,
Form 1**

Released Position   Applied Position

**(b) Eccentric Operated,
Form 2**

GENERAL NOTE: Double eccentrics are provided for heavy duty.

Diagram 2.17.5.1   Instantaneous Safety, Roller- and Eccentric-Operated

wrench to release the safety rewinds the rope on the safety drum. While the rope is being rewound on the safety drum, great care must be used to maintain tension on the drum rope or the rope may jam on the safety drum causing subsequent failure of the safety mechanism. [See Diagram 2.17.5.2(a).]

A Type B flexible guide clamp safety is a safety in which the final force is derived from a spring in the jaw assembly, which is compressed or further compressed as the device is being applied. Because of the presence of the spring member, variations in guide-rail thicknesses or a bad guide-rail joint produce comparatively small variations of the pressure on the rail. The mechanism consists of a pair of tapered wedges with sets of rollers between each wedge and a spring-backed inclined surface. When pulled into contact with the guide rail by a governor rope-operated trip rod, the rollers permit these wedges to deflect the spring until the wedges reach a stop after which they slide on the guide rail with substantially constant pressure. Once the wedges are in contact with the rail, the device is self-actuating, the operating force being derived from the mass and motion of the car. In some applications, a U-shaped spring is used to furnish the pressure directly to rollers. In other applications, a coil spring and pivoted arms are furnished. Because the pressure on the guide rail is determined by the deflection of the spring or springs, the retardation is essentially independent of the speed at which the governor operates and of the tension in the governor ropes. Spring tension must be sufficient to handle reasonable overloads allowing for wearing of the parts. This type of safety is normally released by lifting the car or counterweight. See Diagram 2.17.5.2(b).]

Another type of mechanism consists either of a roller or of a roller-operated wedge, which operates between a spring-backed tapered surface and the guide rail. Once the roller is in contact with the guide rail, the device is self-actuating, the operating force being derived from the mass and motion of the car. The governor rope acts only to bring the roller into contact with the guide rail after which the governor rope continues to pull through the governor jaws. Characteristics are similar to the

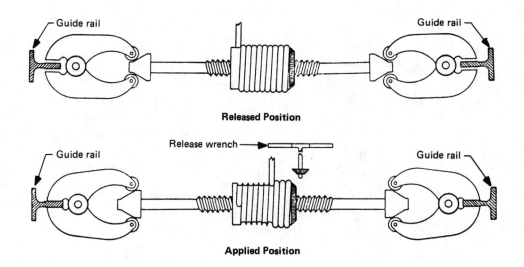

Diagram 2.17.5.2(a)   Type B Wedge Clamp, Drum-Operated Safety

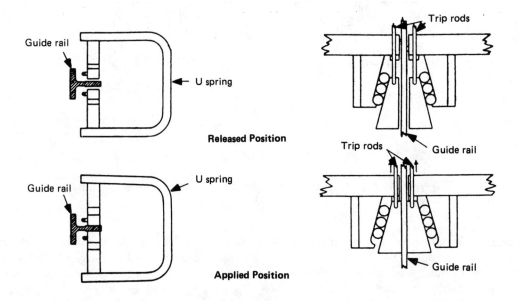

GENERAL NOTE: The U-spring is replaced with clamp and coiled spring for heavy duty.

Diagram 2.17.5.2(b)   Type B Flexible Guide-Clamp Safety, Wedge-Operated

wedge-operated type described above without the follower wedge. This safety tends to produce higher retardation rates with a lightly loaded car. To release the car safety, a lever wrench is inserted in the slot in the car floor and is used to compress and reset the spring. See Diagram 2.17.5.2(c).

2.17.5.3 Type C Safeties (Type A With Oil Buffers). The Type C safety develops retarding forces during the compression stroke of one or more oil buffers interposed between the lower members of the car frame and a governor operated Type A auxiliary safety plank applied on the guide rails. The stopping distance is equal to the effective stroke of the buffer. The safety plank in the car sling is independently guided. Due to the inherent design of the Type A safety described previously, the stopping distance of the auxiliary safety plank is very short, providing an operating platform for the oil buffer or buffers with a minimum of car travel. The car is retarded and brought to a stop by an oil buffer or buffers having a stroke calculated for the application to provide smooth retardation. This type of safety is normally released by lifting the car. See Diagram 2.17.5.3.

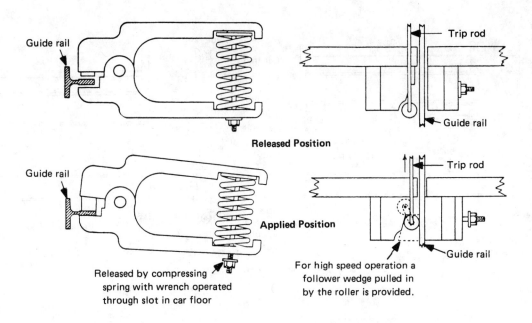

Released Position

Applied Position

Released by compressing
spring with wrench operated
through slot in car floor

For high speed operation a
follower wedge pulled in
by the roller is provided.

Diagram 2.17.5.2(c)   Type B Flexible Guide-Clamp Safety, Contact-Roller-Operated

2.17.7 Governor-Actuated Safeties and Car-Safety-Mechanism Switches Required

2.17.7.2 If this switch was not provided and the safety was set, severe damage could be caused if the drive machine continued to run. Cases are known where the ropes were severed and the drive sheave worn through due to the conditions described.

2.17.8 Limits of Use of Various Types of Safeties

See Handbook commentary on 2.17.5 for discussion on various types of safeties.

2.17.9 Application and Release of Safeties

2.17.9.1 Means of Application. Electric, hydraulic, or pneumatic devices are not a reliable positive means for actuation or holding safeties in the retracted position and, thus, they are prohibited by this requirement.

2.17.9.2 Level of Car on Safety Application. If the car platform went out of level more than the tolerances specified in this requirement, the load on the car could shift, causing structural damage to the car and possibly causing injury to the passengers or freight handlers in the car.

2.17.10 Minimum Permissible Clearance Between Rail-Gripping Forces of Safety Parts

See Diagram 2.17.10.

2.17.11 Maximum Permissible Movement of Governor Rope to Operate the Safety Mechanism

The basis for the allowable governor pull-out distance specified by 2.17.11, which decreases as the car speed increases, is based on the premise that an overspeed could occur in conjunction with a high acceleration such as a freely falling car (broken rope) or an electrical malfunction. If, under such a case, the governor applied when the car speed was at governor tripping speed, the final speed of the car at the time the safeties set and started to apply stopping force would be at a higher value. This would be equal to the sum of the governor tripping speed plus an additional speed increment due to the acceleration acting over the governor pull-out distance.

A flexible guide clamp safety classified as a Type F.G.C. safety under the provisions of the 1937 edition of this Code is now classified as a Type B safety. This type of safety does not require the continual unwinding of the safety drum rope to fully apply the safety. Instead, the safety jaws are quickly applied to the guide rail upon activation of the governor as the rollers move up the rather steep wedge and on to the flat or nearly flat surfaces. The final pressure on the rails results from the spring associated with the safety jaws. It is not necessary that three turns of safety rope be left on the drum after the overspeed test of the safeties. This was recognized in the Elevator Maintenance Bulletin Number 5 published by the American Society of Mechanical Engineers in October 1945. The Inspectors' Manual also states that all rope may be pulled from this type of safety.

111

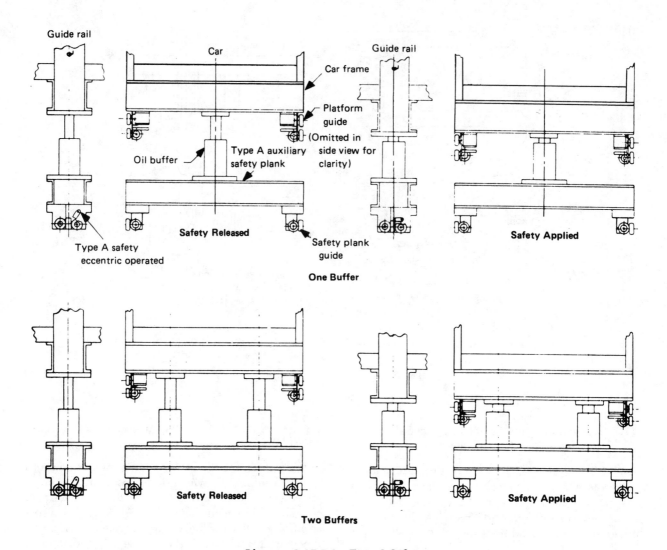

Diagram 2.17.5.3   Type C Safety

The three turns of the safety rope required to be left on the drum after the safety application, ensure a margin of safety against the accidental pulling of all the rope off the drum before the safety is fully applied. Also, see Handbook commentary on 2.17.5.

2.17.12 Minimum Factors of Safety and Stresses of Safety Parts and Rope Connections

2.17.12.4 Phosphor-bronze hoisting rope has historically been used successfully for this rope application. For such application where the rope is subject to limited flexing over sheaves and drums, galvanized rope is recommended by rope manufacturers. Plain iron wire rope would not satisfy the requirements. For further rope information, a supplier of elevator wire rope should be contacted.

Tiller rope construction employs 6 strands of 42 wires each for a total of 252 wires. Each of the 6 strands is actually a complete 6 × 7 rope containing a fiber center. The 6 strands are closed around a central fiber core.

This construction produces one of the most pliable ropes available. The wires, however, are extremely fine, making the rope susceptible to wear, abrasion and crushing.

The Code prohibits the use of Tiller rope construction for use as a safety connection because it has low wear resistance, low abrasion resistance, and low crushing resistance.

2.17.13 Corrosion-Resistant Bearings in Safeties and Safety Operating Mechanisms

The requirement for corrosion-resistant bearings is because the safety is under the car and is apt to be exposed to a corrosive atmosphere from a wet or damp pit. The bearings will be stationary most of their life and actually will be called upon to operate only in an emergency when the safety must be applied.

2.17.15 Governor-Rope Releasing Carriers

A governor-rope releasing carrier is a mechanical device to which the governor rope may be fastened,

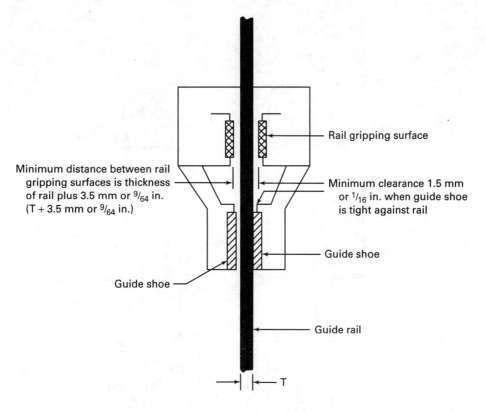

Diagram 2.17.10   Safety Jaw Clearances
(Courtesy Zack McCain, Jr., P.E.)

calibrated to control the activation of a safety at a prede-termined tripping force. A releasing carrier is used to minimize the possibility of the inertia of the governor rope tripping the safety. A governor-rope releasing car-rier is not required, but if such a device is provided, it must conform to the requirements of 2.17.15. If the governor-rope releasing carrier could be adjusted beyond the limits specified in this requirement, it might accidentally be over-tightened. Under this condition, the governor could trip from an overspeed condition, and the governor rope would not release from the releasing carrier and would pull through the governor jaws, thus not allowing the safety to be actuated. The purpose of these requirements is to limit the magnitude of the ten-sile load imparted to the governor rope to a safe working value in order to prevent overloading the rope. See Dia-gram 2.17.15.

2.17.17 Compensating Rope Tie-Down

These requirements are met by the use of compensat-ing ropes, which run from the underside of the car, down to the pit, around a compensating rope sheave, and then up to the bottom of the counterweight. The compensat-ing rope sheave is installed in a frame, which limits its vertical movement.

Tie-down compensation requirements were incorpo-rated in A17.1 during the 1930s when it was found that under certain conditions that elevator safety stops may occur at greater than $1g$. In order to prevent this from happening on high speed elevator systems, tie-down compensation was introduced. The purpose is to transfer energy from the counterweight to the elevator car and actually speed the car up decreasing the deceleration rate to less than $1g$. This can happen when either car or counterweight stops on safety and could happen at the terminals when either is approaching the overhead. Also, see the Handbook commentary on 2.20.3.

SECTION 2.18
SPEED GOVERNORS

2.18.1 Speed Governors Required and Location

Over the years, many different types of governors have been developed and used. The following is a description of five of the most common types:

(a) flyball governor: A flyball governor is one operated by a pair of flyballs attached to and driven by a vertical shaft. Links attached to the flyball arms lift a collar or sleeve operating against an adjustable compression

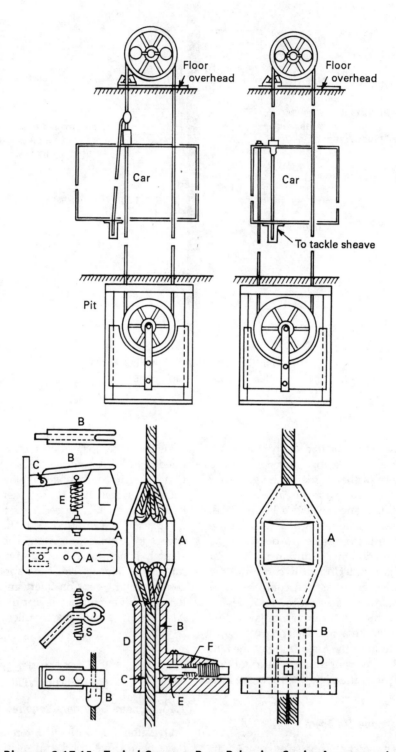

Diagram 2.17.15   Typical Governor-Rope Releasing Carrier Arrangements

spring on the shaft. The vertical shaft is driven through a pair of bevel gears by a sheave, which in turn is driven by a governor rope attached to the car. Various gear ratios are generally available to take care of various tripping speeds. When a predetermined speed is reached, the collar is lifted far enough to trip the rope grips. In older elevators they may consist of a pair of grooved, arched pivots pivoted on opposite sides of the down-running side of the governor rope. These jaws grip and stop the rope when they are tripped. Gear teeth are provided to ensure equal travel of the jaws. Where pull-through is desired, one of the grips is spring-backed.

(b) *overhead flyball:* This is a modification of the usual flyball type. The flyballs are mounted with the points of support below the plane of rotation. When at rest, the flyballs lie inside the lines through their supports. A disk lifted by a pair of levers attached to the ball arms trips the rope grip; because the outward travel of the balls is aided by gravity, small masses may be used.

(c) *horizontal shaft governor:* A horizontal shaft governor is one with a shaft perpendicular to the plane of the sheave. Pivoted masses, spring controlled, operate a lift rod by means of short arms attached to the masses. By varying the tension of the springs, a considerable range of speeds may be covered.

(d) *centrifugal governor:* A centrifugal governor (disk governor, bail-type governor, knockout governor) consists of a sheave containing two or more eccentrically pivoted weights normally held by springs within the periphery of the sheave. For a bail or arm carrying a wedge in line with the governor rope, the bail is mounted eccentrically to the sheave. When the speed of the governor rope reaches a predetermined value, the weights are driven outward by centrifugal force until the bail or arm is engaged or moves in a direction of rotation until the wedge member engages the rope and locks it against the sheave. Ordinarily, this type of governor has no provision for a pull-through but when used for moderate or high speeds is arranged with parallel spring-backed jaws, which permit pull-through. Jaw grips are tripped by a link connecting them to a notched disk, normally stationary, which is operated on overspeed by a dog or lug on the inner end of either of the pivoted weights. See Diagram 2.18.1.

(e) *jawless governor:* This type of governor does not have jaws to grip the governor rope. When the sheave of this type of governor is stopped due to the detection of an overspeed condition, the traction between the sheave and the governor rope also causes the governor rope to stop and therefore actuates the safety mechanism. When these types of governors are used, a switch on the governor tail sheave is required to assure that the tension in the governor rope is maintained.

2.18.3 Sealing and Painting of Speed Governors

If a governor is painted, extreme care must be taken that all bearings and pivoting surfaces are not painted. Governors that have been painted may fail to trip at their required speed and in some cases may fail to trip at any speed. If painted, the governor tripping speed should be tested.

A broken or missing seal indicates that the governor tripping speed mechanism and/or rope retarding means provided for the rope pull-through force (tension) may have been altered. If a broken seal is found, the governor tripping speed and/or rope pull-through force should be tested, making any readjustments that are necessary, and then the governor resealed.

The sealing means required is expressed in performance terms to afford design latitudes in accomplishing the sealing. Regardless of the specific type or method used, the sealing means should be capable of:

(a) providing visual indication that readjustment has not been made; and

(b) withstanding the environment, such as temperature and humidity extremes, in which the governor is installed, without being affected. If the sealing means meet the above requirements, it would comply with these requirements. Two means employed are the lead-wire seal, which has been employed for many years, and sealing wax.

2.18.4 Speed-Governor Overspeed and Car-Safety-Mechanism Switches

2.18.4.1 Where Required. The governor must be equipped with a switch, which will cause the power to be removed from the driving machine motor and brake. Prior to ASME A17.1–2000, the switch was only required for governors that operated Type B and Type C safeties where the rated speed was 0.76 m/s (150 ft/min) and on all governors where static controls were provided. Inherent risk of overspeed exists regardless of speed or type of control (e.g., the motor is always connected to the line but can also have a failure of contactor to the main line). The switch provides protection against overspeed in either direction. A second set of contacts may be provided to regulate the speed of the motor. At times, an additional set of contacts may be provided to limit speed when approaching terminal landings.

Additionally, every car must have a switch located on the car and operated by the safety mechanism, which will cause the power to be removed from the driving machine motor and brake either at the time of or before the safeties engage (2.17.7.2). This additional switch is not practical for counterweight safeties as there typically is no traveling cable to the counterweight.

2.18.4.2 Setting of Speed-Governor Overspeed Switches. The maximum tripping speeds required for speed-governor overspeed switches provide an additional degree of safety by electrically causing an

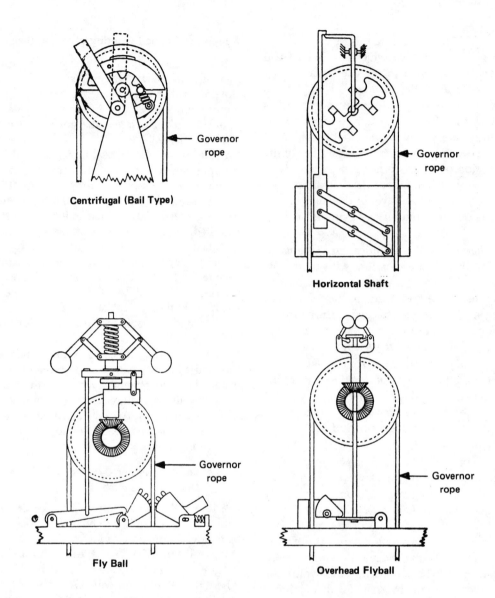

Diagram 2.18.1   Types of Speed Governors

overspeeding elevator to stop or reduce speed before the car safety is applied. Also, see Chart 2.18.4.2. This chart only gives the maximum governor tripping speed for elevator car governors. The maximum tripping speed for counterweight governors is higher. There is also a minimum governor tripping speed specified in 2.18.2.2. See Table 2.28.2(a) in the Inspectors Guide, ASME A17.2 for minimum governor tripping speeds.

On static control elevators, additional safety provisions are required to protect against equipment failures, such as:

(*a*) the power contactor/relay (2.26.9.5.6) closes (makes), the brake lifts, but the power converter/inverter (2.26.9.5.6) fails to conduct; or

(*b*) the power converter/inverter (2.26.9.5.6) combination fails "full-on" in a direction opposite to the dictated direction.

To protect against the above types of equipment failure, it was agreed that a motion detection sensing device be required to stop the car if car movement does not agree with motion dictation.

The type of governor would be one where the safety contact opens for an overspeed condition independent of the direction of car travel.

2.18.4.3 Setting of the Counterweight Governor Switch. This requirement was added in ASME A17.1–2000. It requires the switch to open prior to governor tripping speed and not to open at rated speed.

2.18.4.4 Type of Speed-Governor Overspeed Switches, Speed-Reducing Switches, and Car-Safety-Mechanism Switches Required. If an overspeed or speed-reducing switch is tripped, the reason for it being tripped should be determined. The elevator is not allowed to resume

Chart 2.18.4.2   Governor Overspeed Switch Settings

Type Contol	Rated Speed	% Governor Trip Speed — Table 2.18.2.1		
		Down		Up
		Without Speed Reducing Switch	With Speed Reducing Switch	
Static	All	90	100	90
Other	≤ 500	90	100	100
	> 500	95	100	100

normal operation without the tripped switch being manually reset. If this was not the case and the tripped switch was allowed to automatically reset, the malfunction might not even be observed.

The same logic holds true for the car safety mechanism switch. An elevator should not automatically resume normal operation after operation of the car safety without an inspection of the car safety, as a partially applied safety is probable. If the safety is not properly reset, it may accidentally apply and/or result in equipment damage.

2.18.5 Governor Ropes

2.18.5.1 Material and Factor of Safety. Governor ropes may be preformed or nonpreformed as long as they are of regular lay construction and are iron, steel, monel metal, phosphorus bronze, or stainless steel. Tiller rope and lang-lay rope construction is prohibited. See Handbook commentary on Tiller rope 2.17.12.4.

2.18.5.2 Speed-Governor-Rope Clearance. If the governor rope rubs against the governor jaws, rope guards, or other stationary parts during normal operation of the elevator, the life of the rope would be drastically reduced and it could part when called upon to actuate the safety.

2.18.5.3 Governor Rope Tag. The governor rope data tag gives the inspector the information necessary to assure that the governor rope is compatible with the design of the governor jaws. This information should be checked against the information on the speed governor marking plate (see 2.18.8 and 8.6.3.4).

2.18.6 Design of Governor-Rope-Retarding Means for Type B Safeties

The governor rope for Type B safeties usually has a considerable mass, which could result in damage to the wires or strands of the rope when the governor is actuated. The jaws must permit the pull-through of the governor rope at a predetermined tension. This pull-through value, once set, should remain constant over a period of years. Governor jaws typically do not wear appreciably from stopping the governor rope and safety application or in testing safeties. See Diagram 2.18.6.

2.18.6.5 The term "tripped by hand" was intended to convey a performance requirement that the governor be designed to allow for manual activation, by such means as a finger, hand, cable, lever, cam, electromechanical actuation, etc. It was also intended that the method of hand tripping the numerous designs of governors in the marketplace be done safely and without risk of causing equipment damage or danger to personnel performing the test. It was intended that hand tripping be applied to a stationary or relatively slow moving components and not to any components rotating at the same speed as the governor sheave.

2.18.7 Design of Speed-Governor Sheaves and Traction Between Speed-Governor Rope and Sheave

The governor rope must properly fit into the groove of the governor sheave in order for it to drive the governor at the same speed as the rope. The governor rope should not slip when running in the governor rope sheave. The governor sheave groove must have machined finished surfaces to eliminate excessive wear on the governor rope. The pitch diameter of the governor sheave and the governor tension sheave is regulated in order to reduce the damage caused by a sharp bend in a rope, which will reduce its life.

2.18.9 Speed Governor Marking Plate

This information should be checked against the information on the governor rope data tag to assure that the proper governor rope has been installed (see 2.18.5.3 and 8.6.3.2.2). Lubrication added to governor ropes can prevent the governor jaws from retarding the governor rope sufficiently to operate the car or counterweight safety.

SECTION 2.19
ASCENDING CAR OVERSPEED AND UNINTENDED CAR MOVEMENT PROTECTION

See Chart 2.19.

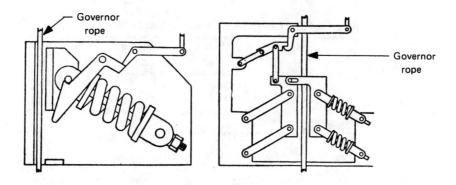

Diagram 2.18.6   Types of Pull-Through Governor Jaws

2.19.1 Ascending Car Overspeed Protection

See Diagram 2.19.1.

2.19.1.1 Purpose. As contradictory as it may sound, elevators can fall up! Traction elevators can overspeed in the up direction because the empty car weighs less than the counterweight. When they do, the results can be extremely serious.

Various failure modes can occur on the elevator driving machine assembly as follows:

(*a*) failure of brake component(s), such as, pins, shoes, links, brake arms, etc.;

(*b*) loss of friction coefficient due to fluid getting on brake surfaces, such as, machine lubrication or water from a sprinkler system, or brake drum and lining overheating due to dragging;

(*c*) failure of machine component(s), such as, shafts, gearing, bearings, etc.; and/or

(*d*) failure in the control system causing the car to be powered into the overhead. This may be caused by failure of the suicide circuit, field protection relay circuit, etc.

When such a failure occurs, the drive sheave starts rotating in a direction dictated by the hoistway masses. If the car loading exceeds balanced load, the car will descend and once it reaches governor tripping speed, the safety will apply and safely retard and stop the car.

However, if the car loading is less than balanced load, being heavier the counterweight will accelerate downward. Since the suspension ropes are intact, the car will be accelerated upward. If there is occupied space below the hoistway, the counterweight is required to have a safety (see 2.6.1), and as soon as the counterweight reaches governor tripping speed, its safety will apply, thus stopping the counterweight. The car will stop also during the safety-stopping interval. If the counterweight has no safety, it will continue to accelerate downward until it strikes its pit buffer.

2.19.1.2 Where Required and Function. The requirements provide protection against ascending car overspeed and/or collision of the elevator with the building overhead by requiring detection of ascending

overspeed of the car and operation of an emergency brake.

To ensure the safe functioning of the detection means, redundancy is required so that no single specified fault will lead to an unsafe condition. Checking of this redundancy is also required. Once the detection means is activated, a manual reset of the detection means is required.

2.19.2 Protection Against Unintended Car Movement

See Diagram 2.19.2.

2.19.2.1 Purpose. The purpose of the requirements is to detect unintended movement of the elevator away from the landing with the car door and hoistway door open.

2.19.2.2 Where Required and Function. The requirements provide protection against unintended motion of the elevator with the doors open by requiring detection of the unintended motion of the car and operation of an emergency brake.

To ensure the safe functioning of the detection means, redundancy is required so that no single specified fault will lead to an unsafe condition. Checking of this redundancy is also required. Once the detection means is activated, a manual reset of the detection means is required.

The reason for specifying that power be removed from the drive motor of the motor-generator set is because of the concern for protection against the failure of the "suicide circuit." An important consideration in formulating this requirement is the fact that it will prevent regeneration back to the AC line, and in so doing, may extend the braking distance.

2.19.3 Emergency Brake

(*a*) The emergency brake is a mechanical device independent of the braking system used to hold, retard, or stop an elevator. It only applies should the car overspeed or move in an unintended manner. Such devices include, but are not limited to, those that apply braking force on one or more of the following:

(*1*) car rails;

Chart 2.19 Traction Elevator Brake Type, Function, and Performance

Brake Type	Location	Normal Operation Function	Emergency Operation Function	Performance (Minimum Required)	
				Normal	Emergency
Driving machine brake (see 1.3 and 2.24.8.3)	Electric driving machine (see 1.3 and 2.24.8.1)	To hold car stationary at floor [Note (1)] [see 2.24.8.3(a) and (b), and 2.26.8]	Retard car during emergency stops [see 2.24.8.3(c), 2.26.8.3(c) and (d)]	Hold 125% rated load [Note (2)] [see 2.24.8.3(a)]	Retard empty car in up direction [see 2.24.8.3(c)]
Braking system (see 1.3 and 2.24.8.2)	Not specified	Note (1) (see 2.26.8)	Retard car during emergency stops, [see 2.24.8.2 and 2.26.8.3(c) and (d)]	Note (1)	Retard 125% rated load car in down direction from rated speed (see 2.24.8.2)
Emergency brake (see 1.3 and 2.19.3)	Electric driving machine, hoist ropes, compensation ropes, car, or counterweight (see 2.19.3.2)	Not permitted [see 2.19.3.2(c)]	Retard car during ascending car overspeed and unintended movement, independently of the braking system [see 2.19.1.2(b) and 2.19.2.2(b)]	Not applicable [see 2.19.3.2(c)]	Retard empty car in the up direction [see 2.19.3.2(a)] up to 110% of governor tripping speed [see 2.19.1.2(a)] Stop unintended motion: 125% rated load down or empty car up [see 2.19.2.2(b) and Note (2)]

GENERAL NOTE: See 1.3, 2.19, and 2.24.8.

NOTES:
(1) It is permitted that the braking system, or the driving machine brake function in normal retardation of the elevator car.
(2) For freight elevators not authorized to carry passengers, 100% rated load (see 2.16.8).

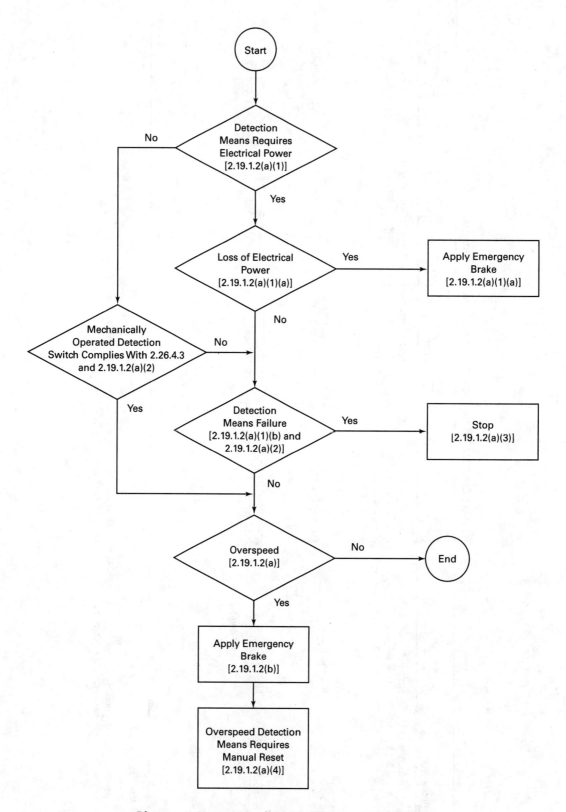

Diagram 2.19.1   Ascending Car Overspeed Protection

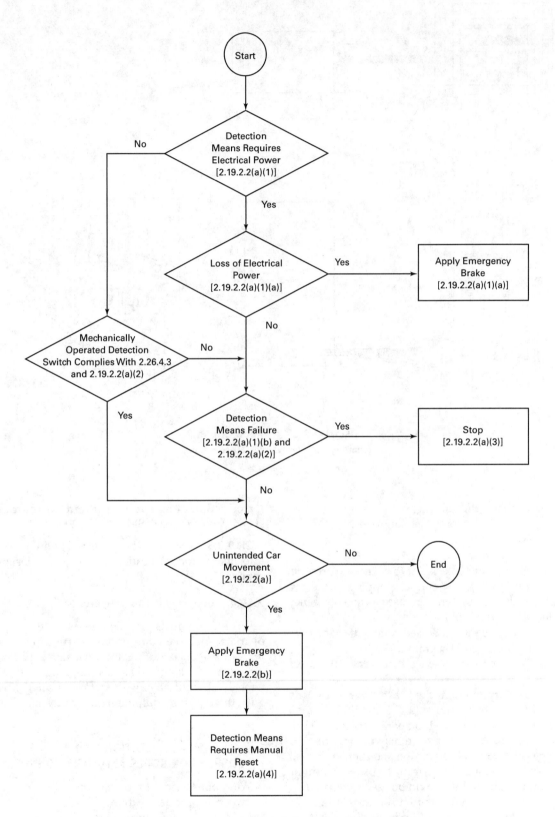

Diagram 2.19.2 Unintended Car Movement Protection

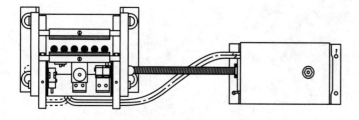

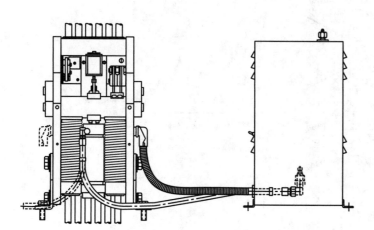

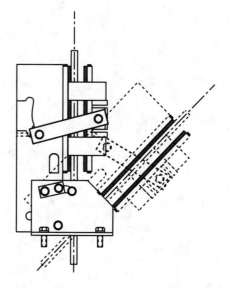

Diagram 2.19.3   Rope Gripper
(Courtesy Hollister-Whitney Elevator Corp.)

(2) counterweight rails;

(3) suspension or compensation ropes;

(4) drive sheaves; and

(5) brake drums.

(b) The emergency brake is not to be used to provide, or assist in providing, the normal stopping of the car. Examples of devices that may be used as an emergency brake include, but are not limited to:

(1) counterweight safety (see commentary below);

(2) rope brake. See Diagram 2.19.3.

(3) sheave jammer, a device that acts on the drive sheave to stop;

(4) secondary brake, if the brake drum is directly connected to the drive sheave [2.19.3.2(a)(5)].

These devices are intended to operate should the primary device fail. A single device may be provided for ascending car overspeed protection and unintended car movement or separate devices may be provided. In the above examples item (b)(1) could be used for ascending car overspeed, but would not provide protection against unintended car movement in the down direction.

Requirements for the elevator braking system (2.24.8) are new, and the elevator driving machine brake requirements (2.24.8.3) were revised concurrently with the

introduction of ascending car and uncontrolled low speed protection in ASME A17.1–2000.

2.19.3.3   The marking plate is required to provide information for verification of the correct design application.

2.19.4 Emergency Brake Supports

The requirements provide criteria for the safe design of the respective components/structures subject to forces developed during the retardation phase of the emergency braking and all other loading acting simultaneously, if applicable, with factors of safety consistent with the types of equipment/structures involved.

SECTION 2.20
SUSPENSION ROPES AND THEIR CONNECTIONS

Wire ropes used with traction elevators are primarily in two general classifications:

(a) 8 × 19 Class, which contains 8 metal strands around a fiber core;

(b) 6 × 19 Class, which contains 6 metal strands around a fiber core.

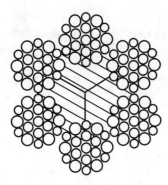

(a) 6 × 19 Suspension Rope, Warrington-Type Strand [Note (1)]

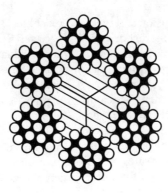

(b) 6 × 19 Suspension Rope, Filler-Wire-Type Strand [Note (2)]

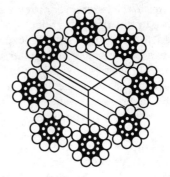

(c) Extra Flexible 8 × 19 Suspension Rope, Seale-Type Strand [Note (3)]

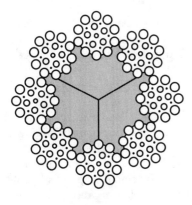

(d) Extra Flexible 8 × 21 Suspension Rope, Filler-Wire-Type Strand [Note (4)]

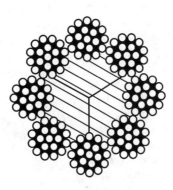

(e) Extra Flexible 8 × 25 Compensating and Governor Rope, FIller-Wire-Type Strand [Note (5)]

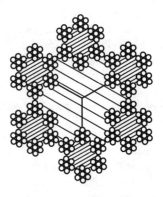

(f) Tiller-Rope Construction [Note (6)]

GENERAL NOTE: Sketches (a) through (f) show ropes with fiber core. Steel cores may be used in the ropes in sketches (a) through (e).

NOTES:

(1) Outer layer of strand is comprised of alternating large and small wires.

(2) Each strand includes six filler wires that are considered as not load bearing.

(3) 8 x 19 construction is more flexible than 6 x 19 and may be used over smaller sheaves.

(4) 8 x 21 construction is more flexible than 6 x 19 [sketches (a) and (b)], also more flexible than 8 x 19 [sketch (c)]. The 8 x 21 is especially designed for basement single-wrap and double-wrap traction machines, also basement drum machines, and under-slung-car installations.

(5) Each strand includes six filler wires that are considered as not load bearing.

(6) Used for hand-operating ropes. Not permitted for suspension rope or governor rope, except for replacement on old elevators with governors designed for it. This type of rope is also prohibited to connect the safety to the governor rope (2.17.12).

Diagram 2.20   Typical Wire Rope Constructions

Each strand can have 19, 21, 25, or 26 individual wires in its configuration. The fiber core is a tightly wound three-strand rope. It may be made of vegetable or synthetic fibers.

In an elevator wire rope, the arrangement of the wires in the strands, which determine its specific label, is of four types. See Diagram 2.20.

Taken in the order of predominant usage, the most common is the Seale construction, where each strand consists of a comparatively heavy center wire around which is a layer of nine smaller wires. The outer layer is supported in the valleys formed by the first layer; hence, both layers have the same number of wires. By simple wire count, this strand is designated 1-9-9. The large wires in the outer layer provide abrasion and wear resistance to this type. The filler wire constructions are designed to give more strand flexibility. The 6 small filler wires laid in between the main wires of the inner layer provide a series of valleys into which wires of the outer layer fit nicely. There are twice as many wires in the outer layer as there are wires in the inner layer. By wire count, this type is 1-6-6-12. The Warrington construction is less commonly used today. It is made up of 2 layers of wire about a center wire. There are 6 wires in the inner layer and 12 in the outer layer, the outer layer of wires being alternately large and small. The wire count designation here is 1-6-12.

From the foregoing it will be seen that 8 × 19 Seale simply means a wire rope of 8 strands of Seale construction having 19 wires each for a total of 152 wires. Generally speaking, the more wires per rope, the more flexible the rope. The large outer wires of the Seale construction strand make it comparatively stiffer than the filler wire strand, but eight-strand rope is more flexible than six-strand. Therefore, a 6 × 25 filler wire has more flexible strand construction, but having 6 strands reduces overall rope flexibility. The 8 × 25 filler wire combines both flexibility factors and has 200 total wires (8 × 25).

The term rope lay is used to designate the distance along the rope's length that one strand requires for one complete helical turn around the rope. It is approximately equal to 6.5 times the nominal rope diameter.

Several grades of steel are used in the manufacture of elevator ropes. The so called "iron rope" is made from a relatively soft, low-carbon steel. It is a low-strength rope used mostly for compensating ropes. Traction steel contains somewhat more carbon and has a tensile strength of at least 1 113 MPa (160,000 psi). This is the steel used for general elevator rope application.

In six-strand traction steel rope, all of the wires are made from traction steel, but eight-strand traction steel rope is made from a combination of traction steel wires in the outer layer and special high strength steel wires for the center wire and the inner layer. The reason for this is that eight-strand rope has a smaller cross-sectional steel area than six-strand rope, but both six- and eight-strand traction steel ropes have the same breaking strength.

In line with the construction of high-rise office buildings in the late 1950s and 60s, the need for higher strength rope was apparent and the wire rope industry responded with extra high-strength traction steel (EH rope). The rope strength was increased 25% to 40% and wire tensile strength was increased to at least 1 517 MPa (220,000 psi).

Whereas 12.7 mm (½ in.) and 15.9 mm (⅝ in.) diameter ropes were former common sizes, now 17.5 mm ($^{11}/_{16}$ in.), 19 mm (¾ in.), 20.6 mm ($^{13}/_{16}$ in.), and 22.2 mm (⅞ in.) diameter suspension rope are not uncommon.

The strength of a rope is determined by taking a specimen of new rope and pulling it apart on a tensile testing machine. The factor of safety is defined as the ratio between the minimum load necessary to break the rope, and the load imposed on it in service. However, after a period of service, changes occur in the rope consisting principally of wear in the outer wires and the formation of microscopic fatigue cracks, which develop into broken wires.

The strength of a new rope then is not the permanent measure of safety. Safety in ropes can only be assured by continued inspection and prompt removal at the appearance of signs of approaching failure. The visible signs are the number of breaks of outer wires on the worst lay length of the rope, as well as the distribution of these breaks over the various strands. Another is minimum rope diameter when it is reduced due to crown wire wear and core deterioration. See 8.11.2.1.3(cc)(1) for wire rope replacement criteria and the Guide for Inspection of Elevators, Escalators, and Moving Walks, ASME A17.2 for inspection recommendations.

The service life of a rope is affected by many things, but there are two in particular that should be mentioned: rope lubrication and sheave diameter. Sheave diameter can change due to wear.

In lubricating traction drive elevator ropes, a lubricant that will excessively reduce the coefficient of friction between the ropes and sheave must be avoided. An extreme pressure lubricant or one containing graphite, for instance, must not be used for this reason. The ideal lubricant should penetrate the strands of the rope and thicken up enough to remain in place to protect them from corrosion, but should not produce a greasy or slippery surface. In general, lubrication of ropes is recommended when they become dry to the touch.

When a rope runs over a sheave, there is sliding friction between the different wires of the rope, between the strands and the hemp center, and usually between the outside wires and the surface of the sheave. This action causes abrasion of the wire and a breaking down of the hemp center. Considerable difficulty was experienced in this respect some years ago, but improvement in lubrication practice, both in rope manufacture and in elevator maintenance, has acted to reduce these troubles.

As sheaves wear, not all grooves in the sheave wear at the same rate. This results in a sheave having grooves of different diameters. The individual ropes must therefore travel different distances in a revolution of the sheave, resulting in ever changing tensions and slipping of some ropes. This accelerates the wear of the rope(s).

As ropes wear into the grooves of a sheave, and the rope diameter shrinks, this often results in grooves that "match" the profile of an undersized rope. When new ropes are installed without regrooving the sheave (making all the groove profiles proper), the "ridge" worn by the previous rope will be a pressure point of wear on the new rope(s).

Requirement 2.24.2 specifies that for hoist ropes, no sheave shall be less than 40 times the rope diameter. The most obvious advantage of a large sheave is that it reduces the strain due to bending the wires of the rope. Another advantage, not quite so obvious, is that the larger the sheave the less the radial pressure between rope and sheave.

While a large sheave is beneficial from the standpoint of sheave wear and rope life, it also increases the cost of the machine. In a geared machine, it results in more load on the gear, and in a gearless machine, it requires more torque from the motor. Thus, the size of the sheave must be a compromise between these conflicting influences.

Recent elevator wire rope application considerations are: preforming; synthetic fiber cores; and prestretching.

Preforming is a wire rope fabrication technique whereby the helical shape of the strands in their position in the finished rope are permanently formed before the strands and fiber core are assembled into rope. The non-preformed wire rope does not have its wires and strands constrained in their final position in the rope; therefore, the ends must be held with wire bands called seizings. In nonpreformed wire rope, severed wires or strands will tend to straighten out. In preformed rope, they do not. The advantages of preformed rope are increased flexibility and longer useful life. Disadvantages are increased rope stretch and greater difficulty in detecting broken wires during service inspection.

Wire ropes with synthetic fiber cores have had successful application in oil fields due to their resistance to moisture and environment conditions detrimental to natural fibers. They have been used on some elevator installations, primarily as compensating ropes.

Construction stretch is permanent stretch and primarily due to progressive embedding of the wires into the fiber core in the kneading action as new ropes run over the sheaves. This process is experienced fairly rapidly in the early stages after installation of the ropes and diminished with time over a period, which would be from a few months to over a year. The amount of this stretch can be estimated fairly accurately and the new rope should be cut short to allow for this anticipated stretch. The manufacturer's values for estimated stretch of wire rope with fiber core are as follows:

(a) 6 × 19, $\frac{1}{2}$% to $\frac{3}{4}$% — 152 mm (6 in.) to 229 mm (9 in.) per 30.5 m (100 ft)

(b) 8 × 19, $\frac{3}{4}$% to 1% — 229 mm (9 in.) to 305 mm (12 in.) per 30.5 m (100 ft)

For preformed ropes, these figures are increased by 50%.

Prestretching is finish treatment of rope, which removes most of the initial (construction) stretch before the ropes are installed. It is utilized to eliminate the need for shortening the ropes, especially in high-rise installations. See Diagram 2.20.

2.20.1 Suspension Means

See Handbook commentary on 8.7.2.21 as to why new ropes are required.

2.20.2 Wire Rope Data

The wire rope data plate is installed by the manufacturer to advise maintenance and inspection personnel what wire rope the equipment was designed for. The rope data tag should be checked against the rope data plate at the time of inspection to ensure that the proper ropes are installed. It should also be checked at the time of re-roping to ensure that the proper ropes are being placed on the elevator. Wire rope lubrication could be detrimental to the system, for the reasons discussed in the Handbook commentary on 2.20. Requirement 2.20.2.2(j) was incorporated in the 2000 edition of A17.1 to advise if lubrication is acceptable, and if acceptable the criteria for lubrication.

2.20.2.2 On Rope Data Tag. The month and year the ropes were installed [2.20.2.2(d)] provides criteria for the replacement of a single damaged suspension rope.

(a) Ideally, the replacement of a single suspension rope should come from the same master reel. However, the likelihood is small that this can be done. Notwithstanding, the replacement rope must come from the same rope manufacturer and be of the same construction and material. The data for the replacement rope must correspond to the data for the entire set of ropes, as given by the Rope Data Tag.

(b) If the suspension ropes have been shortened once, it is doubtful whether the replacement rope can be made to perform uniformly with the original ropes, since the two components of rope stretch, i.e., construction and elasticity, will affect the performance of the new rope. Whereas, the original ropes will only be affected by elastic stretch.

(c) Diameter reduction and crown wire wear must be considered.

(d) Rope tension will have to be monitored closely, particularly when the replacement rope is first put into service. Severe sheave wear can occur in a matter of days on high-speed elevators.

2.20.3 Factor of Safety

The factors of safety noted in 2.20.3 are based on static load conditions. The Code does not address the dynamic condition caused at safety application or buffer engagement during which a condition known as car or counterweight jump occurs. After the ascending car or counterweight jumps, it falls back, taking up the slack in the suspension ropes, and induces a dynamic tension in the ropes that retards and stops the falling mass. The smaller the jump, the higher the rope factor of safety during fall back.

2.20.5 Suspension Rope Equalizers

If the operation of the equalizers is determined to be proper when inspected as required by 8.11.2.1.3(bb), load fluctuations will not be large enough to cause stress reversals and therefore induce fatigue failures.

2.20.7 Spare Rope Turns on Winding Drums

This requirement assures that the ropes will not be pulled off the drum when the car reaches its maximum overtravel in the down direction. Experience has shown that one turn is adequate.

2.20.9 Suspension Rope Fastening

2.20.9.1 Type of Rope Fastenings. The prohibition of the use of U-bolt-type rope clips was intended to preclude fastenings where the U-bolt clip would be directly involved in the attachment of the suspension rope to the rope fastening device. U-bolt-type fastenings have a tendency to damage the rope due to over tightening, thus reducing the rope strength. They also have a tendency to become loose during normal use. U-bolt-type fastenings are not prohibited for fastening of the governor rope.

The current requirements provide for approval of rope fastenings other than individual, tapered, babbitted rope sockets or wedge rope sockets on the basis of adequate tensile engineering tests (see 1.3, Definitions) by a qualified laboratory. Fatigue is not a problem on rope sockets because the actual variations in stress due to the changing loads are very small in comparison to the stresses to which the sockets are continuously subjected. Moreover, the rope sockets are not subjected to cyclical reversing load.

Since the elevator code specifies a very high factor of safety for the hoist ropes and the hoist ropes loading cycle never reaches a zero load condition, the cycles necessary to reach fatigue failure are many more times the cycles for rope replacement by wear or wire breaks. The design of elevator rope is basically forgiving to fatigue problems at the fastening because of the multiple strand/wire design and the elasticity of steel wire. Specification of the rope fastening device design and materials in the Code means that each device design will have a proper review by the related code committees before it is accepted.

Hoist rope wear testing at Westinghouse Elevator Co. and Bethlehem Steel's, Homer Research Center, never resulted in any fatigue failures of the rope fastening even though testing was conducted continuously for many years. The rope crown wear and wire breaks, which normally occur on elevator hoist ropes, caused the testing to be discontinued before any rope fastening fatigue failure could occur.

Any fatigue test that would be performed on the rope and fastening connection at or near the rope, full strength rating would not in any way be similar to the actual elevator system operating conditions.

2.20.9.2 Adjustable Shackle Rods. Adjustable shackle rods assure a means for adjusting the tension in all suspension wire ropes. Suspension ropes on traction elevators must be set at equal tension to ensure that each rope carries its share of the load. If a set is out of balance, the ropes under lesser tension are under little load and may soon show excessive wear from the creeping action in the sheave grooves and will probably cause premature wear on the sheave grooves. Unequal tension can result from improper adjustment at the adjustable shackle rods, but can also be caused by sheave grooves of unequal depth. If the ropes and sheave grooves are not at the same depth, the length of travel of all ropes will be unequal and thus the loading will be unequal.

2.20.9.3 General Design Requirements

2.20.9.3.5 Thread specifications control the tolerance of mating manufactured parts to ensure that the strength of the threaded connection will meet the design requirements.

2.20.9.7 Method of Securing Wire Ropes in Tapered Sockets

2.20.9.7.2 Seizing of Rope Ends. Alternate methods of seizing the ends of wire ropes to be socketed (other than using annealed iron or wire), such as banding, are acceptable if they serve to prevent the untwisting of the strands making up the rope beyond that necessary to make the fastening. The seizing should be removed after the fastening has been completed, as it serves no further purpose.

2.20.9.7.9 Inspection of Socket After Completion. See Diagram 2.20.9.7.9.

2.20.9.8 Antirotation Devices. This requirement prevents loss of rope lay and rope tension due to rotation of the ropes.

**SECTION 2.21
COUNTERWEIGHTS**

2.21.1 General Requirements

2.21.1.1 Frames. Where a counterweight employs filler weight material that is only retained in sealed containers, the structural design of each container must be adequate to withstand the lateral forces induced by the loose material and must be adequately sealed to ensure that there will be no loss of filler (weight) material during normal service, safety engagement, or buffer engagement.

2.21.1.2 Retention of Weight Sections. Where tie rods are used as the retention means, two tie rods are required to pass through all weight sections. The tie rods are not required to be secured to the frame. For example, two

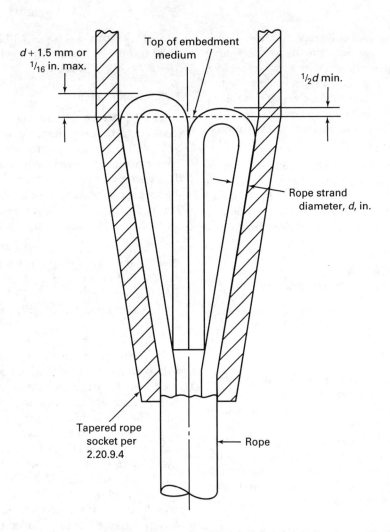

Diagram 2.20.9.7.9   Cross-Section Through Tapered Rope Socket Showing Maximum and Minimum Projection of Loops Above Embedment Medium

tie rods pass through all the counterweight sections and the bottom member of the frame only.

In lieu of using tie rods, some companies weld the counterweight subweight section to the frame. Angle retainers at the top of the counterweight are also utilized. When properly designed, these methods conform to the Code requirements.

2.21.1.3 Guiding Members. Counterweight frames are guided by sliding guide shoes or roller guides.

2.21.2 Design Requirements for Frames and Rods

2.21.2.3 Factor of Safety

2.21.2.3.3   The requirements provide criteria for the safe design of the respective components/structures, subject to forces developed during the retardation phase of the emergency braking and all other applicable loading acting simultaneously, with factors of safety consistent with the types of equipment/structures involved.

2.21.3 Cars Counterbalancing One Another

One car counterbalancing another is prohibited due to the probability that both cars cannot be level with a landing at the same time. Rope stretch due to loading and unloading will vary on each car. Releveling one car will make the other not level with its landing.

2.21.4 Compensation Means

The requirements for "compensating means" are included under 2.21, which is the general section for counterweights due to the common understanding that compensation is related to counterbalancing or counterweighting.

The purpose of compensation is to partially or fully compensate for the weight of the suspension ropes. This requirement does not dictate that compensation must be either chain or rope, and it is not the intent to restrict or specify the type of compensation permitted. The intent is to specify that the compensation means (rope, chain, etc.) be fastened structurally to the counterweight and car and that, if ropes are used with a tension sheave, a means of individual rope adjustment be provided.

Compensation means other than rope or chain may be used. Any method of fastening that transfers the load due to the weight of the compensation to the counterweight or car frame is permissible.

2.21.4.1 The requirements cover the whole compensation system. There is no safety justification for requirements that the means must be fastened to the frame of the car and/or counterweight, provided the point of fastening conforms to the strength requirements.

SECTION 2.22
BUFFERS AND BUMPERS

NOTE: The author expresses his appreciation to George W. Gibson for his extensive contributions to Section 2.22.

2.22.1 Type and Location

2.22.1.1 Spring, Oil, or Equivalent Buffers. A buffer is a device designed to provide a controlled stop of a descending car or counterweight beyond its normal limit of travel, by storing or by absorbing and dissipating the kinetic energy of the car or counterweight. An oil buffer uses oil as a medium, which absorbs and dissipates the kinetic energy of the descending car or counterweight. A spring buffer utilizes a spring to absorb and store the energy of the descending car or counterweight. See Diagram 2.22.1.1.

A Type C safety incorporates an oil buffer between the safety plank and the lower member of the car frame, thus eliminating the need for a separate oil or spring buffer located in the pit. Type C safeties utilize solid bumpers in the pit rather than buffers.

2.22.1.2 Location. Normally, car and counterweight buffers or bumpers are located between a pair or car or counterweight rails. They need not be positioned in this location as long as the car frame and the counterweight frame are designed accordingly.

2.22.2 Solid Bumpers

A bumper is a device, other than an oil or spring buffer, designed to stop a descending car or counterweight beyond its normal limits of travel by absorbing the impact. Solid bumpers are normally constructed of a solid block or blocks of wood or rubber.

Solid bumpers are only permitted for elevators with Type C safeties (2.22.11), or hydraulic elevators (3.22.1.5), and screw-column elevators (4.2.13), which have maximum speeds in the down direction of 0.25 m/s or 50 ft/min. The striking speed for an electric elevator with a rated speed of 0.25 m/s or 50 ft/min could be much greater based on the requirements of 2.18.8.1 and

2.18.2.1. For example, Table 2.18.2.1 permits a 0.25 m/s or 50 ft/min car to have a governor tripping speed of 0.89 m/s or 175 ft/min.

2.22.3 Spring Buffers

2.22.3.1 Stroke. The spring buffer stroke is the distance contact ends of the spring can move under a compressive load until all coils are essentially in contact. The stroke may be reduced by a fixed stop built into the assembly. In this case, the stroke will be the distance the spring can be compressed until the stop is contacted. This is described as the distance the spring can be compressed until it is a solid unit.

2.22.3.2 Load Rating. The Code contains three requirements for spring buffers and all three requirements must be fulfilled in order to comply fully with the requirements:

(*a*) A minimum buffer stroke is required for a given car speed.

(*b*) The spring buffer must be capable of supporting, without being compressed solid, a minimum static load of two times the combined weight of the car and its rated load, for car buffers and two times the weight of the counterweight for counterweight buffers.

(*c*) The spring must be compressed solid with a static load of three times the weight of the car and its rated load, for car buffers and three times the weight of the counterweight for counterweight buffers.

This rating specified in 2.22.3.2.1 is equivalent to the dynamic load, which is equal to the car weight plus its rated load when it is decelerated from rated car speed to zero car speed at the deceleration rate equal to $1g$.

(Imperial Units)

$$R = W + \frac{Wa}{g}$$
$$a = g$$
$$R = 2W$$

where

R = load rating of buffer, kg (lb)

a = deceleration rate, m/s^2 (ft/s^2)

g = gravity acceleration (or deceleration) rate, m/s^2 (ft/s^2)

W = car weight + rated capacity, kg (lb)

To better illustrate how the three spring buffer requirements are interrelated and interdependent, the following examples are given.

(*a*) *Example A.* Rated car speed = 1.02 m/s (200 ft/min or 3.333 ft/s)

Assume a spring buffer having the following parameters:

R = load rating = 2 (car weight + capacity) = 2 268 kg (5,000 lb)

K = spring constant = 25.0 kg/mm (1,400 lb/in.)

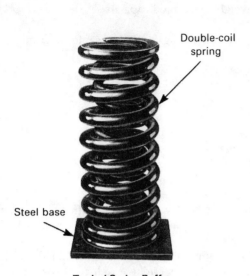

Typical Spring Buffer

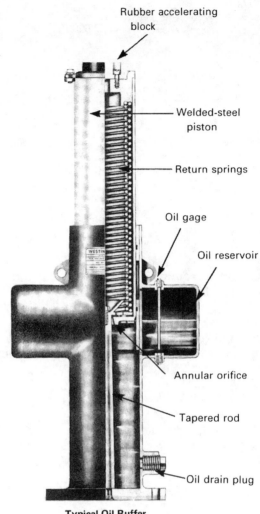

Typical Oil Buffer

Diagram 2.22.1.1   Typical Spring and Oil Buffer
(Courtesy Schindler Elevator Co.)

s = stroke = 102 mm (4 in.) = the minimum permitted by the A17.1 Code for a car speed of 1.02 m/s (200 ft/min)

(1) The spring buffer deflection, d, is

(Imperial Units)

$$d = \frac{R}{K} = \frac{5{,}000}{1{,}400} = 3.57 \text{ in. or } 0.2975 \text{ ft}$$

(SI Units)

$$d = \frac{R}{K} = \frac{2\,268}{25} = 90.7 \text{ mm}$$

This meets the requirement of 2.22.3.2.1, i.e., the buffer supports the load without being compressed solid.

(2) The average deceleration rate, $-a$, from rated car speed [1.02 m/s (3.333 ft/s)] to zero car speed over a distance of 90.7 mm (0.2975 ft) is

(Imperial Units)

$$-a = \frac{v^2}{2s} = \frac{(3.333)^2}{2(0.2975)}$$
$$= 18.67 \text{ ft/s}^2 \text{ or } 0.56g$$

(SI Units)

$$-a = \frac{v^2}{2s} = \frac{(1.02)^2}{2(0.0907)} = 5.7 \text{ m/s}^2$$

(3) The spring buffer deflection, d', at three times car weight plus rated capacity is

(Imperial Units)

$$P_3 = \frac{3R}{2} = \frac{3(5,000)}{2} = 7,500 \text{ lb}$$

(SI Units)

$$P_3 = \frac{3R}{2} = \frac{3(2\ 268)}{2} = 3\ 402 \text{ kg}$$

(Imperial Units)

$$d' = \frac{P_3}{K} = \frac{7,500}{1,400} = 5.3 \text{ in. (theoretically)}$$

(SI Units)

$$d' = \frac{P_3}{K} = \frac{3\ 402}{25.0} = 136 \text{ mm (theoretically)}$$

This meets the requirements of 2.22.3.2.2, i.e., the buffer became fully compressed (solid) at $d' = 4$ in.

(b) Example B. Rated car speed = 1.02 m/s (200 ft/min or 3.333 ft/s).

Assume a spring buffer with a spring constant double that of the spring buffer in Example A under the same dynamic load [2 268 kg (5,000 lb)]

$$R = 2\ 268 \text{ kg (5,000 lb)}$$
$$K' = 2K = 2 \ (1,400)$$
$$= 50.0 \text{ kg/mm (2,800 lb/in.)}$$

    (1) The spring buffer deflection, d, is

(Imperial Units)

$$d = \frac{R}{K'} = \frac{5,000}{2,800} = 1.786 \text{ in. or 0.149 ft}$$

(SI Units)

$$d = \frac{R}{K'} = \frac{2\ 268}{50} = 45.4 \text{ mm}$$

This meets the requirement of 2.22.3.2.1, i.e., the buffer supports the load without being compressed solid.

    (2) The average deceleration rate, a, from rated car speed (3.333 ft/s) to zero car speed over a distance of 0.149 ft is:

(Imperial Units)

$$-a = \frac{V^2}{2s} = \frac{(3.333)^2}{2(0.149)}$$
$$= 37.28 \text{ ft/s}^2 \text{ or } 1.158g$$

(SI Units)

$$-a = \frac{V^2}{2s} = \frac{(1.02)^2}{2(0.0454)} = 11.4 \text{ m/s}^2$$

The deceleration rate is double that under Example A.

    (3) The spring buffer deflection, d', at three times car weight plus rated capacity is:

$$P_3 = 3\ 402 \text{ kg (7,500 lb) (same as Example A)}$$

(Imperial Units)

$$d' = \frac{P_3}{K'} = \frac{7,500}{2,800} = 2.69 \text{ in.}$$

(SI Units)

$$d' = \frac{P_3}{K'} = \frac{3\ 402}{50} = 68 \text{ mm}$$

This does not compress the buffer solid [stroke = 102 mm (4 in.)] and, therefore, does not comply with 2.22.3.2.2.

Now, assume that a fixed stop is added and located such that the buffer stroke is reduced to 64 mm (2½ in.); the buffer now will also comply with 2.22.3.2.2; however, it will not comply with 2.22.3.1, which specifically states that the minimum stroke be not less than 102 mm (4 in.) for a rated car speed of 1.02 m/s (200 ft/min). Therefore, the addition of a fixed stop at a location that will satisfy the compression requirements will not satisfy the stroke requirement of 2.22.3.1.

The above Examples indicate that the spring constant will be essentially the same with either the fixed stop design or the full compression design in order to comply fully with the buffer stroke requirement and the compression requirements of the Code.

2.22.3.2.3 Where there is occupiable space below the hoistway the rating of the spring buffers is increased to reduce the load that can be transmitted to the building structure. This reduces the probability of building damage in the event the buffers are struck by the car or counterweight.

2.22.3.3 Marking Plate. For many years predating the 1980s, spring buffers typically consisted of a single helical spring buffer. In the early 1980s, it was apparent that many manufacturers were employing spring buffer designs containing one or more springs per spring buffer assembly. A spring buffer may consist of an assembly of one or more springs, some of which could be removable. By providing a marking plate for each spring buffer assembly showing the assembly load rating, stroke, and number of springs, an inspector can determine that no springs have been removed. If the springs can be removed, they must be identified and the identification indicated on the marking plate. This is often a number stamped into each spring at the top or bottom and shown on the marking place.

It is the intent of the requirements to require a marking plate that has been permanently and legibly marked and that this marking plate be provided with the spring buffer assembly. A loose marking plate fastened to the

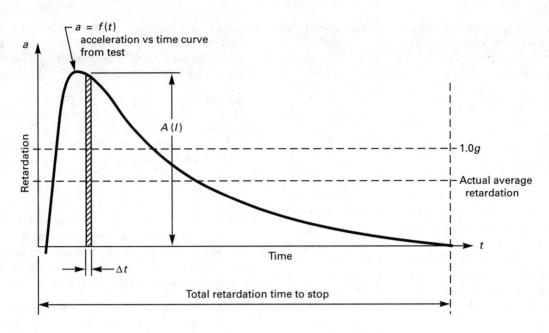

Diagram 2.22.4   Oil Buffer Retardation vs. Time

spring buffer assembly by a wire tie would comply with this requirement.

2.22.4 Oil Buffers

The lower the buffer striking speed, the lesser the jump of the ascending car or counterweight. One advantage to the use of reduced stroke oil buffers (2.22.4.1.2) at reduced engagement speeds is that the car or counterweight jump is less than that obtained with full stroke buffers at buffer-rated strike speeds.

The oil buffer retardation is required to be an average of 9.81 m/s² or 32.2 ft/s² (1 *g*) or or less if struck at 115% of rated speed. A typical oil buffer instrumented when tested would show the general graph of the retardation versus time as indicated in Diagram 2.22.4.

The area under the actual retardation curve is usually found by either a direct integration method or by numerical integration as follows:

(a) Direct Integration

(Imperial Units)

$$\text{Average retardation } a_{\text{avg}} = \frac{\int_{o}^{t} a\,dt}{t}$$

(b) Numerical Integration

(Imperial Units)

$$\text{Area} = \sum_{l=1}^{N} A(l)\,\Delta t$$

$$\text{Average retardation } a_{\text{avg}} = \frac{\sum_{l=1}^{N} A(l)\,\Delta t}{N \Delta t}$$

where

N = number of elemental areas under the curve

In order to comply with this requirement, it is necessary that:

$$a_{\text{avg}} \leq 1.0\,g$$

See Diagram 2.22.4.

2.22.4.1 Stroke. The oil buffer stroke is the oil displacement movement of the buffer plunger or piston excluding the travel of the buffer plunger accelerating device. Requirement 2.22.4.1.2 refers to what is commonly known as a reduced stroke buffer. A reduced stroke buffer is commonly found where the pit is not deep enough, such as on a high-speed installation where the standard oil buffer stroke is extremely long. A reduced stroke buffer can be used on any installation even if the pit is deep enough for the installation of a full-stroke buffer.

Over 50 years ago, it was recognized by the Code developers that the era of urban development would necessitate higher speed elevators than had existed before and that product safeguards still had to be assured. Of necessity, safeguards had to be compatible with the state of the art and be practical. One concern related to safety of the passengers as the car approached the terminal landings. The results of early study by the developers of the A17.1 Code were the requirements for oil buffers that would absorb the kinetic energy of the descending mass (car or counterweight) whose speed exceeded contract running speed but was less than the speed at which the safeties would apply. This terminal stop would be regulated so that an average of a 9.81 m/s^2 or 32.2 ft/s^2 ($1g$) retardation was maintained.

As car speeds increased, required buffer stroke sizes had to increase and new factors entered the picture, notably: the lack of commercially available buffer sizes for higher speeds; building conditions precluded the use of pit depths required to suit the longer buffer strokes; and the lack of sufficient overhead height above the top floor.

The 1931 Edition of the A17.1 Code addressed this issue wherein normal stroke oil buffers could not be used. The Committee concluded that reduced stroke buffers could be used provided that:

(a) there is a device which would detect that the speed of the descending mass (car or counterweight) had been reduced to a predetermined speed that could be retarded with an average of not more than 9.81 m/s^2 or 32.2 ft/s^2 ($1g$) on the reduced stroke buffer; and

(b) an emergency terminal speed-limiting device, which is independent of the normal terminal stopping device, be provided which automatically reduced the speed as the car approached a terminal landing.

This premise has been the basis of numerous Code requirements (e.g., 2.22.4.1, 2.25.4.1), which have been in the Code for 50 years, covering thousands of safe elevator installations all over the country and throughout the world. Clearly, the premise has withstood the test of time.

The use of reduced stroke buffers requires the inclusion of additional control circuits and devices. Commercially, the designation is called "reduced stroke buffers and potential switch slowdown," which the Code calls "emergency terminal speed limiting device."

The reduced stroke buffer/potential switch slowdown feature has no effect on the elevator operation, provided the slowdown of the elevator at the terminal is normal. Simply, the car speed is checked by means of a switch mounted on the car overspeed governor and a switch located in the hoistway or in the machine room, which is activated when the car is a predetermined distance from the buffer engagement point. If the car speed is correct, a normal terminal slowdown occurs. If, however, the car is going too fast at this predetermined checkpoint in the hoistway, the switch indicates this condition and the emergency terminal speed limiting device goes into

operation, removing power from the driving machine motor and brake, thereby assuring that the car strikes the buffer at that speed, which can be accommodated by the reduced stroke buffer.

2.22.4.5 Plunger-Return Requirements. These requirements assure that an oil buffer that has been compressed will return to its fully extended position in case the car or counterweight strikes the buffer again. A spring-return oil buffer is one employing a mechanical spring, not a gas spring.

Requirement of 2.22.4.5 specifically addresses spring-return buffers, which utilize a gas spring rather than a mechanical spring. A switch is required by 2.22.4.5(c) to shut the elevator down if there is a gas leak and the buffer does not return to its fully extended position after being compressed. The gas is only used to extend the buffer after it has been compressed.

2.22.4.6 Means for Determining Oil Level. During the buffer engagement, glass sight gages could shatter, resulting in buffer failure. Additionally, glass is subject to breakage for many reasons and, thus, glass sight gages on buffers are prohibited.

Most oil buffers are provided with two ports: one at the minimum oil level line, and the other at the maximum oil level line. A dipstick is another method employed to comply with this requirement. Where only one port is provided, the oil must be maintained at the level of that port.

2.22.4.7 Type Tests and Certification of Oil Buffers. This requirement permits the testing of only one of a series of buffers of the same type and design as a basis for approval of other buffers in the same series of oil buffers. Also, see the commentary on 8.3.2.

2.22.4.8 Compression of Buffers When Car is Level With Terminal Landings. An elevator may be arranged to use reduced stroke oil buffers subject to the requirements of 2.22.4.1.2 and at the same time utilize up to 25% precompression of the buffer stroke as permitted by this requirement. If precompression is utilized, the buffer must be of the mechanical spring-return type. Gas spring-return-type buffers cannot be used with precompression.

The Code does not address the simultaneous application of 2.22.4.8 and 2.25.4 since they cover independent conditions described as follows:

(a) Under normal elevator operation, the elevator comes into the bottom terminal landing at normal slowdown speeds for which precompression of the buffer stroke is permitted. Under this case, the emergency terminal speed limiting devices are inoperative.

(b) Under an emergency condition, the emergency terminal speed limiting devices, as required by 2.25.5, will function to reduce the speed of the car to a level that can be retarded by a reduced stroke buffer not in excess of 9.81 m/s² or 32.2 ft/s² (1 *g*). Under this case, precompression of the buffer is not utilized.

2.22.4.11 Buffer Marking Plate. The oil buffer is an energy dissipating device. The marking plate requires that information, which is necessary to describe the capacity of the buffer, namely, the maximum load and the maximum speed, which it has been, designed to safely withstand. Inherent in the speed rating is the minimum stroke. The viscosity of the oil will affect the buffer performance. The oil viscosity used during the type testing is required to be stated on the marking place.

The stroke of the buffer is necessary information for the inspector in determining compliance with 2.22. In some cases, when the buffer is fully compressed, there is still 25 mm (1 in.) to 102 mm (4 in.) of the plunger protruding from the body of the buffer, and therefore, the stroke cannot be readily measured.

The gas composition requirement for the marking plate in 2.22.4.11(f) is analogous to the buffer oil requirement specified in 2.22.4.11(a) and (c).

SECTION 2.23
CAR AND COUNTERWEIGHT GUIDE RAILS, GUIDE-RAIL SUPPORTS, AND FASTENINGS

NOTE: The author expresses his appreciation to George W. Gibson for his extensive contributions to Section 2.23.

2.23.1 Guide Rails Required

Guide rails are machined guiding surfaces installed vertically in a hoistway, with the exception of inclined elevators (see 5.1 and 5.4). They direct and restrain the course of travel of an elevator car or counterweight, when provided, within a predetermined path. They also provide the support for the car during safety application.

2.23.2 Material

This requirement applies to all load-bearing components of the guide-rail assembly. It does not apply, nor does the Code address, materials used for other minor components such as thin gage kicker shims used for adjusting rails or brackets.

The term "select wood" means wood that is clear and knot free, with a straight, close grain structure, without any visible defects such as splits and checks. Wood types that comprise the "Select(ed)" category as defined in the wood technology field, would comply with this requirement.

2.23.2.2 Requirements for Metals Other Than Steel. Cast iron is an alloy of iron, carbon, and silicon that is cast in a mold and is hard, nonmalleable, and brittle. Being brittle, it will not withstand the forces that could be applied to a guide rail.

2.23.3 Rail Section

When carrying an eccentric load, the car is prevented from tilting by the guide shoes or rollers pressing on the rails. The rail acts as a beam supported by the brackets and must have sufficient strength to carry these forces and also sufficient stiffness to keep the front edge of the platform level with the landing as loads enter or leave the car. The properties of the rail involved are the section modulus and moments of inertia about both axes. In the case of passenger elevators where the eccentric loads are small, these properties are not as important as with freight elevators where eccentric loads are usually very large. When an elevator or counterweight stops due to a safety application, the guide rails become columns and must be designed accordingly.

Shapes other than the T shape are allowed as long as they meet the requirements in 2.23.3. Other shapes used are the round rails (see Diagram 2.23.3) and the omega rails. Since round sections have uniform properties, they are ideally suited as columns. The round rail has been used on hydraulic elevators and counterweights where a safety would not be applied.

2.23.4 Maximum Load on Rails in Relation to the Bracket Spacing

2.23.4.1 With Single Car or Counterweight Safety. At safety application, the rail acts as a column and thus cross-sectional areas increase as the load increases. The weight per foot of the guide rail is directly proportional to the cross-sectional area. In addition to cross-sectional area, a column must be supported at certain intervals or it tends to buckle if the points of support are too far apart. The distance between the points of support to prevent buckling depends on the moment of inertia of the rail.

A safety application on a pair of rails also tends to spread the rails and thereby cause one of the safeties to pull away from the rail. If supports cannot be provided at sufficiently close intervals or if these supports become so numerous as to be impractical, the section of the rail must be reinforced between brackets. This is accomplished by means of channels or other structural sections fastened to the back of the rail. See Diagram 2.23.4.1.

While the strength and stiffness of the rail, with or without backing, is important, the strength and stiffness of the structure to which the rails are fastened is equally important. This may be brick or concrete walls, steel framing, or other means, depending on the building construction.

It frequently happens that the structural supports of the hoistway enclosure are a considerable distance from the desired location of the rail, in which case it may be necessary to provide columns of sufficient strength to resist bending as well as torsion due to the force on the side of the rail. This tends to twist the column with the torsional (twisting) moment.

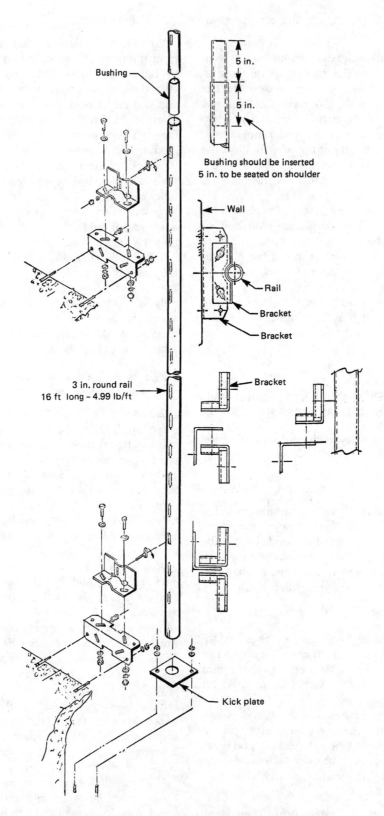

Bushing

5 in.

5 in.

Bushing should be inserted
5 in. to be seated on shoulder

← Wall

Rail

Bracket

Bracket

Bracket

3 in. round rail
16 ft long – 4.99 lb/ft

Kick plate

Diagram 2.23.3   Typical Round Rail Arrangement
(Courtesy Otis Elevator Co.)

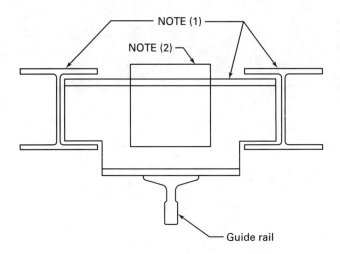

— Guide rail

GENERAL NOTE: Maximum deflection of rail column supports not to exceed 3 mm or $\frac{1}{8}$ in.

NOTES:
(1) Vertical rail column supports and cross-tie members are required and provided by other than the elevator supplier when rated load exceeds 3 500 kg or 8,000 lb. The size of the rail columns are determined by others from rail forces furnished by the elevator supplier.
(2) Alternate method of rail column support, as shown, by other than the elevator supplier when rated load is 3 500 kg or less. The size of the columns are determined by others from rail forces furnished by the elevator supplier.

Diagram 2.23.4.1   Vertical Rail Column Supports
(Courtesy Building Transportation Standards and Guidelines, NEII-1 2000, © 2000, National Elevator Industry, Inc., Teaneck, NJ)

Attendant to the general subject of guides and guide rails is the performance parameter referred to as car riding quality. This is one of the important criteria by which an elevator's performance is judged. The principal factors, which influence car ride, are those, which produce horizontal excitations (accelerations and retardations), which act upon the car, described as follows:

(a) Curvature and twist of the rails is caused by misalignment at installation or is due to building compression. Long, gentle curvatures to the rails occurring over several spans do not represent as big a problem as do distortions occurring in localized spans; however, it should be recognized that, in any case, when the deflection of the guide rail exceeds the available float embodied in the roller guide, an excitation will be imparted to the car.

(b) Being manufactured items, the rails are subject to tolerances, which may produce a step (displacement) due to this buildup of tolerances at the tongue and groove joint.

(c) Guide shoe forces due to any cause produce rail deflections whose magnitudes increase and decrease as a function of car position, which, in turn, is a function of the square of the car speed. These forces occur typically as a result of an eccentric live load in the car, which

is the most frequent load condition that occurs due to the random manner in which passengers distribute themselves. In addition, unbalanced cars produce guide shoe forces at all positions in the hoistway. Evenly balanced cars also produce guide shoe forces as they move away from the center of their rise.

(d) Transverse rope vibrations included in the hoist ropes and/or compensating ropes caused by hitch points moving sideways cause a transverse excitation to the end of the ropes due to the horizontal deflection of the rails or by hoistway wind disturbances due to stack effects or building sway or due to rope oscillations acting on short rope lengths as the car approaches the upper terminal.

(e) The displacement of air around a single car produces forces against the side of the car. In the case of multiple cars, the turbulence produced by cars passing each other in adjacent common hoistways causes side loads on the cars which, in turn, set up horizontal forces. Unevenness in air pressure is produced by projections extending into the hoistway, which also produces horizontal forces on the car.

The Code does not set a maximum spacing for rail brackets rather it specifies the maximum load on the rails in relation to bracket spacing. Bracket spacing is permitted to exceed those shown in Figs. 2.23.4.1-1 and 2.23.4.1-2, subject to the requirements of 2.23.5 and the requirement that the rail deflection caused by the application of the safety, shall not exceed 6.4 mm or $\frac{1}{4}$ in. per rail.

The term "reinforcement" refers to structural additions to a member in order to increase those properties, which increase rigidity such as moment of inertia, section modulus, cross-sectional area, radius of gyration, etc. There are many types of reinforcement that are utilized in a safe and satisfactory guide-rail design.

This requirement and Figs. 2.23.4.1-1 and 2.23.4.1-2 do not apply to a hydraulic elevator unless it is equipped with a safety.

2.23.5 Stresses and Deflections

This requirement also applies to hydraulic elevators with or without safeties. See Diagram 2.23.5.

2.23.5.3 Allowable Stress Due to Emergency Braking. This requirement provides criteria for the safe design of the respective components/structures subject to forces developed during the retardation phase of the emergency braking and all other applicable loading acting simultaneously, with factors of safety consistent with the types of equipment/structures involved.

2.23.8 Overall Length of Guide Rails

The extreme positions of travel are the highest and lowest positions that the car or counterweight could physically reach under any circumstance.

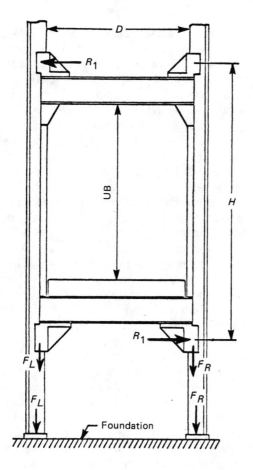

W = gross load to be stopped (lb)
(car wgt. + live load)
a = retardation (ft/s^2)
F_L = stopping force – LH safety (lb)
F_R = stopping force – RH safety (lb)
k = percent load carried by RH safety

$$F_L + F_R = W\left(1 + \frac{a}{g}\right)$$

$$F_R = W\left(1 + \frac{a}{g}\right)k$$

$$F_L = W\left(1 + \frac{a}{g}\right)(1 - k)$$

Guide Rail Force

$$R_1 = \frac{WD}{2H}\left(1 + \frac{a}{g}\right)(2k - 1)$$

Diagram 2.23.5   Guide-Rail Forces at Safety Application
(Courtesy Otis Elevator Co.)

The compressed car buffers establish the lowest position. The highest position is dependent upon the type of elevator system involved. In the case of a traction elevator, the highest position would be a function of the counterweight buffer stroke of the descending mass and a function of the gravity stopping distance of the ascending mass. In the case of a winding drum machine, the highest position would be the underside of the overhead structure. In the case of a hydraulic elevator, the highest position would be determined by the location of the plunger stops.

2.23.9 Guide-Rail Brackets and Building Supports

See Diagrams 2.23.9(a) and 2.23.9(b).

2.23.9.2 Bracket Fastenings. Lag bolts (screws) are an acceptable means of attaching guide-rail brackets to a wood building structure, provided the attachment complies with the stress and deflections specified in the Code.

2.23.10 Fastening of Guide Rails to Rail Brackets

This requirement defines the type of fastenings, which must be used, and not the method in which they must be installed. Sound engineering practice requires the designer to arrange the fastenings between the guide rails and their brackets in such a manner as to effectively transfer the loads through the connections with the allowable stresses and deflections while maintaining the accuracy of rail alignment. The loads include those induced at loading, running, safety application, and those caused by support movement, if applicable.

When guide rails are welded to their brackets, the welding procedure and design must comply with 8.8.2. The welding is required to be done by qualified welders in accordance with the requirements of the American Welding Society (see 8.8). The manufacturer may qualify the welder or, at his option, have the welder qualified by a professional engineer or recognized testing laboratory. The usual method of fastening guide rails to their brackets is with rail clips. Sliding rail clips are used to compensate for building compression. See Diagram 2.23.10.

SECTION 2.24
DRIVING MACHINES AND SHEAVES

NOTE: The author expresses his appreciation to George W. Gibson for his extensive contributions to Section 2.24.

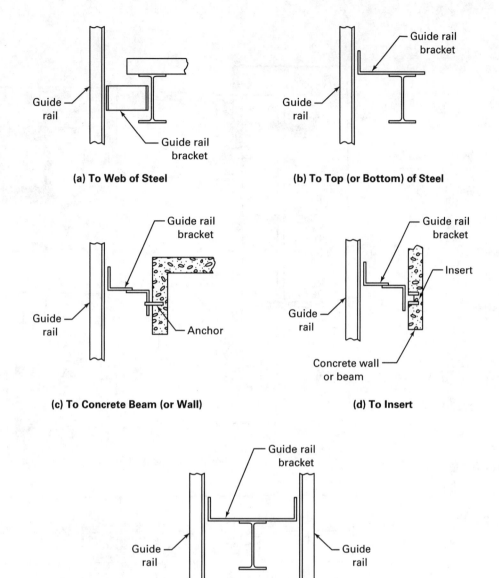

(a) To Web of Steel

(b) To Top (or Bottom) of Steel

(c) To Concrete Beam (or Wall)

(d) To Insert

(e) To Divider Beam

Diagram 2.23.9(a)   Guide-Rail Bracket Fastening Details
(Courtesy Building Transportation Standards and Guidelines, NEII-1 2000, © 2000, National Elevator Industry, Inc., Teaneck, NJ)

2.24.1 Type of Driving Machines

A traction machine is an electric machine in which the friction between the hoist rope and machine sheaves is used to move the elevator car and counterweight. A winding drum machine is a gear-driven machine in which the suspension ropes are fastened to and wind on a drum in order to move the car. See Diagrams 2.24.1(a) through 2.24.1(d).

There was significant concern by the A17 Committee over the inherent safety problems related to drum machine installations which predated A17.1–1955. There is no single written record that completely documents

the Committee's deliberations on this subject over 45 years ago; however, the following safety-related points guided the A17 Committee in prohibiting the use of drum machines for passenger elevators due to a history of accidents.

The car or counterweight, where provided, could be pulled into the building overhead due to a failure of the normal and final terminal stopping devices, resulting in a breakage of the suspension ropes or rope shackles, thus dropping the car or counterweight. Counterweighted cars could hang up in the guide rails causing slack suspension ropes and a collision between the car

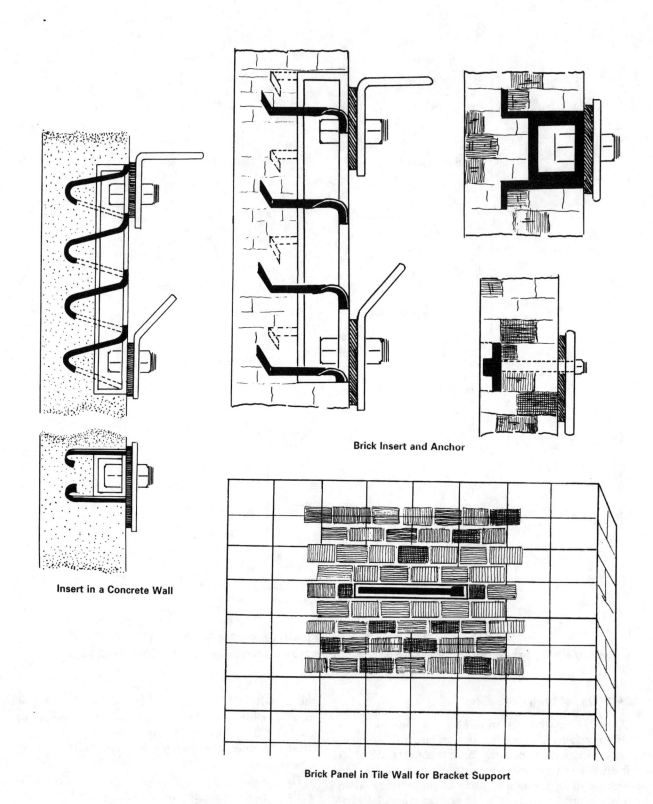

Insert in a Concrete Wall

Brick Insert and Anchor

Brick Panel in Tile Wall for Bracket Support

Diagram 2.23.9(b)   Typical Guide-Rail Bracket Inserts
(Courtesy National Elevator Industry Educational Program)

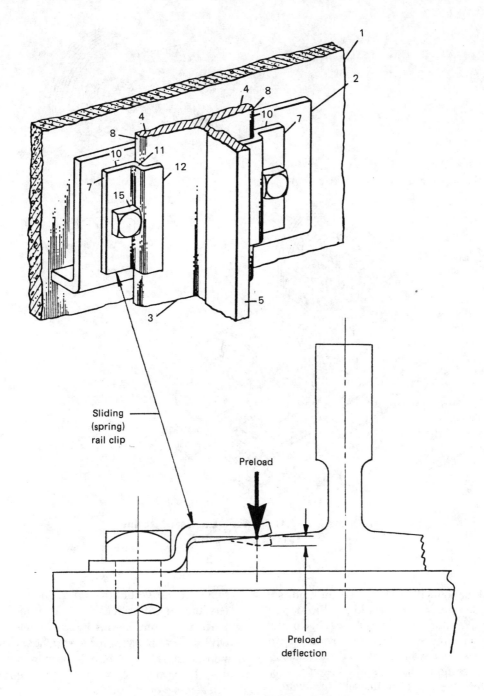

Sliding (spring) rail clip

Preload

Preload deflection

Diagram 2.23.10 Typical Sliding Rail Clip
(Courtesy Otis Elevator Co.)

counterweight and drum counterweight. Light duty drum machine elevators usually were arranged without a counterweight. On heavier duties, counterweighting was often provided to obtain economies of reduced machine size and reduced operating cost. The counterweighting was generally done in two ways, namely, by providing a drum counterweight or by providing both a drum counterweight and a car counterweight. The latter method reduced the total load on the drum machine. When used, both the drum and car counterweights would run in the same guide rails. The car counterweight suspension ropes terminated on the car. The counterweight had to weigh less than the empty car to permit acceleration and retardation under all conditions without slacking the suspension ropes. The weight of the car counterweight plus the weight of its

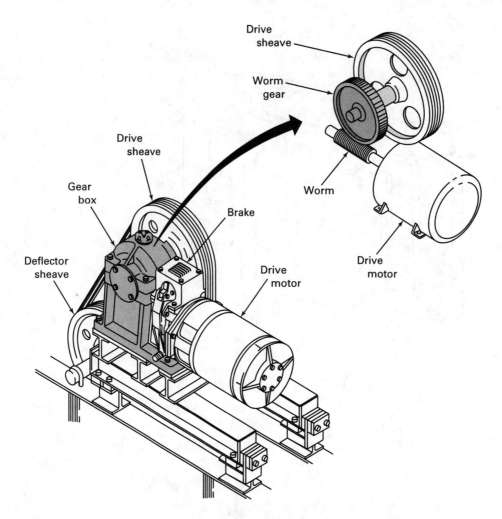

Diagram 2.24.1(a)   Worm and Gear Machine
(Courtesy Otis Elevator Co.)

ropes, in combination with the friction of the guide shoes and dry wooden guide rails, would produce too light a car condition, resulting in a hung-up car. This condition would cause slack ropes on the car side and the drum counterweight section would run into the stalled car counterweight section. This was the reason for disallowing counterweighted drum machine installations.

Upon release of the car safety, cars often fell taking up slack rope. While some manufacturers had various types of safeties, such as the roll type and flexible guide clamp, both of which were generally released after safety application by lifting the car, many early-type safeties were the drum-operating type or the block and tackle type wedge safety. These latter types produced slack suspension ropes at safety application. To release these types of safeties, the mechanic would have to get inside the car and reset them through a hole in the car platform floor. On many occasions, the mechanic would forget to first take up the slack in the ropes before getting on the car and releasing the safety. When he did release the

safety, the car would free-fall 1.5 m (5 ft) or 1.8 m (6 ft), resulting in broken legs or other serious injuries.

There were numerous injuries to elevator personnel while they were trying to take up the slack in the suspension ropes. The ropes have to be manually reseated back into their proper drum grooves. The mechanics and inspectors would reach out to help guide the ropes into their grooves while the drum was rotating, which frequently resulted in the amputation of fingers and hands.

Rope damage occurred due to slack ropes running over themselves as they wind on the drum because of the fleet angle generated as the ropes enter or leave the drum.

Suspension ropes frequently broke due to fatigue of the rope at its termination on the car or counterweight hitch. The hoist ropes will be vertical in one position of the car in its travel. For all other positions, there will be a side pull on the rope which fatigues the section of rope at its entry to its termination, necessitating periodic

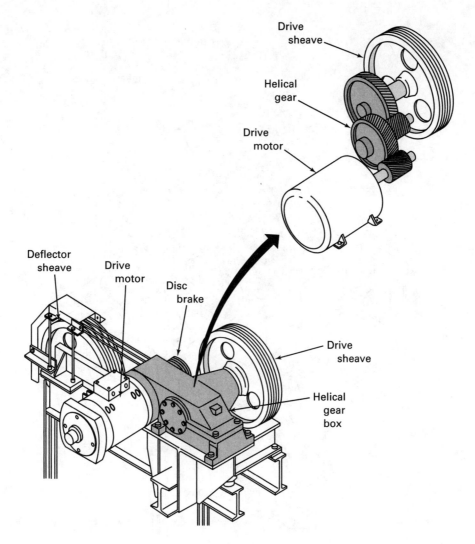

Diagram 2.24.1(b)   Helical Gear Machine
(Courtesy Otis Elevator Co.)

reshackling unless auxiliary rope fastenings are provided. In addition, the ropes tend to twist as they are loaded and unloaded, thus imposing a rope torsion on the rope sections already under fatigue by side forces. Failure of the ropes at the drum clamp was occasionally cited, but this was less prevalent because of the reserve or spare turns of rope provided on the drum.

The chances of dropping an elevator because of worn or corroded ropes were increased because the number of suspension ropes was limited to one or two to keep the drum size within reasonable limits. The limited number of suspension ropes, caused excessive car bounce at starting and stopping.

Drum spokes often fractured. Because it was necessary to have access to the rope termination or clamp inside the drum, the drum embodied a spoke construction at its ends.

In an effort to overcome some of the above problems, the A17.1 Code introduced certain safety features as follows.

(a) A final stop motion device, also known as a "machine automatic," or "machine limit" which is a mechanical device incorporated with the drum machine, mechanically monitored the position of the suspension ropes on the drum in order to prevent the car or counterweight, where provided, from being pulled into the building overhead. This device was arranged with mainline contacts. The operation of these contacts at either terminal opens the mainline circuits to the motor and brake and stops the car from full speed within the top and bottom overtravel.

(b) A slack rope switch monitors the suspension ropes to detect whether the suspension ropes slacken and, if so, stop the machine.

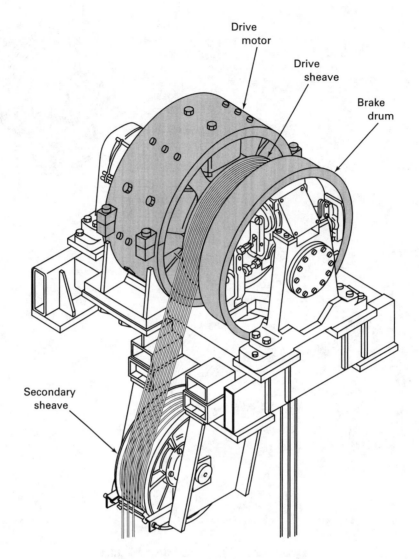

Drive
motor

Drive
sheave

Brake
drum

Secondary
sheave

Diagram 2.24.1(c)   Gearless Machine
(Courtesy Otis Elevator Co.)

(c) A potential switch was arranged with the power contacts located in the main power lines to the drive motor and the control circuits. Its purpose was to remove all power from the drive motor and control circuits when activated by either the slack cable device or the final contact in the machine automatic.

Even with these safety features, the basic safety to the riding public as well as to personnel responsible for servicing and inspecting drum machine installations could not be assured, and most of the hazards described above continued, as did the record of accidents.

At the September 18, 1940 meeting of the A17 Executive Committee, the following motion was passed: "Voted that the next edition of the Code shall prohibit drum type machines except for certain specified uses."

The A17 Committee approved a proposal by the elevator manufacturers that disallowed the use of drum machines for elevators except uncounterweighted,

freight elevators with rise less than 12.5 m or 40 ft and car speed not exceeding 0.25 m/s or 50 ft/min. This change appeared in the 1955 edition of the A17.1 Code and has continued unchanged since that time.

2.24.2 Sheaves and Drums

The sheaves or drum grooves must be metal to assure that they will not disintegrate in case of fire and possibly allow a car to fall. They must be finished to eliminate excessive wear on the ropes. The pitch diameter is regulated to assure that the ropes do not have too sharp a bend, which in turn would reduce their life expectancy. Grooves of sheaves can be lined with nonmetallic material, provided they meet the requirements of 2.24.2.3.1. Liners may be installed to increase traction and to prolong the life of the ropes. See Diagram 2.24.2.

Diagram 2.24.1(d)   Typical Winding-Drum Machine
(Courtesy Otis Elevator Co.)

2.24.2.3 Traction. There are no requirements establishing a maximum permissible amount of rope slippage since there is no way to establish such a limitation, which would be suitable for all designs, all capacities, and all speeds of elevators. Some rope movement relative to the drive sheaves is normal for all traction elevators; in fact, the ability to lose traction in the event of either the car or counterweight over-traveling at the lower terminals and compressing its buffer is a fundamental safety characteristic of traction elevators. Of course excessive slippage can result in accelerated sheave wear and other problems. The Guide for the Inspection and Testing of Elevators, Escalators, and Moving Walks mentions slippage but, like the Code, does not provide specific guidance as to an acceptable amount.

The following analysis documents a method of determining the available traction developed in a drive sheave by considering the frequency of oscillation of the suspended mass.

Consider the simple spring-mass system shown in Diagram 2.24.2.3(a), consisting of a mass, M, spring with constant, k, and a drive sheave.

The drive sheave will be rotated in each direction and the resulting motions studied. Let Q = available traction.

(a) Case 1 — Clockwise Rotation. The free-body diagrams of each element in the system are shown in Diagram 2.24.2.3(b).

δ_0 = static deflection of spring to support static load = $r\theta_0$
δ = deflection of spring due to moving system = $r\theta$
r = drive sheave radius
k = spring constant of spring
α = angular acceleration

For the weight, $(\Sigma V = 0)$. Therefore,

(Imperial Units)

$$T_m = W\left(1 - \frac{a}{g}\right) = W\left(1 - \frac{r}{g}\alpha\right) \qquad (1)$$

When the rope is slipping, the ratio of tensions in the ropes is equal to the available traction. This is expressed as, $T_s/T_m = Q$, from which

(Imperial Units)

$$T_s = QT_m \qquad (2)$$

The load acting on the spring is,

(Imperial Units)

$$T_s = k(\delta_0 + \delta) \qquad (3)$$

Equating (2) to (3) and substituting the angular relationships yields,

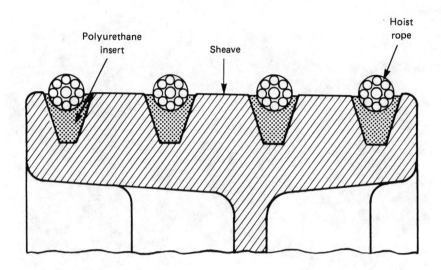

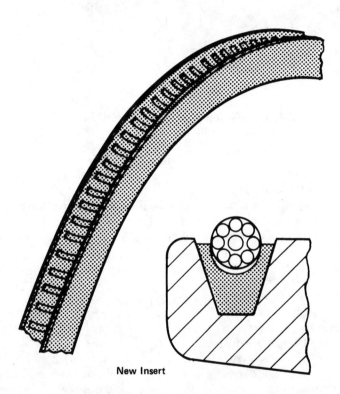

Diagram 2.24.2 Typical Sheave Groove Liners
(Courtesy Otis Elevator Co.)

(Imperial Units)

$$T_m = \frac{kr}{Q}(\theta_o + \theta) \qquad (4)$$

Equating (1) to (4) and substituting $\alpha = d^2\theta/dt^2$ in Eq. (1) yields

(Imperial Units)

$$\frac{Wr}{g}\left(\frac{d^2\theta}{dt^2}\right) + \frac{kr\theta}{Q} + \frac{kr\theta_o}{Q} - W = 0 \qquad (5)$$

In the static equilibrium position, Eq. (1) becomes

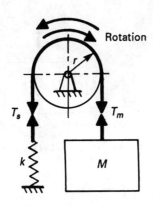

Diagram 2.24.2.3(a)

(Imperial Units)

$$T_m = W \qquad (6)$$

and Eq. (4) becomes

(Imperial Units)

$$T_m = \frac{kr\theta_o}{Q} \qquad (7)$$

Equating (6) to (7) yields

(Imperial Units)

$$W = \frac{kr\theta_o}{Q} \qquad (8)$$

Substituting Eqs. (8) and (5) yields

(Imperial Units)

$$\frac{d^2\theta}{dt^2} + \frac{k/Q}{W/g}\theta = 0 \qquad (9)$$

Equation (9) has the general simple harmonic motion form

(Imperial Units)

$$\frac{d^2\theta}{dt^2} + p^2\theta = 0 \qquad (10)$$

where

(Imperial Units)

$$p^2 = \frac{k/Q}{W/g} \qquad (11)$$

therefore

(Imperial Units)

$$p = \sqrt{\frac{k/Q}{W/g}} = \sqrt{\frac{k}{QM}} \qquad (12)$$

The frequency, f_1, equals

(Imperial Units)

$$f_1 = p/2\pi = \frac{1}{2\pi}\sqrt{\frac{k}{QM}} \text{ (cps)} \qquad (13)$$

The circular frequency, ω_1, equals

(Imperial Units)

$$w_1 = 2\pi f_1 = p = \sqrt{\frac{k}{QM}} \text{ (rad/s)} \qquad (14)$$

(b) *Case 2 — Counterclockwise Rotation.* The free-body diagrams of each element are shown in Diagram 2.24.2.3(c).

For the weight, ($\Sigma V = 0$). Therefore,

(Imperial Units)

$$T_m = W(1 + a/g) = W\left(1 + \frac{r}{g}\alpha\right) \qquad (15)$$

When the rope is slipping, the ratio, $T_m/T_s = Q$, from which

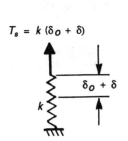

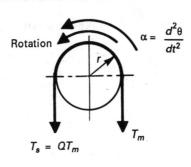

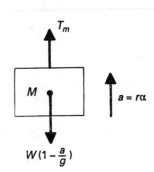

Diagram 2.24.2.3(b)

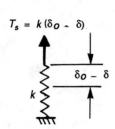 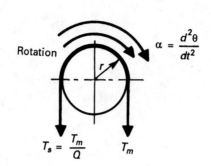 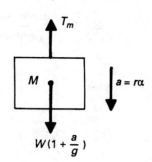

Diagram 2.24.2.3(c)

(Imperial Units)

$$T_s = T_m/Q \tag{16}$$

The load acting on the spring is

(Imperial Units)

$$T_s = k\,(\delta_o - \delta) \tag{17}$$

Equating (16) to (17) and substituting the angular relationships yields

(Imperial Units)

$$T_m = Qkr\,(\theta_o - \theta) \tag{18}$$

Equating (15) to (18) and substituting $\alpha = d^2\theta/dt^2$ in (15) yields

(Imperial Units)

$$\frac{Wr}{g}\left(\frac{d^2\theta}{dt^2}\right) + Qkr\theta + W - Qkr\theta_o = 0 \tag{19}$$

Noting again that, in the static equilibrium position, Eq. (15) becomes

(Imperial Units)

$$T_m = W \tag{20}$$

and Eq. (18) becomes

(Imperial Units)

$$T_m = Qkr\theta_o \tag{21}$$

Equating (20) to (21) yields

(Imperial Units)

$$W = Qkr\theta_o \tag{22}$$

Substituting (22) in (19) yields

(Imperial Units)

$$\frac{d^2\theta}{dt^2} + \frac{Qk}{W/g}\,\theta = 0 \tag{23}$$

Equation (23) has the same form as the simple harmonic motion equation given by (10), i.e.,

(Imperial Units)

$$\frac{d^2\theta}{dt^2} + p^2\theta = 0$$

where

(Imperial Units)

$$p^2 = \frac{Qk}{W/g} \tag{24}$$

therefore

(Imperial units)

$$p = \sqrt{\frac{Qk}{W/g}} = \sqrt{\frac{Qk}{M}} \tag{25}$$

The frequency, f_2, equals

(Imperial units)

$$f_2 = p/2\pi = \frac{1}{2\pi}\sqrt{\frac{Qk}{M}} \text{ (cps)} \tag{26}$$

The circular frequency, ω_2, equals

(Imperial Units)

$$w_2 = 2\pi f_2 = p = \sqrt{\frac{Qk}{M}} \text{ (rad/s)} \tag{27}$$

Taking the ratio of circular frequencies, ω_2/ω_1, equals,

(Imperial Units)

$$\frac{w_2}{w_1} = \sqrt{\frac{Qk/M}{k/QM}} = \sqrt{\frac{Qk}{M}} = \sqrt{\frac{QM}{k}} = \sqrt{Q^2} \tag{28}$$

from which the available traction

(Imperial Units)

$$Q = \frac{w_2}{w_1} \tag{29}$$

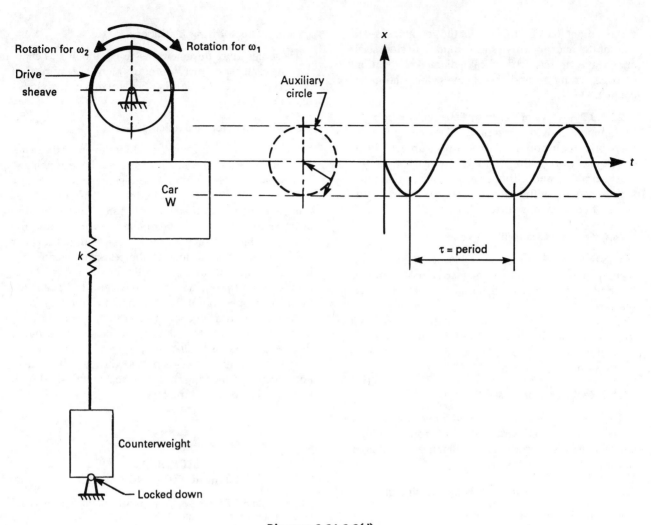

Diagram 2.24.2.3(d)

The preceding equation can be applied to actual measurements taken in an elevator system by locking one mass in place, say, the counterweight, driving the drive sheave in one direction, then the other, and measuring the frequency of the car, i.e., time period for car to move from one extreme to another. [See Diagram 2.24.2.3(d).]

2.24.2.4 Minimum Groove-Bottom Diameter. The minimum depth of metal at the bottom of the drive sheave groove is the numerical value established by the designer of the sheave based upon compliance with the minimum factor of safety for the specific sheave material specified in 2.24.3. Other design criteria may also be used to establish this minimum depth. This requirement applies only to drive sheaves.

The Code does not require that sheaves be manufactured to allow regrooving. The number of permissible regroovings would relate to the minimum groove bottom diameter established by the sheave designer and the amount of groove material removed at each successive regrooving.

2.24.3 Factor of Safety for Driving Machines and Sheaves

The factor of safety referenced, applies to the ultimate strength of the materials for stress calculations. The factor of safety relates to all the machine components subject to load including the motor, brake, driving sheave, deflector and secondary sheaves, gearing, shafting, etc. The high factor of safety is intended to account for impact, stress concentrations, and stress reversals which cause fatigue. In all cases, however, the responsibility of the designer goes beyond the simple application of a minimum allowable factor of safety. It is necessary to consider the application, shock loading, fatigue service, manufacturing variations, and the possibility of overloading, to ensure that the equipment will provide reliable and safe operation throughout its life.

The method of computation of the resultant sheave shaft load is addressed in 2.24.3.1.2. The magnitude of this resultant load is a function of the geometric relationship of the machine drive sheave and deflector/secondary sheave, the respective sheave diameters, the number of rope wraps around the sheaves, and the areas of contact.

2.24.3.2 Factors of Safety at Emergency Braking. The requirement provides criteria for the safe design of the respective components/structures subject to forces developed during the retardation phase of the emergency braking and all other applicable loading acting simultaneously with factors of safety consistent with the types of equipment structures involved.

2.24.4 Fasteners Transmitting Load

The bolt or other member when in place should be so tight that there will be no relative motion between the bolt and the involved member when the driving machine is operated. There is no requirement in the Code for shrink-fitting all members that are connected together to transmit torque. Neither is it the practice of all manufacturers to always use a shrink-fit for such connections.

2.24.6 Cast-Iron Worms and Worm Gears

The use of cast iron is prohibited for worms and worm gears as it is a commercial alloy of iron, carbon, and silicon that is hard, brittle, and nonmalleable, which thus has a tendency to break easily.

2.24.8 Braking System and Driving-Machine Brakes

Machine brake system requirements were completely revised in ASME A17.1–2000, concurrently with the introduction of ascending car/uncontrolled low speed protection (see 2.19). Prior to this edition of the Code there was no requirement for the brake to have the capability to decelerate the car to buffer striking speed. The brake is also required to hold the car at rest with rated load or 125% of rated load for passenger elevators and freight elevators permitted to carry passengers (see 2.16.8) and 100% of rated load for freight elevators. See Diagram 2.24.8.

2.24.8.5 Marking Plates for Brakes. The plate contains vital information required to maintain brakes in safe operating condition.

2.24.8.6 Driving Machine Brake Design. The contact area between the brake and drum determines the ability of the brake to absorb energy and, therefore, affect the stopping capacity of the brake.

2.24.8.9 Means of Manual Release. A manual brake release is not required by the Code, but is permitted. Performance requirements allow for the use of manual brake release tools. These requirements apply only if a manual brake release is provided. If provided, provisions must be made to prevent accidental activation.

2.24.9 Indirect-Driving Machines

These design requirements assure that belt and chain driving machines provide equivalent safety to that which is provided by the other types of driving machines permitted by the Code.

2.24.10 Means for Inspection of Gears

Means of access to geared machines in order that vital components may be visually inspected is very helpful. This is particularly valid for worm-gear machines where the gear is usually manufactured from nonferrous materials such as bronze and the worm is steel. Early warning of impending failure can be obtained by inspecting the contact surfaces of such worm wheels. With helical gear machines and similar devices, the gears are usually less likely to show visible degradation before failure; however, even in these cases it can be useful to have some means of inspection of the gearing and lubricant.

With modern inspection probes such as Borascopes, it is not necessary that the inspection access provide direct visual access to all contact or wear surfaces. However, it is helpful if access is provided such that surfaces not directly visible may still be inspected if desired with instruments such as Borascopes.

SECTION 2.25
TERMINAL STOPPING DEVICES

NOTE: The author expresses his appreciation to Ralph Droste for his extensive contributions to 2.25.

When in doubt about any of the Code and safety requirements, check the definitions in 1.3 and the NEC® 620-2. The following are of particular interest when reviewing the electrical requirements:

    braking, electrically assisted
    certified
    certified organization
    control, motion
    control, operation
    control system
    controller
    controller, motion
    controller, motor
    controller, operation
    hoistway access switch
    labeled/marked
    landing zone
    leveling
    leveling device
    leveling zone
    listed/certified
    normal stopping means
    operating device
    operation, inspection

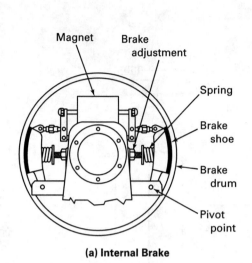

(a) Internal Brake

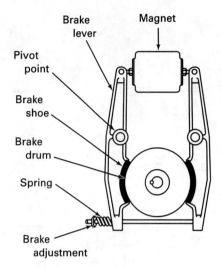

(b) External Brake

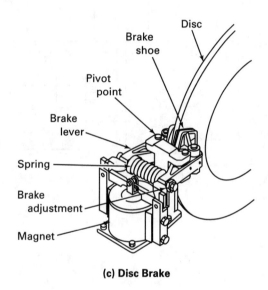

(c) Disc Brake

Diagram 2.24.8   Typical Brake Arrangements
(Courtesy Otis Elevator Co./Ralph Droste)

signal equipment (NEC®)
static switching
terminal speed-limiting device, emergency
terminal stopping device, emergency
terminal stopping device, final
terminal stopping device, machine final (stop-motion switch)
terminal stopping device, normal

Wiring Diagram 2.25/2.26 captures the essential Code requirements of 2.25 and will be referred to in the Handbook commentary. The following legends are used in the wiring diagrams and referenced in the Handbook commentary.

C1	Potential Relay #1
C2	Potential Relay #2
CD1	Car Door Relay #1
CD2	Car Door Relay #2
LD1	Landing Door Relay #1
LD2	Landing Door Relay #2
NOR1	Normal Operation Relay #1
NOR2	Normal Operation Relay #2
AUD	Access Switch Operation
3	Car Door Bypass #1
4	Car Door Bypass #2
5	Hoistway Door Bypass #1

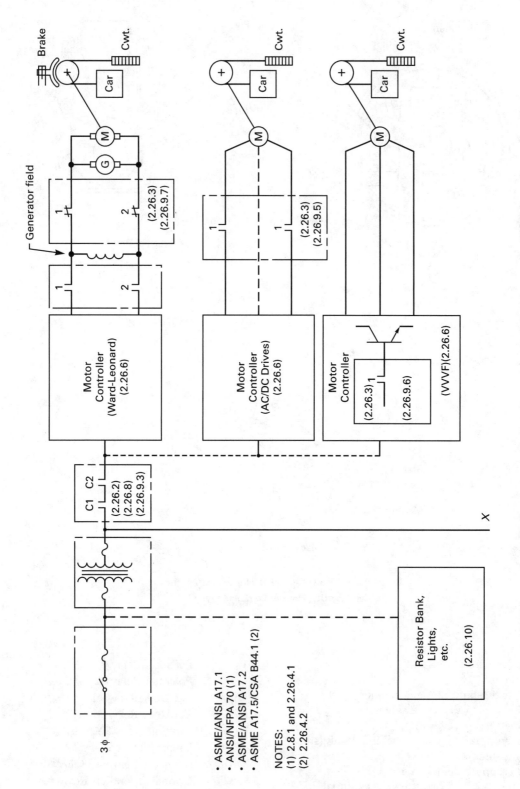

Diagram 2.25/2.26   Wiring Diagram — Summary of Essential Safety Requirements
(Courtesy Otis Elevator Co./Ralph Droste)

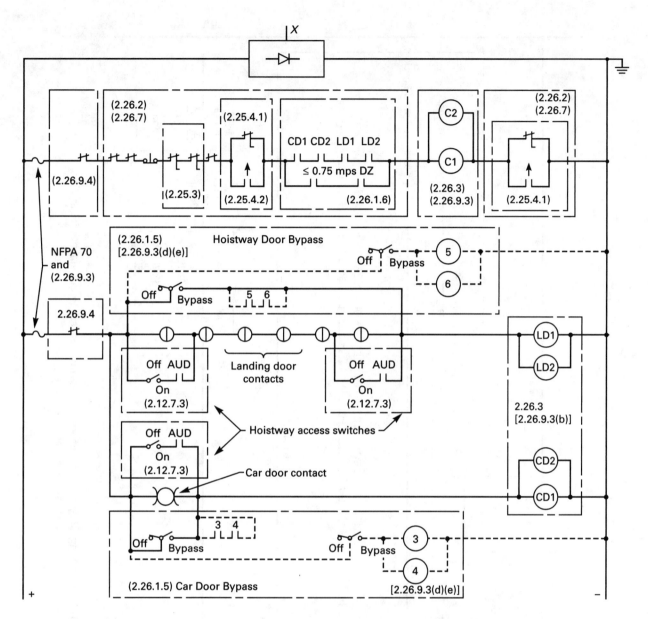

Diagram 2.25/2.26   Wiring Diagram — Summary of Essential Safety Requirements (Cont'd)
(Courtesy Otis Elevator Co./Ralph Droste)

6   Hoistway Door Bypass #2

See introduction to Handbook commentary on 2.26.

2.25.1 General Requirements

The first paragraph of this requirement indicates that mechanically or magnetically operated optical or solid-state devices for determining car position and speed are permitted to be used for normal terminal stopping devices, emergency terminal stopping devices, and emergency terminal speed limiting devices. The second paragraph indicates that mechanically operated switches are required for the final terminal stopping devices.

The revision incorporated in A17.1–1996 permitted the use of mechanically and magnetically operated optical- or static-type devices to determine car position for emergency terminal speed limiting devices required by 2.25.4.1. Prior to A17.1–1996, the Code permitted the use of nonmechanical-type switches for determining car speed only. The revision has no impact on safety performance since 2.25.4.5 and 2.25.4.1.9 cover the failure protection requirements for the nonmechanical devices. The revision permitted the same level of performance language afforded to the requirements for the speed monitoring devices and is more appropriate in today's elevator motor controller.

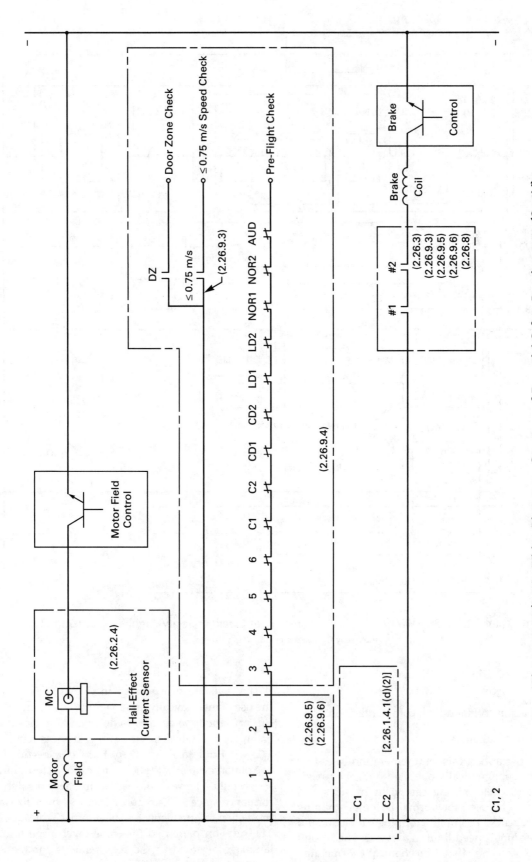

Diagram 2.25/2.26 Wiring Diagram — Summary of Essential Safety Requirements (Cont'd)
(Courtesy Otis Elevator Co./Ralph Droste)

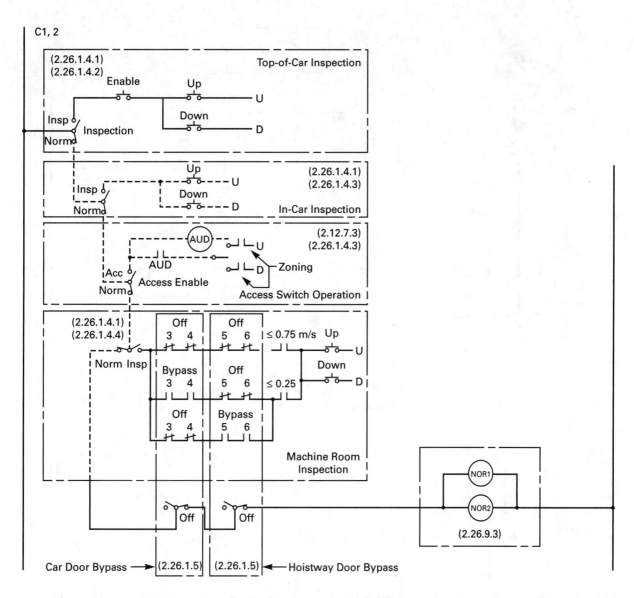

Diagram 2.25/2.26   Wiring Diagram — Summary of Essential Safety Requirements (Cont'd)
(Courtesy Otis Elevator Co./Ralph Droste)

2.25.2 Normal Terminal Stopping Devices

Nonautomatic elevators, such as those with constant pressure push-button or car switch control, require normal terminal slowdown devices to stop the car at or near the terminal floor, in the event the operator forgets to initiate a stop.

Automatic elevators, on the other hand, have a Position transducer (e.g., floor controller, selector or other device) to automatically slow down and stop the car at landings where a car and/or hall call is registered. A failure of the normal stopping means at an intermediate floor might be an inconvenience, but a failure at a terminal landing could allow the car to strike the buffer at full speed. For this reason, the Code has required the stopping devices at the terminal floor to meet certain safety requirements.

Variable voltage elevators were almost universally controlled by resistors and relay contacts in series with the generator field. These relays could easily be deenergized by the normal terminal stopping devices.

When the term *device* is used in 2.25.2, the intent of the text is to permit devices other than mechanical switches (i.e., magnetic, optical, and solid state devices). See Diagrams 2.25.2(a) and 2.25.2(b).

2.25.2.1 Where Required and Function. The "normal terminal stopping device" is required to function independently of the operation of the "normal stopping means" and the "final terminal stopping device," but

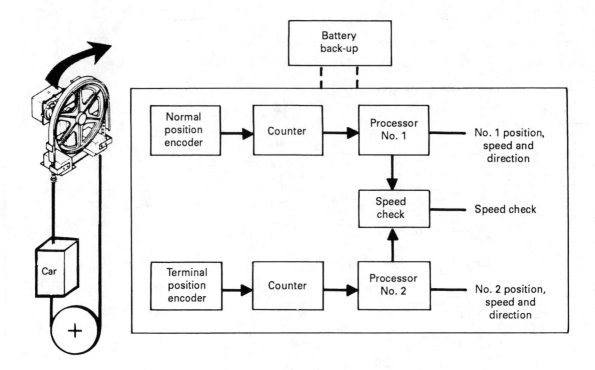

Diagram 2.25.2(a)   Functional Diagram for Position Transducer
(Courtesy Otis Elevator Co./Ralph Droste)

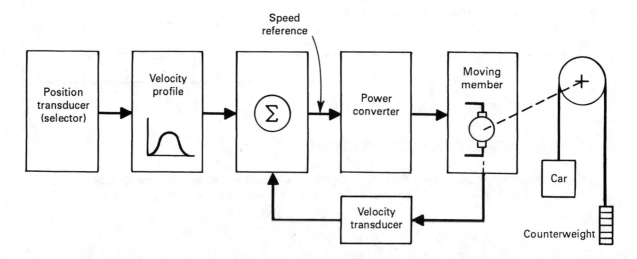

Diagram 2.25.2(b)   Functional Diagram for Normal Speed/Position
(Courtesy Otis Elevator Co./Ralph Droste)

the "normal terminal stopping device" may be used as the "normal stopping means" on elevators with rated speeds of 0.75 m/s or 150 ft/min or less.

A "normal stopping means" is that portion of the operation, which initiates stopping of the car in normal operation at landings in response to an operating device, e.g., car switch, car and/or hall buttons.

A "normal terminal stopping device" is the "device(s) to slow down and stop an elevator, dumbwaiter, or material lift car automatically at or near a terminal landing,

independently of the functioning of the 'normal stopping means'." (See 1.3 and Diagram 2.25.2.1.)

2.25.3 Final Terminal Stopping Devices

At the time the 1955 edition of the Code was being prepared, it was felt that there was a need to eliminate car contact with the overhead structure and for continuous functioning of the top final terminal stopping switch.

In complying with the minimum bottom counterweight runby requirement in 2.4.2 and the minimum

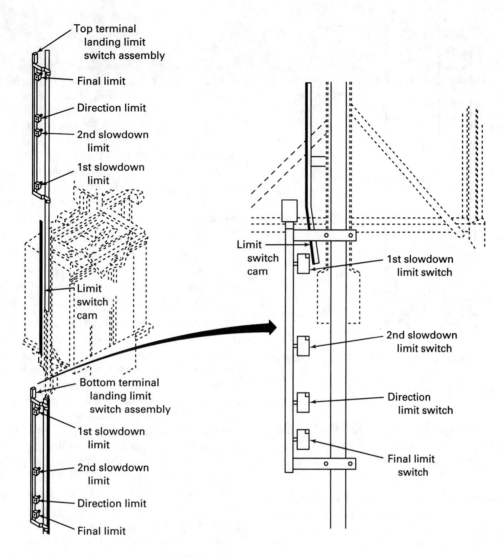

Diagram 2.25.2.1 Normal Terminal Stopping Device
(Courtesy Otis Elevator Co./Ralph Droste)

stroke requirements specified for spring and oil buffers in 2.22.3 and 2.22.4 for various rated car speeds, the distance of the top final terminal stopping switch operation is determined by the sum of the counterweight runby plus $1\frac{1}{2}$ times the buffer stroke requirement.

To ensure that the top final terminal stopping device will continue to function as required, for performance and operation, it was necessary in view of the reduced distance possibility cited above to provide a requirement that it shall be operative for not less than 600 mm or 2 ft. This minimum dimension was established only after full consideration was given to the minimum top car clearance, which is required to be provided. This assures that the device will remain operative if the car overtravels the terminal landing for a car travel of not less than 600 mm or 2 ft above the landing as specified in the present requirement.

The final terminal stopping device automatically causes the power to be removed from an electric elevator driving machine motor and brake independent of the function of the normal terminal stopping device, the operating device, or any emergency terminal speed-limiting device after the car has passed a terminal landing.

This requires that separate and independent final limit switches be provided to complete the driving machine motor and brake circuit in either direction of travel. It is difficult to verify by field inspection that when the final limit is operated, it directly causes the switch to drop out, and that a solid-state device is not interfacing between the two operations. On some controllers, this may be verified by observing the hard wiring from the terminal studs to the final limit and tracing the hard wiring directly to the coil of the contactor required. See Diagram 2.25.3.

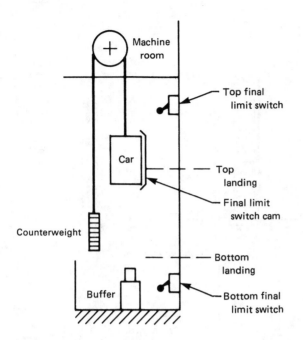

Diagram 2.25.3   Typical Final Terminal Stopping Device Location
(Courtesy Otis Elevator Co./Ralph Droste)

2.25.3.2 Where Required and Function. Winding drum machines must have a final terminal stopping switch located in the hoistway and a machine final, which is also referred to as a stop-motion switch.

2.25.3.5 Additional Requirements for Winding Drum Machines. This requirement refers to the machine final, which is a final terminal stopping device operated directly by the driving machine. It is used on winding drum machines and is also referred to as a stop-motion switch. See Diagram 2.25.3.5.

2.25.4 Emergency Terminal Stopping Means

Provisions are provided for emergency terminal speed limiting devices used with reduced stroke buffers and emergency terminal stopping devices for static control elevators with rated speeds over 1 m/s or 200 ft/min. The confusion caused in previous editions of ASME A17.1 by the arbitrary rated speed where emergency terminal stopping means had been required has been resolved by specifying a speed where oil buffers are required. See Diagram 2.25.4.1.

2.25.4.1 Emergency Terminal Speed-Limiting Device. The emergency terminal speed-limiting device is a device that automatically reduces the car speed, by removing the power from the driving machine motor and brake, as it approaches a terminal landing, independently of the functioning of the operating device and normal terminal stopping device if the latter fails to slow down the car as intended. This device is only installed when a reduced stroke buffer is used. It causes the car

to slow down to at least the speed at which the buffer is rated.

As an example, for an elevator with a contact speed of 8.12 m/s (1,600 ft/min) and a reduced stroke buffer rated for 5.08 m/s (1,000 ft/min), when the emergency terminal speed-limiting device is activated, the car speed would be reduced from 8.12 m/s (1,600 ft/min) to 5.08 m/s (1,000 ft/min) at or just before contacting the buffers. See Diagrams 2.25.4.1 and 2.26.2.

The present requirement permits the use of mechanically operated, magnetically operated, optical, or solid-state devices for determining both car position and speed. Requirements 2.25.4.1.5 and 2.25.4.1.9 cover the failure protection requirements to assure safety.

2.25.4.1.9 The failure protection requirements were added in A17.1–1996 when magnetically operated, optical or solid-state devices are used for determining car position and speed. See Diagram 2.25.4.1.9.

2.25.4.2 Emergency Terminal Stopping Device. The requirements are performance oriented which permits latitude in compliance. The operation of the emergency terminal-stopping device has to be entirely independent of the normal terminal stopping device and the normal speed control system. A test procedure for this type of operation must be prepared by the installer or manufacturer and submitted to the enforcing agency. Acceptable devices include, but are not limited to:

(a) a tachometer on the drive sheave as long as it is not part of the normal control;

(b) a tachometer driven by the governor;

(c) tooth wheel or tapes or a device that converts pulses to speed;

(d) governor overspeed switches.

These are all acceptable speed-indicating devices and can be verified by physical observation on the job site. The elevator company shall demonstrate that both normal stopping means and normal terminal stopping devices have been disabled. The car shall then stop as a result of the emergency terminal stopping device. Note that there is nothing to prevent the car from starting again to attempt to get to terminal. See Diagram 2.25.4.1.

SECTION 2.26
OPERATING DEVICES AND CONTROL EQUIPMENT

NOTE: The author expresses his thanks to Ralph Droste for his extensive contributions to 2.26.

The A17.1 Code along with the referenced standards (e.g., ANSI/NFPA 70, CSA B44.1/ASME A17.5, ICC/ANSI A117.1, ADAAG, ...) are concerned with the safe design, installation, use, operation, and accessibility of the elevator over the life of the equipment. The following are the primary safety concerns: operation at the terminals (2.25); overspeed (2.26.9.8, 2.26.10, 2.18, and 2.19); leveling with open car and landing doors (2.26.1.6);

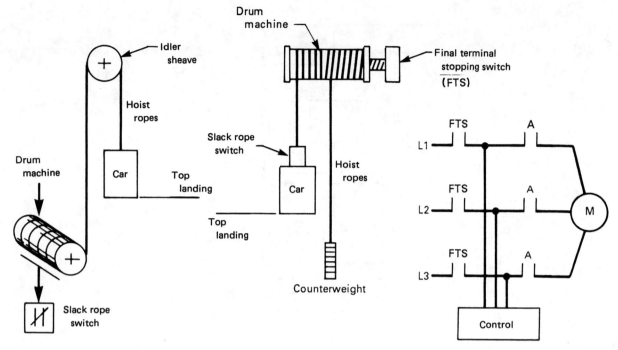

where
L1, L2, L3 = mainline power
A = driving machine motor contactor
FTS = final terminal stopping switch

Diagram 2.25.3.5   Typical Final Terminal Stopping Device, Winding Drum Machine
(Courtesy Otis Elevator Co./Ralph Droste)

equipment failures (hardware and software) (2.26.9.3, 2.26.9.4, and 2.19); and fire and shock hazards (2.26.4.1 and 2.26.4.2).

The principles on which elevator control systems are based have evolved over more than a century. The elevator control system has to be designed so that any hardware/software failure (e.g., 2.26.9.3) or combination of such failures will cause the elevator to either continue its prescribed safe operation or revert to a more restrictive safe state.

It goes without saying that system safety is based on the principal that the safest state achievable is when the car is stopped. Therefore, elevator system design requires positive actions (signals) to allow the elevator system to enter any state beyond the stopped state. Generally speaking, elevator safety is achieved by following proven methods, rules, procedure, etc., which have evolved from past system design. A control algorithm is generally proven to be safe by analyzing all the possible hazards and showing by adherence to industry-accepted practices such as conducting FMEA's, FTA, etc., that each hazard is successfully mitigated. Where it is not possible to predict every failure mode (e.g., large scale integrated circuits and programmable electronics), other methods such as using checked redundancy, N-version programming, diversity, etc., can be used to demonstrate that all hazards have been successfully mitigated.

System safety requires that from the initial conceptual design, through the actual hardware and/or software design, construction, installation, and maintenance of the elevator installation, safety will not be compromised.

With the availability and most frequent use of high performance, fast response time solid-state devices [see 1.3 definitions for static control and motor control and Diagram 2.26(a)] in the 1960s, questions were raised about the adequacy and/or suitability of the A17.1 Code that was in existence at the time.

With the first complete rewrite and reorganization of the A17.1–1955 edition behind them, the Code writers were concerned about the potential hazards peculiar to this type of equipment. In particular, the code writers were concerned with the failure of static power devices and the resulting hazardous conditions that could exist should the failure occur while the car was in the process of leveling and/or was at the landing with open doors.

(a) Doing a risk analysis, the Code writers (i.e., Solid-State Subcommittee) identified a number of potential hazardous conditions that could be caused by catastrophic static control failures. For example:

(1) a full-on power amplifier failure while the car is at floor level with the doors open, could subject the car to a high acceleration rate, and forcing it to move away from the floor;

(2) a failure while the car is leveling into the floor

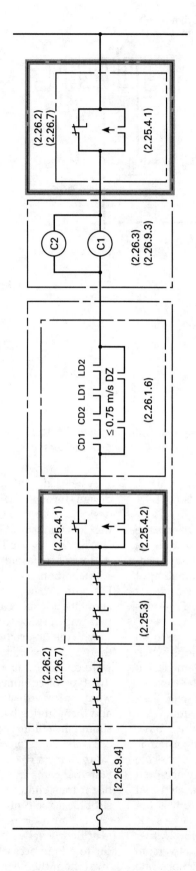

Diagram 2.25.4.1 Typical Safety Circuit for Emergency Terminal Stopping Means
(Courtesy Otis Elevator Co./Ralph Droste)

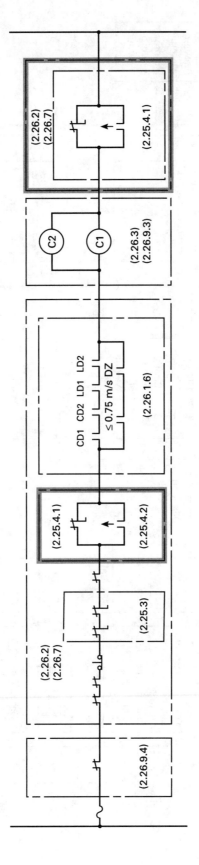

Diagram 2.25.4.1.9   Typical Safety Circuit for Emergency Terminal Speed-Limiting Device
(Courtesy Otis Elevator Co./Ralph Droste)

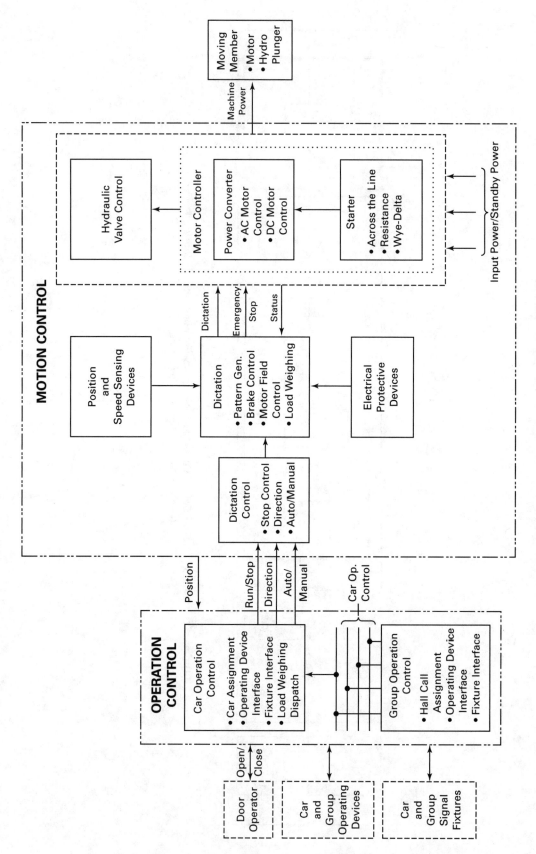

Diagram 2.26(a)   Operation Control and Motion Control

could cause it to travel past the floor with the doors open without stopping;

(3) a failure could cause the car to come to a stop within the leveling zone, but considerably short of the floor, and/or;

(4) a failure could cause the car to stop short of the floor, reverse direction and move away from the floor with the doors open.

(b) As a result of this risk analysis, ASME A17.1(e)–1975 was published and included:

(1) new definitions for "static control", "solid-state device" and "static switching" [Section 3 (1.3)];

(2) restricted power-opening of car and hoistway doors to within 12 inches of the floor for static control systems [Rule 112.2 (2.13.2)];

(3) speed governor overspeed switches were required to operate in both directions to provide motion (direction) sensing to protect against the condition where the motor contactor closes, the brake lifts, and the drive (motor controller) fails to conduct [Rule 206.4 (2.18.4)];

(4) inner landing zone and speed monitoring for leveling operation [Rules 210.1e(6) and (7) (2.26.1.6.6 and 2.26.1.6.7)];

(5) requirements for AC variable voltage drives [Rule 210.9(d) (2.26.9.5)].

(c) Additional revisions were published in ASME A17.1–1981 and included:

(1) a new technology section was added to the Preface advising the AHJ to exercise good judgment in granting exceptions to new products and/or new systems where it is equivalent to that intended by the present Code requirements.

(2) the "motor field sensing means" requirements were revised to allow the use of drives that use field-switching causing the motor shunt field current to pass through zero [Rule 210.2(d) (2.26.2.4)];

(3) single-failure requirements were revised to include solid-state devices [Rule 210.9(c) (2.26.9.3)];

(4) DC drive requirements were added for the first time [Rule 210.9(d) (2.26.9.5)].

Subsequently, requirements were added for VVVF drives [A17.1–1984, Rule 210.9(e)] (2.26.9.6) and the entire control system architecture was redefined in the A17.1–1996 edition, Section 3 (1.3).

With the A17.1–2000 edition the following additional safety requirements were added, which will be discussed in more detail elsewhere:

(a) expanded section on inspection requirements (2.26.1.4);

(b) expanded single failure requirements including software system failures and cyclical checking for failures (2.26.9.3 and 2.26.9.4);

(c) EMI immunity requirements (2.26.4.4);

(d) contactor and relays in critical operating circuits (2.26.3);

(e) ascending car over speed and unintended car movement protection [protection against control, machine and brake failures (2.19)].

Finally, whenever in doubt about the meaning of any of the Code and safety requirements, please check the definitions in 1.3.

The following are of particular interest when reviewing the electrical requirements:

> braking, electrically assisted
> certified
> certified organization
> control, motion
> control, operation
> control system
> controller
> controller, motion
> controller, motor
> controller, operation
> hoistway access switch
> labeled/marked
> landing zone
> leveling
> leveling device
> leveling zone
> listed/certified
> normal stopping means
> operating device
> operation, inspection
> signal equipment (NEC®)
> static switching
> terminal-speed limiting device, emergency
> terminal-stopping device, emergency
> terminal-stopping device, final
> terminal-stopping device, machine final (stop-motion switch)
> terminal stopping device, normal

Wiring Diagram 2.25/2.26 captures the essential Code requirements of Sections 2.25 and 2.26 and will be referred to in the Handbook commentary:

C1	Potential Relay #1
C2	Potential Relay #2
CD1	Car Door Relay #1
CD2	Car Door Relay #2
LD1	Landing Door Relay #1
LD2	Landing Door Relay #2
NOR1	Normal Operation Relay #1
NOR2	Normal Operation Relay #2
AUD	Access Switch Operation
3	Car Door Bypass #1
4	Car Door Bypass #2
5	Hoistway Door Bypass #1
6	Hoistway Door Bypass #2

See Diagram 2.26(b).

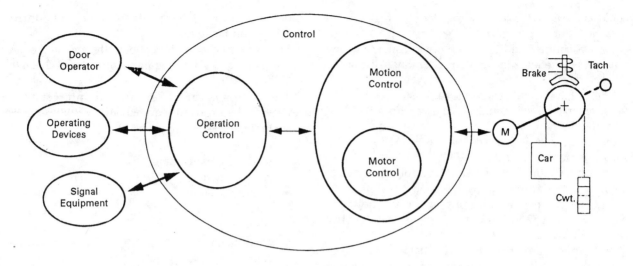

Diagram 2.26(b)   Definitions Control System
(Courtesy Otis Elevator Co./Ralph Droste)

2.26.1 Operation and Operating Devices

2.26.1.1 Types of Operating Devices. An operating device is a car switch, push-button, lever, or other manual device used to actuate a control. All operating devices must be of the enclosed electric type to protect against accidental contact (i.e., prevent electric shock), which is a requirement of the NEC®, 2.26.4.1 (i.e., NFPA 70, Section 620-4) and ASME A17.5/CSA B44.1. See Diagrams 2.26(a) and 2.26(b) and definitions for operating devices in Section 1.3.

2.26.1.2 For Car-Switch Operation Elevators. If the switch did not return to the stop position when the operator's hand was removed, the car would continue in its direction of travel until the final terminal limit was engaged and thus possibly trap the passengers in a stalled elevator. The normal stop position of the handle of a lever-type operating device is the center position. A lever-type operating device that meets these requirements is commonly called a self-centering switch.

2.26.1.3 Additional Operating Devices for Elevators Equipped to Carry One-Piece Loads Greater Than the Rated Load. The operating device for one-piece (safe lift) operation has to be of the continuous pressure type, located near the driving machine with the speed restricted to 0.75 m/s or 150 ft/min. See Handbook commentary on 2.16.7.

2.26.1.4 Inspection Operation. Inspection operation is provided for the use of construction, repair, maintenance, inspection, and rescue personnel. When the elevator is being moved, the operator must have complete control over the elevator. Continuous-pressure operating means afford the operator complete control over the movement of the car. As soon as pressure is removed, the car stops and when pressure is restored,

the car moves in the desired direction. The speed is limited to a speed considered safe when operating from the top of the car.

Portable devices on top of the car allow operators to stand where necessary to perform their duties, which may be desirable on some larger cars. If the device could be unplugged or removed from the car, it may not be available when needed; thus, the requirement that the device cannot be removed. The location of the device on the crosshead assures safety as the user will always know where to find it and take control before he gets on top of the car. If the means of transferring the control to the car top operating station were other than on top of the car, it would not be convenient and there could be a tendency to ride the top of the car without actuating the top of the car operating device. The requirements for inspection operation were extensively revised and reorganized in ASME A17.1–2000.

ASME A17.1–2000 established inspection operation priorities. The order of the priorities are as follows:

(a) top-of-car inspection
(b) in-car inspection
(c) hoistway access operation
(d) machine room inspection
See Handbook commentary on 2.26.1.5.

2.26.1.4.1 General Requirements. The general requirements for inspection operation and transfer switch have been grouped under the heading "General Requirements." The transfer switch labeling requirements were added in ASME A17.1–2000 and the functional requirements are defined for both the "INSP" and "NORMAL" positions. The requirement that "positively opened contacts" are only required when transferring from "NORMAL" to "INSP" and not from "INSP" to

"NORM." Requirements also clearly specify that automatic power door opening and closing shall be disabled anytime that the inspection transfer switch is in the "INSP" position. ASME A17.1–2000 added requirements for inspection operating devices, including labeling requirements and separated "operation" from "operating devices" requirements. The requirements also clarify that car doors may be held closed electrically on inspection operation.

2.26.1.4.2 Top-Of-Car Inspection Operation. All of the additional requirements for top-of-car inspection operation are grouped together in this requirement. Requirements were added in A17.1–2000 for permanent location of transfer switch, stop switch, operating devices including the B44 requirements that the transfer switch be designed to prevent accidental transfer from "INSP" to "NORM" position. ASME A17.1–2000 added requirements for an "ENABLE" device to protect against accidental activation or failure of the inspection operating devices. The inspection-operating device is permitted to be portable as long as the "ENABLE" and "STOP" switch (button) are included within the portable unit. Personnel safety has been adequately addressed by requiring the stop switch to be readily accessible from the hoistway entrance.

2.26.1.4.3 In-Car Inspection Operation. All of the additional requirements for in-car inspection operation are grouped in this requirement. A separate switch or switch position is required to enable hoistway access switches.

2.26.1.4.4 Machine Room Inspection Operation. All of the additional requirements for machine room inspection operation are grouped in this requirement. Machine room inspection operation is the lowest priority inspection operation. See Diagrams 2.26.1.4(a) and 2.26.1.4(b) and Charts 2.26.4(a) and 2.26.4(b). See also Handbook commentary on 2.26.9.3(d)

2.26.1.5 Inspection Operation With Open Door Circuits. The purpose of these requirements is to eliminate the use of jumpers to bypass door and gate contacts, and to replace jumpers with devices that are safer and more convenient. A proposal was originally made by the International Union of Elevator Constructors (IUEC) to the B44 and A17 Main Committee on November 5, 1995 and May 25, 1988, respectively. A meeting of A17.1 and B44 representatives was held on December 8, 1988 to address the IUEC proposal. The general consensus was that:

(a) jumpers should be prohibited;

(b) the bypass means should be located in the controller;

(c) the speed should be limited to inspection speed or less;

(d) a single failure of the bypass equipment should not permit the elevator to revert to automatic operation; and

(e) if the bypass is left "ON" after the problem is corrected, the car shall not be able to revert to automatic operation.

Instances where jumpers have been used to bypass car and landing door electrical contact circuits and inadvertently left in place after servicing the elevator, have resulted in some serious accidents. It has become evident that some means should be provided to inhibit the use of jumpers across door contact circuits.

There are two main hazards to using jumpers:

(1) They can be left on and still be effective when the car is returned to service (high speed operation and carrying passengers).

(2) They can be connected to the wrong terminals and cause any number of unsafe conditions.

The requirements are that a permanently wired bypass switch be effective only on inspection operation. Since it is not always possible to enter or gain access to the top-of-car in some locations (or release passengers who are trapped between floors), inspection operation from the machine room with car or hoistway door contacts bypassed is permitted under the prescribed conditions, but not required.

See Diagrams 2.26.1.4(b), 2.26.1.5, and 2.26.1.5.10.

2.26.1.6 Operation in Leveling or Truck Zone. Generally speaking, all modern controlled elevators have "leveling" and are designed to stop and maintain a certain accuracy level to the floor as passengers enter and leave the car. Although A17.1 is silent on the question of leveling accuracy, the ICC/ANSI A117.1-1998 Accessibility Standard says that each automatic elevator "shall be equipped with a self-leveling feature that will automatically bring and maintain the car at floor landings within a tolerance of 13 mm ($\frac{1}{2}$ in.) under rated loading to zero loading conditions."

The term "leveling" has been around the elevator industry for a very long time. In the early days (1910s to 1920s) automatic leveling was more important on freight elevators where the platform had to be flush with the floor for industrial fork lift loading, cart loading and tailboard loading of trucks. This was done first by an auxiliary elevator machine coupled to the drive machine to turn it at slow leveling speed, called micro machine leveling.

Following auxiliary micro machine leveling, main machine leveling was introduced in the 1920s to 1930s, where the hoist motor was designed to have sufficient torque to operate at leveling speeds.

Passenger elevators, on the other hand, were operated by attendants using car switch control, who brought elevators into the vicinity of the floor. The skill of a passenger car attendant was limited to car speeds up to 1.5 m/s (300 ft/min), above which the attendant could not react fast enough, resulting in the missing of landings and poor stopping accuracy.

On passenger elevators, automatic leveling was first added with resistance control, where the attendant initiated the stop, but the actual slow down and leveling

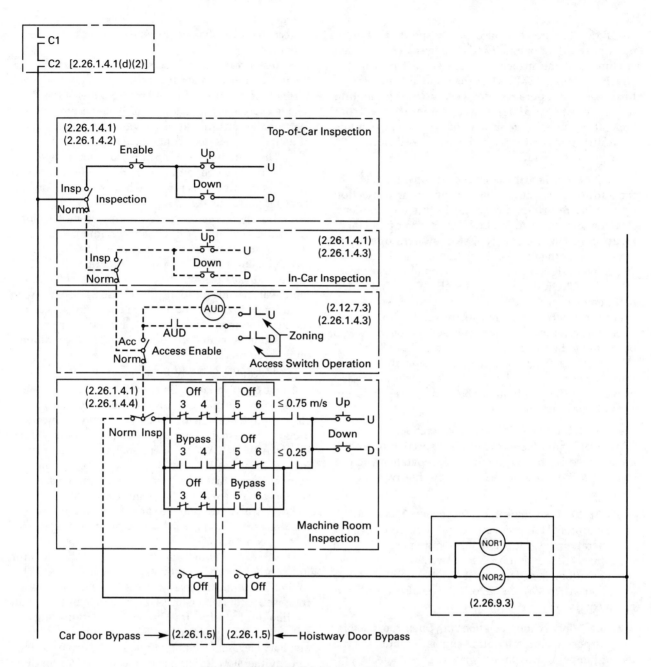

Diagram 2.26.1.4(a)   Inspection Operation
(Courtesy Otis Elevator Co./Ralph Droste)

was completely automatic. When first introduced it was known as "automatic floor stop" and/or "flying stop" operation.

With the introduction of "signal control" operation and the use of generator field control (a.k.a., variable voltage or Ward-Leonard Control) all the attendant had to do was press the "START" or "DOOR CLOSE" button and the elevator automatically accelerated and decelerated in response to car and hall calls ("operating devices") and leveled into the landing with an accuracy of approximately ± 13 mm (± ½ in.).

Ward-Leonard control (i.e., generator field control) was the control system of choice until the 1970s for high-speed installations (and even for the lower car speeds and especially for freight elevators) where leveling accuracy was particularly important. The Vertical Transportation Handbook by George Strakosch suggests that for a quality installation, the elevator should come to a stop at a floor within a prescribed leveling tolerance of ± 7 mm (± ¼ in.) and maintain that tolerance during releveling as passengers enter and leave the car.

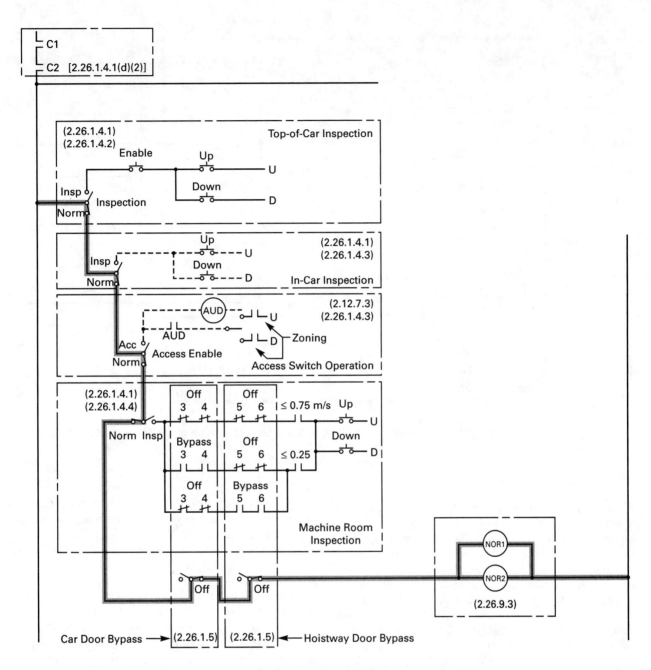

Diagram 2.26.1.4(b)   Inspection Operation and Single Failures
(Courtesy Otis Elevator Co./Ralph Droste)

Automatic leveling (or releveling) occurs when the control of the elevator position is taken off main operation and control of leveling speed and final stop is by a leveling device (see 1.3). See Handbook comments on 2.25, 2.26, and 2.13.2.1.2.

2.26.1.6.6 Two general categories of systems can be designed to meet these requirements. One type of system monitors the car speed with a comparator-type device. The other is a speed-saturated system in which the drive system, under worst case fault conditions, would be incapable of moving the car at a speed

exceeding 0.75 m/s or 150 ft/min. The safety requirements of systems employing the speed monitoring technique would be such that a malfunction of a component involved in the operation of the car (e.g., a tachometer "failing open" causing the car speed to increase) would not result in a corresponding failure in the speed monitor, which employed the tachometer as a speed indicating device.

An independent means is required to limit the leveling speed even if the rated speed is 0.75 m/s or 150 ft/min or less.

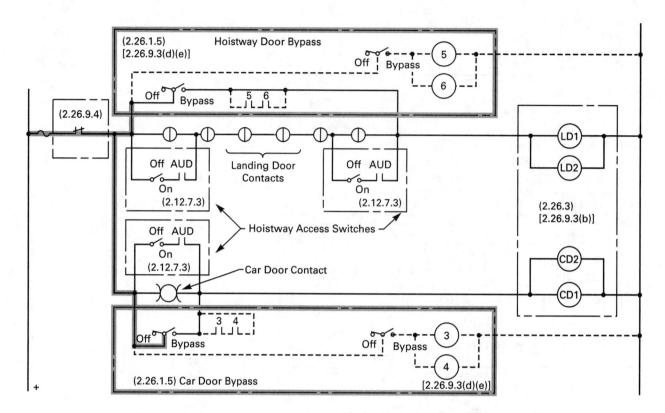

Diagram 2.26.1.5   Inspection Operation With Open Door Circuits and Single Failures (2.26.9.3)
(Courtesy Otis Elevator Co./Ralph Droste)

See Diagram 2.26.1.6.

2.26.1.6.7 This requirement provides protection in the event the car stops with the doors open within the leveling zone, but outside the "inner zone." In this instance, the car would not be permitted to relevel. The car would be permitted to restart if the doors were closed and all other conditions for the initiating circuits were satisfied. In the event that the car stops with the doors open within the leveling zone, but outside the inner zone, and reverses direction, the car would be stopped and not permitted to restart.

2.26.2 Electrical Protective Devices

Where a malfunction of an elevator component and/or device can create a serious hazard for elevator users, the operation of such components and/or devices has to be checked by an "electrical protective device (EPD)." See Diagram 2.26.2.

For example, doors not closing [2.26.2.14 or 2.26.2.15] or an elevator not making a normal stop at the terminals [2.26.2.16 or 2.26.2.11], etc.

Also, an EPD may be necessary to ensure safety of elevator mechanics, such as the stop switches on the car top, in the pit, or machinery spaces.

New wording is used to establish general requirements for EPDs in 2.26.4.3 that they have contacts that are positively opened. Where positively opened contacts

are not required (i.e., where solid state and/or programmable electronics is permitted) the electrical protective devices have to be designed to protect against single failures (e.g., 2.26.2.12, 2.26.2.29, and 2.26.2.30).

All EPDs have the same response when "activated (operated, opened)," i.e., cause power to be removed from the elevator driving machine motor and brake, thereby reverting to a known safe state, meaning that the elevator comes to a stop.

The phrase " . . . remove power from the driving machine motor and brake . . ." is used extensively throughout the Code. The word "power" is used in the sense of "driving power." Requirement 2.26.2.2 provides further clarification for a particular class of control systems by stating that " . . . It is not required that the electrical connections between the elevator driving machine motor and the generator be opened in order to remove power from the elevator motor." The preferred way to make a fast and controlled stop on a DC machine is to maintain the motor field excitation while removing power from the motor armature. The clear intent is to require that the elevator stops in the most expeditious and safe manner. The language is general and performance oriented, and as such covers a wide variation of elevator drive control systems.

System safety is based on the principal that the safest state achievable is when the elevator(s) are stopped.

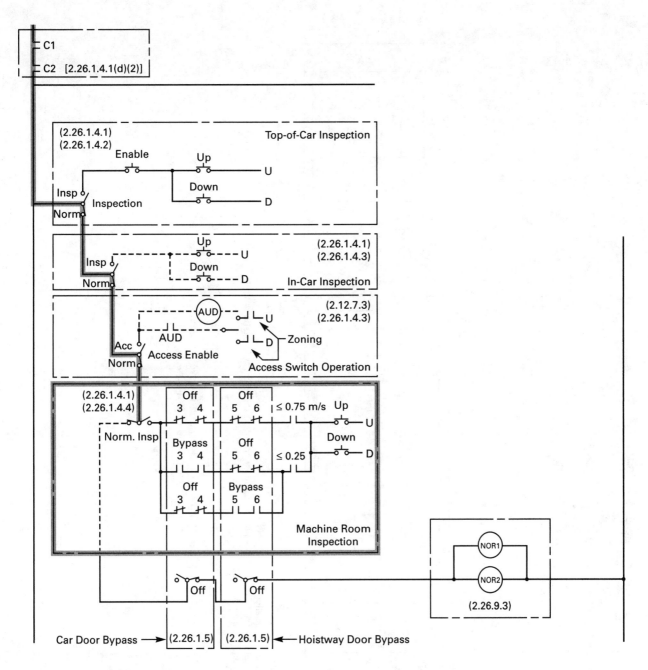

Diagram 2.26.1.5.10   Machine Room Inspection With Open Door Circuits
(Courtesy Otis Elevator Co./Ralph Droste)

Therefore, to allow the car to move beyond the stopped state requires positive action (energize "Potential Relay #1 and #2" in Diagram 2.26.2). To allow the elevator system to enter any state beyond the stopped state requires that all EPDs must not be in the "activated, operated, or opened" state.

The following is an overview of significant revisions incorporated in the 2000 edition of ASME A17.1.

(a) A number of sections had editorial changes made for clarity. Also deleted "cause power to be removed from the elevator driving machine motor and brake,"

since this requirement is now covered in the opening paragraph of 2.26.2.

(b) Requirement 2.26.2.6 — was clarified since broken ropes, tape or chain switch is the same EPD regardless of the function of the rope, tape, or chain.

(c) Requirement 2.26.2.28 — car door interlock added to harmonize with CSA B44 requirements.

(d) Requirement 2.26.2.23 — car access panel locking device added to harmonize with CSA B44 requirements.

(e) Requirement 2.26.2.32 — hoistway access opening

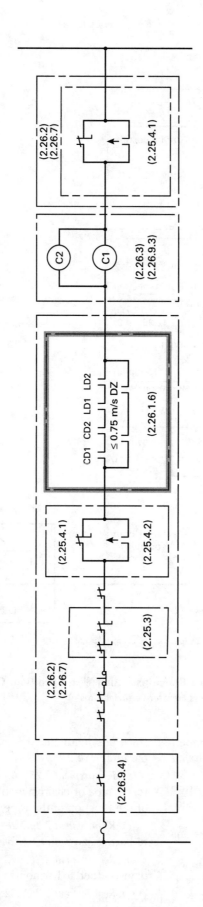

Diagram 2.26.1.6 Operation in Leveling Zone
(Courtesy Otis Elevator Co./Ralph Droste)

GENERAL NOTES:
(a) Requirement 2.26.1.6.3 [Rule 210.1e(3)] – leveling zone ±450 mm (18 in.)
(b) Requirement 2.26.1.6.6 [Rule 210.1e(6)] – ≤ 0.75 m/s (150 ft/min) max. and independent speed check for static control.
(c) Requirement 2.26.1.6.7 [Rule 210.1e(7)] – inner door zone ± 75 mm (3 in.) for static control.

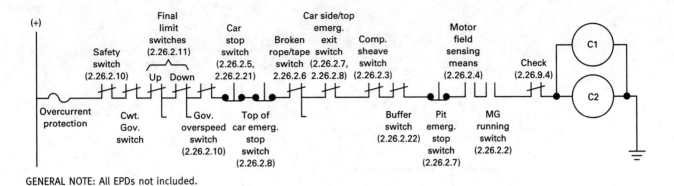

GENERAL NOTE: All EPDs not included.

Diagram 2.26.2 Typical Safety Circuit, Electrical Protective Devices Circuit
(Courtesy Otis Elevator Co./Ralph Droste)

locking device added to harmonize with CSA B44 requirements.

(f) Requirement 2.26.2.29 — ascending car overspeed protection device, to provide protection against ascending car overspeed to prevent collision of the elevator with the building overhead and to harmonize with CSA B44 requirements. To ensure the safe functioning of the detection means, redundancy is required so that no single component fault (hardware or software) will lead to an unsafe condition.

(g) Requirement 2.26.2.30 — unintended car movement device; to prevent movement of the elevator away from the landing when the car and hoistway doors are open and to harmonize with B44 requirements. To ensure the safe functioning of the detection means, redundancy is required so that no single component fault (hardware and/or software) will lead to an unsafe condition.

2.26.2.1 Slack-Rope Switch. See Diagram 2.25.3.5.

2.26.2.3 Compensating-Rope Sheave Switch. See commentary on 2.17.17 and Diagram 2.26.2.3.

2.26.2.4 Motor Field Sensing Means. See Diagram 2.26.2.4.

2.26.2.5 Emergency Stop Switch. The original need for the in-car emergency stop switch is easily understood. When passenger elevators were equipped with perforated car doors, or no doors at all, pinching hazards and the possibility of crushing between hoistway walls and sills existed. However, on today's modern passenger elevators with solid enclosures, these hazards have been eliminated.

Although accidents have been prevented by the use of emergency stop switches, evidence is overwhelming that more accidents and problems are caused by their misuse.

(a) Due to misuse, the states of Wisconsin and Florida were the first to prohibit the emergency stop switch, and they report nothing but positive results. Some examples of misuse reported are as follows:

(1) Robbery, muggings, and other violent crimes are perpetrated after actuating the emergency stop switch, stopping the car between floors.

(2) The emergency stop switch is actuated while the car is leveling, causing an unsafe condition where people can trip while going into or out of the car.

(3) The stop switch is operated while the doors are open, possibly negating stretch-of-cable leveling and causing the same tripping hazard.

(4) The emergency stop switch is pushed inadvertently in a crowded car and passengers remain trapped, thinking there is an equipment failure.

(5) Passengers have stopped the car between floors and opened the hoistway door, jumping to the floor below. This sometimes results in the passenger falling back into the hoistway, causing serious injury or death.

(6) Operating an emergency stop switch on a high-speed elevator at full speed could result in a severe abrupt stop, which may injure the passengers.

(b) In today's modern elevators, the passenger is protected by many new safety requirements added to the Elevator Code over the last decade. Examples are as follows:

(1) Requirement 2.11.6. Passenger elevator hoistway doors shall be so arranged that they may be opened by hand from within the elevator car only when the car is within the unlocking zone. (See 1.3, Definitions.)

(2) Requirement 2.13.2. Power opening can occur only when the car is at rest or leveling. On elevators with static control, power cannot be applied to open car doors until the car is within 300 mm or 12 in. of the landing.

(3) Requirement 2.26.1.6.6. This limits the leveling speed to a maximum of 0.75 m/s or 150 ft/min with doors open.

(4) Requirement 2.26.1.6.7. Requires an inner landing zone extending not more than 75 mm or 3 in. above and 75 mm or 3 in. below the landing. A car shall not move if it stops outside the inner landing zone unless the doors are fully closed.

(5) Requirement 2.26.9.3(b). Prevents the car from

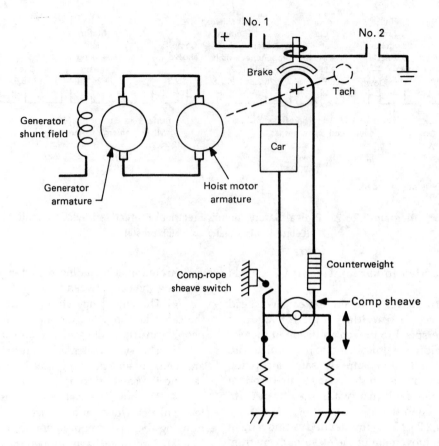

Diagram 2.26.2.3 Typical Compensating-Rope Sheave Switch Arrangement
(Courtesy Otis Elevator Co./Ralph Droste)

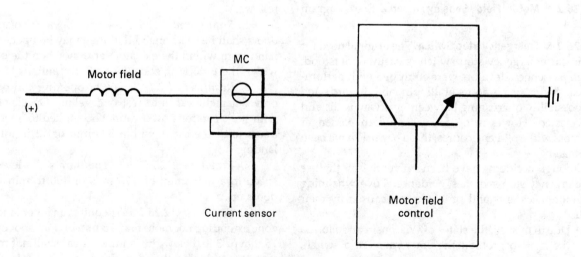

Diagram 2.26.2.4 Typical Motor Field Sensing Means Circuit
(Courtesy Otis Elevator Co./Ralph Droste)

starting to run if any hoistway door interlock is unlocked or hoistway door or car gate electric contact is not in the closed position.

(6) *Requirement 2.26.9.5*. Requires two devices to remove power independently from the driving machine motor each time the car stops.

All of these requirements not only add to safety, but also directly protect the entrance when the car is at or away from the floor.

The European Code CEN EN81 has prohibited the in-car emergency stop switch basically for the same reasons. Moreover, other countries have reported frequent misuse of the emergency stop switch.

Thus, the A17.1 Code prohibits publicly accessible emergency stop switches in passenger elevators and requires a stop switch in the car that is either key operated or located behind a locked cover (2.26.2.21). The in-car stop switch is comparable in function to the stop switch in the pit or on the car top; it is useful for inspectors and maintenance mechanics, and therefore should be required, but in a form inaccessible to the public.

An emergency stop switch is required in a freight elevator as perforations are permitted in freight car enclosures. An emergency stop switch also needs to be accessible to stop an elevator if the load shifts due to improper stacking.

When the emergency stop switch is placed in the "STOP" position, power must be removed from the elevator driving machine motor and brake. The failure of any spring used in the switch must not cause a switch in the "STOP" position to revert to the "RUN" position. Emergency stop switches conforming to these requirements are also required in overhead machinery spaces in the hoistway (2.7.3.5), in the pit (2.2.6), and on top of the car (2.26.2.8).

2.26.2.9 Car-Safety Mechanism Switch. See Handbook commentary on 2.14.7, 2.18.4.1, and 2.18.4.3 and Diagram 2.26.2.9.

2.26.2.10 Speed-Governor Overspeed Switch. See Handbook commentary on 2.18.4.1, 2.18.4.2 and 2.18.4.3 and Diagram 2.26.2.10.

2.26.2.11 Final Terminal Stopping Devices. See Handbook commentary on 2.25.3 and Diagram 2.25.3.

2.26.2.19 Motor-Generator Overspeed Protection. See Diagram 2.26.2.19.

2.26.2.21 In-Car Stop Switch. See Handbook commentary on 2.26.2.5.

2.26.2.22 Buffer Switches for Gas Spring Return Oil Buffers. See Diagram 2.26.2.22.

2.26.3 Contractors and Relays in Critical Operating Circuits

New requirements in ASME A17.1–2000 for electromechanical operated contactors and relays used to satisfy the requirements of 2.26.8.2 and 2.26.9.3 through 2.26.9.7 are considered to be used in critical operating circuits.

The key element of the requirement is that the monitoring contact should not change state if the contact in the critical circuit has not changed state. One method to assure this is to use a relay/contactor that satisfies the following requirements:

(a) if one of the break contacts (normally closed) is closed, all the make contacts are open; and

(b) if one of the make contacts (normally open) is closed, all the break contacts are open.

Springs may be used, but spring failures shall not cause monitoring contact(s) to indicate an open critical circuit contact if the critical circuit contact is not open.

2.26.4 Electrical Equipment and Wiring

2.26.4.1 See also Diagram 2.26.4.1, and Charts 2.26.4(a), 2.26.4(b), 2.26.4.1(b), and 2.26.4.1(c).

The Handbook commentary on 2.26.4.1 is broken into three sections:

(a) General Commentary on ANSI/NFPA 70, Article 620; and

(b) Highlights of the New and Revised Requirements in ANSI/NFPA 70-1996.

(c) Highlights of the New and Revised Requirements in ANSI/NFPA 70-2002.

General Commentary on ANSI/NFPA 70, Article 620

620-2. Definitions

Clearly defines the terms used in Article 620.

New technology has expanded the traditional meaning and use of the terms "controller" and "control system." This stems from the use of the microprocessor and other micro electronics which now permits a control system's and controller's functions to be physically distributed in different locations. The use of this new micro technology also changes the "safety" aspects of the use of such equipment when their voltage levels are low and their power levels are typically measured in micro and mill watts. The new definitions recognize this and separate the control system and controller into its functional parts, while maintaining the traditionally defined controller in the newly defined "Motor Controller." It is within this "Motor Controller" that the high voltage and high power portion of the controller is contained, and with which the major traditional safety concerns exist. The 1996 NEC® rewrite of Article 620 utilizes these new definitions to address and specify the proper safety concerns of these controllers, "Motor," "Motion," and "Operation."

620-3. Voltage Limitations

The voltage limitation is to provide protection for the passenger. The greatest danger is energizing some

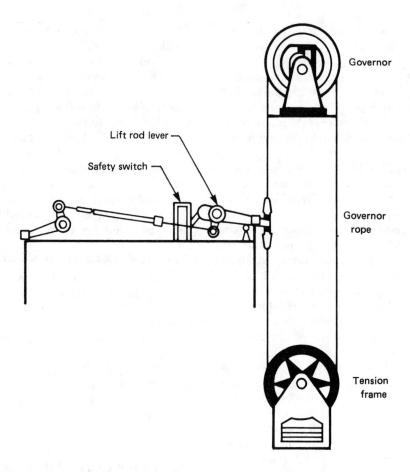

Diagram 2.26.2.9   Typical Car Safety Mechanism Switch
(Courtesy Otis Elevator Co./Ralph Droste)

conductive, noncurrent carrying part of the car/landing devices (i.e., operating devices) through a fault. There is adequate provision for safety in the NEC® to protect trained mechanics when working on equipment. In Section 110-27(a), the requirements for guarding live parts against accidental contact are stated only for systems operating at 50 V or more, which seems to indicate that all voltages at 50 V or more, are considered hazardous. Since the 1925 edition of the A17.1 Safety Code for Elevator and Escalators, the maximum nominal voltage permitted for operating device circuits was limited to 300 V to ground. The first edition of the A17.1 Elevator Safety Code (1921) limited the voltages to 750 V.

620-3(a) Elevator drive technology is using voltage levels above 600 V within the drive cabinet. Voltages over 600 V should be permitted as long as the equipment conforms to the appropriate CSA and ANSI standard established under the provisions of the Canadian and National Electrical Codes. All elevator/escalator and related electrical equipment has to be certified under the provisions of CAN/CSA B44.1-ASME/ANSI A17.5, a harmonized North American Standard; therefore, the fire and shock concerns of the Canadian and National Electrical Codes are addressed.

620-3(b) and (c) Include requirements for lighting, heating, and air-conditioning circuits.

620-5. Working Clearances

The terms used in 620-5(a) are correlated with the definition requirements in 620-2. The equipment in 620-5(a) may be moved so that the intended requirements of Section 110-26(a) are complied with. The conditions described in 620-5(b), (c), and (d) guard against electrical shock under the conditions where only qualified persons will examine, adjust, service, and maintain the equipment. Also, see 670-5, 110-26(a), and 110-26(a)(1) Exception No. 2. Working clearances are applicable to all electrical equipment in this article, not only those in the machine room.

620-11. Insulation of Conductor

Covers hoistway door interlock wiring, traveling cables, and other related wiring. Also, recognizes flame-retardant, limited smoke, low toxicity, and low corrosivity insulating materials. The FPN indicates minimum acceptable requirements for flame-retardant insulation.

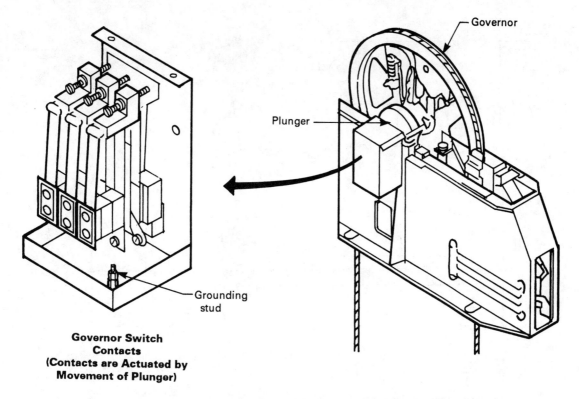

**Governor Switch
Contacts
(Contacts are Actuated by
Movement of Plunger)**

Diagram 2.26.2.10   Typical Speed-Governor Overspeed Switch Arrangement
(Courtesy Otis Elevator Co./Ralph Droste)

620-12. Minimum Size of Conductors

Clarifies that conductors smaller that No. 14 AWG, up to and including No. 20 AWG, may be used in parallel. 620-12(b) was revised to permit smaller than No. 24 AWG if listed for the purpose. Shielded cables interconnecting microprocessors in an elevator distributive system come in sizes smaller than No. 24 AWG.

620-13. Feeder and Branch Circuit Conductors

620-13(a) Due to the Duty Cycle nature of the equipment, the actual current for short periods of time, may exceed the nominal nameplate value, but there are also operational periods of time when actual current levels are well below the nameplated value, as well as periods of time when the actual current is zero. Table 430-22(b) is based on equivalent rms heating current, that is to say, a current that will cause the equivalent heating in the conductors as a Continuous Duty Cycle. The addition of the FPN to 620-13(a) is to remove the confusion as to the values in Table 430-22(b), what it is based on, as well as to point out that currents in excess of the nominal rating will be seen during the duty cycle.

620-13(b) The heating of conductors depends on the rms current value, which may be reflected by the nameplate current rating of the motor controller. The nameplate current rating of a motor controller for a duty cycle may be determined by the elevator manufacturer based on the design or application. See also FPN in Table 620-14. The following calculation will illustrate this concept.

EXAMPLE: The elevator manufacturer decides to rate his controller on the full load up (worst case) load condition. Assume an elevator using an adjustable speed drive of the SCR converter type has the input power at 460-3-60 and the elevator DC motor has the following duty cycle currents:

(a) Motor accelerating full load up = 150 A DC for 2 s
(b) Motor running full load up = 75 A DC for 3.5 s
(c) Motor decelerating full load up = 55 A DC for 2 s

There are 240 starts per hour and the elevator operates at 50% duty cycle (i.e., half time on, half time off). The controller power transformer is located within the controller cabinet and has a 460 V primary and a 300 V secondary. The controller contains power supplies for the machine brake, motor field, and internal drive controls, which draw 10 A (assumed continuous) from the AC line at the primary of the input transformer.

The motor RMS current over the duty cycle is calculated as:

$$I_{rms} = \sqrt{\frac{I_a^2\,T_a + I_r^2\,T_r + I_d^2\,T_d}{T}}$$

where

$$T = T_a + T_r + T_d + T_i$$

T_a = acceleration time
T_r = running time
T_d = deceleration time
T_i = idle time

For 240 starts per hour, the interval between starts is 15 s. For 50% duty cycle, the idle time T_i = 7.5 s.
Solving for I_{dcrms}

173

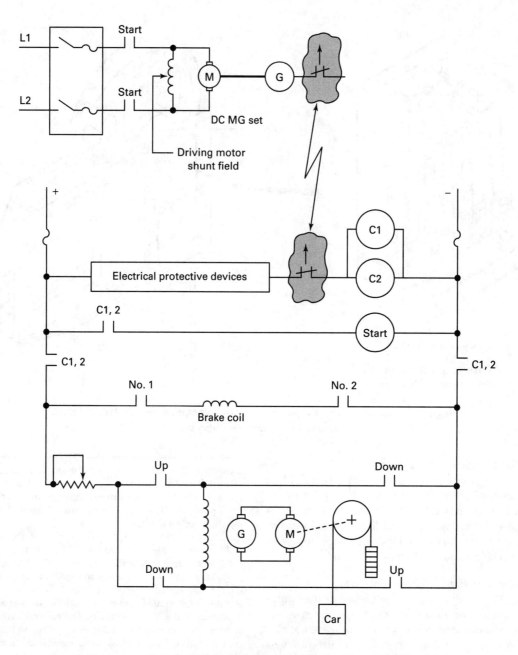

Diagram 2.26.2.19 Typical Motor-Generator Overspeed Switch
(Courtesy Otis Elevator Co./Ralph Droste)

$$I_{dcrms} = \sqrt{\frac{(150^2 \times 2 + 75^2 \times 3.5 + 55^2 \times 2)}{15}} = 68.7 \text{ A}$$

Reflecting this to the AC line input side of the controller transformer,

$$I_{acrms} = 0.816 \times I_{dcrms} \times \left(\frac{V_{sec}}{V_{prim}}\right)$$

$$= 0.816 \times 68.7 \times \left(\frac{300}{460}\right) = 36.6 \text{ A} = 37 \text{ A}$$

Combining this RMS current with the constant drive power supply current of 10 A, $I_{Total} = 37 \text{ A} + 10 \text{ A} = 47 \text{ A}$. This is the nameplate rating the elevator manufacturer has chosen for the controller. $I_{Rated} = 47 \text{ A}$.

620-13(c) A motor controller may be supplied by a power transformer, which changes the line voltage (primary) to the input voltage (secondary) required by the motor controller. The conductors supplying the power transformer carry current at line voltage to supply the

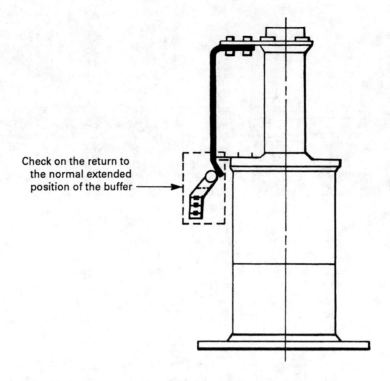

Check on the return to
the normal extended
position of the buffer →

Diagram 2.26.2.22   Gas Spring-Return Oil Buffer Switch
(Courtesy Otis Elevator Co./Ralph Droste)

motor controller through the power transformer. The input (primary) nameplate current of the power transformer includes the motor controller load on the transformer secondary (referred to the transformer primary) and the transformer losses.

620-13(d) Combined 1993 NEC® Sections 620-13 and 620-14 to clarify motor feeder and branch circuit requirements. Table 430-22(b) contains factors to adjust the nameplate current rating of elevator (intermittent duty) rotating equipment based on the time rating of such equipment, i.e., 30 min, 60 min, or continuous. These factors "normalize" the current for intermittent duty application to continuous duty application, which Article 430 is primarily concerned with.

The 125% factor in the 1993 and previous editions of Section 620-13(b) applied to an elevator system with multiple motors. This applies to the feeder conductors supplying a bank of elevators. If the substantiation for (a)(b) and (c) apply to individual components of a single elevator, then it must follow that a feeder conductor, which supplies several cars described in (a)(b) and (c), will be subject to the sum of the currents of the equipment it directly supplies as determined in (a)(b) and (c). In the 1990 NEC® and previous editions, Section 430-24 required 125% of the largest motor current rating for both continuous and noncontinuous duty motors. With the 1993 NEC® change to Section 430-24, there appears to be no justification for the 125% factor that used to be in 620-13(b) for intermittent duty application. Examples 9 and 10 in Appendix D illustrate requirements. See Diagram 2.26.4.1(d).

620-15. Motor Controller Rating
This rule recognizes that when an elevator application does not require the full nameplate rating of the motor in conjunction with a motor controller, which limits the available power to the motor, such as an adjustable speed drive system, that a motor controller of smaller rating than the motor nameplate rating is suitable.

620-21. Wiring Methods
See Diagrams 2.26.4.1(e) through 2.26.4.1(h).

620-22. Branch Circuits for Car Lighting, Receptacle(s), Ventilation, Heating, and Air Conditioning
See Diagram 2.26.4.1(i).

620-23. Branch Circuit for Machine Room/Machinery Space Lighting and Receptacle(s)
Harmonized with the Canadian Electrical Code CSA C22.1 Section 38-052 and ASME/ANSI A17.1–2000, Requirement 2.7.5. A safety problem could be created for persons if the lighting were connected to the load side of GFCI receptacles (i.e., person trying to exit space or reset receptacles without lighting).

620-24. Branch Circuit for Hoistway Pit Lighting and Receptacle(s)

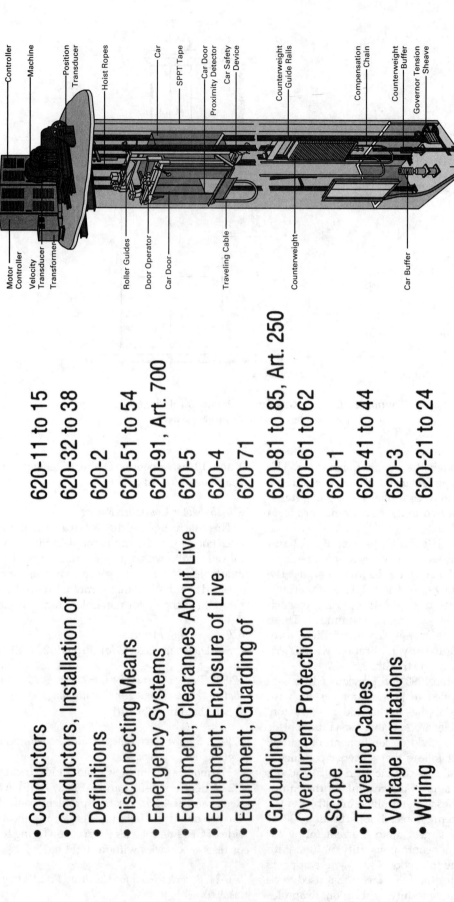

- Conductors — 620-11 to 15
- Conductors, Installation of — 620-32 to 38
- Definitions — 620-2
- Disconnecting Means — 620-51 to 54
- Emergency Systems — 620-91, Art. 700
- Equipment, Clearances About Live — 620-5
- Equipment, Enclosure of Live — 620-4
- Equipment, Guarding of — 620-71
- Grounding — 620-81 to 85, Art. 250
- Overcurrent Protection — 620-61 to 62
- Scope — 620-1
- Traveling Cables — 620-41 to 44
- Voltage Limitations — 620-3
- Wiring — 620-21 to 24

Diagram 2.26.4.1 National Electrical Code® Requirements (NFPA 70)
(Courtesy Otis Elevator Co./Ralph Droste)

Chart 2.26.4(a)   Elevator Requirements National Electrical Code ® Versus Canadian Electrical Code
(Courtesy Otis Elevator Co. / Ralph Droste)

ANSI/NFPA 70-1999		CAN-CSA C22.1-1998	Subject
Part	Article/Section	Section/Rule	
A. General	620-1	38-000	Scope
	620-2	–	Definitions
	620-3	38-002	Voltage Limitations
	620-4	38-004	Live Parts Enclosed
	620-5	38-044	Working Clearances
B. Conductors	620-11	38-006; 008	Insulation of Conductors
	620-12	38-010	Minimum Size of Conductors
	620-13	38-010(3)(4)	Feeder and Branch-Circuit Conductors
	620-14	38-010(4)	Feeder Demand Factor
	620-15	–	Motor Controller Rating
C. Wiring	620-21	38-014; 016; 018-020	Wiring Methods
	620-22	38-012	Branch Circuits for Car Lighting, Receptacle(s), HVAC
	620-23	38-052	Branch Circuit for MR/Machinery Space Lighting and Receptacle(s)
	620-24	38-054	Branch Circuit for Hoistway Pit Lighting and Receptacle(s)
D. Installation of Conductors	620-32	12-1014, Table 8	Metal Wireways and Nonmetallic Wireways
	620-33	12-1014, Table 8	Number of Conductors in Raceways
	620-34	38-024	Supports
	620-35	12-1900, 1902, 1904	Auxiliary Gutters
	620-36	38-022	Different Systems in One Raceway or Traveling Cable
	620-37	38-014; 12-014	Wiring in Hoistways and Machine Rooms
	620-38	–	Electric Equipment in Garages and Similar Occupancies
E. Traveling Cables	620-41	38-028	Suspension of Traveling Cables
	620-42	38-030	Hazardous (Classified) Locations
	620-43	38-032	Location of and Protection for Cables
	620-44	38-014	Installation of Traveling Cables
F. Disconnecting Means and Control	620-51	38-034	Disconnecting Means
	620-52	38-034, Appendix B	Power from More than one Source
	620-53	–	Car Light, Receptacle(s) and Ventilation Disconnecting Means
	620-54	–	Heating and Air Conditioning Disconnecting Means
G. Overcurrent Protection	620-61	38-040; 038	Overcurrent Protection
	620-62	38-034(3)	Selective Coordination
H. Machine Room	620-71	38-042	Guarding Equipment
J. Grounding	620-81	38-046	Metal Raceways Attached to Cars
	620-82	38-048	Electric Elevators
	620-83	–	Nonelectric Elevators
	–	38-050	Bonding
	620-84	–	Escalators, Moving Walks, Wheelchair Lifts and Stairway Chair Lifts
	620-85	–	Ground-Fault Circuit Interrupter Protection for Personnel
K. Emergency and Standby Power Systems	620-91	38-036	Emergency and Standby Power Systems

Chart 2.26.4(b)   Elevator Requirements Canadian Electrical Code Versus National Electrical Code ®

(Courtesy Otis Elevator Co. / Ralph Droste)

CAN-CSA C22.1-1988 Section/Rules	ANSI/NFPA 70-1999 Article/Section/Part		Subject
38-000	620-1	A	Scope
–	620-2	A	Definitions
38-002	620-3	A	Voltage Limitation
38-004	020-4	A	Isolation of Live Parts
38-006	620-11	B	Insulation of Conductors
38-008	620-11(b)	B	Traveling Cables
38-010	620-12, 13, 14	B	Conductor Sizes
–	620-15	B	Motor Controller Rating
38-012	620-22	C	Branch Circuits for Car Lighting, Accessories, Heating and Air Conditioning
38-014	620-21	C	Wiring Methods in Hoistways
	620-44	E	Machine Rooms and Escalator Wellways
38-016	620-21	C	Wiring Methods on Cars
38-018	620-21	C	Wiring Methods Between Motors, Machine Brakes, Valves, Generators, and Control Panels
38-020	–		Wiring Methods on Sidewalk Elevators
38-022	620-36	D	Grouping of Conductors
38-024	620-34	D	Raceway Supports
–	620-38	D	Electric Equipment in Garages
38-026	–		Fittings
38-028	620-41	E	Suspension of Traveling Cables
38-030	620-42	E	Hazardous Locations
38-032	620-43	E	Mechanical Protection
38-034	620-51, 52	F	Disconnecting Means
	620-62	G	
–	620-53	F	
–	620-54	F	
38-036	620-91	K	Emergency Power
38-038	620-61(b)	G	Overload Protection for Motors
38-040	020-01(a)	G	Overcurrent Protection of Operating, Control, and Signal Circuits
38-042	620-71	H	Installation of Machines
38-044	620-5	A	Installation of Control Panels
38-046	620-81	J	Bonding of Raceways to Car
38-048	620-82	J	Bonding of Equipment
–	620-83	J	Non-electric Elevators
38-050	–		Methods of Bonding
–	620-84	J	Escalator
–	620-85	J	GFCI
38-052	620-23	C	Branch Cicuits and Lighting for Machine Rooms
38-054	620-24	C	Branch Circuits and Lighting for Hoistway Pits

Harmonized with the Canadian Electrical Code CSA C22.1 Section 38-054 and ASME/ANSI A17.1–2000, Requirement 2.2.5. A safety problem could be created for persons if the lighting were connected to the load side of GFCI receptacles (i.e., person trying to exit space or reset receptacles without lighting).

620-37. Wiring in Hoistways and Machine Rooms

620-37(b) If a lightning protection system grounding "down" conductor(s), located outside of the hoistway, is within a critical horizontal distance of the elevator rails, bonding of the rails to the lightning protection system grounding "down" conductor(s) is required by NFPA 780 to prevent a dangerous side flash between the lightning protection system grounding "down" conductor(s) and the elevator rails. A lightning strike on the building air terminal will be conducted through the lightning protection system grounding "down" conductor(s), and if the elevator rails are not at the same potential as the lightning protection system grounding "down" conductor(s), a side flash may occur. Equipment and wiring not associated with the elevator is prohibited from being installed in elevator machine rooms and

Chart 2.26.4.1(b) NEC® Section 620-3 — Voltage Limitations
(Courtesy Otis Elevator Co./Ralph Droste)

Section	Type Circuits	Voltage (Max.)
(a) Power circuits	• Motor controllers • Driving machine motors • Machine brakes • Motor-generator sets • Door operator controller • Door motors	600
	• Internal to power conversion	> 600
(b) Lighting circuits	• All	Per Article 410
(c) Heating and air conditioning on car	• All	600

GENERAL NOTE: 300 V max. between conductors unless permitted by (a), (b), and (c).

Chart 2.26.4.1(c)
NEC® Section 620-21(a)(b)(c) — Wiring Methods
(Courtesy Otis Elevator Co./Ralph Droste)

• Rigid Metal Conduit (Article 346)
• Intermediate Metal Conduit (Article 345)
• Electrical Metallic Tubing (Article 348)
• Rigid Nonmetallic Conduit (Article 347)
• Metal and Nonmetallic Wireways (Article 362)
• Liquidtight Flexible Nonmetallic Conduit (LFNC-B)(Article 351)
• MC Cable (Article 334)
• MI Cable (Article 330)
• AC Cable (BX) (Article 333)

hoistways (620-37). Only electrical equipment and wiring used directly in connection with the elevator may be installed inside the hoistway. (Also, see ASME A17.1–2000, requirement 2.8.1.

620-38. Electric Equipment in Garages and Similar Occupancies

Local code authorities have referenced this Section as a basis for requiring explosion-proof elevator equipment in garages used for parking or storage where no repair work is done. As indicated in Section 511-2, parking garages used for parking or storage and where no repair work is done except exchange of parts and routine maintenance requiring no use of electrical equipment, open flame, welding, or use of volatile flammable liquids are not classified, and there is therefore no hazardous location involved. The requirement in edition 511-2 first appeared in 1975 edition of the Code as part of Section 511-1. However, the requirement in Section 620-38 first appeared in the 1971 edition of the Code, at a time when there was no exemption from hazardous location classification for some types of garages.

620-41. Suspension of Traveling Cables
See Diagram 2.26.4.1(j).

620-44. Installation of Traveling Cables
See Diagram 2.26.4.1(k).

620-51. Disconnecting Means

620-51(a) Requirements are coordinated with Part J of Article 430. See Diagrams 2.26.4.1(l) and 2.26.4.1(m).

Passenger safety, comfort, and elevator maintenance is enhanced by separating the main power supply conductors from the car lighting, et al, supply conductors.

620-51(b) ASME A17.1–2000, requirement 2.8.2.3 requires that if sprinklers are installed in hoistways, machine rooms, or machinery spaces, a means has to be provided to automatically disconnect the main line power supply to the affected elevator(s) prior to the application of water. Water on elevator electrical equipment can result in hazards such as uncontrolled car movement (wet machine brakes), and movement of elevator with open doors (water on safety circuits bypassing car and/or hoistway door interlocks). See Diagram 2.26.4.1(n).

620-51(c) Correlates requirements with Definitions in 620-2 and provides for mechanic's safety when equipment is located in remote machinery spaces by requiring a disconnecting means to open all ungrounded main power supply conductors. See Diagrams 2.26.4.1(o) and 2.26.4.1(p).

620-51(d) See Diagram 2.26.4.1(q).

620-52 Power From More Than One Source
See Diagrams 2.26.4.1(r) and 2.26.4.1(s).

620-53. Car Light, Receptacle(s), and Ventilation Disconnecting Means

To assist elevator mechanics to troubleshoot elevators. Signs to identify the location of the supply side overcurrent protective device is to assist in troubleshooting in case of power loss.

620-54. Heating and Air-Conditioning Disconnecting Means
See commentary on 620-53.

620-61. Overcurrent Protection

Operating devices, control, signaling, and power-limited circuit conductor overcurrent protection requirements are cited in the referenced Section.

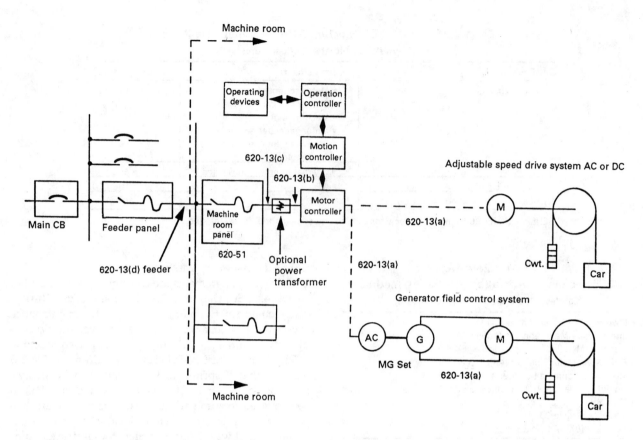

Diagram 2.26.4.1(d) NEC® Section 620-13 — Feeder and Branch Circuit Conductors
(Courtesy Otis Elevator Co./Ralph Droste)

Sections of Article 725 have been combined, simplified, and relocated to position the "installation" requirements before the "listing" requirements. The NEC® Code Making Panel responsible for Article 725 concluded that it is no longer practical to evaluate Class 2 and 3 equipment in the field. Both the testing and construction evaluation require special knowledge of product standards, materials, and procedures that are commonly available only through qualified testing laboratories. Based on this conclusion, the listing requirements for Class 2 and 3 power sources, including Tables 725-31(a) and (b) and Sections 725-31 to 725-36 in 1993 and older editions have been moved to Chapter 9 to separate the installation requirements from the listing requirements so as to provide direction for testing laboratories properly equipped and qualified to evaluate these products. Definitions were also added (725-2) to better understand the Class 1, 2, and 3 circuits.

620-62. Selective Coordination
To assist elevator mechanics to troubleshoot and help located supply side overcurrent protective devices in case of power loss. See Diagram 2.26.4.1(t).

Single elevator installations are exempted because they are typically located in low-rise structures and the electrical service is close to the elevator equipment and

would not create a problem for elevator maintenance personnel in locating supply side overcurrent devices. Also, simplifies compliance by use of all available types of overcurrent devices.

Section IX. Grounding. See Diagrams 2.26.4.1(u) and 2.26.4.1(v).

620-82. Electric Elevators
See Diagram 2.26.4.1(w).

620-85. Ground-Fault Circuit-Interrupter Protection for Personnel
To clarify that each receptacle is to be of the GFCI type, except in machine rooms and machinery spaces. Receptacles on the elevator car tops, pits, and in escalator and moving walk wellways have to be of the GFCI type. It is not in the best interest of safety to introduce an unnecessary action, such as climbing off the top of the elevator car and potentially falling, in order to reset the only device that provided protection in the first place. The receptacle for the sump pump is exempted. See Diagrams 2.26.4.1(x) and 2.26.4.1(y).

620-91. Emergency and Standby Power Systems
620-91(a) A new paragraph (a) explains that for a regenerative elevator, some means are required to either

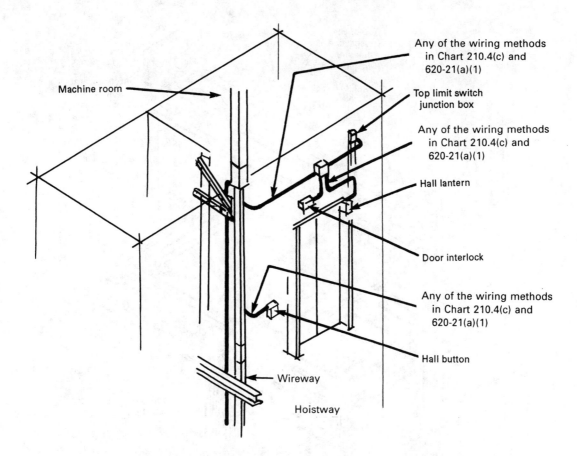

Machine room

Any of the wiring methods
in Chart 210.4(c) and
620-21(a)(1)

Top limit switch
junction box

Any of the wiring methods
in Chart 210.4(c) and
620-21(a)(1)

Hall lantern

Door interlock

Any of the wiring methods
in Chart 210.4(c) and
620-21(a)(1)

Hall button

Wireway

Hoistway

Diagram 2.26.4.1(e) NEC® Section 620-21(a)(1) — Hoistway Wiring Methods
(Courtesy Otis Elevator Co./Ralph Droste)

absorb the regenerative power or prevent the elevator from overspeeding. This may be the power system itself, the emergency or standby power generator, if the prime mover is sufficiently large, a separate load bank provided by the emergency or standby generator system. See Diagram 2.26.4.1(z).

620-91(b) Other building loads may be considered as an absorption means required in 620-91(a) as long as they are automatically connected to the emergency or standby power source (i.e., cannot be inadvertently omitted) and are large enough to absorb the regenerative energy without causing the elevator from attaining a speed equal to the governor tripping speed or a speed in excess of 125% of the elevator rated speed, whichever is the lesser.

620-91(c) The disconnecting means shall disconnect the elevator from both the emergency or standby and the normal power source. This requires that the power transfer switch between the emergency or standby and the normal power systems, as well as the regenerative power absorption means is to be located on the supply side of the elevator disconnect switch. See Diagrams 2.26.4.1(aa) and 2.26.4.1(bb).

The requirements in the second paragraph of (c) were added to harmonize with the Canadian Electrical Code requirements in Section 38-036(3). The auxiliary contact prevents operation of the elevator when the disconnecting means is open.

Highlights of the New and Revised Requirements in ANSI/NFPA 70-1996

The following are highlights of the new and revised requirements in the 1996 National Electrical Code® (NEC®), ANSI/NFPA 70 that are of interest to the elevator industry.

There were 4,259 proposals submitted to change the NEC® and 3,246 public comments were made in response to the actions of the Code Making Panels (CMPs) on the proposals.

As part of an effort to enhance the usability of the NEC® many of the Articles within the Code were reorganized to make them more user friendly and much of the material formerly expressed within exceptions

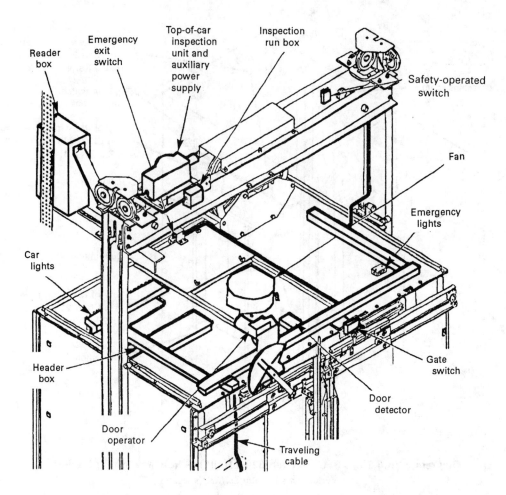

GENERAL NOTE: Wiring — Any method in Chart 2.26.4.1(c) and 620-21(a)(2).

Diagram 2.26.4.1(f)   NEC® Section 620-21(a)(2) — Car Wiring Methods, Top of Car
(Courtesy Otis Elevator Co./Ralph Droste)

were integrated into the main body of the Code. Article 620 was extensively revised in the 1996 NEC® to achieve the same objective.

(a) Article 620 — Elevators, Dumbwaiters, Escalators, Moving Walks, Wheelchair Lifts, and Stairway Chair Lifts

(1) The "Fine Print Notes (FPN)" in Sections 620-1 through 620-3, 620-24, 620-51, and 620-91 were changed to reflect the latest editions of A17.1, A17.5/B44.1, and B44 Codes.

(2) The opening paragraph of Section 620-5(a) was revised to clarify the intent of Section 620-5(a)(1) through (4).

(3) In the interest of NEC®/CEC harmonization Sections 620-21(a)(1)b, 620-21(a)(2)c, 620-21(a)(3)c, 620-21(b)(2) and 620-21(c)(2) were revised to delete the voltage ranges for Class 2 circuits since the values in the CEC and NEC® are slightly different.

(4) A new "Exception" was added to Section 620-21(a)(2)(a) (cars), Section 620-21(a)(3)(a) [machine room

and machinery spaces], Section 620-21(b)(1) [escalators], and Section 620-21(c)(1) (wheelchair lifts and stairway chair lifts) to permit "liquidtight flexible nonmetallic conduit (LFNC-B)" in unlimited lengths as described in Section 351-22(2), i.e., "a smooth inner surface with integral reinforcement within the conduit wall, designated as type LFNC-B."

The wording in this new exception was not supported by NEC® Code Making Panel 12, however, it received NFPA Electrical Section member support on the floor of the 1998 NFPA Annual Meeting and through both the NEC® Technical Correlating Committee and the NFPA Standards Council.

(5) New Section 620-21(a)(2)d was added for driving machine wiring when located on the car.

(6) Section 620-22 was harmonized with CEC Rule 38-012(3), i.e., the over-current protection for the car light and/or heating and air conditioning branch circuits

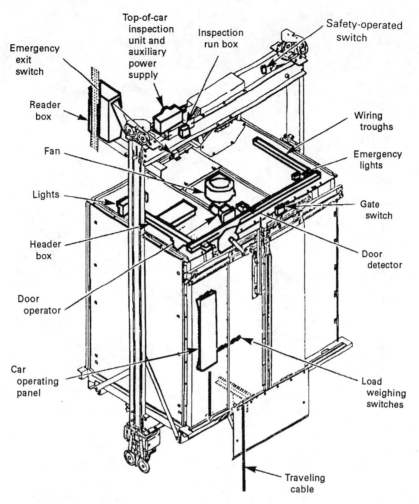

Emergency exit switch

Reader box

Fan

Lights

Header box

Door operator

Car operating panel

Top-of-car inspection unit and auxiliary power supply

Inspection run box

Safety-operated switch

Wiring troughs

Emergency lights

Gate switch

Door detector

Load weighing switches

Traveling cable

GENERAL NOTE: Wiring — Any method in Chart 2.26.4.1(c) and 620-21(a)(2).

Diagram 2.26.4.1(g)   NEC® Section 620-21(a)(2) — Car Wiring Methods
(Courtesy Otis Elevator Co./Ralph Droste)

has to be located in the elevator machine room/machinery space. This new requirement will facilitate ease of maintenance and trouble shooting.

(7) Sections 620-23 and 620-24 were revised to indicate that the required lighting in the machine room and hoistway pit are not permitted to be connected to the load side of a GFCI because the lighting could be de-energized during a fault condition.

(8) Section 620-36 was revised to clarify that power-limited and non-power-limited fire alarm conductors are permitted in the same raceway or traveling cable as power and other types of signaling conductors; however, all power, signaling, and fire alarm conductors have to be insulated for the maximum voltage applied to any conductor in the raceway or cable.

(9) Sections 620-53 and 620-54 were revised to indicate that where there is no machine room, the disconnecting means for the car lights and/or heating and air-conditioning branch circuits have to be located in the machinery space for the respective elevator car.

(10) Section 620-85 was revised to permit ground fault circuit interrupter protection (in lieu of a GFCI-type receptacle) not only in machine rooms, but in machinery spaces as well.

(11) Section 620-91(c) was harmonized with CEC Rule 38-036, i.e. when an auxiliary lowering device is provided, the auxiliary contact on the disconnecting means has to be positively opened mechanically and not be solely dependent on springs.

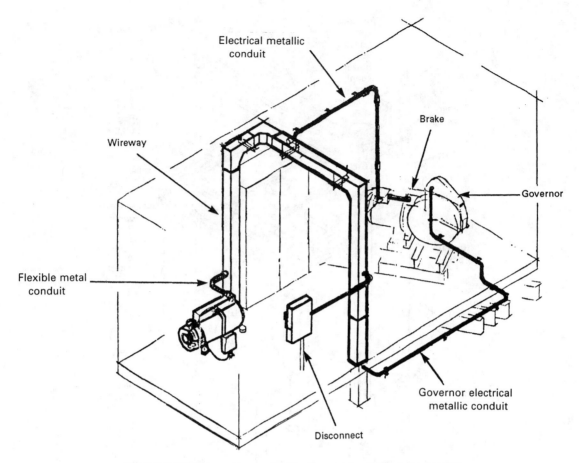

GENERAL NOTE: Wiring — Any method in Chart 2.26.4.1(c) and 620-21(a)(2).

Diagram 2.26.4.1(h) NEC® Section 620-21(a)(3) — Machine Room and Machinery Space Wiring Methods
(Courtesy Otis Elevator Co./Ralph Droste)

(12) Feeder/ branch Circuit Calculations in Examples 9 and 10 are now located in Appendix D.

(b) Article 110 — Requirements for Electrical Installations

(1) Section 110-14(c) [temperature limitations of electrical connections] is an example of a significant format change found in many sections of the 1999 NEC®. Changes made in the 1993 NEC®, to address temperature ratings of electrical connections, did not specifically address motor connections; therefore, language was added to clarify the temperature selection requirements for type B, C, D, and E design motors. Motors designed to NEMA standards are evaluated based on the 750°C compacity rating of conductors rather than 60°C rating, even for equipment rated 100 amperes or less.

(2) Section 110-26 [spaces about electrical equipment] was extensively revised. Section 110-26(a) [working space] was formerly Section 110-16(a). The text of this section was divided into three headings:

(a) depth of working space (no change);

(b) width of working space; and

(c) height of working space.

Section 110-26(a)(2) was revised to make it clear that a work space greater than 30 in. is necessary where the equipment is wider than that dimension. Section 110-26(a)(3) was revised to allow equipment to extend up to 6 in. further into workspace in front of equipment.

(c) Article 200 — Use and Identification of Grounded Conductors

(1) Sections 200-6 and 200-7 (means of identifying grounded conductors and use). In addition to "white" or "natural gray" insulation for a grounded conductor (often the neutral) it is now also permitted to use "three continuous white stripes on other than green insulation along its outline length (conductors No. 6 and smaller)" or "distinctive white marking at its termination (conductors larger than No. 6)."

(d) Articles 210, 215, and 220 were extensively revised to provide a more logical format for circuit requirements and load calculations.

(1) Article 210 - Branch Circuits. A complete listing of the sections that were relocated either from Article 220 or from within Article 210 are shown below. There

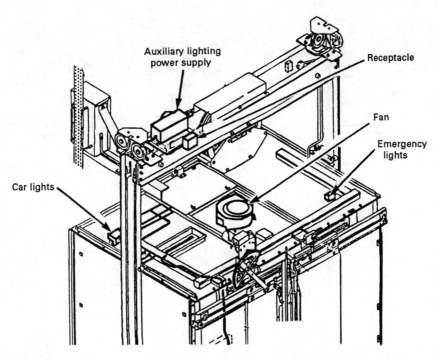

Auxiliary lighting power supply

Receptacle

Fan

Emergency lights

Car lights

GENERAL NOTE: Separate branch circuit on each elevator car to supply car lights, receptacle(s), auxiliary lighting power source, and ventilation.

Diagram 2.26.4.1(i)   NEC® Section 620-22(a) — Branch Circuit for Car Lighting
(Courtesy Otis Elevator Co./Ralph Droste)

are no substantive changes in the application of the rules. [See Sections 620-13, 620-22, 620-23, and 620-24 for applicable branch-circuit requirements.]

(2) Article 210 Reorganization

1999 NEC® Section	Section Heading or Description	1999 NEC® Section
220-4	Branch Circuits Required	210-11
220-4(a)	Number of Branch Circuits	210-11(a)
220-4(d)	Load Evenly Proportioned Among Branch Circuits Dwelling Units	210-11 (b)
		210-11(c)
220-4(b)	Small Appliance Branch Circuits	210-11(c)(1)
220-4(c)	Laundry Branch Circuits	210-11(c)(2)
210-52(d) (part)	Bathroom Branch Circuits	210-11(c)(3)
210-19, 220-3(a), 210-22(c),	Conductors-Minimum Ampacity and Size	210-19
210-19 (part)	General (Continuous and Noncontinuous Loads)	210-19(a)
210-19(a) (part)	Multioutlet Branch Circuits	210-19(b)
210-19(b)	Household Ranges and Cooking Appliances	210-19(c)
210-19(c)	Other Loads	210-19(d)
210-20	Overcurrent Protection	210-20
210-22(c), 220-3(a)	Continuous and Non-continuous Loads	210-20(a)
210-20(1)	Conductor Protection	210-20(b)
210-20(2)	Equipment	210-20(c)
210-20(3)	Outlet Devices	210-20(d)

(e) Article 215 — Feeders. Article 215 was reorganized to include material relocated from Article 220. See listing below. The reorganization did not result in any substantive changes to the way feeders are sized or installed (See Section 620-13).

(1) Article 215 Reorganization

1996 NEC® Section	Section Heading or Description	1999 NEC® Section
215-2	Minimum Rating and Size	215-2
220-10(b) (part)	General	215-2(a)
215-2(a)	For Specked Circuits	215-2(b)
215-2(b)	Ampacity Relative to Service Entrance Conductors	215-2(c)
215-2 (part)	Individual Dwelling Unit or Mobile Home Conductors	215-2(d)
220-10(b) (part)	Over-current Protection	215-3

(f) Article 220 — Branch Circuit, Feeder, and Service Calculations. Article 220 was extensively revised to correlate with changes in Articles 210 and 215. See listing below. Article 220 is the base for all load calculations covered in the NEC® except for some applications such as elevators. See Section 620-13 and Examples D9 and D10 in Appendix D.

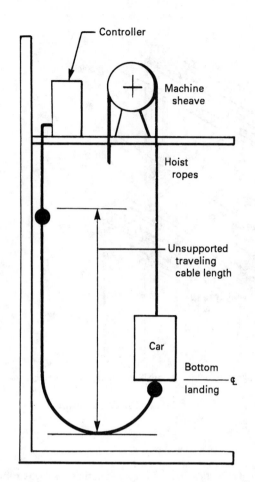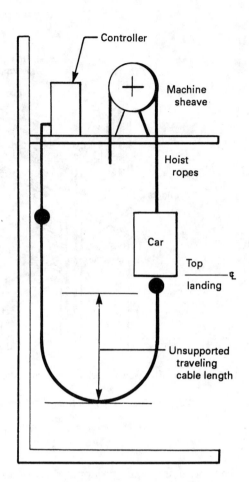

Diagram 2.26.4.1(j)   NEC® Section 620-41 — Unsupported Length of Traveling Cable
(Courtesy Otis Elevator Co./Ralph Droste)

(1) Article 220 Reorganization

1996 NEC® Section	Section Heading or Description	1999 NEC® Section
220-3(b)	Lighting Load for Specified Occupancies	220-3(a)
220-3(c)	Other Loads — All Occupancies	220-3(b)
220-3(c)(1)	Specific Appliances or Loads	220-3(b)(1)
220-3(c) Exc. No. 2 & No. 5	Electric Dryers and Household Electric Cooking Appliances	220-3(b)(2)
220-3(c)(2)	Motor Loads	220-3(b)(3)
220-3(c)(3)	Recessed Lighting Fixtures	220-3(b)(4)
220-3(c)(4)	Heavy Duty Lampholders	220-3(b)(5)
220-3(c)(5)	Track Lighting (Deleted-See 410-102)	220-12(b)
220-3(c)(6)	Sign and Outline Lighting	220-3(b)(6)
220-3(c) Exc. No.3	Show Windows	220-3(b)(7)
220-3(c) Exc. No. 1	Fixed Multi-outlet Assemblies	220-3(b)(8)
220-3(c)(7)	Other Outlets	220-3(b)(9)
220-3(d)	Loads for Additions to Existing Installations	220-3(c)
Table 220-3(b)*Note	General Use Receptacles In Dwellings and Guest Rooms of Hotels and Motels	220-3(b)(10)

1996 NEC® Section	Section Heading or Description	1999 NEC® Section
Table 220-3(b)	General Lighting Loads by Occupancies	Table 220-3(a)
210-22	Maximum Loads	220-4
210-22(a)	Motor Operated and Combination Loads	220-4(a)
210-22(b)	Inductive Lighting Loads	220-4(b)
210-22(c) (part)	Range Loads	220-4(c)

(g) Article 240 — Overcurrent Protection

(1) Section 240-3(d) (small conductor overcurrent protection). The "obelisk" note information from the ampacity Tables 310-16 and 310-17 was relocated to Section 240-3(d) to improve NEC® usability. The practical application of this rule has not changed (i.e., overcurrent protection for conductors sizes No. 14, 12, and 10 AWG). Elevator/Escalator requirements are covered in Section 620-61(a) and by reference to Sections 725-23 and 725-24.

(2) Section 240-3(a) (tap conductor definition). The term "tap" is used in many ways, both in the NEC® and in the field. While not an Article 100 definition, the

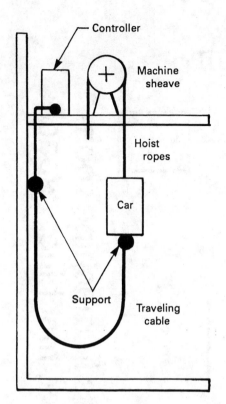

- 6 ft (1.83 m) length limitation without use of raceway or conduit

- Suitably supported and protected from damage

Diagram 2.26.4.1(k) NEC® Section 620-44 — Installation of Traveling Cables
(Courtesy Otis Elevator Co./Ralph Droste)

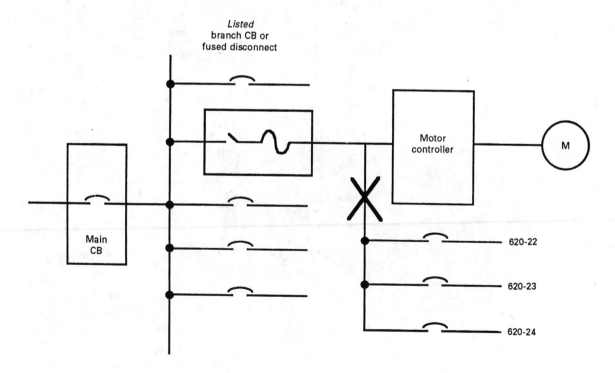

Diagram 2.26.4.1(l) NEC® Section 620-51 — Disconnecting Means
(Courtesy Otis Elevator Co./Ralph Droste)

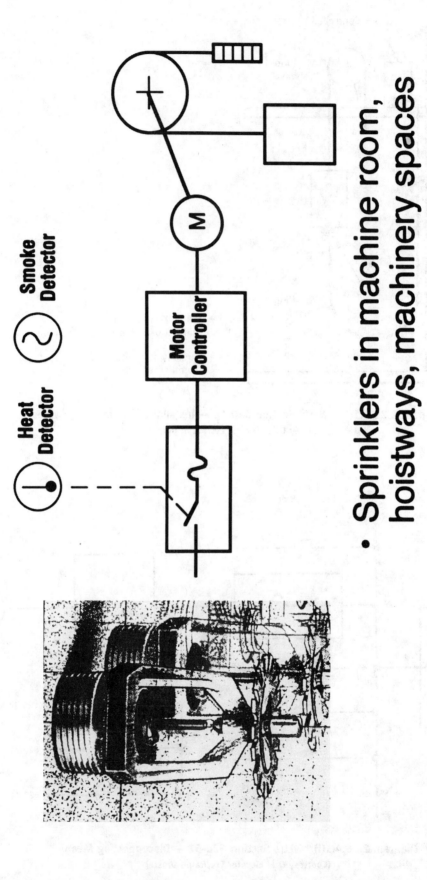

• Sprinklers in machine room, hoistways, machinery spaces

Diagram 2.26.4.1(m) NEC® Section 620-51(b) — Operation Sprinklers
(Courtesy Otis Elevator Co./Ralph Droste)

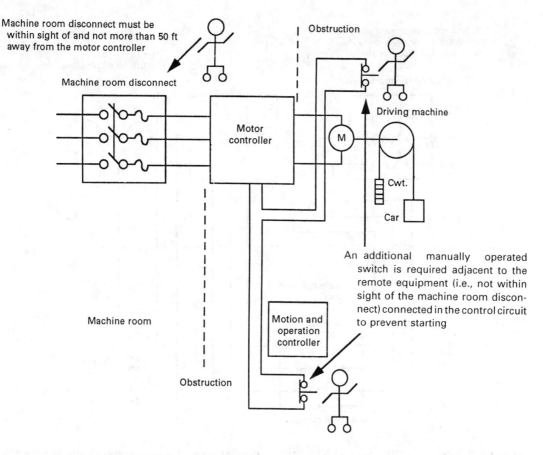

Machine room disconnect must be within sight of and not more than 50 ft away from the motor controller

Machine room disconnect

Obstruction

Driving machine

Motor controller

M

Cwt.

Car

An additional manually operated switch is required adjacent to the remote equipment (i.e., not within sight of the machine room disconnect) connected in the control circuit to prevent starting

Machine room

Motion and operation controller

Obstruction

Diagram 2.26.4.1(n)   NEC® Section 620-51(c) — Disconnecting Means Location
(Courtesy Otis Elevator Co./Ralph Droste)

revision assigns a clear meaning to branch circuit, feeder, transformer secondary, and motor circuit taps.

(3) Section 240-3(g) (over-current protection rules) was re-organized with many references to other articles, including Article 725 [see Section 620-61(a)]. The change should make it easier to use Section 240-3.

(4) Section 240-4(b)(3) [field-assembled extension cord sets] was revised so that if a cord set (conductor size No. 16 or larger) is assembled from listed components in the Field, it is permitted to be protected by 20 A over-current devices. This was not previously permitted.

(5) Section 240-21 tap conductor rules have been reorganized, with rules for feeder taps separated from rules for transformer secondary conductors (a usability issue).

The outline of revised Section 240-21 is:

240-21	Location in Circuit
240-21(a)	Branch Circuit Conductors
240-21(b)	Feeder Taps
240-21(b)(1)	Taps Not over 10 ft Long
240-21(b)(2)	Taps Not over 25 ft Long
240-21(b)(3)	Taps Supplying a Transformer
240-21(b)(4)	Taps over 25 ft Long

240-21(b)(5)	Outside Taps of Unlimited Length
240-21(c)	Transformer Secondary Conductors
240-21(c)(1)	Protection by Primary Overcurrent Device
240-21(c)(2)	Transformer Secondary Conductors Not over 10 ft Long
240-21(c)(3)	Secondary Conductors Not over 25 ft Long
240-21(c)(4)	Outside Secondary Conductors
240-21(c)(5)	Secondary Conductors from a Feeder Tapped Transformer
240-21(d)	Service Conductors
240-21(e)	Busway Taps
240-21(f)	Motor Circuit Taps
240-21(g)	Conductors from Generator Terminals

(h) Article 250 — Grounding. (See NEC® Appendix E for cross-reference between old and new sections).

(1) Article 250 was completely revised to provide a more logical approach to grounding and, therefore, make it easier to apply. Grounding performance information in the Fine Print Notes (FPNs) was rewritten in mandatory language and combined with former Section

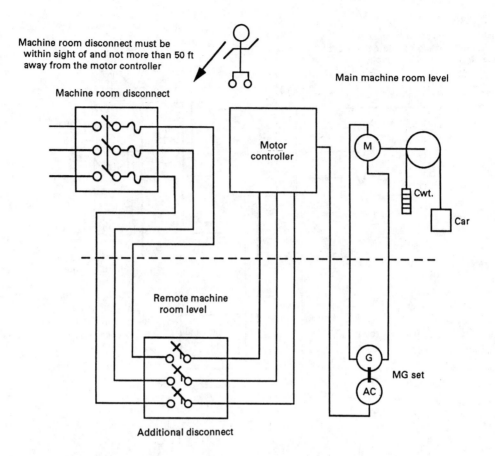

Machine room disconnect must be within sight of and not more than 50 ft away from the motor controller

Machine room disconnect

Main machine room level

Motor controller

M

Cwt.

Car

Remote machine room level

G

AC

MG set

Additional disconnect

Diagram 2.26.4.1(o)   NEC® Section 620-51(c) — Disconnecting Means Location With Remote MG
(Courtesy Otis Elevator Co./Ralph Droste)

250-51 as a new Section 250-2. The old Section 250-2 was renumbered as Section 250-4.

(2) Section 250-106 (lightning protection system) now requires only that the ground terminals for the lightning protection system be bonded to the building or structure grounding electrode system (Section 250-46 of the 1996 NEC®). The previous bonding requirement of electrical equipment within 6 ft of lightning protection system conductors has been deleted. Bonding of the ground terminal (grounding electrode) of the lightning protection system to the grounding electrode system for the structure electrical system is required. Also, see Section 620-37(b).

(i) *Article 300 — Wiring Methods*

(1) Section 300-14 (length of free conductors at outlets, junctions, and switch points) was revised to clarify how to determine free conductor length and to address the length that extends beyond the enclosure.

(j) *Article 310 — Conductors for General Wiring*

(1) Section 310-15. The ampacity tables and notes were reorganized. The notes are now incorporated into Section 310-15 (see listing below).

1996 NEC® Section	Section Heading or Description	1999 NEC® Section
310-15	Ampacities for Conductors Rated 0-2000 Volts	310-15
310-15 (part)	General-Tables or Engineering Supervision	310-15(a)(1)
310-15 (FPN)	Tables do not account for voltage drop	310-15(a)(1) (FPN No. 1)
Tables Note 7 FPN	Type MTW Machine Tool Wire Ampacities	310-15(1) (FPN No. 2)
310-15(c)	General-Selection of Ampacity	310-15(a)(2)
310-15(a)(part)	Tables	310-15(b)
310-15(a) (FPN)	Load based on Article 220-ampacity considerations	310-15(b) (FPN)
Table Note 1	Tables-General-insulation-type letters	310-15(b)(1)
Tables Note 8(a)	Adjustment Factors and Table	310-15(b)(2)(a)
Tables Note 8(b)	More Than One Conduit, Tube, or Raceway	310-15(b) (2)(b)
Tables Note 5	Bare or Covered Conductors	310-15(b)(3)
Tables Note 10	Neutral Conductor-when counted as current carrying	310-15(b)(4)
Tables Note 11	Grounding or Bonding Conductor	310-15(b)(5)

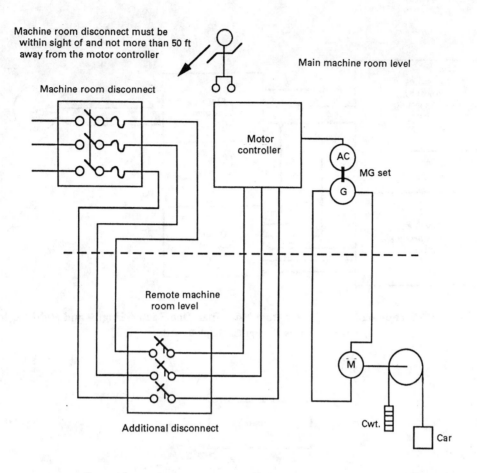

Diagram 2.26.4.1(p) NEC® Section 620-51(c) — Disconnecting Means Location With Remote Driving Motor
(Courtesy Otis Elevator Co./Ralph Droste)

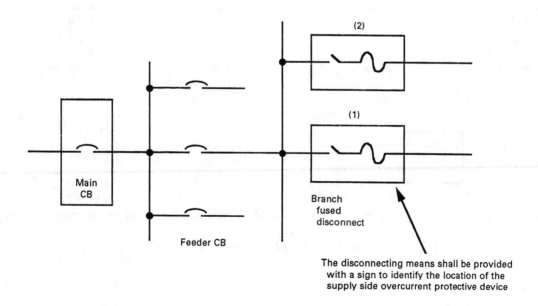

Diagram 2.26.4.1(q) NEC® Section 620-51(d) — Disconnecting Means Identification and Signs
(Courtesy Otis Elevator Co./Ralph Droste)

191

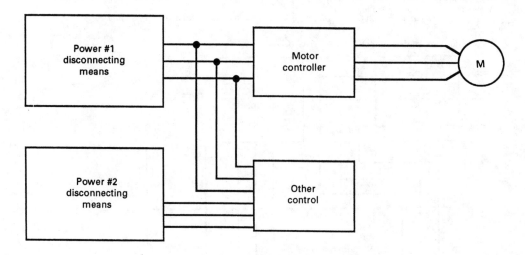

Diagram 2.26.4.1(r)   NEC® Section 620-52 — Power From More Than One Source Single- and Multi-Car Installations
(Courtesy Otis Elevator Co./Ralph Droste)

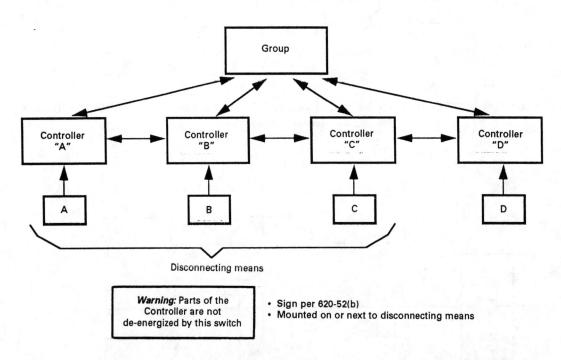

Diagram 2.26.4.1(s)   NEC® Section 620-52 — Power From More Than one Source Interconnection Multi-Car Controllers
(Courtesy Otis Elevator Co./Ralph Droste)

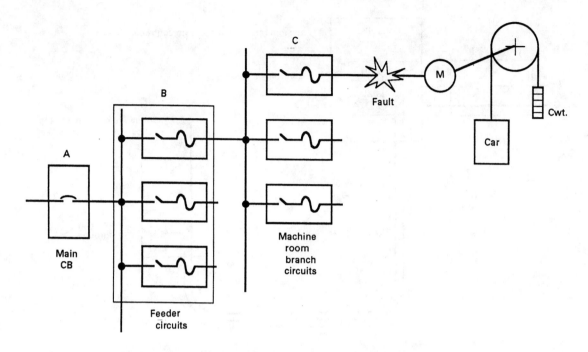

Diagram 2.26.4.1(t)   NEC® Section 620-62 — Selective Coordination
(Courtesy Otis Elevator Co./Ralph Droste)

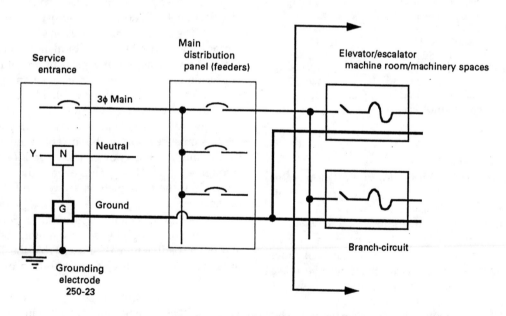

Diagram 2.26.4.1(u)   NEC® Sections 620-81, 82, 83, and 84 — Grounding
(Courtesy Otis Elevator Co./Ralph Droste)

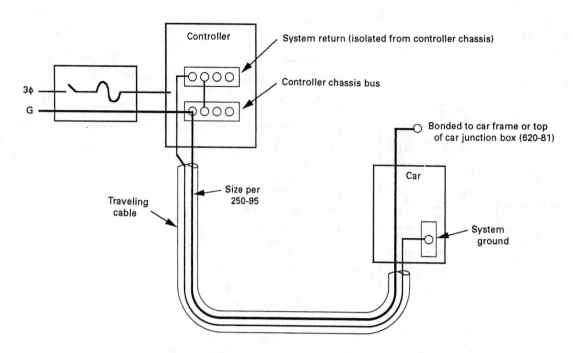

Diagram 2.26.4.1(v)   NEC® Sections 620-81, 82, 83, and 84 — Grounding
(Courtesy Otis Elevator Co./Ralph Droste)

1996 NEC® Section	Section Heading or Description	1999 NEC® Section
Tables Note 3	Dwelling Units feeder and service conductor ampacities	310-15(b)(6)
Tables Note 6	Mineral-Insulated, Metal-Sheathed Cable	310-15(b)(7)
310-15(b)	Engineering Supervision	310-15(c)
Tables Note 9	Overcurrent Protection	Deleted-See 240-3
Obelisk Notes at bottom of Tables 310-16 & 310-7	Limitations on overcurrent protection for small conductors	240-3(d)

(k) Article 333 — AC Cable

(1) Section 333-7 (cable ties for support of AC cable) now permits use of cable ties to secure type AC cable.

(l) Article 362 — Wireways

(1) Section 362-6 (raceway entries to wireways) was revised to require that the distance between raceway entries, where a conductor enters a wireway in one raceway and exits through another raceway, must be at least six times the trade diameter of the largest raceway in order to ensure adequate bending space.

(2) Section 362-26 (bonding of raceways connected to nonmetallic wireways) was revised to require an equipment grounding conductor (or be bonded) between the metal raceway extension and nonmetallic wireway.

(m) Article 400 — Flexible Cords and Cables

(1) Section 400-8(2)(5) was revised to prohibit flexible cords or cables going through holes in walls, structural ceilings, suspended ceilings, dropped ceilings, or floors since damage or deterioration of the cord or cable would not be readily observed.

(n) Article 430 — Motors, Motor Circuits, and Controllers. Section 430-22(a) (single motor)

(1) The first sentence of Section 430-22(a) (i.e., single motors) was revised to read: "branch-circuit conductors that supply a single motor used in a continuous duty application (e.g., escalator) shall have an ampacity not less than 125% of the motor full-load current rating as determined by Section 430-6(a)(1)."

Section 430-6(a)(1) requires that the ampacity of conductors must be based on the current ratings determined by the marked horsepower rating and Tables 430-147 through 430-150. This applies even if the current rating marked on the nameplate is known and is of a lower rating.

(2) A similar change was made in Section 430-24 for conductors supplying several motors, or a motor(s) and other load(s), to make it clear that the Table values must also be used for the calculation of conductor ampacity.

(3) Note that Section 430-22, Exception No. 1 in the 1996 NEC® became Section 430-22(b) in the 1999 NEC® (see Section 620-13).

(o) Article 450 Transformers and Transformer Vaults

(1) Section 450-3 (transformer overcurrent protection rules) was rearranged into two tables to make the selection of overcurrent protection for transformers easier; however, the requirements have not been changed from the 1996 NEC®.

(p) Articles 725 and 800

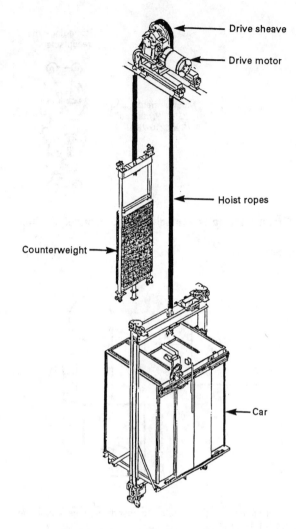

Diagram 2.26.4.1(w)   NEC® Sections 620-82 and 250-58(b) — Metal Car Frames Equipment Considered Effectively Grounded
(Courtesy Otis Elevator Co./Ralph Droste)

(1) Section 725-61 (application of listed class 2, class 3, and PLT cables) and Section 800-53 (application of listed communication wires and cables, and communication raceways) on cable substitutions were revised and clarified for remote control, signaling, and communication. Numerous permitted substitutions were eliminated, including the permission to substitute fire alarm cables for CL2 and CL3 cables. See Tables 725-61 and 800-53.

Highlights of the New and Revised Requirements in ANSI/NFPA 70-2002

There were 28 requirement changes to Article 620 in the 2002 NEC®. The majority of the changes were revisions to present requirements to include the machine-room-less type of elevator installations. In addition, there were 38 "editorial" changes proposed by the NFPA Editorial Committee such as changing to dual metric/ (imperial) designations and adding titles to Section subparagraphs and to figures in Article 620.

The following is a summary of the 28 changes for Article 620 — Elevators, Dumbwaiters, Escalators, Moving Walks, Wheelchair Lifts, and Stairway Chair Lifts. The rationale for each change can be found in the NFPA 70 Report on Proposals (Jan. 10-22, 2000 NEC Panel meetings) and the Report on Comments (Dec. 4-16, 2000 NEC Panel meetings).

(a) Section 620-5(a) received a "wording" change to use the accepted code language phrase "permitted to be provided with," while leaving the basic Section intact.

(b) Section 620-11(d) "editorially" dropped the reference to an "LS" marking on wire insulations that meet limited smoke criteria.

(c) Section 620-12 received change to add "AWG" to the numerical wire number designations in the Section.

(d) Section 620-21 was revised to add the words

- Pits, elevator car tops, and escalator and moving walk wellways
- Each 125 V, single-phase, 15 and 20 A receptacle

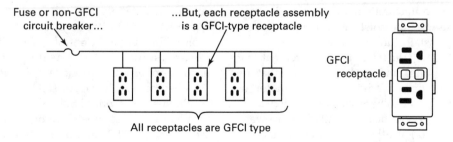

Diagram 2.26.4.1(x)   NEC® Section 620-85 — Ground Fault Circuit Interrupter Protection for Personnel
(Courtesy Otis Elevator Co./Ralph Droste)

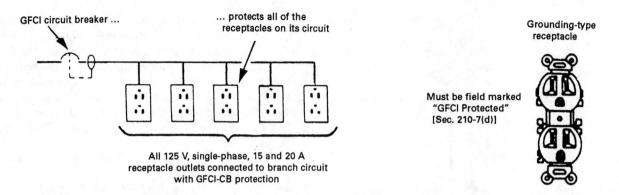

Diagram 2.26.4.1(y) NEC® Section 620-85 — Ground Fault Circuit Interrupter Protection for Personnel Machine Rooms
(Courtesy Otis Elevator Co./Ralph Droste)

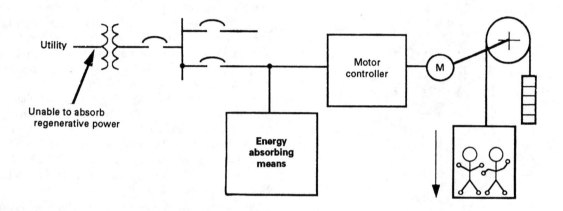

GENERAL NOTE: For elevator systems that regenerate power back into the power source, which is unable to absorb the regenerative power under overhauling elevator load conditions, a means shall be provided to absorb this power.

Diagram 2.26.4.1(z) NEC® Section 620-91 — Emergency and Standby Power Systems Regenerative Power
(Courtesy Otis Elevator Co./Ralph Droste)

"Machinery Spaces, and Control Spaces." This is to address the wiring methods for machine-room-less type elevator installations.

(e) Section 620-21(a)(1)(c) is a new requirement that permits flexible cords and cables that are part of listed equipment and used in circuits not exceeding 30 vrms or 42 vdc to be used in the hoistway in lengths up to 6 ft, without being installed in raceway.

(f) Section 620-21(a)(1)(d) is a new requirement that permits flexible conduit, and cords and cables that are part of listed equipment, a driving machine motor, or a driving machine brake to be used in the hoistway in lengths up to 6 ft, without being installed in raceway.

(g) Section 620-21(a)(3) title was revised to add the words "Control Rooms, and Control Spaces." This is to address the machine-room-less type elevator installations.

(h) Section 620-21(a)(3)(a) Exception, was changed to

add a Metric Designator 12 as the equivalent to ³⁄₈ Trade Size.

(i) Section 620-22(a) was revised to add the words "Control Rooms, and Control Spaces." This is to address the car light branch circuit for machine-room-less type elevator installations.

(j) Section 620-22(b) was revised to add the words "Control Rooms, and Control Spaces." This is to address the air conditioning and heating branch circuit for machine-room-less type elevator installations.

(k) Section 620-23 was revised to add the words "Control Rooms, and Control Spaces." This is to address the machine-room-less type elevator installations.

(l) Section 620-23(a) was revised to add the words "Control Rooms, and Control Spaces." This is to address the lighting branch circuit for machine-room-less type elevator installations.

(m) Section 620-23(b) was rewritten to include the

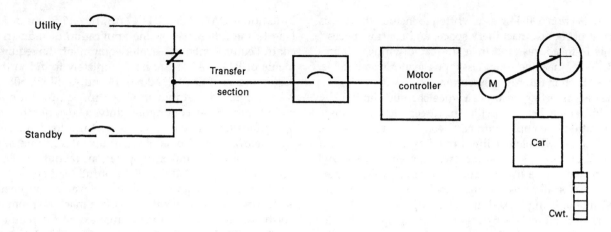

GENERAL NOTE: The disconnecting means required by Section 620-51 shall disconnect the elevator from both the emergency or standby power system and the normal power system.

Diagram 2.26.4.1(aa)   NEC® Section 620-91 — Emergency and Standby Power Systems Disconnecting Means
(Courtesy Otis Elevator Co./Ralph Droste)

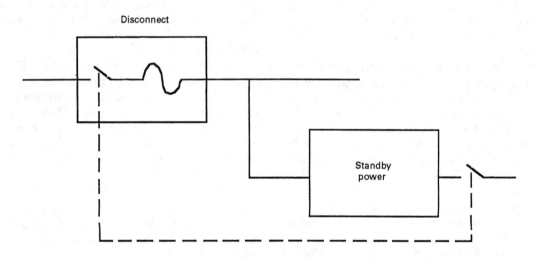

Diagram 2.26.4.1(bb)   NEC® Section 620-91(c) — Emergency and Standby Power Systems/Disconnecting Means
(Courtesy Otis Elevator Co./Ralph Droste)

words "Control Rooms, and Control Spaces." This is to address the lighting switch location for machine-room-less type elevator installations.

(n) Section 620-23(c) was rewritten to include the words "Control Rooms, and Control Spaces". This is to address the receptacle requirement for machine-room-less type elevator installations.

(o) Section 620-25 is a new requirement that permits branch circuits for other utilization equipment that is not permitted to be connected to the car light, air conditioning, machine room, hoistway, and pit lighting branch circuits. The limitations as to the type of utilization

equipment are covered by the scope of Article 620, Section 620-1 that restricts the equipment to that 'used in connection with the elevators."

(p) Section 620-37 has a change to the Section title that now includes the words "Control Rooms, Machinery Spaces, and Control Spaces." This is to address the wiring requirements for machine-room-less type elevator installations.

(q) Section 620-37(a) was rewritten to include the words "control rooms, machinery space, and control spaces." This is to address the wiring requirements for machine-room-less type elevator installations.

(r) Section 620-44 was rewritten to include the words "control rooms, machinery space, and control spaces." This is to address the traveling cable wiring requirements for machine-room-less type elevator installations.

(s) Section 620-51(a) Exception is a new exception for the disconnecting means of a wheelchair lift supplied by an individual branch circuit. The disconnecting means is permitted to comply with NEC 430-109(c) which states that, if the wheelchair lift is rated at 2HP or less and 300 V or less, a general use switch, a general use AC snap switch or a listed manual motor controller is permitted. For all cases, the disconnect must be listed and capable of being locked in the open position.

(t) Section 620-51(b) was rewritten to include the words "control rooms, and control spaces." This is to address the use of sprinklers and the disconnection of power prior to the application of water requirements for machine-room-less type elevator installations.

(u) Section 620-51(c)(1) was revised to add the words "remote machinery room." This is to address the disconnecting means used for both machine room type and machine-room-less type elevator installations.

(v) Section 620-51(c)(2) was revised to add the words "remote machinery room." This is to address the disconnecting means used for machine-room-less type elevator installations.

(w) Section 620-53 was rewritten to include the words "control room." This is to address the location of the disconnecting means for the car light branch circuit in machine-room-less type elevator installations.

(x) Section 620-54 was rewritten to include the words "control room." This is to address the location of the disconnecting means for the heating and air conditioning branch circuit in machine-room-less type elevator installations.

(y) Section 620-55 is a new requirement for the utilization equipment branch circuits (see item 15) disconnecting means.

(z) The title of Part H was changed to include "Control Room, Machinery Spaces, and Control Spaces." This is to address the machine-room-less type elevator installations.

(aa) Section 620-71 was rewritten to change the word "enclosure" to "space." This is to address the location and securing of the equipment in machine-room-less type elevator installations.

(bb) Section 620-85 was revised to require Ground Fault Circuit Interrupter protection for personnel in hoistways. This is to address the use of hoistway-mounted receptacles for machine-room-less type elevator installations.

2.26.4.2 Drive-machine controllers, logic controllers, and operating devices necessary thereto for starting, stopping, regulating, controlling, or protecting electric motors, generators, or other equipment has to be listed/certified and labeled/marked to the requirements of the bi-national CAN/CSA-B44.1/ASME A17.5 standard. Due to the differences in the application of industrial control equipment and elevator equipment, the requirements of B44.1/A17.5 are not completely found in the requirements of CSA C22.2 No. 14 and ANSI UL 508. It was recognized that industrial control equipment normally operates at continuous duty, a low number of operations (about 3000/year), and at full load current; while elevator control equipment operates at intermittent duty, a high number of operations (about 500,000/year) and at up to 200% to 250% of full load current in order to accelerate a mass. Further, elevator equipment is always protected by either a locked machine room or hoistway. However, to a very large extent the requirements of B44.1/A17.5 are based on CSA C22.2 No. 14 and ANSI UL 508. Where there were differences between the UL and CSA standards, the more stringent requirements were used. This will minimize the risk of electricity as a source of electric shock and as a potential ignition source of fires.

Approval of equipment is the responsibility of the elevator and electrical inspection authority and many "approvals" are based on tests and listings of testing laboratories such as Underwriters Laboratories (UL), Canadian Standards Association (CSA), ETL Testing Laboratories (ETL), etc.

Revisions incorporated in ASME A17.1–2000 include:

(a) Adopted the language in the scope of A17.5/B44.1.

(b) Clarifies what electrical equipment has to be listed/certified and labeled/marked to the requirements of CAN/CSA-B44.1/ASME - A17.5, 1996 edition. Previously requirements said that all electrical equipment had to be certified to A17.5/B44.1. Also, see definitions for listed/certified and labeled/marked that were added to harmonize the terms in the US and Canada.

2.26.4.3 ASME A17.1–2000 adopted the CSA B44 language in Clause 3.12.4.3, which consolidates the requirements for "electrical protective devices" under one requirement.

2.26.4.4 This new requirement addresses reported incidences of EMI causing an unsafe condition (e.g., car movement initiated by two-way radios and other forms of RF transmissions). When control equipment is exposed to interference levels at the test values specified for "safety circuits" in EN12016: 1998 (European Standard on immunity for elevators, escalators and passenger conveyors), the interference can not cause any of the conditions in 2.26.9.3(a) through (e) or cause the car to move while on inspection operation. See Diagram 2.26.4.4.

NOTE: See EMI/EMC Comparison of Worldwide EMS Standards Report. ISO/DTR 16744.

Warning sign requirements are given for the case where doors or suppression equipment must not be removed in order to meet the immunity requirements.

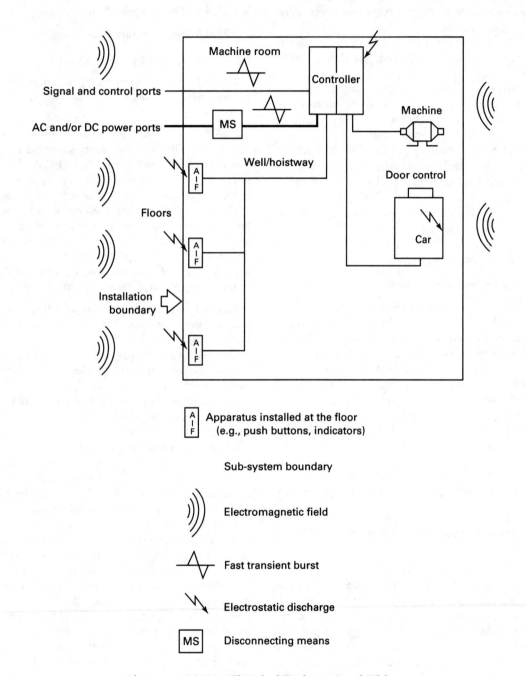

Signal and control ports

AC and/or DC power ports

Floors

Installation boundary

Machine room

Controller

Machine

Door control

Car

Well/hoistway

MS

| A I F | Apparatus installed at the floor (e.g., push buttons, indicators) |

Sub-system boundary

Electromagnetic field

Fast transient burst

Electrostatic discharge

MS   Disconnecting means

Diagram 2.26.4.4   Electrical Equipment and Wiring
(Courtesy Otis Elevator Co./Ralph Droste)

2.26.5 System to Monitor and Prevent Automatic Operation of the Elevator With Faulty Door Contact Circuits

The requirement only applies to installations with power-operated car doors that are mechanically coupled with the landings doors.

The intent is to monitor the position of the car doors while the car is in the landing zone (see definition) in order to:

(a) prevent operation of the car, (except when on access switch operation, leveling or bypass switch operation) if the car door is not closed with the car and/or landing contacts closed or open; and

(b) prevent power closing of doors if the car door is fully open and any of the following conditions exists:

(1) car-door contact is made (closed) or bypassed (jumped);

(2) landing door contact is made (closed) or bypassed (jumped); and

(3) car and landing door contacts are made (closed) or bypassed by a single jumper.

2.26.6 Phase Protection of Motors

If phase rotation is in the wrong direction, the motor rotation would be in the wrong direction and thus the elevator car or elevator door(s) would be traveling in an opposite direction, as indicated by the controller. This would allow the car to operate without protection of slowdowns and normal limits or could cause the door(s) to move in the wrong direction when a reopening device operates. The use of reverse-phase protection prevents the driving machine motor from being activated under these conditions.

This type of protection is required only on those drives where a phase-reversal of the incoming alternating current power to the elevator could cause an unsafe condition.

2.26.8 Release and Application of Driving-Machine Brakes

Partially energizing the brake coil of a car at a floor level that is about to close its doors and start away from the floor does not violate the intent of this requirement. In fact, while the doors are closing, passengers may still be entering or leaving the car and thus, may produce enough rope stretch to move the car away from the floor enough to activate releveling. The brake would then normally be fully energized and power applied to the elevator motor to bring the elevator back to floor level even though the doors would not be fully closed.

When the elevator is at the floor, the brake must be set. If loading or unloading causes the car leveling device to be active, then the brake can be electrically released in accordance with 2.26.1.6. The brake cannot be connected across the armature as residual magnetism (voltage) remaining after the elevator has moved could hold the brake in the open position. Two devices are required by 2.26.9.3, 2.26.9.5, and 2.26.9.6 to remove power from the brake, thus all power feed lines do not need to be opened. See Diagrams 2.26.8, 2.26.8.2, and 2.26.8.5.

2.26.8.2 See Diagram 2.26.8.2.

2.26.8.5 See Diagram 2.26.8.5.

2.26.9 Control and Operating Circuits

2.26.9.1 The reason for this requirement is that springs in compression are less likely to fail than springs in tension.

2.26.9.2 This requirement can be traced back to the first edition (1921) of the safety standard for elevators. It was not until the third edition (1931) of the elevator safety code that the restriction for interrupting power to the motor and brake "at the terminal" was added.

The circa 1931 committee left no known reason, except the recognized need for redundant safe guards at the terminal landing. The intent of 2.26.9.2 is that the contactors used to interrupt the power to the elevator driving machine motor or brake at the terminal shall not be in the energized state. See Diagram 2.26.9.2.

2.26.9.3 The principles on which elevator control system safety are based have evolved since the days of Elisha Otis.

Elevator safety requires that the elevator control system be designed such, that under any hardware or software system failure as described in 2.26.9.3, the elevator

(a) continues its prescribed safe operation; or

(b) reverts to a more restrictive known safe operating state; or

(c) comes to a complete stop.

The nature of electrical equipment failures and the occurrence of accidental grounds on the safe operation of elevators were recognized in the very first edition of A17.1–1921 and have evolved ever since.

The forerunner of the present day language dates back to the A17.1–1955 edition. It recognized the failure mechanism of magnetically operated switches, contactors and relays to release in the intended manner, thereby implying that when ever such devices were used in controller design, redundancy (i.e., two switch protection) was required to prevent the elevator from moving away from a floor with open car and hoistway landing doors because of this type of failure.

Over the years, 2.26.9.3 was revised to include static control failures (A17.1–1981), and in ASME A17.1a-1988 supplement the requirement that the failures listed could not make the EPDs ineffective was added.

With the A17.1–2000 edition a number of new concepts are being introduced.

First, the concept of software system failures and the effect on elevator safety and secondly, the concept of single failures and their effects on elevator safety when on hoistway access operation, inspection operation, door

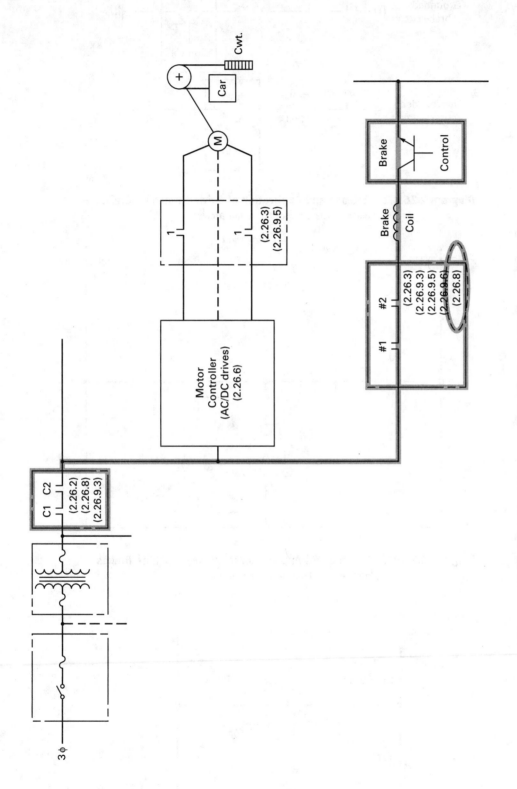

Diagram 2.26.8   Release and Application of Driving Machine Brakes
(Courtesy Otis Elevator Co./Ralph Droste)

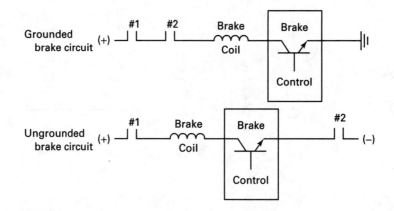

Diagram 2.26.8.2   Release and Application of Driving Machine Brakes
(Courtesy Otis Elevator Co./Ralph Droste)

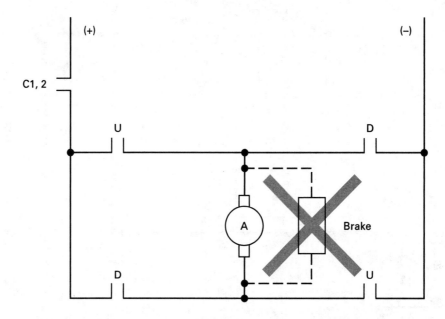

Diagram 2.26.8.5   Release and Application of Driving Machine Brakes
(Courtesy Otis Elevator Co./Ralph Droste)

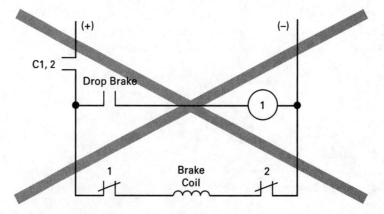

Diagram 2.26.9.2   Control and Operating Circuits
(Courtesy Otis Elevator Co./Ralph Droste)

bypass operation, and leveling operation. See Diagrams 2.26.9.3(a) through (e).

The use of programmable electronics (software) is increasingly becoming more pervasive in today's elevator control systems. The software may be involved directly or indirectly with the elevator safety. Such failure must result in a "nondangerous" state of elevator operation as listed in 2.26.9.3(a) through (e).

The Code also recognizes that the construction, installation, and maintenance of the elevator system must not compromise the basic design safety requirements in 2.25 and 2.26.

Currently work is underway in both the A17.1 Electrical Committee and ISO/TC178/WG8 to prepare a standard for programmable electronic system in safety-related applications for elevators. IEC Standard #61508 covers basic (generic) aspects when electrical/electronic/programmable electronic systems (E/E/PESs) are used to carry out safety functions, but it does not specify elevator specific requirements that could have an impact on the safety of the person, equipment, and/or the environment.

2.26.9.4 This requirement was added in ASME A17.1–2000. Requires cyclical checking of redundancy. Does not consider the possibility of a second failure occurring before the elevator comes to a stop. Where software is used, removal of power from the elevator driving machine motor and brake cannot be solely dependent on software. System safety is based on the principal that the safest state achievable is when the elevator is stopped.

"Checked prior to each start of the elevator from a landing," means checking for failures anytime since the last start from a landing. See Diagram 2.26.9.4.

2.26.9.5 The requirements cover both AC and DC drives (motor controller). Requires two devices to remove power independently from the drive motor. Since a single short circuit must not nullify the safe operation of the brake, the operation of the brake has to be made subject to an additional switch. It also provides protection in case the motor contactor welds close when the elevator prepares to make a stop at a landing. See Handbook commentary for 2.26.8.

Both the motor and brake contactor have to be subject to the car door and landing door safety contacts (i.e., EPDs). Further, the operation of the initiating circuits (i.e., to permit elevator to run) is dependent on the prior drop out of the motor and brake contactors.

2.26.9.5.1 This requirement allows more than one contactor and less than one pole per power line to the motor. A four-pole contactor would be needed if one pole for the brake and one pole for each of the AC lines is used. This would be a special contactor. Requirement 2.26.2 does not specify the number of lines to be opened in the requirement for the "removal of power" for non-static systems. ANSI/NFPA 70 Code Section 430-84 says

that all conductors need not be opened. See Diagram 2.26.9.5.

2.26.9.5.3 This requirement is consistent with the requirements in 2.26.9.6.3 to assure that the brake is dropped while there is no drive torque applied to the motor. The difference in language permits the use of an auxiliary device to open the brake circuit. This is analogous to "electrical protective devices" in a safety circuit operating relays, contactors, and/or devices, which in turn prevent the brake from being lifted if the safety circuit is not closed. See also Diagram 2.26.9.5.

2.26.9.6 The only difference from 2.26.9.5 is that instead of a motor contactor, VVVF drives can use an electromechanical relay to independently inhibit the flow of AC current through the inverter that connects the DC power source to the AC driving motor. The thinking was that a full on failure of the inverter power devices, applies DC to the AC motor, thereby providing a braking torque; the speed governor overspeed switches (see commentary for 2.18.4) provide motion sensing to protect against the condition where the electromechanical relay closes, the brake lifts, and the inverter fails to conduct. See also Handbook commentary for 2.26.9.5 and Diagram 2.26.9.6.

2.26.9.7 Suicide Circuit Protection. The magnetic properties of all electromagnetic apparatus have a certain amount of "memory." That is, after a magnetic field has been established in the iron of a generator, for instance, by passing current through its field windings, the removal of that current does not necessarily mean that the flux will drop to zero value immediately. This is expressed as residual magnetism.

The residual magnetism in the generator field, after removal of the excitation of the leveling field will permit the series field windings to excite the generator and in turn develop some armature current. The armature current, in turn, will cause the elevator motor to move, particularly if brake is not clamped tight. It is essential, therefore, to avoid this sequence by reversing the direction of the generator field, and connecting it across the generator armature to buck the residual flux and kill its own generated voltage. Hence, the name "suicide." See Diagram 2.26.9.7.

2.26.9.8 See Diagrams 2.26.9.8(a) and 2.26.9.8(b).

2.26.10 Absorption of Regenerated Power

If the normal power source is incapable of absorbing the energy generated by an overhauling load, a separate means such as a resistor bank must be provided on the load side of each elevator power supply line disconnecting means to absorb the regenerated power. For an emergency power source, the requirements of 2.27.2 apply. See Section 620-91 of ANSI/NFPA 70. See Diagram 2.26.10.

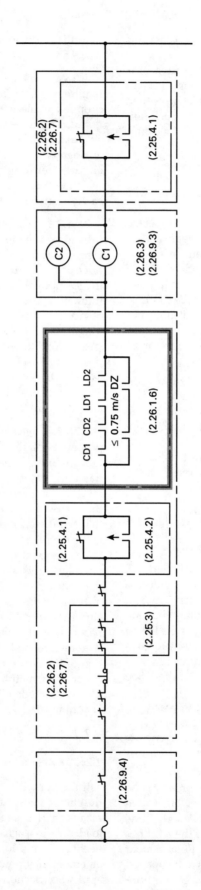

Diagram 2.26.9.3(a) and (b)
Single Failure
(Courtesy Otis Elevator Co./Ralph Droste)

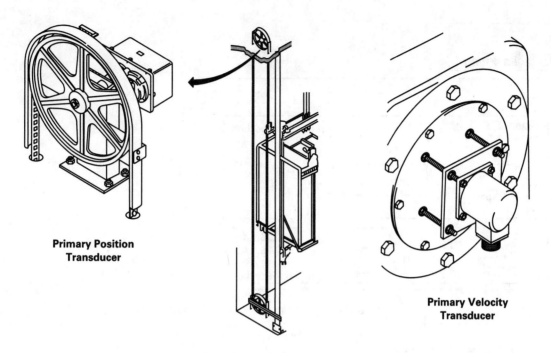

**Primary Position
Transducer**

**Primary Velocity
Transducer**

Diagram 2.26.9.3(c)   Single Failure
(Courtesy Otis Elevator Co./Ralph Droste)

2.26.11 Car Platform to Hoistway Door Sills Vertical Distance

It establishes leveling accuracy where ICC/CABO A117 or ADAAG is not applicable (e.g., Canada). It was added previously to B44 at the request of the Association of Canadian Chief Elevator Inspectors because they wanted a criteria for assessing leveling performance.

2.26.12 Operating Device Symbols

The operating device symbols were developed to eliminate the language barrier and to assist the visually impaired. The American National Standard for Accessible and Usable Buildings and Facilities (CABO/ANSI A117.1), The Americans with Disabilities Act Accessibility Guidelines (ADAAG), and the Uniform Federal Accessibility Standard (UFAS) require the symbols to be adjacent and to the left of the controls on a contrasting color background. These standards also require that letters, numbers, and symbols be a minimum of 15.86 mm ($^5/_8$ in.) high and raised or recessed 0.76 mm (0.030 in.).

In the early 1990s NEII became aware that despite the ASME A17.1 requirements for car control button symbols, further clarification was required. The ASME A17.1 Standards allowed the designer too much leeway. Complaints were being raised by the blind that the symbols in some cases were not clear when read tacitly. NEII formed a task group to work with the National Federation of the Blind. The findings of the task group were presented to the ASME A17 and ANSI A117 Committees. Both committees have adopted those findings.

The identical requirements for control button identification now appear in ASME A17.1–2000 and ICC/ANSI A117.1-1998.

The emergency stop switch symbol refers to the emergency stop switch marking only, and not the "STOP" and "RUN" position labeling required by 2.26.2.5(c).

SECTION 2.27
EMERGENCY OPERATION AND SIGNALING DEVICES

A note was added in the 2000 edition of ASME A17.1 to recognize that additional requirements may be found in the building codes. An example is the requirement in the IBC for fire department communication system in high-rise building (IBC Section 907.2.12.3).

2.27.1 Car Emergency Signaling Devices

All elevators must be provided with an audible signaling device. A means of two-way communications to an accessible point outside the hoistway is required in elevators with a travel of 18 m or 60 ft or more. See Diagram 2.27.1 for a diagrammatic description of these requirements. The two-way communications system does not have to be provided with a means of activation from within the car. Most communications systems, excluding telephones, are activated in the main lobby or building emergency control center. The alarm bell is a means that is provided to signal that there is a problem. The means of two-way communications is provided to assure occupants that their signal has been received and

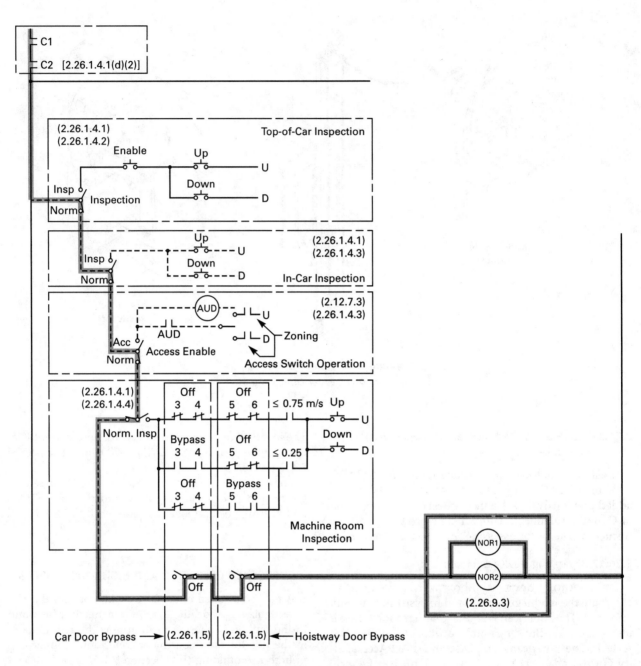

Diagram 2.26.9.3(d)   Single Failure
(Courtesy Otis Elevator Co./Ralph Droste)

measures are being instituted to get them out of the stalled car.

One telephone could be used to meet both the requirements of 2.27.1.1.2 and 2.27.1.2, provided that it is capable of reaching someone outside of the hoistway who can take appropriate action.

The intent of 2.27.1.1.2 is that the telephone or intercom is to be connected to a point where two-way communications can be established by emergency personnel with the occupants of the car, assuring them that help is on the way. A "readily accessible point" is a location

that is accessible to emergency personnel. The intent of this requirement is that emergency personnel have the ability to establish a communications link from within the building to the car. It was determined that when the travel is less than 18 m or 60 ft that communications could be established from outside the hoistway to the occupants of a car without the need of a two-way communications device. The exact location is determined on a local basis. The emergency signaling system is used to signal a problem and the communication system is then used to communicate with the occupants of the car,

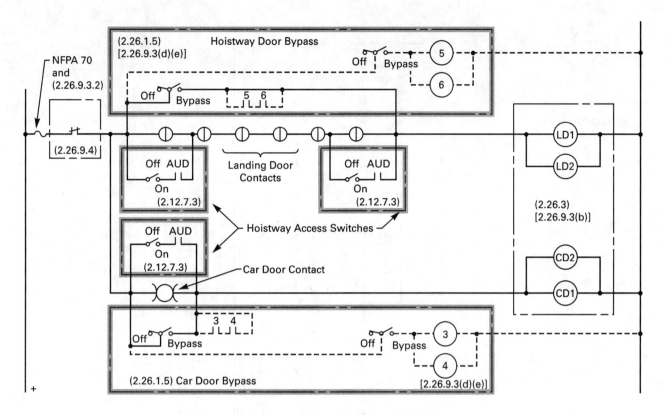

Diagram 2.26.9.3(e)   Single Failure
(Courtesy Otis Elevator Co./Ralph Droste)

to alleviate fears, prevent panic, and prevent dangerous attempts to exit the car without assistance.

A means of signaling or communicating with a service, which can take appropriate action in an emergency, is required by 2.27.1.2 in buildings where authorized personnel (see 1.3) are not continuously available. If a building is occupied, at off hours, by a single person (e.g., watchman, building engineer) that person may use the elevator and 2.27.1.2 would be applicable. See also 8.6.10.4 for written emergency evacuation procedures plan that must be provided in every building. It should be noted that a means of only signaling is acceptable.

Requirements for emergency communications can also be found in the Americans with Disabilities Act Accessibility Guidelines (ADAAG) ICC/ANSI A117.1 and Uniform Federal Accessibility Standards. All three standards state, "the emergency intercommunication system shall not require voice communication." These requirements address the needs of those with hearing and speech impairments.

The use of a light indicating the call has been received and responded to will provide a system that meets the intent for the hearing impaired. The light on the hands-free phone panel should only be illuminated or blink when activated by the recipient of the call. Those who are blind will be able to use the phone to hear that help is on the way. The instructions on the use of the light(s),

need not be raised or in Braille as this feature is not being provided for those with a vision impairment. Finally, although an illuminated alarm button is required by ASME A17.1, it does not by itself meet the full intent of ADAAG or ICC/ANSI A117.1.

In addition to these requirements, the building codes require that all elevators in high-rise buildings (see building code for definition) have a communication system from the elevator lobby, car, and machine rooms to the building's central control station. See commentary on car lighting in 2.14.7.1.

2.27.2 Emergency or Standby Power System

Emergency or standby power for an elevator is not required by this Code. If provided, then it must comply with 2.27.2. The building codes typically require standby power for at least one elevator that can travel to each floor in a high-rise building (see definition of building code in 1.3). One elevator does not have to stop at every floor, but every floor must be served by at least one elevator that is supplied with standby power. As an example, in a 20-story building elevator group A serves floors 1 through 10 and elevator group B serves floors 1 and 11 through 20. Standby power would have to be supplied to one elevator in group A and one elevator in group B. If the same 20-story building had an elevator

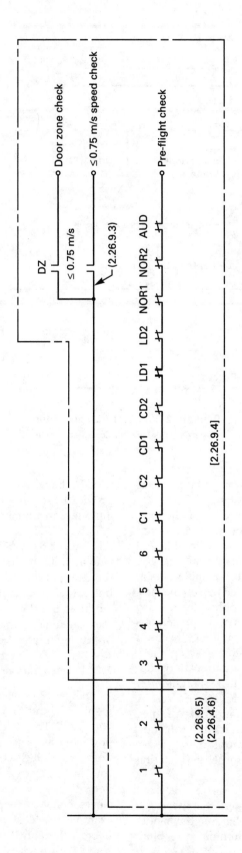

Diagram 2.26.9.4   Single Failure/Checking
(Courtesy Otis Elevator Co./Ralph Droste)

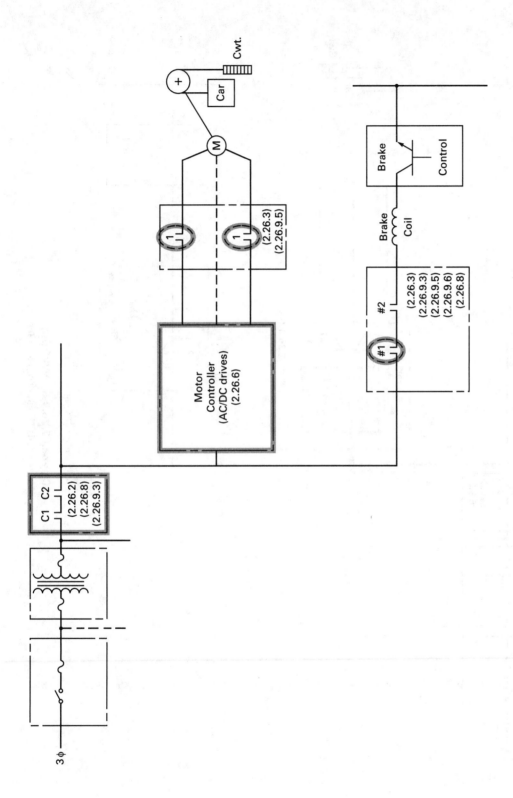

Diagram 2.26.9.5   AC/DC Drives
(Courtesy Otis Elevator Co./Ralph Droste)

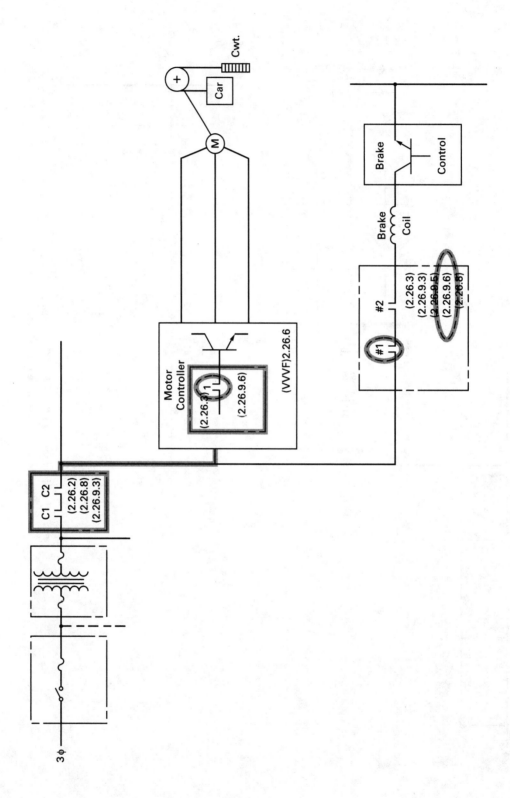

Diagram 2.26.9.6 VVVF Drives
(Courtesy Otis Elevator Co./Ralph Droste)

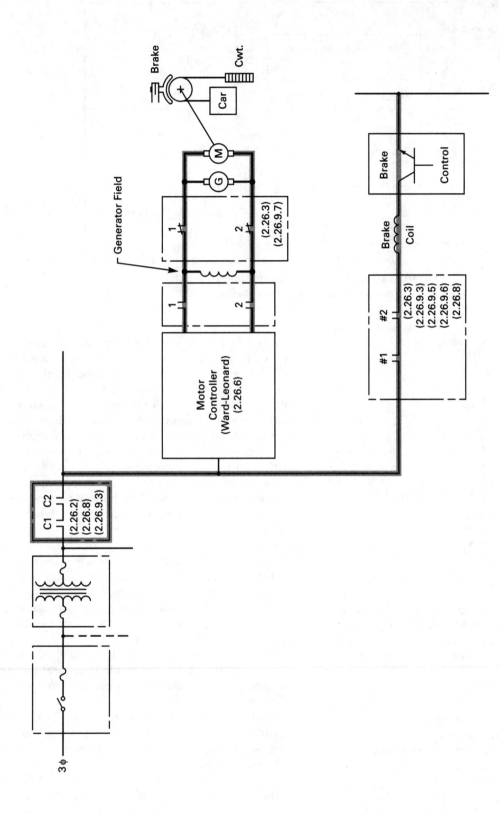

Diagram 2.26.9.7   Generator Field Control
(Courtesy Otis Elevator Co./Ralph Droste)

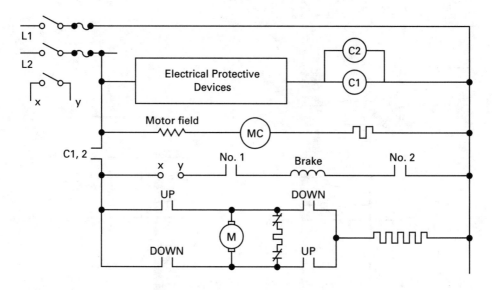

Diagram 2.26.9.8(a)   Overspeed in Down Direction
(Courtesy Otis Elevator Co./Ralph Droste)

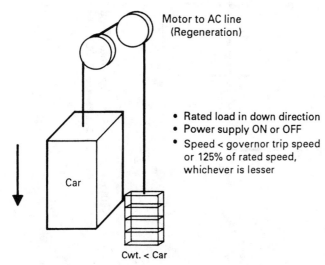

Motor to AC line
(Regeneration)

- Rated load in down direction
- Power supply ON or OFF
- Speed < governor trip speed or 125% of rated speed, whichever is lesser

Cwt. < Car

Diagram 2.26.9.8(b)   Typical Control and Operating Circuit for Controlling Overhauling Load in Down Direction
(Courtesy Otis Elevator Co./Ralph Droste)

that served all floors 1 through 20, then standby power could be supplied to that elevator only.

The US building codes addresses the need to provide accessible means of egress during a fire. Accessible means of egress typically include elevators, operating on Phase II emergency operation. When accessible means of egress include elevators, they shall be provided with standby power.

The National Electrical Code (NEC®) has requirements for both legally required standby power systems (Article 701) and optional standby power systems (Article 702). Legally required standby power systems provide electric power when normal power is interrupted to aid fire fighting, rescue operations, control of health hazards, and similar operations. Optional standby power systems provide electric power when normal power is interrupted to eliminate physical discomfort, interruption of an industrial process, damage to equipment, or disruption of business.

Emergency power systems are those power systems that are essential for safety to human life and must conform to the requirements of the NEC® Article 700. For additional information, see the NFPA 110 standard for emergency and standby power systems.

Legally required standby power systems have requirements that are very similar to emergency power systems. At the loss of normal power, the legally required standby power system must be able to supply power within 60 s whereas emergency power system must be able to supply power within 10 s. Wiring for legally required standby power systems can be installed in the same raceway, cables, and boxes as other general wiring. Emergency power system wiring must be kept entirely independent of all other wiring.

Elevators are normally connected to legally required standby power systems and not emergency power systems. Some hospital elevators are hooked into emergency power systems.

Requirement 2.27.2.1 facilitates the availability of power for the elevator system. It speeds up the evacuation process, minimizes entrapments, and allows the use of the elevator by firefighters.

Requirement 2.27.2.4 requires automatic sequence operation. Automatic sequencing must be so arranged

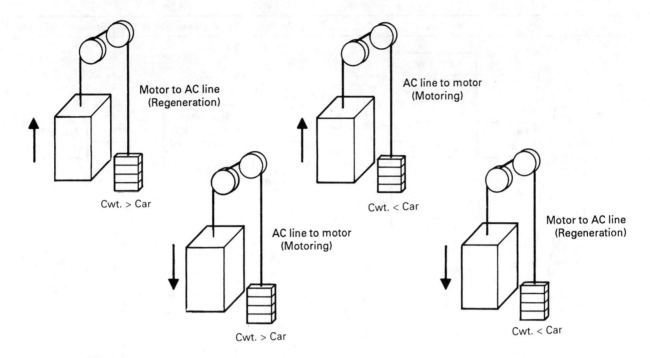

Diagram 2.26.10 Regenerated Power
(Courtesy Otis Elevator Co./Ralph Droste)

that it can be overridden by the manual selection switch. This switch must not stop the elevator when in motion; power will be transferred only after the elevator is stopped.

See also the commentary on 2.26.10 and Section 620-101 of the National Electrical Code®.

2.27.3 Firefighters' Emergency Operations — Automatic Elevators

NOTE: The author expresses his appreciation to John J. O'Donoghue for his extensive contribution to 2.27.3.

In 1998, the National Fire Protection Association reported 4,035 fire deaths in the United States. Of these deaths, 5 were the result of fires in high-rise buildings. Since 1980, The NFPA has recorded 184 fatal fires in high-rise buildings in the United States and Canada. These fires resulted in 783 deaths of the public and firefighters. The last available set of figures showing a comparison of the two countries is from 1995. The total fire incidents in the United States were 1,965,500 with 4,585 deaths resulting. In Canada, the comparative figures were 64,300 fire incidents, resulting in 399 deaths. An equally important set of figures is the total number of injuries, in both the United States (25,775) and Canada (2,455). When these are examined, only then will we get a true picture of the fire problem that we are facing today. Between both countries, it is estimated that in the past 20 years, there have been thousands of new high-rise buildings (22.9 m/75 ft) built. In the past, office occupancies had been classified as a low-risk fire hazard.

This was true in older, compartmentalized high-rise buildings, with no multi-floor HVAC systems, and with office furnishings made of wood or metal. However, the modern high-rise building is a different creature. Fire and smoke spreading to other floors of the structure, or to adjoining ones (i.e., World Trade Center bombing of 1993), is the true nightmare facing the firefighting forces of both countries today. Other factors in this fire problem are the furnishings. The vast majority of today's building interiors are made up of plastic in one form or another. Among firefighters, plastic is referred to as frozen gasoline. As plastic is heated from exposure to fire, it goes through a decomposition process that emits toxic, disabling gases (e.g., hydrogen chloride) and flammable hydrocarbon gases, which readily spread and cause a rapid, lightning-fast propagation of the fire across large areas. The smoke from these products of combustion is dense and black, causing trapped occupants to have little or no visibility, encounter choking smoke, and face heat of temperatures well above what a human being can endure. Today's high-rise buildings present new and different problems to fire suppression personnel and techniques. Yet the cause of fires in high-rise buildings and the materials used in them, including furniture and fixtures, are not any different from those used in conventional low-rise structures. However, if a fire breaks out in the top story of a high rise, the fire service must transport their firefighters and equipment to the upper floors via elevators operating on firefighters' in-car operation. Some fires have necessitated the use of stairs to

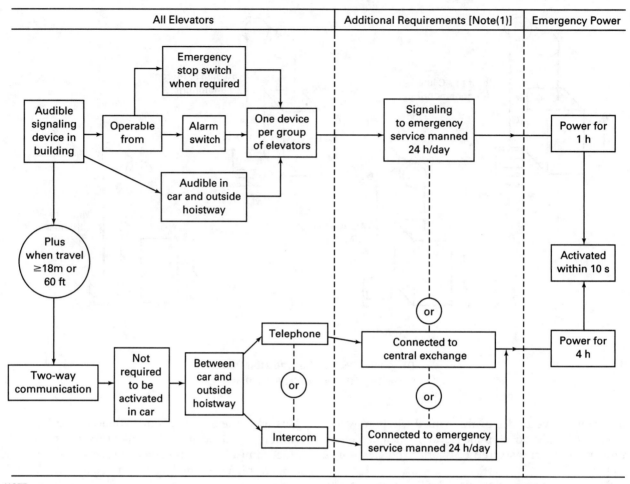

| All Elevators | Additional Requirements [Note(1)] | Emergency Power |

Diagram 2.27.1 reflects:

- Audible signaling device in building
- Emergency stop switch when required
- Operable from
- Alarm switch
- Audible in car and outside hoistway
- One device per group of elevators
- Plus when travel ≥18m or 60 ft
- Two-way communication
- Not required to be activated in car
- Between car and outside hoistway
- Telephone
- Intercom
- Signaling to emergency service manned 24 h/day
- Connected to central exchange
- Connected to emergency service manned 24 h/day
- Power for 1 h
- Activated within 10 s
- Power for 4 h

NOTE:
(1) In a building in which a building attendant, building employee, or watchman is not continuously available to take action when an emergency signal is operated.

Diagram 2.27.1   Emergency Signaling Devices

reach the fire floor. Keep in mind that the firefighters, with wearing all of their protective gear, weigh an additional 65 lb, and they are carrying long lengths of hose and attachments that weigh an additional 50 lb to 65 lb. Using the stairs to gain access to upper floors in a high-rise structure is the last resort. Elevators must be made a reliable tool to be utilized by the firefighters in the performance of their duties; however, elevators cannot be relied upon during a fire in a building. A firefighting commander would take a fire any day of the week, which is located on the top floor of a building, rather than one on a lower floor. Why? Because the life hazard is on the fire floor, with only the roof and the sky being exposed. In contrast, a fire on the 10th floor of a 34-story building has a life hazard on all floors above the fire, as well as the fire floor, requiring additional staffing to accomplish the tasks of search, rescue, and fire extinguishment. Firefighters have immediate concerns relating to smoke and heat spread, stack effect, and the mass, uncontrolled evacuation down the same stairways that

firefighters are trying to use to move up to locate, surround, and extinguish this fire. Remember a few points: At the Meridian Plaza fire in Philadelphia, there was a total failure of all building systems early in the fire, and secondly, the World Trade Center explosion and resulting fire was below grade, and it took 11 h to complete the evacuation.

Let's review some of the reasons why elevators are unsafe in a fire and what led to the Code requirements for firefighters' operation of elevators. Elevators are unsafe in a fire because:

(a) person may push a corridor button and have to wait for an elevator that may never respond; valuable time to escape is lost;

(b) elevators respond to car and corridor calls; one of these may be at the fire floor;

(c) elevators cannot start until the car and hoistway doors are closed. This could lead to overcrowding of an elevator and the blockage of the doors, and thus prevent closing.

(d) power failure during a fire can happen at any time and thus lead to passenger entrapment.

Fatal delivery of the elevator to the fire floor can be caused by any of the following:

(1) an elevator passenger pressing the car button for the fire floor;

(2) one or both of the corridor call buttons may be pushed on the fire floor;

(3) heat may melt or deform the corridor push button or its wiring at the fire floor;

(4) normal functioning of the elevator, such as high or low reversal, may occur at the fire floor.

(5) heat from the fire or loss of air conditioning in the machine room or hoistway may have a detrimental effect on solid-state control equipment, resulting in erratic elevator operation.

The A17.1 Code recognized all of these conditions and has reacted by mandating elevator recall, more commonly referred to as Phase I Emergency Recall Operation. The building code also requires a sign in elevator lobbies similar to that shown in Diagram 2.27.3(a) to advise building occupants not to use elevators in a fire.

The features described in 2.27.3 are also known as Firefighters' Service or special emergency service [SES] features. The Code Committee has strived to standardize the operation to ensure that there are no variations thereby eliminating confusion of firefighters during an emergency. The automatic or manual return of elevators to the designated level (see 1.3 for definition) is referred too as Phase I Emergency Recall Operation. The provision to allow emergency personnel to operate the elevator from within the car on emergency in-car operation is commonly referred to as Phase II Emergency In-Car Operation.

The requirements for firefighter emergency operation are detailed in the flow chart shown in Diagrams 2.27.3(b)(1) through 2.27.3(b)(15).

NOTE: Diagram 2.27.3(b)(1) shows the logic needed inside the fire alarm panel to generate the three signals for the elevator system.

Diagram 2.27.3(b)(2) shows the initiation of fire service for a car on Inspection.

Diagram 2.27.3(b)(3) shows the initiation of fire serive for a car on Hospital Service.

Diagrams 2.27.3(b)(4) and (b)(5) show the initiation of fire service for a car on Designated Attendant Service.

Diagrams 2.27.3(b)(6) through (b)(8) show the Phase I recall.

Diagrams 2.27.3(b)(9) shows the car at the lobby after Phase I recall.

Diagrams 2.27.3(b)(10) through (b)(13) show Phase II operation.

Diagrams 2.27.3(b)(14) and (b)(15) show the car going off of Phase II operation.

2.27.3.1 Phase I Emergency Recall Operation. A17.1b–1989 through A17.1b–1992 required Phase I for an elevator with a travel of 7.62 m (25 ft) or more. Under earlier editions of the Code, an elevator could have nearly 15.24 m (50 ft) of travel [just less than 7.62 m (25 ft) above and below the designated landing], and still not have

been required to have Phase I and Phase II operation. The term "designated level" (see 1.3) refers to the main floor or other level that best serves the needs of emergency personnel for firefighting and rescue purposes. The term "alternate landing" (see 1.3) refers to a floor level identified by the building code or fire authority, other than the designated landing. The term "recall level" (see 1.3) refers to the designated or alternate level that the car returns to when Phase I Emergency Recall Operation is activated. These requirements apply for all automatic elevators except when the hoistway or a portion thereof is not required to be constructed of fire-resistive construction (2.1.1.1), the travel does not exceed 2 000 mm or 80 in., and the hoistway does not penetrate a floor. An example of this would be an elevator that would allow a person to be transported from one level in a lobby to a second level not penetrating the next floor or fire barrier. This can be seen in department stores or malls, where this exception has been applied. Where Firefighters' Emergency Operation is provided voluntarily, (e.g., Canada) the requirements in 2.27.3 apply.

A three-position key-operated switch must be provided in the designated level lobby for each single elevator or for a group of elevators. The location of the three-position Phase I Emergency Recall switch has been specified for standardization and to make sure that it is located where all of the elevators will be within sight and readily accessible from this location. The key is to be removable only in the "ON" and "OFF" positions. The specified key positions standardize a clockwise rotation to reach the "ON" position similar to the requirements for the Phase II Emergency In-Car Operation switch. Prior to the 2000 edition of ASME A17.1 the three-position Phase I key switch included the "BYPASS" position. This allowed building or emergency personnel to return elevators to normal service in event that the activated (alarmed) fire alarm initiating device (e.g., smoke detector) could not be cleared. However, the great drawback to this feature was that because one faulty smoke detector could not be cleared, the entire Phase I Emergency Recall system would be disabled when the "BYPASS" was activated. The fire alarm initiating devices that are used today are far superior to the ones that were first used. Today's systems can be monitored, maintained, and cleared from their control panel. In the 1996 edition of the ASME A17.1 Code, the expertise of NFPA 72, Chapter 3 was recognized as the proper authority to determine the number and location of fire alarm initiating devices in a building. In continuing with this transition, in ASME A17.1–2000 the "BYPASS" position was replaced with the "RESET" position on the three-position Phase I Emergency Recall Operation switch. The "RESET" feature will be utilized by emergency personnel (1.3) to reset the elevators returning them to normal service, after Phase I Emergency Recall Operation has been activated. If the fire alarm initiating

In Case Of Fire
Elevators Are Out Of Service

Use Exit

White

Black

Red

Diagram 2.27.3(a)   Elevator Corridor Call Station Pictograph

device has not been cleared before using the "RESET" feature, then the Phase I Emergency Recall Operation will continue in effect until the device has been replaced, repaired, or bypassed by the fire alarm system. The switch shall be labeled "FIRE RECALL" and its positions marked "RESET," "OFF," and "ON" (in that order), with the "OFF" position as the center position. An additional key-operated "FIRE RECALL" switch, with two positions, marked "OFF" and "ON" (in that order), is permitted only at the building fire control station. Keys shall be removable in the "OFF" and "ON" positions only.

Phase I operation ensures that the elevator is not available to the general public when automatic elevator operation may be hazardous during a fire emergency. When an elevator is out of service during a fire, the public is unable to use it as a means of exiting from the building. Depending on the building fire plan, occupants will be directed to either the stairwells, places of refuge, or to be active participants in the emergency plan developed

for their safety by the building owners/operators. Earlier editions of the ASME A17.1 Code addressed only automatic door operations. The 1981 and later editions of the ASME A17.1 Code cover the operation of vertically sliding doors, doors controlled by constant pressure buttons, and manual doors. The only time an automatic elevator will not return upon activation of Phase I is when that car is at a landing with its door(s) open and the in-car stop switch, emergency stop switch (in-car, top-of-car, pit, etc.), or some other electrical protective device is activated. Requirement 2.27.3.1.6(b) recognizes that an elevator at a landing with the in-car stop switch or emergency stop switch in the "STOP" position should not be captured, as this is not a normal condition. The stop switch may have been activated to facilitate an inspection or maintenance and capturing the elevator could be a hazard to the elevator personnel. The in-car stop switch is disabled, just like the in-car emergency stop switch, so that the car cannot be inadvertently taken

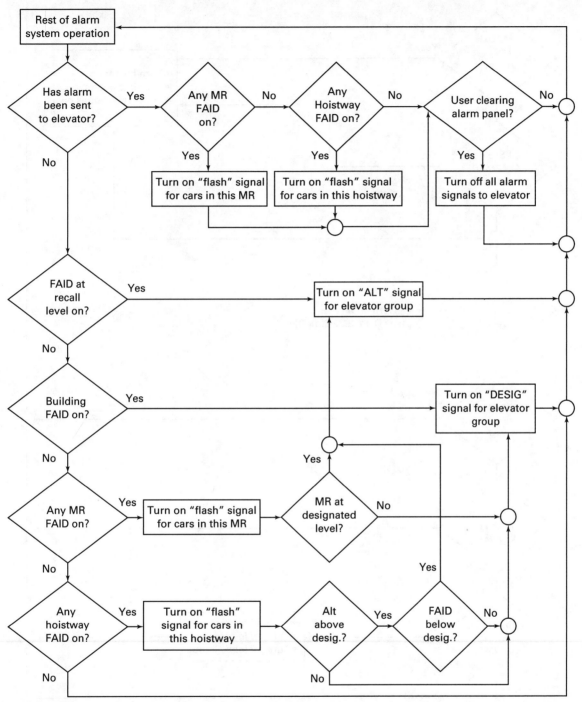

GENERAL NOTE: Interface consists of one signal per group to command a designated level recall, one per group to command an alternate level recall, and several signals (min. = 1, max. = # of cars) to indicate the in-car lamp should flash. The number depends on how many cars share motor rooms and how many share hoistways. FAID = fire alarm initiating device. MR = machine room.

Diagram 2.27.3(b)(1) Fire Alarm System Logic to Generate Elevator Input Signals (Courtesy Matt Martin)

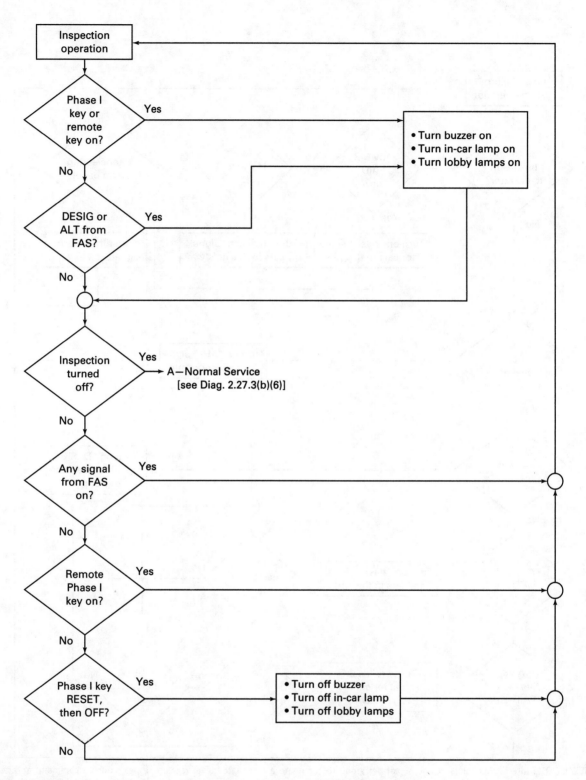

GENERAL NOTES:
(a) Anytime "FLASH" signal comes from FAS, change in-car lamp to flashing. FAS = fire alarm system. If there is no remote Phase I switch, assume it is OFF.
(b) Generally single-letter references (e.g., A, B, C) are related to Phase I operation and double-letter references (e.g., BB, CC, GG) are related to Phase II operation.

Diagram 2.27.3(b)(2) Fire Service Phase I Recall When on Inspection Operation (Courtesy Matt Martin)

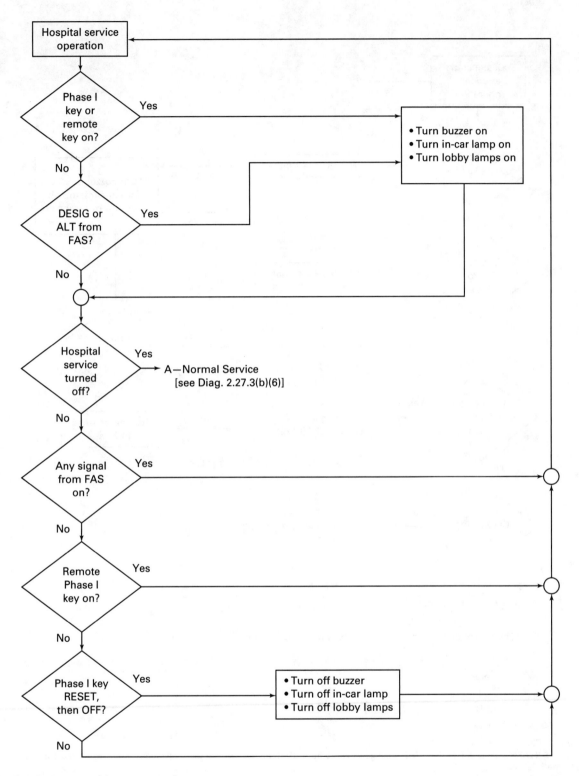

GENERAL NOTES:
(a) Anytime "FLASH" signal comes from FAS, change in-car lamp to flashing. FAS = fire alarm system. If there is no remote Phase I switch, assume it is OFF.
(b) Generally single-letter references (e.g., A, B, C) are related to Phase I operation and double-letter references (e.g., BB, CC, GG) are related to Phase II operation.

Diagram 2.27.3(b)(3) Fire Service Phase I Recall When on Hospital Service Operation (Courtesy Matt Martin)

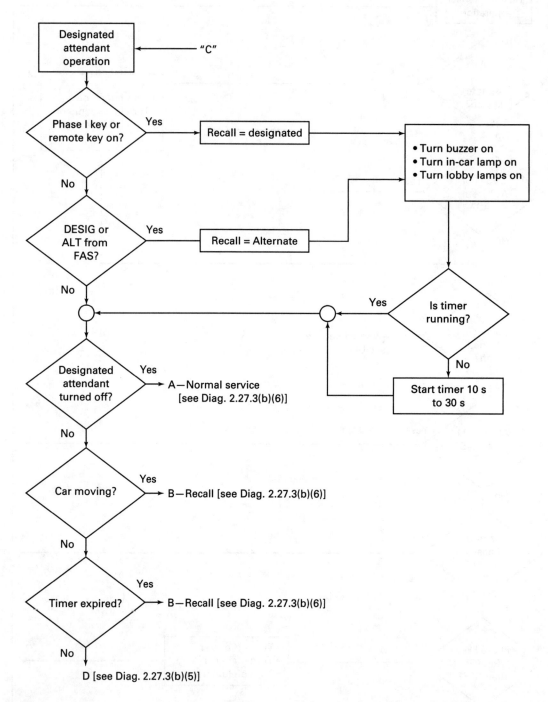

Diagram 2.27.3(b)(4) Fire Service Phase I Recall When on Designated Attendant Operation (Courtesy Matt Martin)

GENERAL NOTES:
(a) Anytime "FLASH" signal comes from FAS, change in-car lamp to flashing. FAS = fire alarm system. If there is no remote Phase I switch, assume it is OFF.
(b) Generally single-letter references (e.g., A, B, C) are related to Phase I operation and double-letter references (e.g., BB, CC, GG) are related to Phase II operation.

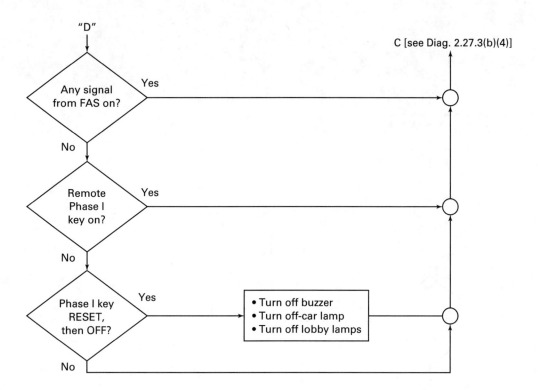

GENERAL NOTES:
(a) Anytime "FLASH" signal comes from FAS, change in-car lamp to flashing. FAS = fire alarm system. If there is no remote Phase I switch, assume it is OFF.
(b) Generally single-letter references (e.g., A, B, C) are related to Phase I operation and double-letter references (e.g., BB, CC, GG) are related to Phase II operation.

Diagram 2.27.3(b)(5) Fire Service Phase I Recall When on Designated Attendant Operation (Courtesy Matt Martin)

off of recall operation. A person with access to the key for the in-car stop switch, after the doors are closed and the elevators start such as a building cleaning person might be inclined to activate the switch while the elevator is being recalled. This would create an unsafe condition, as the firefighters' would have to search for the car immediately, and the occupant could be putting himself or herself into great danger. For passenger safety, 2.27.3.1.6(e) requires elevators to close their doors at a slower speed when a door-reopening device is rendered inoperative. Mechanically actuated door-reopening devices are not sensitive to smoke or flame and can remain operative. Flame is the glowing, gaseous, visible part of a fire. Smoke or flame can register a signal, whereas direct flame impingement by the fire will destroy. In actuality, unless the fire was started within the car's contents, the cars would all have responded to the recall level well before in advance of the fire growing to that stage. The reference to 2.13.5 recognizes "nudging" and, therefore, it is not unsafe to disconnect mechanically actuated door-reopening devices.

Requirements 2.27.3.1.6(f) and (g) allow full control of those doors that may have to be closed from the corridor. Also, the automatic closing of vertical slide doors requires an active "OPEN" or "STOP" button on the corridor. The corridor "DOOR OPEN" button must be operative to cover the case where a manual gate is open with the hoistway door closed. A firefighter may need the "DOOR OPEN" button to open the door to see if anyone is in the car. Operating hall position indicators [2.27.3.1.6(f)] may convey a message that the elevators may be used. Only where elevator location is important to firefighters, such as the designated level, the alternate level, and the building fire control station are the hall position indicators to remain in operation. Car position indicators are always required by the firefighters utilizing the elevators, thus these devices are to remain operative.

Unless the car is at the fire floor, it is safer to keep the doors open. Arriving firefighting forces will be able to immediately determine that all cars have answered the recall, and have the knowledge that no passengers are trapped in cars within the hoistway. The visible and audible signals [2.27.3.1.6(h)] alert passengers in an automatically operated elevator of the emergency and can minimize any apprehension that the passengers may have while the elevator is returning to the main floor. In an attendant-operated elevator, this signal alerts the

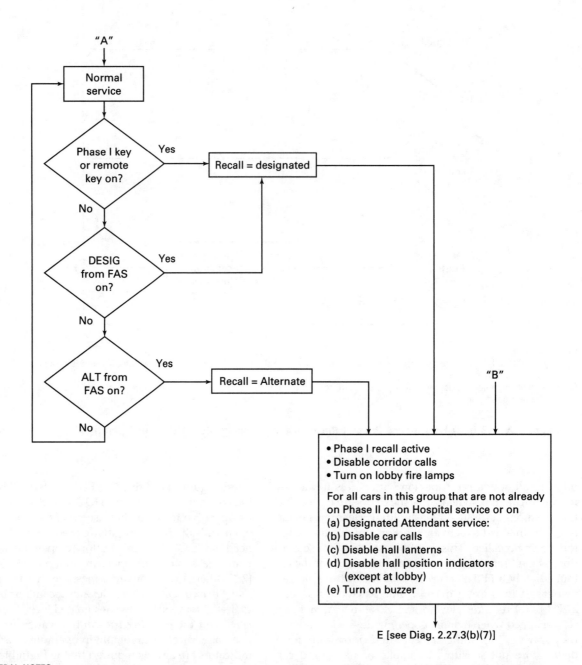

"A"

Normal service

Phase I key or remote key on? — Yes → Recall = designated

No

DESIG from FAS on? — Yes → Recall = designated

No

ALT from FAS on? — Yes → Recall = Alternate

No

"B"

- Phase I recall active
- Disable corridor calls
- Turn on lobby fire lamps

For all cars in this group that are not already on Phase II or on Hospital service or on
(a) Designated Attendant service:
(b) Disable car calls
(c) Disable hall lanterns
(d) Disable hall position indicators (except at lobby)
(e) Turn on buzzer

E [see Diag. 2.27.3(b)(7)]

GENERAL NOTES:
(a) Anytime "FLASH" signal comes from FAS, change in-car lamp to flashing. FAS = fire alarm system. If there is no remote Phase I switch, assume it is OFF.
(b) Generally single-letter references (e.g., A, B, C) are related to Phase I operation and double-letter references (e.g., BB, CC, GG) are related to Phase II operation.

Diagram 2.27.3(b)(6) Fire Service Phase I Recall When on Normal Operation   (Courtesy Matt Martin)

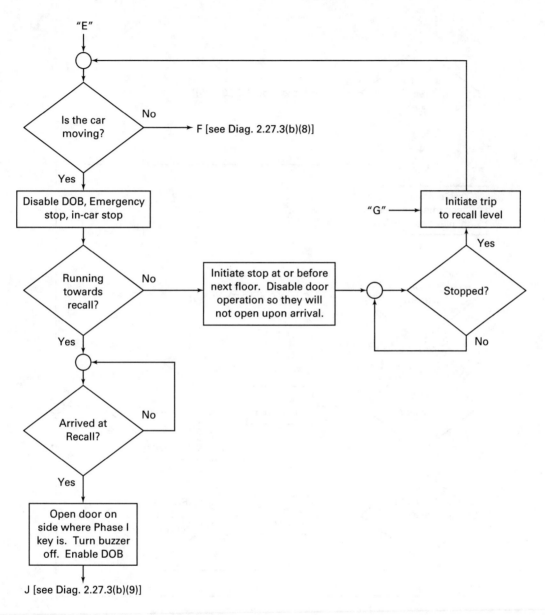

"E"

Is the car moving? — No → F [see Diag. 2.27.3(b)(8)]

Yes ↓

Disable DOB, Emergency stop, in-car stop

Running towards recall? — No → Initiate stop at or before next floor. Disable door operation so they will not open upon arrival.

Stopped? — Yes → Initiate trip to recall level ← "G"

Stopped? — No

Yes ↓

Arrived at Recall? — No

Yes ↓

Open door on side where Phase I key is. Turn buzzer off. Enable DOB

J [see Diag. 2.27.3(b)(9)]

GENERAL NOTES:
(a) Anytime "FLASH" signal comes from FAS, change in-car lamp to flashing. DOB = "DOOR OPEN" button.
(b) Generally single-letter references (e.g., A, B, C) are related to Phase I operation and double-letter references (e.g., BB, CC, GG) are related to Phase II operation.

Diagram 2.27.3(b)(7) Fire Service Phase I Recall When on Normal Operation (Courtesy Matt Martin)

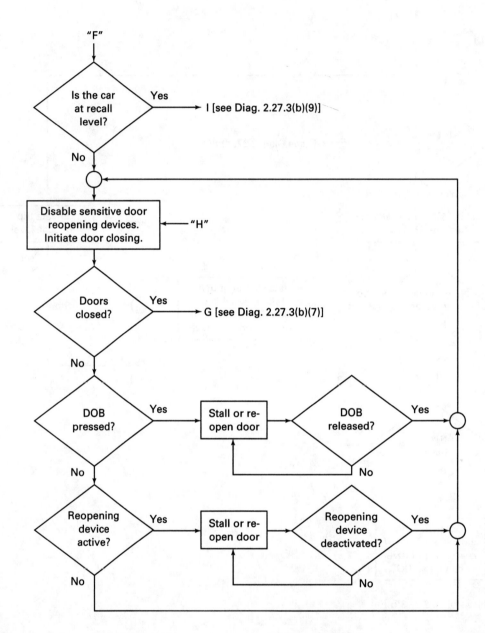

GENERAL NOTES:
(a) Anytime "FLASH" signal comes from FAS, change in-car lamp to flashing. DOB = "DOOR OPEN" button
(b) Generally single-letter references (e.g., A, B, C) are related to Phase I operation and double-letter references (e.g., BB, CC, GG) are related to Phase II operation.

Diagram 2.27.3(b)(8) Fire Service Phase I Recall When on Normal Operation (Courtesy Matt Martin)

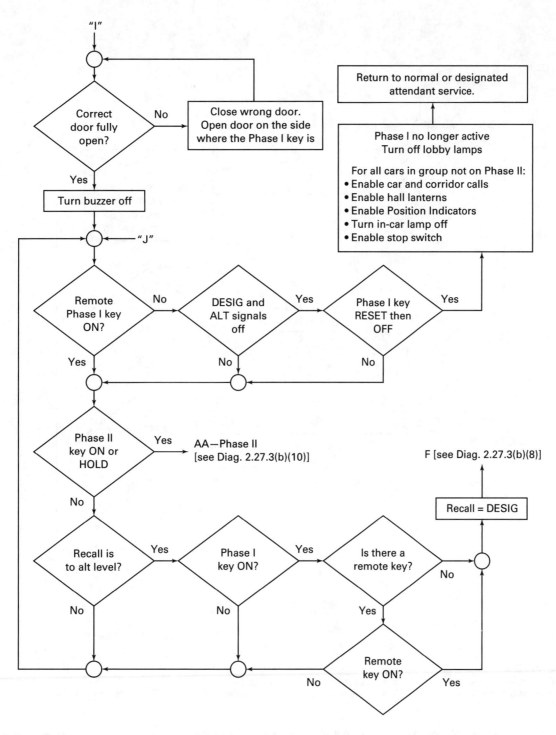

GENERAL NOTES:
(a) Anytime "FLASH" signal comes from FAS, change in-car lamp to flashing. DOB = "DOOR OPEN" button
(b) Generally single-letter references (e.g., A, B, C) are related to Phase I operation and double-letter references (e.g., BB, CC, GG) are related to Phase II operation.

Diagram 2.27.3(b)(9) Fire Service Phase I Recall When on Normal Operation (Courtesy Matt Martin)

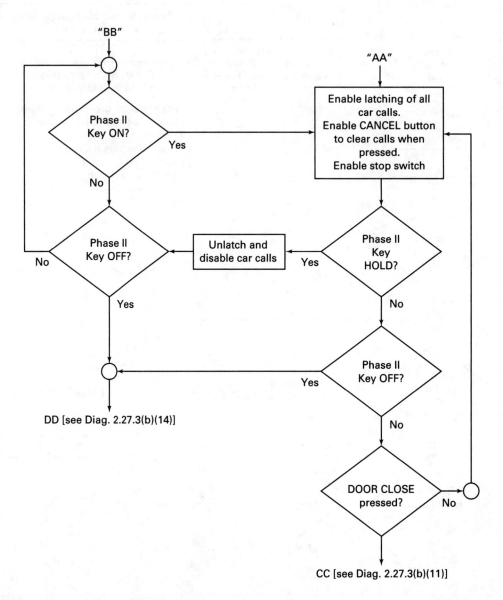

"BB"

"AA"

Enable latching of all car calls.
Enable CANCEL button to clear calls when pressed.
Enable stop switch

Phase II Key ON? — Yes

No

Phase II Key OFF? ← Unlatch and disable car calls ← Yes — Phase II Key HOLD?

No — No

Yes

No

Phase II Key OFF?

Yes

No

DD [see Diag. 2.27.3(b)(14)]

DOOR CLOSE pressed? — No

CC [see Diag. 2.27.3(b)(11)]

GENERAL NOTES:
(a) Anytime "FLASH" signal comes from FAS, change in-car lamp to flashing.
(b) Generally single-letter references (e.g., A, B, C) are related to Phase I operation and double-letter references (e.g., BB, CC, GG) are related to Phase II operation.

Diagram 2.27.3(b)(10) Phase II "HOLD," Phase II "ON" Parked With Doors Open   (Courtesy Matt Martin)

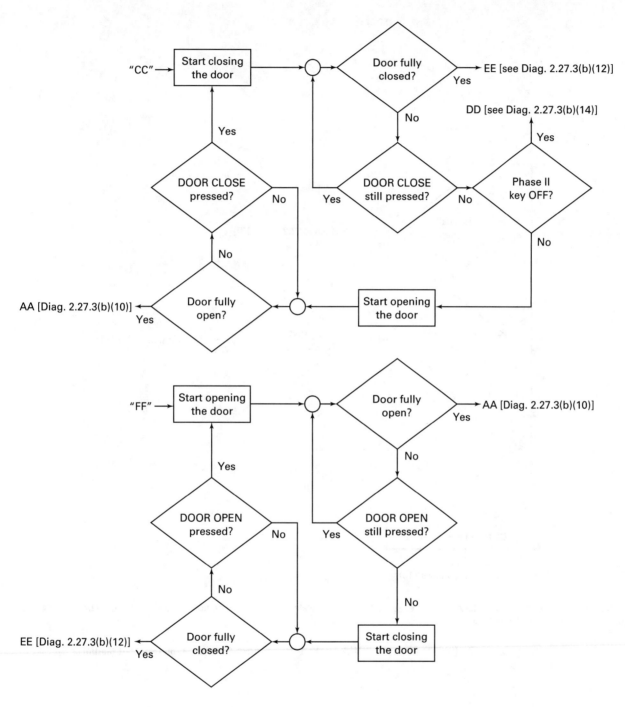

GENERAL NOTE: Generally single-letter references (e.g., A, B, C) are related to Phase I operation and double-letter references (e.g., BB, CC, GG) are related to Phase II operation.

Diagram 2.27.3(b)(11) Phase II "ON," Opening and Closing the Doors (Courtesy Matt Martin)

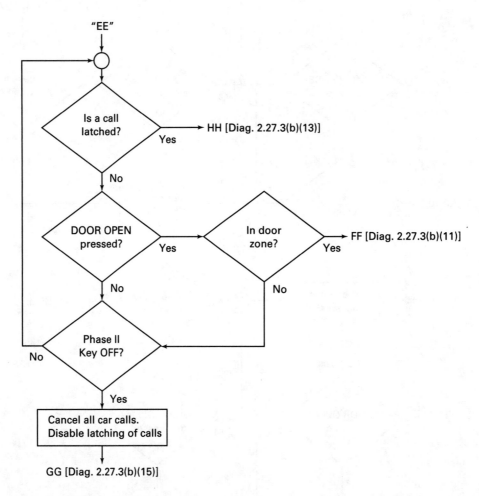

GENERAL NOTE: Generally single-letter references (e.g., A, B, C) are related to Phase I operation and double-letter references (e.g., BB, CC, GG) are related to Phase II operation.

Diagram 2.27.3(b)(12) Phase II "ON," Parked With Doors Closed   (Courtesy Matt Martin)

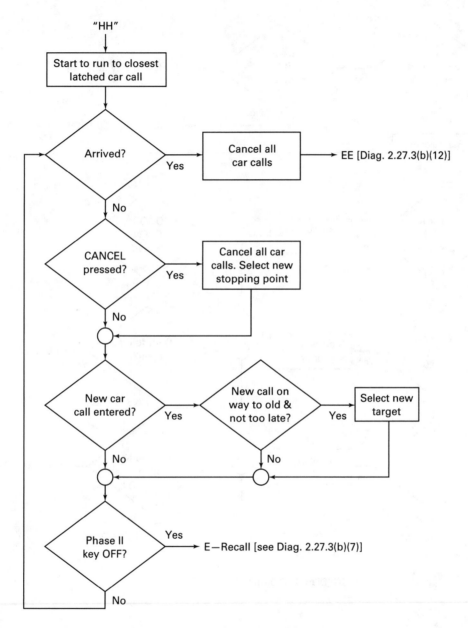

"HH"

Start to run to closest latched car call

Arrived? — Yes → Cancel all car calls → EE [Diag. 2.27.3(b)(12)]

No

CANCEL pressed? — Yes → Cancel all car calls. Select new stopping point

No

New car call entered? — Yes → New call on way to old & not too late? — Yes → Select new target

No / No

Phase II key OFF? — Yes → E—Recall [see Diag. 2.27.3(b)(7)]

No

GENERAL NOTE: Generally single-letter references (e.g., A, B, C) are related to Phase I operation and double-letter references (e.g., BB, CC, GG) are related to Phase II operation.

Diagram 2.27.3(b)(13) Running on Phase II   (Courtesy Matt Martin)

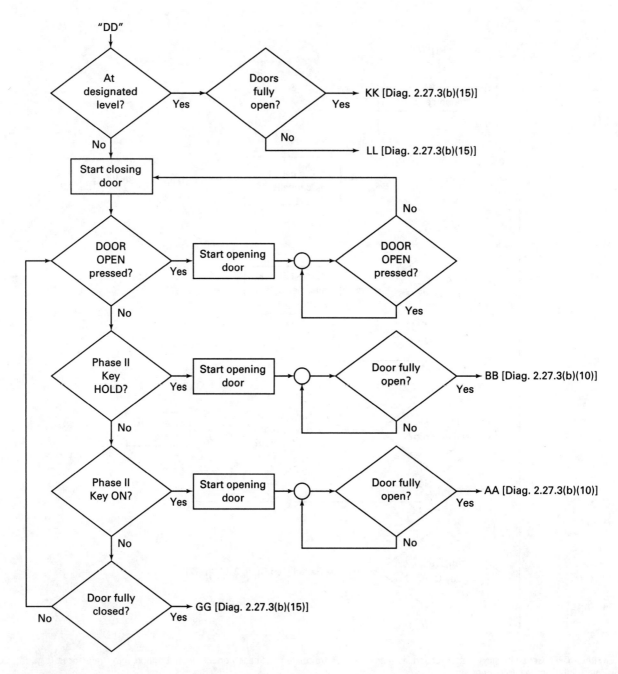

GENERAL NOTE: Generally single-letter references (e.g., A, B, C) are related to Phase I operation and double-letter references (e.g., BB, CC, GG) are related to Phase II operation.

Diagram 2.27.3(b)(14) Going Off Phase II — Closing the Doors (Courtesy Matt Martin)

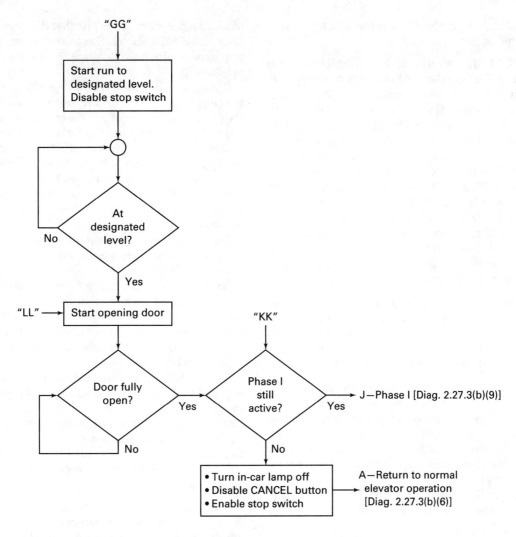

"GG"

```
        Start run to
        designated level.
        Disable stop switch
```

At designated level? — No

Yes

"LL" → Start opening door   "KK"

Door fully open? — Yes → Phase I still active? → Yes → J—Phase I [Diag. 2.27.3(b)(9)]

No   No

• Turn in-car lamp off   A—Return to normal
• Disable CANCEL button   →   elevator operation
• Enable stop switch   [Diag. 2.27.3(b)(6)]

GENERAL NOTE: Generally single-letter references (e.g., A, B, C) are related to Phase I operation and double-letter references (e.g., BB, CC, GG) are related to Phase II operation.

Diagram 2.27.3(b)(15) Going Off Phase II   (Courtesy Matt Martin)

attendant of the emergency and warns him to return immediately to the designated level. When on inspection operation, the inspector or maintenance personnel are also alerted to the emergency by this signal.

Requirement 2.27.3.1.6(j) recognized that if the smoke detector at the designated level is activated, turning the additional Phase I switch to the "ON" position will not override the fire alarm initiating device sending the car to the designated level. This switch may be at a location where the condition of the designated level lobby cannot be determined. Where an additional "FIRE RECALL" switch is provided, both "FIRE RECALL" switches must be in the "ON" position to recall the elevator to the designated level if the elevator was recalled to the alternate level (2.27.3.2.4). To remove the elevators from Phase I, the "FIRE RECALL" switch shall first be rotated to the "RESET" position, and then to the "OFF" position.

If a second recall switch is provided, it must be in the "OFF" position to remove the elevator from Phase I Emergency Recall Operation. Means used to remove elevators from normal operation, other than as specified in this code, shall not prevent Phase I Emergency Recall Operation. Plainly, this means that the many uses of elevators by service personnel, movers, locked out, etc., will no longer prevent the firefighters from having all the available elevators at their disposal during a fire. This feature has been included at the request of the fire service community, who on too many occasions, have arrived at a fire, only to find that most of the elevators were not accessible for their use.

2.27.3.2 Phase I Emergency Recall Operation by Fire Alarm Initiations Device. An initiation device is defined by NFPA 72 as a system component that originates transmission of a change-of-state condition, such as in a smoke detector, etc.

2.27.3.2.1 The reference to the National Fire Alarm Code®, NFPA 72 is to a standard with expertise to specify the type and installation of automatic initiating devices. NFPA 72 has been revised, at the request of the ASME A17 Committee, to address fire alarm systems in all building types. The expertise for determining when conditions, due to a fire, require automatic elevator recall are within the jurisdiction of the National Fire Alarm Code®, NFPA 72. See Chart 2.27.3.2.1 for excerpts from the NFPA Fire Alarm Code® Handbook. Beginning with ASME A17.1b–1997 a fire alarm initiating device must be provided at all floors. ASME A17.1 recognizes that devices other than smoke detectors may be more appropriate under some conditions. Those conditions are specified within NFPA 72.

2.27.3.2.2 This requirement recognizes that only manual Emergency Recall Operation is required by the NBCC. Note that within this requirement the terms "smoke detector" and "elevator lobby" have been utilized because of requirements in the NBCC.

2.27.3.2.4 The following is the basis for alternate floor recall. It is not preordained that the designated level has the lowest fuel load of any other floor in the buildings. This may be the case in some major, high-rise office buildings, but it certainly is not applicable to many other buildings, such as apartments, hotels, showrooms, or buildings with elaborate reception areas. However, even if it were a fact, a firebomb can suddenly provide an enormous fuel load on an otherwise fire-resistant floor.

The preponderance of buildings have elevators without an express zone. In the event that there are express zones, it is safer to park elevators away from any potential fire floor. It is feared that if the mandatory alternate floor requirement is repealed and made permissive, then many buildings would ultimately revert to the early Code requirements that only required return to the designated level.

What is often overlooked is the fact that if the elevators are returned to an alternate level, the firefighters have not lost control of the elevators. If conditions dictate that the designated level can provide safe egress, there is no reason why the firefighters cannot exercise the option of calling the elevators to the designated level by turning the required three-position keyed switch to the "ON" position. The key switch overrides the alternate floor recall operation and will return all elevators to the designated level, even though the elevators may be parked at a floor above an express zone.

Typically, the designated level is also the location of the central command station. It would be difficult to effectively utilize a designated level central command station if the designated level is engulfed in a rapidly spreading fire such as the one that destroyed the main floor of the MGM Hotel. Elevators that are returned to, or parked at, the main floor are of no value if the result is loss of life.

It may also be assumed that sprinklers will reduce the probability of a large fire, but one cannot rule out the possibility that smoke in dangerous quantities may be produced. Although sprinkler manufacturers say that sprinklers have been proven effective in stopping fire in a large number of buildings, smoke control advocates note that smoke, not the flames themselves cause the majority of fire deaths and say that sprinklers allow too much smoke to develop before the sprinklers are activated.

The effectiveness of smoke detectors for the recall function has also been questioned because of the possibility that smoke may be present on floors above and/ or below the fire floor. The Code has addressed this in 2.27.3.2.5 by indicating that the first smoke detector activation determines the recall level (see 1.3). This is based on the fact that it is highly improbable that the smoke detectors on floors other than the fire floor would be activated beforehand.

As a corollary to smoke detectors that are now required to initiate elevator recall, it has been cited that the water flow switch associated with the on-floor sprinkler system is more positive. Smoke detectors were chosen in order to initiate elevator recall, as soon as possible in order to prevent the elevators from being used by building occupants during a fire when their use may be hazardous.

2.27.3.2.5 This requirement specifically states that the elevator only need to respond to the first detector, which was activated. The likelihood of two simultaneous fires is infinitesimal. It is assumed that the smoke detector at the fire floor will be the first one that is activated. Subsequent alarms would most likely occur due to smoke migration and should not affect the choice of the recall floor.

2.27.3.3 Phase II Emergency In-Car Operation. Phase II firefighters' operation is for the benefit of firefighters. Some of the input received from firefighters is as follows:

Firefighters need the elevator for their use. The Fire Services have long complained of elevators not being available to them upon arrival at the fire building. This will provide the maximum number of elevators available to them. Firefighters' will take command during a fire, and they will determine whether and how many elevators are to be used. Firefighters' are willing to accept the risks that are associated with running elevators during a fire. It is standard operating procedure for the firefighters to use the elevators not only to carry equipment for firefighting or evacuation purposes, but to also disperse fire personnel to non-fire involved floors. The presence of a firefighter reduces occupant fears, and firefighters can direct occupant movement

Chart 2.27.3.2.1
NFPA National Fire Alarm Code® Handbook

3-9.3 Elevator Recall for Fire Fighters' Service

3-9.3.1* System-type smoke detectors or other automatic fire detection as permitted by 3-9.3.5 located in elevator lobbies, elevator hoistways, and elevator machine rooms used to initiate fire fighters' service recall shall be connected to the building fire alarm system. In facilities without a building fire alarm system, these smoke detectors or other automatic fire detection as permitted by 3-9.3.5 shall be connected to a dedicated fir alarm system control unit that shall be designated as "elevator recall control and supervisory panel," permanently identified on the control unit and on the record drawings. Unless otherwise required by the authority having jurisdiction, only the elevator lobby, elevator hoistway, and the elevator machine room smoke detectors or other automatic fire detection as permitted by 3-9.3.5 shall be used to recall elevators for fire fighters' service.

Elevator lobby, elevator hoistway, and elevator machine room smoke detectors are the only smoke detectors required to initiate elevator recall in accordance with ANSI/ASME A17.1-1995, *Safety Code for Elevators and Escalators,* which requires recall of elevators to the designated or alternate recall level when these detectors are actuated.

Buildings that do not have and are not required to have a fire alarm system, use a type of control unit designated and permanently labeled as the "Elevator Recall Control and Supervisory Panel," that serves the smoke detectors that initiate elevator recall.

A-3-9.3.1 In facilities without a building alarm system, dedicated fire alarm system control units are required by 3-9.3.1 for elevator recall in order that the elevator recall systems be monitored for integrity and have primary and secondary power meeting the requirements of this code.

The control unit used for this purpose should be located in an area that is normally occupied and should have audible and visible indicators to annunciate supervisory (elevator recall) and trouble conditions; however, no form of general occupant notification or evacuation signal is required or intended by 3-9.3.1.

The elevator recall control and supervisory unit should be placed in an area that is constantly attended for monitoring, especially when it is installed as a stand-alone control. See Exhibit 3.24 for an example of an elevator recall control and supervisory system.

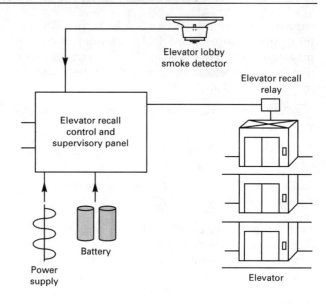

Exhibit 3.24   Elevator Recall System. (Source: FIREPRO Incorporated, Andover, MA)

3-9.3.2 Each elevator lobby, elevator hoistway, and elevator machine room smoke detector or other automatic fire detection as permitted by 3-9.3.5 shall be capable of initiating elevator recall when all other devices on the same initiating device circuit have been manually or automatically placed in the alarm condition.

Generally, unless the required smoke detectors are installed on individual fire alarm-initiating device circuits without any other fire alarm devices installed on those circuits, the smoke detectors should be powered separately from the initiating device circuit. Smoke detectors installed on signaling line circuits will not be affected.

3-9.3.3 A lobby smoke detector shall be located on the ceiling within 21 ft (6.4 m) of the centerline of each elevator door within the elevator bank under control of the detector.

Exception: For lobby ceiling configurations exceeding 15 ft (4.6 m) in height or that are other than flat and smooth, detector locations shall be determined in accordance with Chapter 2.

Paragraph 3-9.3.3 covers a new requirement for the 1999 edition of the code. Chapter 3 requires a smoke detector to be within 0.7 times the selected spacing of the detector. On smooth ceilings under 15 ft (4.6 m), the selected spacing is typically permitted to be 30 ft (9.1 m). This

(continued)

Chart 2.27.3.2.1
NFPA National Fire Alarm Code® Handbook (Cont'd)

ensures that a smoke detector will be within 21 ft (6.4 m) of the elevator door. High or nonsmooth ceilings may require a different spacing. There is no requirement that the smoke detector be located immediately adjacent to the elevator doors.

3-9.3.4 Smoke detectors shall not be installed in elevator hoistways.

Exception 1: Where the top of the elevator hoistway is protected by automatic sprinklers.

Exception 2: Where a smoke detector is installed to activate the elevator hoistway smoke relief equipment.

Even though 2-1.4.2 mentions elevator hoistways as one of the areas where detection must be installed where total coverage is specified, smoke detectors installed in elevator hoistways require continuous maintenance and are a source of numerous false or nuisance alarms. For this reason, the code has specifically excepted them from the elevator hoistway unless the top of the hoistway is protected by an automatic sprinkler system. If sprinklers are installed at the top of the hoistway, then the smoke detector is needed to provide the recall feature before the heat detector or waterflow switch on the hoistway sprinkler system actuates. (See Exhibit 3.23, Rule 211.3b of ANSI/ASME A-17.1a-1997.) Section 5-13.6.1 of NFPA 13, *Standard for the Installation of Automatic Sprinkler Systems*, determines if the hoistway is required to be sprinklered. See in Exhibit 3.25 the excerpt from NFPA 13.

Exception No. 2 was added to the 1999 edition of the code to allow for smoke relief equipment, such as a smoke hatch.

3-9.3.5 If ambient conditions prohibit installation of automatic smoke detection, other automatic fire detection shall be permitted.

The intent of paragraph 3-9.3.5 is to prevent nuisance alarms from smoke detectors installed in areas that are inappropriate for their use such as unheated areas. Where the designer, the authority having jurisdiction or another code requires a detector in areas where the ambient conditions are unsuitable for a smoke detector, 3-9.3.5 allows the use of any other type of detector that would be stable and still provide necessary detection.

3-9.3.6 When actuated, each elevator lobby, elevator hoistway, and elevator machine room smoke detector or other automatic fire detection as permitted by 3-9.3.5 shall initiate an alarm condition on the building fire alarm system and shall visibly indicate, at the control

---

5-13.6.1* Sidewall spray sprinklers shall be installed at the bottom of each elevator hoistway not more than 2 ft (0.61 m) above the floor of the pit.

Exception: For enclosed, noncombustible elevator shafts that do not contain combustible hydraulic fluids, the sprinklers at the bottom of the shaft are not required.

A-5-13.6.1 The sprinklers in the pit are intended to protect against fires caused by debris, which can accumulate over time. Ideally, the sprinklers should be located near the side of the pit below the elevator doors, where most debris accumulates. However, care should be taken that the sprinkler location does not interfere with the elevator toe guard, which extends below the face of the door opening.

ASME A17.1, *Safety Code for Elevators and Escalators*, allows the sprinklers within 2 ft (0.65 m) of the bottom of the pit to be exempted from the special arrangements of inhibiting waterflow until elevator recall has occurred.

5-13.6.2* Automatic sprinklers in elevator machine rooms or at the tops of hoistways shall be of ordinary- or intermediate-temperature rating.

A-5-13.6.2 ASME A17.1, *Safety Code for Elevators and Escalators*, requires the shutdown of power to the elevator upon or prior to the application of water in elevator machine rooms or hoistways. This shutdown can be accomplished by a detection system with sufficient sensitivity that operates prior to the activation of the sprinklers (see also NFPA 72, *National Fire Alarm Code®*). As an alternative, the system can be arranged using devices or sprinklers capable of effecting power shutdown immediately upon sprinkler activation, such as a waterflow switch without a time delay. This alternative arrangement is intended to interrupt power before significant sprinkler discharge.

5-13.6.3* Upright or pendent spray sprinklers shall be installed at the top of elevator hoistways.

A-5-13.6.3 Passenger elevator cars that have been constructed in accordance with ASME A17.1, *Safety Code for Elevators and Escalators*, Rule 204.2a (under A17.1a-1985 and later editions of the code) have limited combustibility. Materials exposed to the interior of the car and the hoistway, in their end-use composition, are limited to a flame spread rating of 0 to 75 and a smoke development rating of 0 to 450.

Exhibit 3.25 Excerpt from NFPA 13, Standard for the Installation of Sprinkler Systems, 1999 edition.

---

unit and required remote annunciators, the alarm initiation circuit or zone from which the alarm originated. Actuation from elevator hoistway and elevator machine room smoke detectors or other automatic fire detection as permitted by 3-9.3.5 shall cause separate and distinct visible annunciation at the control unit and required annunciators to alert fire fighters and other emergency

(continued)

Chart 2.27.3.2.1
NFPA National Fire Alarm Code® Handbook (Cont'd)

personnel that the elevators are no longer safe to use. Actuation of these detectors shall not be required to actuate the system notification appliances where the alarm signal is indicated at a constantly attended location.

The intent of paragraph 3-9.3.6 is to ensure that the area, or zone of alarm (floor, room, etc.) be indicated on the fire alarm control unit and the remote annunciator. The code requires that the elevator hoistway smoke detector (if one is present) and the elevator machine room smoke detector(s) be connected to the fire alarm control unit and remote annunciator as a separate zone or point of alarm indication. The detectors located in the elevator hoistway and elevator machine room actuate Phase I elevator recall, but are not required to sound the building evacuation alarm. Their annunciation, however, must conform to the requirements as stated.

Exception: If approved by the authority having jurisdiction, the elevator hoistway and machine room smoke detectors shall be permitted to initiate a supervisory signal.

The exception to 3-9.3.6 is provided to minimize the nuisance alarms from smoke detectors in these areas. The elevator recall system would still operate, but the fire alarm signal would not sound. The option should be used only where trained personnel are constantly in attendance and can immediately respond to the supervisory signal and investigate the cause of the signal. Means should be provided for initiating the fire alarm signal if the investigation of the cause of the supervisory signal indicates that building evacuation is necessary.

In addition to having trained personnel constantly in attendance, it is recommended that if the supervisory signal is not acknowledged within a given period of time (3 minutes to 10 minutes), the fire alarm system will automatically and immediately initiate an alarm.

3-9.3.7* For each group of elevators within a building, three separate elevator control circuits shall be terminated at the designated elevator controller within the group's elevator machine room(s). The operation of the elevators shall be in accordance with Rules 211.3 through 211.8 of ANSI/ASME A17.1, *Safety Code for Elevators and Escalators*. The smoke detectors or other automatic fire detection as permitted by 3-9.3.5 shall actuate the three elevator control circuits as follows:

(a) The smoke detector or other automatic fire detection as permitted by 3-9.3.5 located in the designated elevator recall lobby shall actuate the first elevator control circuit. In addition, if the elevator is equipped with front and rear doors, the smoke detectors in both lobbies at the designated level shall actuate the first elevator control circuit.

(b) The smoke detectors or other automatic fire detection as permitted by 3-9.3.5 in the remaining elevator lobbies shall actuate the second elevator control circuit.

(c) The smoke detectors or other automatic fire detection as permitted by 3-9.3.5 in elevator hoistways and the elevator machine room(s) shall actuate the third elevator control circuit. In addition, if the elevator machine room is located at the designated level, its smoke detector or other automatic fire detection as permitted by 3-9.3.5 shall also actuate the first elevator control circuit.

Three elevator control circuits are needed for proper operation of the recall sequence and for safe use of the elevators by the fire department. The first circuit is needed to prevent recalling the elevators and discharging passengers to the designated floor when the designated floor is the fire location, and to provide for an alternate recall location (determined by the authority having jurisdiction) when the designated floor is reporting a fire condition.

The second circuit configuration [part (b)] provides for standard recall to the designated floor when any other elevator lobby smoke detector is in alarm.

The operation of the third circuit, [part (c)], although not defined in the code, is intended for the safety of fire fighters who may be using the elevators to bring equipment to staging areas in a high-rise building. The third circuit's feature is intended to recall the cab(s) to the designated level during Phase I operation and to warn fire fighters of a fire in the hoistway or machine room during Phase II operation. The third circuit will sound a warning in the elevator cab to notify the fire department personnel using the elevators on Phase II Operation during the fire to immediately move to a safe floor and exit the elevator. See Exhibit 3.26 for an illustration of a connection of the fire alarm system to the elevator controller.

A-3-9.3.7 It is recommended that the installation be in accordance with Figures A-3-9.3.7(a) and (b). Figure A-3-9.3.7(a) should be used where the elevator is installed at the same time as the building fire alarm system. Figure A-3-9.3.7(b) should be used where the elevator is installed after the building fire alarm system.

(continued)

Chart 2.27.3.2.1
NFPA National Fire Alarm Code® Handbook (Cont'd)

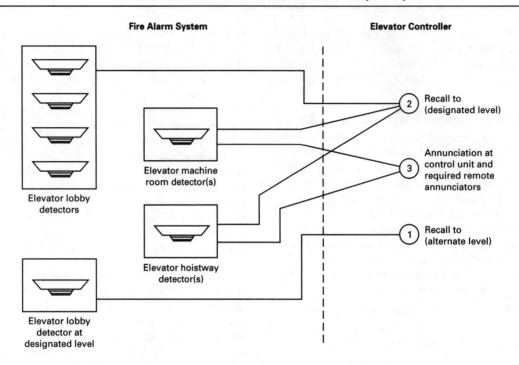

Exhibit 3.26 Connection to Third Circuit for Firefighter Notification. (Source: Bruce Fraser, Simplex Time Recorder Company, Gardner, MA)

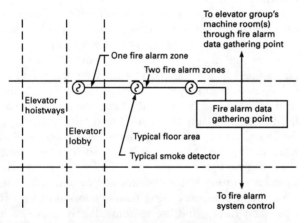

Figure A-3-9.3.7(a) Elevator Zone — Elevator and Fire Alarm System Installed at Same Time

Reprinted with permission from NFPA 72, *National Fire Alarm Code® Handbook,* Copyright© 1999, National Fire Protection Association, Quincy, MA 02269. This reprinted material is not the complete and official position of the National Fire Protection Association on the referenced subject which is represented only by the standard in its entirety.

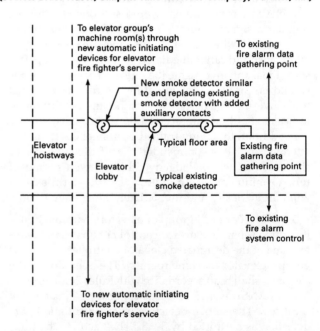

Figure A-3-9.3.7(b) Elevator Zone — Elevator Installed After Fire Alarm System

strategy since they are in constant communication with the fire command post.

Many firefighters stated that there are times when they cannot afford the luxury of using personnel to operate an elevator on a return trip to the main floor. They requested placing the Phase II switch in the elevator to the "OFF" position to automatically cause the elevator to return to the main floor for use by later arriving firefighters. However, standard operating procedures dictate that whenever staffing does permit, a firefighter with a radio be dedicated as the operator of any car being placed onto Phase II operation. That firefighter would be responsible for the shuttling of firefighters and their equipment to the discharge floor, usually a minimum of two stories below the fire floor. This would be their task until relieved by another firefighter assigned that position. There also is a requirement that when the car is on Phase II operation, turning the switch in the car to the "HOLD" position at a floor will permit the firefighter to remove the key, and leave the car without the danger of an unauthorized person taking the car to another floor.

Beginning with the 1981 edition of ASME A17.1, Phase II operation is required whenever Phase I operation is provided. The previous A17.1 requirement was predicated on the needs of emergency personnel only. The current requirement takes into account the need for evacuating the disabled during an emergency by firefighters. Disabled persons are always a concern, but the term "handicapped" in the normal context is no longer applicable; under fire conditions, even a firefighter can be considered disabled, especially when near exhaustion or if their compressed air supply is gone. Further, able-bodied occupants can become disabled from smoke, from walking up or down steps, or from hysteria. Therefore, when you hear that provisions must be made for the disabled during a fire, expand your overall picture, because even normal ambulatory persons can suddenly become non-ambulatory.

The A17 Code Committee was aware of the need to evacuate the disabled during a fire. A17.1a–1992 and later editions require firefighters' service on all elevators. The building codes, Life Safety Code (NFPA 101) and American with Disabilities Act Accessibility Guidelines (ADAAG) envision the use of elevators operating on Phase II as a principal means of evacuating the disabled during a fire. The use of Phase II is certainly a major step in providing the fire service with the necessary tools to accomplish this task, but the elevator industry must continue its research and development into making the elevator a reliable part of the picture. We are constructing buildings today, and placing those who will need this assistance into higher parts of the building, with no guarantee that we will have a means to get them down via the elevator. The firefighters will always go after anyone in danger, but the industry must provide the necessary equipment for the firefighter to do their job.

Some major changes have been added to 2.27.3.3. First, the key-operated switch is now called the "FIRE OPERATION" switch, labeled "OFF," "HOLD" and "ON," in that order. It shall be rotated clockwise, to go from "OFF" to "HOLD" to "ON." Removable only in the "OFF" and "HOLD" position, it shall not change the mode of operation within Phase II Emergency In-Car Operation until the car is at a landing with its doors in the normal open position. By preventing the key from being removed from the fire operation switch, when in the "ON" position, reduces the potential of leaving a car stranded after a fire emergency. It also reduces the possibility of unauthorized personnel entering an empty car, and taking it to another floor, as it is unlikely a firefighter would not remove their fire operations key when exiting a car.

Firefighters are allowed to reverse their direction at the next available landing by turning Phase II switch to "OFF" (2.27.3.3.4). This change has been requested by the firefighters. The car shall remain on Phase II operation, and, without going through a door open and close sequence, will return to the designated level. This provides the firefighter with a means of aborting an upward flight when conditions warrant their return to the designated level where they can review their options.

Means used to remove elevators from normal operation, other than as specified in this Code, shall not prevent Phase II Emergency In-Car Operation.

Appropriate requirements for cars with more than one entrance are established [2.27.3.3.1(d)]. Requirement 2.27.3.3.1(e) recognizes that on large vertically sliding doors with continuous pressure closing, reopening is impractical and could introduce a delay factor that may impair safety. Requirement 2.27.3.3.1(g) recognizes that door-reopening devices are not necessary since constant pressure operation is required.

For standardization, 2.27.3.3.3 and 2.27.3.3.4 define Phase II operation in the "OFF" position and the car is not at the designated level.

2.27.3.3.6 To prevent water from automatic sprinklers and/or fire department hose streams from causing an occurrence of an accidental ground or short circuit in elevator electrical equipment that is located on the landing side of the hoistway enclosure, once it has been activated, Phase II shall not be deactivated.

2.27.3.4 Interruption of Power. This requirement clarifies that even in the case of power interruption upon restoration of power, Phase I or Phase II would continue to remain in effect.

2.27.4 Firefighters' Emergency Operation — Non-Automatic Elevators

A "designated attendant" (see 1.3) is where the elevator is controlled solely by authorized personnel (see 1.3) such as attendant service, independent service, hospital service, and similar operations.

This Rule establishes the operation requirements of elevators, which are not covered by 2.27.3. A standard sign is required to alert the operator as to which floor the elevator should be returned to. See commentary on 2.27.3.

2.27.5 Firefighters' Emergency Operation — Automatic Elevators With Designated Attendant Operation

A car could be left at the fire floor, exposing the hoistway to fire, when it could be recalled. The delay is to give ample warning prior to recall. "Hospital Service" (see 1.3) is a special case of operation by a designated attendant (see 1.3) used only for medical emergencies. Hospital service has been excluded, since there may be valid reasons for not returning the car.

2.27.6 Firefighters' Emergency Operation — Inspection Operation

This requirement recognizes that taking the elevator control away from elevator personnel may be hazardous.

2.27.7 Firefighters' Emergency Operation — Operating Procedures

Operating procedures must be incorporated with or be adjacent to the Phase I and Phase II key-operated switches to assure that during an emergency, emergency personnel have quick access to available instructions for their use. Since firefighters' service is now standardized, the Code specifies signs with simple and clear wording that explain the operation of the elevators. An explanation of the "RESET" position is not necessary on the Phase I sign because this is a building function, not a firefighters' operation.

Requirement 2.27.7.4 recognizes a requirement in NBCC.

2.27.8 Switch Keys

A Note is included to suggest that local jurisdictions may legislate uniform keyed lock box requirements to assure the availability of emergency keys when building personnel are not available. It is not within the jurisdiction of A17 to make such a requirement so it is only suggested through the use of the Note. The use of lock boxes is one means of assuring that the fire service community has access to the keys to operate this system. In some jurisdictions, a common key has been designated as the Fire Service Key. The distribution of the uniform key can be controlled through laws and training.

The required switch keys for Phase I and Phase II operation must be the same to assure a speedy response to the emergency situation. A key is required for all Phase I and Phase II key switches, to allow for simultaneous operation of all key switches.

SECTION 2.28
LAYOUT DATA

2.28.1 Information Required on Layout Drawing

All the necessary layout data has been consolidated in this requirement. The purpose is to require that the elevator company provide the building designer with the forces that the elevator system will impart on the building structure and foundation. With this information, the authority having jurisdiction is also in a position to confirm that the elevator is designed to conform to the requirements of the Code.

SECTION 2.29
IDENTIFICATION

2.29.1 Identification of Equipment

This requirement, in conjunction with The National Electrical Code® requirement NEC® Section 620-51(d) that disconnecting means be numbered when there is more than one driving machine in the machine room, provides elevator personnel and emergency personnel with the information necessary to remove power from only one machine without interrupting the power to other equipment.

2.29.2 Floor Numbers

The floor numbers are necessary for elevator personnel when on top of an elevator car.

PART 3[1]
HYDRAULIC ELEVATORS

SCOPE

Roped-hydraulic elevators were not covered in this Part in the 1955 Edition through A17.1–1988 of the A17.1 Code. In the early part of the twentieth century, many multireeved roped-hydraulic elevators were installed. Many were operated by city water pressure. Many had hydraulic driving machines. Cylinders were either mounted horizontally, as shown in Diagram 8.6.3.3(l), or vertically, as shown in Diagram 8.6.3.3(k).

Elevators reeved with ratios as high as 12:1 were quite common. One foot of plunger travel resulted in 3.66 m (12 ft) of car travel. If the plunger traveled at 0.305 m/s (1 ft/s), the resulting car speed would be 3.6 m/s (720 ft/min). These installations were plagued with leaks. At this high reeving ratio, just a small leak could cause the car to have excessive downward drift if parked for just a short period of time.

In those very early days, true interlocks were not required. There were many accidents resulting from people opening the hoistway doors and stepping through the entrance, but the car was not there.

Most of these installations had wedge clamp safeties on the car. There were a number of accidents involving these. When the car was on the safety, a mechanic could get on the car with his tee wrench, remove the cover in the floor, and release the safety, not realizing the piston had retracted and the hoist ropes were slack.

The directional valves were mechanically operated by an abundance of levers, cams, and other devices that were connected to a lever in the car by a rope or cable. These valves and the operating linkages had a tendency to freeze, causing the car to attempt to exceed its extreme limits of travel, which resulted in a catastrophic accident. Many of these valves were later converted to electric solenoid operation, but they still had a tendency to freeze up due to the corrosive action of the water.

On many of these installations, there were also speed regulating valves installed in the circulating piping that were supposed to regulate the amount of water exiting the hydraulic system so the down car speed would not reach the governor trip speed and set the car safety, regardless of how far the operator pushed the control lever in the car. These valves, which consisted of a group of levers, cams, and other parts, had a tendency to become gummed up, so they would not open properly and thus slowed the car down considerably. A car that was supposed to be operating at 3.05 m/s (600 ft/min) would sometimes be slowed down to less than 1.52 m/s (300 ft/min). These valves were very difficult and expensive to maintain, so in many cases all the internal workings were removed from them, leaving only the housing. When this happened, only the operator had control of the down speed of the elevator car, and it depended upon his skill at operating the lever in the car. If the operator lacked the necessary skill, he could cause an unnecessary stop on the car safety or run the car into the pit at a car speed much greater than the designed striking speed of the pit buffers.

It has also been determined that during the time the requirements for the 1955 edition of the Code were being drafted, not a single manufacturer of roped hydraulics had any intention of continuing the product line. The Committee saw no need to spend its time writing requirements for something that was no longer being manufactured. In the mid 1980s, a renewed interest was shown in roped-hydraulic elevators. The A17 Committee was aware of the problem associated with earlier equipment of this design as described above. They were also aware that roped-hydraulic elevators had been safely used in Europe for many years. A study was undertaken and requirements for roped-hydraulic elevators were incorporated in the 1989 Addenda to the A17.1 Code.

SECTION 3.1
CONSTRUCTION OF HOISTWAYS AND HOISTWAY ENCLOSURES

Hoistways must be enclosed as required by the local building code, or if there is no local building code, one of the following model building codes:

(a) National Building Code (NBC)
(b) Standard Building Code (SBC)
(c) Uniform Building Code (UBC)
(d) National Building Code of Canada (NBCC)

Where fire-resistive construction is required by the building code, the hoistway must comply with the requirements of 2.1.1.1. When non-fire-resistive construction is permitted by the building code, the hoistway must conform to the requirements of 2.1.1.2. Partially enclosed hoistways must conform to the requirements

---

[1] Throughout Part 3, references are made to other requirements in this Code. To gain a complete understanding of a requirement, the reader should review the Handbook commentary on all referenced requirements.

of 2.1.1.3. After determining the type of hoistway construction required, see commentary on referenced requirement. The International Building Code (IBC) is the heir apparent of the NBC, SBC, and UBC, which are no longer being updated. The IBC, Section 3006.4 requires "machine rooms and machinery spaces to be enclosed with construction having a fire resistance rating not less than the required rating of the hoistway enclosure served by the machinery." In some jurisdictions, this may be a departure from past practices.

The number of elevators permissible in a hoistway (2.1.1.4) shall be in accordance with the local building code, or if there is no local building code, one of the model building codes listed above. Requirements are given in 2.1.1 for the strength of the enclosure, construction at the top and bottom of the hoistway, and requirements for the protection of the hoistway in case of fire.

3.1.1 Strength of Pit Floor

The loads of both the cylinder and buffer have to be considered. As roped-hydraulic elevators are roped 1:2 and have the dead end of the hoist ropes terminated in the pit, rope up-pull also has to be considered.

SECTION 3.4
BOTTOM AND TOP CAR CLEARANCES AND RUNBYS FOR CARS AND COUNTERWEIGHTS

3.4.1 Bottom Car Clearance

The intent of this requirement is to provide safe refuge space for a person under a car. Trenches and depressions or foundation encroachments permitted by 3.4.1.4 (2.2.2) are not considered in determining bottom car clearance. When the car rests on its fully compressed buffer, no part of the car or any equipment attached thereto is permitted to strike any part of the pit or any part of the equipment located therein. It is not the intent of this requirement to consider the hydraulic supply pipe, the cylinder support, or the buffer support channels as the pit floor reference.

The intent of 3.4.1.3 is that there be no encroachment on the floor area of refuge space. However, in situations where it is necessary for the pipe or beam to encroach the refuge area, the volume is measured from the top of the pipe or beam. The minimum refuge area specified must be provided.

3.4.1.6 Marking of the spaces where refuge clearance is not provided throughout the entire cross-sectional area of the pit is required. The marking is the color for danger. This will assist elevator personnel identify where the refuge space is located. A sign is also required to notify elevator personnel of the condition.

3.4.2 Minimum Bottom and Top Car Runby

The bottom car runby is the distance between the car buffer striker plate and the striking surface of the car buffer when the car floor is level with the bottom terminal landing. Top car runby for direct-plunger hydraulic elevators is the distance the elevator car can run above its top terminal landing before the plunger strikes its mechanical stop.

The minimum runby is based upon the operating speed in the down direction with rated load in the car. These distances allow adequate space for an elevator to stop after passing its top terminal landing without the plunger striking its mechanical stop, and for the car to run past the bottom terminal landing without striking its buffers. See Diagram 3.4.2.

3.4.3 Car Top and Bottom Maximum Runby

The bottom car runby is the distance between the car buffer striker plate and the striking surface of the car buffer when the car floor is level with the bottom terminal landing. The top car runby, for direct-plunger hydraulic elevators, is the distance the elevator car can run above its top terminal landing before the plunger strikes its mechanical stop. The maximum distance for bottom and top car runby assures that, if the elevator passes its terminal landings, there is adequate space to evacuate the passengers from the stalled elevator through the car entrance. See Diagram 3.4.2.

3.4.4 Top Car Clearance

The top car clearance for a hydraulic elevator is the shortest vertical distance within the hoistway between the horizontal plane described by the top of the car enclosure and the horizontal plane described by the lowest part of the overhead structure or other obstruction when the car floor is level with the top terminal landing. See Diagram 3.4.4.

3.4.5 Equipment Projecting Above the Car Top

The intent of these requirements is to avoid the possibility of a crushing hazard to a mechanic or inspector on top of the car. When a car travels above the top terminal landing and the plunger strikes its mechanical stop, all equipment exclusive of the guide shoe assemblies and guide post for vertically sliding gates attached to and projecting above the car top, must have at least 150 mm or 6 in. clearance to any part of the overhead structure or equipment located in the hoistway. The clearance above the crosshead is to prevent a crushing hazard to the chest or body. Due to the central location of the crosshead, it is foreseeable that a person could be leaning over the crosshead. Guide-shoe assemblies or the guide post for vertically sliding gates shall not strike any overhead structure or equipment located in the hoistway when the car is in this position, but the 150 mm or 6 in. clearance is not required. Guide-shoe assemblies and gate posts for vertically sliding gates are usually the highest item on the elevator. They are off to the side of the car top where personnel normally do not

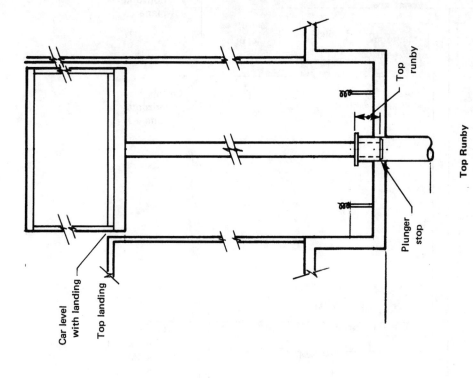

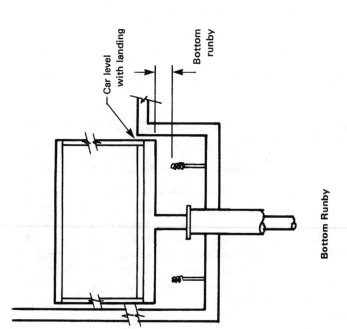

GENERAL NOTES:
(a) Minimum runby:
  (1) not less than 75 mm or 3 in. for rated speeds up to 0.50 m/s or 100 ft/min;
  (2) increased from 75 mm or 3 in. to 150 mm or 6 in. in proportion to the increase in rated speed from 0.50 m/s or 100 ft/min to 1.00 m/s or 200 ft/min; and
  (3) a minimum of 150 mm or 6 in. for rated speeds exceeding 1.00 m/s or 200 ft/min.
(b) Maximum runby: 600 mm or 24 in.

Diagram 3.4.2   Hydraulic Elevator Runby
(Courtesy W. Banister)

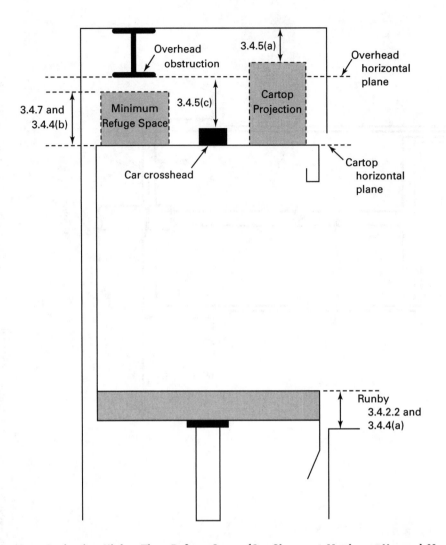

Diagram 3.4.4 Projection Higher Than Refuge Space (Car Shown at Maximum Upward Movement)

stand during inspection or maintenance. Thus, no specified clearance is required above this remote equipment. See Diagram 3.4.5.

3.4.6 Top Clearance and Bottom Runby of Counterweight

When a counterweight is provided and the car has reached its uppermost limit of travel (the plunger has struck its mechanical stop), there must be at least 150 mm or 6 in. of clearance between the bottom of the counterweight and its striking surface. The 150 mm or 6 in. clearance allows for rope stretch. A counterweight buffer is prohibited by 3.22.2. If a counterweight was allowed to strike anything, including a buffer, a slack rope condition could be created.

When the car is on its fully compressed buffer, there must be at least 150 mm or 6 in. of clearance between the top of the counterweight and any part of the overhead structure. If a counterweight was allowed to be pulled into the overhead structure, the ropes could separate

from the counterweight, allowing it to fall. See Diagrams 3.4.6.1 and 3.4.6.2.

3.4.7 Refuge Space on Top of Car Enclosure

The required refuge area provides a clear, unobstructed space on which service personnel and inspectors can stand when on top of the car. The area is wide enough to accommodate full shoulder width with a height to accommodate a large individual in a crouched position. The maximum upward movement is when the plunger strikes its mechanical stop. See Diagrams 2.4.12 and 3.4.4.

SECTION 3.6
PROTECTION OF SPACES BELOW HOISTWAY

The requirements are intended to assure that the occupiable portions of a building below a hydraulic elevator are safe. This requirement is to be complied with when

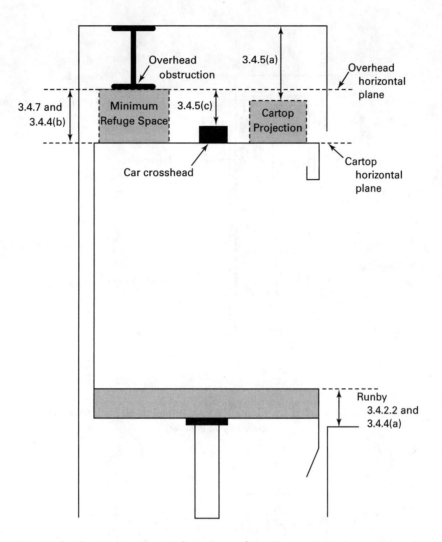

Diagram 3.4.5   Projection Lower Than Refuge Space (Car Shown at Maximum Upward Movement)

any of the spaces on each succeeding floor directly below the hoistway are not permanently secured against access. The operating speed attained in the down direction is the maximum speed with rated load under normal operating conditions. See definition in 1.3.

The building structure must be adequate to support the jack. If a counterweight is used, a counterweight safety must be provided. Spring buffers must not be fully compressed by a fully loaded car striking at maximum speed in the down direction. Energy dissipation buffers are always required. The buffer support structure must resist the impact load.

SECTION 3.7
MACHINE ROOMS AND MACHINERY SPACES

3.7.1 Location of Machine Rooms

Spaces outside the hoistway enclosure, which contain hydraulic machinery and electric control equipment, must be enclosed with noncombustible material

extending to a height of not less than 2 000 mm or 79 in. Openwork material, if used, must reject a ball 50 mm or 2 in. in diameter. Access doors to machine rooms must be 750 mm or 29½ in. wide by 2 030 mm or 80 in. high. They must be

(a) self-closing;

(b) self-locking;

(c) provided with a spring-type lock arranged to permit the door to be opened from the inside without a key; and

(d) kept closed and locked.

Access doors to machinery spaces must comply with the requirements of 2.7.3.4. In addition to these requirements, the building code should be checked as it may require the machine room to be enclosed with fire-resistive construction.

The International Building Code (IBC) may require that the machine room be of fire-resistive construction. If the IBC requires a fire-resistive hoistway enclosure; the same would apply to the machine room and machinery space. Spaces containing pumps, motors, valves, and

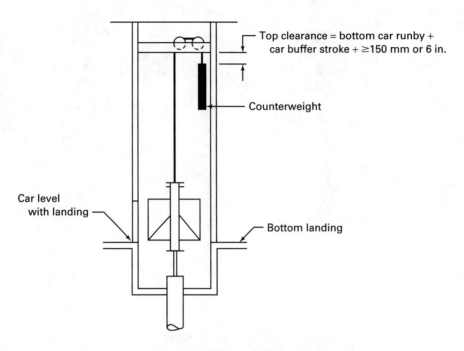

Diagram 3.4.6.1   Top Car Clearance Counterweighted Hydraulic Elevator
(Courtesy W. Banister)

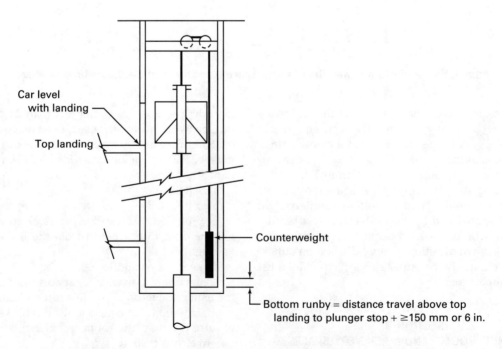

Diagram 3.4.6.2   Bottom Runby Counterweighted Hydraulic Elevator
(Courtesy W. Banister)

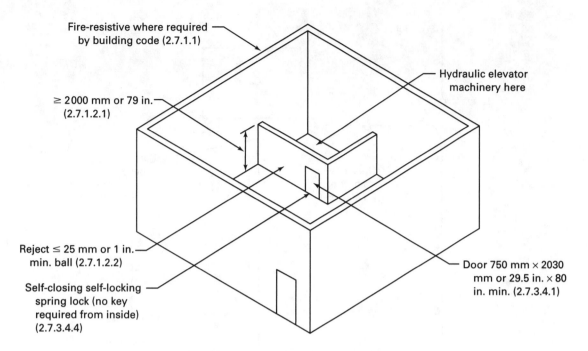

Fire-resistive where required
by building code (2.7.1.1)

≥ 2000 mm or 79 in.
(2.7.1.2.1)

Hydraulic elevator
machinery here

Reject ≤ 25 mm or 1 in.
min. ball (2.7.1.2.2)

Self-closing self-locking
spring lock (no key
required from inside)
(2.7.3.4.4)

Door 750 mm × 2030
mm or 29.5 in. × 80
in. min. (2.7.3.4.1)

Diagram 3.7 Hydraulic Elevator Machine Room Requirements

electrical control equipment can also contain other machinery, provided it is separated from the other machinery by a metal grille enclosure a minimum of 2 000 mm or 79 in. high, which will reject a ball 50 mm or 2 in. in diameter and is provided with a self-closing and self-locking door. See Diagram 3.7. The electrical clearance requirement specified in the National Electrical Code® would be taken from the metal grille enclosure or room enclosure, whichever is closer to the equipment.

SECTION 3.11
PROTECTION OF HOISTWAY-LANDING OPENINGS

3.11.1 Emergency Doors

Requirement 2.11.1 provides for emergency access to a single blind hoistway in order to evacuate a stalled elevator.

When a car safety is not provided, a stalled elevator can be lowered with a manual-lowering valve by trained elevator personnel to a lower landing and the car and hoistway doors opened to allow normal egress from the elevator. Hydraulic elevators are not normally provided with either a car safety or a side emergency exit, but are provided with a manual lowering valve.

SECTION 3.14
CAR ENCLOSURES, CAR DOORS AND GATES, AND CAR ILLUMINATION

A stalled hydraulic elevator with a manual-lowering valve and without car safeties can be lowered by elevator

personnel to a lower landing and the car and hoistway doors opened to allow normal egress from the elevator.

SECTION 3.15
CAR FRAMES AND PLATFORMS

The car frame of a roped-hydraulic elevator is suspended so it must conform to all the requirements in 2.15 and 8.2.2.

3.15.1 Requirements

The car frame of a direct-acting hydraulic elevator may be omitted provided that:

(a) the platform frame can withstand any eccentric loads and transmit them safely to the plunger designed to withstand the additional loads;

(b) only one guide shoe per rail is provided on platform frame; and

(c) car safeties are not used.

SECTION 3.17
CAR AND COUNTERWEIGHT SAFETIES AND PLUNGER GRIPPER

3.17.1 Car Safeties

Car safeties are not required or normally provided on direct-acting hydraulic elevators. A roped-hydraulic elevator is suspended rather than supported, thus car safeties are required. A slack rope device removes power to stop the jack from lowering if slack develops in the rope. The car must move in the up direction to release

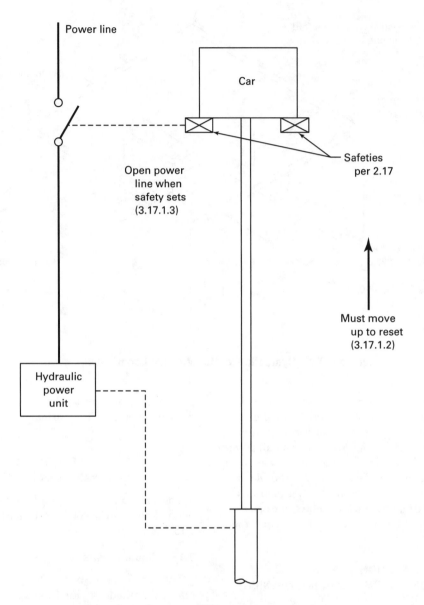

Diagram 3.17.1   Hydraulic Elevator Car Safeties
(Courtesy W. Banister)

the safety. This assures there is no slack rope or loss of pressure in the hydraulic system, which would allow the car to fall after the safety was released. When car safeties are provided, see the commentary on 2.17, and Diagram 3.17.1.

The Code gives performance criteria such that the car must be raised to reset the safety. The design of the device to do this is up to the manufacturer. Hand pumps may be used since they are readily available and designed for this purpose. Other means such as an auxiliary power-driven pump are permitted.

Roped-hydraulic elevators equipped with pistons have an indirect coupling means, usually ropes, passing through a seal fixed in the cylinder. A rope break is theoretically possible such that there is no rope contained in the seal and there is an open hole and a serious high pressure oil leak.

A function of the slack rope device is to set the safety for a broken rope condition as well as the overspeed condition. The additional slack rope device is permitted to set the safety, but not required, for roped-hydraulic elevators using hydraulic jacks equipped with plungers.

3.17.2 Counterweight Safeties

A hydraulic elevator normally would not attain sufficient overspeed to operate a safety governor, hence the requirement to operate the safety as a result of the breaking or slackening of the counterweight suspension ropes.

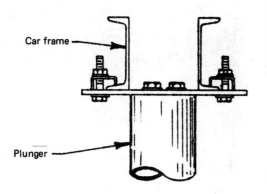

Diagram 3.18.1.1 Typical Driving Machine Connection
(Courtesy Dover Elevator Co.)

Counterweight safeties are required only when there is occupiable space below the counterweight runway. Counterweights are very seldom found on hydraulic elevators.

SECTION 3.18
HYDRAULIC JACKS

3.18.1 Hydraulic Jack and Connections

3.18.1.1 Direct-Acting Hydraulic Elevators. A direct-acting hydraulic elevator is an elevator having a plunger or cylinder directly attached to the car frame or platform. This requirement does not mandate any particular design, but does specify the design criteria of the connection of the driving member to the car frame or car platform. See Diagram 3.18.1.1 for a typical connection.

3.18.1.2 Roped-Hydraulic Elevators. Requirements are specified for the suspension members and sheaves.

The loss of one hydraulic jack is essentially equivalent to loss of one rope where there are two or more ropes per driving machine and provides for alternate designs. The pressure is limited to 150%, equivalent to the maximum relief valve pressure. The requirement that the remaining hydraulic jacks be capable of supporting the load without overloading the frame is to protect these items in the unlikely event of a rope failure.

Rope retaining means are always required whether or not seismic conditions are applicable. See commentary on referenced Sections and Requirements. See Diagram 3.18.1.2.

3.18.1.2.2 The references to 2.15.1.3 and 2.20 provide the necessary requirements for suspension ropes and their connections.

3.18.1.2.5 The reference to 2.24.2 requires that the sheave have a pitch diameter of 40 times the rope diameter. The reference to 2.24.3 requires a safety factor of 8 for a steel sheave. The reference to 2.24.5 requires that fillets be provided at any change in diameter to prevent stress concentration.

3.18.2 Plungers

3.18.2.3 Plunger Connection. The plunger must be capable of carrying in tension the weight of the plunger. That the weight of the plunger is included in the buffer calculations where applicable, such as on conventional (noninverted) direct-acting hydraulic jacks. This is the reason 3.16.1(b) requires the plunger weight to be specified on the data plate. The plunger weight would typically not be included in roped-hydraulic elevator buffer designs.

3.18.2.4 Plunger Joints. See Diagram 3.18.2.4 for a typical plunger joint.

3.18.2.7 Plunger-Follower Guide. The intent of this requirement is to retain the L/R ratio required by 8.2.8.1 and further require that the power circuit be opened when this requirement is not met.

The intent of this requirement is to adjust the free length based on the use of a plunger follower. If one follower is positioned at the halfway point, the free length is half of the full length measured from the bearing point in the cylinder head to the bearing point at the car frame. The plunger-follower guide is to be designed to be within the car frame stresses tabulated in 2.15 since the follower functions as an auxiliary car plank to provide lateral stability.

If the maximum free length is exceeded, returning the car to the lowest landing and rendering the elevator inoperative will maintain the elevator in a safe condition.

3.18.3 Cylinders

3.18.3.3 Clearance at Bottom of Cylinder. The intent of this requirement is to protect a cylinder from the impact forces and a possible rupture caused by the plunger striking the bottom of the cylinder.

3.18.3.4 Safety Bulkhead. See Diagram 3.18.3.4 for typical safety bulkhead design.

A slip casing (sleeving) does not meet the requirements for a double-wall cylinder.

3.18.3.5 Cylinder Packing Heads. See Diagram 3.18.3.5.

3.18.3.6 Closed Cylinder and Plunger Heads. Cylinders supported at the bottom and not installed below ground, are exempted by 3.18.3.6.1 as they are not subject to corrosion and are readily available to visual inspection, from the requirements of 3.18.3.4.

3.18.3.7 Collection of Oil Leakage. This requirement is normally complied with by providing a collector ring with tubing running to a container in the pit or to the hydraulic oil storage tank. The container is normally a 4 to 19 L or 1 to 5 gal can that is manually emptied by the service mechanic. The limit of 19 L or 5 gal is imposed to prevent an unnecessary fire hazard and provide assistance in identifying a leak.

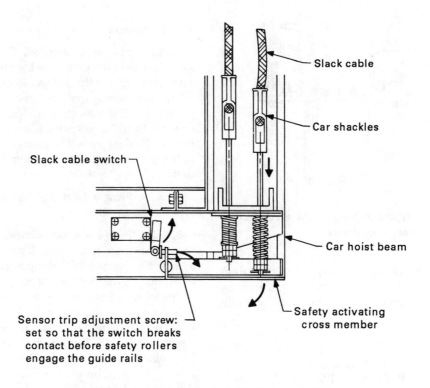

Slack cable

Car shackles

Slack cable switch

Car hoist beam

Sensor trip adjustment screw:
set so that the switch breaks
contact before safety rollers
engage the guide rails

Safety activating
cross member

Diagram 3.18.1.2   Typical Slack Rope Switch
(Courtesy Cemco Lift, Inc.)

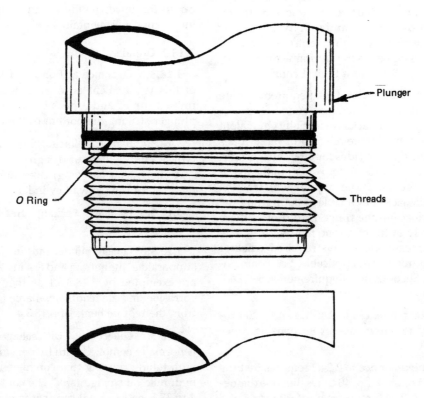

Plunger

O Ring

Threads

Diagram 3.18.2.4   Typical Plunger Joint
(Courtesy Dover Elevator Co.)

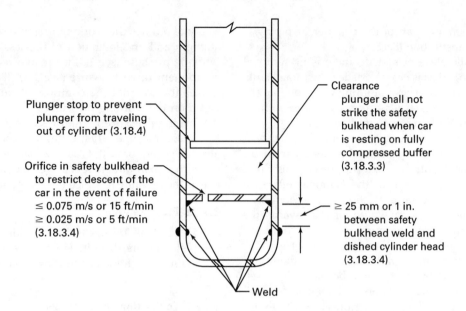

Plunger stop to prevent — plunger from traveling out of cylinder (3.18.4)

Orifice in safety bulkhead — to restrict descent of the car in the event of failure ≤ 0.075 m/s or 15 ft/min ≥ 0.025 m/s or 5 ft/min (3.18.3.4)

Clearance plunger shall not strike the safety bulkhead when car is resting on fully compressed buffer (3.18.3.3)

≥ 25 mm or 1 in. between safety bulkhead weld and dished cylinder head (3.18.3.4)

Weld

Diagram 3.18.3.4   Typical Cylinder, Plunger Head, and Plunger Stop

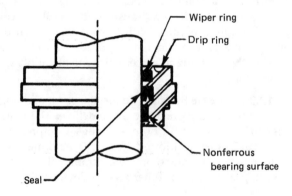

Wiper ring

Drip ring

Nonferrous bearing surface

Seal

Diagram 3.18.3.5   Typical Cylinder Packing Head

3.18.3.8 Cylinders Buried in the Ground. In the summer of 1987 the United States Environmental Protection Agency (EPA) released the regulations on underground storage tanks for public comments. At the time, the EPA included in the regulations underground cylinders of hydraulic elevators. The final regulations did not include requirements for hydraulic elevators. The elevator industry presented the EPA with statistics showing the reliability of underground cylinders. The information provided also indicated that a leak would not go undetected.

The average travel of these hydraulic elevators is about 7.92 m (26 ft) with an average cylinder volume of 11.4 L/m (3 gal/ft) or a total of 291.5 L (77 gal) of oil encased by the cylinder below ground when the elevator is in the uppermost position. When the elevator car is lowered, the volume of oil in this cylinder would be 159 L (42 gal).

The total oil in the average elevator system installed in the 1980s is about 363.4 L (96 gal), including 56.8 L (15 gal) of oil reserved over and above that required for moving the car to the top floor and that oil in the oil line connecting the power unit to the cylinder.

The hydraulic cylinder in a hydraulic elevator system should be considered a process tank enabling a reversible flow system to power the plunger for car movement and not viewed as a well or storage tank.

The hydraulic elevator system is filled at the time of installation and the same oil is normally used in the system throughout the life of the elevator. Assuming it is not contaminated such as by water, etc. It is often filtered to remove particulate contaminants.

The elevator seal through which the plunger passes is made of neoprene providing a dry wipe of the micro-finished jack without oil loss. In the event of damage to the jack seal, which could cause a leak, an oil recovery ring is provided on the jack head. In addition, the ASME A17 Code requires that a container, not to exceed 19 L or 5 gal be installed in the elevator pit. This container would, of course, be checked during routine maintenance. An oil usage log must also be maintained (8.6.5.7). Hydraulic elevators are also required to be regularly inspected by Certified QEI Inspectors (8.11.1.1). A substantial loss of oil, such that the pump sucks air as the elevator tries to go to the top landing, produces a significant noise. A leak should not go undetected for long.

This inspection would include an examination of the record of oil usage, the power unit in the machine room, the oil piping, and the 19 L (5 gal) container required by the Code in the pit for oil leakage. A written record of the quantity of hydraulic fluid added and emptied is required when any part of the cylinder or piping is not visible for inspection (8.6.5.7).

In a hydraulic elevator system, there is no removal and filling up of oil as is characteristic to a storage tank. The need for an elevator in a building will mandate the replacement of a cylinder if a leak occurs since without the elevator the building may not be usable.

In order to conform with a requirement in the National Electrical Code® a timer is furnished as an industry standard on each controller. The purpose of this timer is to meet the requirement in the National Electrical Code® for phase reversal protection and 3.26.5. Since the hydraulic elevator pump would turn backwards, the car would not move, and of course the motor could run to a point of overheating. With the timer set for 20 s more than the trip time from bottom to top, the motor is then cut off and the timer must be manually reset, thereby requiring a qualified person to make a determination of the cause. This protection would likewise announce the loss of oil in the system since, generally only 56.8 L (15 gal) of reserve is installed in the system. If there were an oil loss either above or below ground exceeding this amount, the car, in attempting to go to the top floor, would stall at some point but the motor would continue to run until the expiration of the preset time and then the car would reverse, lower to the bottom floor, and wait for a qualified person to reset the time. See 3.26.9.

The EPA was satisfied with this information but requested that corrosion protection be provided on all underground cylinders and piping. The A17 Committee developed these requirements in 3.18.3.8 and in exchange, the EPA excluded elevator cylinders from the final regulations.

The requirements are performance oriented and the manufacturer has the option of choosing one or more, whichever best suit its needs.

3.18.4 Plunger Stops

On a new installation, it is important to verify that the stop ring has been installed. This can be accomplished by running the car up to engage the stop ring at inspection speed, with or without a load, until the car stops and the relief valve operates. Running the car up to engage the stop ring at full speed is unnecessary and could damage the stop ring. See Diagram 3.18.3.4 for a typical plunger stop.

SECTION 3.19
VALVES, PRESSURE PIPING, AND FITTINGS

3.19.1 Materials and Working Pressures

3.19.1.2 Working Pressures. All supply piping materials, valves, and fittings, with the exception of flexible hydraulic hose and fitting assemblies and flexible couplings, must be of ductile material and have a factor of safety of 5 times the working pressure based on tensile strength and an elongation of not less than 10%. The working pressure (see 1.3) is not allowed to exceed the component-rated pressure (see 1.3). The requirement makes it clear that the component-rated pressure is based on the working pressure.

Working pressure is the pressure measured at the hydraulic machine when lifting car and its rated load at rated speed, or with Class C2 loading when leveling up with maximum static load. Component-rated pressure is the pressure to which the component is subjected.

3.19.2 Pressure Piping

3.19.2.1 Wall Thickness. The working pressure for a completed oil line is limited by the lowest rated element in the line, which includes fittings, valves, and connections.

3.19.3 Connections and Fittings

3.19.3.2 Grooved Pipe Fittings. Fittings must be of the type designed and constructed for use on grooved pipe. There is also available a type of coupling for use on nongrooved pipe called a "roustabout" coupling. This type of coupling is fitted with threads that, when the coupling is tightened in place, presses into the pipe securing the coupling in place. This type of coupling is prohibited.

3.19.3.3 Flexible Hydraulic Connections. All markings required must be permanent and indelible. The markings can be either on the hose or fittings. Compression-type fittings are prohibited. A mechanical means is required to prevent the complete separation of a connection if a gasket, seal, or other sealing device fails.

Flexible hoses must:
   (a) be designed for 10 times working pressure;
   (b) be tested at 5 times working pressure;
   (c) have the date of test permanently marked;
   (d) not be used in the hoistway; and
   (e) not project into or through any wall.

This requirement does not apply to a flexible hose installed between the pump and tank.

3.19.4 Valves

See Diagram 3.19.4.

3.19.4.1 Shutoff Valve. The relief valve pressure should be set using the shutoff valve prior to the entire system being subjected to the relief pressure. This will protect the system from excessive pressure if the relief valve does not open and the system is over pressured. The shutoff valve in the machine room provides an additional degree of safety for personnel entering and working in a pit of a hydraulic elevator. It is safer to locate the shutoff valve outside the hoistway rather than the pit. If an additional shutoff valve is provided in the pit, as is often the case, it must meet the pressure-rating

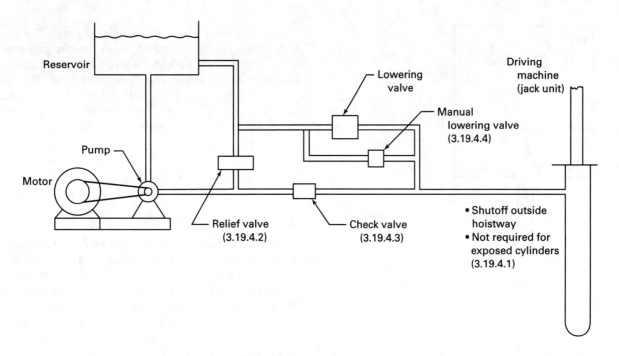

Diagram 3.19.4 Relief, Check, and Manual Lowering Valves
(Courtesy W. Banister)

requirements. Brass valves used in plumbing systems usually do not meet the pressure and safety factor requirements for hydraulic elevators. The shutoff valve also does not require a permanent handle, as it is not provided for emergency use.

3.19.4.2 Pump Relief Valve. The pump relief valve is to be set to allow full flow at a pressure not greater than 150% of the working pressure (see 1.3). The opening pressure is usually lower; however, it is very difficult to accurately measure. The requirement allows large pumps or groups of pumps to use more than one relief valve to achieve the required flow capacity so pressure never goes more than 50% above working pressure. See Diagram 3.19.4.

3.19.4.5 Pressure Gage Fittings. The location specified allows static pressure to be read at the hydraulic machine; to determine the pressure at the hydraulic jack, the pipe pressure drop must be determined and subtracted from the hydraulic machine pressure. A shutoff valve is used to be sure a leaky or broken fitting or gage can prevent oil loss.

3.19.5 Piping Buried in the Ground

See Handbook commentary on 3.18.3.8.

3.19.6 Welding

Welding of supply piping, valves, and fittings must be done by welders qualified in accordance with the requirements of the American Welding Society. The

welders may be qualified, at the option of the manufacturer, by one of the following: the manufacturer, a professional consulting engineer, or a recognized testing laboratory.

**SECTION 3.22
BUFFERS AND BUMPERS**

3.22.1 Car Buffers or Bumpers

Bumpers can be used when the car is equipped with a Type C safety or on hydraulic elevators having a maximum speed in the down direction with rated load in the car of 0.25 m/s or 50 ft/min.

3.22.2 Counterweight Buffers

An up-traveling hydraulic elevator will be stopped when its stop ring is struck; therefore, a counterweight buffer is not needed. The hydraulic system is designed for counterweight load always being effective. The need for the counterweight is never removed. Removal of the counterweight with car at the top places a buckling load on the plunger, which reduces the design safety factor. The removal of the counterweight load also creates excess system hydraulic pressure on valves, fittings, and supply piping. Counterweight buffers could also create a slack rope condition with rope stretch.

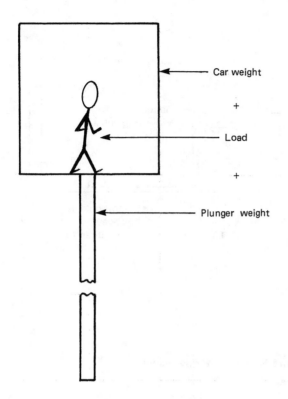

Diagram 3.23.1 Rail Bracket Spacing, Hydraulic Elevator With Safeties
(Courtesy W. Banister)

SECTION 3.23
GUIDE RAILS, GUIDE-RAIL SUPPORTS, AND FASTENINGS

3.23.1 Direct-Acting Hydraulic Elevators

See Diagram 3.23.1.

3.23.2 Roped-Hydraulic Elevators

Car safeties are required on roped-hydraulic elevators. Therefore, all of 2.23 is applicable. A means of keeping the traveling sheave, when provided, in alignment is required.

SECTION 3.24
HYDRAULIC MACHINES AND TANKS

3.24.2 Tanks

3.24.2.2 Minimum Level Indication. See commentary on 3.24.3.3.

3.24.3 Atmosphere Storage and Discharge Tanks

3.24.3.1 Covers and Venting. Tanks must be covered to prevent foreign materials from entering the oil system and reduce a fire hazard.

3.24.3.3 Means for Checking Liquid Level. The most common means of complying with these requirements is to provide a dipstick in the top of the tank similar to the type used in an automobile engine. Another method is to provide a sight glass on the side of the tank. Both methods require the car be at its lowest landing in order to obtain a correct reading of the oil level.

SECTION 3.25
TERMINAL STOPPING DEVICES

3.25.1 Normal Terminal Stopping Devices

3.25.1.1 Where Required and Function. The normal terminal stopping device is required for the purpose of stopping the car at the terminal landing when the normal stopping means fails to stop the car. The normal terminal stopping device is required to function until:
(a) at the top terminal, the stop ring is engaged;
(b) at the bottom terminal, the car is resting on its fully compressed buffer(s).

During the recycling operation for telescopic plungers, the normal terminal stopping device can be bypassed if the requirements of 3.26.7 are met.

3.25.1.3 Requirements for Stopping Switches on the Car or in the Hoistway. When the switch is located on the car or in the hoistway, it must be of the enclosed type. It must be securely mounted so that horizontal movement of the car does not affect its operation. Operating cams must be metal, and the switch contacts must be directly opened mechanically. Opening of contacts must not depend on a spring or gravity or a combination thereof. Stopping switches are not emergency stop switches.

3.25.2 Terminal-Speed Reducing Devices

The term "solid limit of travel" used in this requirement refers to the stop ring (plunger stop). In the up direction the car speed is to be reduced to 0.25 m/s or 50 ft/min, with a retardation rate not greater than gravity, before it strikes the stop (3.18.2.5), if the normal terminal stopping device fails to slow down the car. Two independent control means are required. At least one of which is operated by the terminal-speed reducing device, the other or both by the terminal stopping device. The terminal-speed reducing device is permitted to be mechanically operated.

3.25.3 Final Terminal Stopping Devices

Final terminal stopping devices are not necessary since the stop ring would stop the elevator if the normal devices do not function. The run timer would prevent running of the motor after the plunger is against the stop ring.

SECTION 3.26
OPERATING DEVICES AND CONTROL EQUIPMENT

3.26.1 Operating Devices and Control Equipment

By referencing all of 2.26, the following requirements are mandatory:

(a) inspection operation, 2.26.1.4;

(b) inspection operation with open door circuits 2.26.1.6;

(c) contactors and relays in critical operating circuits 2.26.3;

(d) electrical equipment and wiring including listing and labeling requirements and EMC immunity, 2.26.4;

(e) monitoring of door circuits 2.26.5;

(f) making electrical protective devices inoperative 2.26.7; and

(g) redundancy and checking of redundancy.

3.26.3 Anticreep and Leveling Operation

The anticreep operation remains operational when the in-car stop switch or the emergency stop switch is in the stop position (3.26.4). It was common practice to shut an elevator down for the night or over a weekend using the in-car emergency stop switch, and also turning the lights off. If the anticreep operation was not functioning and the door closing device did not function as required by 2.11.3, the car could descend to the pit, leaving the hoistway door in the open position. A person could inadvertently step into the hoistway thinking he was entering a dark car.

Anticreep operation may be bypassed during the recycling operation of telescopic plungers (3.26.7).

The anticreep operation is required for the purpose of preventing the car from drifting away from a floor. The anticreep operation needs to operate in the up direction only for electrohydraulic elevators. The requirements do not apply to a car when it is in the process of responding to a call, or to a malfunction of a switch.

3.26.5 Phase-Reversal and Failure Protection

Since there is no immediate danger, as there is with electric elevators of running in the wrong direction, the need for immediate action in case of a phase-reversal as called for in 2.26.6 is not required. Means should be provided to prevent the pump from running backward for any extended period of time and to remove the drive motor from the power supply in the event of a single-phase condition. The "run" indication timer used by some manufacturers on hydraulic controllers to protect against pump failure (i.e., loss of oil) also provides protection against phase reversal (i.e., oil is not delivered to jack).

Thermistors, bimetallics temperature sensors, and phase failure, and phase loss relays are options that can be utilized. Requirement is performance oriented.

3.26.6 Control and Operating Circuits

Springs in switches, relays, etc., to stop elevators must be in compression. Power should not be required to stop the elevator.

3.26.7 Recycling Operation for Multiple or Telescopic Plungers

Telescopic plungers are normally found on holeless hydraulic elevator installations and may be vertically realigned periodically. Realignment must be accomplished in accordance with these requirements. If a two-stage plunger is not lowered for realignment, a micro-leveling pump will typically be utilized.

3.26.8 Pressure Switch

If a car hangs up mechanically for any reason and the lowering valve is opened, the oil is free to flow back to the tank leaving the cylinder void or partially void of oil. Experiments have shown that present seals are not effective in maintaining a vacuum in the cylinder. See Diagram 3.26.8.

SECTION 3.27
EMERGENCY OPERATION AND SIGNALING DEVICES

The firefighter community has advised that in their opinion passengers should not be restricted from exiting an elevator even if the elevator is not capable of reaching the recall level. It is safer to exit even at a potential fire floor since the fire is most likely in its early stages. If passengers are not allowed to exit, they may be overcome by smoke. The low oil indicator means the car cannot reach all floors, but it may still be able to reach the fire recall level; it should make the attempt. A car with low oil or only emergency lowering will not be useful to the firemen, after Phase I recall, and may sink away from the floor, so once the passengers are out at the recall level the doors should close.

Once firefighters are in the car on Phase II, it should give them as much control of the cars as possible. Firefighters are provided with a visual indication [see 3.27.3(c)] when an elevator should not be used.

SECTION 3.28
LAYOUT DATA

3.28.1 Information Required on Layout Drawing

All the necessary layout data has been consolidated in this requirement. The intent is to require that the elevator company provide the building designer with the necessary forces that the elevator will impart on the building and other needed data for the designer. With this information the authority having jurisdiction is also in a position to confirm that the elevator is designed to conform with the requirements of the Code.

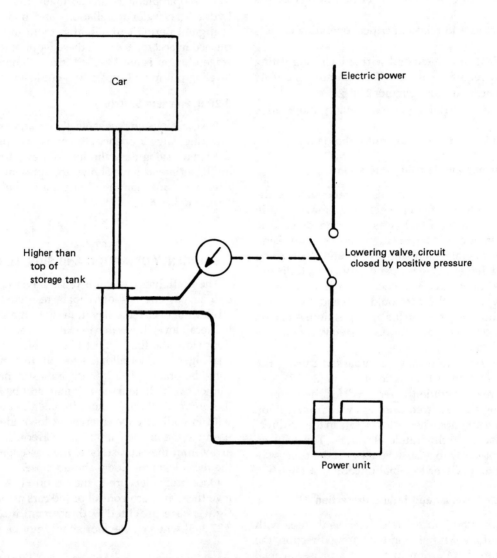

Diagram 3.26.8 Pressure Switch
(Courtesy W. Banister)

PART 4
ELEVATORS WITH OTHER TYPES OF DRIVING MACHINES

SECTION 4.1
RACK AND PINION ELEVATORS

NOTE: Throughout Section 4.1, references are made to other requirements in this Code. To gain a complete understanding of a requirement, the reader should review the Handbook commentary on all referenced requirements.

This Part covers rack and pinion elevators with the driving machine located on the car with one or more fixed racks and with one or more rotating pinions. Rack and pinion elevators conforming to the requirements in this Section can be used for passenger and freight service. See Diagram 4.1.

4.1.2 Machinery Rooms and Machinery Spaces

Rack and pinion driving machines are located on the car, in the hoistway, and these requirements cover those conditions. If a separate machine room is provided, it must conform to the requirements in 2.7.

4.1.3 Equipment in Hoistways or Machine Rooms

The requirements are similar to those for electric elevators except that the installation of the main electrical feeder inside the hoistway is permitted. This is necessary since the driving machine is located on the car. Design requirements of rack and pinion machines necessitate that the drive assembly be located on the car, usually requiring a 480 V main electrical feeder to the motor. This requirement has been an industry standard reflected in the ANSI A10.4, Safety Code for Personnel Hoists, since 1973.

4.1.4 Supports and Foundations

These requirements are similar to those for electric elevators.

4.1.5 Emergency Doors

These requirements are similar to those in 2.11.1.

4.1.6 Car Enclosures, Car Doors and Gates, and Car Illumination

The requirements are the same as for an electric elevator.

4.1.7 Car Frames and Platforms

The requirements are the same as for an electric elevator.

4.1.8 Capacity and Loading

The requirements are the same as for an electric elevator.

4.1.9 Car Safeties and Speed Governor

The basic requirements are equivalent to those for an electric elevator. However, it does allow rack and pinion safeties and governors to be an integral unit mounted on the car. The stopping distances in Table 4.1.9.1 are derived from combining the requirements in 2.17.3 and 2.17.11. Most manufacturers require that integral rack and pinion governor/safety system be returned to the factory periodically for rebuilding. This typically is every 5 years. This is in addition to the periodic inspection and test required by 8.11.5.11. See Diagram 4.1.9.

4.1.10 Counterweights

Counterweights are not required, but if used, they must conform to the same requirements as an electric elevator.

4.1.11 Car Buffers

The requirements are the same as for electric elevators except the rotational kinetic energy must be added to the translative kinetic energy of the car.

4.1.12 Guide Rails, Guide-Rail Supports, and Fastenings

See Diagram 4.1.12.

4.1.13 Rack and Pinion Driving Machine

The safety requirements for the driving machine and the rack are unique to rack and pinion elevators. However, many of the requirements parallel those for other types of elevators. See Diagrams 4.1.13(a) and 4.1.13(b).

4.1.14 Terminal Stopping Devices

4.1.14.1 Normal and Final Terminal Stopping Devices.
The requirements are basically the same as those for an electric elevator. Typically, the limit switches are on the car as illustrated in Diagram 4.1.14.1.

4.1.15 Operating Devices and Control Equipment

4.1.15.1 Applicable Requirements.
The requirements are generally the same as for an electric elevator with exceptions to those, which are not applicable due to the design of the equipment.

255

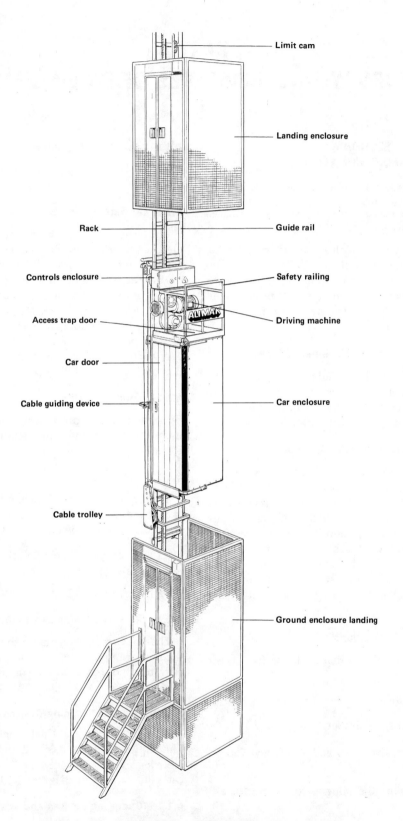

Limit cam

Landing enclosure

Rack

Guide rail

Controls enclosure

Safety railing

Access trap door

Driving machine

Car door

Cable guiding device

Car enclosure

Cable trolley

Ground enclosure landing

Diagram 4.1   Typical Rack and Pinion Elevator Arrangement
(Courtesy Alimak, Inc.)

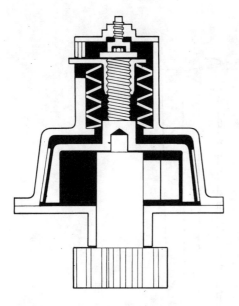

Released Position

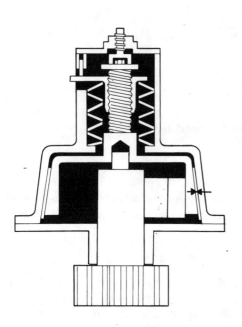

Applied Position

Diagram 4.1.9   Typical Rack and Pinion Safety
(Courtesy Alimak, Inc.)

4.1.15.2 Permitted Voltage. Control and operating circuit voltages are not permitted above 300 V. See also Handbook commentary on 4.1.3.

4.1.16 Emergency Operation and Signal Devices

The requirements are equivalent to those for electric elevators.

4.1.17 Information Required on Layout Drawings

The requirements are the same as for electric elevators except for two additional requirements, which address needs unique to rack and pinion elevators.

SECTION 4.2
SCREW-COLUMN ELEVATORS

NOTE: Throughout Section 4.2, references are made to other requirements in this Code. To gain a complete understanding of a requirement, the reader should review the Handbook commentary on all referenced requirements.

Prior to the 1983 Supplement to the ASME A17.1 Code, screw machines were covered by Rule 208.9.

4.2.1 Hoistways, Hoistway Enclosures, and Related Construction

Fire-resistive hoistway construction is typically required by the building code for all screw-column elevators except for:

(a) elevators that are entirely within one story or which pierce no solid floors; and

(b) observation elevators, which are located adjacent to a building wall without penetrating the fire-resistive areas of the building. The requirements for the following components, construction, or operations are the same as for electric elevators: guards for exposed auxiliary equipment, pits, protection of hoistway landing openings, hoistway door locking devices, car door or gate electric contacts, hoistway access switches, elevator parking devices, power operation, power opening, power closing of hoistway doors and car doors or gates.

4.2.2 Vertical Clearance and Runby for Cars

The bottom and top clearance requirements ensure that there will be no damage to the elevator equipment or the building hoistway construction in the event the elevator should malfunction resulting in overtravel in either the up or down direction. The bottom and top clearances also provide sufficient space for a maintenance mechanic or inspector to take refuge and avoid injury if trapped either on top of the car or in the pit when the car reaches its extreme limit of travel.

4.2.3 Horizontal Car Clearance

The requirements for horizontal car clearance in relation to the hoistway enclosure are the same as for electric elevators. Since screw-column elevators have no counterweights, the clearance requirements are the same for all sides not used for loading and unloading.

4.2.4 Protection of Spaces Below Hoistway

In the event the elevator is located above a space to which persons have access, the supporting structure for the screw column, guide rails, and buffers are required

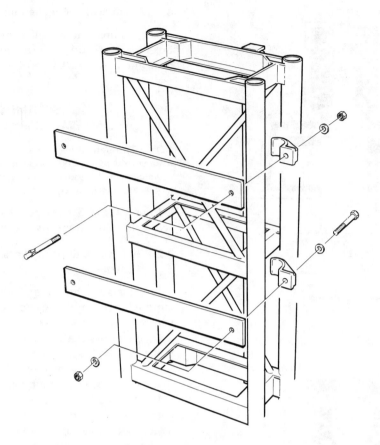

Diagram 4.1.12   Typical Guide Rail Support Tower
(Courtesy Alimak, Inc.)

to be strong enough to withstand the most severe impact that might occur due to a failure of the driving nut, a safety application, or a buffer engagement by the car while carrying rated load at maximum down speed. The structure must withstand this impact without permanent deformation, thus eliminating hazards to persons who might be underneath the elevator at the time such an impact occurs.

4.2.5 Machine Rooms and Machinery Spaces

Depending on the design of the screw-column elevator, the screw machine may be located in a machinery space within the hoistway, on the elevator car, in the elevator pit, or in a separate machinery room or machinery space. Where the screw machine is located in the pit, the normal means of access to the pit is considered adequate. See Diagram 4.2.5.

4.2.6 Equipment in Hoistways and Machine Rooms

Main line feeders are permitted in the hoistway. However, there must be no intermediate access to the conductors between the disconnecting means and the termination at the motor or controller. This arrangement

eliminates any need for other building personnel to enter the hoistway to inspect, service, or alter any electrical connections.

4.2.7 Supports and Foundations

The design of the supports and foundations is required to be in accord with the building code (see 1.3). Since the elevator may transmit impact loading into its supports and the building structure, the maximum normal loading is doubled when calculating the unit stresses, to compensate for impact and accelerating stresses.

4.2.8 Car Enclosures, Car Doors and Gates, and Car Illumination

The requirements for car enclosures, car doors and gates, and car illumination are identical to those for electric elevators.

4.2.9 Car Frames and Platforms

The requirements for car frames and platforms are substantially the same as for electric elevators.

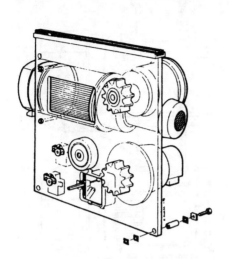

Simple Drive Unit

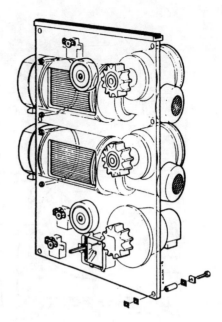

Double Drive Unit

GENERAL NOTE: A drive unit is a compact unit with one or two pinions engaging into the rack of the mast. The pinion is mounted on a key joint and fitted to the secondary shaft of the worm gear, which is driven by a direct started, squirrel-cage induction motor with built-in electromagnetic disc brake.

Diagram 4.1.13(a)   Typical Rack and Pinion Driving Machine
(Courtesy Alimak, Inc.)

4.2.10 Capacity and Loading

The capacity and loading requirements are identical to those for electric elevators.

4.2.11 Car Safeties and Speed Governor

The requirements for the car safeties and speed governors are identical to those for electric elevators, except that either one of two specified alternate safety devices may be used for slow speed (not exceeding 0.37 m/s or 75 ft/min) screw-column elevators that are driven by a substantially constant speed alternating current motor. The alternate safety devices provide safety either by:

(a) limiting the down speed of the elevator car with rated load to 0.87 m/s or 175 ft/min in the event of failure of the driving means; or

(b) limiting the fall to a distance not exceeding 13 mm or ½ in. in the event of failure of the driving nut.

Since these alternate safety devices cannot be readily tested in the field, engineering-type tests of such devices are required before use in the field.

4.2.12 Safety Nut and Data Tag

A safety nut is required as a backup to the drive nut on all screw machines, which use a drive nut made of material other than metal. Further, a safety nut may be used on all screw machines. The safety nut travels parallel with the driving nut but is not subject to any load

and thus has no wear. The metal data tag provides a record at the machine as to when the driving nut and safety nut were installed. It also records the initial spacing between the driving and safety nuts. The inspector can get an indication of when the driving nut is beginning to wear by determining any reduction from the original spacing between the two nuts.

4.2.13 Car Buffers

Solid bumpers are permitted on cars having a maximum down speed of 0.25 m/s or 50 ft/min or less provided with a safety nut. The option of providing solid bumpers is not permitted on elevators equipped with a speed limiting device or a conventional governor and safety. Either of these safety devices might allow the car to approach the lower limit of travel at up to 0.87 m/s or 175 ft/min in the event of failure of the driving nut, a speed considered excessive for solid bumpers. The option for the use of solid bumpers parallels the requirements for hydraulic elevators.

4.2.14 Guide Rails, Guide-Rail Supports, and Fastenings

The requirements for guide rails, guide-rail supports, and fastenings are substantially the same as for electric elevators.

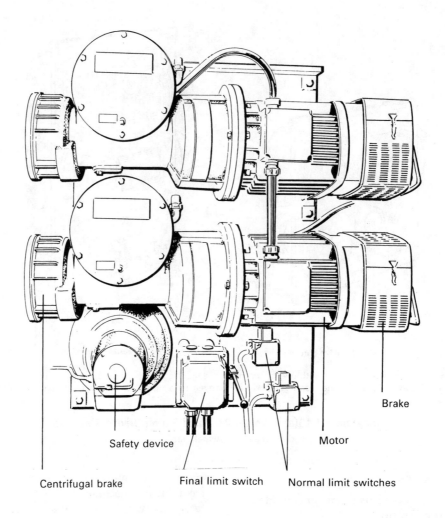

Rack and Pinion Driving Machine

Diagram 4.1.13(b)   Typical Rack and Pinion Driving Machine
(Courtesy Alimak, Inc.)

4.2.15 Driving Machine and Screw Column

The elevator car is directly supported by the screw column, which may be either in tension or compression. The design formulas for screw columns in compression are similar to those for the plunger of a hydraulic elevator. Design guidance is included for screw columns in tension to ensure a safe design.

Means are required to permit the release of passengers who may be trapped in the elevator car because of a power failure. For elevators available to the general public, either manual means to move the car from a position outside the car or standby emergency power is required. For elevators not available for use by the general public, i.e., private residence elevators and special purpose personnel elevators, means for moving the car may be the same as those for elevators available to the general public or, as an exception, such elevators may be provided only with means to permit the trapped passenger to move the car to a landing manually. Should the trapped passenger be handicapped and unable to move the car to a landing, the car emergency signal devices are available to summon help. See Diagram 4.2.15.

4.2.16 Terminal Stopping Devices

4.2.16.1 Normal Terminal Stopping Devices. The requirements for normal terminal stopping devices are identical to those for electric elevators.

4.2.16.2 Final Terminal Stopping Devices. Final terminal stopping devices identical to those for electric elevators are required for screw-column elevators having a rated speed exceeding 0.50 m/s or 100 ft/min. Screw-column elevators having a rated speed of 0.50 m/s or 100 ft/min or less have less potential for severe impact in the event of overtravel at a terminal, and thus final

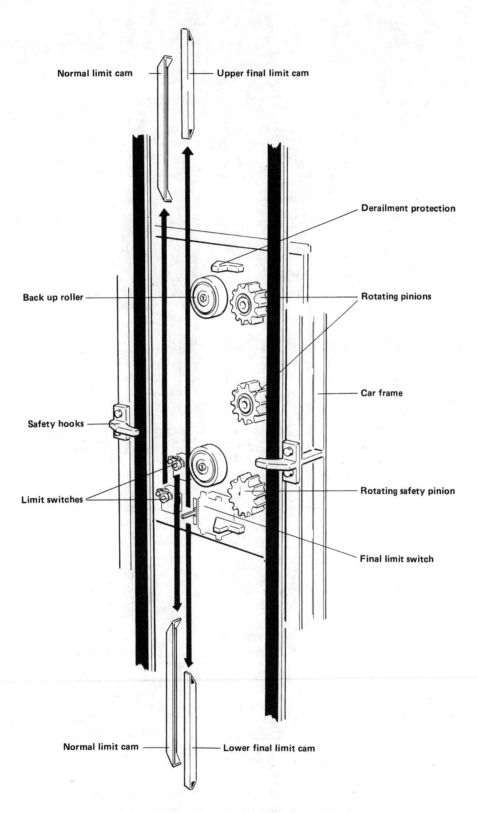

Diagram 4.1.14.1 Typical Terminal Stopping Device Arrangement
(Courtesy Alimak, Inc.)

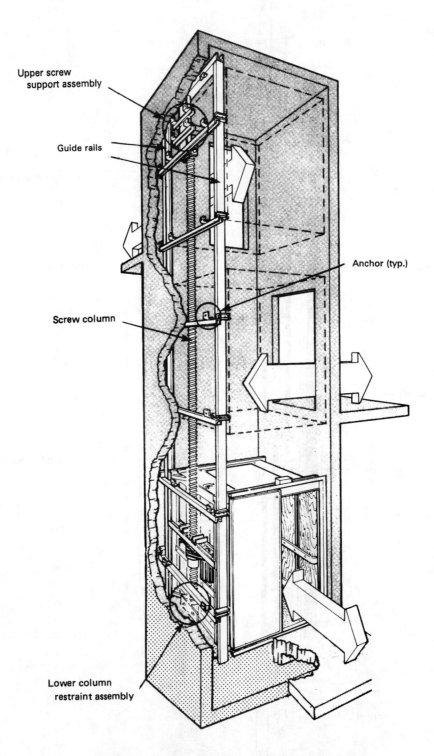

Diagram 4.2.5   Typical Screw-Column Elevator Schematic
(Courtesy Urban Mass Transportation Administration)

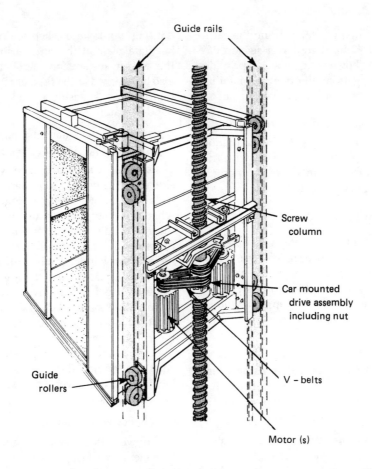

Diagram 4.2.15 Detail of Car-Mounted Drive Assembly, Nut, and Screw Column
(Courtesy Urban Mass Transportation Administration)

terminal stopping devices are not required. Instead, the elevator car must be brought to a mechanical stop without damage in the event of overtravel at either terminal due to a malfunction.

4.2.16.3 Emergency Terminal Speed Limiting Devices. The requirements for emergency terminal speed limiting devices, where utilized, are identical to those for electric elevators.

4.2.17 Operating Devices and Control Equipment

4.2.17.1 Applicable Requirements. The listed requirements for operating devices and control equipment are identical to those for electric elevators.

4.2.18 Emergency Operation and Signal Devices

The requirements for emergency operation and signal devices are identical to those for electric elevators.

4.2.19 Layout Data

The layout information required for electric elevators is supplemented by information unique to screw-column elevators. The complete layout information will permit the authority having jurisdiction to check the structural adequacy of the building to support the elevator, prior to approval for installation.

4.2.20 Welding

The requirements for welding are identical to those for electric elevators.

SECTION 4.3
HAND ELEVATORS

NOTE: Throughout Section 4.3, references are made to other requirements in this Code. To gain a complete understanding of a requirement, the reader should review the Handbook commentary on all referenced requirements.

A hand elevator is an elevator using manual energy to move the car. New installations of hand elevators are very rare, and these are usually installed in private residences or where unusual circumstances exist. Few companies, if any, are currently manufacturing hand elevators.

4.3.6 Hoistway Entrances

Hoistway entrances cannot be locked closed when the car is not at the landing, as the lifting rope must be accessible from the landing in order to bring the car to same. The 1 070 mm or 42 in. high, lower section of a dutch door [4.3.6.1(b)] or hoistway gate (4.3.7) provides protection at the opening from the danger of falling into the hoistway when operating the elevator from the landing.

If the hoistway is required by the building code (see 2.1.1.1) to be enclosed with fire-resistive construction, then all hoistway openings must be provided with fire-resistive protective assemblies with an hourly rating as specified in the building code (see 1.3).

4.3.7 Hoistway Gates for Landing Openings

This requirement assures that the gate in the extended position cannot be forced towards the hoistway and thus interfere with the running clearances of the car. This requirement applies when a single-section swinging door is provided, where as 4.3.6.1(b) applies when a two-section (Dutch type) swinging door is provided.

4.3.16 Suspension Means

4.3.16.4 Securing of Drum Ends and Turns on Drum. This requirement assures that the rope will not be pulled off the drum when the car or counterweight reaches its maximum overtravel in the down direction.

4.3.16.5 Suspension-Member Data. The suspension-member data plate is installed by the manufacturer to advise maintenance and inspection personnel of the design characteristics of the suspension members. The

data tag should be checked against the suspension-member data plate at the time of inspection to ensure that the proper suspension member is installed. It should also be checked at the time of replacement to assure that the proper suspension member is being placed on the hand elevator.

4.3.18 Guide Rails and Fastenings

4.3.18.3 Extension of Guide Rails at Top and Bottom of Hoistway. Car and counterweight guide rails must extend at the bottom to a point that the car or counterweight will not become disengaged from its guide rail when it is resting on its fully compressed buffer(s). The car guide rails must extend to the top of the hoistway to a point that the car will not become disengaged when the counterweight is on its fully compressed buffer plus a distance equal to one-half the gravity stopping distance calculated using the governor tripping speed when spring buffers are used. The counterweight guide rails must extend to the top of the hoistway to a point that the counterweight will not become disengaged when the car is on its fully compressed buffer plus a distance equal to one-half the gravity stopping distance calculated using the governor tripping speed when spring buffers are used.

4.3.20 Power Attachments

If power attachments are added to a hand elevator, that elevator must meet all of the applicable requirements for electric elevator (Part 2) or hydraulic elevator (Part 3) of the ASME A17.1 Code.

PART 5
SPECIAL APPLICATION ELEVATORS

SECTION 5.1
INCLINED ELEVATORS

NOTE: Throughout Section 5.1, references are made to other requirements in this Code. To gain a complete understanding of a requirement, the reader should review the Handbook commentary on all referenced requirements.

An inclined elevator is defined as an elevator, which travels at an angle of inclination of 70 deg or less from the horizontal. See 5.14 for private residence inclined elevators.

5.1.2 Construction of Hoistway and Hoistway Enclosures

5.1.2.2 Non-Fire-Resistive Enclosures. These requirements apply where non-fire-resistive construction is permitted by the building code. An example is the condition where an inclined elevator is located between two escalators and the perimeter walls surrounding the escalator meet the fire-resistive construction requirements. However, hoistway guarding would be required between the inclined elevator and the escalator. Examples include areas where people may come in contact with the elevator, such as a passageway, stairway, elevator landing, walkway, etc.; it must be enclosed as specified by the requirements. See Diagrams 5.1.2.2(a), 5.1.2.2(b), and 5.1.2.2(c).

5.1.3 Pits and Working Spaces

5.1.3.1 Work Space Dimensions. See Diagram 5.1.3.1.

5.1.5 Clearances for Cars and Counterweights

5.1.5.1 Bottom Car Clearances. The purpose of this requirement is to provide refuge space below the car even when it is resting on its fully compressed buffer. Bottom car clearance is defined as the clear vertical distance from the pit floor to the lowest structural or mechanical part, equipment, or device installed beneath the car platform except guide shoes or rollers, safety device assemblies, and platform aprons or guards when the car rests on its fully compressed buffer.

5.1.5.2 Top Car Clearance for Uncounterweighted Inclined Elevators. Top car clearance is defined as the shortest vertical distance between the top of the car crosshead or between the top of the car when no crosshead is provided and the nearest part of the overhead structure or any other obstruction when the car floor is level with the top terminal landing.

5.1.6 Protection of Spaces in Line With the Direction of Travel

Requirements provide for the angular direction of travel.

5.1.8 Protection of Hoistway Openings

5.1.8.2 Landing Sill Guards. See Diagram 5.1.8.2.

5.1.9 Restricted Opening of Hoistway or Car Doors

The maximum unlocking zone is reduced due to the horizontal movement of the car. Note that the maximum unlocking zone is measured in the direction of travel.

5.1.10 Access to Hoistways for Inspection, Maintenance, and Repairs

5.1.10.3 Special Operating Requirements

5.1.10.3.1 Slower speed is required due to the horizontal component of travel.

5.1.11 Car Enclosures

5.1.11.1 Top Emergency Exits. Above 49 deg, it is considered safe to access the car top from a landing opening. At 49 deg and below, a top emergency exit is not required. Emergency exiting is taken into account by the requirements of 5.1.11.1.2 and 5.1.11.1.3.

5.1.11.1.2(e) The uphill emergency exit is the preferred location for inspection of the hoistway.

5.1.12 Car Frames and Platforms

5.1.12.2 Platform Guards (Aprons). The purpose of the guard is to reduce the shearing action at the edge of the car sill and to block the opening from the landing to reduce the possibility of falling into the hoist.

5.1.13 Capacity and Loading

5.1.13.1 Benches or Seats. This provision allows the installation of seats without restricting standing capacity.

5.1.16 Suspension Ropes and Their Connections

5.1.16.1 Protection of Ropes. See Diagram 5.1.16.1.

5.1.17 Car and Counterweight Buffers

The basis of the tables and formulas for buffer strokes and safety stopping distances was the preliminary acceptance of $\frac{1}{2}g$ by enforcing authorities familiar with

Diagram 5.1.2.2(a)   Hoistway Enclosure Non-Fire-Resistive Construction

Diagram 5.1.2.2(b)   Hoistway Without Enclosure

inclined elevators. Additionally, reference was made to a study performed some years ago in Japan regarding braking on passenger trains.

Allowable maximum acceleration and deceleration during normal operation is not addressed in this Code. Each manufacturer should evaluate their control system as is done with vertical elevators and then set a standard that would be acceptable using $\frac{1}{2}g$ as a maximum.

5.1.18 Car and Counterweight Guide Rails, Guide-Rail Supports, and Fastenings

5.1.18.1 Guide-Rail Section. See Diagram 5.1.18.1.

5.1.18.4 Safety Guide Rail. On many inclined elevators using other than "T" rails for main guides, the historical method for application of the safety is to use a centrally located "T" rail.

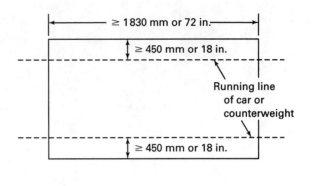

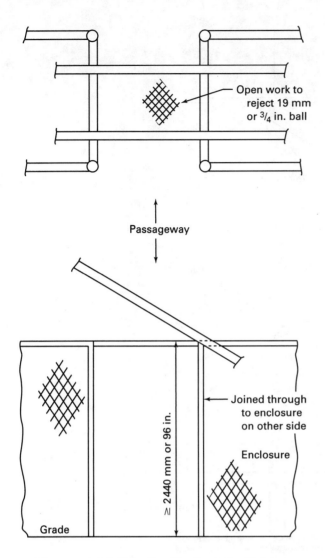

Diagram 5.1.2.2(c) Non-Fire-Resistive Hoistway Enclosure Details

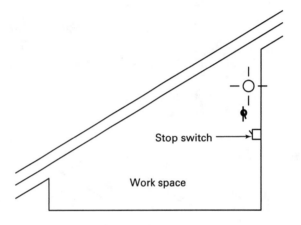

Diagram 5.1.3.1 Work Space Dimensions

5.1.22 End-Loading Incline Elevators

5.1.22.2 Speed. Slow speeds are required due to the combined horizontal/vertical movement of the car.

5.1.22.3 Buffers. The referenced requirement did not accomplish the intent. As these buffers are directional and not intended to stop all operation, the requirements in 5.1.22.3 were developed.

SECTION 5.2
LIMITED-USE/LIMITED-APPLICATION ELEVATORS

NOTE: Throughout Section 5.2, references are made to other requirements in this Code. To gain a complete understanding of a requirement, the reader should review the Handbook commentary on all referenced requirements.

A limited-use/limited-application (LU/LA) elevator is intended to provide vertical transportation for people with physical disabilities. The Code restrictions on size, capacity, speed, and rise will regulate where this type of elevator is permitted.

At this time, when the commentary is being written, a LU/LA elevator does not comply with the requirements in ADAAG. ADAAG Section 4.1.3 states that in new construction if a building or facility is eligible for the exemption for providing an elevator and one is nevertheless planned, that elevator shall meet all the requirements of ADAAG Section 4.10. A LU/LA elevator cannot meet the minimum size required by ADAAG Section 4.10.9. Presumably, a LU/LA elevator could be provided in an existing building eligible for the elevator exemption specified in ADAAG Section 4.1.3. A LU/LA elevator can also be provided in buildings not subjected to the requirements in ADA as an example churches, lodges, etc. In those occupancies the building code should be reviewed. If the building code requires the elevator to comply with ICC/ANSI A117.1 then a LU/LA elevator must comply with A117.1, Section 407.4 in addition to the requirements in ASME A17.1, Section 5.2.

A proposal has been submitted and tentatively accepted by the Architectural Transportation Barriers Compliance Board (ATBCB) to specify accessibility requirements for LU/LA elevators in the next edition of ADAAG.

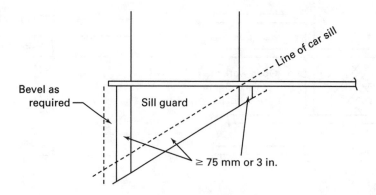

Diagram 5.1.8.2   Landing-Sill Guards

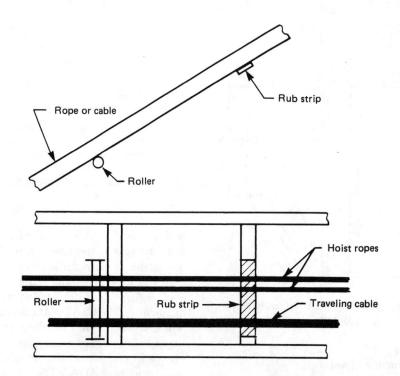

Diagram 5.1.16.1   Example of Protection of Ropes and Traveling Cable

Said requirements are the same as in ICC/ANSI A117.1–1998, except their use would be permitted only in a building where an elevator is not required by ADA. LU/LA elevators complying with ICC/ANSI A117.1 require inside car dimensions sufficient to carry, but not turn around in a wheelchair. The car door is required to be on the narrow side of the enclosure. Power-0perated swinging doors are permitted. They shall be low energy power operated complying with ANSI/BHMA A156.19 and shall be required to remain open a minimum of 20 s.

The reader would be well advised to check the Federal and Local regulations before providing a LU/LA elevator, to meet accessibility requirements.

5.2.1 Electric Limited-Use/Limited-Application Elevators

5.2.1.1 Construction of Hoistway and Hoistway Enclosures. Due to the differences in intended use and code requirements a LU/LA elevator is not permitted in a hoistway with other elevators. A separate machine room is not envisioned for a LU/LA elevator. The floor loading in 2.1.3.3 is excessive for the light machines used for LU/LA elevators. The requirement also recognizes that wood floors may be throughout the building including the machine room.

5.2.1.3 Location and Guarding of Counterweights. The use of warning chains in lieu of counterweight guards is acceptable due to the slow speed of the elevator.

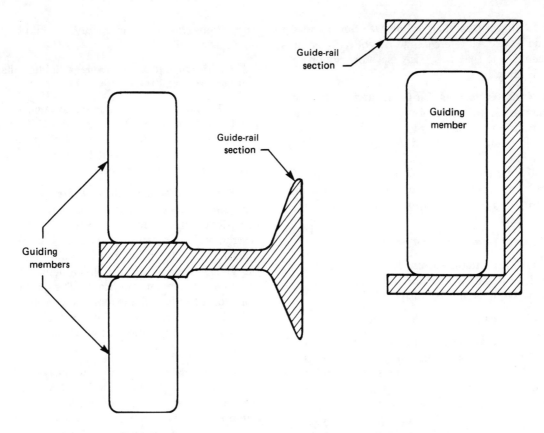

Diagram 5.1.18.1   Examples of Typical Guide-Rail Sections

5.2.1.4 Vertical Clearances and Runbys for Cars and Counterweights. It is assumed that persons will have to be on top of the car and in the pit for maintenance and inspections. Requirements 5.2.1.4.2 and 5.2.1.4.4 allow the use of an alternative mechanical means to provide a safe area of refuge on top of the car and in the pit. A key provision in both requirements is that this means be capable of being activated without complete bodily entry into the hoistway or pit.

5.2.1.7 Machine Rooms and Machinery Spaces. To control the location of mechanical equipment. Electrical equipment location and guarding is subject to the requirements of ANSI/NFPA 70 or CSA C22.1.

5.2.1.7.1 A separate machinery space is not necessary.

5.2.1.7.2 Space where equipment can be utilized will not always permit construction of 7-ft high machine rooms. Equipment will be small and compact, such that reduced ceiling height is considered adequate.

5.2.1.7.3 Necessary provisions are provided to prevent accidental contact with equipment.

5.2.1.7.4 Requirements permit safe servicing of equipment.

5.2.1.7.5 Requirements are applicable if a separate machine room is provided.

5.2.1.7.6 Requirements are applicable if a separate machine room is provided.

5.2.1.7.7 Requirements permit safe servicing of equipment in hoistway.

5.2.1.7.8 See 5.2.1.7.1. The NEC only requires 78 in. headroom.

5.2.1.7.9 Requirements are applicable if a separate machine room is provided.

5.2.1.7.10 Requirements permit safe servicing of equipment in hoistway.

5.2.1.7.12 While the equipment provided on a LU/LA elevator will be small and compact, the shape of the car platform may create a hoistway of such width or depth that all of the components within the machinery space cannot be easily accessed from outside the hoistway. These requirements provide a platform without necessitating a separate machine room.

5.2.1.11 Protection of Hoistway Landing Openings. All references to freight requirements have been deleted, as a LU/LA elevator cannot be a freight elevator.

5.2.1.12 Hoistway Door Locking Devices and Electric Contacts and Hoistway Access Switches. All references to freight requirements have been deleted, as a LU/LA elevator cannot be a freight elevator.

5.2.1.13 Power Operation of Hoistway Doors and Car Doors and Gates. All references to freight requirements have been deleted, as a LU/LA elevator cannot be a freight elevator.

5.2.1.14 Car Enclosures, Car Doors, and Car Illumination. The provision in 5.2.1.14 is subject to the requirements for manual lowering operation, similar to 5.3.1.16.2(i), being incorporated in the driving machine requirements.

Typically, the top of the car is too small to allow for the installation of a top emergency exit. This alternative allows for the rescue of trapped passengers from a stalled elevator.

5.2.1.15 Car Frames and Platforms

5.2.1.15.2 Platform Guards. This allows the use of toe guards of less than 1 200 mm or 48 in. [2.15.9.2(a)], which may be necessary because of shallow pits.

5.2.1.16 Capacity, Loading, Speed, and Rise

5.2.1.16.1 Rated Load and Platform Area. Limiting the size of the equipment, limits its use and lessens the chance that it will be misused for transporting freight. A LU/LA elevator by definition is a "passenger elevator" (1.3). Also, three standard ISO car sizes [320 kg (8.75 ft^2), 400 kg (11 ft^2), and 630 kg (15 ft^2)] would fit within the 1.67 m^2 or 18 ft^2 limitation.

5.2.1.16.5 Maximum Rise. Limiting the speed and rise will reduce the use of the elevator and allow the use of alternate provisions for some of the requirements in Parts 2 or 3.

5.2.1.20 Suspension Ropes and Their Connections

5.2.1.20.1 Suspension Means. The requirements in 5.2.1.20.1(b) are based on aircraft cable fully tested to a Military Specification and successfully utilized in aircraft. Aircraft cable has been permitted to be used for private residence elevator with a long history of displaying no safety concerns. These elevators run at speeds higher than allowed for LU/LA elevators. They also are allowed up to 15.2 m or 50 ft of rise, which is twice as high as the 7.6 m or 25 ft, of rise allowed for LU/LA elevators. Residence elevators are currently allowed up to 340 kg (750 lb) capacity. A LU/LA elevator is allowed up to 630 kg (1,400 lb) and based on the limited use/limited application, lower speed, lower rise, tight cable specification versus no specification, and information package provided should allow this cable.

5.2.1.21 Counterweight

5.2.1.21.1 Independent Car Counterweight. See Handbook commentary on 5.2.1.23.1.

5.2.1.22 Buffers and Bumpers

5.2.1.22.1 Bumpers. Forces on the car frame on impact should be a definable maximum. Substantiating

calculations from the manufacturer should be available on request.

5.2.1.23 Car and Counterweight Guide Rails, Guide-Rail Supports, and Fastenings

5.2.1.23.1 Use of Common Guide Rails. It is common practice for small elevators such as LU/LA elevators, private residence elevators, special purpose personnel elevators, and dumbwaiters to have the car and counterweight run on a common rail.

5.2.1.23.2 Guide-Rail Sections. Allows for rails with smaller section modulus and moment of inertia consistent with cars and counterweights of smaller size, weight, and rated capacity. Does not dictate the size of bolts required for fastening the guide rail to the rail brackets; however, 2.23.10.1 which defines the strength requirements remains applicable. The guide-rail deflection must comply with the requirements in 2.23. In addition the deflection must be limited during safety application such that the safeties will not disengage from the rail.

5.2.1.24 Driving Machines and Sheaves. Some of the conditions in 2.24 allow winding drum machine use beyond that generally stipulated for LU/LA elevators. Although somewhat repetitive, the conditions have been restated here so that winding drum machines are no longer allowed to exceed the general stipulation via a reference.

In addition, the condition that multiple layers of wrap are not intended has been clearly stated.

5.2.1.24.1 Types of Driving Machines. This prohibition on the use of a counterweight with a winding drum machine is a condition, which is the same as required by 2.24.1.

5.2.1.24.2 Material and Grooving

5.2.1.24.2(b) The Elevator Wire Rope Technical Board stated that 6 × 19 WSC (7 × 19) aircraft cable is made as a cross-laid strand design, and therefore should not be subjected to crushing pressure in application. Thus, no V-type or undercut grooves are permitted; only finished "U" grooves.

5.2.1.25 Terminal Stopping Devices. Requirements for emergency terminal stopping means are not applicable since the speed is limited to 0.15 m/s or 30 ft/min. Due to the limited travel of this device, a redundant final stopping device is required.

5.2.1.27 Emergency Operations and Signaling Devices. Due to the limited rise, a communication system serves no useful purpose. Until recently, elevators up to 7.6 m or 25 ft rise did not require firefighters' service; 0.15 m/s or 30 ft/min is too slow for effective elevator use by firefighters. Firefighters would use the stairs rather than take an elevator for 7.6 m or 25 ft.

5.2.1.28 Manual Operation. The top of a LU/LA elevator typically will not be large enough to provide a top emergency exit. These requirements provide an alternative means for evacuating a passenger from a stalled elevator.

5.2.2 Hydraulic Elevators

See Handbook commentary on referenced requirements.

SECTION 5.3
PRIVATE RESIDENCE ELEVATORS

NOTE: Throughout Section 5.3, references are made to other requirements in this Code. To gain a complete understanding of a requirement, the reader should review the Handbook commentary on all referenced requirements.

Section 5.3 applies to elevators installed in or at a private residence. It also applies to similar elevators installed in buildings other than private residences as a means of access to a private residence in that building, provided the elevator is installed so that it is not accessible to the general public or to other occupants of the building. A private residence is defined in this Code as "a separate dwelling or separate apartment in a multiple dwelling which is occupied by the members of a single family unit."

This Code does not permit the installation of private residence elevators in locations other than described above even if they are key operated.

The rationale for 5.3 is best described by the introduction to ASA A17.1.5-1953 when requirements for private residence elevators were first added to the Code.

"This Part of the Code has been developed in response to demands for a separate section of Rules to cover the installation in a private residence of a small electric power passenger elevator or inclined lift which serves only the members of a single family. These elevators or inclined lifts are installed primarily for the convenience of elderly persons or invalids who are unable to walk up or down stairs. In such cases, equipment of very low speed and small capacity meets the requirements. Such elevators and lifts, while usually installed in single-family private residences, may be installed within a separate apartment in a multiple dwelling.

"It is frequently necessary to install such elevators in open stairwells, as the construction of the building does not provide space to permit installing a standard type of enclosed hoistway inside the building.

"Due to their limited size, speed, load, and travel, and the fact that their use is limited to the members of a single family and is under the control of the head of the family, adequate safety of operation can be secured without requiring that such equipment meet the standards set up in other parts of this Code for equipment installed in buildings of other types which is used by the general public and is thus subjected to much more severe and frequent service. The Rules of this Part of the Code have, therefore, been developed to set up minimum standards for equipment of the type described which, if conformed to, should result in adequate safety of operation and use.

"It should be noted that the Rules of this Part of the Code do not apply to all power elevators installed in private residences, but only to those which meet the definition for Private Residence Elevator as given in this Code and which are installed in a Private Residence as defined in this Code. All other elevators in private residences, even when the latter terms conform to the definition in this Code, are required to comply with all the Rules of the other Parts of this Code."

See Diagram 5.3 of a typical traction private residence elevator.

5.3.1 Private Residence Electric Elevators

5.3.1.1 Construction of Hoistway and Hoistway Enclosures

5.3.1.1.1 The under-platform pressure device is commonly found on a private residence elevator and normally consists of sensitive switches covered by an operating surface of metal or plywood for the full width and depth of the platform. See Diagram 5.3.1.1.

5.3.1.1.3 When 90 N or 20 lbf force is placed on the cover, the springs or cushioning materials are compressed sufficiently to activate the electric contacts, which stop the car from moving in the up direction. This also can be accomplished by a lift frame with switch gear located on the car top, which will stop the upward motion of the car when the lift frame attempts to lift the hatch with the 90 N or 20 lbf. See Diagram 5.3.1.1.3.

The electric contact [5.3.1.1.3(c)] can be located on the car or on the cover supports.

5.3.1.2 Pits. A pit is normally not provided with a private residence elevator, but if one is utilized, it must be in compliance with 5.3.1.2. A pit can be created by a depression in the floor or a ramp providing access to the car.

5.3.1.3 Top Car Clearance. The top car clearance is measured between the top of the car crosshead or between the top of the car when no crosshead is provided and the nearest part of the overhead structure and any other obstruction when the car floor is level with the top terminal landing. Two examples of top car clearance are:

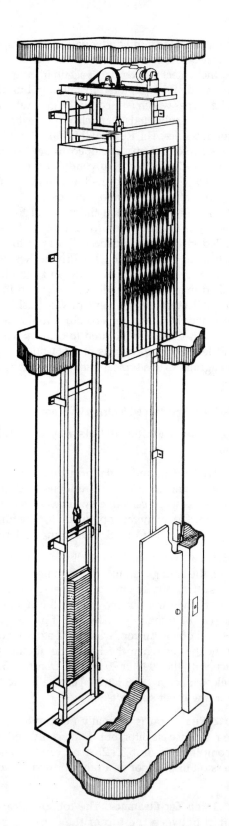

Diagram 5.3   Typical Private Residence Elevator With Traction Driving Machine
(Courtesy Sedgwick Lifts, Inc.)

(a) for a car with a rated speed of 0.13 m/s or 25 ft/min; 152 mm or 6 in.;

(b) for a car with a rated speed of 0.25 m/s or 50 ft/min; 305 mm or 12 in.

Where the machine or its controls are located on the top of the car, a refuge space is required on the car top.

5.3.1.5 Pipes in Hoistways. The user of the private residence elevator is confined within the car and would not be able to immediately retreat from the effects of a ruptured pipe; therefore, pipes are prohibited in hoistways of private residence elevators.

5.3.1.6 Guarding of Suspension Means

5.3.1.6.2 Suspension Means Immediately Adjacent to a Stairway. See Diagram 5.3.1.6.2 for an illustration of ropes or chains, which are guarded by their location within a guide or track.

5.3.1.7 Protection of Hoistway Openings

5.3.1.7.2 Clearance Between Hoistway Doors or Gates and Landing Sills and Car Doors or Gates. Diagram 5.3.1.7.2 illustrates the required clearances between the hoistway doors or gates and landing sills and car doors or gates.

5.3.1.7.4 Locking Devices for Hoistway Doors and Gates. A single contacted lock, which only detects the keeper and does not stop the car within 150 mm or 6 in. of the landing if the door lock fails to lock, does not conform to this requirement.

5.3.1.7.5 Opening of Hoistway Doors or Gates. The requirements, as illustrated in Diagram 5.3.1.7.2, ensure that a pinching hazard is not present when opening or closing the hoistway door or gate.

5.3.1.7.6 Hangers and Stops for Hoistway Sliding Doors. Door stops are required to prevent the door from traveling beyond the length of the door track.

5.3.1.8 Car Enclosures, Car Doors and Gates, and Car Illumination

5.3.1.8.1 Car Enclosures

5.3.1.8.1(d) Car Top Mounted Machine or Controller. Where the machine or controls are located on the car top, the car top must then be designed to support the weight of the workman, provide a means of operating the car from the car top, and to ensure that the machine or controls are properly guarded.

5.3.1.8.3 Light in Car. The minimum illumination level assures adequate lighting for safe ingress to and egress from the car.

5.3.1.9 Car Frames and Platforms

5.3.1.9.1 Car Frames. Cast iron is an alloy of iron, carbon, and silicon that is cast in a mold and is hard, nonmalleable, and brittle with low tensile strength. Being brittle, it will not withstand the forces applied to the car frame or platform.

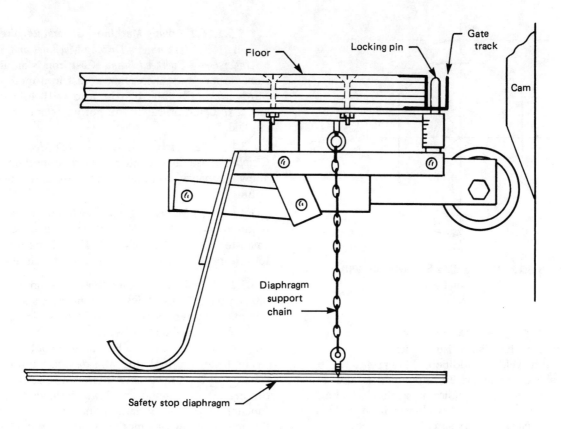

Diagram 5.3.1.1   Typical Detail of Under-Platform Pressure Switch

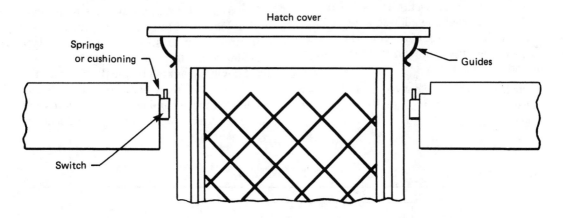

Diagram 5.3.1.1.3   Typical Detail of Hatch Cover Arrangement

5.3.1.10 Capacity, Loading, Speed, and Rise. If any of these parameters are exceeded, the installation must comply in its entirety with Parts 2, 3, or Section 4.2 of this Code.

5.3.1.10.1 Capacity. The permitted net inside platform area was increased in ASME A17.1–2000 to 1.4 m^2 or 15 ft^2. Prior to the 2000 edition, the maximum platform area was limited to 1.1 m^2 (12 ft^2). A 1.1 m^2 (12 ft^2) inside net area platform is not adequate space to accommodate today's wheelchair, especially if an attendant is present. History has shown that private residence elevators

designed for a floor loading of 195 kg/m^2 (40 lb/ft^2) is sufficient for platform sizes up to and including 1.1 m^2 (12 ft^2). When a platform area exceeds 1.1 m^2 (12 ft^2), the committee used the requirements in 2.16.1 to determine that the rated load should be based on 305 kg/m^2 or 62.5 lb/ft^2.

5.3.1.11 Safeties and Governors. See Diagrams 5.3.1.11(a) and 5.3.1.11(b) of typical private residence counterweight safety arrangements.

5.3.1.11.3 Application of Safeties. These requirements address guide-rail shapes other than T-shaped.

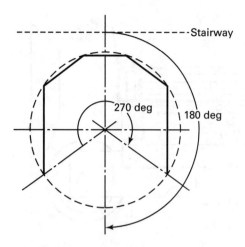

Diagram 5.3.1.6.2   Guarding by Enclosing Within Guide or Track

5.3.1.12 Suspension Means. Welded link chain is not permitted, as there is no visual indication of impending failure at a weld. In addition, there appears to be no practical way to detect an impending failure due to fatigue. Hairline cracks indicating a fatigue situation can appear at any moment. Any chain failure could result in a life threatening exposure.

5.3.1.15 Car and Counterweight Guide Rails and Guide Fastenings. Where guide rail sections other than those specified in 2.23.3 are used, the allowable deflection of the guide rail is limited by these requirements to prevent the safety device from disengaging from the rail during application with full load.

5.3.1.16 Driving Machines, Sheaves, and Their Supports. See Diagram 5.3.1.16 of typical private residence hydraulic elevator.

5.3.1.16.2 Driving Machines: General Requirements

5.3.1.16.2(c) Fastening of Driving Machines and Sheaves to Underside of Overhead Beams. Cast iron is an alloy of iron, carbon, and silicon that is cast in a mold and is hard, nonmalleable, brittle, and with low tensile strength. Being brittle, it will not withstand the forces applied.

5.3.1.16.2(e) Friction Gearing, Clutch Mechanisms, or Couplings. Use of other than a solid connection would allow a runaway condition in the case of a connection failure.

5.3.1.16.2(i) Manual Operation. Manual operation is required on private residence elevators as the car is too small for a top emergency exit. There is no other way for a trapped passenger to be removed from the car.

5.3.1.17 Terminal Stopping Devices. A normal terminal stopping device slows down and stops an elevator automatically at or near a terminal landing independent of the function of the car switch, push button, lever, or other manual device used to actuate a control.

The final terminal stopping device automatically causes the power to be removed from an electric elevator driving machine motor and brake independent of the functioning of the normal terminal stopping device or the operating device after the car has passed a terminal landing.

5.3.1.17.1 Stopping Devices Required. Alternate and independent but equivalent means for stopping a winding drum or sprocket and chain-type driving machine at the terminal landings is required. Machine limit devices operated by the driving machine are difficult to set and reset because of cable stretch directly related to capacity and loading. Pit depths and overhead space requirements in a private residence are minimal,

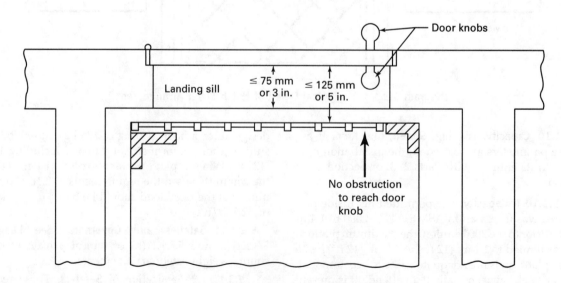

Diagram 5.3.1.7.2   Clearances at Landing Sill

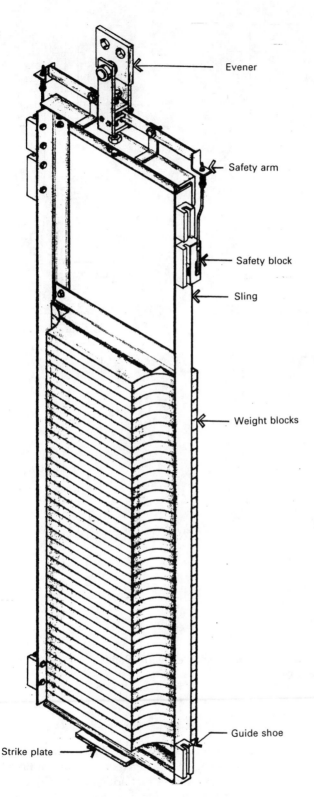

**Diagram 5.3.1.11(a)   Typical Private Residence Elevator
Counterweight With Safety**
(Courtesy Sedgwick Lifts, Inc.)

requiring the final terminal stopping devices to be precisely located.

The stopping device operated by the driving machine may be indirectly connected to the machine by a belt, rope, or chain. This type of connection has been provided on private residence elevators for more than 60 years.

5.3.1.18 Operating Devices and Control Equipment

5.3.1.18.2 Control and Operating Circuit Requirements

5.3.1.18.2(b) A tension-type spring, if broken, could cause the device to stay in the made position.

5.3.1.18.2(d) A majority of single-phase AC motors will continue to rotate in the same direction upon rapid operation of the reversing control as caused by maintaining the opposite direction control device in the made position when stopping at a landing.

5.3.1.18.4 Electrical Equipment and Wiring

5.3.1.18.4(b) The specified electrical equipment is required to be certified as complying with CSA B44.1/ASME A17.5. This will minimize the risk of electricity as a source of electric shock and as a potential ignition source of fires.

5.3.2 Private Residence Hydraulic Elevators

See Diagram 5.3.2.

5.3.2.2 Driving Machines, Sheaves, and Supports for Direct-Plunger and Roped-Hydraulic Driving Machines

5.3.2.2.2   A pressure switch is required. This device adds a means of protection should the plunger be obstructed in its descent. See also commentary on referenced requirements.

5.3.2.3 Terminal Stopping Devices. Compliance with 3.25 for the requirements for terminal stopping devices is required.

Private residence elevators are limited to a maximum speed of 0.20 m/s or 40 ft/min, thus the exception to the requirements for terminal speed reducing devices.

5.3.2.4 Anticreep Leveling Devices. The requirements are similar to those specified in 3.19.3 and 3.26.4. They have been modified and are compatible with the electrical protective devices required on a private residence elevator.

SECTION 5.4
PRIVATE RESIDENCE INCLINED ELEVATORS

NOTE: Throughout Section 5.4, references are made to other requirements in this Code. To gain a complete understanding of a requirement, the reader should review the Handbook commentary on all referenced requirements.

Section 5.4 applies to elevators installed in or at a private residence. It also applies to similar elevators installed in buildings other than private residences as a

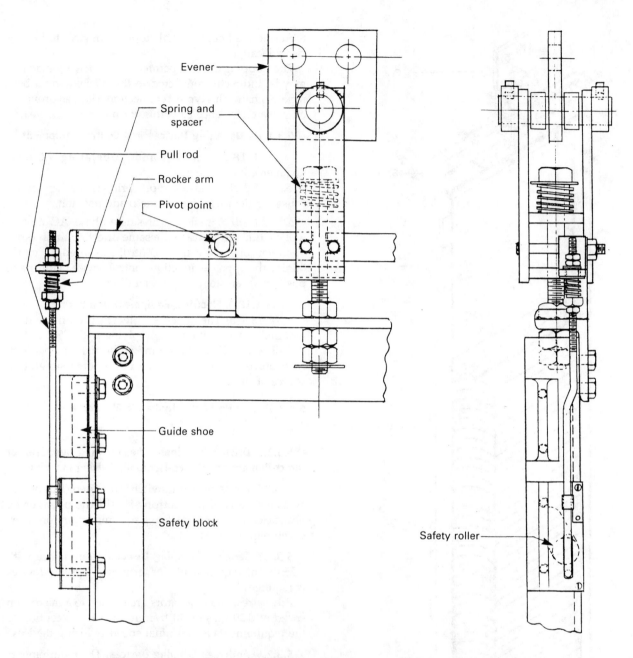

Evener

Spring and spacer

Pull rod

Rocker arm

Pivot point

Guide shoe

Safety block

Safety roller

Diagram 5.3.1.11(b) Typical Private Residence Elevator Counterweight Safety
(Courtesy Sedgwick Lifts, Inc.)

means of access to a private residence in that building provided the elevator is installed so that it is not accessible to the general public or to other occupants of the building. A private residence is defined in this Code as "a separate dwelling or separate apartment in a multiple dwelling which is occupied by the members of a single family unit."

This Code does not permit the installation of a private residence elevator in locations other than described above even if the elevator is key operated.

The rationale for 5.4 is best described by the introduction to ASA A17.1.5-1953 when requirements for Private Residence Elevators were first added to the Code.

"This Part of the Code has been developed in response to demands for a separate section of Rules to cover the installation in a private residence of a small electric power passenger elevator or inclined lift which serves only the members of a single family. These elevators or inclined lifts are installed primarily for the convenience of elderly

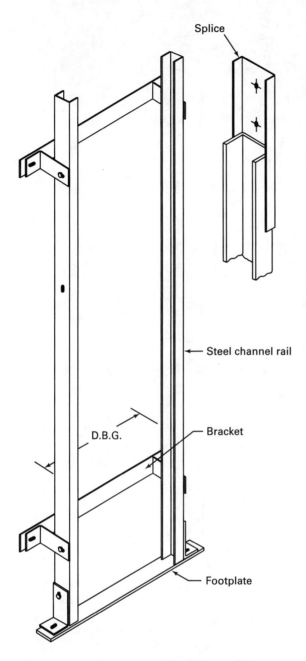

Splice

Steel channel rail

D.B.G.

Bracket

Footplate

**Diagram 5.3.1.16 Typical Private Residence Elevator
Guide-Rail System**
(Courtesy Sedgwick Lifts, Inc.)

persons or invalids who are unable to walk up or down stairs. In such cases, equipment of very low speed and small capacity meets the requirements. Such elevators and lifts, while usually installed in single-family private residences, may be installed within a separate apartment in a multiple dwelling.

"It is frequently necessary to install such elevators in open stairwells, as the construction of the building does not provide space to permit installing a standard type of enclosed hoistway inside the building.

"Due to their limited size, speed, load, and travel, and the fact that their use is limited to the members of a single family and is under the control of the head of the family, adequate safety of operation can be secured without requiring that such equipment meet the standards set up in other parts of this Code for equipment installed in buildings of other types which is used by the general public and is thus subjected to much more severe and frequent service. The Rules of this Part of the Code have, therefore, been developed to set up minimum standards for equipment of the type described which, if conformed to, should result in adequate safety of operation and use.

"It should be noted that the Rules of this Part of the Code do not apply to all power elevators installed in private residences, but only to those which meet the definition for Private Residence Elevator as given in this Code and which are installed in a Private Residence as defined in this Code. All other elevators in private residences, even when the latter terms conform to the definition in this Code, are required to comply with all the Rules of the other Parts of this Code."

5.4.1 Runway Protection

The requirements protect against, or move pinch and snag points outside the reach of passengers. The building code should also be consulted, as there may be a requirement for a fire-resistive hoistway (runway) enclosure.

5.4.2 Landing Enclosures and Gates (Where Required)

5.4.2.1 Landing Enclosures. Provides protection from falling or entering the elevator runway.

5.4.2.4 Clearance Between Hoistway Doors or Gates and Landing Sills and Car Doors or Gates. Diagram 5.3.1.7.2 illustrates the required clearances between the hoistway doors or gates and landing sills and car doors or gates.

5.4.3 Machinery Beams and Supports

5.4.3.3 Fastening of Driving Machines and Sheaves to Underside of Beams. Cast iron is an alloy of iron, carbon, and silicon that is cast in a mold and is hard, nonmalleable, and brittle with low tensile strength. Being brittle, it will not withstand the forces applied to supporting members.

5.4.4 Car Enclosures, Car Doors, and Gates

5.4.4.2 Car Doors or Gates

5.4.4.2.4 Latching of Swinging Gates. This requirement is intended to prohibit movement of the car when

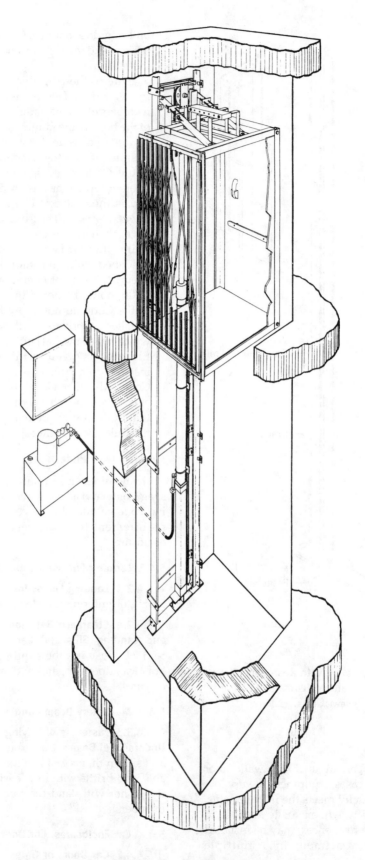

Diagram 5.3.2   Typical Hydraulic Private Residence Elevator
(Courtesy Sedgwick Lifts, Inc.)

the car door is open, as the open door would result in severe damage to the elevator.

5.4.4.3 Use of Glass, Plastics, or Acrylics. See Handbook commentary on 2.14.1.8.

5.4.6 Area, Rated Load, and Rated Speed

If any of the requirements are exceeded the entire installation must comply with all the requirements in 5.1.

5.4.8 Suspension Means

Welded link chain is not permitted, as there is no visual indication of impending failure at a weld. In addition, there appears to be no practical way to detect an impending failure due to fatigue. Hairline cracks indicating a fatigue situation can appear at any moment. Any chain failure could result in a life threatening exposure.

5.4.14 Terminal Stopping Devices

5.4.14.1 Terminal Stopping Devices. A normal terminal stopping device slows down and stops an elevator automatically at or near a terminal landing independent of the function of the car switch, push button, lever, or other manual device used to actuate a control.

The final terminal stopping device automatically causes the power to be removed from an electric elevator driving machine motor and brake independent of the functioning of the normal terminal stopping device or the operating device after the car has passed a terminal landing.

SECTION 5.5
POWER SIDEWALK ELEVATORS

NOTE: Throughout Section 5.5, references are made to other requirements in this Code. To gain a complete understanding of a requirement, the reader should review the Handbook commentary on all referenced requirements.

A sidewalk elevator is an elevator of the freight type for carrying material and operating between a landing of a sidewalk or other areas exterior to a building and floors below a sidewalk or grade elevation. Being an elevator of the freight type, it must be used primarily for carrying freight in which only the operator and persons necessary for unloading and loading the freight are permitted to ride. Passengers are not permitted to ride on a sidewalk elevator.

See Diagram 5.5.

5.5.1 Electric Sidewalk Elevators

5.5.1.1 Construction of Hoistway Enclosures. Sidewalk elevators must comply with the requirements in 2.1, except as modified by 5.5.1.1(a) through (e).

The requirements of the local building code should also be reviewed since they may have additional requirements. If the hoistway is required by the building code to be enclosed with fire-resistive construction, then all hoistway openings in that construction must be provided with fire-resistive protective assemblies with an hourly rating as specified in the building code (see 1.3, Definitions).

5.5.1.2 Pits. Due to the nature of their operation, sidewalk elevator hoistways are exposed to the weather, therefore, means must be provided to automatically remove water from the pit.

5.5.1.4 Vertical Clearances and Runbys. These requirements reflect the possible use of a car top and incorporate the necessary clearances when provided. Requirements are also included for vertical lifting covers when provided.

5.5.1.5 Horizontal Car and Counterweight Clearances. These requirements are similar to those for electric elevators with clarification of clearances where adjacent openings are provided.

5.5.1.8 Equipment in Hoistways and Machine Rooms. Sidewalk elevator hoistways and pits are subjected to water, thus the requirement to locate switches a minimum of 600 mm or 24 in. above the floor. For the same reason, the use of electrical metallic tubing (EMT) is prohibited. Water corrodes EMT quickly, allowing the wire to get wet and deteriorate, resulting in grounding.

5.5.1.11 Protection of Hoistway Landing Openings

5.5.1.11.1 Vertical Openings. Requirement 2.11.2.1 is not applicable as it applies only to passenger elevators. Hoistway entrance panels are not allowed to be perforated.

5.5.1.11.2 Horizontal Openings in Sidewalks and Other Areas Exterior to the Building. The requirements in 5.5.1.11.2(a) are intended to provide a safe area for pedestrian traffic when the elevator is at the upper landing.

The requirements in 5.5.1.11.2(c) assure that an area which could entrap pedestrians is not created.

The public is protected by requiring hinged metal doors or vertical lifting covers [5.5.1.11.2(d)]. When closed, they must be capable of safely supporting a static load of not less than 1 460 kg/m^2 or 300 lb/ft^2, which is the normal load that would be encountered with traffic moving over the door or cover. When opened, the doors or covers will form a barricade around the opening. Requirement 5.5.1.11.2(e) addresses a potential tripping hazard. Requirement 5.5.1.11.2(g) addresses the need to keep as much water as possible from entering the hoistway. Requirement 5.5.1.11.2(h) assures that doors or covers of sidewalk elevators, which might be exposed to vehicular traffic, are designed to support anticipated loads. See Diagram 5.5.1.11.2.

5.5.1.11.3 Hinged-Type Swing Sidewalk Doors. See Diagram 5.5.1.11.3.

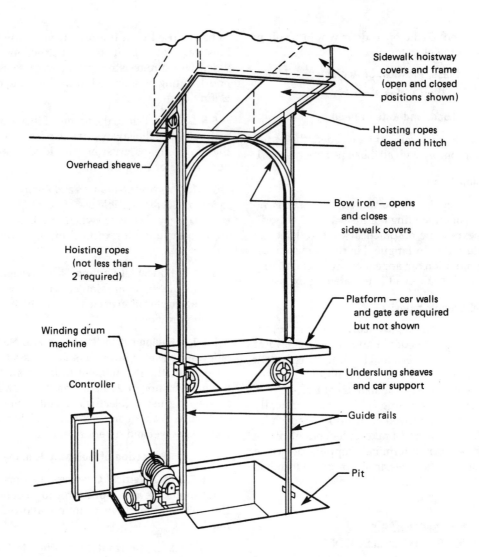

Diagram 5.5   Typical Sidewalk Elevator Layout

5.5.1.11.4 Vertical Lifting Sidewalk Covers. See Diagram 5.5.1.11.4.

5.5.1.12 Hoistway Door Locking Devices. The modifications to the requirements in 2.12 are to provisions, which only pertain to passenger elevators.

5.5.1.14 Car Enclosures, Car Doors and Gates, and Car Illumination. Car tops are permitted for sidewalk elevators subject to the limitations of this requirement and 5.5.1.4, 5.5.1.11.4, and 5.5.1.25.3, which assure the necessary safety requirements.

Lighting fixtures are usually located on the bow iron and can be damaged when it strikes the door. Requirement 5.5.1.14.3 permits the lighting to be exterior to the car, provided that the same illumination levels are provided.

5.5.1.15 Car Frames and Platforms. Requirements are included for conditions that are unique to sidewalk elevators.

5.5.1.16 Capacity and Loading. Requirements in 2.16 on carrying passengers are not referenced, as passengers are not permitted on sidewalk elevators.

5.5.1.23 Driving Machines and Sheaves. The smaller diameter has proved satisfactory for sidewalk elevators.

5.5.1.25 Operating Devices and Control Equipment

5.5.1.25.3 Top-of-Car Operating Devices and Stop Switch. Requirements reflect the safest possible use of a car top.

5.5.1.25.4 Maximum Rated Speed. When the car is fully enclosed, there is no need to limit car speed. Also, see requirements 5.5.1.25.2 and 5.5.1.25.3 for additional speed limitation requirements.

5.5.1.26 Car Emergency Signaling Devices. If the travel is less than 7.6 m or 25 ft, an audible signal must be provided, and if the building is not occupied 24 hours a day, a means of communicating with, or signaling to,

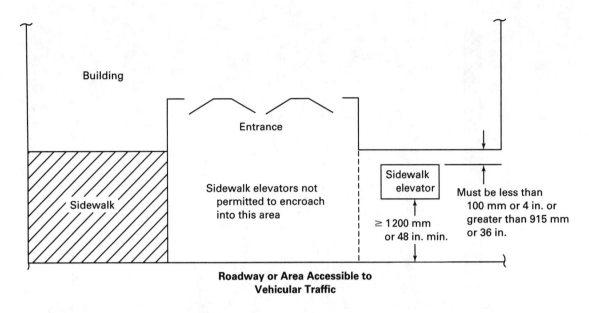

Building

Entrance

Sidewalk elevators not permitted to encroach into this area

Sidewalk

Sidewalk elevator

Must be less than 100 mm or 4 in. or greater than 915 mm or 36 in.

≥ 1200 mm or 48 in. min.

Roadway or Area Accessible to Vehicular Traffic

Diagram 5.5.1.11.2   Sidewalk Elevator Location

a service for assistance must be provided. When the travel is more than 7.6 m or 25 ft, a communication system must also be provided in the sidewalk elevator.

SECTION 5.6
ROOFTOP ELEVATORS

NOTE: Throughout Section 5.6, references are made to other requirements in this Code. To gain a complete understanding of a requirement, the reader should review the Handbook commentary on all referenced requirements.

This Section was developed in response to a request for elevators to bring passengers and freight to rooftop heliports. Where a roof is used for a heliport, the FAA places severe restrictions on obstruction such as elevator hoistways.

5.6.1 Electric Rooftop Elevators

5.6.1.2   Pits. A means must be provided to automatically remove water from a pit as the hoistway is exposed to the weather when the car is at the top landing. Also, see Handbook commentary on 5.5.1.2.

5.6.1.4   Vertical Clearances and Runbys. The requirements in 2.4 that do not apply, address clearances between the top of car and the overhead building structure. Rooftop elevators have a door in a horizontal opening at the top of the hoistway.

5.6.1.4(g) There will be no need for a person to ride on top of the car when travel is less than or 6.1 m or 20 ft, thus no need for a refuge space. The refuge space is measured when the bow iron or stanchion is in contact with the door to assure that adequate refuge space is provided.

5.6.1.8   Electrical Equipment, Wiring, Pipes, and Ducts in Hoistways and Machine Rooms. Additional requirements are included because the elevator equipment is exposed to the weather. Also, see Handbook commentary on 5.5.1.8.

5.6.1.11   Protection of Hoistway Landing Openings

5.6.1.11.2   Horizontal Openings in Rooftops. See Handbook commentary on 5.5.1.11.2.

5.6.1.11.3   Hinged-Type Rooftop Doors. See Handbook commentary on 5.5.1.11.3. Additional requirements are incorporated to address ramps provided for handicapped wheelchair and hospital gurney access to the elevator.

5.6.1.11.4   Vertical Lifting Rooftop Covers. See Handbook commentary on 5.5.1.11.4. Additional requirements are incorporated to address ramps provided for handicapped access to the elevator.

5.6.1.11.5   Setting of the Door. The intent of this requirement is to locate the roof door in such a way that any water on the roof flows away from the opening.

5.6.1.12   Hoistway Door Locking Devices and Electric Contacts and Hoistway Access Switches. See Handbook commentary on the 5.5.1.12.

5.6.1.14   Car Enclosures, Car Doors, Gates, and Car Illumination. Weatherproof electric equipment is required as the elevator is exposed to the elements when at the top landing.

5.6.1.15   Car Frames and Platforms

5.6.1.15.1   Platforms. These requirements provide for the safe operation of equipment arranged to travel

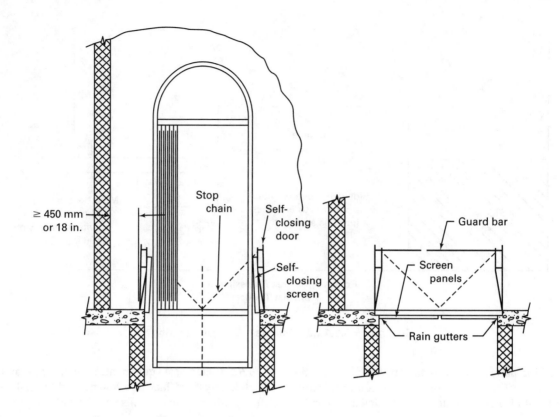

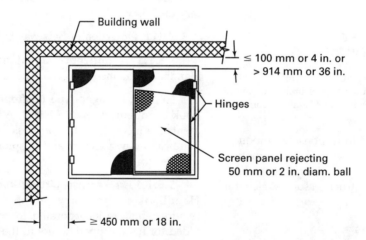

Diagram 5.5.1.11.3   Requirements for Hinged-Type Sidewalk Doors
(Courtesy Gillespie Corp.)

above the level of the roof for freight loading and unloading. See Handbook commentary on 5.5.1.15.1.

5.6.1.15.2 Bow-Irons and Stanchions. See Handbook commentary on 5.5.1.15.2.

5.6.1.17 Safeties. See Handbook commentary on 5.5.1.17.

5.6.1.23 Driving Machines and Sheaves. See Handbook commentary on 5.5.1.23.

5.6.1.24 Terminal Stopping Devices. Requirements provide protection of the equipment from the elements.

5.6.1.25 Operating Devices and Control Equipment. Requirements 5.6.1.25.1 and 5.6.1.25.2 require operation of the car by an operator located on the rooftop to ensure the roof area is clear. Requirements are added for protection of the equipment from the elements [5.6.1.25.2(c)].

5.6.1.25.4 Maximum Rated Speed. The operating speed is not limited until the bow-iron or stanchion is in contact with the door.

5.6.1.25.5 Landings Served. The equipment is limited to serving the rooftop and one floor below. In the case of a hospital with a heliport on the roof, the floor

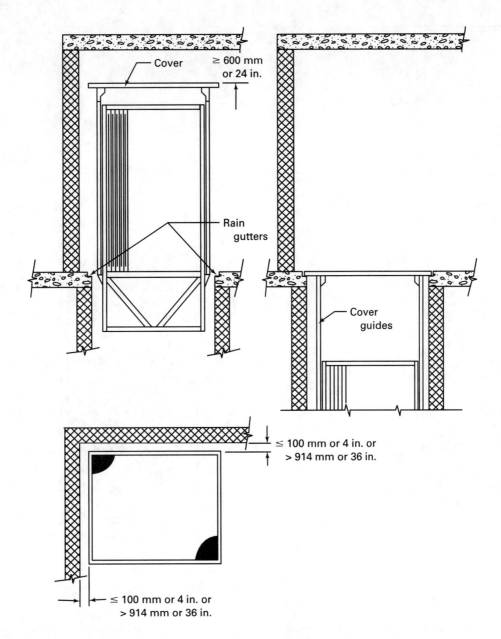

Diagram 5.5.1.11.4 Requirements for Vertical Lifting Sidewalk Covers
(Courtesy Gillespie Corp.)

below that may be served as the floor with the emergency room. Since this is a relatively new application, the requirements were written for equipment within a limited scope until more experience has been gained.

SECTION 5.7
SPECIAL PURPOSE PERSONNEL ELEVATORS

NOTE: Throughout Section 5.7, references are made to other requirements in this Code. To gain a complete understanding of a requirement, the reader should review the Handbook commentary on all referenced requirements.

Elevators covered by this Section are limited to rated loads of 454 kg or 1,000 lb or less, rated speeds of 0.76 m/s or 150 ft/min or less, and a platform area of 1.208 m² or 13 ft² or less (see 5.7.12.2). By definition, special purpose personnel elevators are to be used only by authorized personnel (1.3).

5.7.1 Construction of Hoistways and Hoistway Enclosures

5.7.1.1 Hoistways and Hoistway Enclosures. Where a hoistway is adjacent to an area where people may come in contact with the elevator, such as a passageway, stairway, elevator landing, walkway, etc., the hoistway must

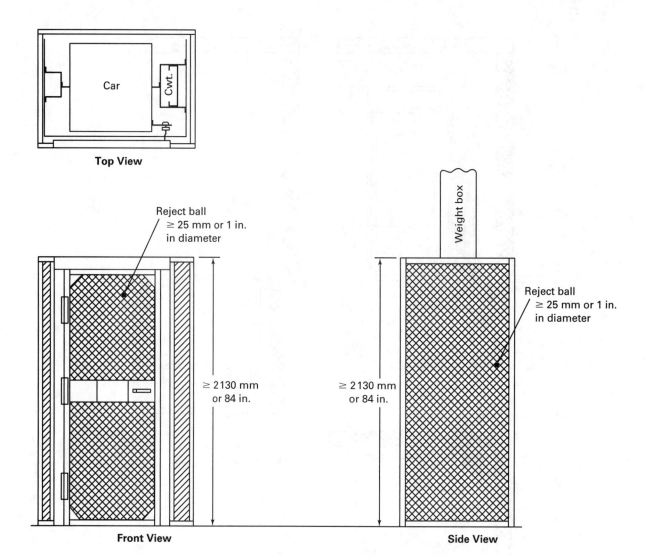

Diagram 5.7.1.1   Typical Hoistway Enclosure
(Courtesy Sidney Manufacturing Co.)

be enclosed to a height of 2 130 mm or 84 in. above the floor or stair tread. Above that height, the hoistway can be unenclosed. A landing may be comprised of nothing more than a metal grating a few feet wide that is adjacent to the hoistway. If a person is expected to be on this platform, the hoistway must be guarded at that location. The enclosure up to 2 130 mm or 84 in. may be of the openwork type as long as it rejects a ball 25 mm or 1 in. in diameter (see Diagram 5.7.1.1). The local building code or one of the following model building codes should be complied with if a fire-resistive enclosure (floor or wall) is penetrated, as it is quite likely that a fire-resistive enclosure would be necessary for the entire hoistway.

    (a) National Building Code (NBC)
    (b) Standard Building Code (SBC)
    (c) Uniform Building Code (UBC)
    (d) National Building Code of Canada (NBCC)

It is noted that the International Building Code (IBC) issued in the first quarter of 2000 is the successor code for the NBC, SBC, and UBC.

5.7.2 Pits

A pit must be provided. The Code does not require a minimum bottom car clearance, only the runby is addressed. The small cross-sectional area of the hoistway precludes providing a refuge area in the pit under a car resting on its fully compressed buffers.

5.7.3 Location and Enclosing of Counterweights

5.7.3.1 Counterweight Coming Down to Floors or Passing Floors or Stairs. A counterweight runway adjacent to a floor or stairway must be enclosed to a height of 2 130 mm or 84 in. above that floor or stairway tread by a solid or openwork enclosure. A landing may be comprised of nothing more than a metal grating a few feet wide that is adjacent to the hoistway. If a person is expected to be on this platform, the hoistway must be guarded at that location. The openwork enclosure must reject a ball 25 mm or 1 in. in diameter and be at least 100 mm or 4 in. between the outside of the enclosure and the closest member of the counterweight assembly. Above 2 130 mm or 84 in. the counterweight runway need not be enclosed. The local building code or one of the following model building codes should be consulted if the counterweight runway passes through a fire-resistive enclosure (floor or wall).

(a) National Building Code (NBC)

(b) Standard Building Code (SBC)

(c) Uniform Building Code (UBC)

(d) National Building Code of Canada (NBCC)

It is noted that the International Building Code (IBC) issued in the first quarter of 2000 is the successor code for the NBC, SBC and UBC.

5.7.3.2 Access to Enclosed Counterweights and Ropes. This requirement refers to counterweights that may be enclosed throughout their full run and to counterweights that are partially enclosed throughout a significant portion of the counterweight runway. The electrical contact on the access door assures that the elevator will not run when servicing or inspecting the counterweight. The self-locking key tumbler lock assures that only authorized personnel can get into the counterweight runway for inspection or maintenance purposes.

5.7.4 Vertical Clearances and Runby

5.7.4.2 Top Car Clearance. A rack and pinion elevator will only travel in the up position to the point where the rack teeth are terminated.

5.7.7 Overhead Machinery Beams and Supports

5.7.7.1 Securing of Machinery Beams and Type of Supports. Cast iron is an alloy of iron, carbon, and silicon that is cast in a mold and is hard, nonmalleable, and brittle. Being brittle, it will not withstand the forces applied.

5.7.8 Hoistway Doors and Gates

5.7.8.1 Where Required. Each entrance must be protected by a door or gate to its full height or 2 030 mm or 80 in.; whichever is less. Openwork entrances are permissible as long as they reject a ball 25 mm or 1 in. in diameter. The local building code should also be checked to see if a fire-resistive entrance is required.

5.7.8.3 Access to Hoistways for Emergency and Inspection Purposes. Hoistway door unlocking devices are required at the top and bottom landing for access to the hoistway for emergency, inspection, and maintenance purposes. These devices must be of a design to prevent unlocking of the door with common tools, such as with a screwdriver. They must be kept on the premises in a location readily available only to qualified or emergency personnel.

5.7.8.6 Distance Between Hoistway Doors or Gates and Landing Sills and Car Doors or Gates. Compliance assures that no one can stand between the car door or gate and hoistway door or gate, and that no one can stand on the landing sill with the hoistway door or gate closed and the car at a remote location in the hoistway.

5.7.9 Locking Devices for Hoistway Doors or Gates

There is no difference in the exposure of a special purpose elevator compared to a passenger elevator in terms of travel or number of openings. The use of mechanical locks and contacts where there is no electrical confirmation of the locking action should not be permitted except under very limited conditions. The limitations cover freight elevators not permitted to carry passengers, with manual doors, a travel not over 4 570 mm or 15 ft or for the lowest landing where the pit is not more than 1 525 mm or 60 in. from the landing sill. These limitations provide a short fall distance in the event of inadvertent access to the open hoistway. With mechanical locks and contacts, door opening is available as an elevator in flight passes a landing during the period the unlocking cam rides on the mechanical lock roller arm. In the event the locking arm does not return to the closed position after the elevator leaves the landing, the door will remain unlocked and can be opened with the car at any position. In which case, a dangerous falling hazard exists.

5.7.10 Car Enclosures

5.7.10.4 Top Emergency Exits. As this type of elevator is not used by the general public, an emergency exit is not required. If one is provided, it must conform to these requirements. The exit size must be large enough for use and the exit cover must be equipped with a switch or contact, which will not allow the car to move if the exit cover is open. A written emergency evacuation procedure is required by 8.6.10.4.

5.7.12 Capacity and Loading

5.7.12.2 Limitation of Load, Speed, and Platform Area. As this elevator is permanently installed in a wide variety of structures and locations to provide vertical transportation for authorized personnel and their equipment only, such elevators are limited in load, speed, and platform area.

The speed and capacity provisions have been increased since this type of equipment was recognized by the Code. The maximum allowed speed was increased because it is recognized that on tall structures it would be more efficient to decrease travel time. The maximum capacity was also increased. As the elevator is used to move equipment, the Committee recognized a need to provide a means of moving heavy equipment safely. It was determined the previous limit was not practical.

If these requirements cannot be met, then the elevator must conform to the applicable requirements in Part 2 or 3 or Section 4.1 or 4.2 of this Code.

5.7.13 Car Safeties and Governors

5.7.13.1 Car Safeties and Governors for Traction and Winding Drum Type Elevators. See Diagram 5.7.13.1.

5.7.13.2 Car Safeties and Governors for Rack and Pinion-Type Elevators. A rack and pinion-type elevator typically utilizes a safety device controlled by a pinion, which is permanently engaged in the rack. In the event of overspeed, a centrifugal device engages and screws a cone brake into action, bringing the car to a gradual stop. These safeties are usually sealed and require that they be returned to the manufacturer for calibration and maintenance at scheduled intervals. See 8.6.6.1 for maintenance requirements. Also, see Handbook commentary on 4.1.9.

5.7.13.3 Opening of Brake and Motor Control Circuits on Safety Application. In addition to these requirements, the speed governor should be located where it can be easily serviced.

5.7.14 Suspension Means

See Diagram 5.7.14 for examples of typical traction and winding drum roping arrangements.

5.7.14.1 Types Permitted. Only wire rope that is specifically designed and constructed for suspending elevator cars and counterweights can be used. Suspension means other than wire rope, such as chain, are prohibited.

5.7.14.5 Arrangement of Wire Ropes on Winding Drums. This requirement assures that when the elevator or counterweight has reached its limit of travel, the wire rope will not become separated from the winding drum.

5.7.14.6 Lengthening, Splicing, Repairing, or Replacing Suspension Means. The design strength of the wire rope is severely weakened when spliced. Splicing also increases the diameter of the rope at the splice. This increase in diameter would adversely affect the relationship between the sheave groove and rope.

All ropes must be replaced at one time to ensure uniform rope life. If one rope has excessive wear, the cause of the wear should be determined before replacing a full set of ropes. Replacement of the entire set of ropes when any single worn or damaged rope has to be replaced is required.

5.7.15 Counterweight Guiding and Construction

5.7.15.1 Guiding. Counterweights, if used, must be guided by the use of guide rails, which must conform to the requirements of 5.7.17.

5.7.16 Car and Counterweight Buffers

5.7.16.2 The rotational kinetic energy is additive to the translative kinetic energy of the car.

5.7.17 Car Guide Rails and Guide-Rail Fastenings

Guide rails other than the T-shaped type may be used on special purpose personnel elevators. Some other types of shapes used are:
  (a) round rail;
  (b) angle rail; and
  (c) rack rail.

5.7.18 Driving Machines and Sheaves

Winding drum machines are required to conform to the travel and speed limitations specified in 2.24.1 (see 5.7.12.2). Special purpose personnel elevators, with winding drum machines, are permitted to transport authorized personnel.

5.7.20 Operation

5.7.20.1 Types of Operation. Continuous pressure operation is operation by means of buttons or switches, in the car and at the landing, any one of which may be used to control the movement of the car as long as the button or switch is manually maintained in the actuating position.

Single automatic operation is automatic operation by means of one button in the car for each landing served and one button at each landing, so arranged that if any car or landing button has been actuated, the actuation of any other car or landing button will have no effect on the operation of the car until its response to the first button has been completed.

SECTION 5.8
SHIPBOARD ELEVATORS

NOTE: Throughout Section 5.8, references are made to other requirements in this Code. To gain a complete understanding of a requirement, the reader should review the Handbook commentary on all referenced requirements.

5.8.1 Electric Shipboard Elevators

5.8.1.1 Hoistway Enclosures. This requirement addresses the need to construct a hoistway with sufficient strength to resist being struck by a moving object as a result of ship movement. The requirement also addresses the different construction materials utilized in maritime construction. The use of an expanded metal hoistway enclosure when the hoistway is within one compartment can be compared to an observation elevator hoistway with an atrium.

5.8.1.3 Protection of Space Below Hoistway. Counterweight safeties must be provided because the area below the counterweight could be either occupiable or the ship's hull.

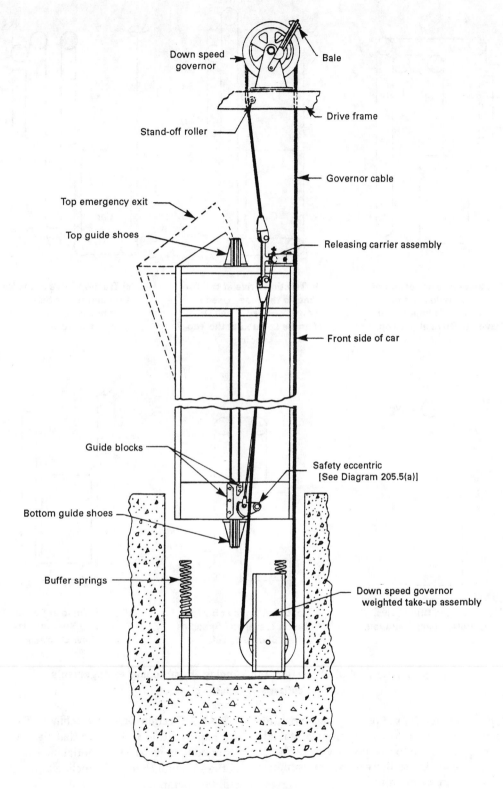

Down speed governor

Bale

Drive frame

Stand-off roller

Governor cable

Top emergency exit

Top guide shoes

Releasing carrier assembly

Front side of car

Guide blocks

Safety eccentric
[See Diagram 205.5(a)]

Bottom guide shoes

Buffer springs

Down speed governor
weighted take-up assembly

Diagram 5.7.13.1   Cutaway View Governor, Releasing Carrier, and Type A Safety
(Courtesy Sidney Manufacturing Co.)

287

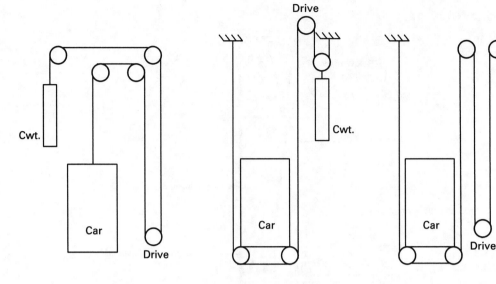

(a) Traction Drive at Bottom or Some Intermediate Point— Used Where it Is Impractical to Have the Drive at the Top.

(b) Traction Drive at the Top and to the Side. Used for Low Overhead When Drive Can Be Located at the Top Landing.

(c) Traction Drive at the Bottom and to the Side. Similar but With the Drive at the Bottom.

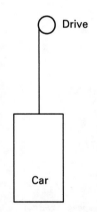

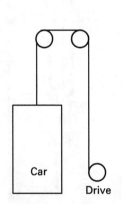

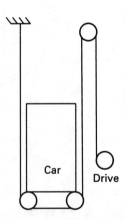

(d) Overhead Drum Drive. No Counterweight Required.

(e) Drum Drive at Bottom. Same Travel and Speed as Shown to Left.

(f) Drum Drive at Bottom With Secondary Sheaves. Used for Low Overhead.

Diagram 5.7.14   Typical Traction and Winding Drum Roping Arrangements
(Courtesy Viola Industries)

5.8.1.5 Top Emergency Exits. Special provisions are made to allow a passenger to escape from a stalled elevator in a ship using the top emergency exit. These are provided as a person in a stalled elevator in a sinking ship must be given an escape route.

5.8.1.6 Illumination of Cars. The National Electrical Code® does not cover all of the electrical requirements that are deemed necessary on a ship nor does the referenced standard IEEE-45. In addition to IEEE-45, compliance with NFPA 70 or CSA C22.1 is required.

5.8.1.7 Traction Driving Machines. These provisions are required due to the fact that the elevator may be required to move at a substantial incline. The movement of the ship will change the incline at which the elevator will be operating.

5.8.1.9 Special Conditions. These requirements, in addition to those in 2.14 through 2.28, apply to shipboard elevators. Elevators on a ship must be capable of operating safely as the ship moves.

5.8.1.10 Handrails. As the ship moves, the elevator moves. The handrail gives the passengers a means of steadying themselves.

5.8.1.11 Flooring. As the ship moves the elevator moves. Floors must be slip resistant to allow a secure foothold.

5.8.2 Hydraulic Shipboard Elevators

5.8.2.1 Storage Tanks. These requirements address the potential of an oil spill, which would be dangerous on a ship. Storage tanks must be capable of withstanding the moving environment encountered on a ship.

5.8.2.2 Special Conditions. These requirements, in addition to those in Part 3, apply to hydraulic shipboard elevators. Elevators on a ship must be capable of operating safely as the ship moves.

5.8.2.3 Handrails. As the ship moves the elevator moves. The handrail gives the passengers a means of steadying themselves.

5.8.2.4 Flooring. As the ship moves the elevator moves. Floors must be slip resistant to allow a secure foothold.

5.8.3 Rack and Pinion Shipboard Elevators

5.8.3.1 Special Conditions. These requirements, in addition to those in Part 4.1, apply to rack and pinion shipboard elevators. Elevators on a ship must be capable of operating safely as the ship moves.

5.8.3.2 Handrails. As the ship moves the elevator moves. The handrail gives the passengers a means of steadying themselves.

5.8.3.3 Flooring. As the ship moves the elevator moves. Floors must be slip resistant to allow a secure foothold.

SECTION 5.9
MINE ELEVATORS

NOTE: Throughout Section 5.9, references are made to other requirements in this Code. To gain a complete understanding of a requirement, the reader should review the Handbook commentary on all referenced requirements.

A mine elevator is an elevator installed in a mine hoistway (see 1.3, Definitions) used to provide access to a mine for personnel, materials, equipment, and supplies.

To be considered a mine elevator the elevator components must be designed and installed in conformance with Part 2, except as those requirements are modified by 5.9. Although mine elevators are very similar to high-rise, high-speed elevators, the mine environment is more hostile to the equipment. Fire and emergencies are handled differently. The equipment is also regulated by Title 30 Code of Federal Regulations. See Diagram 5.9.

5.9.1 Construction of Hoistways and Hoistway Enclosures

5.9.1(a) Fire-resistive construction and other specific construction requirements are not applicable.

5.9.1(b) Projections, recesses, and setbacks are not applicable, (see 1.3, definition of *mine hoistway*).

5.9.2 Pits

It is important to be aware of water accumulation in the elevator pit. Elevator operation is inhibited if the power to this alarm is interrupted to prevent a silent failure. In some conditions, the best solution is to position the bottom landing above the mine level and locate the pit equipment at the mine floor level.

5.9.5 Horizontal Car and Counterweight Clearance

Mine hoistways are often round or present an irregular surface (Diagram 5.9.5). Fascia plates are not practical because of high airflow and a car door interlock is provided. In the event the car runs on to the compressed buffer, entrapped miners may force the car doors open to evacuate the car. Fascia is provided at the bottom landing to prevent them from falling into the pit. Elsewhere in the hoistway, the car top exit panel would be utilized to evacuate the car. Also, see Handbook commentary on 5.9.12.

5.9.7 Machine Rooms and Machinery Spaces

Fire-resistive construction is not applicable.

5.9.8 Equipment in Hoistways and Machine Rooms

5.9.8.1 Hoistway and Car Wiring. The requirements address the conditions that may be encountered in a mine hoistway.

5.9.8.2 Sprinklers are not used in mine hoistways. Pipes, ducts, etc. must be run in the elevator hoistway, since it is usually the only vertical shaft available for this purpose.

5.9.11 Protection of Hoistway Openings

Glass doors are not suitable in a mine environment. Mine hoistways are not required to be fire resistive; therefore, fire-rated doors are not required.

5.9.12 Hoistway-Door Locking Devices and Electric Contacts, and Hoistway Access Switches

Cars are provided with a car door interlock, modified to suit the environment, conforming to 2.14.4.2. Therefore, anytime the car is outside the unlocking zone, the car doors are locked. However, entrapped miners may force the car doors open to evacuate the car. See also Handbook commentary on 5.9.5.

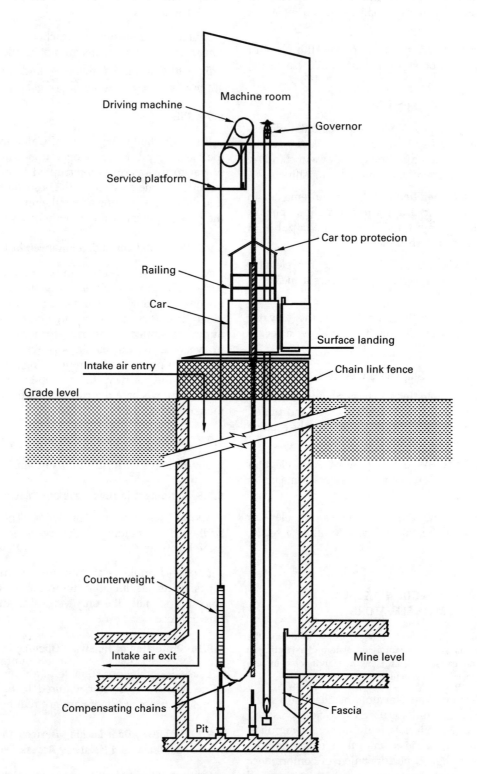

Diagram 5.9 Typical Mine Elevator Layout
(Courtesy US DOL MSHA/Thomas Barkland)

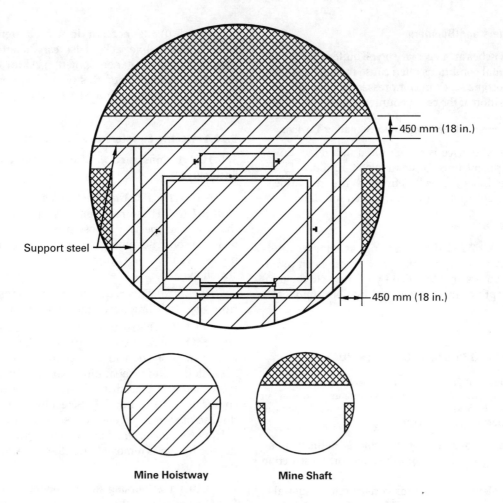

450 mm (18 in.)

Support steel

450 mm (18 in.)

Mine Hoistway   **Mine Shaft**

Diagram 5.9.5   Mine Shaft and Mine Elevator Hoistway
(Courtesy US DOL MSHA/Thomas Barkland)

5.9.14 Car Enclosures, Car Doors and Gates, and Car Illumination

Environmental conditions in a mine require frequent inspections of the hoistway to ensure that safe operating conditions exist.

Access to the car top by passengers (miners) is necessary for emergency evacuation.

Mine hoistways are often round or irregular. A car door interlock is necessary since environmental conditions make fascia plates impractical. A car door interlock prevents the car door from being opened when the car is between landings.

5.9.14.5 Miners wear cap lamps that will provide auxiliary lighting when normal car lighting fails.

5.9.15 Car Frames and Platforms

5.9.15.1 Corrosion Protection. Environmental conditions in a mine are corrosive.

5.9.17 Car and Counterweight Safeties

Mine environmental conditions such as accumulation of ice on the guide rails, or ground movement-causing misalignment of the guide rails, can cause inadvertent actuation of the counterweight safeties. Many times the rail surfaces are exposed to high humidity and rail surfaces develop a coating of "rust" that may dramatically impact safety slide distance.

5.9.17.4 Access to the elevator is not readily available to release the car safeties when the car is in the "blind" mine hoistway.

5.9.17.5 Access to the car or counterweight is not readily available to release the safeties in the "blind" hoistway.

5.9.17.6 Pit conditions, such as falling ice or debris, could foul compensating ropes or sheaves, thus making tie down impractical.

5.9.18 Speed Governors

5.9.18.1 Governor Rope Tension Sheaves. Vast changes in temperature and humidity occur in a mine hoistway. This can result in both stretching and shrinkage of the governor rope. Means must be provided to ensure the tension frame operates as intended.

5.9.22 Buffers and Bumpers

Buffer switches are necessary on oil buffers, since mine environmental conditions often cause the buffers to corrode and become stuck in compressed or partially compressed position, if the car or counterweight overtravels.

5.9.27 Emergency Operations and Signaling Devices

Firefighters' service is not required in a mine. Emergency equipment must be available at all times. Emergency procedures are well defined. A fire in a mine is a very different problem from that in a building.

5.9.29 Identification

Requirement 2.29.2 does not apply, as mine elevators are typically two-stop installations. Landings on multiple stop elevators are identified by reference to the mining level depth of the landing, not by floor numbers.

SECTION 5.10
ELEVATORS USED FOR CONSTRUCTION

NOTE: Throughout Section 5.10, references are made to other requirements in this Code. To gain a complete understanding of a requirement, the reader should review the Handbook commentary on all referenced requirements.

An elevator used for construction is defined as an elevator being used temporarily only for construction purposes.

Existing elevators used for construction must also comply with the requirements of 5.10. The requirements of 5.10 are applicable throughout the period the elevator is used for construction. After that period, the requirements for alterations (8.7) would apply.

Section 5.10 does not apply to the design, construction, installation, operation, inspection, testing, maintenance, alteration, or repair of structures and hoists, which are not a permanent part of the building, are installed inside or outside of buildings during construction, alteration, or demolition, and are used to raise and lower workers and other personnel connected with or related to a building project and/or for the transportation of materials. The requirements relating to that type of equipment are contained in:

(a) American National Standard Safety Requirements for Personnel Hoists, ANSI A10.4; or

(b) American National Standard Safety Requirements for Material Hoists, ANSI A10.5.

Elevators used for construction must be located where the general public will not come in contact with the equipment. This elevator is for use of construction personnel and not the general public. Elevator entrances must open into lobbies separated from areas accessible to the general public. This is especially important in buildings that are undergoing renovation, and new buildings that are partially occupied.

Section 5.10 does not, nor do any requirements in this Code apply to line jacks, false cars, shafters, moving platforms, and similar equipment used for installing an elevator. Information on this type of equipment can be located in the Elevator Industry Field Employees' Safety Handbook.

5.10.1 Electric Elevators Used for Construction

5.10.1.1 Construction of Hoistways and Hoistway Enclosures

5.10.1.1.1 Hoistway Enclosures

5.10.1.1.1(a) These requirements are similar to those that apply to special purpose personnel elevators. Fire-resistive construction is not required because the building is usually open from floor to floor during construction.

5.10.1.1.1(b) This requirement provides protection from objects falling down the hoistway.

5.10.1.1.1(c) This requirement protects personnel in the adjacent hoistways while they are working and affords a means of protection to the occupants of the elevator. Screening used for separation may remain in place after the elevator is placed in general service (i.e., elevator conforms to Parts 1 and 3). Screening must not restrict the use of side emergency exits when provided.

However, the original intent of the requirement is that the separation is removed once the elevator is placed in general use.

5.10.1.1.2 Working Requirements in the Hoistway. Conformance with these requirements will protect the occupants of the elevator from material falling in the hoistway and/or interfering with the necessary running clearances.

5.10.1.3 Location and Guarding of Counterweights. Counterweights located in the hoistway of the elevator they serve facilitate the inspection and maintenance of the counterweight and its ropes. When counterweights are located in a separate hoistway, as permitted by 2.3.3, the requirements provide access for inspection and maintenance. Requirement 5.10.1.3.3 also recognizes that a separate counterweight hoistway can be temporary as long as it complies with the requirements of 5.10.1.1.1.

5.10.1.5 Horizontal Car and Counterweight Clearances. A 100 mm or 4 in. clearance between the car landing sill and hoistway landing sill is permitted because the permanent sills may not be installed.

5.10.1.6 Protection of Spaces Below Hoistways. The securing of the space below the hoistway can be provided by the use of a temporary fence or partition as long as it will not be occupied when the elevator is in use.

5.10.1.7 Machine Rooms and Machinery Spaces. This requirement provides for a machinery area protected from the elements. It also requires that the machinery area be secured to protect nonelevator personnel from accidental exposure to the machinery.

5.10.1.7.2 Machine Room and Machinery Space Floors. A floor is not required at the top of the hoistway when the elevator machine is located below or at the side of the hoistway, and where elevator equipment can be serviced and inspected from outside the hoistway or from the top of the car. When an elevator machine is located over the hoistway, a floor is not required below secondary deflecting sheaves if the elevator equipment can be serviced and inspected from the top of the car or from outside of the hoistway. A floor provides a platform that makes servicing the elevator equipment less hazardous. The requirement also recognizes that wood floors may be necessary, as the machinery room as well as the building floors may not be in place at this time.

5.10.1.8 Machinery and Sheave Beams, Supports, and Foundations. The Note recognizes the temporary use of this equipment and provides for same.

5.10.1.9 Hoistway Doors and Gates. Every landing needs a door or gate. The doors may be constructed of wood and have a wire mesh vision panel. Mechanical means are acceptable (see 5.10.1.9.5) for a slower speed car (≤1.75 m/s or ≤350 ft/min) to lock hoistway doors or gates in the closed position. It is the responsibility of the operator to close and latch the door before moving the car. On high-speed (> 1.75 m/s or >350 ft/min) applications, the exposure is greater and interlocks or combination mechanical locks and electric contacts are required.

The requirement also has a provision, which allows the elevator hoistway to be locked from the landing when the car is taken out of service.

5.10.1.10 Car Enclosure, Car Doors and Gates, and Car Illumination

5.10.1.10.1 Enclosures Required. Car enclosure requirements permit both temporary and permanent enclosures. See Diagram 5.10.1.10.1.

5.10.1.10.4 Emergency Exits

5.10.1.10.4(e) This requirement recognizes that all loads cannot be carried totally within the car. The term "authorized personnel" refers to persons who have been instructed in the operation of the equipment and designated to use the equipment.

5.10.1.10.7 Car Emergency Signal. The signaling device could be an audible signal, i.e., an alarm bell or a two-way means of communication such as a permanently or temporarily installed intercom, walkie-talkie, etc.

5.10.1.10.8 Car Doors or Gates. The car door or gate could be either permanent or temporary.

5.10.1.11 Car Frames and Platforms. This requirement recognizes that temporary car platforms may be used and allows them to be constructed of wood without protection against fire.

5.10.1.12 Rated Load and Speed

5.10.1.12.1 Rated Load. Capacity is calculated on 90 kg or 200 lb per passenger to allow for the fact that most passengers will be carrying tools and/or other equipment.

5.10.1.12.3 Speed. The limitation on car speed is due to the temporary nature of the hoistway, car enclosure, hoistway door locks, etc. Elevators with higher speeds are required to comply with Part 1, 2, or 4 of this Code.

5.10.1.15 Ascending Car Overspeed and Unintended Car Movement Protection. It is not deemed necessary to add ascending car overspeed and unintended car movement protection to an existing elevator unless the elevator has been or is in the process of being altered and said requirements will be mandated by 8.7.

5.10.1.16 Suspension Means. The factor of safety of suspension wire ropes is the same as freight elevators since they are not used by the general public. Elevators used for construction are typically used as freight elevators to carry construction material.

5.10.1.21 Operating Devices and Control Equipment

5.10.1.21.1 Applicable Requirements. The requirement recognizes that not all of the requirements of 2.26 can practically be enforced on elevators used for construction. Requirement 5.10.1.21.1(c) recognizes that elevators used for construction are not going to be used by the disabled, thus the landing accuracy required by 2.26.11 is not required.

5.10.1.21.3 Electrical Equipment and Wiring. This requirement recognizes and allows temporary wiring as long as it conforms to the requirements of Article 305 of the National Electrical Code® or Section 76 of CSA-C22.1. Permanent wiring must also conform to the National Electrical Code® or CSA-C22.1.

Permanent electric elevator equipment is required to be certified as complying with CSA B44.1/ASME A17.5. This will minimize the risk of electricity as a source of electric shock and as a potential ignition source of fires.

5.10.1.23 Marking Plates or Signs. This requirement recognizes that a temporary data and capacity plate may be provided. The permanent (final) data and capacity plate may not contain correct information on the elevator used for construction; hence the requirement for the temporary plates.

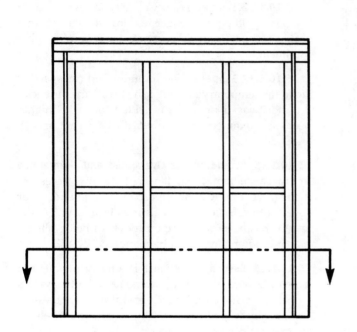

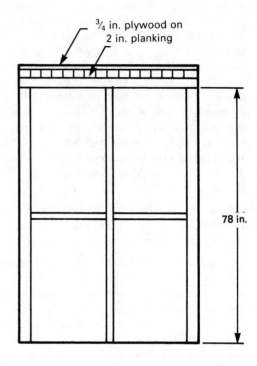

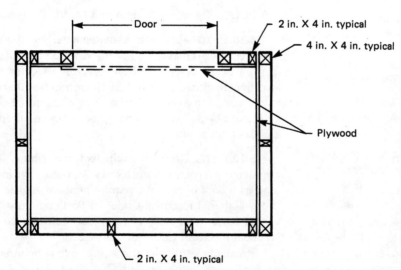

GENERAL NOTE: Attachment of temporary car to platform is not shown.

Diagram 5.10.1.10.1   Typical Temporary Car Enclosure

294

PART 6
ESCALATORS AND MOVING WALKS

SECTION 6.1
ESCALATORS

Escalators are designed and installed for transporting people only. They are not to be used for moving heavy, bulky, crude objects, which may result in overloading, or damage that could contribute to the failure of parts, scratching of skirts, congestion, or jam-ups that could cause an unsafe condition. The use of a stationary escalator as a stairway is a violation of the building codes. Escalators typically have a 203 mm (8 in.) riser and 406 mm (16 in.) tread. The building codes requirements for a stair specify that the greatest riser height within a flight of stair shall not exceed the smallest by more than 9.5 mm ($^3/_8$ in.). Escalator steps in transition do not comply with the uniform riser requirement.

NOTE: For educational information on safely using escalators, contact the Elevator/Escalator Safety Foundation at (205) 479-2199, or www.eesf.org.

6.1.1 Protection of Floor Openings

6.1.1.1 Protection Required. Escalators passing through fire-rated floor assemblies must have their openings protected against the passage of flame, heat, and/or smoke. If not, a fire would have a clear path of travel from one floor to another. Not all escalator openings pass through fire-rated floor assemblies and in these cases, no protection of the opening is required. An example of this type of construction is an escalator that serves two floors that are within an atrium.

The building code is an ordinance, which sets forth requirements for building design and construction. Where such an ordinance has not been enacted, one of the following model codes is observed:

(a) National Building Code (NBC)
(b) Standard Building Code (SBC)
(c) Uniform Building Code (UBC)
(d) National Building Code of Canada (NBCC)

It is noted that the International Building Code (IBC) issued in the first quarter of 2000 is the successor code for the NBC, SBC, and UBC.

Most building codes recognize that it is not necessary to fully enclose an escalator in the same way as one would enclose a stair since escalators are normally not part of a required means of egress. Chart 6.1.1.1 (excerpts from *NFPA 101 Life Safety Code® Handbook*) describes several methods, which are utilized in completely sprinklered buildings, where the escalators are not used as exits, which permit them to be unenclosed from floor to floor.

NOTE: *The Life Safety Code®* is the registered trademark of the National Fire Protection Association, Inc., Quincy, MA 02269.

6.1.2 Protection of Trusses and Machine Spaces Against Fire

6.1.2.1 Protection Required. The sides and undersides of escalator trusses and machinery spaces must be enclosed in noncombustible or limited-combustible materials to protect the trusses from heat during a fire in the vicinity immediately outside the truss. The Life Safety Code®, NFPA 101 defines limited-combustible and noncombustible as follows:

"**Limited-Combustible:** Refers to a building construction material not complying with the definition of noncombustible (see 3.3.131) that, in the form in which it is used, has a potential heat value not exceeding 3500 Btu/lb (8141 kJ/kg), where tested in accordance with NFPA 259, *Standard Test Method for Potential Heat of Building Materials*, and includes (1) materials having a structural base of noncombustible material, with a surfacing not exceeding a thickness of $^1/_8$ in. (3.2 mm) that has a flame spread index not greater than 50; and (2) materials, in the form and thickness used, other than as described in (1), having neither a flame spread index greater than 25 nor evidence of continued progressive combustion, and of such composition that surfaces that would be exposed by cutting through the material on any plane would have neither a flame spread index greater than 25 nor evidence of continued progressive combustion."

"**Noncombustible:** Refers to a material that, in the form in which it is used and under the conditions anticipated, does not ignite, burn, support combustion, or release flammable vapors, when subjected to fire or heat. Materials that are reported as passing ASTM E 136, *Standard Test Method for Behavior of Materials in a Vertical Tube Furnace at 750 Degrees C*, are considered noncombustible materials."

Noncombustible or limited-combustible material is not required between multiple escalators in a single wellway. This may also help contain a fire that may start

Chart 6.1.1.1
Excerpts From NFPA 101 Life Safety Code® Handbook

8.2.5.11 Any escalators or moving walks serving as a required exit in existing buildings shall be enclosed in the same manner as exit stairways. *(See 7.2.7)*.

Where used as an exit, an escalator must be completely enclosed with fire-rated construction, including entrance and discharge doors. It is rare to find an escalator enclosed in such a manner. Escalators located within the required means of egress in existing buildings that maintain compliance with the *Code* usually make use of one of the exceptions to 8.2.5.12 to avoid creating an unprotected vertical opening. By doing so they are classified as exit access. Note that 7.2.7 prohibits escalators from constituting any part of the required means of egress in new buildings. Thus, in new construction, an escalator can be installed but is not recognized as satisfying the requirements for exit access, exit, or exit discharge.

8.2.5.12 Moving walks not constituting an exit and escalators, other than escalators in large open areas such as atriums and enclosed shopping malls, shall have their floor openings enclosed or protected as required for other vertical openings.

Exception No. 1: In buildings protected throughout by an approved automatic sprinkler system in accordance with Section 9.7, escalators or moving walk openings shall be permitted to be protected in accordance with the method detailed in NFPA 13, Standard for the Installation of Sprinkler Systems, or in accordance with a method approved by the authority having jurisdiction.

A.8.2.5.12 Exception No. 1 The intent of the exception is to place a limitation on the size of the opening to which the protection applies. The total floor opening should not exceed twice the projected area of the escalator or moving walk at the floor. Also, the arrangement of the opening is not intended to circumvent the requirements of 8.2.5.6.

As with any opening through a floor, the openings around the outer perimeter of the escalators should be considered as vertical openings. The sprinkler draftstop installation is intended to provide adequate protection for these openings, provided that the criteria of NFPA 13, *Standard for the Installation of Sprinkler Systems*, as well as the area criteria described in the preceding paragraph, are met.

Exception No. 2: Escalators shall be permitted to be protected in accordance with 8.2.5.13.

Exception No. 2 to 8.2.5.12 provides that new escalators (which cannot be part of the means of egress) and

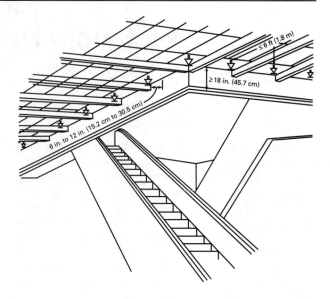

Exhibit 8.21 Sprinklers Around an Escalator Opening

existing escalators not serving as exits (see 8.2.5.11) need not be enclosed if certain provisions are met.

The sprinkler-draft curtain method is detailed in NFPA 13, *Standard for the Installation of Sprinkler Systems*. It consists of surrounding the escalator opening, in an otherwise fully sprinklered building, with an 18-in. (45.7-cm) deep draft stop located on the underside of the floor to which the escalator ascends. This draft stop serves to delay the heat, smoke, and combustion gases developed in the early stages of a fire on that floor from entering into the escalator well. A row of closely spaced automatic sprinklers located outside of the draft stop also surrounds the escalator well. As sprinklers along this surrounding row are individually activated by heat, their water discharge patterns combine to create a water curtain. A typical installation is shown in Exhibit 8.21. In combination with the sprinkler system in the building, this system should delay fire spread effectively and allow time for evacuation.

Prior editions of the *Code* detailed several methods that permitted the use of unenclosed escalators in completely sprinklered buildings where the escalators were not used as exits. In addition to the sprinkler-draft curtain or rolling shutter methods, the authority having jurisdiction might consider one of the following when evaluating existing buildings:
(1) Sprinkler-vent method
(2) Spray nozzle method
(3) Partial enclosure method

(continued)

Chart 6.1.1.1
Excerpts From NFPA 101 Life Safety Code® Handbook (Cont'd)

The following discussion details these three methods.

(a) *Sprinkler-Vent Method.* Under the conditions specified, escalator or moving walk openings are permitted to be protected by the sprinkler-vent method, which consists of a combination of an automatic fire or smoke detection system, an automatic exhaust system, and an automatic water curtain. This combination of fire protection and system design is required to meet the following criteria and be approved by the authority having jurisdiction.

(1) The exhaust system should be capable of creating a downdraft through the escalator or moving walk floor opening. The downdraft should have an average velocity of not less than 300 ft/min (1.5 m/s) under normal conditions for a period of not less than 30 minutes. This requirement can be met by providing an air intake from the outside of the building above the floor opening. The test of the system under "normal" conditions requires that the velocity of the downdraft be developed when windows or doors on the several stories normally used for ventilation are open. The size of the exhaust fan and exhaust ducts must be sufficient to meet such ventilation conditions. Experience indicates that fan capacity should be based on a rating of not less than 500 ft³/min/ft² (8.3 m³/s/m²) of moving stairway opening to obtain the 300-ft/min (1.5-m/s) velocity required. If the building is provided with an air conditioning system arranged to be automatically shut down in the event of fire, the test condition should be met with the air conditioning system shut down. The 300-ft/min (1.5-m/s) downdraft through the opening provides for the testing of the exhaust system without requiring the expansion of air that would be present under actual fire conditions.

(2) Operation of the exhaust system for any floor opening should be initiated by an approved device on the involved story and should use one of the following means, in addition to a manual means, for operating and testing the system:

    a. Thermostats (fixed-temperature, rate-of-rise, or a combination of both)

    b. Waterflow in the sprinkler system

    c. Approved, supervised smoke detection located so that the presence of smoke is detected before it enters the stairway.

(3) Electric power supply to all parts of the exhaust system and its control devices should be designed and installed for maximum reliability. The electric power supply provision of NFPA 20, *Standard for the Installation of Stationary Pumps for Fire Protection*, can be used as a guide to design and installation features that help to ensure maximum reliability.[30]

(4) Any fan or duct used in connection with an automatic exhaust system should be of the approved type and should be installed in accordance with the applicable standards in Chapter 2 and Annex B.

(5) Periodic tests should be made of the automatic exhaust system, at least quarterly, to maintain the system and the control devices in good working condition.

(6) The water curtain should be formed by open sprinklers or by spray nozzles located and spaced to form a complete and continuous barrier along all exposed sides of the floor opening and to reach from the ceiling to the floor. Water discharge for the water curtain should be not less than approximately 3 gal/min/lineal ft (0.6 L/sec/lineal m) of water curtain, measured horizontally around the opening.

(7) The water curtain should operate automatically from thermal-responsive elements of a fixed-temperature type. These elements should be located with respect to the ceiling/floor opening so that the water curtain actuates upon the advance of heat toward the escalator or moving walk opening.

(8) Every automatic exhaust system (including all motors, controls, and automatic water curtain system) should be supervised in an approved manner that is similar to that specified for automatic sprinkler system supervision.

(b) *Spray Nozzle Method.* Under the conditions specified, escalator openings are permitted to be protected by the spray nozzle method, which consists of a combination of an automatic fire or smoke detection system and a system of high-velocity water spray nozzles. This combination of fire protection and system design is required to meet the following criteria and be approved by the authority having jurisdiction.

(1) Spray nozzles should be of the open type and should have a solid conical spray pattern with discharge angles between 45 and 90 degrees. The number of nozzles, their discharge angles, and their location should be such that the escalator or moving walk opening between the top of the wellway housing and the treadway will be completely filled with dense spray on operation of the system.

(2) The number and size of nozzles and water supply should be sufficient to deliver a discharge of 2 gal of water/ft²/min (1.4 L of water/m²/sec) through the wellway, with the area to be figured perpendicularly to the treadway. See Exhibit 8.22.

(continued)

Chart 6.1.1.1
Excerpts From NFPA 101 Life Safety Code® Handbook (Cont'd)

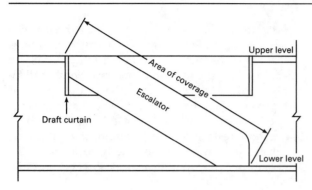

Exhibit 8.22 Area of Coverage for Spray Nozzle Method of Protecting Escalator Openings

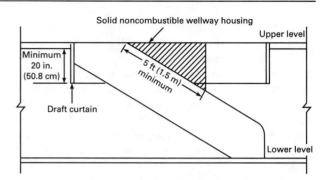

Exhibit 8.23 The Draft Curtain and Wellway Housing Method of Protecting Vertical Openings

(3) Spray nozzles should be located to take full advantage of the cooling and counterdraft effect. They should be positioned so that the centerline of spray discharge is as closely in line as possible with the slope of the escalator or moving walk, not more than an angle of 30 degrees with the top slope of the wellway housing. Nozzles should also be positioned so that the centerline of discharge is at an angle of not more than 30 degrees from the vertical sides of the wellway housing.

(4) Spray nozzles should discharge at a minimum pressure of 25 lb/in.^{2} (172 kPa). Water supply piping is permitted to be taken from the sprinkler system, provided that an adequate supply of water is available for the spray nozzles and that the water pressure at the sprinkler farthest from the supply riser is not reduced beyond the required minimum. Water supply taken from the sprinkler system is designed to provide protection from life hazards to the wellway opening during the exit period but is not to be relied on to provide an effective floor cutoff.

(5) Control valves should be readily accessible to minimize water damage.

(6) A noncombustible or limited-combustible draft curtain should be provided that extends at least 20 in. (50.8 cm) below and around the opening, and a solid noncombustible wellway housing at least 5 ft (152 cm) long, measured parallel to the handrail and extending from the top of the handrail enclosure to the soffit of the stairway or ceiling above, should also be provided at each escalator floor opening. Where necessary, spray nozzles should be protected against mechanical damage or tampering that might interfere with proper discharge. See Exhibit 8.23.

(7) The spray nozzle system should operate automatically from thermal-response elements of the fixed-temperature type and be located with respect to the ceiling/floor opening so that the spray nozzle system actuates upon the advance of heat towards the escalator opening. Supervised smoke detection located in or near the escalator opening is permitted to be used to sound an alarm. The spray nozzle system should also be provided with manual means of operation. It is not desirable to have smoke detection devices activate the spray nozzles; safeguards against accidental discharge must be provided to prevent both panic and property damage.

(8) Control valves for the spray nozzle system and approved smoke detection or thermostatic devices should be supervised in accordance with the applicable provisions of Section 9.6.

(c) *Partial Enclosure Method.* Under the conditions specified, escalator or moving walk openings are permitted to be protected by a partial enclosure, or so-called kiosk, designed to provide an effective barrier to the spread of smoke from floor to floor. This method of fire protection is required to meet the following criteria and be approved by the authority having jurisdiction.

(1) Partial enclosure construction should provide fire resistance equivalent to that specified for stairway enclosures in the same building, with openings therein protected by approved self-closing fire doors. The openings also are permitted to be of approved wired-glass and metal frame construction with wired-glass panel doors.

(2) Fire doors are permitted to be equipped with an electric opening mechanism, which opens the door automatically upon the approach of a person. The

(continued)

Chart 6.1.1.1
Excerpts From NFPA 101 Life Safety Code® Handbook (Cont'd)

mechanism should return the door to its closed position upon any interruption of electric current supply, and it should be adjusted so that the pressures generated by a fire will not cause the door to open.

8.2.5.13 In buildings protected throughout by an approved automatic sprinkler system in accordance with Section 9.7, escalators or moving walk openings shall be permitted to be protected by rolling steel shutters appropriate for the fire resistance rating of the vertical opening protected. The shutters shall close automatically and independently of each other upon smoke detection and sprinkler operation. There shall be a manual means of operating and testing the operation

of the shutter. The shutters shall be operated not less than once a week to ensure that they remain in proper operating condition. The shutters shall operate at a speed not to exceed 30 ft/min (0.15 m/s) and shall be equipped with a sensitive leading edge. The leading edge shall arrest the progress of a moving shutter and cause it to retract a distance of approximately 6 in. (15.2 cm) upon the application of a force not exceeding 20 lbf (90 N) applied to the surface of the leading edge. The shutter, following this retraction, shall continue to close. The operating mechanism for the rolling shutter shall be provided with standby power complying with the provisions of NFPA 70, *National Electrical Code.*

within the truss. It can be met simply by applying gypsum drywall or an equally noncombustible or limited-combustible material on the outside of the truss and machinery space. No hourly rating is required. A truss or machinery space suspended in an approved fire-rated ceiling assembly, where the truss is exposed within that assembly, does not meet this requirement. The truss and machinery space must be separately enclosed within the rated ceiling assembly; also, the sides and undersides must be enclosed. Dampers in the noncombustible or limited-combustible material enclosing the truss must be provided with fire links, which will close the opening when there is a fire.

6.1.3 Construction Requirements

See Diagrams 6.1.3(a) and 6.1.3(b).

6.1.3.1 Angle of Inclination. Limiting the angle of inclination to 30 deg without a tolerance of 1 deg due to field conditions is impractical. Most escalators are designed to be installed at a 30 deg angle of inclination. Construction practices are such that a precise dimension cannot be guaranteed. A maximum 1 deg tolerance is therefore necessary and is not considered unsafe. An angle of inclination above this gives the impression that the escalator is too steep and could cause dizziness in some passengers.

The angle of inclination varies on curved (spiral) escalators depending on where along the step width it is measured. The centerline was selected as the location to measure the angle because it is closest to the location where most passengers stand.

6.1.3.2 Geometry. The handrail clearances reduce entrapment and shear exposure and position the handrail to facilitate its use. These clearances must be maintained wherever the handrail is exposed. Components of the handrail guard are not subject to this requirement. See Diagrams 6.1.3(a), 6.1.3(b), and 6.1.3.2.

6.1.3.3 Balustrades

6.1.3.3.1 Construction. Glass balustrades require gaps of about 5 mm or $\frac{3}{16}$ in. between glass panels to allow for expansion. This condition is allowed. Raised or depressed areas of molding more than 6.4 mm or $\frac{1}{4}$ in. high could snag clothing, packages, and parts of the body.

6.1.3.3.2 Strength. This requirement specifically states that the balustrades shall be designed to resist the forces specified. There is no Code requirement to test the design in the field. Conformance with the design requirement may be verified by examining the manufacturer's design criteria. This requirement simply means that the balustrade has been designed to "resist" the applied load and when the load is removed, there is no permanent deformation or damage.

6.1.3.3.3 Use of Glass or Plastic. See Handbook commentary on 2.14.1.8 for additional information on ANSI Z97.1 and 16 CFR Part 1201.

6.1.3.3.4 Interior Low Deck. The requirements provide a maximum angle of the interior deck profile that assist in preventing a youngster from slipping while attempting to walk on the interior decking.

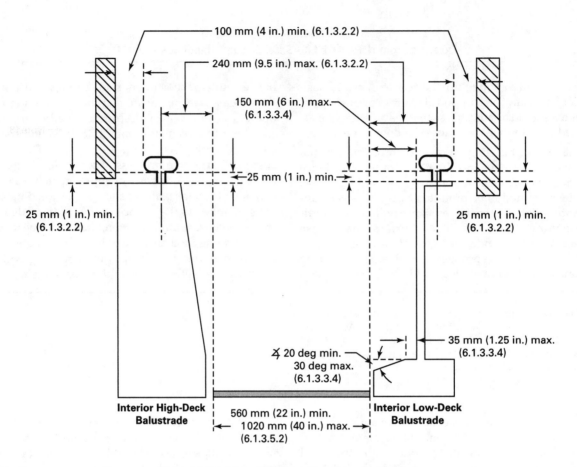

Diagram 6.1.3(a)   Relationship of Escalator Parts

6.1.3.3.5 Loaded Gap Between Skirt and Step. See Handbook commentary on 6.1.3.3.7.

6.1.3.3.6 Skirt Panels. Recent testing has shown that the deflection requirements for the skirt panel contained in this requirement, the loaded gap requirements in 6.1.3.3.5, and the Step/Skirt Index requirements in 6.1.3.3.7 provide the optimum protection against accidental entrapment of a foot, finger, clothing, shoes, sneakers, etc., between the skirt and the steps. See Diagram 6.1.3.3.6.

6.1.3.3.7 Step/Skirt Performance Index. In 1997 National Elevator Industry, Inc. (NEII) contracted with Arthur D. Little (ADL) to develop an Escalator Step/Skirt Performance Index. These requirements are the result of this comprehensive study. The Index, valued from zero to one, represents the relative potential for entrapment of objects in the step-to-skirt gap. A lower Index represents a lower potential.

The Index values were established from the following criteria:

(a) the nominal estimated Index value of the current ASME A17 Code;

(b) the desired Index value for a low entrapment potential for hands;

(c) the desired Index value for a low entrapment potential for leg calf.

The Step/Skirt Index value was based on estimates of the two primary escalator parameters in the ASME A17.1c–1999 and earlier editions of the Code. The loaded gap parameter (the value of the step-to-skirt gap under a spreading force of 110 N or 25 lbf and skirt coefficient of friction parameters were estimated due to ambiguity or nonexistent ASME A17.1 Code requirements. The prior ASME A17.1 Code specification of 4.8 mm ($^3/_{16}$ in.) maximum step-to-skirt clearance does not address additional parameters of step stiffness and step dead band movement identified in the ADL study. These contribute to the loaded gap index parameter, and so were estimated. This additional gap value is nominally 1.27 mm (0.05 in.) resulting in a total gap of 6.1 mm (0.24 in.). An additional gap increase due to the ASME A17.1 Code specified skirt stiffness of 1.6 mm (0.06 in.) at 667 N (150 lbf) or 0.25 mm (0.01 in.) at 111.2 N (25 lbf) increases the total gap to a value of 6.35 mm (0.25 in.) loaded gap.

ASME A17.1c–1999 and earlier editions of the Code specify that the skirt be "made from a low coefficient of friction material or treated with a friction reducing material." A conservative coefficient of friction of 0.4

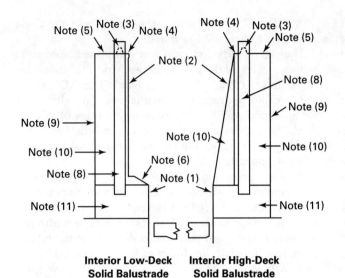

Interior Low-Deck Solid Balustrade **Interior High-Deck Solid Balustrade**

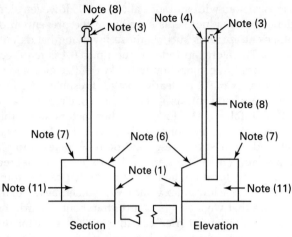

Interior Low-Deck Glass Balustrade

NOTES:
(1) Skirt panel
(2) Interior panel
(3) Handrail stand
(4) High-deck interior
(5) High-deck exterior
(6) Low-deck interior
(7) Low-deck exterior
(8) Handrail
(9) Exterior panel
(10) Newel
(11) Newel base

Diagram 6.1.3(b) Escalator Nomenclature

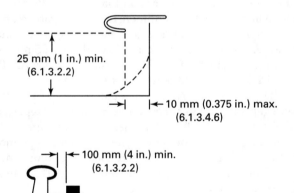

25 mm (1 in.) min.
(6.1.3.2.2)

10 mm (0.375 in.) max.
(6.1.3.4.6)

100 mm (4 in.) min.
(6.1.3.2.2)

Diagram 6.1.3.2 Handrail

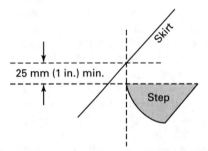

25 mm (1 in.) min.

Skirt

Step

Diagram 6.1.3.3.6 Skirt Panel — Step Nose

between the skirt and polycarbonate (the friction test sample) is estimated based upon ADL tests in both the lab and field.

The estimated loaded gap of 6.35 mm (0.25 in.) and a polycarbonate coefficient of friction of 0.4 results in an Index of 0.4 for escalators that comply with the requirements in ASME A17.1c–1999 and earlier editions of the Code. However, it is also possible that an existing escalator could have an Index as high as 0.7 and still comply with the prior Code requirements.

The ADL study included a series of highly stressed tests to try to induce entrapments of shoes and artificial body parts (referred to as Sawbones parts). These tests indicate that an Index of 0.15 is needed for low entrapment potential of all objects studied, including the leg calf that had the highest incidents of entrapment. One method of complying with the requirements is to provide escalators with an Index of 0.15 or below. However, this low Index value may not realistically be achieved and maintained on all existing and new escalator designs, thus additional design options are needed.

Skirt deflector devices installed in compliance with 6.1.3.3.8, should be effective in the prevention of leg calf entrapments and the proposal makes them mandatory for escalators with an Index above 0.15. However, skirt deflectors may not be as effective in the prevention of other entrapments such as those involving hands and shoes. Therefore, an Index value up to 0.4 is required, when used in conjunction with skirt deflector devices.

An Index of 0.2 is clearly a valid threshold based on Sawbones hand entrapment tests (see ADL report Figure 5-2 and Table 5-1). However, other factors show this Index value to be conservative. First, the Index was derived from severe test conditions with test sample placement into the gap and maximum expected entrapping force applied. Second, the leg calf, entrapments at these low index values were sometimes actually pinches that were classified as entrapments. Third, the object coefficient of friction with stainless steel for the test sample Sawbones hand and calf of 0.8 is significantly higher than that for real skin at 0.5. This makes the entrapment of Sawbones hand samples significantly more likely than real hands at the same loaded gap.

Estimation can be made of the effect of Sawbones sample coefficient of friction as compared to real skin coefficient of friction on the Index. For Sawbones hand tests, the ADL study showed a low entrapment potential below an object Index of 0.4, with object coefficient of friction of 0.8 and loaded gap of 4.1 mm (0.16 in.). The index curves show that an object Index of 0.4 and a low entrapment potential are maintained when the coefficient of friction is decreased from 0.8 (Sawbones skin) to 0.5 (real skin) while the loaded gap is increased from 4.1 mm (0.16 in.) to 5.8 mm (0.23 in.). Thus, the loaded gap (5.8 mm or 0.23 in.) at which real hand entrapment potential diminishes is larger than the loaded gap (4.1 mm or 0.16 in.) associated with diminished Sawbones hand entrapment potential. The Index of an escalator will be measured with a polycarbonate test sample with an assumed coefficient of friction of 0.4. Using this test sample, a Sawbones hand Index of 0.4 with a loaded gap of 4.1 mm (0.16 in.) would be reduced to a measured Index of 0.2. In like manner, a real hand Index of 0.4 with a loaded gap of 5.8 mm (0.23 in.) would be reduced to a measured Index of 0.35. Therefore, an escalator with a loaded gap of 5.8 mm (0.23 in.) and an Index of 0.35 has a low real hand entrapment potential. A real hand would have the same low entrapment potential at an Index of 0.35 as the test Sawbones hand at an Index of 0.2.

Similar analysis can be made for the leg calf. Low entrapment potential exists for the Sawbones calf below object Index of 0.2, with object coefficient of friction of 0.8 and loaded gap of 1.6 mm (0.06 in.). Equivalently low entrapment potential exists for the real calf at object Index of 0.2 with object coefficient of friction decreased to 0.5 and loaded gap increased to 3.3 mm (0.13 in.). The equivalent measured Index is 0.1 for Sawbones and 0.16 for real calves. Entrapment potential is low for real calves at an Index of 0.16 or below.

The above assumptions provided the basis for the index recommendations:

ASME A17.1c–1999 and earlier editions of the Code reflect an Index of 0.4 nominal.

The ADL study indicates low entrapment potential with an Index of 0.2 for Sawbones hand that translates to an Index of 0.35 for a real hand. The ADL study indicates low entrapment potential with an Index of 0.1 for Sawbones calf that translates to an Index of 0.16 for a real calf. These are conservative numbers due to the severity of testing and the conservative classification of calf entrapments.

Because the nominal ASME A17.1 Code Index of 0.4 is reasonably close to the low hand entrapment potential, a maximum Index of 0.4 is required for escalators installed under ASME A17.1d–2000 and earlier editions. A lower entrapment potential is desired for the future. Therefore, a maximum Index of 0.25 is required beginning with ASME A17.1a–2001. This should allow sufficient time for manufacturers to design for and achieve the desired Index. These Index thresholds, when used in conjunction with the required skirt deflector devices, will significantly reduce the entrapment potential on existing and new escalators. An Index of 0.15 or below will allow an escalator to be installed without skirt deflector devices.

Even though the Index provides a comprehensive measure of entrapment potential, a requirement for a loaded gap in the new ASME A17.1 Code requirement is desirable. The loaded gap parameter provides an additional margin of control of escalators that rely heavily upon low coefficient of friction that can be difficult to maintain in the field. In addition, the loaded gap parameter provides a sound means of monitoring step band wear and need for correction. A loaded gap of 5 mm or 0.2 in. is required for escalators installed under ASME A17.1d–2000 and later editions of the Code.

The prior ASME A17.1 Code parameters of step-to-skirt gap and skirt treatment are redundant and replaced by 6.1.3.3.5 and 6.1.3.3.7. If these requirements were maintained, they would be misleading, as compliance with the old requirements would not assure compliance with the new loaded gap and step/skirt performance index requirements.

6.1.3.3.8 Skirt Deflector Devices. Deflector systems are an uncomplicated method of deflecting the feet of riders away from the step/skirt panel interface and can be secured in various manners depending on the design. The requirements prevent the use of common flat or Phillips head screws, which can be easily removed by nonelevator personnel.

6.1.3.3.8(a)(1) This dimension is consistent with the ASME A17 Committee's expressed previous consensus, and the majority, but not all, of what has been safely applied in other parts of the world. The dimension also conforms to what has been used in California without problems. The requirements prevent footwear, etc., from being caught.

6.1.3.3.8(a)(2) Based on historical experience with use of these types of devices, this should provide sufficient clearance and angle to prevent footwear from catching between the step nose and horizontal protrusion. The required clearance is also the value stipulated in European Standard, EN 115 Clause 5.1.5.6, on deflector device application. EN 115 Clause 5.1.5.6 was added a number of years ago specifically for deflectors, thus it would be logical to accept the experience obtained in other parts of the world. The 25 mm or 1 in. dimension is also the historical value deemed necessary in A17.1, requirement 6.1.3.3.6(a) for many years. The 10 deg angle at this clearance is sufficient based on international history for installations at a 25 mm or 1 in. height.

6.1.3.3.8(a)(3) Based on historical experience this should provide sufficient clearance and angle to take care of the change in elevation at the comb and allow passage of footwear.

6.1.3.3.8(a)(4) This is a reasonable value, which could withstand the force exerted by an adult's foot. Test plate area is specified; similar to requirements for the step fatigue test, combplate, etc.

6.1.3.3.8(b)(1) To minimize restriction of usable step area.

6.1.3.3.8(b)(2) The requirement that the flexible part be capable of bending to above the 10 deg line covers the concern of easy withdrawal of a foot.

6.1.3.3.8(b)(3) Although this would only apply to a very small area of contact at the step nose, requirements have been included to prevent damage. For example, to avoid the possibility of a brush providing a flat stepping surface by extending over two or more step cleats.

6.1.3.3.8(b)(4) Any step contact with continuous flexible elements could potentially cause damage to the elements.

6.1.3.3.9 Guard at Ceiling Intersection. A solid guard must be provided in the intersecting angle of each outside top of the balustrade and the point of penetration of the ceiling or soffit. It is not required on high-deck escalators where the side of the penetration at the ceiling or soffit is more than 600 mm or 24 in. from the centerline of the handrail, or on low-deck escalators where the centerline of the handrail is more than 350 mm or 14 in. from the centerline of the soffit. The vertical edge of the guard must be a minimum of 350 mm or 14 in. and be flush with the face of the wellway. The leading edge or the exposed edge of the guard should be vertical and must be rounded. The material used for the guard should be solid without holes or openings passing through it. Glass or plastic can be used for the guards as long as they conform to the requirements of 6.1.3.3.3. This guard protects escalator passengers, if they are leaning off to the side of an up-traveling escalator, from the possibility of trapping their heads or arms in the angle

where the escalator and ceiling or soffit meet. See Diagram 6.1.3.3.9.

6.1.3.3.10 Antislide Devices. Antislide devices, sometimes referred to as baggage stops, discourage pranksters from using the escalator deck as an amusement slide. They also discourage passengers from resting packages on the deck while traveling on the escalator. See Diagram 6.1.3.3.10(a) and 6.1.3.3.10(b).

6.1.3.3.11 Deck Barricades. Deck barricades discourage people, especially small children, from walking or going up or down the low-deck exterior balustrade from the ground level. See Diagram 6.1.3.3.11. Barricade attachment fasteners are required to be tamper resistant to discourage and/or prevent the removal of some or all of the fasteners and/or the barricade itself, which would create a dangerous falling hazard.

6.1.3.4 Handrails

6.1.3.4.1 Type Required. A handrail not moving in the same direction or speed would have a tendency to pull passengers backward or forward causing them to lose their balance. It is not necessary for the handrail to move at exactly the same speed as the step since passengers can easily adjust the position of their hands to compensate for minor differences in speeds. The minimum handrail driving force is necessary to ensure that the handrail continues to operate under a passenger load.

6.1.3.4.2 Extension Beyond Combplates. This requirement allows passengers to grab hold of a handrail before they step on to the escalator and allows them to continue to hold on to the handrail until they have completely stepped off the moving portion of the escalator. The moving handrail extending beyond the exit point also has the tendency to immediately move the passenger away from the exit, thus reducing bunching at this location.

6.1.3.4.3 Guards. Hand or finger guards provide protection for the curious, especially children, who want to explore the area where the handrail enters the balustrade.

6.1.3.4.4 Splicing. Although field splicing of escalator handrails is rare, it is done on occasion.

6.1.3.4.5 Vertical Height. The specification of maximum height assures that handrails will be accessible to a majority of users of the escalator. All manufacturers of escalators provide handrail heights, which are below this value, based on the manufactures internal studies.

6.1.3.4.6 Handrail Clearance. Provides a limitation of a dimension that, if excessive, can permit fingers to become caught. See Diagram 6.1.3.2.

6.1.3.5 Steps

6.1.3.5.1 Material and Type. A step frame is the supporting structure of an escalator step to which the

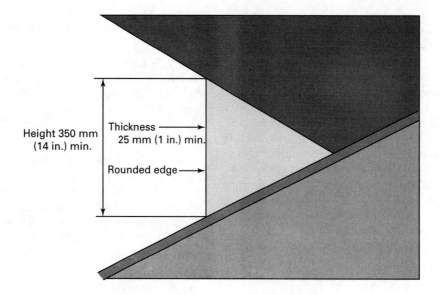

Diagram 6.1.3.3.9   Ceiling or Soffit Guard

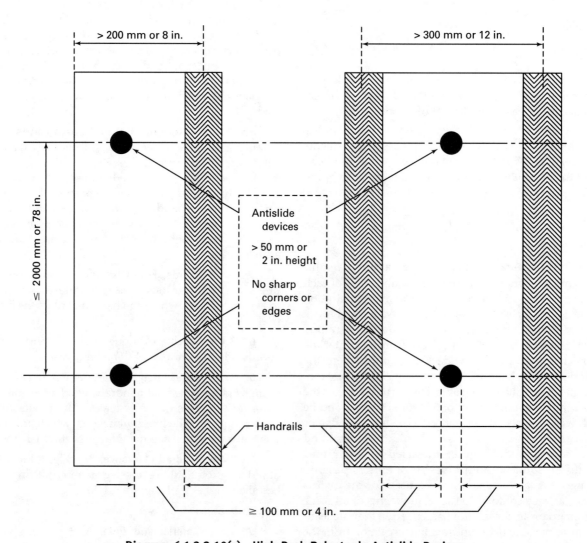

Diagram 6.1.3.3.10(a)   High Deck Balustrade Antislide Devices

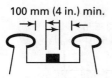

Diagram 6.1.3.3.10(b)   Antislide Device

tread and riser sections are fastened. A concern addressed by this requirement is the potential of non-fire-rated escalator components spreading a fire.

6.1.3.5.1(a) The limit of the step tipping is included as this is a factor that if left uncontrolled, can produce a dangerous condition for the passengers.

6.1.3.5.2 Dimensions of Steps. The minimum dimension of the step tread assures that there is adequate area for persons to stand. The maximum dimension of the step riser assures that a person can walk up the step if necessary. No manufacturer in the United States currently makes a step less than 597 mm (23½ in.) wide. The wider the step, the more room the passenger has to avoid contact with the skirt, especially if an adult and a small child are standing on the same step. See Diagram 6.1.3.5.2.

6.1.3.5.3 Cleated Step Risers. See Handbook commentary on 6.1.3.5.5 and Diagram 6.1.3.5.3.

6.1.3.5.5 Slotting of Step Treads. Slotting of step treads and risers provides protection against entrapment of a foot or finger and articles of clothing such as shoes, sneakers, etc., as the step makes the transition from the incline to the horizontal. See also Handbook commentary on 6.1.3.6.

6.1.3.5.6 Step Demarcation. The dimensions where a result of suggestions from ICC/ANSI A117. A MARTA study found that yellow is the best color, particularly for site-impaired individuals. ICC/ANSI A117 also felt yellow was appropriate.

A marking on the back of the tread is effective in assisting boarding passengers to recognize the separation of adjacent steps. It is in clear view as a passenger approaches the escalator and boards it, but does not provide a distraction from the boarding area as step nose markings do when the steps move away from the landing. This is most prevalent on an up escalator as the strobe effect of the marks emerging from under the comb cause a passengers eyes to seek a more stable environment by following the markings up the incline.

When traveling down, the step's nose markings tend to merge into a solid line of color extending down the incline that can produce a confusing pattern to the passengers looking down the incline and cause them to become disoriented. Neither one of these conditions occur when the marking is at the back of the step since it is hidden from the passengers view on the incline.

It is unsound to attempt to place an insert into the nose of the step. This area plays a critical part in the strength and dimensional stability of the modern one-piece die-cast step that is used today by most escalator manufacturers. Although doable, it would increase the number of areas that could develop problems that could result in a stop failure. Additionally, any insert must have a surface that would provide the same physical and dimensional characteristics as the parent step surface.

The nose marking would most likely be painted on to offset these problems. However, paint on any step surface is subject to wear and dirt that would diminish the intended effect as well as alter the friction characteristics of the step treads. None of these problems are experienced when a marking insert is placed on the back of the step tread.

The addition of this marking on the step nose will do more to reduce safety than promote it. The only condition, for which the markings on the nose are more

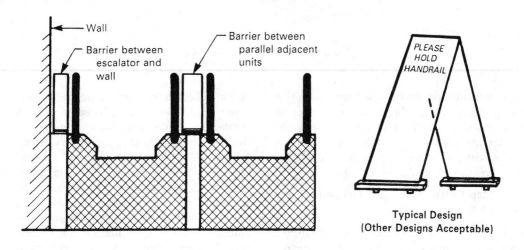

Diagram 6.1.3.3.11   Deck Barricade

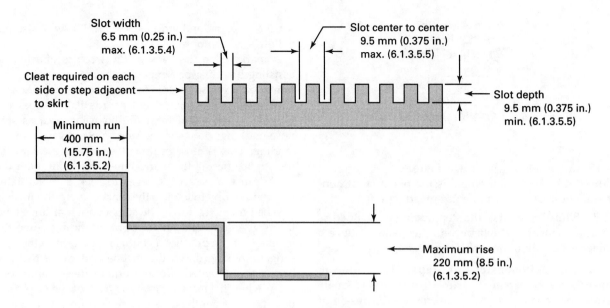

Diagram 6.1.3.5.2   Escalator Step Tread

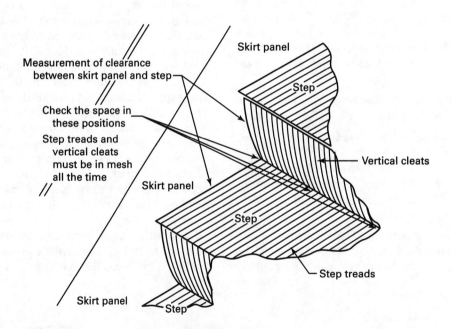

Diagram 6.1.3.5.3   Cleated Riser

effective than markings on the back, is for persons that are walking up or down the escalator. This is a dangerous practice for any person and more so for a handicapped individual. The depth of the step treads and the varying riser heights are difficult to negotiate and cause a walking passenger to stumble and possibly fall.

6.1.3.5.7 Step Fatigue Tests. See Handbook commentary on 8.3.1.1.

6.1.3.5.8 Step Wheels. There have been a number of instances where the trailing wheels have become separated from the step. Although the front of the step is still attached to the step chain, the other end can pivot down into the escalator interior. If the wheels were located inside the step, then when the step is pivoted down it would be stopped by the step wheel track located below. However, if the wheels are located outside the width of the step, the wheel support track is located beyond the edge of the step and there is nothing to prevent the step from pivoting into the escalator interior. If this happens, passengers standing on the step could fall into the escalator interior and be seriously hurt. This requirement is intended to prevent that condition.

6.1.3.6 Entrance and Egress Ends

6.1.3.6.1 Combplates. The combplate is the section of the floor plate nearest the step. It normally has replaceable combs made of brittle material such as plastic or aluminum. Comb teeth must be brittle and break rather than bend. A bent comb tooth could cause a steep wreck piling the steps up at the comb. The comb teeth are so arranged that the cleats of the step treads will pass between them with minimum clearance. This arrangement allows the steps to move under the combplate allowing passengers to move safely from the step onto the floor. Because the combs are not an integral part of the combplate assembly, worn or damaged comb sections can be replaced readily without disturbing the balance of the comb.

6.1.3.6.1(a) A clear distinction between the combs and the combplate is required.

6.1.3.6.1(d) This requirement ensures that the combs and combplate do not drag on the steps when a passenger is standing on them.

6.1.3.6.2 Distinction Between Comb and Step. A significant number of falls occur on boarding an escalator. This requirement is important in that it requires a visual aid for passengers to assist in boarding the moving steps of the escalator. This should reduce the hazard by calling attention to that area.

6.1.3.6.3 Adjacent Floor Surfaces. This requirement addresses a potential tripping hazard. This requirement applies to the difference in elevation between the landing plate and the adjacent building floor, not between the comb and landing plate. There always will be a difference in elevation between the comb and landing plate due to the following factors:

(a) combplate thickness required to meet current deflection criteria;

(b) allowance for combplate deflection;

(c) step tail or nose swing — the tail or nose moves in an arc as the step begins its travel through the turnaround;

(d) a small amount of additional running clearance to allow for manufacturing tolerance and wear-in of parts;

(e) possible increased combplate thickness necessitated by proposed requirements for combplates.

Floor height variations usually occur when buildings are renovated and different floor materials are used than were present when the escalator was originally installed.

6.1.3.6.4 Safety Zone. The safety zone assures there is adequate space for passengers when they exit an escalator. If bunching occurs at the exit of an escalator, there will be no space for passengers exiting the escalator. This presents a serious safety hazard to those passengers on the escalator. See Diagram 6.1.3.6.4.

6.1.3.6.5 Flat Steps. Entering passengers require adequate time to position their feet before the steps

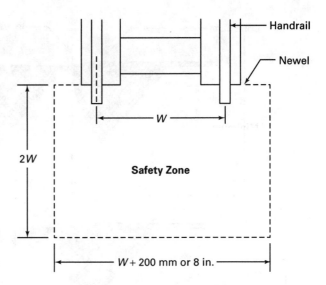

Diagram 6.1.3.6.4   Safety Zone

change their direction of movement from horizontal to the incline. Exiting passengers require adequate time to prepare to step off the escalator; however, excessive time can cause a lack of concentration. Measurement of the flat step distance is from the line of intersection where the top of the comb tooth passes below the top of the step tread, not from the root of the comb teeth. This represents the point where a passenger can first contact the step tread when boarding.

If more than four flat steps are used, when starting an escalator utilizing the key-operated starting switches located on the newels it is very difficult to see if any passengers are either on or boarding the escalator. See Diagram 6.1.3.6.5.

6.1.3.7 Trusses or Girders. The truss is the structural foundation of the escalator. Attached to and/or within the truss are the various components that form the escalator. Structural steel is used to make the truss and may differ from one manufacturer to another since escalator trusses may be made of round tubing, angle iron, square tubing, or in some cases combinations of these materials.

The purpose of the requirement is to prevent any part of the running gear or passengers on the running gear from falling through the truss to the space below. Tension weights are not normally found on current escalator designs.

6.1.3.9 Rated Load. This requirement contains design parameters. There is no Code requirement to test the design in the field. Conformance with these design requirements may be verified by examining the manufacturer's design criteria. The length of the horizontal projection of the truss varies on curved (spiral) escalators as a function of the truss with width. The centerline was selected as the location to measure the length because it represents the average length between the inside minimum and the outside maximum.

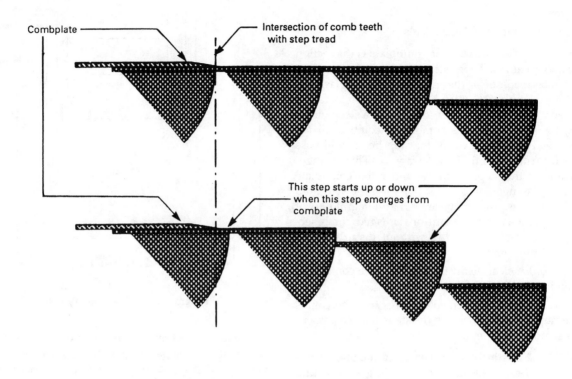

Combplate — Intersection of comb teeth with step tread

This step starts up or down when this step emerges from combplate

Diagram 6.1.3.6.5   Two Flat Steps

6.1.3.10 Design Factors of Safety. Requirements for reinforced concrete structures are not specified in the Code. It is suggested that the normal design criteria for concrete structures be followed in providing supports for escalators. The design of the structure is to be based on static loading.

The intent of the requirements in 6.1.3.10.2 through 6.1.3.10.4 is to ensure that the drive will safely support a loaded escalator. While the limiting mode of failure is ultimate strength and logically the initial design could be tested to ensure the safety factor is achieved, this does not mean that fatigue, etc., should be neglected in the design. Whether the AGMA standards or escalator manufacturers' internal standard is used in drive design, other modes of failure need to be considered. Such standards should include consideration of the endurance limit of a component in regard to the maximum stress that can be varied or reversed indefinitely without producing fracture of the component material.

6.1.3.11 Chains. Cast iron is a commercial alloy of iron, carbon, and silicon that is cast in a mold, is hard, brittle, nonmalleable, has a low tensile strength, and is capable of being hammer welded.

6.1.4 Rated Speed

6.1.4.1 Limits of Speed. A proportion of escalator accidents are related to falls. A problem could occur if the speed was varied during operation without going through a stop and restart sequence. This in fact could be hazardous; due to games played walking or running

against the direction of operation even though this is intensively discouraged. In such a case, misgauging of the speed could result in a fall anywhere in the escalator rise. Variable frequency acceleration to provide controlled starts or Wye-delta motor control is not prohibited.

The tangential speed varies on curved (spiral) escalators as a function of the step width. The centerline was selected as the location to measure the speed for consistency with the other requirements, and because it is closer to the location, where passengers stand while riding the escalator.

6.1.5 Driving Machine, Motor, and Brake

6.1.5.1 Connection Between Driving Machine and Main Drive Shaft. This requirement addresses the connection between the driving machine and the main drive shaft and not the number of driving machines permitted per escalator. This requirement does not say, nor was it ever intended to say, that the escalator shall be driven by only one motor. A fluid coupling is prohibited. A fixed connection is required.

6.1.5.2 Driving Motor. It is not always safe to shut down multiple escalators when circumstances require the stopping of one unit. See Handbook commentary on 6.1.5.1.

6.1.5.3 Brakes. The escalator brake maximum stopping rate provides a smooth stopping action for all passenger loadings. The nameplate indicating the required brake torque will provide a means for periodically checking the brake performance, eliminating the practice of checking the brake by persons carrying weights on a moving escalator, which if not performed carefully, could be unsafe. The maximum deceleration rate is based on a test of high-speed pedestrian conveyors, conducted by the Engineering Physics Department of the Royal Aeronautical Establishment in Farnsborough, England. It also is the normal acceleration and deceleration rate of the San Francisco and the Washington Metro transit vehicles.

Prior to 1937, only an electrically released mechanically applied brake was required on the escalator drive machine (Rule 612a in the 1925 and 1931 editions of the Code). In 1937, recognizing that a dangerous runaway condition could develop should the connection between the drive machine and the main drive shaft part, a new requirement was established. Rule 512f in the 1937 edition required a brake on the main drive shaft to prevent such a runaway condition. When the Code was extensively revised in 1955, Rule 804.3 continued the requirement of the previous Rule 512f, except that the main drive shaft brake was only required when a chain connected the machine to the main drive shaft. This change acknowledged the fact that a damaged or worn gear connection would not produce the runaway condition that a worn or broken drive chain could.

A chain that drives the step chains could permit a runaway condition if they both were to break. A difference exists, however, in the fact that there are two separate chains. Unlike the conventional escalator chain drive, where one link exists that, in failing, immediately produces an unsafe condition, now both independent chains must fail. This introduces a concept of redundancy of equipment to provide safety in operation — a concept widely accepted today. Brake systems in automobiles and control systems in air and space vehicles are examples of its application. If each system in a redundant design is capable of individually fulfilling all the demands imposed on it, and the failure of either system is detected and acted upon, a high degree of safety is achieved. The two drive chains referred to above constitute a redundant design if each can safely support the total escalator loads, and a device is present that would cause the escalator machine brake to stop the unit should either one of the chains break. This would fulfill the original intent by effectively preventing a runaway condition. Safety of operation, the prime objective of the Code, is achieved without the use or need of an auxiliary brake in this particular design.

6.1.5.3.1 Escalator Driving Machine Brake

6.1.5.3.1(a) A braking control system that applies the brake, without intentional delay, with a force that varies as necessary to decelerate the escalator as specified in 6.1.5.3.1(c) but not necessarily the maximum braking force, is an acceptable means of complying with this requirement.

6.1.5.3.1(c) A restriction of deceleration was considered in the up direction, but it was rejected for two basic reasons. First, there was an absence of a history of falls on escalators traveling up that were attributable to the escalator stopping. It is reasonable to assume that up escalators carrying passengers would stop for the same reasons that down running ones would stop. The lack of falls due to the up escalator stopping is the result of the geometry of the escalator steps and handrail and the dynamic interaction of the passenger body, the escalator, and the forces of gravity.

The second reason for not applying a restriction to the up stop deals with the complexity of accomplishing a stop that must extend that which is produced by gravity. Such a stop would require using the motor to bring the escalator to a stop in the up direction. This technology is used in elevator control and is well known to the industry but its complexity and necessary safeguards are not justified by the experience of escalator operation.

The A17 Committee realizes that there is a degree of risk when an up-running escalator is decelerated too quickly and a fall and/or injury could possibly result. The A17 Committee also realizes that when a down-running escalator is decelerated too quickly, there is a much greater risk of fall and more serious injury and an additional risk of causing others to fall. In weighing the consequences of complexity versus degree of risk, the A17 Committee feels that the requirement as written is the proper choice. See Diagram 6.1.5.3.1(c).

6.1.5.3.1(d) The purpose of providing a brake torque nameplate, as required since the stopping requirement outlined in 6.1.5.3.1(d) and 6.1.6.3.6 were established in A17.1–1983, was to replace a potentially dangerous and difficult type test from being performed when a new model escalator was released. This test only confirmed that the basic brake requirement, requiring the fully loaded escalator to stop, was complied with. Since this was only performed for a new model installation, the brake setting was not required to be checked this way during the life of the escalator. Instead, the brake torque was measured from time to time and adjusted to meet the manufacturer's specification, if it was known.

When stopping requirements became more defined in 1983, and maintaining brake torque became more critical, the Code required a brake torque nameplate that displayed the manufacturer's specified brake torque. The brake nameplate was modified in ASME A17.1–2000, to include the minimum distance from the skirt-switch to the combplate, and the minimum stopping distance. Now, inspectors and maintenance personnel alike would know the correct brake torque and be able to periodically check the brake and adjust it if necessary.

Manufacturers develop the brake torque values, for each escalator model, to fulfill the Code requirements

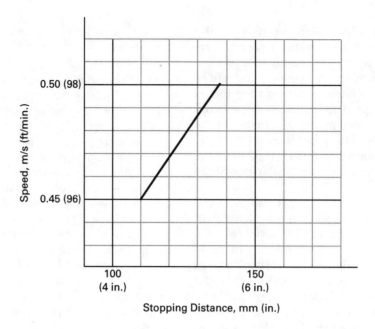

GENERAL NOTE: The above represents the stopping distance of an escalator under a constant deceleration of 0.91 m/s² (3 ft/s²) and does not represent the total stopping distance of the escalator when it is stopped under no load.

Diagram 6.1.5.3.1(c)   Stopping Distances Corresponding to a Deceleration Rate of 0.91 m/s²

through engineering analysis of the data unique to that model, and confirm these results with tests. These tests involve using test weights, encoders, recording devices, and circuit modifications to the controller. This equipment is usually found and best used in a test facility. Because it is both costly and bulky, it does not lend itself to be used periodically on each escalator installation.

When a new model escalator is released, manufacturers can be asked to produce their brake analysis by authorities having jurisdiction over the installation, in the same manner that other design calculations such as truss analyses are requested. This permits verification to the accuracy of the brake calculations. Reliable periodic brake checks can then be made by measuring the existing torque against the nameplate.

6.1.6 Operating and Safety Devices

6.1.6.2 Starting and Inspection Control Switches

6.1.6.2.1 It is important that an individual who is starting an escalator has a clear, unobstructed view of that escalator to ensure that no one is on the steps before the escalator is started. Only key-operated-type starting switches are permissible for starting the escalator. Automatic starting by the use of mats, etc., violates the intent of this requirement. Starting an escalator with passengers on the step is an unsafe practice. This has occurred with automatic starting operations due to misuse of the automatic starting means.

The starting switches must be located such that the escalator steps are within sight. The starting switch must

also be within reach of the emergency stop button (6.1.6.3.1). All persons who start an escalator must also be trained in the proper procedure as required by 8.6.10.5. A closed circuit TV camera used to view the steps does not comply with the intent of the requirements.

6.1.6.2.2 Inspection Control. The intent is to prevent accidents through inadvertent starting/running of escalators by elevator personnel when attempting to position escalators in order to check, adjust, or repair components by providing a switch for use during maintenance and repairs that will prevent the escalator from running except under constant pressure operation.

6.1.6.3 Electrical Protective Devices

6.1.6.3.1 Emergency Stop Buttons. Specifying the exact location of the emergency stop button assures that at the time of an emergency, precious time will not be lost hunting for the stop buttons before the escalator is stopped. The location also gives the user of the emergency stop button a clear, unobstructed view of the escalator. The cover is required to prevent accidental contact. The alarm is to discourage unauthorized/unnecessary use and to alert passengers.

The basic premise of remotely stopping escalators is unsafe. Introducing the human element in the decision process is the weak link. Using TV monitors to supervise the conditions on a remote escalator does not provide for objective and safe control of the unit. Unlike devices that respond to a specific condition and then stop the escalator, a person sitting in a room full of TV monitors and controls is susceptible to fatigue, emotional stress, inattentiveness, carelessness, and other factors that can result in unnecessary or erroneous stops. This type of job is a monotonous one, particularly susceptible to these problems. It is a job sometimes filled by unskilled personnel that may lack motivation.

Although it is true that the major suppliers of escalators have installed units with the capability of providing remote stopping as required by contract specifications, none of the escalators running with this feature are operational. The fact is, that the local authorities have not allowed it, and the suppliers have recommended against its use.

6.1.6.3.2 Speed Governor. The Code does not specifically require a low-slip squirrel cage induction motor, but this type of motor is often used. When an AC induction motor exceeds synchronous speed (an amount equal to its design slip), a braking action takes place electrically that prevents excessive overspeeding.

The distinction between the low and high slip is that the escalator motor must be of low-slip characteristics or loss of speed would occur under load in the up direction. A low-slip motor cannot overspeed beyond 140%. Enough speed cannot be gathered to trip the governor. Thus, an overspeed governor would be superfluous.

6.1.6.3.5 Stop Switch in Machinery Spaces. The stop switch or main line disconnect switch provides the means to ensure that an escalator cannot be started by the starting switch while working in the machinery space.

6.1.6.3.6 Skirt Obstruction Device. In the event that someone's shoe, sneaker, finger, foot, etc., becomes entrapped between the skirt and step, the skirt obstruction device would act to stop the escalator. Skirt obstruction devices are required only in the area where the step approaches the upper and lower combplate areas. This is due to the fact that steps are extending and retracting and have more of a tendency towards entrapment in this area. During the rest of the escalator travel, entrapment usually can only be caused by deliberately forcing an object between the step and skirt. The skirt obstruction devices must be located no less than the distance specified on the brake data plate [6.1.5.3.1(d)(5)] from the combplate. See Diagram 6.1.6.3.6.

The requirement does not establish a specific force value to operate the switch. The performance requirement places the responsibility on the designer to accomplish compliance.

6.1.6.3.7 Escalator Egress Restriction Device. An up-running escalator would bunch people, at a closed rolling shutter or a closed door at the egress end, if the escalator was allowed to continue running. People traveling on the escalator would not be able to come down the escalator because it would still be running in

the up direction. See also the commentary on rolling shutters in Diagram 6.1.1.1.

6.1.6.3.9 Step Upthrust Device. See Diagram 6.1.6.3.9.

6.1.6.3.10 Disconnected Motor Safety Device. This requirement is essential when a non-positive drive connects the escalator motor to the machine. A separate device is required in the event a governor is not used on this equipment.

6.1.6.3.11 Step Level Device. A downward displaced step entering the combplate, as a result of a missing wheel, its tire, or a failure of the end frame where the wheel fastens to the step or pallet, can be hazardous. The step level device must be positioned such that when activated, it will bring the displaced step to a stop before it enters the comb. These devices will prevent the escalator from running under a condition of downward displacement.

6.1.6.3.12 Handrail Entry Device. Objects caught in the area where the handrail reenters the balustrade can produce bodily injury and/or damage to the escalator equipment. A switch in this area should prevent or limit the extent of injury or damage in the event of this occurring. The manual reset is essential to provide the opportunity to examine the handrail return area and handrail entry device for damage before placing the escalator back in service.

6.1.6.3.13 Comb-Step Impact Devices. Escalator step wrecks can occur from a variety of circumstances, some of which may not be detected, or be detectable, with the other safety devices. When these occur, steps continue to be drawn into the pileup until an existing safety device is haphazardly tripped or someone pushes the stop button. This requirement is intended to detect the first step to impact the comb and stop the escalator as a direct result of the impact. Because steps can strike the comb from below as well as entering, both vertical and horizontal impact detection is necessary.

6.1.6.3.14 Step Lateral Displacement Device. The tension in the step chain of curved (spiral) escalators will attempt to straighten the chain out if there is a failure in the side guiding system. This will force the steps against the skirt, producing an unsafe condition if the escalator continues to operate.

6.1.6.4 Handrail-Speed Monitoring Device. There is a need to stop an escalator if the speed of the handrail deviates by 15% or more. While a speed deviation of less than 15% would also not comply with 6.1.3.4.1, it would not be enough to necessitate stopping the escalator right away. A delay is necessary so that the escalator is not stopped every time someone pulls on the handrail.

A handrail not moving in the same direction or speed has a tendency to pull passengers backward or forward causing them to lose their balance. It is not necessary

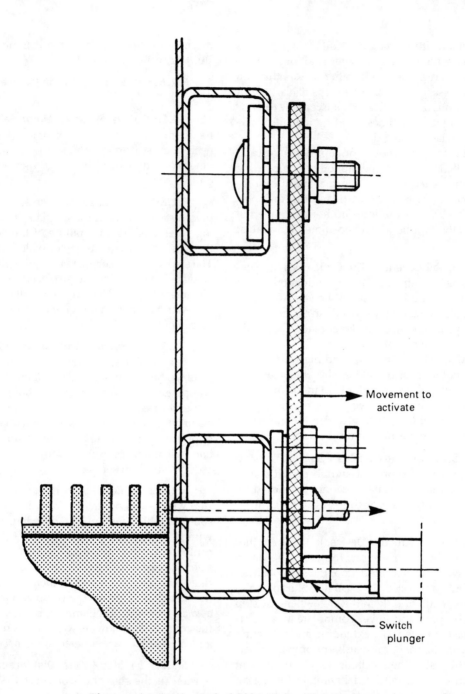

Diagram 6.1.6.3.6   Typical Skirt Obstruction Device

for the handrail to move at exactly the same speed as the step since passengers can easily adjust the position of their hands to compensate for minor differences in speeds.

6.1.6.5 Missing Step Device. This requirement will prevent an escalator from continuing to run if a step is missing. It will prevent a void from a missing step to emerge from under the comb. This requirement was adopted because of the seriousness of the injury that results from stepping into the gap of a missing step, and

the fact that a passenger would not see the gap until starting to step into it.

6.1.6.6 Tandem Operation. See Diagram 6.1.6.6.

6.1.6.7 Step Demarcation Lights. Step demarcation lights warn the escalator passengers that they are approaching the area where they must exit or enter the escalator. The demarcation lights must be green and that may be achieved by using green fluorescent lamps or filters. The demarcation lights are required to show the

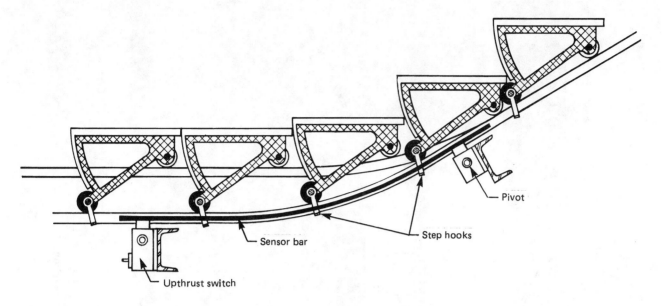

Diagram 6.1.6.3.9 Typical Step Upthrust Device

separation between steps as they emerge from the comb. The intent of the requirement is that the lighting be visible between adjacent step treads when they travel over the demarcation lights. These lights provide assistance to passengers boarding the escalator by highlighting the area where they should place their feet, especially on escalators that are not located in well-lit areas. See 6.1.3.6 for the requirements on the illumination of the step and floor plate.

6.1.6.8 Escalator Smoke Detectors. Smoke detectors are not required in wellways by ASME A17.1. Some building codes do require smoke detectors in wellways and adjacent areas. When a smoke detector is used to stop a moving escalator, these requirements must be adhered to. This requirement was adopted because if smoke detectors were set to stop the escalator, certain events were deemed necessary to permit the escalator to stop safely.

6.1.6.9 Signs

6.1.6.9.1 Caution Signs. The sign (Fig. 6.1.6.9.1) cautions people on how to use an escalator properly and indicates that it is for the transportation of people only. Most accidents are caused by improper use of an escalator. The pictograph reduces the problems caused by language barriers.

6.1.6.9.2 Additional Signs. Additional signs may be provided since during the normal walking flow of people entering an escalator, there is only sufficient time to read and comprehend the required sign. To properly board an escalator, it is important to continue to walk without stopping or otherwise causing abrupt bodily contact with others. For this reason a sign that is well outside the normal boarding traffic flow is permitted.

Additional signs within the boarding area are prohibited. Due to escalators being installed in many different locations, there is no consensus on what this sign should say. Some of the most frequently used messages are: "No Luggage or Carts," "No Bare Feet," "No Strollers," "No Wheelchairs," "Handicapped-Use Elevator."

The Committee has, therefore, provided some broad guidelines and a message to building owners, etc., to consider anything peculiar to the building site that requires further unique messages.

6.1.6.10 Control and Operating Circuits. This requirement provides for protection against singular failures resulting in an unsafe condition.

6.1.6.12 Installation of Capacitors or Other Devices to Make Electrical Protective Devices Inoperative. This mandates that the electrical protective devices always be operative and that they may not be bypassed except as provided for in the Code.

6.1.6.13 Completion or Maintenance of Circuit. The requirement is written in a manner such that failure of any circuit associated with an electrical protective device will cause the device to operate. This is a basic principle of fail-safe design. The intent of 6.1.6.13 is that the contactors used to interrupt the power to the escalator motor and brake shall not be energized to perform this function.

6.1.6.14 Escalator Manual Reset. An external power failure should not result in any safety device requiring manual reset to lose its indication of malfunction.

6.1.6.15 Contactors and Relays for Use in Critical Operating Circuits. This requirement specifies the requirements for single failure protection, i.e., redundancy, in the design of specific control and operating

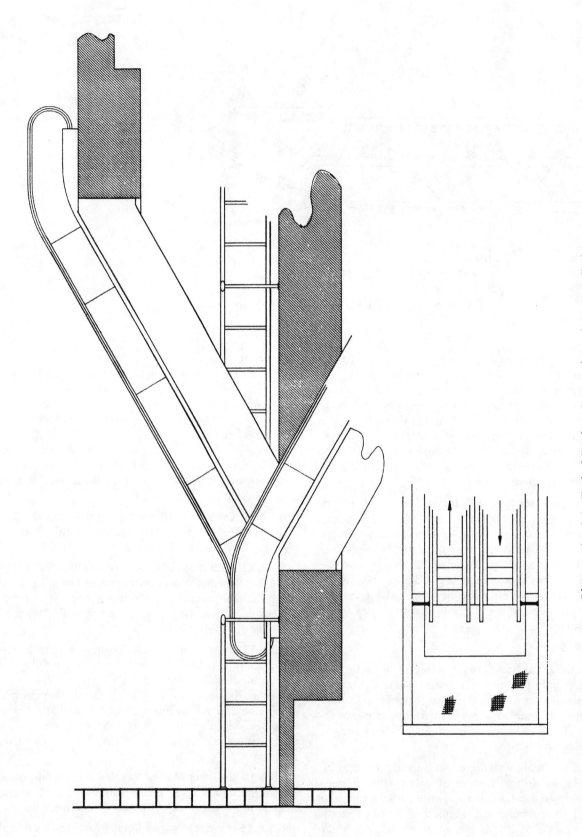

Diagram 6.1.6.6 Typical Tandem-Operated Escalator Installation
(Courtesy Schindler Elevator Co.)

circuits (6.1.6.10.1) and requires the checking of that redundancy (6.1.6.10.2). This may require the use of two relays instead of one in these circuits, and the checking of the operation of these relays.

When electromechanical relays and contactors are being checked for operation, specifically drop out operation, by utilizing contacts on the relays or contactors themselves, care must be taken that the contact used for monitoring the state of the contact used in the critical circuit is a true and correct representation of the critical circuit contact.

The required operation of the monitoring contact is specified, and it is clarified that springs in the design of the relay, contactor, or contact design may be used, but that spring failure shall not cause the monitoring contact to indicate an open critical circuit contact if the critical circuit contact is not open.

6.1.7 Lighting, Access, and Electrical Work

6.1.7.1 Lighting of Machine Room and Truss Interior. The requirements assure that a source of illumination and power is provided in machine rooms for elevator personnel. Machine spaces in the truss, either at the end or on the incline, do not require permanent lighting, as it is impractical to locate a light fixture, which will provide lighting to all work areas.

6.1.7.1.2 Truss Interior. The requirement for a duplex receptacle in the truss interior is to allow for the use of a drop cord light for inspection and maintenance inside the truss.

6.1.7.2 Lighting of Escalator. Step treads must be illuminated properly in order to reduce a tripping hazard.

6.1.7.3 Access to Interior. The removal of escalator steps is considered a reasonable means of access to the interior of an escalator.

6.1.7.3.1 The use of terrazzo, stone, or concrete, and other similar floor coverings that are excessively heavy, and would be a hazard for maintenance and inspection personnel to lift. This type of material also tends to crack, resulting in a tripping hazard. Access plates must be securely fastened to prevent accidental opening.

6.1.7.4 Electrical Equipment and Wiring

6.1.7.4.1 See Handbook commentary on 2.26.4.1.

6.1.7.4.2 The specified electric equipment is required to be certified as complying with CSA B44.1/ASME A17.5. This will minimize the risk of electricity as a source of electric shock and as a potential ignition source of fires. See Handbook commentary on 2.26.4.2.

6.1.7.4.3 See Handbook commentary on 2.26.4.3.

6.1.8 Outdoor Escalators

The Code requires outdoor escalators to be covered (see 6.1.8.2). Experience has shown that outdoor escalators must be protected from the weather. Escalators installed outdoors without any cover were usually shut down when it rained; however, they have since been covered. It was also observed that when it rained or snowed and the escalator was uncovered, passengers ran up the steps, which is considered unsafe.

6.1.8.2 Precipitation. Skirt deflectors and other devices must be protected so they do not present a hazard to the rider. See Diagram 6.1.8.2.

SECTION 6.2
MOVING WALKS

Moving walks are for transporting people only, not for moving heavy, bulky, crude objects that may cause overloading or damage that could contribute to failure of parts, scratching of skirts, congestion or jam-ups, which could cause an unsafe condition.

6.2.1 Protection of Floor Openings

6.2.1.1 Protection Required. Moving walks passing through fire-rated floor assemblies must have their openings protected against the passage of flame, heat, and/or smoke. If not, a fire has a clear path of travel from one floor to another. Typically, moving walks do not pass through fire-rated floor assemblies and in these cases, no protection of the opening is required. An example of this type of construction is a moving walk serving two floors that are within an atrium.

The building code is an ordinance which sets forth requirements for building design and construction, or where such an ordinance has not been enacted one of the following model codes:

(a) National Building Code (NBC)
(b) Standard Building Code (SBC)
(c) Uniform Building Code (UBC)
(d) National Building Code of Canada (NBCC)

It is noted that the International Building Code (IBC) issued in the first quarter of 2000 is the successor code for the NBC, SBC, and UBC.

Most building codes recognize that unlike stairs, moving walks do not need to be fully enclosed, since moving walks are normally not part of a required means of egress. Diagram 6.1.1.1 describes several methods, which are utilized in completely sprinklered buildings, where the moving walks are not used as exits, which permit them to be unenclosed from floor to floor. The reader is cautioned to check for conformance with the building code requirement before proceeding to use any of the methods described in Diagram 6.1.1.1.

6.2.2 Protection of Supports and Machine Spaces Against Fire

6.2.2.1 Protection Required. The sides and undersides of moving walk trusses and machinery spaces must be enclosed in noncombustible or limited-combustible materials to protect the trusses from heat during a fire

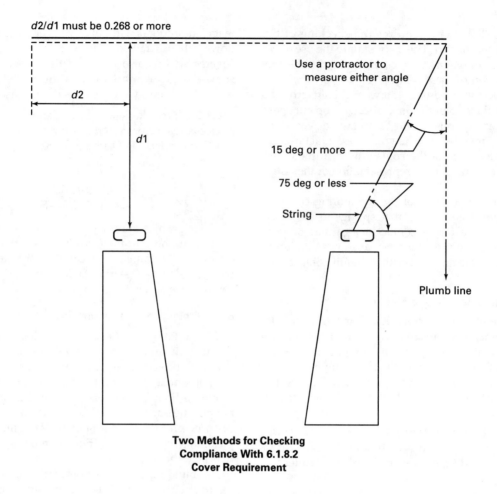

*d*2/*d*1 must be 0.268 or more

*d*2

*d*1

Use a protractor to
measure either angle

15 deg or more

75 deg or less

String

Plumb line

**Two Methods for Checking
Compliance With 6.1.8.2
Cover Requirement**

Diagram 6.1.8.2   Outdoor Escalator Cover

in the vicinity immediately outside the truss. The Life Safety Code®, NFPA 101 defines limited-combustible and noncombustible as follows:

"Limited-Combustible: Refers to a building construction material not complying with the definition of noncombustible (see 3.3.131) that, in the form in which it is used, has a potential heat value not exceeding 3500 Btu/lb (8141 kJ/kg), where tested in accordance with NFPA 259, *Standard Test Method for Potential Heat of Building Materials,* and includes (1) materials having a structural base of noncombustible material, with a surfacing not exceeding a thickness of ⅛ in. (3.2 mm) that has a flame spread index not greater than 50; and (2) materials, in the form and thickness used, other than as described in (1), having neither a flame spread index greater than 25 nor evidence of continued progressive combustion, and of such composition that surfaces that would be exposed by cutting through the material on any plane would have neither a flame spread index greater than 25 nor evidence of continued progressive combustion."

"Noncombustible: Refers to a material that, in the form in which it is used and under the conditions anticipated, does not ignite, burn, support combustion, or release flammable vapors, when subjected to fire or heat. Materials that are reported as passing ASTM E 136, *Standard Test Method for Behavior of Materials in a Vertical Tube Furnace at 750 Degrees C,* are considered noncombustible materials."

Noncombustible or limited-combustible material is not required between multiple moving walks in a single wellway. This may also help contain a fire that may start within the truss. It can be met simply by applying gypsum drywall or an equally noncombustible or limited-combustible material on the outside of the truss and machinery space. No hourly rating is required. A truss or machinery space suspended in an approved fire-rated ceiling assembly, where the truss is exposed within that assembly, does not meet this requirement. The truss and machinery space must be separately enclosed within the rated ceiling assembly; also, the sides and undersides must be enclosed. Dampers in the noncombustible or limited-combustible material enclosing the truss must

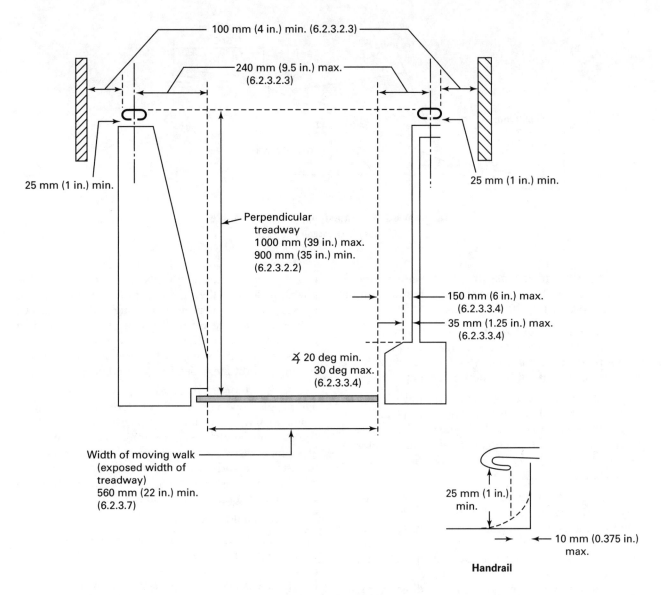

Diagram 6.2.3   Moving Walk Geometry

be provided with fire links, which will close the opening when there is a fire.

6.2.3 Construction Requirements

See Diagram 6.2.3.

6.2.3.1 Angle of Inclination. An angle of inclination above those specified by this Rule is considered too steep for safe use. A larger angle of inclination may cause foot strain and/or dizziness in some riders. See Diagram 6.2.3.1.

6.2.3.2 Geometry. The handrail clearances reduce entrapment and shear exposure and position the handrail to facilitate its use. These clearances must be maintained wherever the handrail is exposed. Components of the handrail guard are not subject to this requirement. See Diagram 6.2.3.

6.2.3.3 Balustrades

6.2.3.3.1 Construction. Glass balustrades require gaps of about 5 mm or $^{3}/_{16}$ in. between glass panels to allow for expansion. This condition is allowed. Raised or depressed areas of molding more than 6.4 mm or $^{1}/_{4}$ in. high could snag clothing, packages, and parts of the body.

6.2.3.3.2 Strength. This requirement specifically states that the balustrades shall be designed to resist the forces specified. There is no Code requirement to test the design in the field. Conformance with the design requirement may be verified by examining the manufacturer's design criteria. This requirement simply means that the balustrade has been designed to "resist" the applied load and when the load is removed, there is no permanent deformation or damage.

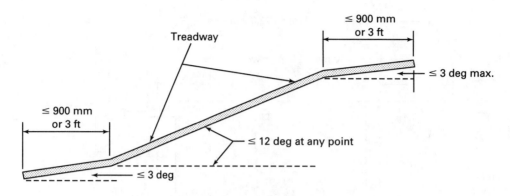

Diagram 6.2.3.1 Maximum Treadway Angle
(Courtesy Zack McCain, Jr., P.E.)

6.2.3.3.3 Use of Glass or Plastic. See Handbook commentary on 2.14.1.8 for additional information on ANSI Z97.1 and 16 CFR Part 1201.

6.2.3.3.4 Interior Low Deck. The requirements provide a maximum angle of the interior deck profile that assist in preventing a youngster from slipping while attempting to walk on the interior decking.

6.2.3.3.5 Skirtless Balustrade. Vertical dimension assures that the balustrade would not have a knife edge. It should be smooth to avoid snagging clothing or footwear. Excessive deflection could contribute to objects getting caught between the balustrade and treadway. See Diagram 6.2.3.3.5.

6.2.3.3.6 Skirt Panels. The deflection requirement for the skirt panel in combination with the clearance requirement provides the best protection against accidental entrapment of a foot, finger, clothing, shoes, sneakers, etc., between the skirt and the pallet. Entrapment incidents rarely occur on a moving walk as the walkway surface is moving in a parallel direction to the skirt's. Vertical dimension assures that the balustrade would not have a knife edge. It should be smooth to avoid snagging clothing or footwear. Excessive deflection could contribute to objects getting caught between the balustrade and treadway. See Diagram 6.2.3.

6.2.3.3.7 Guard at Ceiling Intersection. A solid guard must be provided in the intersecting angle of each outside top of the balustrade and the point of penetration of the ceiling or soffit. It is not required on high deck moving walks where the side of the penetration at the ceiling or soffit is more than 600 mm or 24 in. from the centerline of the handrail or on low deck moving walks where the centerline of the handrail is more than 350 mm or 14 in. from the centerline of the soffit. The vertical edge of the guard must be a minimum of 350 mm or 14 in. and be flush with the face of the wellway. The leading edge or the exposed edge of the guard should be vertical and must be rounded. The material used for the guard should be solid without holes or openings

passing through it. Glass or plastic can be used for the guards as long as they conform to the requirements of 6.2.3.3.3. This guard protects moving walk passengers, if they are leaning off to the side of an up-traveling moving walk, from the possibility of trapping their heads or arms in the angle where the moving walk and ceiling or soffit meet. See Diagram 6.1.3.3.9.

6.2.3.3.8 Deck Barricades. Deck barricades discourage people, especially small children, from walking or going up or down the low deck exterior balustrade from the ground level. See Diagram 6.1.3.3.11. Barricade attachment fasteners are required to be tamper resistant to discourage and/or prevent the removal of some or all of the fasteners and/or the barricade itself, which would create a dangerous falling hazard.

6.2.3.4 Handrails

6.2.3.4.1 Type Required. A handrail not moving in the same direction or speed would have a tendency to pull passengers backward or forward causing them to lose their balance. It is not necessary for the handrail to move at exactly the same speed as the pallet since passengers can easily adjust the position of their hands to compensate for minor differences in speeds. The minimum handrail driving force is necessary to ensure that the handrail continues to operate under a passenger load.

6.2.3.4.2 Extension Beyond Combplates. This requirement allows passengers to grab hold of a handrail before they step on to the moving walk and allows them to continue to hold on to the handrail until they have completely stepped off the moving portion of the moving walk. The moving handrail extending beyond the exit point also has the tendency to immediately move the passenger away from the exit, thus reducing bunching at this location. See Diagram 6.2.3.4.2.

6.2.3.4.3 Guards. Hand or finger guards provide protection for the curious, especially children, who want to explore the area where the handrail enters the balustrade.

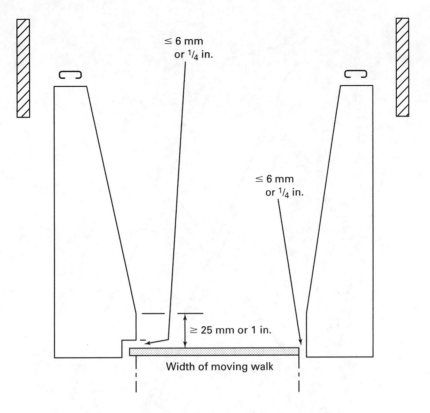

Diagram 6.2.3.3.5   Skirtless Balustrade Clearances

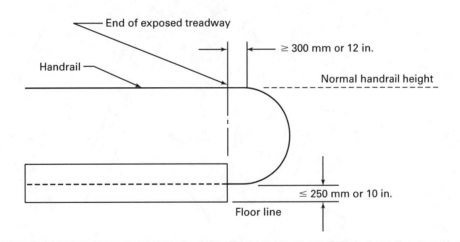

Diagram 6.2.3.4.2   Moving Walk Handrail Extension

6.2.3.4.4 Splicing. Although field splicing of moving walk handrails is rare, it is done on occasion.

6.2.3.4.5 Handrail Clearance. Provides a limitation of a dimension that, if excessive, can permit fingers to become caught. See Diagram 6.2.3.

6.2.3.5 Pallet-Type Treadway. A pallet-type treadway is a treadway consisting of closely coupled intermeshing pallets or small platforms, with the exposed surface of

each being grooved. Diagram 6.2.3.5 illustrates a pallet-type treadway.

6.2.3.5.1 Slotting of Treadway. The safety principles of the slots in the tread are the same for moving walks and escalators. Therefore, the dimensions are the same. See Diagram 6.2.3.5.1.

6.2.3.5.2 Intermeshing Pallets. The intermeshing of adjacent pallet slots protects against entrapment of a

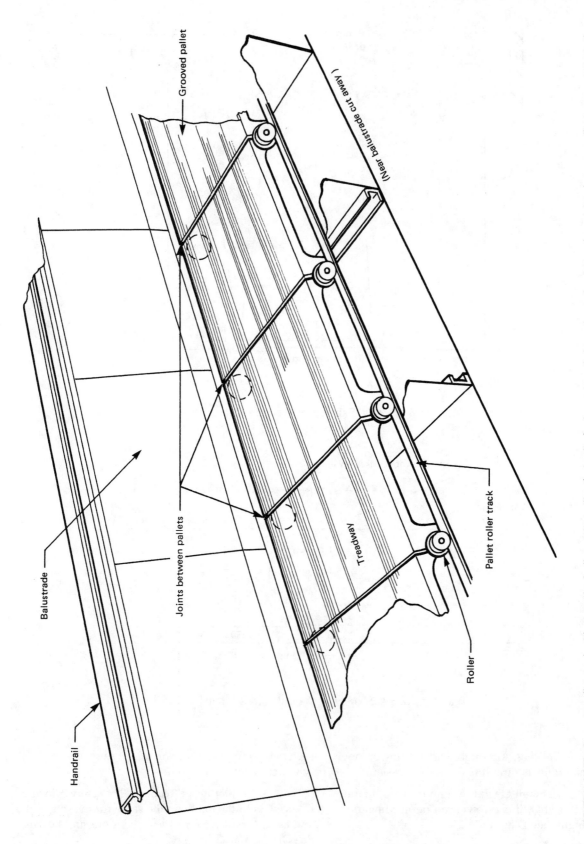

Handrail

Balustrade

Joints between pallets

Grooved pallet

(Near balustrade cut away)

Treadway

Pallet roller track

Roller

Diagram 6.2.3.5   Pallet-Type Moving Walk
(Courtesy Charles Culp)

GENERAL NOTE: Separate pallets each with one pair of rollers at its trailing (or leading) edge are coupled together to form a treadway. Joints between pallets are apparent in the treadway surface.

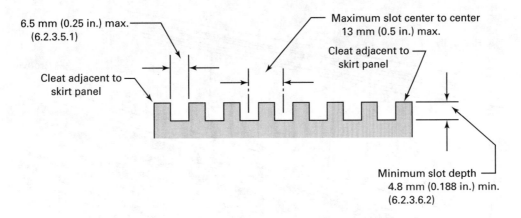

Diagram 6.2.3.5.1   Moving Walk Treadway Slots

foot or finger and articles of clothing such as shoes, sneakers, etc. See also commentary on 6.2.3.8.

6.2.3.5.3 Alignment of Pallet Tread Surfaces. Since people have a tendency to walk on the moving walk, no abrupt changes in the walk surface should occur, which could create tripping hazards.

6.2.3.5.4 Pallet Fatigue Tests. See commentary on 8.3.11.

6.2.3.5.5 Material and Type. A concern addressed by the requirement is the potential of moving walk components spreading a fire.

6.2.3.5.6 Pallet Wheels. There have been a number of instances where the wheels have become separated from the pallet. Although the one end of the pallet is still attached to the pallet chain, the other end can pivot down into the moving walk interior. If the wheels were located inside the pallet, then when the pallet pivoted down it would be stopped by the pallet wheel track located below. However, if the wheels are located outside the width of the pallet, the wheel support track is located beyond the edge of the pallet and there is nothing to prevent the pallet from pivoting into the moving walk interior. If this happens, passengers standing on the pallet could fall into the moving walk interior and be seriously hurt. This requirement is intended to prevent that condition.

6.2.3.6 Belt-Type Treadway. A belt-type treadway is a treadway consisting of an integral belt of uniform width and thickness made or spliced in one continuous piece forming a loop. The particular design determines the belt thickness. A belt-type treadway is illustrated in Diagram 6.2.3.6(a). Some earlier moving walk designs utilized a belt-pallet-type treadway. A belt-pallet-type treadway consists of a continuous integral belt supported on coupled pallets. The pallets may not be ordinarily visible to a person standing on the treadway. A belt-pallet-type treadway is illustrated in Diagram 6.2.3.6(b).

6.2.3.6.2 Slotting of Treadway. Slotting is required for proper combing and safe footing. The fact that the tread is a belt or pallet does not change the dimensional requirements.

6.2.3.8 Entrance and Egress Ends

6.2.3.8.1 Combplates. The combplate is the section of the floor plate nearest the pallet or belt. It normally has replaceable combs made of brittle material such as plastic or aluminum. Comb teeth must be brittle and break rather than bend. A bent comb tooth could cause a pallet wreck piling the pallets up at the comb. The comb teeth are so arranged that the cleats of the pallet or belt treads will pass between them with minimum clearance. This arrangement allows the pallets to move under the combplate allowing passengers to move safely from the pallet onto the floor. Because the combs are not an integral part of the combplate assembly, worn or damaged comb sections can be replaced readily without disturbing the balance of the comb.

6.2.3.8.1(d) This requirement ensures that the combs and combplate do not drag on the pallets when a passenger is standing on them.

6.2.3.8.2 Distinction Between Comb and Pallet. A significant number of falls occur on boarding a moving walk. This requirement is important in that it requires a visual aid for passengers to assist in boarding the moving walk. This should reduce the hazard by calling attention to that area.

6.2.3.8.3 Adjacent Floor Surfaces. This requirement addresses a potential tripping hazard. This requirement applies to the difference in elevation between the landing plate and the adjacent building floor, not between the comb and landing plate. There always will be a difference in elevation between the comb and landing plate due to the following factors:

(a) combplate thickness required to meet current deflection criteria;

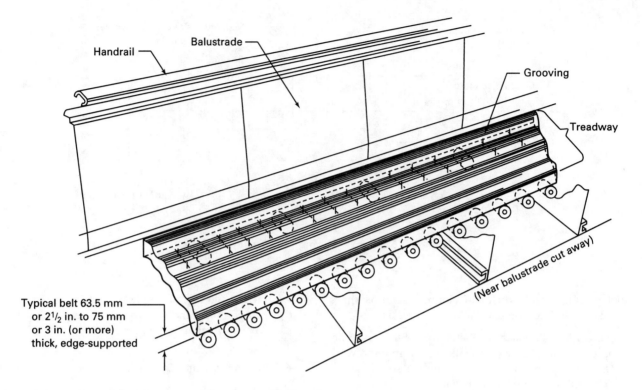

Handrail

Balustrade

Grooving

Treadway

Typical belt 63.5 mm or 2½ in. to 75 mm or 3 in. (or more) thick, edge-supported

(Near balustrade cut away)

Diagram 6.2.3.6(a)   Belt-Type Moving Walk
(Courtesy Charles Culp)

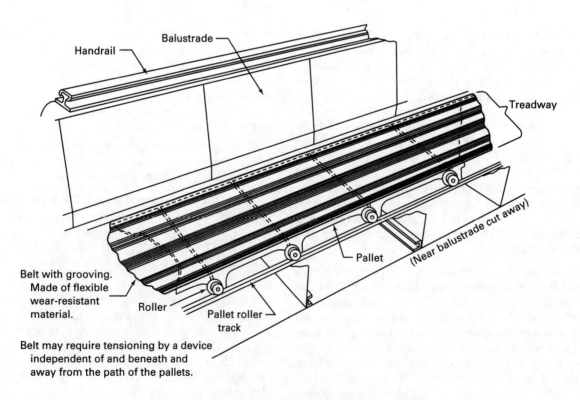

Handrail

Balustrade

Treadway

Belt with grooving. Made of flexible wear-resistant material.

Roller

Pallet roller track

Pallet

(Near balustrade cut away)

Belt may require tensioning by a device independent of and beneath and away from the path of the pallets.

Diagram 6.2.3.6(b)   Belt-Pallet-Type Moving Walk
(Courtesy Charles Culp)

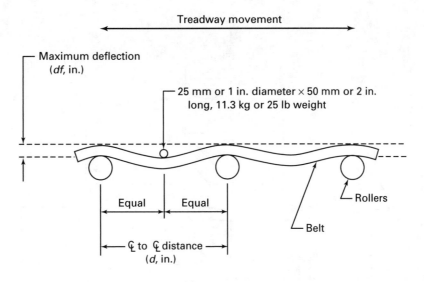

GENERAL NOTE: Maximum Deflection $(df) = 0.094$ in. $+ (0.004 \times d)$

Diagram 6.2.3.9(b)   Belt Deflection With Roller Bed Support

(b) allowance for combplate deflection;

(c) pallet tail or nose swing — the tail or nose moves in an arc as the pallet begins its travel through the turnaround;

(d) a small amount of additional running clearance to allow for manufacturing tolerance and wear-in of parts;

(e) possible increased combplate thickness necessitated by proposed requirements for combplates.

Floor height variations usually occur when buildings are renovated and different floor materials are used than were present when the moving walk was originally installed.

6.2.3.8.4 Safety Zone. The safety zone assures there is adequate space for passengers when they exit a moving walk. If bunching occurs at the exit of a moving walk, there will be no space for passengers exiting the moving walk. This presents a serious safety hazard to those passengers on the moving walk. See Diagram 6.1.3.6.4.

6.2.3.9 Supporting Structure. The truss is the structural foundation of the moving walk. Attached to and/or within the truss are the various components that form the moving walk. Structural steel is used to make the truss and may differ from one manufacturer to another since moving walk trusses may be made of round tubing, angle iron, square tubing, or in some cases combinations of these materials.

The purpose of the requirement is to prevent any part of the running gear or passengers on the running gear from falling through the truss to the space below. Belt deflection requirements are illustrated in Diagrams 6.2.3.9(b) and 6.2.3.9(c).

6.2.3.10 Rated Load. This requirement contains design parameters. There are no Code requirements to test the design in the field. Conformance with these design requirements may be verified by examining the manufacturer's design criteria.

6.2.3.11 Design Factors of Safety. Requirements for reinforced concrete structures are not specified in the Code. It is suggested that the normal design criteria for concrete structures be followed in providing supports for moving walks. The design of the structure is to be based on static loading.

The intent of the requirements in 6.2.3.10.2 through 6.2.3.10.4 is to ensure that the drive will safely support a loaded moving walk. While the limiting mode of failure is ultimate strength and logically the initial design could be tested to ensure the safety factor is achieved, this does not mean that fatigue and notch sensitivity factors should be neglected in the design. Whether the AGMA standards or moving walk manufacturers' internal standard is used in drive design, other modes of failure need to be considered. Such standards should include consideration of the endurance limit of a component in regard to the maximum stress that can be varied or reversed indefinitely without producing fracture of the component material.

6.2.3.12 Chains. Cast iron is a commercial alloy of iron, carbon, and silicon that is cast in a mold. It is hard, brittle, and nonmalleable, and with low tensile strength.

6.2.5 Driving Machine, Motor, and Brake

6.2.5.1 Connection Between Driving Machine and Main Drive Shaft. This requirement addresses the connection between the driving machine and the main drive shaft

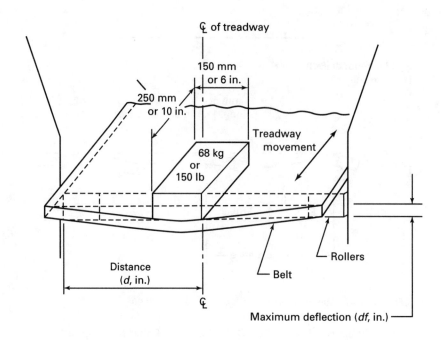

GENERAL NOTE: Maximum Deflection $(df) = d \times 0.03$

Diagram 6.2.3.9(c)   Belt Deflection With Edge Support

and not the number of driving machines permitted per moving walk. This requirement does not say, nor was it ever intended to say, that the moving walk shall be driven by only one motor. A fluid coupling is prohibited. A fixed connection is required.

6.2.5.2 Driving Motor. It is not always safe to shut down multiple moving walks when circumstances require the stopping of one unit. See Handbook commentary on 6.1.5.1.

6.2.5.3 Brakes. The moving walk brake maximum stopping rate provides a smooth stopping action for all passenger loadings. The name plate indicating the required brake torque will provide a means for periodically checking the brake performance and eliminating the practice of checking the brake by persons carrying weights on a moving walk, which if not performed carefully, could be unsafe. The maximum deceleration rate is based on a test of high-speed pedestrian conveyors conducted by the Engineering Physics Department of the Royal Aeronautical Establishment in Farnsborough, England. It is also the normal acceleration and deceleration rate of the San Francisco and the Washington Metro transit vehicles. See also, Handbook commentary on 6.1.5.3.

6.2.5.3.1 Moving Walk Driving-Machine Brake

6.2.5.3.1(a) A braking control system that applies the brake, without intentional delay, with a force that varies as necessary to decelerate the moving walk as specified in 6.2.5.3.1(c) but not necessarily the maximum braking force, is an acceptable means of complying with this requirement.

6.2.5.3.1(c) A restriction of deceleration was considered in the up direction, but it was rejected for two basic reasons. First, there was an absence of a history of falls on moving walks traveling up that were attributable to the moving walk stopping. It is reasonable to assume that up-running moving walks carrying passengers would stop for the same reasons that down-running ones would stop. The lack of falls due to the up-running moving walk stopping is the result of the geometry of the moving walk pallets and handrail and the dynamic interaction of the passenger body, the moving walk, and the forces of gravity.

The second reason for not applying a restriction to the up stop deals with the complexity of accomplishing a stop that must extend that which is produced by gravity. Such a stop would require using the motor to bring the moving walk to a stop in the up direction. This technology is used in elevator control and is well known to the industry but its complexity and necessary safeguards are not justified by the experience of moving walk operation.

The A17 Committee realizes that there is a degree of risk when an up-running moving walk is decelerated too quickly and a fall and/or injury could possibly result. The A17 Committee also realizes that when a down-running moving walk is decelerated too quickly, there is a much greater risk of fall and more serious injury and an additional risk of causing others to fall.

In weighing the consequences of complexity versus degree of risk, the A17 Committee feels that the requirement as written is the proper choice. See Diagram 6.1.5.3.1(c).

6.2.5.3.1(d) The purpose of providing a brake torque nameplate, as required since the stopping requirement outlined in 6.2.5.3.1(d) and 6.2.6.3.6 were established in A17.1–1983, was to replace a potentially dangerous and difficult type test from being performed when a new model moving walk was released. This test only confirmed that the basic brake requirement, requiring the fully loaded moving walk to stop, was complied with. Since this was only performed for a new model installation, the brake setting was not required to be checked this way during the life of the moving walk. Instead, the brake torque was measured from time to time and adjusted to meet the manufacturer's specification, if it was known.

When stopping requirements became more defined in 1983, and maintaining brake torque became more critical, the Code required a brake torque nameplate that displayed the manufacturer's specified brake torque. The brake nameplate was modified in ASME A17.1–2000, to include the minimum distance from the skirt-switch to the combplate, and the minimum stopping distance. Now, inspectors and maintenance personnel alike would know the correct brake torque and be able to periodically check the brake and adjust it if necessary.

Manufacturers develop the brake torque values, for each moving walk model, to fulfill the Code requirements through engineering analysis of the data unique to that model, and confirm these results with tests. These tests involve using test weights, encoders, recording devices, and circuit modifications to the controller. This equipment is usually found and best used in a test facility. Because it is both costly and bulky, it does not lend itself to be used periodically on each moving walk installation.

When a new model moving walk is released, manufacturers can be asked to produce their brake analysis by authorities having jurisdiction over the installation, in the same manner that other design calculations such as truss analyses are requested. This permits verification to the accuracy of the brake calculations. Reliable periodic brake checks can then be made by measuring the existing torque against the nameplate.

6.2.6 Operating and Safety Devices

6.2.6.2 Starting and Inspection Control Switches

6.2.6.2.1 It is important that an individual who is starting a moving walk has a clear, unobstructed view of that moving walk to ensure that no one is on the pallets before the moving walk is started. Only key-operated-type starting switches are permissible for starting the moving walk. Automatic starting by the use of mats, etc., violates the intent of this requirement. Starting a moving walk with passengers on the pallet is an unsafe practice. This has occurred with automatic starting operations due to misuse of the automatic starting means.

The starting switches must be located such that the moving walk pallets are within sight. The starting switch must also be within reach of the emergency stop button (6.2.6.3.1). All persons who start a moving walk, must also be trained in the proper procedure as required by 8.6.10.5. A closed circuit TV camera used to view the treadway does not comply with the intent of the requirements.

6.2.6.2.2 Inspection Control. The intent is to prevent accidents through inadvertent starting/running of moving walks by elevator personnel when attempting to position moving walks in order to check, adjust, or repair components by providing a switch for use during maintenance and repairs that will prevent the moving walk from running except under constant pressure operation.

6.2.6.3 Electrical Protective Devices

6.2.6.3.1 Emergency Stop Buttons. Specifying the exact location of the emergency stop button assures that at the time of an emergency, precious time will not be lost hunting for the stop buttons before the moving walk is stopped. The location also gives the user of the emergency stop button a clear, unobstructed view of the moving walk. The cover is required to prevent accidental contact. The alarm is to discourage unauthorized/unnecessary use and to alert passengers.

The basic premise of remotely stopping moving walks is unsafe. Introducing the human element in the decision process is the weak link. Using TV monitors to supervise the conditions on a remote moving walk does not provide for objective and safe control of the unit. Unlike devices that respond to a specific condition and then stop the moving walk, a person sitting in a room full of TV monitors and controls is susceptible to fatigue, emotional stress, inattentiveness, carelessness, and other factors that can result in unnecessary or erroneous stops. This type of job is a monotonous one, particularly susceptible to these problems. It is a job sometimes filled by unskilled personnel that may lack motivation.

Although it is true that the major suppliers of moving walk have installed units with the capability of providing remote stopping as required by contract specifications, none of the moving walks running with this feature are operational. The fact is, that the local authorities have not allowed it, and the suppliers have recommended against its use.

6.2.6.3.2 Speed Governor. The Code does not specifically require a low-slip squirrel cage induction motor, but this type of motor is often used. When an AC induction motor exceeds synchronous speed, a braking action takes place electrically that prevents excessive overspeeding. The more the motor attempts to exceed synchronous speed, the more the braking action increases, preventing excessive overspeeding.

The distinction between the low and high slip is that the moving walk motor must be of low-slip characteristics or loss of speed would occur under load in the up direction. A low-slip motor cannot overspeed beyond 140%. Enough speed cannot be gathered to trip the governor. Thus, an overspeed governor would be superfluous.

6.2.6.3.5 Stop Switch in Machinery Spaces. The stop switch or main line disconnect switch provides the means to ensure that a moving walk cannot be started by the starting switch while working in the machinery space.

6.2.6.3.6 Moving Walk Egress Restriction Device. An up-running moving walk would bunch people, at a closed rolling shutter or a closed door at the egress end, if the moving walk was allowed to continue running. People traveling on the moving walk would not be able to come down the moving walk because it would still be running in the up direction. See also the commentary on rolling shutters in Diagram 6.1.1.1.

6.2.6.3.8 Disconnected Motor Safety Device. This requirement is essential when a non-positive drive connects the moving walk motor to the machine. A separate device is required in the event a governor is not used on this equipment.

6.2.6.3.9 Pallet Level Device. A displaced pallet entering the combplate, as a result of a missing wheel, its tire, or a failure of the end frame where the wheel fastens to the pallet, can be hazardous. The pallet level device must be positioned such that when activated, it will bring the displaced pallet to a stop before it enters the comb. These devices will prevent the moving walk from running under a condition of downward displacement.

6.2.6.3.10 Handrail Entry Device. Objects caught in the area where the handrail reenters the balustrade can produce bodily injury and/or damage to the moving walk equipment. A switch in this area should prevent or limit the extent of injury or damage in the event of this occurring. The manual reset is essential to provide the opportunity to examine the handrail return area and handrail entry device for damage before placing the moving walk back in service.

6.2.6.3.11 Comb-Pallet Impact Devices. Moving walk pallet wrecks can occur from a variety of circumstances, some of which may not be detected, or be detectable, with the other safety devices. When these occur, pallets continue to be drawn into the pileup until an existing safety device is haphazardly tripped or someone pushes the stop button. This requirement is intended to detect the first pallet to impact the comb and stop the moving walk as a direct result of the impact. Because

pallets can strike the comb from below as well as entering, both vertical and horizontal impact detection is necessary.

6.2.6.4 Handrail-Speed Monitoring Device. There is a need to stop a moving walk if the speed of the handrail deviates by 15% or more. While a speed deviation of less than 15% would also not comply with 6.1.3.4.1, it would not be enough to necessitate stopping the moving walk right away. A delay is necessary so that the moving walk is not stopped every time someone pulls on the handrail.

A handrail not moving in the same direction or speed has a tendency to pull passengers backward or forward causing them to lose their balance. It is not necessary for the handrail to move at exactly the same speed as the pallet since passengers can easily adjust the position of their hands to compensate for minor differences in speeds.

6.2.6.5 Missing Pallet Device. This requirement will prevent a moving walk from continuing to run if a pallet is missing. It will prevent a void from a missing pallet to emerge from under the comb. This requirement was adopted because of the seriousness of the injury that results from stepping into the gap of a missing pallet, and the fact that a passenger would not see the gap until starting to step into it.

6.2.6.6 Tandem Operation. See Diagram 6.1.6.6.

6.2.6.7 Moving Walk Smoke Detectors. Smoke detectors are not required in wellways by ASME A17.1. Some building codes do require smoke detectors in wellways and adjacent areas. When a smoke detector is used to stop a moving walk, these requirements must be adhered to. This requirement was adopted because if smoke detectors were set to stop the moving walk, certain events were deemed necessary to permit the moving walk to stop safely.

6.2.6.8 Signs

6.2.6.8.1 Caution Signs. The sign, (Fig. 6.1.6.9.1), cautions people on how to use a moving walk properly and indicates that it is for the transportation of people only. Most accidents are caused by improper use of a moving walk. The pictograph reduces the problems caused by language barriers.

6.2.6.8.2 Additional Signs. Additional signs may be provided since during the normal walking flow of people entering a moving walk, there is only sufficient time to read and comprehend the required sign. To properly board a moving walk, it is important to continue to walk without stopping or otherwise causing abrupt bodily contact with others. For this reason, a sign that is well outside the normal boarding traffic flow is permitted. Additional signs within the boarding area are prohibited. Due to moving walks being installed in many different locations, there is no consensus on what this sign should say. Some of the most frequently used messages are: "No Luggage or Carts" "No Bare Feet," "No Strollers," "No Wheelchairs," "Handicapped-Use Elevator."

The Committee has, therefore, provided some broad guidelines and a message to building owners, etc., to consider anything peculiar to the building site, which requires further unique messages.

6.2.6.9 Control and Operating Circuits. This requirement provides for protection against singular failures resulting in an unsafe condition

6.2.6.11 Installation of Capacitors or Other Devices to Make Electrical Protective Devices Inoperative. This mandates that the electrical protective devices always be operative and that they may not be bypassed except as provided for in the Code.

6.2.6.12 Completion or Maintenance of Circuit. The requirement is written in a manner such that failure of any circuit associated with an electrical protective device will cause the device to operate. This is a basic principle of fail-safe design. The intent of 6.2.6.12 is that the contactors used to interrupt the power to the moving walk motor or brake shall not be energized to perform this function.

6.2.6.13 Moving Walk Manual Reset. An external power failure should not result in any safety device requiring manual reset to lose its indication of malfunction.

6.2.6.14 This requirement specifies the requirements for single failure protection, i.e., redundancy, in the design of specific control and operating circuits (6.2.6.9.1) and requires the checking of that redundancy (6.2.6.9.2). This may require the use of two relays instead of one in these circuits, and the checking of the operation of these relays.

When electromechanical relays and contactors are being checked for operation, specifically drop out operation, by utilizing contacts on the relays or contactors themselves, care must be taken that the contact used for monitoring the state of the contact used in the critical circuit is a true and correct representation of the critical circuit contact.

The required operation of the monitoring contact is specified, and it is clarified that springs in the design of the relay, contactor, or contact design may be used, but that spring failure shall not cause the monitoring

contact to indicate an open critical circuit contact if the critical circuit contact is not open.

6.2.7 Lighting, Access, and Electrical Work

6.2.7.1 Lighting of Machine Room and Truss Interior. The requirements assure that a source of illumination and power is provided in machine rooms for elevator personnel.

6.2.7.1.2 Truss Interior. The requirement for a duplex receptacle in the truss interior is to allow for the use of a drop cord light for inspection and maintenance inside the truss.

6.2.7.2 Lighting of Treadway. Treadways must be illuminated properly in order to reduce a tripping hazard. Machine spaces in the truss and at either end do not require permanent lighting, as it is impractical to locate a lighting fixture, which will provide lighting to all work areas.

6.2.7.3 Access to Interior. The removal of pallets is considered a reasonable means of access to the interior of a moving walk.

6.2.7.3.1 The use of terrazzo, stone, or concrete, and other similar floor coverings that are excessively heavy, would be a hazard for maintenance and inspection personnel to lift. This type of material also tends to crack, resulting in a tripping hazard. Access plates must be securely fastened to prevent accidental opening.

6.2.7.4 Electrical Equipment and Wiring

6.2.7.4.1 See Handbook commentary on 2.26.4.1.

6.2.7.4.2 The specified electric equipment is required to be certified as complying with CSA B44.1/ASME A17.5. This will minimize the risk of electricity as a source of electric shock and as a potential ignition source of fires. See Handbook commentary on 2.26.4.2.

6.2.7.4.3 See Handbook commentary on 2.26.4.3.

6.2.8 Outdoor Moving Walks

The Code requires outdoor moving walks to be covered (6.2.8.2). Experience has shown that outdoor moving walks must be protected from the weather. Moving walks installed outdoors without any cover were usually shut down when it rained; however, they have since been covered.

PART 7
DUMBWAITERS AND MATERIAL LIFTS

SCOPE

The A17.1 Code defines a dumbwaiter as, "a hoisting and lowering mechanism equipped with a car of limited size which moves in guide rails and serves two or more landings."

The B20.1 Standard defines a conveyor as, "a horizontal, inclined, or vertical device for moving or transporting bulk material, packages, or objects, in a path predetermined by the design of the device, and having points of loading and discharge, fixed or selective."

In response to questions concerning the difference between a "conveyor" and a "dumbwaiter," the B20 Committee responded in Interpretation 7-1 in the following manner:

> "The difference between a vertical reciprocating conveyor and a dumbwaiter are typically found in guides, hoistway construction, platform requirements, controls, and power mechanisms. The distinguishing features of a dumbwaiter are its limitation in capacity, size, door heights, and car clearances."

There are no practical differences between a dumbwaiter and a conveyor, except that a dumbwaiter is limited in size and a conveyor is not. See Inquiry 91-53.

The ASME A17.1 Code defines a material lift as a hoisting and lowering mechanism normally classified as an elevator, equipped with a car which moves within a guide system installed at an angle of greater than 70 deg from the horizontal, serving two or more landings, for the purpose of transporting materials which are manually or automatically loaded or unloaded. Material lifts without an automatic transfer device are Type A or Type B. On Type A material lifts, no persons are permitted to ride. On Type B material lifts, authorized personnel are permitted to ride.

The differences between a freight elevator, reciprocating conveyor conforming to ASME B20.1, Type A material lift, Type B material lift, and a dumbwaiter can be described in Chart 7.

SECTION 7.1
POWER AND HAND DUMBWAITERS WITHOUT AUTOMATIC TRANSFER DEVICES

NOTE: Throughout Section 7.1, references are made to other requirements in this Code. To gain a complete understanding of a requirement, the reader should review the Handbook commentary on all referenced requirements.

7.1.1 Construction of Hoistways and Hoistway Enclosures

Hoistways must be enclosed throughout their height. The fire resistance rating of the hoistway enclosure, including the entrances and protective assemblies in other openings, should not be less than required by the building code, or where there is no local building code, one of the following model codes:

(a) National Building Code (NBC)

(b) Standard Building Code (SBC)

(c) Uniform Building Code (UBC)

(d) National Building Code of Canada (NBCC)

It is noted that the International Building Code (IBC) issued in the first quarter of 2000 is the successor code for the NBC, SBC, and UBC.

Fire-resistive construction retards the spread of fire, hot gases, and smoke from penetrating one floor or area to another. The use of non-fire-resistive construction is usually permitted by the building codes when fire-resistive assemblies are not penetrated by the hoistway. Most local codes and the model building codes generally require that the fire resistance rating of the hoistway enclosure be a minimum of 2 h, with entrances and other protective assemblies in the openings a minimum of $1\frac{1}{2}$ h. In specific types of construction, some codes allow a fire resistance rating of the hoistway enclosure of less than 2 h with entrances and other opening protective assemblies rated accordingly. The term "opening protective assemblies" refers to access openings in the hoistway for maintenance and inspection (see also 2.7.3.4).

7.1.2 Pits

7.1.2.3 Most dumbwaiter pits are shallow, thus ambient illumination is usually acceptable.

7.1.3 Location and Guarding of Counterweights

Requirements of 2.3.1 have been made applicable to facilitate inspection and maintenance of counterweights and ropes. Requirements 2.3.2 and 2.3.3 are not necessary since personnel will not be in the pit.

7.1.4 Vertical Car Clearances and Runbys for Cars and Counterweights

Vertical clearances are not necessary since passengers are not carried.

Chart 7 The Differences Between a Freight Elevator, B20 Reciprocating Conveyor, Type A Material Lift, Type B Material Lift, and Dumbwaiter

	Freight Elevator	B20 Reciprocating Conveyor	Type A Material Lift	Type B Material Lift	Dumbwaiter
Travel	Unlimited	Unlimited	Unlimited	5000 mm or 16 ft 8 in. Penetrate one floor	Unlimited
Speed	Unlimited	Unlimited	Unlimited	0.15 m/s or 30 ft/min	Unlimited
Riders permitted (authorized material handlers)	Yes	No	No	Yes	No
Location	No restrictions	Only in areas with trained personnel [Note (1)]	No restrictions	Not accessible to the general public	No restrictions
Maximum Size	No maximum	1220 mm or 48 in. wide by 2280 mm by 90 in. high. No maximum depth	No maximum. Enclosure walls must be 2030 mm or 80 in. high	1m^2 or 10.75 ft^2 by 1220 mm or 48 in. high	
Operation	Automatic	Automatic	Automatic	Constant pressure push button hall station. Cannot override car station	Automatic
Summary					
Limited Travel	—	—	—	X	—
Limited Speed	—	—	—	X	—
No riders	—	X	X	—	X
Restricted locations	—	X	—	X	—
Limited size	—	—	X	—	X
Restricted operation	—	—	—	X	—
Autorized persons	X	X	X	X	X

NOTE:
(1) ASME B20.1-1996, para. 5.12(c) — Personnel working on or near a conveyor shall be instructed as to the location and operation of pertinent stopping devices.

7.1.5 Horizontal Car and Counterweight Clearances

Requirements for horizontal clearances are specified. Where the rated load is 227 kg or 500 lb or less, shifting loads in the car would not create the need for the same clearances required for elevators. Requirement 2.5.1.5 is for the protection of passengers, and therefore, not applicable.

7.1.6 Protection of Spaces Below Hoistway

Safeties and buffers are required regardless of rated load because of the impact on the hoistway or pit floor. The Code does not require the counterweight frame to be designed or constructed to restrain the weights. There are also no requirements for constraints on guide rails or brackets or the strength of the hoistway wall. Since the dumbwaiter is often in an area accessible to the general public, concern is shown for the fact that even though the pit floor might sustain the impact loads, the falling car or counterweight might cause injury to people adjacent to or beneath the hoistway.

7.1.7 Machine Rooms and Machinery Spaces

7.1.7.1 Separate machinery spaces are not always necessary.

7.1.7.3 Necessary provisions are provided to prevent accidental contact with equipment.

7.1.7.3(c) It is not safe to get in the hoistway below an operable car in order to service it.

7.1.7.4 Dumbwaiters are not required to have a permanent stair or ladder when the machinery spaces are within the hoistway. Obstructions that are easily removable, such as, false ceilings or ceiling tiles, are permitted.

7.1.7.6 A contact is necessary to protect against moving equipment.

7.1.7.10 Control equipment and dumbwaiter machines are allowed to be located in the hoistway.

7.1.8 Electrical Equipment, Wiring, Pipes, and Ducts in Hoistways and Machine Rooms

7.1.8.1 As dumbwaiter use in a fire is not essential, protecting interlock wiring from high temperature is not necessary; thus, SF-2-type wiring is not required.

7.1.8.2 Requirement similar to requirement for electric elevators except for need to remove power prior to or upon application of sprinkler, as there are no passengers on dumbwaiters.

7.1.11 Protection of Hoistway Openings

7.1.11.2 Types of Entrances
7.1.11.2.1(a) See Diagrams 2.11.2.1(a)(1) and 2.11.2.1(a)(2).
7.1.11.2.1(b) See Diagram 2.11.2.1(b).
7.1.11.2.1(c) See Diagram 2.11.2.1(c).
7.1.11.2.1(d) See Diagram 2.11.2.2(e)(2).
7.1.11.2.1(e) See Diagrams 2.11.2.1(d) and 2.11.2.2(f).
7.1.11.2.2(a) See Diagrams 2.11.2.1(d) and 2.11.2.2(f).
7.1.11.2.2(b) See Diagram 2.11.2.2(e)(2).
7.1.11.2.2(c) See Diagram 2.11.2.1(b).

7.1.11.3 Closing of Hoistway Doors. This requirement assures the fire-resistance integrity of the hoistway. If the doors were allowed to remain open in a fire, the fire could enter the hoistway and thus be given a path to other portions of the building.

7.1.11.3.2 Provisions included to close hoistway doors on hand-operated dumbwaiters when activated by smoke sensors in order to be compatible with building code requirements.

7.1.11.8 Hoistway-Door Vision Panels. Requirements have been added for vision panels, if provided, per applicable requirements of 2.11.7.1.

7.1.11.10 Landings and Landing Sills. Requirements for illumination (2.11.10.2) are to provide adequate illumination for operating the equipment, and for hinged landing sills (2.11.10.3), if provided. Hinged car sills are not prohibited by this requirement.

7.1.11.11 Horizontal Slide-Type Entrances. Requirements of 2.11.11.1(b) are deleted since they are for the protection of passengers.

7.1.11.12 Vertical Slide-Type Entrances. Requirements in 2.11.12 for vertical slide-type entrances suitable to dumbwaiter applications are referenced. See Diagram 7.1.11.12.

7.1.11.13 Swing-Type Entrances. Requirements in 2.11.13, which are deleted, are for the protection of riders.

7.1.12 Hoistway-Door Locking Devices, Access Switches, and Unlocking Devices

7.1.12.1 Hoistway-Door Locking Devices for Power Dumbwaiters

7.1.12.1.3 See Diagram 7.1.12.1.3.

7.1.12.3 Hoistway-Door Unlocking Devices. This requirement allows a means for access to the hoistway where access switches are not provided and/or in addition to access switches.

7.1.12.4 Hoistway Access Switches. Hoistway door access switches provide a safe means for inspection and maintenance.

When the travel is less than 7.6 m or 25 ft, either hoistway-door unlocking devices or hoistway access switches must be provided. When the travel is 7.6 m or 25 ft or more, hoistway access switches must be provided. Hoistway-door unlocking devices may be provided, in addition to hoistway access switches, but are not required.

SECTION 7.2
ELECTRIC AND HAND DUMBWAITERS WITHOUT AUTOMATIC TRANSFER DEVICES

NOTE: Throughout Section 7.2, references are made to other requirements in this Code. To gain a complete understanding of a requirement, the reader should review the Handbook commentary on all referenced requirements.

In addition to the requirements in 7.2, the requirements in 7.1 apply to electric and hand dumbwaiters without automatic transfer devices.

7.2.1 Car Enclosures, Car Doors and Gates, and Car Illumination

7.2.1.1 Car Enclosures. Many of the requirements in 2.14 are not applicable because they are for the protection of passengers.

7.2.1.2 Car Doors and Gates. Car doors or gates are required. Car doors prevent the load from projecting from the car.

7.2.1.3 Lighting Fixtures. Where lighting fixtures are provided in the car, conformance to the requirements of 2.14.7.3 and 2.14.7.4 will provide the necessary safety features.

7.2.2 Car Frames and Platforms

The requirements for car frames and platforms reference applicable portions of 2.13. Because of their smaller size, frames for dumbwaiter cars and platforms are not always necessary.

7.2.3 Capacity and Loading

7.2.3.1 Rated Load and Platform Area. The ASME A17.1 Code requirements prior to the 1987 edition specified a maximum rated load of 226.8 kg (500 lb) and a minimum structural load based on the inside net platform area. ASME A17.1–1987 and later editions recognized that, regardless of the posted rated load, a car is capable of being loaded to an extent based on its inside volume. Requiring a minimum rated load, compatible with the structural capacity, is a more logical safety requirement. There is no maximum allowable capacity.

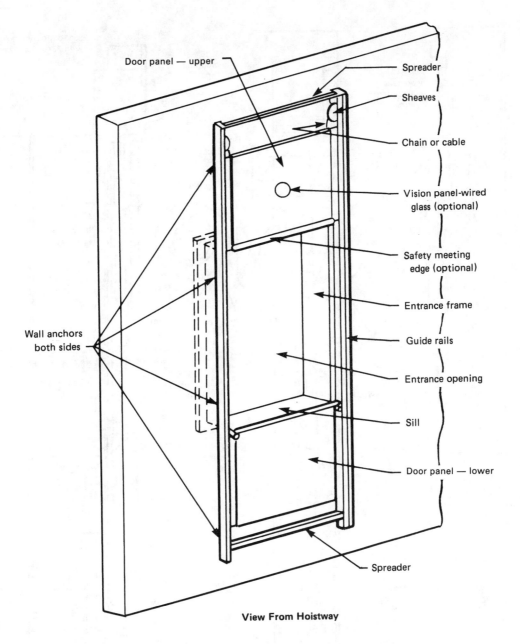

Diagram 7.1.11.12   Typical Dumbwaiter Vertical Biparting Door
(Courtesy The Peelle Co.)

7.2.3.2 Capacity Plate. The capacity plate gives information on safe loading to the user.

7.2.3.3 Data Plate. The data plate is for the use of elevator personnel.

7.2.4 Car and Counterweight Safeties

Car and counterweight safeties are only necessary where the hoistway is over occupiable space (7.1.6). Maximum stopping distances are specified; however, there is no need for minimum stopping distances since there are no passengers. These requirements address guide-rail shapes other than T-shaped. Persons should never be on top of or below a dumbwaiter car without safeties, unless means are provided to secure the car. See safety practices in Elevator Industry Field Employees' Safety Handbook. Procurement information for this document can be found in the Handbook commentary on Part 9.

7.2.5 Speed Governors

Requirements for speed governors, if provided, are referenced. It would be inconsistent to require the governor rope to be 9.5 mm or $\frac{3}{8}$ in. in diameter when the suspension ropes may be less.

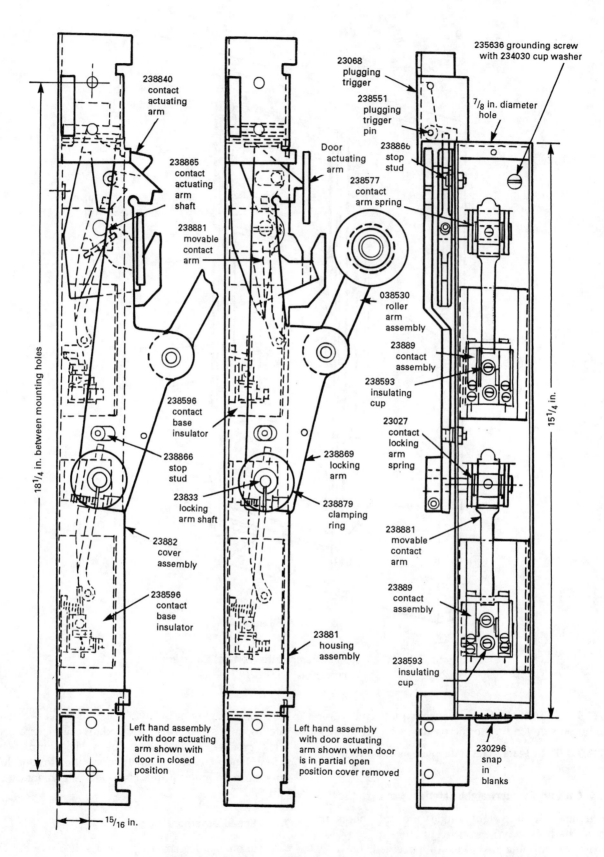

Diagram 7.1.12.1.3   Typical Interlock for Dumbwaiter
(Courtesy The Peelle Co.)

7.2.6 Suspension Means

Welded link chain is prohibited as a suspension means because:

(a) there appears to be no practical way of visually detecting an impending failure at the weld of a welded link proof-tested steel chain;

(b) there appears to be no practical way to detect an impending failure of such a chain due to fatigue. Hairline cracks develop quickly;

NOTE: Items (a) and (b) above are important factors in the safety of users and people, and place an undue burden of responsibility upon an inspector trying to determine when such chain is unsafe.

(c) there is literature published by some chain manufacturers that welded link proof-tested steel chain is not recommended for overhead lifting purposes or where its failure is likely to cause damage to property or persons;

(d) the exposure of a substantial number of people from all walks of life to dumbwaiter usage in their normal everyday work environment subjects them to whatever hazards may exist around and in the entrances of such dumbwaiters. The fact that arms, hands, and even heads encroach into and through the entrance areas makes it imperative that a safe and adequate means of suspension exists at all times. Any chain failure in this situation could be disastrous.

7.2.6.2 The rope data tag is not permitted to be within the car. The crosshead data plate is permitted to be in the car.

7.2.6.8 Fastening of Suspension Means

7.2.6.8.1(b) U-bolt-type rope clips (clamps) are permissible, provided the fastening develops at least 80% of the ultimate breaking strength of the rope.

7.2.8 Car and Counterweight Bumpers and Buffers

Buffer stroke requirements are based on the fact that dumbwaiters do not carry passengers.

7.2.9 Car and Counterweight Guide Rails, Guide-Rail Supports, and Fastenings

Other sizes and shapes for guide rails are permitted, provided they meet the performance requirements of 2.23.

7.2.9.1 Guide-Rail Section. The allowable deflection of 6 mm or $\frac{1}{4}$ in. required by 2.23.5.1.1 relates to T-shaped guide rails and conventional elevator safeties. When other shapes of guide rails are used, the deflection is controlled by this requirement.

7.2.10 Driving Machines and Sheaves

7.2.10.1 Power Dumbwaiters. The sheave and drum diameters are based on what is currently being used, with no history of any problem.

7.2.10.2 Hand Dumbwaiters. Manually applied brakes pose a safety exposure and are prohibited.

7.2.10.2.2 Hand ropes must be located outside the travel path of the car and counterweight when operating the dumbwaiter. Automatic brakes ensure that the brake is applied anytime the operator releases the hand rope.

7.2.10.3 Types of Driving Machines. See Diagrams 7.2.10.3(a) and 7.2.10.3(b).

7.2.11 Terminal Stopping Devices

Requirement 2.25.3.3 is not applicable to dumbwaiters. The final terminal limits must be in the hoistway. The "machine limit" for winding drum machines required in 2.29.3.5 is not required. If the additional stopping device operated by the driving machines is provided, it must comply with 2.29.3.5.

The final terminal stopping device within the confines of the hoistway in addition to the normal stopping means provide an adequate level of redundancy and sufficient overtravel protection on a device designed solely to carry materials.

7.2.12 Operating Devices and Control Equipment

Requirements in 2.26 have been referenced, except for those requirements that are only for the protection of persons in the car. Broken rope, tape, and chain switches used in connection with normal terminal stopping devices are not necessary for dumbwaiters since final terminal stopping devices are now required. If the rope, tape, or chain breaks, the final terminal stopping device will still stop the dumbwaiter. The final terminal stopping device within the confines of the hoistway, in addition to the normal stopping means, provide an adequate level of redundancy and sufficient overtravel protection on a device designed solely to carry materials. Requirements are also added to reference 2.26.6 to prevent the continued rotation of the motor when it is reversed electrically. Where a top of car device is provided, indicating that maintenance or inspection may occur from the car top, car safeties must be provided. Electrical equipment is required to be certified as complying with CSA B44.1/ASME A17.5. This will minimize the risk of electricity as a source of electric shock and as a potential ignition source of fires. Dumbwaiter electrical equipment does not have to comply with EN 12016, Part 2. Other referenced requirements are modified to reflect dumbwaiter conditions.

SECTION 7.3
HYDRAULIC DUMBWAITERS WITHOUT AUTOMATIC TRANSFER DEVICES

NOTE: Throughout Section 7.3, references are made to other requirements in this Code. To gain a complete understanding of a requirement, the reader should review the Handbook commentary on all referenced requirements.

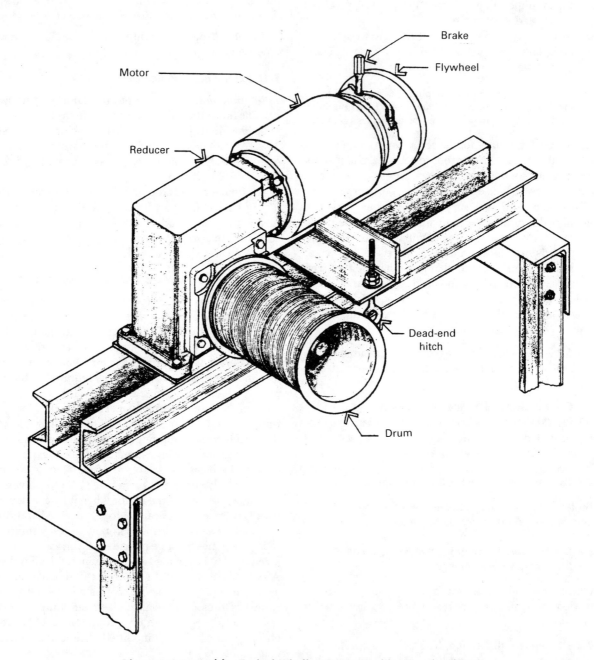

Diagram 7.2.10.3(a)   Typical Winding Drum Machine Dumbwaiter
(Courtesy Sedgwick Lifts, Inc.)

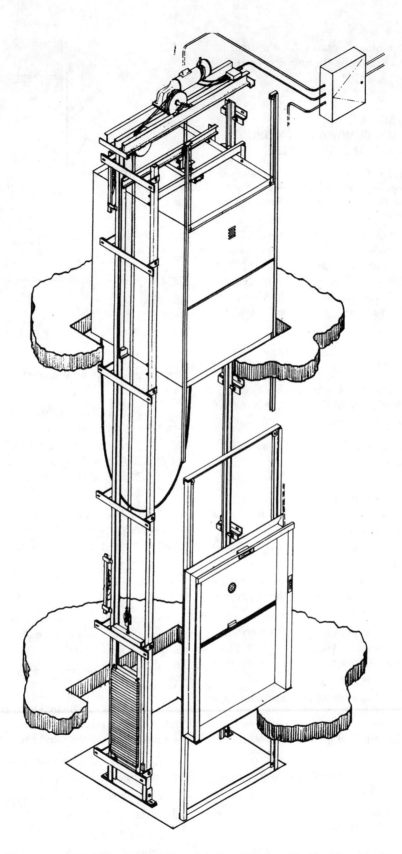

Diagram 7.2.10.3(b)   Typical Floor Loading Traction Machine Dumbwaiter
(Courtesy Sedgwick Lifts, Inc.)

In addition to the requirements in 7.3, the requirements in 7.1 apply to hydraulic dumbwaiters without automatic transfer devices.

SECTION 7.4
MATERIAL LIFTS AND DUMBWAITERS WITHOUT AUTOMATIC TRANSFER DEVICES

NOTE: Throughout Section 7.4, references are made to other requirements in this Code. To gain a complete understanding of a requirement, the reader should review the Handbook commentary on all referenced requirements.

The comments that follow only relate to Type A Material Lifts unless otherwise specified. A new type of material lift called Type B has been introduced in the 2000 edition of A17. This lift was added as part of the harmonization process and closely matches the Canadian Freight Platform Lift contained in Section 15 of CSA B44-94.

7.4.3 Construction of Hoistways and Hoistway Enclosures

The building code (see 1.3, Definitions) will dictate whether the hoistway must be of fire-resistive construction or non-fire-resistive construction.

7.4.6 Vertical Clearances and Runbys for Car and Counterweights

Means are provided for refuge space and in the overhead when appropriate conditions are present. As Type A material lifts do not carry passengers, less restrictive clearances and runbys are permitted.

7.4.6.1.3 and 7.4.6.1.4 Hoistway construction and machinery location in the hoistway may not provide adequate refuge space when the car is at the upper landing. Alternative methods to create the space are permitted

7.4.6.1.4 Tighter tolerances are permitted on these relatively smaller cars on which passengers are not carried. The use of a single rail for both the car and counterweight requires tighter clearances.

7.4.7.4 Beveling and Clearance Requirements for Type B Material Lifts. See Diagram 7.4.7.4.

7.4.10 Electrical Equipment in Hoistways and Machine Rooms

7.4.10.1 Type SF-2 wire is not required as firefighters' service is not required.

7.4.10.2 Provides sufficient protection for material carrying equipment upon sprinkler activation.

7.4.13 Protection of Hoistway Landing Openings

7.4.13.2.1 Emergency access doors to blind hoistways are not required for material lifts that do not

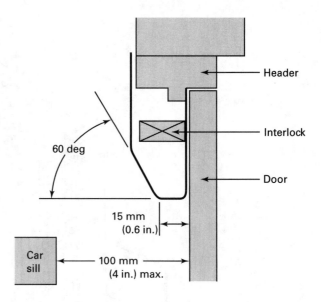

Diagram 7.4.7.4   Beveling and Clearance Requirements

carry people. Opening size is determined by material lift usage.

7.4.13.2.2 The requirements for passenger elevators are not applicable.

7.4.13.2.2(a) See Diagram 2.11.2.1(a)(1).
7.4.13.2.2(b) See Diagram 2.11.2.1(b).
7.4.13.2.2(c) See Diagram 2.11.2.2(c).
7.4.13.2.2(d) See Diagram 2.11.2.2(e)(2).

7.4.13.2.3 Requirement 2.11.3 was modified to suit door styles permitted, with proviso that doors be closed when not being used for loading or unloading.

7.4.13.2.4 Type A material lifts cannot be operated from within the car and passengers are not permitted.

7.4.13.2.5 Since there is no means to operate Type A material lift from the car, a risk of locking someone in the car who is in the process of loading and unloading should not be allowed.

7.4.13.2.6 Hoistway door vision panels serve no safety purpose.

7.4.13.2.7 The inclusion of glass on a device designated to carry materials adds an unnecessary risk that the materials may shift and cause breakage of the glass.

7.4.13.2.8 Landing sill guards are not required on nonpassenger elevators.

7.4.13.2.9 Bridging devices may be desirable to help with the transfer of carts from the landing to the car.

7.4.13.2.10 Pull straps inside the car are not necessary since the doors should not be closed from inside the car.

7.4.13.2.11 Combination horizontally sliding and swinging panels are not allowed.

7.4.14 Hoistway Door Locking Devices and Electric Contacts, and Hoistway Access Switches

7.4.14.1 The requirements for passenger elevators are not applicable.

7.4.14.2 The requirements for passenger elevators are not applicable.

7.4.14.3 Unlocking devices are only needed for access to the overhead and pit for maintenance and inspection purposes.

7.4.14.4 The requirements for passenger elevators are not applicable.

7.4.15 Power Operation of Hoistway Doors and Car Doors and Gates

7.4.15.1 No door or car operating pushbuttons will be in a Type A material lift car.

7.4.15.2 No door or car opening pushbuttons will be in a Type A material lift car. The momentary button called for in 2.13.3.3.2 is placed at the landing.

7.4.15.3 There is no safety reason for sequence operation. There will be no door or car operating pushbuttons in the car.

7.4.15.4 No passengers will be in Type A material lift car.

7.4.15.5 There is no safety reason for sequence operation. No passengers will be in the car.

SECTION 7.5
ELECTRIC MATERIAL LIFTS WITHOUT AUTOMATIC TRANSFER DEVICES

NOTE: Throughout Section 7.5, references are made to other requirements in this Code. To gain a complete understanding of a requirement, the reader should review the Handbook commentary on all referenced requirements.

In addition to the requirements in 7.5, the requirements in 7.4 apply to electric material lifts without automatic transfer devices.

7.5.1 Car Enclosures, Car Doors and Gates, and Car Illumination

7.5.1.1 Car Enclosure

7.5.1.1.1(a) To prohibit the use of a forklift truck when placing a load on a Type A material lift car.

7.5.1.1.2 This requirement contains the essence of the requirement, the additional wordage contained in 2.14.1.2 relates to passenger-type elevators.

7.5.1.1.4 The criteria for design of the enclosure walls is more closely related to the actual load being carried than to a generalized design criteria suitable for a passenger-type application. Type A material lifts may have multiple compartments, fixed and/or removable shelves.

7.5.1.1.5 Emergency exits are not required in non-passenger carrying Type A material lifts.

7.5.1.1.8 The inclusion of glass on a device designed to carry materials adds an unnecessary risk that the materials may shift and cause breakage of the glass.

7.5.1.1.9 This requirement contains the essence of the requirement, the additional wordage contained in 2.14.1.9 relates to passenger-type elevators.

7.5.1.1.10 Emergency exits are not required in non-passenger carrying Type A material lifts.

7.5.1.1.11 Specific requirements for passenger elevators, not applicable to Type A material lifts.

7.5.1.1.12 Grille or perforated construction may be used for the full height and car top (if provided). Type A material lift enclosures may be less than 1 828 mm (72 in.).

7.5.1.1.13 If ventilation grilles are provided, their location is irrelevant in a nonpassenger carrying Type A material lift.

7.5.1.2 Car Doors and Gates

7.5.1.2.2 This requirement ensures that the lift will not run unless the car door or gate is in the closed position.

7.5.1.2.3 As there are no passengers, there is no safety hazard to personnel if a car door or gate is not provided. The purpose of a car door or gate is to assist in retaining the materials carried within a Type A material lift car. This requirement provides minimum design criteria for a car door or gate.

7.5.1.2.4 As there are no passengers, there is no safety hazard to personnel if a Type A material lift car door or gate is not provided. The purpose of a car door or gate is to assist in retaining the materials carried within the car. This requirement provides minimum design criteria for a car door or gate.

7.5.1.2.6 As there are no occupants on a Type A material lift, there is no hazard if the counterweight is located within the car enclosure. Guides or counterweight boxes prevent the weights from swinging freely and causing incidental damage plus provisions for suitable restraint in the event of suspension member failure is included.

7.5.1.3 Car Illumination and Lighting Fixtures. Provides for a minimum level of illumination within Type A material lift car for loading and plus minimum design criteria for the light.

7.5.2 Car Frames and Platforms

7.5.2.1 It is not the intent to dictate the design of the equipment, only to ensure that is designed to meet its intended purpose.

7.5.2.2 Exceptions to specific platform design parameters and specific loading classifications.

7.5.2.3 A Type A material lift does not carry passengers. Protection of the platform against fire is irrelevant.

7.5.2.4 The leveling truck or landing zone plus 75 mm or 3 in. provides adequate shear protection.

7.5.2.5 With an overlap of 75 mm or 3 in. required, the need for the bevel is eliminated.

7.5.2.6 The maximum allowable stresses provide sufficient safety factors and design criteria for a non-passenger carrying Type A material lift.

7.5.3 Capacity and Loading

7.5.3.1 This is not a passenger elevator.

7.5.3.2 Similar to rated load requirements for freight elevators. Provisions are included for the weight of hand truck. Type A material lifts are designed for hand truck use but not intended for forklift use.

7.5.3.3 Riders are not permitted on Type A material lifts.

7.5.4 Car and Counterweight Safeties

7.5.4.1 The objective is to stop the car should a failure occur. Since there are no passengers on a Type A material lift car, the retardation force is of little importance. Type B safeties are permitted.

7.5.4.2 Passengers are not carried.

7.5.4.3 The objective is to stop the car should a failure occur. Since there are no passengers on a Type A material lift car, the retardation force is of little importance. Type B safeties are permitted. When the speed is slow, broken rope safeties are considered sufficient.

7.5.4.4 Provision to allow for rails of different design.

7.5.5 Speed Governors

7.5.5.1 Passengers are not carried.

7.5.5.2 To conform to previous changes in rated speed from 0.75 m/s or 150 ft/min to 1 m/s or 200 ft/min.

7.5.5.3.1 Incorporate provisions for governors ropes smaller than 9.5 mm or $\frac{3}{8}$ in. when the suspension means is less than 9.5 mm or $\frac{3}{8}$ in.

7.5.6 Suspension Ropes

7.5.6.1 Permits the use of chains and wire ropes other than elevator wire rope.

7.5.6.2 Includes requirements for chain data similar to that required for wire ropes.

7.5.6.3 To include an appropriate safety factor for chains.

7.5.6.5 Two ropes provide sufficient redundancy on a nonpassenger carrying Type A material lift. Allows single-bar-type equalizers on all drives when only two ropes are provided.

7.5.6.6 Adds suspension fastening requirements for chains.

7.5.8 Buffers and Bumpers

Retardation force on a nonpassenger carrying Type A material lift can meet less stringent requirements.

7.5.9 Car and Counterweight Guide Rails, Guide Rail Supports, and Fastenings

7.5.9.1 Use of Common Guide Rails. It is common practice for small elevators such as residence elevators, personnel elevators, and dumbwaiters to have the car and counterweight run in a common rail.

7.5.9.2 Guide-Rail Sections. Allows for rails with smaller section modulus and moment of inertia consistent with cars and counterweights of smaller size, weight, and rated capacity. Does not dictate the size of bolts required for fastening the guide rail to the rail brackets; however, strength requirements remain applicable. Adds the requirement that the allowable deflection, in addition to compliance with that specified in 2.23, cannot allow the safety to disengage from the rail.

7.5.10 Driving Machines and Sheaves

7.5.10(a)(1) A material only handling device can meet a lower standard for pitch diameter.

7.5.10(b) Permits the use of chain drives and provides for general design criteria.

7.5.11 Terminal Stopping Devices

A final terminal stopping device within the confines of the hoistway in addition to the normal stopping means provides an adequate level of redundancy and sufficient overtravel protection on a device designed solely to carry materials.

7.5.12 Operating Devices and Control Equipment

7.5.12.1.1 Passengers are not permitted on Type A material lift car.

7.5.12.1.2 One-piece loads greater than the rated load are not permitted.

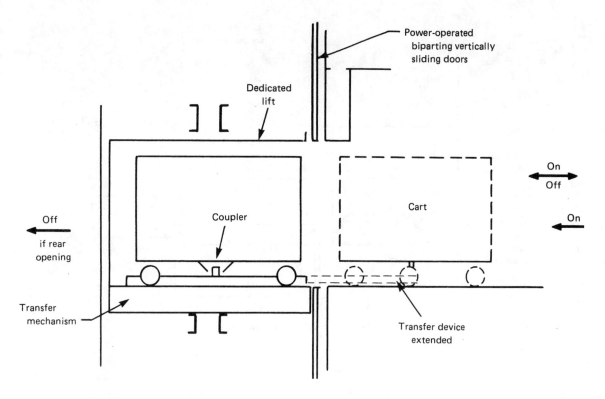

Diagram 7.7(a)   Cart Lift

7.5.12.1.5   Passengers are not permitted on Type A material lift car.

SECTION 7.6
HYDRAULIC MATERIAL LIFTS WITHOUT AUTOMATIC TRANSFER DEVICES

NOTE: Throughout Section 7.6, references are made to other requirements in this Code. To gain a complete understanding of a requirement, the reader should review the Handbook commentary on all referenced requirements.

In addition to the requirements in 7.6, the requirements in 7.4 apply to hydraulic material lifts without automatic transfer devices.

SECTION 7.7
AUTOMATIC TRANSFER DEVICES

There are many varieties of lifts with automatic transfer devices and the following are descriptions of the commonly used types. These descriptions are generalized and are not intended to describe any particular manufacturer's approach.

(a) Cart Lift [see Diagram 7.7(a)]. This type is mounted on a small floor stopping elevator or dumbwaiter and acts to pull the cart from the floor landing onto the lift or to push the cart from the lift onto the floor landing. The sequence is automatic in that the cart is placed in a position in front of the lift and a call or send signal is registered. Lift operation consists of automatic opening of doors, extension of the transfer device, and engaging of the coupler on the underside of the cart with the mating coupler of the transfer device. The transfer device retracts pulling the cart into the lift, the doors automatically close, and the lift travels to the destination. When the reverse cycle takes place, the doors open, the transfer device pushes the cart out, disengages the coupler, retracts, the doors close, and the lift is free to move to service the next demand. Lifts have capacities of from 113.4 kg (250 lb) to 453.6 kg (1,000 lb) and speeds from 0.25 m/s (50 ft/min) to 2.54 m/s (500 ft/min) or more. Dumbwaiter types have a capacity up to 227 kg (500 lb) with a platform area not exceeding 0.84 m^2 (9 ft^2) with speeds up to 2.54 m/s (500 ft/min). They are generally used in hospitals for transporting supplies and meals and in office buildings for mail distribution.

For the cart-lift system, an indexing type floor conveyor may also be installed to allow the queueing of carts at a main dispatch landing. The carts are then automatically moved into position to be loaded, each cart having an encoded address to indicate the destination.

(b) Tote Box Lift [See Diagram 7.7(b)]. This type is a small version of the cart lift, utilizing a counter-stopping dumbwaiter, which provides for the handling of tote boxes. Boxes are placed on a table in front of the lift and the call button is operated. The lift arrives, opens its

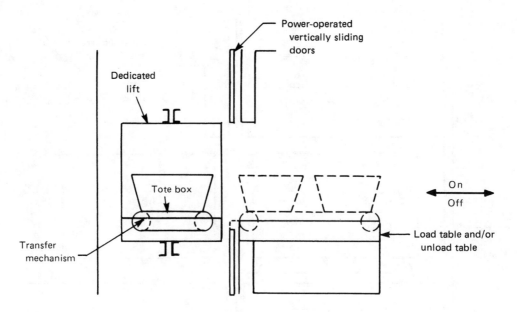

Diagram 7.7(b)   Tote Box Lift

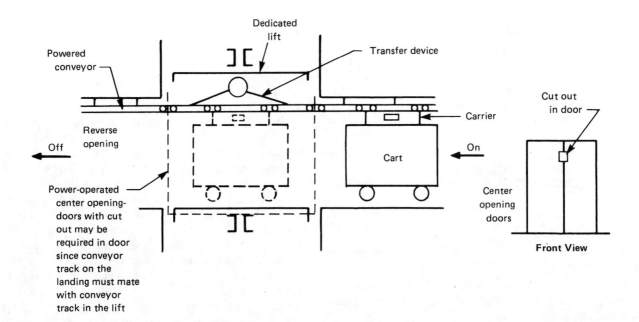

Diagram 7.7(c)   Overhead Chain Conveyor Lift

doors, the transfer mechanism operates to pull a box into the lift, the doors close, and the lift travels to another landing. Ejection is the reverse process. Usually, the tote box lift is used in hospitals for transporting records and small supplies. It is also used in banks for security transfers and in office buildings for mail handling. The capacity ranges from 22.7 kg (50 lb) to 227 kg (500 lb) and speeds vary from 0.25 m/s (50 ft/min) to 2.54 m/s (500 ft/min), with 0.51 m/s (100 ft/min) being the most common speed.

(c) Overhead Chain Conveyor Lift [See Diagram 7.7(c)].

Carriers on a power and free chain conveyor system are driven by the power chain to a position in front of the lift where a destination signal calls the lift. The door opens and a transfer device, either on the lift or at the landing, operates to transfer the carrier onto the lift, which has a mating power conveyor in the overhead. The door closes and the lift travels. Ejection is in a similar manner. The carrier is equipped with a fastening device, which engages a cart, and generally, by means of an inclined power chain, which lifts the cart off the floor. This chain conveyor lift is commonly used in hospitals

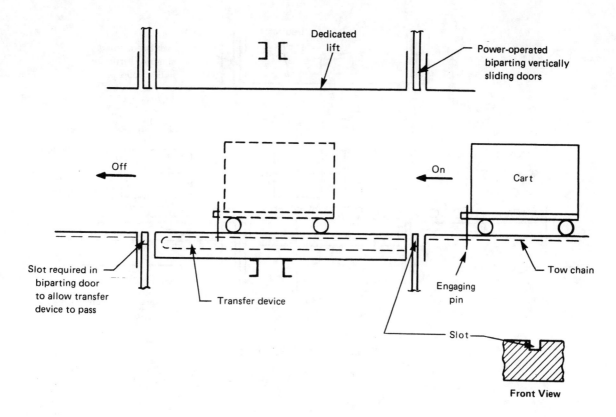

Diagram 7.7(d)   In Floor Tow Chain Conveyor Lift

for transporting food and for supply distribution, in stores for stock distribution, and for industrial use. The capacity in hospitals is usually 272.2 kg (600 lb) and speeds range from 0.25 m/s (50 ft/min) to 2.54 m/s (500 ft/min), with 1.27 m/s (250 ft/min) to 1.52 m/s (300 ft/min) being the most common speeds.

(d) In-Floor Tow Chain Conveyor Lift [See Diagram 7.7(d)]. Carts are propelled along the floor by a powered tow chain engaged by a pin on the cart through a continuous slot in the floor. At the entrance to the lift, the cart is released from the floor tow chain and a signaling device calls the lift. The lift doors open and a transfer mechanism in the floor of the lift extends to engage the pin on the cart and pulls the cart into the lift where it is engaged with a lift tow chain. Ejection is in the reverse order through a rear lift entrance. The cart is then picked up by the floor tow chain and moved to its destination. This lift design is used in warehouses for loads from 1 360.8 kg (3,000 lb) to 2 721.6 kg (6,000 lb) and at speeds from 0.51 m/s (100 ft/min) to 3.05 m/s (600 ft/min).

(e) Pallet Lift [See Diagram 7.7(e)]. Loaded or empty pallets are automatically loaded onto the lift by the use of a powered roller conveyor or conveyors at the landing and powered roller conveyor on the lift. The landing conveyor may be an indexing type where a number of pallets are rolled one at a time into position in front of the lift; the lift automatically opens its doors, and both the last section of the floor conveyor and the first

section of the lift conveyor are actuated. When the pallet reaches a fixed position on the lift, the doors close, the lift travels, and the conveyor operates at the destination to transfer the pallets onto a receiving floor conveyor or conveyors, which may be powered or nonpowered. Pallet lifts are used in warehouses and industrial plants for loads from 1 814.4 kg (4,000 lb) to over 5 443.2 kg (12,000 lb) with speeds from 0.25 m/s (50 ft/min) to 3.05 m/s (600 ft/min). They may be arranged to carry more than one pallet at a time.

(f) Robot Vehicle Lift [See Diagram 7.7(f)]. A self-propelled cart is used to carry a load and/or pull a trailer carrying a load. The cart moves at about 1.61 k/h (1 mi/h) following a guide path on or in the floor. The cart is equipped with a designation-coding device, which will signal an elevator or lift. When the elevator or lift arrives, the car will move onto the elevator or lift. Further signals will indicate the destination and initiate door closing. Variations have included a single cart operating onto a passenger-type elevator shared with pedestrians and equipped for self-propelled cart operation and freight-type elevators accommodating a self-propelled tractor pulling a number of loaded trailers. This system must comply with the requirements of Parts 1 or 2, or 4.2 of this Code when operated as an elevator and Part 7 when operated as a material lift. Robot vehicle lifts are used in hospitals for loads of 907.2 kg (2,000 lb) to 1 587.6 kg (3,500 lb) at speeds from 0.51 m/s (100 ft/

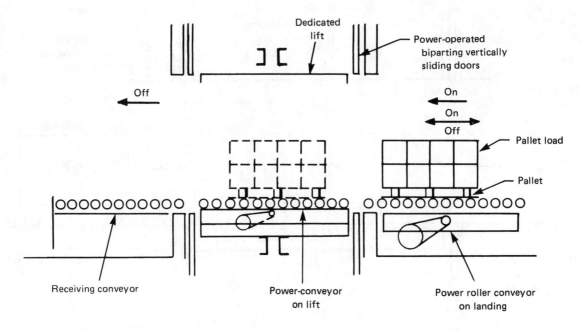

Diagram 7.7(e)   Pallet Lift

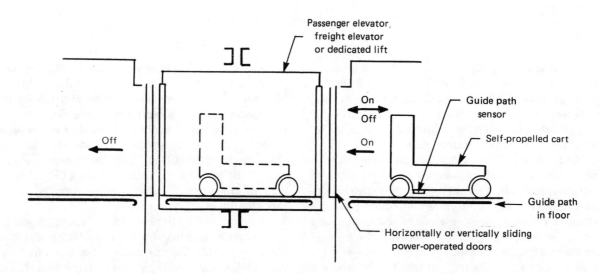

Diagram 7.7(f)   Robot Vehicle Lift

min) to 2.54 m/s (500 ft/min). They are used in industrial plants for loads up to 13 608 kg (30,000 lb) and speeds up to 3.05 m/s (600 ft/min).

7.7.1 General

See Diagram 7.7.1, which shows an automatic load, and unload unit (automatic transfer device) shown projected from the lift preparing to couple with a cart located in front of the lift door. The underside of the cart is equipped with a coupling bar.

SECTION 7.8
POWER DUMBWAITERS WITH AUTOMATIC TRANSFER DEVICES

NOTE: Throughout Section 7.8, references are made to other requirements in this Code. To gain a complete understanding of

Diagram 7.7.1   Typical Automatic Transfer Device
(Courtesy Courion Industries, Inc.)

a requirement, the reader should review the Handbook commentary on all referenced requirements.

7.8.2 Safety Devices

Where the suspended or supported gross load exceeds 700 kg or 1,500 lb, the car is generally large enough for maintenance or inspection personnel to ride on top of the car. This requirement is designed to protect such persons. In addition, a load of this nature falling from any height can damage the enclosing structure and may injure persons in the vicinity of the lowest landing.

7.8.3 Emergency Stop Switch

The emergency stop switch should be so located that a person does not have to reach over a transfer device to operate the emergency stop switch while the material is being transferred to or from the car.

7.8.4 Structural Capacity Load

This area is large enough for a person to stand on top of the dumbwaiter. As such, it must be capable of holding a load of not less than 135 kg or 300 lb.

SECTION 7.9
ELECTRIC MATERIAL LIFTS WITH AUTOMATIC TRANSFER DEVICES

NOTE: Throughout Section 7.9, references are made to other requirements in this Code. To gain a complete understanding of a requirement, the reader should review the Handbook commentary on all referenced requirements.

7.9.1 Hoistways, Hoistway Enclosures, and Related Construction

7.9.1.1 The maximum clearance has been increased from 32 mm or $1\frac{1}{2}$ in. to 125 mm or 5 in. in order to provide for the necessary clearances required by some of the transfer devices, especially those on overhead track conveyors.

7.9.1.5 Roller guides are not permitted since they allow a lateral movement of the car, which could allow the car to strike a protruding transfer device or track.

7.9.1.7 The intent of this requirement is to allow a car to start before its doors are fully closed. A restricted area is an area accessible only to authorized personnel who have been fully instructed in the use and operation of the automatic aspects of the transfer equipment.

7.9.2 Machinery and Equipment

7.9.2.2 Requirement 2.14.3.1 states that car enclosures must be at least 1,825 mm or 72 in. high. Material lift enclosures only need to be as high as required for the material or cart to be transferred plus the height of the transfer device and necessary clearances.

7.9.2.4 The transfer device can form part of the car platform.

7.9.2.12 through 7.9.2.14 People are not permitted to ride material lifts, therefore, the additional impact forces when striking spring buffers at higher speeds than allowed by Section 201 are acceptable and can be tolerated by the equipment.

7.9.2.15 When provided, the car operating device must be in a locked cabinet or consist of key-operated-type switches, assuring that the material lift can be operated from inside the car only by authorized personnel.

7.9.2.18 The location of the emergency stop switch is to assure that a person can operate the emergency stop switch while material is being transferred to or from the lift.

SECTION 7.10
HYDRAULIC MATERIAL LIFTS WITH AUTOMATIC TRANSFER DEVICES

NOTE: Throughout Section 7.10, references are made to other requirements in this Code. To gain a complete understanding of a requirement, the reader should review the Handbook commentary on all referenced requirements.

PART 8
GENERAL REQUIREMENTS

SECTION 8.1
SECURITY

In ASME A17.1d–2000 and earlier editions of the Code, security provisions were addressed throughout the Code. In some areas restrictions were placed on the availability of keys in other areas, the Code was silent. The Committee was concerned that unqualified people were obtaining and using keys, placing themselves and the public in danger. An extensive study was undertaken and the Committee concluded that security levels needed to be established and specified for every requirement where a lock is required by the Code. To assist the building owner/operator, it was determined that 8.1 should be developed rather than specify the specifics for each security level, repeatedly, throughout the Code requirements.

Section 8.1 applies to both new and existing installations (see 1.1.3, Application of Parts). Building owners/operators are required by 8.6.10.2 to control the access to keys as specified in this Section of the Code.

8.1.2 Group 1: Restricted

Elevator personnel (see 1.3) are persons who have been trained in the construction, maintenance, repair, inspection, or testing of equipment. This category includes elevator mechanics and inspectors. It does not include building engineers, building security, electricians, janitors, etc. The Note includes some but not necessarily all of the Code requirements that reference Group 1 Security.

8.1.3 Group 2: Authorized Personnel

Authorized personnel (see 1.3) are persons who have been instructed in the operation of the equipment and designated by the owner to use the equipment. This category includes building personnel who have been properly trained and elevator personnel (see 1.3).

The Note includes some but not necessarily all of the Code requirements that reference Group 2 Security.

8.1.4 Group 3: Emergency Operations

Emergency personnel (see 1.3) are persons who have been trained in the operation of emergency or standby power and firefighters' emergency operation or emergency evacuation. This category typically includes building engineers, fire wardens, etc., who have been

properly trained, firefighters, and elevator personnel (see 1.3).

The Note includes some but not necessarily all of the Code requirements that reference Group 3 Security.

8.1.5 Group 4: Other

This is the lowest security requirement and typically is applicable where there are no training requirements for personnel who have access to and/or use keys assigned to this security group.

SECTION 8.2
DESIGN DATA AND FORMULAS

NOTE: Throughout Section 8.2, references are made to other requirements in this Code. To gain a complete understanding of a requirement, the reader should review the Handbook commentary on all referenced requirements.

This Section of the Code contains design data, formulas, and charts based on the requirements found in Parts 1, 2, and 7 and Section 5.1 of this Code. This information is useful to the designer of elevator equipment.

The author expresses his appreciation to George W. Gibson for his extensive contributions to Section 8.2.

8.2.1 Minimum Rated Load of Passenger Elevators

8.2.1.1 Minimum Rated Load, Passenger Elevators. These requirements and 2.16 establish a minimum rated load for the inside net platform area of various size passenger elevator cars. The Code does not prohibit a rated load greater than the minimum rated load required for a given inside net platform area.

8.2.2 Electric Elevator Car Frame and Platform Stresses and Deflections

In 1.3, the Code defines the following:

car frame: the supporting frame to which the car platform, upper and lower sets of guide shoes, car safety and the hoisting ropes or hoisting rope sheaves, or the hydraulic elevator plunger or cylinder are attached. The car frame is sometimes referred to as the car sling. Specific types include:

car frame, overslung: car frame to which the hoisting rope fastenings or hoisting rope sheaves are attached to the crosshead or top member of the car frame.

car frame, subpost: a car frame all of whose members are located below the car platform.

345

car frame, underslung: a car frame to which the hoisting rope fastenings or hoisting rope sheaves are attached at or below the car platform.

The car frame is a rigid rectangular frame consisting of a crosshead, uprights, and plank, as shown in Diagram 8.2.2(a).

The crosshead is a pair of structural members, generally channel shaped, which form the top of the frame.

The uprights or stiles are the vertical structural members at the side of the car.

The plank is the structural member (or pair of members), similar to the crosshead, forming the bottom of the car frame.

The guide shoes, which are mounted on the four corners of the car frame, serve as the point of riding contact between the car frame and guide rails.

The car safety is the device mounted beneath or within the planks, which retards and stops the car in case of an overspeed.

The car frame hitch is the pickup point on the crosshead where the hoist ropes are fastened.

Bracing members, although not part of the car frame proper, support the corners of the platform and are, in turn, fastened to the car frame.

Except for very large load capacities, the car frame members have historically been rolled structural steel channel sections as produced by the steel mills; however, in recent years, the elevator industry has seen the introduction of car frame members of formed or bent-up design. The primary advantage of this latter type of design is that the metal can be put where it does the most good. In this way, more efficient and lighter weight structures result.

The general design, necessarily modified to suit the variations in sizes, is substantially the same in all cases where the hoisting ropes are attached to the crosshead; therefore, the following design considerations apply to all.

The forces acting on the car frame members are more complex than those acting on any other part of the elevator equipment due to the nature of its design and the fact that it must carry variable loads.

If the loading condition, that is, the manner in which the load enters or leaves the platform, as well as the position it takes while the elevator is running, is unknown, then the forces acting on the car frame cannot be determined.

If the center of gravity of the duty load and car coincided with the center of action of the hoist ropes, there would be mainly tension in the uprights and bending in the crosshead and plank, but this condition cannot exist with a moving load; therefore, the car frame must be able to resist turning moments that produce bending in the uprights and, at times, twisting of the crosshead and plank.

Since we are dealing with loads that move about while entering or leaving the platform, the exact position of the load at any given time is unknown, so we must make certain assumptions regarding the position of the load while the elevator is being loaded and while it is running. We do know, for instance, the clear opening at the entrance and that any load passing through this point cannot be more eccentric from the center of the car than the entrance will permit. On the inside of the car the load can be no more eccentric than its physical size next to the cab wall will permit. Therefore, we have a starting point for designing a car frame. Figure 8.2.2.5.1 of A17.1–2000 shows the specified loading eccentricities specified by 8.2.2.5 as a function of Class of Loading. [See also Diagram 8.2.2(a).]

The calculations are in most cases simple for determining the required sizes of the car frame members, the crosshead and plank. With the uprights, however, the case is somewhat different. Here the sections are more or less limited since they must occupy a minimum amount of space between the rail and the side of the car so as not to encroach too much on the width of the car. The result is that the uprights become limited in strength for bending in the plane of the car frame, and a check is therefore necessary for special cases to make certain they are of sufficient strength and stiffness.

In both cases of loading shown in Diagram 8.2.2(a), the car frame is subjected to an overturning moment within its own plane, the magnitude of which is equal to the load times the eccentricity, i.e., to the center of the car frame. The guide shoes contacting the rails resist this overturning moment and in doing so, guide-shoe forces are impressed upon the shoes and rails. It is these guide-shoe forces, which produce bending of the uprights. As a result, the uprights deflect and cause the platform to sag on one side. A typical example of this is shown in Diagram 8.2.2(b).

Although side braces and trusses are necessary adjuncts to the car frame, they are usually considered as separate parts of the equipment, since either type of platform supports may, under certain conditions, be used with the same type of car frame. When taken in connection with the forces acting on the car frame a distinction must be made, as the stresses induced by the trusses are not the same as those caused by the side braces. Trusses, as a general rule, cause greater stresses than side braces and will be treated in a subsequent section.

The following is a brief outline of the points that must be considered in order to determine the required strength of uprights for the most common cases, during loading and running conditions.

It was noted previously that space beyond the car line is at a premium because it represents lost rentable floor space in the building per floor, and so for that reason, car frame uprights are set such that they occupy a minimum

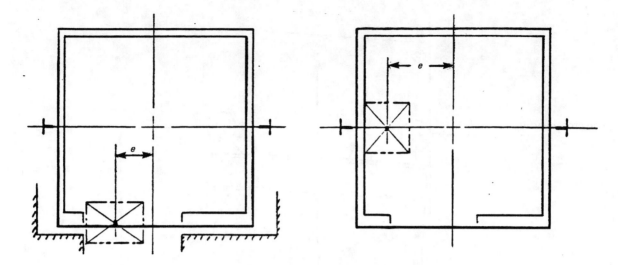

Diagram 8.2.2(a)   Loading Positions
(Courtesy Otis Elevator Co.)

amount of space. However, in so doing, some of the inherent strength of the upright is sacrificed. In this position, the upright will bend about its weak axis. Occasionally, when the duty loads are high, causing large car frame bending moments, and the available hoistway space is small, alternate means of structural reinforcing must be used, such as adding reinforcing members. When the loads become too great for these types of construction, typically occurring on heavy-duty freight cars, double uprights are used and placed in such a manner that they will bend about their strong axis. Hoistway space is generally sufficient to do this.

A simple case of a side-post elevator subject to uniformly distributed loads and arranged with four side braces will be discussed. This loading is seen in Diagram 8.2.2(b) in elevation.

This is the simplest case and the only one where the loading condition is definitely known. The maximum bending moment in the upright occurs when one-half of the duty load is uniformly distributed over one-half of the platform area between the side of the cab and the centerline. The overturning moment is $P/2$, times $E/4$, and is equal to $PE/8$. This moment causes reactions at the guide shoes, which in turn produce bending about the weak axis of the upright and bending of the guide rails. In addition, the uprights are also in tension due to the weight of the duty load, car and compensation. The ends of the platform are supported by the side braces, which are in turn fastened to the upright. When short side braces are used, which is the standard arrangement for passenger elevators, an additional bending moment is produced in the upright due to the brace loads; however, these moments cause bending about the strong axis of the upright and are of very

low magnitude. When long side braces are used, this additional bending does not occur.

Corner-post cars are used where it is necessary to have adjacent openings. In this case, the car frame is located approximately on the diagonal of the platform. The loads imposed on a corner-post car frame are generally more severe than on a side-post car. See Diagrams 2.15.1(a) and 2.15.1(b).

In this case, more than in any of the previous cases, the exact loading condition must be known if the car frame is to be of economical design. The points of application of the load with reference to the entrances both at loading or unloading, as well as its position with reference to the center of the car while the elevator is running, have such a great influence on the size of the required car frame members that, unless the loading is known, the extreme conditions must be assumed for all calculations, resulting in a much heavier construction than may actually be required.

If the platform has considerably more length than width, or vice versa, and if the load enters at the short side, then the turning moment PD, which acts in the weak direction of the uprights, may become quite large, depending on how close to the uprights the center of gravity of the load will set.

The total sag of the corner of the platform is caused by the cumulative effects of:

(a) vertical crosshead deflection due to the direct load on the truss;

(b) crosshead twist, which carries the truss with it, and magnifies the deflection at the end of the rod;

(c) truss deflection;

(d) upright deflection due to the twist of the crosshead;

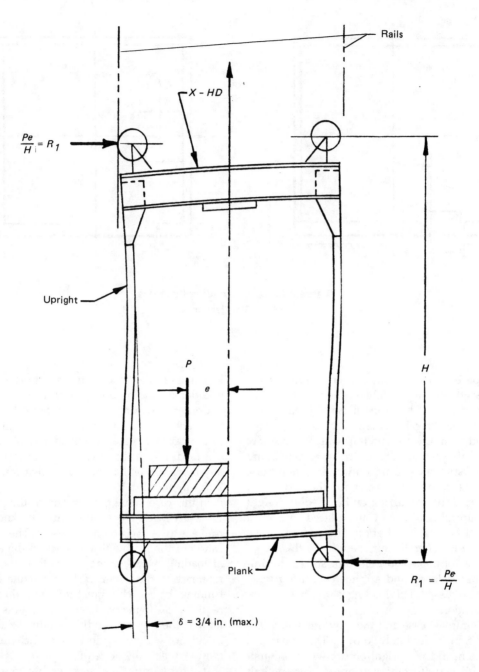

Diagram 8.2.2(b)   Upright Deflection
(Courtesy Otis Elevator Co.)

(e) guide rail deflection.

In the load cases described above, it should be noted that stresses are induced not only because of the live load, but also due to the dead load of the car itself. The objectionable deflections of the platform are those due to the live load only. Deflections due to dead load, such as the weight of the platform, enclosure, truss, etc., can be taken up by adjustment of the truss rods, provided the allowable stresses in the members are not exceeded.

The few cases mentioned above are those most frequently encountered and constitute the greater part of the conditions that must be considered in designing car frame members so far as the loading conditions are concerned. In addition to designing the car frame for normal running and loading conditions, we also must be concerned with emergency conditions such as those caused by safety application and buffer engagement.

During a safety application, the car frame must withstand the forces impressed upon it when the safeties grip the rails. Since the safeties and rails are manufactured items with manufacturing tolerances imposed on them and also since no two guide rails will be lubricated

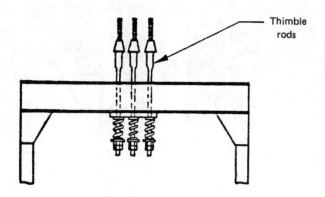

Diagram 8.2.2(c)   1:1 Hitch
(Courtesy Otis Elevator Co.)

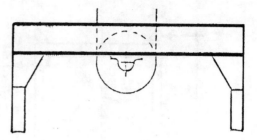

Diagram 8.2.2(d)   2:1 Roping One Sheave
(Courtesy Otis Elevator Co.)

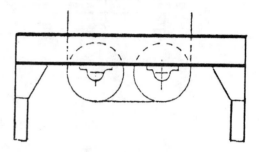

Diagram 8.2.2(e)   2:1 Roping Two Sheaves
(Courtesy Otis Elevator Co.)

the same, a difference in retarding force at the two sides of the car will exist. This difference in forces has the same effect as an eccentric load, and therefore an overturning moment will be introduced. As noted before, this overturning moment produces bending of the uprights. In addition to the bending, the uprights will also be subjected to compression due to that part of the force of retardation, which is transmitted through the side braces.

At buffer engagement, the plank channels are subjected to a severe bending load. The uprights will undergo compression and bending depending on the position of the load in the car.

Two bending moments are set up for corner-post cars, namely, *PC* and *PD*, each of which produces bending of the upright in two planes 90 deg apart. The moment *PC* acts through the truss rod, which transmits it to the truss. The truss then tends to twist the crosshead, which induces a moment at the upper end of the upright; this, in turn, produces a reaction in the guide shoes in contact with the rails. This twisting moment is divided between the two uprights in proportion to its distance from each, that is, equally divided between the uprights for a square car; and for a rectangular car, proportionally divided so that one of the uprights will take a greater part of the moment.

For these emergency conditions or for those times when tests must be conducted, a much higher allowable stress is used as is allowed by 2.15.10. This is justified since the time duration over which these high forces act is very small, and these emergency conditions are not encountered regularly.

In the cases previously discussed, it has been assumed that the car frames were suspended by ropes attached to the center of the crosshead which is the case of 1:1 roping [Diagram 8.2.2(c)], 2:1 roping with one sheave [Diagram 8.2.2(d)], and 2:1 hitch when two sheaves are used, provided all of the ropes lead around the two sheaves [Diagram 8.2.2(e)].

8.2.2.5 Car Frame Uprights (Stiles). The evaluation of the stresses due to bending of the car frame uprights is based on the following load distribution and assumptions (see Diagram 8.2.2.5).

(a) Each corner of the frame carries one-fourth the total moment *K* taken at the intersection of the centroids of the members.

(b) Free length bending of the upright occurs between the lowest fastening in the crosshead gusset-to-upright connection and the highest fastening in the plank-to-upright fastening.

(c) Each upright carries equal shear loads since both uprights have equal stiffness.

Due to car frame overturning moment *K*, caused by an eccentric live load, guide-shoe forces are induced to magnitude

(Imperial Units)

$$R_1 = \frac{K}{H} \text{ (lbf)}$$

The moment acting at each end of the upright is

(Imperial Units)

$$M = \frac{KL}{4H} \text{ (in.-lbf)}$$

For the upright member shown, the expression for the rotation of the tangent to the elastic curve at the ends is found from the standard beam formula

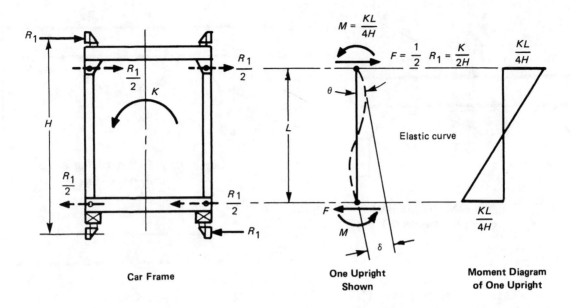

Diagram 8.2.2.5   Car Frame Stress (Imperial Units)

(Imperial Units)

$$\theta = \frac{ML}{6EI} \text{ (radians)}$$

The deflection of one end of the upright relative to the other end is equal to the distance between the end tangents as

(Imperial Units)

$$\delta = L\theta \ (in.)$$

Combining these equations and substituting 0.019 m ($^3/_4$ in.) for δ yields the equation in 8.2.2.5.3.

(Imperial Units)

$$I = \frac{KL^3}{18EH} \ (in.^4)$$

Substituting dimensional units yields,

(Imperial Units)

$$I = \frac{(in.\text{-}lb) \times (in.^2)}{\dfrac{lb}{in.^2} \times (in.)} = (in.^4)$$

8.2.2.6 Freight Elevator Platform. The construction of freight platforms typically consists of a welded channel frame with intermediate stringer channels welded to the outer frame. A steel floor plate is then welded to the stringers and frame. For purposes of design analysis, the stringers are supported by at least three points, namely, the front and rear end channels (members) and at the center by the car frame plank. Depending upon

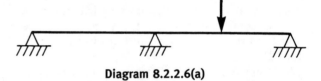

Diagram 8.2.2.6(a)

the load ranges and the specific manufacturer, there may be additional intermediate supports. The platform end channels are supported by the side braces.

The load-carrying capacity of the floor plate between the stringers depends upon the plate thickness; the capacity of the stringers depends upon their strength in combination with an effective width of floor plate.

When hoistway conditions preclude the use of a one-piece platform, two-piece or multipiece sections, bolted together at erection, are commonly furnished.

As noted above, the stringers are supported at a minimum of three points (front, center, and rear). Therefore, they are designed as beams continuous over three supports, as shown in Diagram 8.2.2.6(a), and acted upon by a concentrated load at any one point. Since the stringers are spaced at certain intervals, depending upon the strength and rigidity of the floor plate, a certain amount of the duty can be concentrated at one point. For regular freight loading (Class A), we consider one-fourth of the duty acting at one point. For truck loadings (Class C), an entire wheel load may occur directly over a stringer.

The end channels are treated as simple beams, supported at the ends by the side braces, under the action of a load or loads applied at any point in accordance with the provisions of 2.16.2.2 on classes of loading. For regular freight service (Class A loading), as an example, one-fourth of the duty load would be considered as

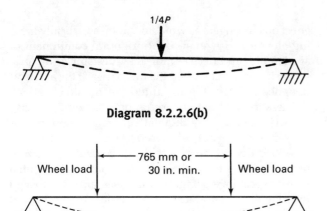

Diagram 8.2.2.6(b)

Wheel load | ← 765 mm or 30 in. min. → | Wheel load

Diagram 8.2.2.6(c)

acting at the center of the end channel as shown in Diagram 8.2.2.6(b).

For Class C (industrial truck loading), the beam would be acted upon by two wheel loads, placed in such a manner so as to produce maximum bending, as shown in Diagram 8.2.2.6(c).

Corner-post freight platforms are used when openings occur on adjacent sides of the car. The method of construction is basically the same as for side post except that the frame channel usually borders the periphery of the platform. The design requirements are somewhat different, especially for truck loading elevators, since it is possible to have two wheel loads acting upon one stringer at the same time.

8.2.2.6(b) The 15 400 kg or 34,000 lb limit is set as this is the maximum tandem axle load by the United States Department of Transportation "Bridge Gross Weight Formula."

8.2.8 Hydraulic Machines and Piping

8.2.8.1 Plunger Design. The maximum free length of the plunger is determined by the unsupported length of the plunger when fully extended. This can be expressed as the distance between the top of the cylinder head and the top of the plunger with the stop ring engaging the stop collar. A plunger follower guide can be used to reduce the maximum free length.

The safety factors used on hydraulic elevators pertain to several different concerns such as bending, compression, tension, and pressure.

(a) Bending

(1) In 8.2.8.1(b) the factor of safety given by formula is approximately 3 times the limit of the value given by Euler's equation for various grades of steel. Requirement 8.2.8.1(d) merely restates that this value of three times to be applied to plungers at varying cross sections. However, there is no formula given since how the plungers will vary on a special application is not known.

(2) In 3.18.2.3, it is required that the connection

transmitting the eccentric bending moment into the plunger have a factor of safety of 4. This is slightly higher than the factor of safety of 3 used for elastic stability and is considered adequate.

(b) Car Hung in Hatch

(1) In 3.18.1.2 the situation of the car being hung in the hatch intentionally or unintentionally is covered so that if this occurs, the plunger will not separate from the car. A factor of safety of 4 is considered adequate for this condition.

(2) Requirement 3.18.2.4(a) considers the same condition except covers plunger joints, if any. The same factor of safety of 4 is also used.

(c) Compressions. Requirements 3.18.2.4(b) and 3.18.2.1 are for the actual loading based on gross load. The gross load allowed, W is defined in 8.2.8.1. Both requirements are consistent with a factor of safety of 5 being used. An appearance of redundancy of 3.18.2.1(a) is not a valid conclusion since as noted in 8.2.8.1 some plungers are also subjected to external pressure (factor of safety also 5), which needs to be considered.

(d) Pressures (Tensile Stress)

(1) Requirements 8.2.8.2, 8.2.8.3, and 8.2.8.4 all cover pressure and all state that one-fifth of the ultimate strength shall be used in the formula. In the case of mild steel, the ultimate strength at 60,000 psi has already been divided by 5 to give a value of 12,000 psi to make it easier for the Code user since this is the material usually used. This division by 5 gives a factor of safety of 5 for items subject to pressure.

(2) Requirement 3.18.2.1 covering material is consistent with the safety factor of 5 and includes the cylinder connecting couplings.

(3) Requirement 3.19.1.1 also extends the safety factor of 5 to valves and fittings.

(4) For this special case of flexible hose and fitting assemblies, 3.19.3.3 specifies an even higher factor of safety of 10.

8.2.8.1.1 Plungers Not Subject to Eccentric Loading. This formula results from the application of the Euler equation for the allowable column load for a pin-ended column of constant cross section and a factor of safety on buckling equal to 3, as follows:

(Imperial Units)

$$W = \frac{\pi^2 E l}{F_s L^2}$$

where F_s = factor of safety.
Substituting $l = AR^2$ yields

(Imperial Units)

$$\frac{W}{A} = \frac{\pi^2 E}{F_s \left(\frac{L}{R}\right)^2}$$

$$t = \frac{pd}{2s}$$

FOS = 5

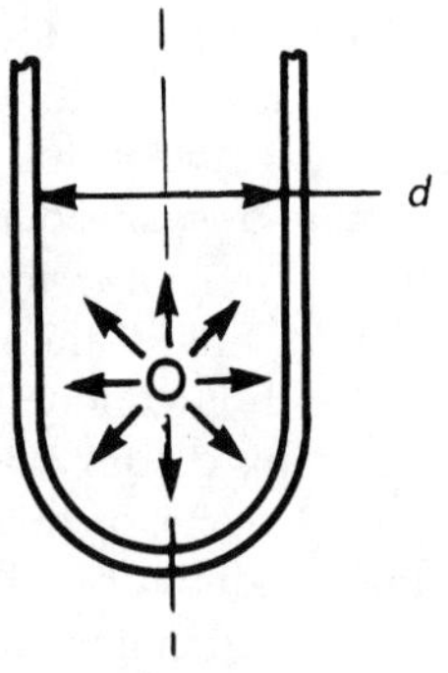

GENERAL NOTE:
t = wall thickness
s = material strength

Diagram 8.2.8.2   Cylinder Design

Substituting $E = 29,000,000$ psi (2.000×10^6 MPa) and $F_s = 3$ yields

(Imperial Units)

$$\frac{W}{A} = \frac{95,000,000}{\left(\frac{L}{R}\right)^2}$$

8.2.8.2 Cylinder Design. See Diagram 8.2.8.2.

8.2.8.3 Cylinder and Plunger Heads. See Diagram 8.2.8.3.

8.2.8.4 Wall Thickness of Pressure Piping. The allowable pressure for an oil line is limited by the lowest rated element in the line, which includes, but is not limited to, threads, grooves, fittings, valves, connections, etc.

8.2.11 Stopping Distances for Car and Counterweight Safeties for Inclined Elevators

(*a*) All elevators are subject to accelerations and decelerations acting in the direction of motion. Under normal elevator operation, the car must be accelerated from its rest, or stopped, position to contract, or rated speed, and decelerated (retarded) in going from rated speed to its stopped position.

(*b*) Unlike vertically-moving elevators where the resultant acceleration/deceleration is directed vertically, inclined elevators are subject to a resultant acceleration/deceleration, a, acting in the direction of motion, which is resolved into two components, one acting vertically, a_v, the other acting horizontally, a_h, as shown in Diagram 8.2.11. While these normal acceleration/deceleration values in an inclined elevator system will be accompanied by a horizontal component, its magnitude will be of sufficiently small order so as to be a nonissue for ambulatory passengers.

(*c*) The vertical component of acceleration and deceleration for an inclined elevator system has the same effect on passengers as would occur on a vertically moving elevator. However, the horizontal component of acceleration and deceleration may destabilize the passenger(s) depending upon the rate of rise of the acceleration/deceleration, the magnitude of the acceleration/deceleration and the time duration over which it acts.

(*d*) Concern for the destabilizing retardation, principally horizontal but whose effects on passengers are amplified due to the simultaneous presence of vertical retardation, has to be viewed in the context of potential injury to passengers if the elevator was subjected to high retardations at an emergency electrical or mechanical stop and the resulting allegations of design defect.

(*e*) Whatever retardations are generated, the onus will be on the elevator designer, manufacturer, installer, and maintainer to show that a rational basis was used to quantify the magnitude of these retardations to which passengers would be subjected.

(*f*) There are several types of stops that can occur and produce horizontal retardations of varying magnitudes:

    (*1*) Normal elevator operation

    (*2*) Emergency electrical stop (Electrical protective device tripped). Some examples include:

        (*a*) power failure;

        (*b*) open electrical protective device;

        (*c*) weakened motor field;

        (*d*) strengthened generator field; and

        (*e*) feedback failure.

    (*3*) Emergency mechanical stop (Safety application and buffer engagement)

    (*4*) Initialization stop

(*g*) Superior car-riding quality assurance is accomplished during normal elevator operation by the reduction of stimuli to a level below the threshold of passenger perception. However, in the event that an inclined elevator is subject to an emergency electrical or mechanical stop, the goal will be to impress retardations upon the elevator to stop it quickly, on the one hand, but at a rate of change of speed and acceleration that can be withstood by passengers without destabilizing them.

(*h*) The riding public can be expected to behave passively during an elevator ride because of its perceived functionality rather than for its amusement value, and its acceptance as a routine event. Most passengers treat the elevator ride as a place of momentary respite, and while they tend not to communicate with each other during the ride, their posture is generally relaxed as opposed to being ready for a horizontal disturbance at the floor. Any attempt to counter an upsetting or destabilizing event will be largely intuitive.

(*i*) In the erect position, the human body is a multidegree of freedom system with a complex transfer function and nonconstant damping coefficient, i.e., a rather complex nonlinear transducer for acceleration/deceleration excitations. An extensive analysis is required to assess the effects of external retardations impressed upon such

Type	Illustration	Formula (Imperial Units)

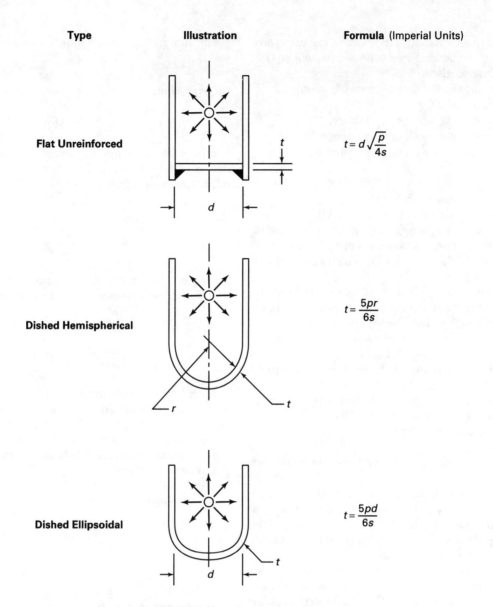

Flat Unreinforced — $t = d\sqrt{\dfrac{p}{4s}}$

Dished Hemispherical — $t = \dfrac{5pr}{6s}$

Dished Ellipsoidal — $t = \dfrac{5pd}{6s}$

Diagram 8.2.8.3 Cylinder Head Design

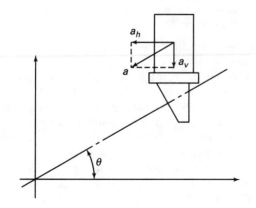

Diagram 8.2.11 Inclined Elevator Horizontal Retardations

a body. Notwithstanding, the passenger response to these suddenly-occurring motions is difficult at best to predict due to the variables involved at the onset of the horizontal retardation, which include the following:

(1) Passenger posture/stance

(2) Ability/mobility to make compensatory body adjustments:

   (a) to shift the center of gravity of the body

   (b) to stabilize the body

(3) Agility of the passenger

(4) Alignment of the body axes with respect to the direction of motion

(5) Strength and stability of the musculoskeletal structure

(6) Passenger restraint

353

(j) Horizontal excitation to the elevator can cause any or many of the following effects on the passenger(s):

  (1) feeling of discomfort

  (2) difficulty in the maintenance of balance

  (3) alarm/anxiety

  (4) destabilizing and loss of balance

  (5) lateral impact with other passenger(s)

  (6) lateral impact with elevator interior

  (7) fall

  (8) injury

(k) Body disturbances include

  (1) minor, i.e., negligible horizontal effect

  (2) moderate, i.e., staggering/stumbling potential

  (3) major, i.e., destabilizing, resulting in falls and/or impact with elevator walls, handrails and car operating panel

(l) The following types of passengers are likely to have difficulty in maintaining their stability on a horizontally accelerating/decelerating surface:

  (1) persons with slow physical reactions to an upsetting or destabilizing effect

  (2) persons with slow mental reactions to an upsetting or destabilizing effect

  (3) persons able to provide only a small restoring force to counteract a destabilizing effect

  (4) persons carrying baggage

  (5) young children

  (6) ambulatory handicapped people: walkers; crutches

(m) Research has shown that the upsetting effect of acceleration/deceleration depends not only on the level of the acceleration deceleration but also on the time taken to reach that level, i.e., the rate of change of acceleration/deceleration, which is referred to as the Jerk.

(o) The overall upsetting effect during an inclined elevator retardation is applicable to the entire retardation pattern, i.e., during the onset of the retardation (constant jerk phase) followed by the constant retardation phase, followed by the constant jerk phase at the end of the retardation.

(p) Since the time duration over which the emergency stopping retardation occurs is small, issues relating to tolerance to vibration should be deemed secondary to the upsetting or destabilizing effects of the retardation.

The safety stopping distances are derived from the general equation, which assumes a constant retardation

(Imperial Units)

$$v^2 = 2aS$$

Letting $a = kg$ and solving for the stopping distance yields

(Imperial Units)

$$S = \frac{V^2}{2kg}$$

Substituting the correct dimensions for the variables yields

(Imperial Units)

$$S = \frac{V^2}{231,840k}$$

For the minimum stopping distance, a maximum retardation of $1.0g$ is assumed. Therefore, $k = 1.0$ and the minimum stopping distance

(Imperial Units)

$$S' = \frac{V^2}{231,840}$$

For the maximum stopping distance, a minimum retardation of $0.35g$ is assumed. Substituting this value yields

(Imperial Units)

$$S = \frac{V^2}{231,840\,(0.35)} = \frac{V^2}{81,144}$$

An additional increment equal to 0.84 ft [approximately 255 mm (10 in.)] was added, presumably to account for the length of the wedge plus an additional amount to account for the vertical distance over which the safety retarding force builds up. The maximum stopping distance becomes

(Imperial Units)

$$S = \frac{V^2}{81,144} + 0.84$$

SECTION 8.3
ENGINEERING TESTS, TYPE TEST AND CERTIFICATTON

NOTE: Throughout Section 8.3, references are made to other requirements in this Code. To gain a complete understanding of a requirement, the reader should review the Handbook commentary on all referenced requirements.

The tests specified in 8.3 are not performed in the presence of an inspector employed by the enforcing authority. The tests are performed only on a sample of a product of the same design and are not performed on every assembly manufactured. The tests subject a specimen to a controlled exposure in order to evaluate its performance during the tests. These tests are intended to develop data, which will enable enforcing authorities to determine whether the product conforms to the requirements of this Code. Any variation from the construction or conditions under which the sample products were tested may substantially change the performance characteristics of the item tested. Changes to the product shall be made only as permitted by 8.3.1.4.

The readers should familiarize themselves with the following terms, which are defined in 1.3, and are used throughout 8.3.

engineering test: a test carried out by or witnessed by a registered or licensed professional engineer, testing laboratory, or certifying organization to ensure conformance to Code requirements.

type test: a test carried out by or witnessed by a certifying organization concerned with product evaluation and the issuing of certificates to ensure conformance to Code requirements.

certifying organization: an approved or accredited, independent organization concerned with product evaluation that maintains periodic inspection of production of listed/certified equipment or material and whose listing/certification states whether that equipment meets appropriate standards or has been tested and found suitable for use in a specified manner.

NOTE: For the purpose of this definition, *accredited* means that an organization has been evaluated and approved by an Authorized Agency to operate a Certification/Listing program, and is designated as such in a publication of the Authorized Agency.

labeled/marked: equipment or material to which has been attached a label, symbol, or other identifying mark of an approved or accredited independent certifying organization, concerned with product evaluation, that maintains periodic inspection of production of labeled/marked equipment or material, and by whose labeling/marking the manufacturer indicates compliance with appropriate standards or performance in a specified manner.

NOTE: For the purpose of this definition, *accredited* means that an organization has been evaluated and approved by an Authorized Agency to operate a Certification/Listing program, and is designated as such in a publication of the Authorized Agency.

8.3.1 General Requirements for Tests and Certification

The intent is to apply procedural and administrative rules to certification of elevator, escalator, and moving walk components.

8.3.1.4 Changes to Listed/Certified Components or Equipment. As the tested product is a representative sample of the finished product that is listed/certified, it must represent the tested product. However, it is not practical to assume that minor changes may not be made to a product after it has been tested. The requirements in 8.3.1.4 outline the procedures that must be followed to assure the product continues to be acceptable to the certification organization to be listed/certified. The certifying organization may make an engineering evaluation of the change or require a new test.

8.3.2 Type Tests of Car and Counterweight Oil Buffers

8.3.2.4 Installation of Buffer and Preparations for Tests

8.3.2.4.4 Filling Buffer With Oil. The intent of these requirements is to duplicate actual field conditions. The oil level is to be at the lowest level recommended by the manufacturer to assure that the buffer performs as required when subjected to a drop test under the most severe conditions. The specified procedure assures there is not any trapped air in the buffer.

8.3.2.5 Buffer Tests. The oil buffer is subjected to conditions, which duplicate in a laboratory the most severe conditions that could be encountered during actual service. The test speeds are equal to the maximum overspeed encountered before a car safety would be engaged. The plunger return test is made to ascertain the ability of the oil buffer to return to its normal position, fully extended. A malfunctioning elevator could compress the oil buffer, travel to an upper landing, and then strike the oil buffer again for a second time, etc. The plunger return test is to confirm the ability of an oil buffer to be ready for sequential compressions.

8.3.3 Type Test of Interlocks, Combination Mechanical Locks and Electric Contacts, and Door or Gate Electric Contacts

These tests subject hoistway door interlocks, hoistway door combination mechanical locks and electric contacts, and hoistway door and car door or gate electric contacts to laboratory conditions that represent field conditions to which the product will be subjected during use. Products tested prior to August 1, 1996, can be tested to the engineering test specified in ASME A17.1a–1994, Section 1101 [2.12.4.1(a)]. Products that have successfully completed these tests, have a history of reliability without serious malfunctions during actual field use.

UL 104 provides guidance and useful information that the designer will find helpful. Compliance to UL 104 is not a requirement.

8.3.3.3 General Requirements

8.3.3.3.1 Connections for and Test of Electrical Parts. The test current and voltage is related to the rated voltage and current, not an arbitrary value.

8.3.3.4 Required Tests and Procedures. The device is subjected to a total of 1,000,000 cycles by requiring the use of one device for all tests. Devices used for private residence elevators are not subjected to the same use, thus the 50,000 cycles specified for this application. Because of the number of cycles, lubrication is permitted in accordance with the manufacturers' instructions. The number of cycles is based on a survey, which concluded the number of stops per year ranged from 100,000 in a quiet apartment building to 600,000 in a hotel.

8.3.3.4.2 Current Interruption Test

8.3.3.4.2(a) The test current and voltage is related to the rated voltage and current.

8.3.3.4.2(c) Evidence of insulation breakdown and deterioration provides objective criteria for passing the test.

8.3.3.4.4 Test in Moist Atmosphere. The test described requires the device to be exposed to atmosphere saturated with a 3.5% salt solution for 72 h. Similar testing conducted by Underwriters Laboratory (UL) is the Rust Resistance and Corrosion Resistance Tests. The Rust Resistance Test is conducted for 24 h and is intended to demonstrate a minimum level of resistance to corrosion for steel enclosures intended for indoor use. The Corrosion Resistance Test is conducted for 200 h and is intended for steel enclosures for outdoor use. Both tests are conducted with a 5% by weight solution and are based on the Salt Spray (Fog) Testing Standard, ASTM B117-1985 for corrosion testing.

The salt spray test is an effective accelerated corrosion-aging test that determines a minimum level of a material's corrosion-resistant properties.

UL conducts the salt spray tests in a sealed chamber, which keeps the NaCl solution in a semiliquid state. As such, the only time we have experienced salt deposits is when the test device is removed from the chamber and the liquid droplets of the solution are allowed to evaporate, leaving (solid) salt deposits.

As far as the percentage of saturation of the NaCl solution, UL's testing is from 5% to 20%. The origin of the saturation range or where it was developed is not known.

8.3.4 Entrance Fire Tests

Elevator entrance assemblies are subjected to identical testing as that which a standard fire-rated door must endure. The only variation is the test acceptance criteria for the horizontal slide-type entrance assembly. The two test standards, UL 10B, and NFPA 252, are similar.

The lobby side of passenger elevator entrances is exposed to the furnace during the test. The hoistway side of freight elevator entrances is exposed to the furnace. On a passenger elevator, it is assumed the fire is in the lobby. On a freight elevator, it is assumed the fire is on the car or in the lobby and the door is open. In either case, it is assumed, the fire will be stopped by a door. On a passenger elevator, the closed lobby door will stop the fire from entering the hoistway. On a freight elevator, the fire will enter the hoistway and would be stopped by a closed entrance at upper floors. In either case, the fire would be contained from extending through the hoistway to upper floors in the building.

These tests are intended to provide a method of evaluating fire door assemblies when exposed to laboratory fire conditions. Entrance assemblies include the door panel(s), frame, and related hardware. The effectiveness of the opening protection is directly related to the entire assembly; thus, all necessary parts must be in place during testing. Once the complete assembly is successfully tested, the testing laboratory may label the components, panel, frame, and hardware separately. They may further allow one or more of the components from one test to be used with components from another test, which, combined together, make an approved fire-rated entrance assembly.

Fire-resistance ratings are assigned to indicate that the component has successfully conformed to the requirements for a period of 3, $1\frac{1}{2}$, 1, $\frac{3}{4}$, $\frac{1}{2}$, or $\frac{3}{4}$ h. Labels may also have a letter designation of A, B, C, D, or E. The alphabetical letter designation was one method employed to classify the opening for which the fire door is considered suitable. Traditionally, the relationship between the alphabetical designation and its use was as follows:

(a) Class A: Openings in firewalls and in walls that divide a single building into fire areas.

(b) Class B: Openings in enclosures of vertical communications through buildings and in 2 h rated partitions providing horizontal fire separations.

(c) Class C: Openings in walls or partitions between rooms and corridors having a fire-resistance rating of 1 h or less.

(d) Class D: Openings in exterior walls subject to severe fire exposure from outside of the building.

(e) Class E: Openings in exterior walls subject to moderate or light fire exposure from outside of the building.

This designation is not part of the test standard but is used by many codes and standards. Elevator entrances have historically been $1\frac{1}{2}$ h (B) labeled. Recently some codes are also recognizing 1 h (B) and $\frac{3}{4}$ h (B) labeled elevator entrances. Labeling agencies may also allow an entrance assembly to be down-labeled, e.g., from $1\frac{1}{2}$ h to 1 h or less.

The test requires that a complete entrance assembly be installed in a wall, duplicating the type of construction (masonry, gypsum drywall, etc.) for which it is intended. This specimen is then placed in a furnace. The applicable standard provides details on the operating characteristics and temperatures in the test furnace. A temperature-time relationship is detailed in the standard and is represented in a curve (T - t). This curve was adopted in 1918 and is representative of a severe building fire.

The assembly is subjected to the fire test for a time duration equal to that for which it is intended to be used, i.e., $1\frac{1}{2}$ h, 1 h, etc. Immediately following the fire test, the complete assembly is removed from the furnace and the door panel is then sprayed with a stream of water from a 63.5 mm ($2\frac{1}{2}$ in.) hose with a 28.5 mm ($1\frac{1}{4}$ in.) nozzle. This produces a stress in the assembly and is a gauge of its structural capabilities. The duration, location of the hose, and area sprayed are detailed in the standard.

The standard specifies the condition of acceptance not only during the fire test and hose stream test, but immediately following same. The most important conditions are:

(a) the door remains in the opening during the fire endurance test and the hose stream test;

(b) no openings develop anywhere in the assembly during the fire endurance test or hose stream test except small portions of the glass may dislodge during the hose stream test;

(c) the door shall not release from its guides and the guides shall not loosen from their fastenings;

(d) separations of meeting edges shall be within the tolerances specified in the standard.

The acceptance criteria for swing doors is developed based on the door fitting into a frame with rabbets and the fire impinging on the side of the door that swings into the furnace. The door panel, therefore, moves away from the frame under test while the horizontal slide elevator door overlap moves against the frame. Pairs of elevator doors have a greater allowable tolerance for separation at the meeting edge parallel to the panel surface than do pairs of swing doors. From a fire penetration point of view, they are less able to sustain such penetration than swing doors and may be considered weaker. From a practical point of view, the swing door pairs are restrained at their respective hinges but the horizontal slides are not and, therefore, more liable to separate under the test condition.

The atmospheric pressure in the furnace is as nearly equal to the atmospheric pressure as possible. In recent years the Uniform Building Code (UBC) and International Building Code (IBC) have changed that requirement for fire doors and require the natural pressure level in the furnace be established at 1 016 mm (40 in.) above the sill. This change is intended to more appropriately reflect actual fire conditions. However, elevator doors have been excluded as they have a reliable track record of stopping the advancement of a fire past closed labeled doors. One need only review the findings of investigations of two recent major fires to confirm this record. At both the MGM Grand in Las Vegas and Meridian Plaza in Philadelphia, elevators were found at the fire floor. Upon opening the doors, only smoke damage and incidental melting of plastic fixtures was observed. At the MGM Grand, two cars, side by side, in the same hoistway were at the fire floor. One had its doors open and was consumed by the fire. The adjacent car suffered only smoke damage. There was no evidence of flame spread inside the car with closed doors.

Vertical and horizontal slide elevator doors as presently designed would not pass the new test standards (UL 10C). It is doubtful they could even be redesigned to pass the revised test criteria and meet the additional criteria mandated by the elevator code. That criteria controls kinetic energy and door closing forces, which is critical for the safety of the riding public.

With the neutral pressure level established at 1 016 mm (40 in.) above the floor and elevator door with openings between panels and/or frame, would in all probability fail the test due to flaming on the unexposed surface of the test as specified in the test standards. In actual use, the neutral plane of a fire, either in or migrating through an elevator hoistway, is well above the top of the elevator door due to stack effect in the hoistway. In addition, elevator hoistways have minimal fire loading and except for the floor where the elevator car is located, there is an empty hoistway adjacent to the fire-rated door.

8.3.7 Vertical Burn Test

Prior to ASME A17.1a–1985, the use of napped, tufted, woven, looped (i.e., carpet), and similar material on car enclosure walls was prohibited. This material is now permitted as long as it meets the acceptable criteria based on a vertical burn test as described in this section. The vertical burn test is based on applicable portions of a Federal Aviation Administration (FAA) test for aircraft interiors. A specimen, duplicating an end-use configuration, is placed in a draft-free cabinet 19 mm or $\frac{3}{4}$ in. above a bunsen burner with a flame length of 38 mm or $1\frac{1}{2}$ in. applied for 12 s, then removed. After the burner is removed, flame time, burn length, and flaming time of the drippings is recorded. A minimum of three specimens must be tested. The material is acceptable if:

(a) average burn length does not exceed 203 mm or 8 in. or;

(b) average flame time after the removal of the burner does not exceed 15 s; and

(c) the drippings do not flame for more than 5 s.

This requirement is in addition to the test of the enclosure material [see 2.14.2.1.1], as napped, etc., material is more susceptible to accidental ignition.

8.3.8 Test Method for Evaluating Room Fire Growth Contribution of Textile Wall Covering

Materials are fire tested in an 2.4 m (8 ft) × 3.7 m (12 ft) × 2.4 m (8 ft) high room/corner utilizing a product mounting system, including adhesive, representative of actual installation. Prior to testing, the sample shall be conditioned at 21°C (70°F) ± 2°C (5°F) and at a relative humidity of 50% ± 5% until the sample reaches a rate of weight change of less than 0.1% per day. The product shall be exposed to a flame from a gas diffusion burner for 15 min. The fire exposure shall be 40 kW ± 1 kW for the first 5 min, followed by an exposure of 150 kW ± 5 kW for an additional 10 min. Such tests shall demonstrate that a product will not spread fire to the edge of the specimen or cause flashover in the test room. For additional information, see NFPA 265, Fire Test for Evaluation Room Fire Growth Contribution of Textile Wall Coverings.

8.3.10 Engineering Tests — Safety Nut and Speed-Limiting Device of Screw-Column Elevators

8.3.10.2 Test of Safety Nut. The engineering test simulates the most severe condition to which the safety nut might be subjected in the field. Satisfactory performance during the engineering test indicates that the safety nut should function to prevent the elevator car from falling, or any damage to the elevator equipment in the event of failure of the driving nut on an elevator equipped with this type of safety nut.

8.3.10.3 Test of Speed-Limiting Device. The engineering test simulates the most severe condition to which the speed-limiting device might be subjected in the field. Satisfactory performance during the engineering test indicates that the speed-limiting device should function to control the descent of the car within safe limits and prevent injury to passengers or damage to the elevator equipment in the event of failure of the driving means of an elevator equipped with this type of speed-limiting device.

8.3.11 Engineering Test of Escalator Steps and Moving Walk Pallets

This test subjects an escalator step or moving walk pallet to a fatigue loading in order to evaluate the structural integrity of the step or pallet. This is not a field test and can be performed at either the manufacturer's facility or in a testing laboratory. The test is performed on a sample only, which duplicates the product installed in the field. Each step or pallet width supplied is subjected to this test. A copy of the test report, certified by a registered professional engineer, should be supplied for each installation and/or model installed.

SECTION 8.4
ELEVATOR SAFETY REQUIREMENTS FOR SEISMIC RISK ZONE 2 OR GREATER

NOTE: Throughout Section 8.4, references are made to other requirements in this Code. To gain a complete understanding of a requirement, the reader should review the Handbook commentary on all referenced requirements.

Earthquake requirements are probably just as important as the requirements for firefighters' service except people tend to discount the severity unless they are in an active earthquake zone. The San Fernando earthquake of 1971 caused extensive damage and provided a good base of statistics to consider in the development of the following requirements:

(a) counterweights out of their rails, 674;

(b) counterweights out of their rails with damaged cars, 109;

(c) rail brackets broken or damaged, 223;

(d) roller guides broken or loose, 286;

(e) compensating cables out of their grooves or damaged, 100;

(f) governor cable hung up, 20;

(g) hoisting ropes damaged or out of their sheave grooves, 7.

The International Building Code (IBC) and later editions of the National Building Code (NBC) and Standard Building Code (SBC) do not refer to "seismic risk zones." These codes specify earthquake risk in terms of Peak Horizontal Ground Acceleration, A_v.

The National Building Code of Canada (NBCC) specifies earthquake risk in terms of Peak Horizontal Ground Acceleration, Z_v.

The correlation of A_v and Z_v to seismic risk zones can be found in 8.4.13.

8.4.1 Horizontal Car and Counterweight Clearances

8.4.1.1 Between Car and Counterweight and Counterweight Screen. Greater clearances are required because the moving equipment can be displaced from its normal operating plane during a seismic disturbance.

8.4.2 Machinery and Sheave Beams, Supports, and Foundations

8.4.2.1 Beams and Supports. The intent of these requirements is to inhibit movement during a seismic disturbance.

8.4.2.2 Overhead Beams and Floors. These requirements ensure the use of adequate equipment fastening means. Requirement 2.9.3.1.2 does not apply; therefore, it is required that isolated equipment also be restrained.

8.4.2.3 Fastenings and Stresses. This requirement specifies minimum design parameters to assure adequate strength of the fastenings. Substantially more strength is required for fastenings subject to impact loading, and this requirement is contained in 8.4.2.3.2. The maximum combined stress allowed under seismic conditions is contained in 8.4.2.3.3 and is one-third greater than that allowed under normal conditions.

8.4.3 Guarding of Equipment

8.4.3.1 Rope Retainer Guards. These requirements provide a means for retaining each wire rope in its sheave groove during seismic activity.

8.4.3.2 Guarding of Snag Points. These requirements provide a means for preventing the snagging of ropes, chains, and traveling cables when they are displaced from their normal location as a result of a seismic disturbance.

8.4.4 Car Enclosures, Car Doors and Gates, and Car Illumination

8.4.4.1 Top Emergency Exits

8.4.4.1.1 This requirement provides a means whereby the hoistway can be inspected from within the car for earthquake damage. The Exception allows the use of standard access switch keys so that one type of key is all that is necessary for the elevator installation.

8.4.4.1.2 This requirement prevents normal operation of the car while the top emergency exit is open and yet allows slow-speed operation of the car for hoistway inspection following an earthquake.

8.4.5 Car Frames and Platforms

8.4.5.1 Guiding Members and Position Restraints. This requirement provides a means, other than the normal guiding means, of keeping the car within the plane of the guide rails and requires that the additional restraint means are located to the best advantage of the guide-rail system. The last sentence allows the restraint to be integral with the guiding member.

8.4.5.2 Design of Car Frames, Guiding Members, and Position Restraints

8.4.5.2.2 This requirement provides the clearances necessary between the restraint and the guide rail so that, under normal conditions, the restraints will not interfere with the riding comfort of the elevator.

8.4.6 Car and Counterweight Safeties

8.4.6.1 Compensating Rope Sheave Assembly. This requirement assures that the tension sheave equipment is adequately designed to withstand, without damage, the specified seismic forces and thereby not cause unnecessary shutdowns at lower levels of seismic activity. The third paragraph provides a means of shutting down the elevator when a compensating sheave assembly is not located in its normal frame of operation for any reason.

8.4.7 Counterweights

8.4.7.1 Design

8.4.7.1.1 This requirement addresses the design where the height of the weight stack occupies less than two-thirds of the height of the counterweight frame.

8.4.7.1.2 This requirement assures that not more than one-third of the total weight of the counterweight appears as a reactive force on the guide rail at the lower position restraint location, when the counterweight is acted upon by a seismic acceleration rate of not more than $g/2$.

8.4.7.1.4 This requirement assures that the major-portion of the side thrust force of the counterweight assembly is transferred to the building structure through a main guide-rail bracket after the guide rails have deflected more than 13 mm or $\frac{1}{2}$ in. at the upper and lower position restraint locations as a result of a $g/2$ seismic acceleration force.

8.4.7.2 Guiding Members and Position Restraints

8.4.7.2.1 This requirement provides a means, other than the normal guiding means, of keeping the

counterweight assembly within the plane of the guide-rail system and requires that the additional restraint means are located to the best advantage of the guide-rail system. The last sentence takes care of the situation where the position restraint may be an integral part of the guiding member.

8.4.7.2.3 This requirement provides the clearance necessary between the restraint and the guide rail so that under normal conditions, the restraints will not interfere with the riding comfort of the elevator.

8.4.8 Car and Counterweight Guide-Rail Systems

The requirements are prescriptive in that the specific design detail tends to obscure the intent of selecting design solutions based upon the application of indeterminate structural analysis. In order that a guide-rail system be selected, which will safely withstand the seismic loading condition, the installed guide-rail system must perform in a manner consistent with the theoretical basis upon which the requirements are premised. The term "guide-rail system," as used in 8.4.8, includes the guide rails and reinforcement, where used, fishplates, rail clips, rail clip bolts, main brackets, and intermediate tie brackets, where used.

8.4.8.2 Seismic Load Application. This requirement assures that an adequate size guide rail is used for a given bracket spacing and a given weight of counterweight or a given car weight plus 40% of its rated load, when the car or counterweight is acted upon by a seismic force of $\frac{1}{2} g$ horizontally. The Exception to this requirement allows a degree of flexibility in achieving an acceptable guide-rail system.

The guide rail selection curves shown in Figs. 8.4.8.2-1 through 8.4.8.2-7 are the graphical solutions to the equations given in 8.4.12, which are based on continuous beam design theory in which the guide rails are designed to develop continuity at their supports and joints, i.e., rail brackets and fishplates, respectively. In order that this structural theory comprise a valid analytical model, the guide rail joints must be capable of transferring the bending moments, shears, and axial forces, induced by the seismic loading condition, from the end of guide rail to the adjacent one through the connection.

The curves in Figs. 8.4.8.2-1 through 8.4.8.2-7 depict the allowable total weight of the counterweight for a given rail size and a given bracket spacing based on the number of intermediate tie brackets used. Under these conditions, an adequate size guide-rail system will be capable of withstanding, without damage, a seismic force of $\frac{1}{2} g$ acting horizontally.

The curve in Fig. 8.4.8.2-8 takes care of the condition where the rail span is much greater than the height of the counterweight and provides the modification necessary to assure that an adequate guide-rail system is provided.

8.4.8.2.4 Use of intermediate tie brackets assures the strength of the guide-rail system located between main guide brackets and, therefore, heavier counterweights can be used. The Exception to this requirement allows some flexibility in the design of the guide-rail system.

8.4.8.3 Guide Rail Stress. This requirement specifies a maximum allowable stress under seismic conditions, which is one-third larger than the stress allowed under normal conditions.

8.4.8.4 Brackets, Fastenings, and Supports. For purposes of clarification, the guide rail bracketry comprises the structural system intermediate between the guide rail and the building structure. It is intended to include the guide rail brackets and fastenings, bracket reinforcements, where provided, and intermediate tie brackets, where provided, including any other structural supports exclusive of the building structure. The building structure is the foundational structure to which the guide rail bracketry is fastened. Examples of the building foundational structure include hoistway separator beams, hoistway perimeter structural framing or load-bearing walls, etc.

In the development of the requirements for structural deflections, it was intended to differentiate between the deflections of the guide rail bracketry at or in close proximity to the point of load versus the deflection of the foundational structure, i.e., at the location of load transfer from the rail and/or its bracketry to the building.

A comprehensive analytical study of guide rail loading conditions showed that the most adverse loading condition occurred when the resultant line of action of seismic acceleration forces coincided with the plane of the guide rails, i.e., in a direction so as to cause bending of the guide rails about their own x-x axis, as shown in Fig. 8.4.8.9. If the resultant line of action of the seismic acceleration forces is at an oblique angle with respect to the plane of the guide rails, a load distribution occurs to both rails, thus reducing the overall effect to one rail.

The method of structural analysis used for the guide-rail system subject to seismic loading is based upon continuous beam theory in which the guide rail supports are considered to be relatively rigid, i.e., nondeflectable. Continuous beam theory is valid for continuous beams with flexible supports with relatively small deflections compared to the maximum deflections in the rail span. Therefore, it was decided to limit the deflection of the support to a value equal to 2.54 mm or 0.10 in., which is approximately one order of magnitude lower than the maximum deflections allowed for the guide rails. (See Table 8.4.12.2.)

The note subordinate to 8.4.8.7, advises that the design variations between specific designs of guide rail bracketry and methods of attachment to the building structure require that the maximum stresses and deflections be analyzed to suit the specific designs. Requirement 8.4.8.7 is intended to limit the maximum operating stresses and deflections of the guide rail bracketry to the values specified in Table 8.4.8.7.

The allowable deflection of 2.54 mm or 0.10 in. given in Table 8.4.8.7 applies only to the guide rail bracket and reinforcement, if used. The following rationale was used to establish the 2.54 mm or 0.10 in. value:

(a) The maximum allowable deflection at an elastic, i.e., nonrigid, support will be limited by the precedent value of 3.2 mm or $\frac{1}{8}$ in. deflection per 3 m or 10 ft 0 in. span length, which had been used historically for the primary elevator structures, i.e., car frames and platforms. This is expressed as $\frac{1}{960}$ of the span length (see 2.15.11).

(b) The distance between two unyielding rail brackets is taken as equal to two spans of 2.4 m or 8 ft 0 in. length, which is deemed to be the smallest practical value corresponding to 2.4 m or 8 ft 0 in. floor heights.

(c) The incremental deflection of the rail bracket is measured at the point of application of the load in a direction coincident with the line of action of the load.

(d) The total deflection at the rail bracket equals the incremental deflection of the rail bracket itself plus the deflection of the building (foundational) support plus distortion of the fastening. The incremental deflection of the rail bracket itself is assumed to be no greater than one-half of the total.

(e) The limiting incremental deflection of the guide rail bracket together with any reinforcement, if used, is equal to:

(Imperial Units)

$$\delta_{bkt} = \frac{\text{span}}{960}\left(\frac{1}{2}\right) = \frac{2 \times 96}{960 \times 2} = 0.10$$

In the development of these requirements the A17 Earthquake Safety Committee noted in the Minutes of Meeting of October 30, 1980 that the total deflection of 3.2 mm (0.125 in.) at the point of support as originally specified in ASME A17.1b–1992 and earlier editions, Appendix F, Rule F200.5b is the combined total deflection of the guide rail brackets and fastenings plus their foundational supports, such as, building beams and walls. However, the text of this statement did not appear in the official ballot comments. Its exclusion is attributed more to oversight than a change of intent, since there was no reassessment of the appropriateness of this value until 1986.

In the subsequent development the A17 Earthquake Safety Committee agreed at the June 10, 1986 Committee meeting that the aggregated deflection limit of 3.2 mm ($\frac{1}{8}$ in.) for bracketry, bolts, and building foundational supports contained in ASME A17.1b–1992 and earlier editions Appendix F, Rule F200.5b was too stringent and unrealistic.

The simple proof of this conclusion is seen from the mathematical statement of Appendix F, Rule F200.5b:

(Imperial Units)
$$\delta_{bkt} + \delta_{bolts} + \delta_{bldg} \le 0.125$$

If the 2.54 mm or 0.10 in. limiting deflection of the bracket is assumed, and a rail bracket-to-building support connection utilizes either a welded or body-fit bolted fastening, then,

(Imperial Units)
$$0.10 + 0 + \delta_{bldg} \le 0.12$$

from which the allowable deflection of the building foundational support is

(Imperial Units)
$$\delta_{bldg} \le 0.025$$

Clearly, an allowable deflection of the building foundational supports not exceeding 6.4 mm or $\frac{1}{4}$ in. was too stringent and unrealistic.

In order to have a rational basis for the selection of a more realistic value, the Committee considered that the California Code allowed a deflection of 6.4 mm or $\frac{1}{4}$ in. for the building foundational support. Section 3030(k)(1)(C) of the California Code of Regulations, Title 8, requires that the building construction forming the foundational supports for the guide rail brackets not deflect more than 6.4 mm or $\frac{1}{4}$ in. under a horizontal seismic acceleration force due to 0.5 g with the car and/or counterweight located in their most adverse position in relation to any rail bracket.

However, the Committee concluded that, under no circumstances, should the total deflection at the face of the guide rail adjacent to its bracket exceed 6.4 mm or $\frac{1}{4}$ in. Therefore, 8.4.8.4 allows a cumulative deflection not to exceed 6.4 mm or $\frac{1}{4}$ in., thus, superseding the previous limit of 9.125 mm or $\frac{1}{8}$ in., which was contained in ASME A17.1–1995 Appendix F, Rule F200.5b. The mathematical equivalent of 8.4.8.4 is as follows:

(Imperial units)
$$\delta_{bkt} + \delta_{bolts} + \delta_{bldg} \le 0.25$$

In summary, the deflection value of 6.4 mm or $\frac{1}{4}$ in. specified in 8.4.8.4 limits the cumulative deflections of the rail bracket, fastenings, and building foundational supports taken with respect to a datum at the face of the guide rail, whereas, the deflection limit of 2.54 mm or 0.10 in. specified in Table 8.4.8.7 pertains to the incremental deflection of the guide rail bracket with respect to a datum at the face of the guide rail.

8.4.8.6 Design and Construction of Rail Joints

8.4.8.6.1(e) The fishplate requirement assures a rail system without the discontinuity (strength wise) that the standard type fishplate would introduce if it were used. The structural connection between adjacent rail sections, i.e., fishplate and fastenings, must have the strength and rigidity necessary to develop continuity at the supports. Literal compliance with 8.4.8.6.1(e) ensures that structural continuity is provided at the rail joint. However, compliance with the fundamental intent of structural continuity at the joint is ensured by meeting 8.4.12.1 and 8.4.12.2. In other words, if the individual section modulus and moment of inertia of the guide rail and fishplate, each taken singly, comply with formulas for the minimum section modulus and moment of inertia as specified in 8.4.12.1 and 8.4.12.2, respectively, then the guide rail and fishplate design comply with the intent of 8.4.8.6.1(e) and the literal test of 8.4.8.6.2.

8.4.8.7 Design and Strength of Brackets and Supports.

These design requirements assure that guide-rail brackets of adequate strength are provided and that bracket designs are compatible with the specific building structure to which they are attached.

Table 8.4.8.7 provides a visual description of the bracket requirements contained in 8.4.3, including all design parameters.

8.4.8.8 Type of Fastenings.

This requirement makes certain that fastening bolts of sufficient strength are furnished for all conditions of loading including seismic loading.

8.4.8.9 Information on Elevator Layouts.

The orientation of the principal axes of the guide rail section is shown in Fig. 8.4.8.9. Forces acting to cause bending about the y-y axis of the rail are resisted by the fishplate, which would be subjected to bending about its strong axis. This orientation of the fishplate attached to the back of the rail section inherently provides structural continuity at the supports and joints for bending about the y-y axis, so that the rail can be analyzed and designed as a continuous beam over multiple supports. The forces acting normal to the face of the rail, i.e., causing bending about the x-x axis, must be withstood by the joint connection, normally referred to as a full strength fishplate, in order that the rail functions as a continuous beam over multiple supports. If a full strength fishplate were not used, the curves shown in Figs. 8.4.8.2-1 through 8.4.8.2-7 would not be valid.

8.4.9 Driving Machines and Sheaves

8.4.9.1 Seismic Requirements for Driving Machine and Sheaves.

This requirement assures that the driving machine and its supports can withstand the dynamic effects of their own masses as a result of the seismic disturbances specified without sustaining serious damage.

8.4.10 Emergency Operation and Signaling Devices

8.4.10.1 Operation of Elevators Under Earthquake Emergency Conditions

8.4.10.1.2 Equipment Specifications. This requirement specifies the desired parameters of the devices and defines the most convenient location for maintenance and inspection accessibility and includes a specific recommended location best suited for responding to a seismic disturbance if such location is available.

8.4.10.1.2(d) This requirement provides a safe environment for people who work on or around the exposed equipment in the hoistway.

8.4.10.1.3 Elevator Operation. See ASME A17.1 Fig. 8.4.10.1.3.

8.4.10.1.3(a) and (b) The purpose of these requirements is to get the elevator safely to a floor as quickly as possible so that the passengers can exit the car and not be trapped in the elevator between floors.

8.4.10.1.3(c) This requirement assures that elevators open their doors when they arrive at the floor. The Exception assures that the doors will not open automatically at a fire-involved floor.

8.4.10.1.3(d) This requirement assures that the elevator will remain at the floor and that it opens its doors. The Exception assures that the doors will not open automatically under fire conditions.

8.4.10.1.3(e) This requirement assures that the elevator remains at the floor.

8.4.10.1.3(f) This requirement allows the use of an elevator at slow speeds following a seismic disturbance.

8.4.10.1.3(g) This requirement prevents the use of an elevator when seismic damage has occurred to the counterweight or its rail system.

8.4.10.1.3(h) This requirement prevents the loss of earthquake condition status when power fails for any reason.

8.4.10.1.3(i) This requirement provides a means for restoring normal operation.

8.4.10.1.3(j) This requirement prevents bypassing critical safety circuits in the control system, which could lead to serious injury or a fatality.

8.4.10.1.4 Maintenance of Equipment. This requirement assures that the equipment is serviced, maintained, and accurately calibrated at prescribed intervals so that it remains operational at all times.

8.4.11 Hydraulic Elevators

8.4.11.1 Machinery Rooms and Machinery Spaces. This requirement prevents locating a machine room and its associated hoistway on opposite sides of an expansion joint. Seismic activity could cause enough joint separation to rupture the oil supply line to the hydraulic cylinder.

8.4.11.2 Overspeed Valve. The overspeed valve provides a means to stop and hold a fully loaded car if loss of hydraulic pressure occurs between the cylinder and the hydraulic pump due to a rupture in the oil supply line. The overspeed valve is to be located where it can monitor as much of the oil supply line as possible and to assure that the loss of electrical power will not affect its operation should a rupture occur. The exception applies where car safeties are provided and states that the safety will act to stop and hold the car if the oil supply line ruptures. The loss of pressure will allow the car to attain sufficient speed to cause the safety to be actuated.

8.4.11.3 Pipe Supports. This requirement provides sufficient support for hydraulic piping to inhibit its movement during seismic activity. Such movement, if not restrained, could result in rupturing the hydraulic oil supply line.

8.4.11.5 Guide Rails, Guide-Rail Supports, and Fastenings. This specifies requirements that apply to hydraulic elevators and the procedures for determining maximum load.

8.4.11.6 Support of Tanks. This requirement provides sufficient support to inhibit tank movement during seismic activities.

8.4.11.7 Information on Elevator Layouts. This requirement provides sufficient information so that proper supports can be furnished.

8.4.12 Design Data and Formulas for Elevators

8.4.12.1 Maximum Weight Per Pair of Guide Rails. The formulas in 8.4.12.1.1 indicate the allowable vertical loads as a function of the seismic zone acceleration factor, guide rail section modulus, span length and numerical coefficient relating to multiple spans in the continuous beam model in which the loads are appropriately located a certain distance from the supports in order to produce the maximum loading condition. These formulas can be restated to solve for the minimum section modulus as a function of the known vertical load, seismic zone, span length, and numerical coefficient. The formulas in 8.4.12.2 provide the minimum allowable moment of inertia of the guide-rail section as a function of the known vertical load, span length, modulus of elasticity, and numerical coefficient. Therefore, for any elevator and/or counterweight application, where a known vertical load, seismic zone and bracket spacing are given, the application of the formulas in 8.4.12.1.1 and 8.4.12.2 provide the minimum section modulus and moment of inertia of the guide rail, continuous over multiple supports.

8.4.13 Ground Motion Parameters

The International Building Code (IBC) and later editions of the National Building Code (NBC) and Standard Building Code (SBC) do not refer to "seismic risk zones."

These codes specify earthquake risk in terms of Peak Horizontal Ground Acceleration, Z_v.

SECTION 8.5
ESCALATOR AND MOVING WALK SAFETY REQUIREMENTS FOR SEISMIC RISK ZONE 2 OR GREATER

NOTE: Throughout Section 8.5, references are made to other requirements in this Code. To gain a complete understanding of a requirement, the reader should review the Handbook commentary on all referenced requirements.

The International Building Code (IBC) and later editions of the National Building Code (NBC) and Standard Building Code (SBC) do not refer to "seismic risk zones." These codes specify earthquake risk in terms of Peak Horizontal Ground Acceleration, Z_v.

The National Building Code of Canada (NBCC) specifies earthquake risk in terms of Peak Horizontal Ground Acceleration, Z_v.

The correlation of A_v and Z_v to seismic risk zones can be found in 8.4.13.

8.5.1 Balustrade Construction

This requirement allows manufacturers and building owners a choice of options to alleviate a potential hardship while still providing for safety.

Examples:

(a) where the balustrades are constructed of laminated panels entirely of glass/clear plastic layers, documentation can be presented to support that the seismic loads can be entirely supported by the plastic layer(s);

(b) stationary external guard rails a minimum of 1 067 mm (42 in.) in height vertically above the step nose line are provided outboard of the moving handrail. The guardrails are permitted to be of metal or plastic materials with supporting calculations to indicate they can resist a lateral load 730 N/m or 50 lbf/ft in seismic zones 2 or higher and 584 N/m or 40 lbf/ft in lower seismic zones.

(c) where steel or other metal sandwich panels are alternated with clear glass panels and the sandwich panels can support the entire seismic loads.

8.5.2 Truss Members

Since escalator trusses follow AISC calculation in the same manner as buildings, the method of calculation should be similar. The formula proposed is similar to that used in the Uniform Building Code for elements of structures and their attachments. Friction from gravity loads is not considered to provide resistance to seismic forces.

8.5.2.1.1 Z is a factor, not a percent of gravity. The actual values of Z are: for Zone 2A, $Z = 0.15$; for Zone 2B, $Z = 0.20$; for Zone 3, $Z = 0.30$; and for Zone 4, $Z =$

0.40; however, values of Z have been rounded for ease of calculation. The affect on F_p is slightly conservative.

8.5.2.1.2 The value of l follows the considerations listed in the building codes. Values of C_p are in accordance with UBC Table 16-0 Items 2 and 3 for nonstructural components and equipment and 1630.2, which includes allowances for components below ground level.

8.5.2.2 To treat escalator vertical force requirements in a manner similar to that used for elevators in 8.4.

8.5.2.3 Truss Calculations. The AISC Section A5-2 for wind and seismic loading as well as the allowable stress section in AISC, which is referenced by the ASME A17.1 Code only for escalators and moving walks, modifies the value of stress used in calculations to something other than the yield stress that will cause permanent deformation.

8.5.3 Supporting Connections Between the Truss and the Building

8.5.3.1 The requirements are written to clearly indicate that the escalator truss is not a structural element of the building and cannot resist building member movement creating forces several magnitudes above normal escalator loads. The supports do require the building to support seismic forces, which act on the escalator or moving walk. Friction from gravity loads is not considered to provide resistance to seismic forces.

8.5.3.2 If the fixed end is free, enough to rotate slightly, forces transferred to the truss will be reduced. Also, some small allowance for longitudinal and transverse motions at both ends can help accommodate this rotation and minimize any damage to either the escalator truss or the supporting structure (balustrade) for the moving handrails.

8.5.3.2.2 The most conservative value from the 1997 National Earthquake Hazards Reduction Program (NEHRP) Recommended Provisions for Seismic Regulations for New Buildings, Table for Allowable Story Drifts, was selected to be used for all types of buildings and seismic hazard exposure groups.

8.5.4 Earthquake Protective Devices

8.5.4(b) Setting the seismic switch above that advised by the manufacturer it is possible that some displacement or damage will be incurred by the escalator or moving walk. Turning the escalator off will allow it to be moved back into position and/or repaired to ensure safe operation.

SECTION 8.6
REPAIRS, REPLACEMENTS, AND MAINTENANCE

NOTE: Throughout Section 8.6, references are made to other requirements in this Code. To gain a complete understanding of a requirement, the reader should review the Handbook commentary on all referenced requirements.

This Section applies to existing installations. The requirements of 8.6 assure that alteration, maintenance, repairs, or replacements of damaged, broken, or worn parts are done in such a way that the safe operation of the equipment is not affected. Under no circumstances should the safety provisions for an existing elevator, escalator, etc., be reduced or eliminated due to alteration, maintenance, repair, or replacement of damaged, broken, or worn parts. Due to the intricacies of altering all types of existing equipment, the details of which cannot be covered by any code, the enforcing authority and elevator contractor are advised to keep in mind the provisions of 1.2, which states in part, "The provisions of this Code are not intended to prevent the use of systems, methods, or devices of equivalent or superior quality, strength, fire resistance, effectiveness, durability, and safety to those prescribed by this Code, provided that there is technical documentation to demonstrate the equivalency of the system method or device. The specific requirements of this Code may be modified by the authority having jurisdiction based upon technical documentation or physical performance verification to allow alternative arrangements that will assure safety equivalent to that which would be provided by conformance to the corresponding requirements of this Code."

8.6.1 General Requirements

8.6.1.1 Maintenance, Repair, and Replacement. The requirements in 8.6 are limited to maintaining equipment and components required by the Code in compliance with the applicable requirements (8.6.1.1.2) performance issues (e.g., car start delay, door opening time, ride quality) are not addressed. Detailed maintenance checklists can be found in the Building Transportation Standards and Guidelines, NEII-1-2000. See Handbook commentary on Part 9 for ordering information.

8.6.1.1.2 This requirement for maintenance (8.6.1.1.2) and period inspections and tests (8.11.1.2) are intended to assure the level of safety is equal to what was required when the equipment was installed and/or altered. This requirement does not refer to the ASME A17.1 Code in effect at the time of installation and/or alteration. It refers to the code the Authority Having Jurisdiction was enforcing at the time. That may be a version of ASME A17.1 with or without modifications or a locally developed elevator code. If no code was in effect at the time of installation and/or alteration, the ASME A17.1 Code in effect (see Preface to ASME A17.1 for the effective date of editions, supplements, and addenda) could be used for guidance. For example, if at the time of installation the equipment was required to conform to A17.1–1995, then that is the Code with which the equipment must comply. If at the time of

installation the equipment was required to conform to A17.1–1931 and in 1978 the brake on an existing driving machine was replaced, then the equipment is to be maintained in compliance with A17.1–1931 and A17.1–1978, Rules 208.8 and 210.8. The reason for maintaining the brake in compliance with A17.1–1978, Rules 208.8 and 210.8 is that at the time the new brake was installed, it was considered by the Code to be a major alteration (see A17.1–1978, Rule 1200.2).

If the equipment is required to conform to the requirements of ASME A17.3, Safety Code for Existing Elevators and Escalators, compliance with that document must also be confirmed during maintenance. For example, an elevator installed under the A17.1–1955 edition of the Code would have to comply with the requirements of that Code and any additional or more stringent requirements in the A17.3 Code, such as firefighters' service.

8.6.1.2 General Maintenance Requirements

8.6.1.2.2 Minimum or maximum maintenance intervals are not specified as there is no engineering justification for universally applied intervals for all equipment. A maintenance control program (mcp) should take into account the conditions listed in 8.6.1.2.2 and specify appropriate intervals for specific maintenance items. There are a few exceptions where it was felt that it was better to error on the side of safety. A couple of examples are the monthly test of Firefighters' Service (8.6.10.1) and Escalator and Moving Walk Startup (8.6.10.5).

8.6.1.3 Maintenance Personnel. Elevator personnel are defined (see 1.3) as persons who have been trained in the construction, maintenance, repair, inspection, or testing of equipment. The purpose of this requirement is to prevent unqualified persons from performing maintenance that may render the equipment unsafe.

8.6.1.4 Maintenance Records. Records are an important part of maintaining equipment in a safe operation condition and are necessary for scheduling maintenance.

8.6.1.4.2 Record Availability. Records must be available to elevator personnel (see 1.3) including elevator mechanics and inspectors. The requirement is written in performance terms. It can be met by a number of methods such as: on site storage of records or electronic records. The key is that the records must be kept and be made available when requested.

8.6.1.5 Code Data Plate. Requirement 8.11.1.2 requires that inspections and tests be performed to confirm the equipment complies with the Code under which it was installed and/or altered. This information should be readily available at the site for inspection and maintenance personnel. This requirement recognizes that existing equipment, which currently must comply with the requirements of 8.11.1.2 may have been installed and/or altered prior to the adoption of A17.1 in the jurisdiction. See Handbook commentary on 8.7.

8.6.1.6 General Maintenance Methods and Procedures

8.6.1.6.1 Making Safety Devices Inoperative or Ineffective. Making safety devices inoperative is permitted only after the installation is removed from public use. The reader is cautioned that jumpers unintentionally left on safety circuits have been the cause of fatalities and serious injury. The reader should be familiar with and adhere to the jumper procedures in the latest edition of the Elevator Industry Field Employees' Safety Handbook.

8.6.1.6.2 Lubrication. There should be a lubrication schedule including type and grade for every installation. Excessive lubrication is a fire hazard and if allowed to collect on the floor it poses a fall hazard. For example, heavy buildup of lubricant in motor can prevent proper cooling.

Excess lubricant also collect dirt and can inhibit safe operation of equipment. Lubricant on governor jaw can prevent operation of the safety. Excess lubricant on the selector can also foul electrical contacts.

8.6.1.6.3 Controllers, Wiring, and Wiring Diagrams

8.6.1.6.3(a) A complete wiring diagram is not required. However, wiring diagrams necessary for diagnostics and repair of code-required safety circuits are required.

8.6.1.6.3(c) and (d) The requirements prohibit the use of jumpers, switches or contact blocks to return the equipment to service in lieu of correcting the problem in the proper manner. Jumper wires stored in the machine room, hoistway, or pit cannot be accounted for and are subject to misuse. Storage of jumpers would also not comply with the jumper control procedures in the Elevator Industry Field Employees' Safety Handbook.

8.6.1.6.5 Fire Extinguishers. Extinguishers are labeled with a classification system so that users may quickly identify the classes of fires for which a particular extinguisher may be used. The classification system is contained in the extinguisher standard (NFPA 10), which gives the applicable class symbol or symbols with supplementary wording to recall the meaning of the letters. Color coding is also used. UL requires a classification system on its labels. The fire extinguisher must be inside the machine space. It is intended only for first response. If it is necessary to leave the machine room where a fire exists, the room should not be re-entered. The fire fighting should be left to professional firefighters. In Canada this issue is addressed in NBCC.

8.6.1.6.6 Workmanship. Manufactures design tolerances should be adhered to whenever they are available. Good workmanship is necessary to assure safety is not compromised. For example, the improper tightening of a bolt can result in a failure.

8.6.2 Repair

8.6.2.1 General Requirements. These requirements ensure that repairs provide an equivalent level of safety to that originally provided.

When a component in a labeled product is replaced, it must be replaced with an identical component manufactured under the original labeling service (certifying agency). Substitution of a different component is only permitted where it is equivalent to that which was tested, as determined by the certifying agency (8.6.3.7).

8.6.2.2 Welding and Design. Welding where properly designed and performed by qualified welders is permitted. The requirements referenced address both the design and welders qualifications.

8.6.2.3 Repair of Speed Governors. The requirements ensure that the speed governor will operate as required subsequent to a repair.

8.6.2.5 Repair of Ropes. The ropes are critical in the operation of the elevator and 8.7.2.21 ensures proper replacement procedures. Splicing of ropes weakens them and causes poor fit in sheave grooves.

8.6.3 Replacement

8.6.3.2 Replacement of a Single Suspension Rope. This provision applies to replacement of a single rope only, which has been damaged during installation or acceptance testing and prior to the replacement set of ropes being placed in service. Ideally, the replacement of a single suspension rope should come from the same master reel. However, the likelihood is small that this can be done. Notwithstanding, the replacement rope must come from the same rope manufacturer and be of the same construction and material. The data for the replacement rope must correspond to the data for the entire set of ropes, as given by the Rope Data Tag (2.20.2.2).

If the suspension ropes have been shortened once, it is doubtful whether the replacement rope can be made to perform uniformly with the original ropes, since the two components of rope stretch, i.e., construction and elastic, will affect the performance of the new rope, whereas, the original ropes will only be affected by elastic stretch. Diameter reduction and wear of crown wires must be considered. Rope tension will have to be monitored closely, particularly when the replacement rope is first put into service. Severe sheave wear can occur in a matter of days on high speed elevators.

8.6.3.3 Replacement of Ropes Other Than Governor Ropes. Over the years, many different roping arrangements have been used. See Diagrams 8.6.3.3(a) through 8.6.3.3(l).

8.6.3.3.1 Replaced suspension ropes must conform to the same safety requirements as those established for new equipment. Refer to the various rules specified and the commentary on them.

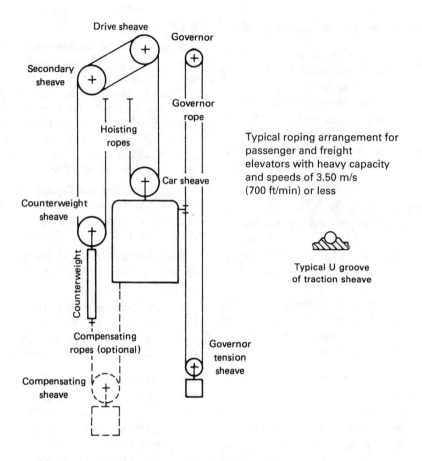

Diagram 8.6.3.3(a)   2:1 Double-Wrap Overhead Traction Machine

Requirement 8.6.3.3.1(b) specifically prohibits the installation of used ropes.

There are several reasons that justify using only new ropes on either new or existing installations including but not limited to the following:

(a) Prior to the planned reuse of a set of ropes, they must be removed from an elevator installation, during which the ropes could be kinked or twisted during coiling, outer wires abraded due to contact with floor surfaces or other parts of the building structure, and contaminated by dirt and debris.

(b) Once a set of ropes has been removed from its original elevator, some of the data recorded on the rope data tag required by 2.20.2.2 may no longer be valid or the accuracy verifiable, and may be discarded or lost during the removal, storage, and reinstallation process. Without rope data verification, compliance with 2.20.2 cannot be verified and 2.20.2.2 will not be met.

(c) Requirement 2.20.3 requires that the factor of safety of the suspension ropes conform to the values given in Table 2.20.3, which is directly related to the manufacturer's rated breaking strength of the rope in its new or unused condition. Depending upon the length of time a rope has been in use, its loading, frequency

and direction of rope bends, rope maintenance and lubrication, the breaking strength of any used rope will undoubtedly be less than the corresponding value when the rope was new due to the presence of internal and/or external wire breaks.

While the rope manufacturers have published data on the approximate reserve strength of a rope as a function of the visible wire breaks, the use of this data would require an intensive inspection of the complete rope length and still only yield an approximation of the actual breaking strength.

Without destructive testing of the rope(s), there is no verifiable method of determining the actual breaking strength of a used rope. Therefore, it would not be possible to verify compliance with 2.20.3. Depending upon the reduction in rope strength due to some or all of the causes noted above, the used ropes cannot be assumed to comply with 2.20.3.

8.6.3.3.2 Rope Fastenings and Hitchplates. This requirement is to ensure that replacement rope fastenings do not interfere with each other. It also permits alternate fastenings to be installed at one connection end only when ropes are being shortened, resocketed, or

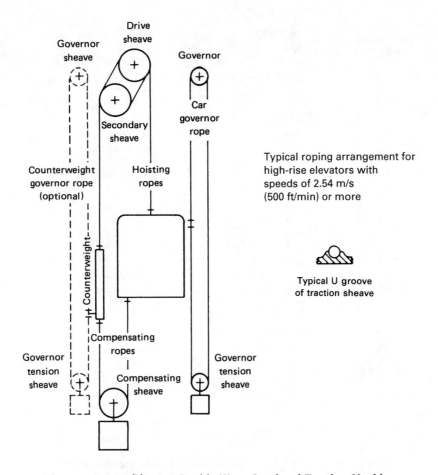

Diagram 8.6.3.3(b) 1:1 Double-Wrap Overhead Traction Machine

where the length of existing ropes or space conditions do not permit the installation of wedge sockets.

8.6.3.4 Replacement of Governor or Safety Rope

8.6.3.4.1 Reference to 8.7.2.19 implies that installation of ropes of different material or construction constitutes an alteration, and test specified in 8.7.2.19(b) must be performed.

8.6.3.5 Belts and Chains. This requirement establishes the procedure for replacements.

8.6.3.7 Labeled and Listed Devices. When a component in a labeled product is replaced, it must be replaced with an identical component manufactured under the original labeling service (certifying agency). Substitution of a different component is only permitted where it is equivalent to that which was tested, as determined by the certifying agency. Failure to use an approved replacement component will negate the listing/certification of the device.

8.6.3.8 Replacement of Door-Opening Device. If a mechanical door-opening device is removed and a non-contact door-opening device installed, the requirements for firefighters' service will require the closing kinetic

energy not exceed the requirements specified in 2.13.4.2.1(c).

8.6.4 Maintenance of Electric Elevators

8.6.4.1 Suspension and Compensating Wire Ropes

8.6.4.1.2 The lubrication of a wire rope applied during its manufacture may not last the full life of the rope and the rope may have to be relubricated periodically. Proper lubrication of suspension ropes will prolong rope life by reducing abrasive action of wire on wire or strand on strand and will retard deterioration of the fiber cores, eliminate distortion of the rope, and retard corrosion by providing a moisture-repellent coating. As a practical guide to the need for lubrication, a finger wiped in a sheave groove should show a faint smudge and have a slightly oily feel. If this test leaves the finger dry and clean, lubrication is advisable.

Excessive or improper lubricants may, in the case of traction elevators, seriously reduce the available traction and cause rope slippage. The lubricants and the amount used should be limited to those supplied or approved by established elevator or wire rope manufacturers. Slide of the ropes during acceleration or retardation may be an

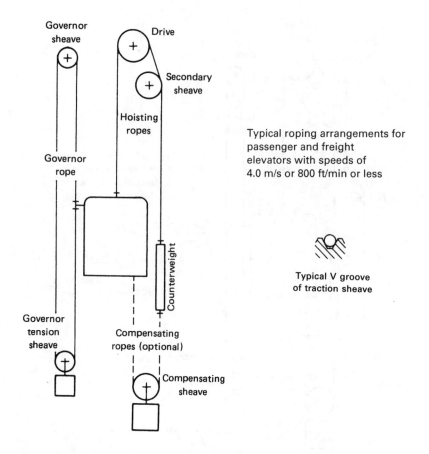

Typical roping arrangements for
passenger and freight
elevators with speeds of
4.0 m/s or 800 ft/min or less

Typical V groove
of traction sheave

Diagram 8.6.3.3(c)   1:1 Single-Wrap Overhead Traction Machine

indication that the lubrication is excessive. To determine this, it will usually be necessary to observe the ropes where they pass around the driving machine sheave during acceleration and retardation. Some rope creepage is normal.

In the case of winding drum machines, excessive lubrication does not create a hazardous condition but could interfere with the proper inspection of ropes.

8.6.4.2 Governor Wire Ropes. While it is evident there is lubrication in a new rope and that periodic lubrication may extend the rope life, the question of proper lubrication and identification of governors designed and set for lubricated rope versus those designed and set for dry rope presents a problem.

The Committee on many occasions has reviewed the subject of lubrication of governor ropes after installation and came to the following conclusions.

(*a*) While it is evident, there is lubrication in a new rope and that periodic lubrication may extend the rope life, the question of proper lubrication and identification of governors designed and set for lubricated rope versus those designed and set for dry rope presents a problem.

(*b*) There is no reported history of governor rope failure or shortened rope life due to lack of lubrication.

(*c*) The lubrication would not only have to be controlled (amount, type, etc.) and checked, but all of the existing governors that have not been set up for lubricated ropes would have to be identified in the same manner or reset.

(*d*) The use of lubricated ropes would require increasing the governor jaw rope gripping forces to ensure that the governor rope will still activate the safeties, thus exerting higher crushing forces on the rope than as now in practice.

(*e*) Unlike suspension ropes, where excess lubrication would manifest itself in the ropes sliding over the drive sheave when the car started or stopped, excess governor rope lubrication might not be detected, with catastrophic potential.

It was thus determined that lubrication of governor ropes should continue to be prohibited.

A governor rope that has been lubricated and subsequently cleaned to remove the applied lubrication, may not have the same surface characteristics as the same rope in its original nonlubricated state. The intent of this requirement can only be met if the governor rope in its end-use state is reasonably similar to the condition of the governor rope as originally tested during the acceptance

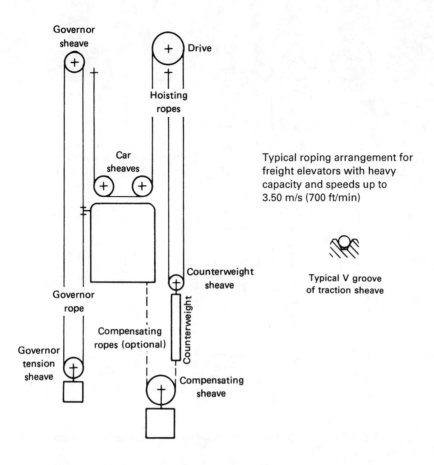

Diagram 8.6.3.3(d) 2:1 Single-Wrap Overhead Traction Machine

test. The removal of applied lubrication to a governor rope is acceptable if tests of the governor and safety are performed in accordance with the referenced requirements.

8.6.4.3 Lubrication of Guide Rails. Requirements specify guide rail lubricant specifications and refer to other safety requirements when using solvents. It is inadequate to simply require a nonflammable or high flash point solvent. Regulations apply when any solvent is used.

Special precautions must be taken when removing excess rail lubricants. Nontoxic, nonflammable, water soluble solvents, cleaners, or degreasers should be the only type cleaners used for cleaning guide rails, fishplates, brackets, door tracks, overhead sheaves, car tops, slings, pit equipment, escalator tracks, and escalator truss pans. Toxic solvents, cleaners, or degreasers should not be used in any confined area where toxicity may induce nausea or asphyxiation. Flammable solvents should never be used in areas where the possibility of vapor ignition exists. Mineral spirits and solvents with a minimum flash point of 37.8°C (100°F) should be used for cleaning and flushing machine bearings, housings, and gear casings. Mineral spirits and solvents should not be used for cleaning electric motors, field windings,

armature windings, commutators, brushes, or electrical control components. Flames and sparks should be kept away from flammable liquids and their vapors. Smoking should be prohibited in areas where such materials are used or stored. Approved safety containers must be used to promptly and safely dispose of used saturated cleaning rags. Carbon tetrachloride should never be used or stored at the job site. Nonflammable chlorinated solvents used as a substitute for carbon tetrachloride should be used for cleaning electrical motor field windings, armature windings, commutators, brushes, and electrical control components. Flammable and/or toxic cleaning liquids should only be used in well-ventilated areas. The Material Safety Data Sheet (MSDS) for the cleaning material should be reviewed and all precautions therein complied with. Appropriate protective equipment must be worn if the composition of the material is harmful to the skin, lungs, or eyes. Read and comply with all precautionary instructions on the label. If any part of the clothing becomes saturated with solvent, it should be changed at once.

8.6.4.4 Oil Buffers. The oil buffer marking plate (2.22.4.1.1) will indicate the permissible range in viscosity of the buffer oil to be used stated in Saybolt Seconds Universal (SSU) at 37.8°C or 100°F, the viscosity index

369

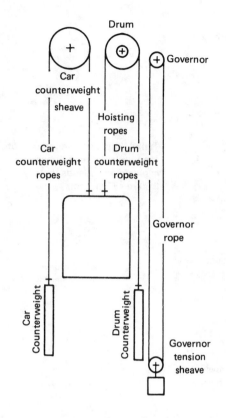

Typical roping arrangement for
low-rise passenger and
freight elevators with speeds
of 2.03 m/s (400 ft/min) or less

Diagram 8.6.3.3(e)   1:1 Overhead Winding-Drum Machine With Counterweight

number of the oil to be used, and the pour point in degrees Fahrenheit of the oil to be used. The viscosity is a measure of the oil resistance to flow at a specific temperature. The viscosity index number is determined by the rate of change of viscosity with temperature. The higher the index number the less the viscosity change with temperature. A viscosity index between 95 and 100 is usually specified. A viscosity index number may be as high as 150 for synthetic oils.

8.6.4.5 Safety Mechanisms

8.6.4.5.2 This requirement is included to emphasize the need for maintaining clearance between the safety jaws and rail.

8.6.4.6 Brakes

8.6.4.6.1 Those items are deemed critical to ensure proper brake operation where listed. The list is not intended to be all-inclusive.

8.6.4.7 Cleaning of Hoistways and Pits. Requirements prevent unauthorized persons from entering the pit. Allowing blocks and/or pipe stands to be safely stored in the pit for ready use, will make them available when needed. This will improve both efficiency and safety. Absolutely no other material, supplies, tools, or equipment is allowed to be stored in the pit.

8.6.4.8 Machine Rooms and Machinery Spaces. Use of the machine room for storage not necessary for the elevator operation creates a hazard for both personnel and using public. Access to the machine room is classified as Security Category 2 restricting keys to authorized personnel. This is to prevent improper use of the space and hazards to personnel.

8.6.4.10 Refastening or Resocketing of Car-Hoisting Ropes and Winding-Drum Machines

8.6.4.10.1 General. The intervals for refastening on winding drum driving machines with 1:1 roping are based on fatigue developing in the section of rope adjacent to the rope socket due to the bending encountered by the wires as a result of vibration being dampened at this location or due to stress of a torsional nature that occurs at this point as a result of changes in tension on the ropes. This problem is more apt to be encountered where a rope socket is employed but can also develop with attachments of any type. If the hoisting ropes are not refastened at the interval specified, a dangerous situation could occur. Wedge rope sockets complying with 2.20.9.5 are permitted.

On winding drum driving machines with 2:1 roping, there is no requirement to refasten the hoisting ropes periodically as there is no fatigue developed in the rope section adjacent to the rope socket.

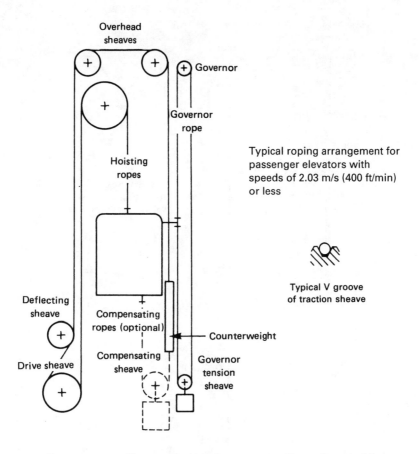

Diagram 8.6.3.3(f)   1:1 Single-Wrap Basement Traction Machine

8.6.4.10.2 Procedure. Diagrams 8.6.3.3(a) through 8.6.3.3(l) show the different roping arrangements that can be found on elevators.

An unsafe condition can occur if the rope ends that extend beyond their clamps protrude from the drum, which can result in damage to the drum spokes, bearing stands, and bearings, as well as injury to personnel working around the machine.

A source of additional information on wire ropes is available from the American Iron and Steel Institute, 1000 16th Street, NW, Washington, D.C. 20036. Request the "Wire Rope Users Manual."

8.6.4.11 Runby. If a reduction in runby is observed, it is important that the top clearance not be reduced below that which was required at the time of installation. The final terminal stopping devices must also remain operational. They should be tested as specified in 8.11.2.2.5. When the counterweight runby has been reduced to zero, the ropes must be shortened or replaced. When spring-return oil buffers are provided and compression of the buffer with the car at their terminals was part of the original design (2.4.2 and 2.22.4.8) it may continue. Under no circumstances should compression exceed 25%.

8.6.4.13 Door Systems

8.6.4.13.1 General. Those items are deemed critical to safe door operation where listed. This is not an all-inclusive list.

8.6.4.13.2 Kinetic Energy and Force Limitations for Automatic-Closing, Horizontal Sliding Car, and Hoistway Doors or Gates. As a significant number of accidents are attributed to being struck by closing doors, it is critical that these requirements be adhered to.

8.6.5 Maintenance of Hydraulic Elevators

8.6.5.5 Gland Packings and Seals

8.6.5.5.2 Oil Leakage Collection. Provides for the removal of leakage oil from the pit in order to reduce the fire loading. The container is required to be covered for the same reason.

8.6.5.6 Flexible Hoses and Fittings. ASME A17.1–1996 revised Rule 303.3c(1)(e). The replacement date for the flexible hose was eliminated when the rupture valve requirements was added. This requires replacement of the flexible hose assembly when a rupture valve is not installed. Requirement for rupture valves was added in

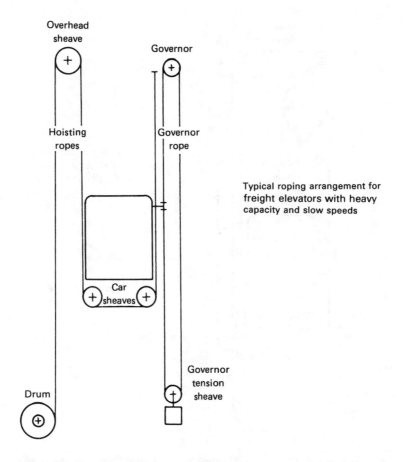

Typical roping arrangement for
freight elevators with heavy
capacity and slow speeds

Diagram 8.6.3.3(g)   2:1 Basement Winding-Drum Underslung Machine

1987, B44 Supplement. Before that, 6-year replacement was required.

8.6.5.7 Record of Oil Usage. The annual cylinder leak test required by 8.11.3.2.2 is intended to detect any leak in the underground cylinder or piping and to identify potential cylinder failure. However, a leak could begin the day after this annual test and not be detected for an extended time period. The logging of the hydraulic fluid, added and compared with the oil collected from visible leaks, will provide early detection of any leak and prevent hazardous operation as well as possible ground pollution. When the oil loss cannot be accounted for, the specified tests must be performed. An inspector is not required to witness this test.

The log is required to be kept in the machine room and available to elevator personnel (1.3).

8.6.5.8 Safety Bulkhead. Single-bottom cylinders generally have been in the ground for more than 30 years. The potential for catastrophic failure resulting from the separation of the single bottom from the cylinder increases as the cylinder ages. This requirement relates to maintenance, repair, and replacement of a component that has exceeded its life span.

8.6.5.9 Relief-Valve Setting. If the seal is not intact, the specified test must be performed. This test is not required to be witnessed by an inspector.

8.6.5.11 Cylinder Corrosion Protection and Monitoring. If monitored cathodic protection is provided, it is critical it be correctly maintained. If it is not, cylinder corrosion can be accelerated and failure may result in a very short time.

8.6.8 Maintenance of Escalator and Moving Walks

8.6.8.2 Step-to-Skirt Clearance. See Handbook commentary on 6.1.3.3.5.

8.6.8.3 Step/Skirt Performance Index. See Handbook commentary on 6.1.3.3.7.

8.6.8.5 Escalator Skirt Panels. Manufacturers know the characteristics of their new equipment and how best to apply lubricant on existing escalators.

Experience has indicated that reducing the friction on the skirt panels significantly reduces skirt-step entrapment incidents. If a spray is used to coat the skirt panel, care must be taken not to introduce the friction-reducing agent into the air and onto adjacent surfaces as this could result in a slipping hazard on the steps and/or

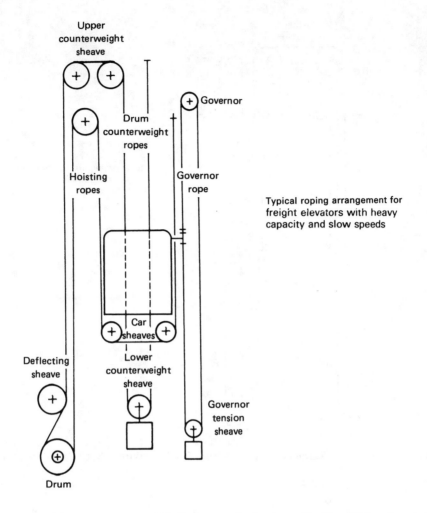

Typical roping arrangement for freight elevators with heavy capacity and slow speeds

Diagram 8.6.3.3(h)   2:1 Basement Winding-Drum Underslung Machine With a Counterweight

adjacent walkways. Escalators installed under ASME A17.1–2000 and later editions are required to have a friction-reducing agent incorporated into the construction of the skirt panel.

8.6.8.14 Cleaning. The interior of escalators and moving walks need to be cleaned periodically to reduce the amount of combustible material, which collects in this area. The annual examination to determine cleaning is a minimum. In some environments, examinations that are more frequent may be necessary.

8.6.8.15 Entrance and Egress Ends. See Handbook commentary on 6.1.6.3.4.

8.6.10 Special Provisions

The provisions in 8.10 typically apply to the equipment owner/operator not the maintenance company. However, ASME A17.1 does not assign responsibility as stated in the Preface. Responsibility, to comply with the Code, is typically assigned by the authority having jurisdiction. The owner/operator is ultimately responsible.

8.6.10.1 Maintenance of Firefighters' Service. The exercising of the circuitry and relays should be done on a periodic basis. This is not an inspection or test as covered in 8.10 and 8.11. It is especially important with relay control, as relays are only called on to work when firefighters' service is activated. The relays may be dirty and not work properly in an emergency, which could result in a hazardous condition. The requirement does not specify who is to perform the operation — maintenance company, owner, operator — only that it is performed and that a written record of the findings is to be kept on the premises.

8.6.10.2 Access Keys. To ensure keys are available and properly stored to permit ready access by personnel in the assigned group, while preventing their use by the general public.

8.6.10.3 Cleaning Inside Hoistway. This requires that only authorized personnel clean observation elevator glass. The Code defines "authorized personnel" (see 1.3) as persons who have been instructed in the operation

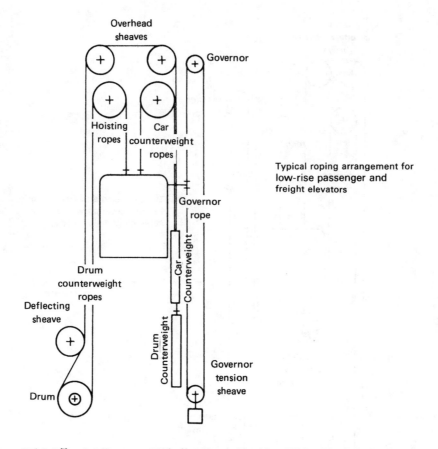

Overhead
sheaves

Governor

Hoisting
ropes

Car
counterweight
ropes

Typical roping arrangement for
low-rise passenger and
freight elevators

Governor
rope

Drum
counterweight
ropes

Deflecting
sheave

Car
Counterweight

Drum
Counterweight

Governor
tension
sheave

Drum

Diagram 8.6.3.3(i)   1:1 Basement Winding-Drum Machine With a Back-Drum Counterweight

and/or maintenance of the equipment and designated by the owner to use or maintain the equipment. The written procedures should assist in identifying the hazards and training of cleaning personnel to assure that they clean glass in a manner that will not expose them to danger.

8.6.10.4 Emergency Evacuation Procedures for Elevators. The procedure can be as simple as providing authorized elevator and emergency personnel written instructions to communicate with trapped passengers and to call the maintenance company or emergency services, or as detailed as written instructions on how to perform an evacuation. The possible unavailability of elevator personnel, due to wide-spread power outages, seismic activities, etc., must also be taken into account when preparing emergency evacuation procedures. The readers should familiarize themselves with ASME A17.4.

8.6.10.5 Escalator or Moving Walk Startup. The requirements are intended to ensure that unsafe equipment is removed from service and maintenance personnel are made aware of equipment deficiencies.

SECTION 8.7
ALTERATIONS

NOTE: Throughout Section 8.7, references are made to other requirements in this Code. To gain a complete understanding of a requirement, the reader should review the Handbook commentary on all referenced requirements

The author expresses his appreciation to Albert Saxer for his extensive contributions to Section 8.7.

This Section applies to existing installations. The requirements of this Section assure that alterations (a.k.a., modernizations) are done in such a way that the safe operation of the equipment is not affected. Under no circumstances should the safety provisions for an existing elevator, escalator, etc., be reduced or eliminated due to an alteration. Detailed information and helpful checklists for alterations can be found in the Building Transportation Standards and Guidelines, NEII-1-2000. Examples of the checklists are given in Charts 8.7(a) and 8.7(b). See Part 9 for procurement information. Because of the intricacies of altering all types of existing equipment, the details of which cannot be covered by any

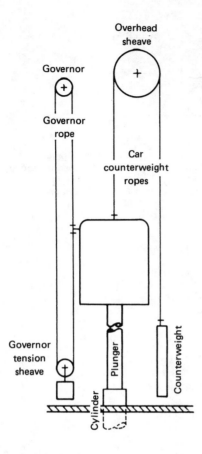

Governor

Governor
rope

Overhead
sheave

Car
counterweight
ropes

Typical roping arrangement for
low-rise passenger and
freight hydraulic elevators

Governor
tension
sheave

Plunger

Cylinder

Counterweight

Diagram 8.6.3.3(j)   Hydraulic Elevator With a Counterweight and Governor

Code, the authority having jurisdiction and elevator contractor are advised to keep in mind the provisions of 1.2 of this Code, which states in part: "The provisions of this Code are not intended to prevent the use of systems, methods, or devices of equivalent or superior quality, strength, fire resistance, effectiveness, durability, and safety to those prescribed by this Code, provided that there is technical documentation to demonstrate the equivalency of the system method or device. The specific requirements of this Code may be modified by the authority having jurisdiction based upon technical documentation or physical performance verification to allow alternative arrangements that will assure safety equivalent to that which would be provided by conformance to the corresponding requirements of this Code."

A comprehensive list of possible alterations and the requirements with which they must comply are included in Appendix L. In developing these requirements, three major points were considered.

(a) When an alteration is made, all affected safety requirements must be complied with.

(b) Compliance with safety requirements, which are not affected by the alteration, should not be required since they may discourage an equipment owner from making the alteration.

(c) The requirements should encourage alterations

that improve the safety of the equipment and not discourage safety improvement due to unrelated improvements that increase the overall cost.

In ASME A17.1d–2000 and earlier editions, an alteration required that the entire installation comply with ASME A17.3. That requirement does not appear in this edition of the ASME A17.1 Code. ASME A17.3, Code for Existing Elevators and Escalators, was developed to establish minimum safety requirements to ensure a reasonable level of safety on existing elevators for the riding public. While ASME A17.3 was under development, Section 8.7, Part XII at the time, was revised to eliminate major and minor alterations and clarify the various alteration, repair, and maintenance requirements. A new class of work, replacement, was defined to permit obsolete equipment to be replaced with modern equipment that performs the same function. The Committee also recognized that replacement components may not be an exact duplicate of the original, as the original manufactured item may no longer be available.

When an alteration had been made to an installation, Rule 1200.1 of ASME A17.1d–2000 and earlier editions requires compliance with the entire ASME A17.3 Code, as a minimum. There was a tendency to interpret replacements as alterations. The advocates of this trend claim that every opportunity should be explored to

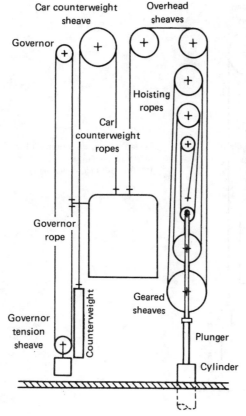

GENERAL NOTE: Requirement 3.18.1.2 requires the rope ratio not to exceed 1:2.

Diagram 8.6.3.3(k)   Roped-Hydraulic Elevator With a Vertical Driving Machine

enhance safety. They claim this is best ensured by compliance with ASME A17.3.

ASME A17.3 requires firefighters' service, top of car inspection, door restrictors, etc. Alterations that should be classified as a repair or replacement were not being made due to the economic burden of total compliance with ASME A17.3. This results in an increased safety exposure for existing equipment, rather than the enhanced safety that the addition of modern components would add. It is very evident the present requirements for compliance with ASME A17.3 for "any" alteration results in owners putting off safety improvements to existing equipment.

The Committee concluded that the requirement was becoming such a financial burden that it was having a negative impact on the adoption of ASME A17.1 by authorities having jurisdiction. It was undermining rather than enhancing safety.

8.7.1 General Requirements

Before starting an alteration, a list of the items to be altered should be made. The requirements in 8.7 that apply can be identified using Appendix L.

An alteration is any work done on an elevator, escalator, or moving walk as defined in 1.3. Appendix L is an index of alterations for which there are specific requirements in 8.7. In addition, good practice suggests that when any alteration is made, the safety requirements of ASME A17.3 should be considered even if A17.3 has not been adopted by the authority having jurisdiction. The requirements of A17.3 are minimum safety requirements that should exist on all equipment in service.

A17.1d–2000 and earlier editions of this Code required full compliance with ASME A17.3 before any alteration was made. This requirement was deleted in the 2000 edition to prevent replacements from being interpreted as alterations. The requirements of 8.7 are often more stringent than the requirements of ASME A17.3.

If an existing installation was, or is now required to meet more stringent requirements than appear in 8.7, it is required to continue to meet those requirements.

Diagrams 8.7(a) through 8.7(l) break down the various alteration Code requirements based on specific areas. Alteration or modernization affects various parts of an elevator system. In many cases, the work is limited to a few items. The flow diagrams are arranged by specific

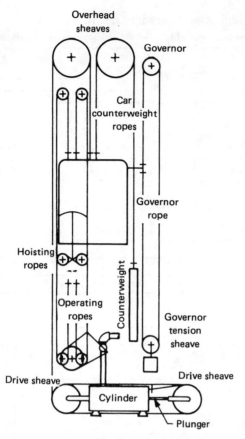

Overhead sheaves

Governor

Car counterweight ropes

Typical roping arrangement for passenger roped hydraulic elevator of intermediate speed

Governor rope

Hoisting ropes

Counterweight

Operating ropes

Governor tension sheave

Drive sheave

Drive sheave

Cylinder

Plunger

GENERAL NOTE: Not permitted by 3.18.1.2.

Diagram 8.6.3.3(l) Roped-Hydraulic Elevator With a Horizontal Driving Machine

areas, therefore, some repetition occurs. For example, reuse of buffers, safeties, and governor requirements appear in Diagram 8.7(e) but also appears in Diagrams 8.7(b) and 8.7(d).

8.7.1.2 Items Not Covered in This Part. If the item altered is not mentioned, the alteration shall not affect any other item. The alteration, replacement of parts, repair, or maintenance must be analyzed thoroughly so that the level of safety is not diminished below that specified in 8.7.1.1 and 8.6.1.1.2.

8.7.1.3 Testing. All requirements for testing, including those required as a result of the alteration, can be found in 8.10 and 8.11 of the Code. When an alteration is made to a part that requires a periodic or acceptance test, that test must be performed before the equipment is placed back into service. See also testing requirements 8.10.2.3.2, 8.10.3.3.2, and 8.10.4.2.

8.7.1.4 Welding. Welding, in most cases, is now permitted. These requirements permit welding where properly designed by a licensed professional engineer and performed by a qualified welder.

Over the years, the design of traction machines dramatically changed. The original machines that were produced were of the combination spider and drive sheave. The latest design of machines utilizes removable traction sheaves for obvious reasons. Through the years, the non-demountable sheaves have been replaced by removing the worn outer surface and replacing it with a new drive sheave ring. The usual practice is to have the basic casting support the ring, and then the ring is held in place with a minimum of six mounting bolts. The diameter and grade of the bolts being adequate for the forces acting on the drive sheave. The design of this alteration does not have to be prepared nor certified by a licensed professional engineer, however, where the alteration is made in the driving machine sheave, 8.7.2.25.1 requires that the driving machine sheave conform to the requirements of 2.24.2, 2.24.3, and 2.24.4.

8.7.1.5 Design. See Handbook commentary on 8.7.1.4.

8.7.1.6 Temporary Wiring. When an alteration (modernization) is in process, temporary wiring is necessary to continue elevator service in the building. The necessary safety precautions are given in this requirement.

Chart 8.7(a) Pre-Modernization Checklist Electric Elevators
(Courtesy Building Transportation Standards and Guidelines, NEII-1-2000, © 2000, National Elevator Industry, Inc., Teaneck, NJ)

BUILDING: _____ EQUIPMENT #: _____
Manufacturer: _____ Model: _____
Evaluation Performed By: _____

INTRODUCTION:
The following checklist of items should be considered when evaluating existing elevators for modernization. It is not intended to replace the complete hoistway and machine room survey that is usually required to modify or prepare a proper layout. Given the scope, variety and complexity of the equipment, a small item that is overlooked, may prove extremely costly when the job is in progress. The format has been arranged to follow the sequence of Sections 8.10 and 8.11 of the ASME A17.1 Code. The engineering section is intended to aid the Engineering Office in their evaluation. Insert "NA" in the note column where the item does not apply.

"Notes" reference individual user comments on attached separate sheet on factors that may affect compliance with contract specifications.

Preface each question with "Have I checked"

Item #	Details	YES	NOTES
1.0	**INSIDE CAR**		
1.1	For compliance with codes, clearances and space including out of plumb conditions?		
1.2	Car and car doors?		
1.2.1	Car inside dimension vs. capacity?		
1.2.2	Condition (enclosure, floor, hang on panels, support rails)?		
1.3	Existing operator will meet performance specification?		
1.3.1	For ability to add nudging?		
1.3.2	For heavy doors (limit open and close speeds)?		
1.3.3	Sound produced vs. specifications?		
1.3.4	That reversal device is suitable to be reused or is replacement required? Note: Certain light rays or equal require reduced energy closing.		
1.3.5	Door restrictors?		
1.4	Firefighters' Service to suit applicable code?		
1.5	Communication system per code (fire phone) (fire speaker)?		
1.6	Fixtures?		
1.6.1	Compliant with accessibility regulations?		
1.6.2	Technical interface?		
1.7	Emergency lighting?		
1.8	Car light fixture is code compliant (light bulbs or tubes properly protected and retained)?		
2.0	**MACHINE ROOM**		
2.1	Elevator equipment rooms for access, lighting, space, guards, and clearances? (also see item #1.1)		
2.2	For temperature and humidity ranges as required (HVAC)?		
2.3	And gathered completed data for drive motor and MG?		
2.4	Standby (emergency) power characteristics and requirements?		
2.5	Size and configuration of disconnect, feeders, fuses, or circuit breakers?		
2.7	Machine support?		
2.8	Hoist beams; door or trap door permit equipment access?		
2.9	Basement machine support?		
2.10	Basement machine sheave support (overhead)?		
2.11	Sprinklers?		
2.12	Railings, ladders, steps in machine rooms?		
2.13	Fire alarm initiating devices (smoke detectors)?		
2.14	Number of units?		

Chart 8.7(b)   Pre-Modernization Performance Evaluation Form
(Courtesy Building Transportation Standards and Guidelines, NEII-1-2000, © 2000, National Elevator Industry, Inc., Teaneck, NJ)

ELEVATOR IDENTIFICATION NUMBER: _____

SYSTEM TYPE: _____   BUILDING: _____

ADDRESS: _____ CITY: _____ STATE: _____

CAPACITY (kg)(lb): _____ SPEED (m/s)(ft/min): _____ STOPS: _____ OPENINGS: _____

RISE (m)(ft-in.): _____ DOOR TYPE: _____ DOOR OPENING (mm)(ft-in.): _____

POWER: V _____ A _____ PHASE: _____ Hz: _____ NUMBER OF WIRES: _____

CAR INSIDE (ft-in.) (mm) WIDTH: _____ DEPTH: _____

	EXISTING MEASUREMENT	STANDARD SPECIFICATION	DEVIATION	LIMITING FACTOR	NOTES
MOTION:					
ACCELERATION (m/s^2) or (ft/s^2)					
DECELERATION (m/s^2) or (ft/s^2)					
MAX. JERK (m/s^3) or (ft/s^3)					
CONTRACT SPEED REGULATION (±%)					
VERTICAL VIBRATION (milli-g)					
HORIZONTAL VIBRATION — S/S (milli-g)					
HORIZONTAL VIBRATION — F/B (milli-g)					
STOPPING ZONE (mm) or (in.)					
TIMING:					
PERFORMANCE TIME (s)					
DOOR OPENING TIME, NOMINAL (s)					
DOOR CLOSING TIME (s)					
SOUND:					
DOOR SOUND — OPENING (dBA)					
DOOR SOUND — CLOSING (dBA)					
DOOR SOUND — REVERSAL (dBA)					
SOUND IN CAR AT RATED SPEED (dBA)					
SOUND IN STOPPED CAR, DOOR CLOSED, FAN ON (dBA)					
SOUND IN MACHINE ROOM (dBA)					

NOTES:
1. Use separate form for each elevator.
2. Typical measurements taken with a maximum of two people in the car.
3. Circle unit of measurement used in completing this form.
4. "Notes" reference individual user comments on attached separate sheet on factors that may affect compliance with contract specifications.
5. Also complete "Pre-Modernization Checklist."

Name   Date

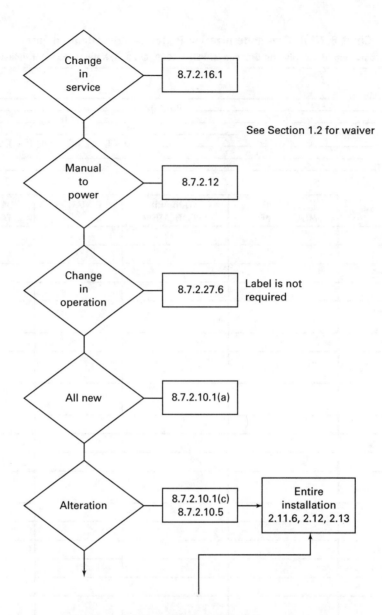

Diagram 8.7(a)   Alterations That Affect Entrances
(Courtesy Albert Saxer)

8.7.1.7 Repairs and Replacements. The requirements for this type of work have been moved to 8.6.

8.7.1.8 Code Data Plate. The data plate will list the various Code editions under which the equipment was installed or altered, to enable the inspector to do a complete and objective inspection.

See Handbook commentary on 8.9, which contains a checklist to assist in preparing a new code data plate when the equipment has been altered.

8.7.2 Alterations to Electric Elevators

8.7.2.1 Hoistway Enclosures

8.7.2.1.1 Hoistway Enclosure Walls. Alterations considered include increasing the height of the building,

elimination of entrances, addition of entrances, enclosure of openwork hoistway, and repair of the enclosure. The building code (1.3, Definitions) should also be reviewed for the applicable appropriate requirements for existing buildings.

8.7.2.1.2 Addition of Elevator to Existing Hoistway. Elevators may be added to existing hoistways if the number of elevators in the multiple hoistway complies with the building code and the required clearances are maintained. See also Handbook commentary on 2.1.1.4.

8.7.2.1.3 Construction at Top of Hoistway. Construction to maintain fire integrity, smoke venting, and safe clearances is necessary. See also Handbook commentary on 2.1.2.1, 2.1.2.2, and 8.7.2.4.

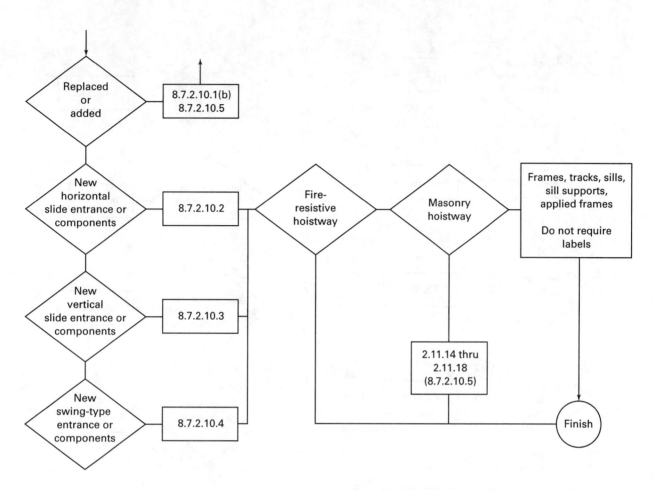

Diagram 8.7(a)   Alterations That Affect Entrances (Cont'd)
(Courtesy Albert Saxer)

8.7.2.1.4 Construction at Bottom of Hoistway. Construction to maintain proper clearances is necessary. See also Handbook commentary on 2.2 and 8.7.2.4.

8.7.2.1.5 Control of Smoke and Hot Gases. Alterations are not permitted to adversely affect the control of smoke in the hoistway. See also Handbook commentary on 2.1.4.

8.7.2.2 Pits. Safety requirements are specified for the pit and the bottom of the hoistway. See also Handbook commentary on 2.2.

8.7.2.3 Location and Guarding of Counterweights. Newly installed counterweights must meet the requirements for new equipment. A counterweight safety is necessary if the space below is accessible. See also Handbook commentary on 2.3.1 and 2.6.

8.7.2.4 Bottom and Top Car and Counterweight Clearances and Runbys. If the travel or rated speed is increased, it is necessary to comply with vertical clearance requirements for new equipment. In other cases, existing clearances may be maintained but not reduced. See also Handbook commentary on 2.4, 8.7.2.25.2, 8.7.2.17.1, and 8.7.2.17.2.

8.7.2.5 Horizontal Car and Counterweight Clearances. With an increase in rated speed, the horizontal clearance must comply with those for new equipment. In other alterations, existing clearances may be maintained, but not reduced. See also Handbook commentary on 8.7.2.17.2.

8.7.2.6 Protection of Spaces Below Hoistways. Compliance with 2.6 becomes necessary if the building is altered such that space below the hoistway becomes accessible. See also Handbook commentary on 2.6.

8.7.2.7 Machine Rooms and Machinery Spaces. New machine room enclosures are required to meet the requirements for new construction, including the clearance requirements. See also Handbook commentary on 2.7.1.1, 2.7.2, 2.7.3.2, 2.7.3.3, 2.7.3.4, 2.7.3.5, 2.7.5.2, and 2.7.6.

8.7.2.8 Electrical Equipment, Wiring, Pipes, and Ducts in Hoistways and Machine Rooms. Wiring, pipes, and ducts in hoistways and machine rooms, if altered or newly installed, must meet the requirements for new equipment. This requirement does not require existing wiring, pipes, or ducts to be removed or to conform to

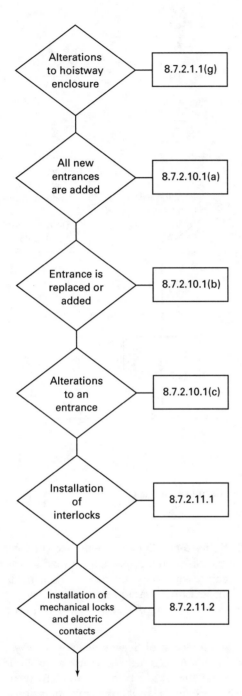

Diagram 8.7(b)   Door Locking Alterations
(Courtesy Albert Saxer)

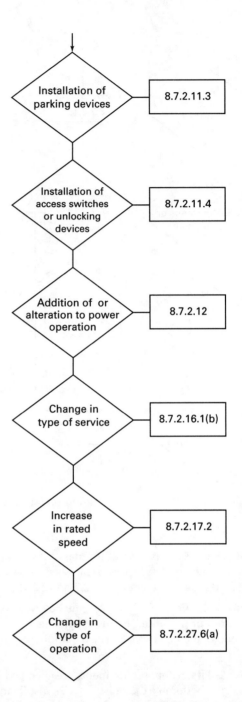

Diagram 8.7(b)   Door Locking Alterations (Cont'd)
(Courtesy Albert Saxer)

the latest edition of NFPA 70 or CSA-C22.1, whichever is applicable, if they are not altered. See also Handbook commentary on 2.8.2.

8.7.2.9 Machinery and Sheave Beams, Supports, and Foundations. Allowance is made in order to make changes in cars, door operators, and similar equipment. In the event there are major changes, the adequacy of the building structure must be verified. See also Handbook commentary on 2.9.

8.7.2.10 Hoistway Entrances. The primary concerns in the application of entrances are their fire rating, interlocks, door operation, and unlocking zone, whether new or used. The reuse of entrance components while still maintaining the integrity of the assembly for fire safety is addressed in 8.7.2.10.5.

8.7.2.10.1 General Requirements
8.7.2.10.1(a) If all new entrances are installed, they

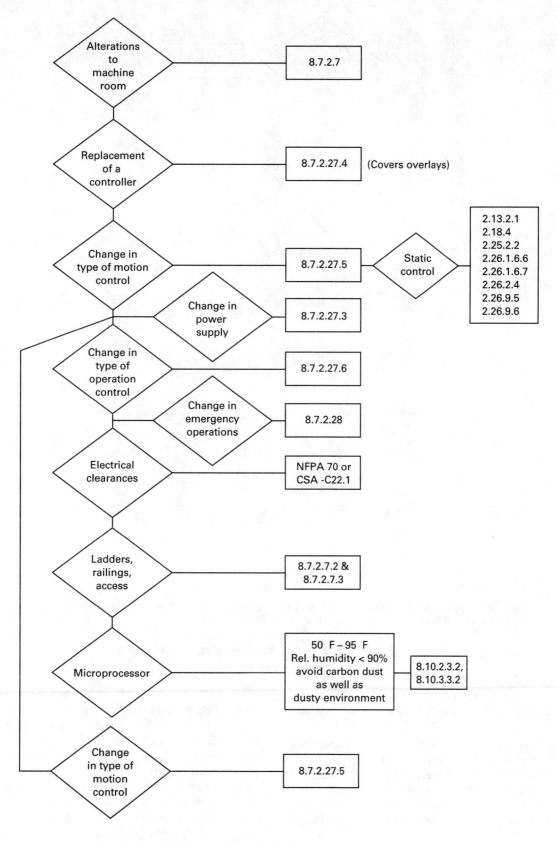

Diagram 8.7(c) Machine Room Alterations
(Courtesy Albert Saxer)

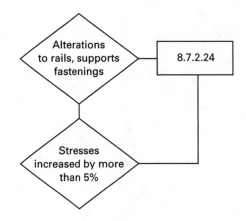

Load Changes

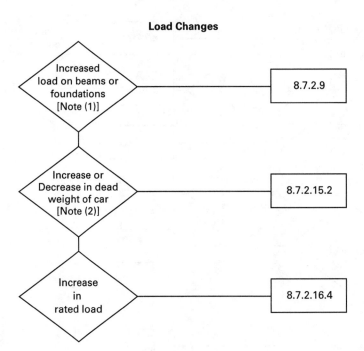

NOTES:
(1) Weight of machine and other loads can be used in calculating the 5%, thereby allowing larger than the 5% of the suspended load.
(2) If weight decreases, the buffer rating for minimum load must be verified.

Diagram 8.7(d)   Alterations to Rails, Fishplates, and Brackets
(Courtesy Albert Saxer)

must comply with the current requirements. See also Handbook commentary on 2.11, 2.12, and 2.13.

8.7.2.10.1(b) When an entrance is replaced or added, the entrance must conform to current requirements in 2.11.1, 2.11.8, and 8.7.2.10.5. It is also necessary to comply with the type of entrance as described in 8.7.2.10.2, 8.7.2.10.3, and 8.7.2.10.4. Requirement 8.7.2.10.5 allows reuse of certain hardware and frames when installed in masonry.

This requirement also allows for the replacement of the door panels and reuse of the frames, if in masonry, and reuse of door tracks, sills, and sill support. It is

expected that the entrance hardware be replaced by hardware that is labeled or listed as required by 2.11.15.

When 2.11.6 is applied along with 2.12 and 2.13, the entire system is made safe. Modern interlocks and door operation are required, and 2.12.5 requires an interlocking zone (door restraint) when the car is outside the unlocking zone. These requirements apply whether it involves only one door or the complete system.

8.7.2.10.1(c) The Handbook commentary on 8.7.2.10(b) applies to this requirement as well. Compliance with 2.11.2 and 2.11.4 is not required since the entrances are in place. Compliance is also required to the type of

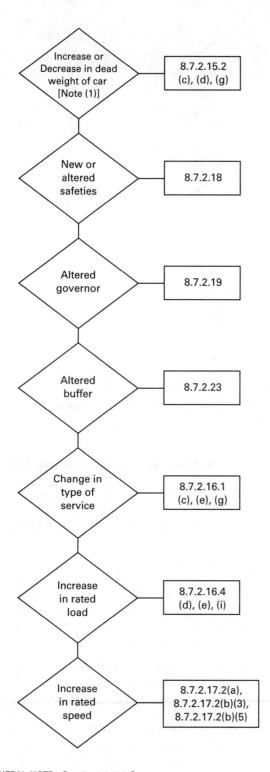

GENERAL NOTE: See 8.10.2.3.2 for tests.

NOTE:
(1) If weight is decreased, the buffer rating for minimum load must be verified.

Diagram 8.7(e) Reuse of Buffers, Safeties, and Governors

(Courtesy Albert Saxer)

entrance as described in 8.7.2.10.2, 8.7.2.10.3, and 8.7.2.10.4.

8.7.2.10.1(d) See Handbook commentary on 2.11.1.

8.7.2.10.2 Horizontal Slide-Type Entrances. See Handbook commentary on 2.11.10 and 2.11.11. Notice that 8.7.2.10.2(e) will allow a 13 mm or $\frac{1}{2}$ in. door lap.

Detailed requirements are specified for conditions when only certain components as well as when a complete entrance are installed.

Door safety retainers conforming to 2.11.11.8 are required.

8.7.2.10.3 Vertical Slide-Type Entrances. See Handbook commentary on 2.11.10 and 2.11.12.

8.7.2.10.4 Swing-Type Entrances. See Handbook commentary on 2.11.10 and 2.11.13.

8.7.2.10.5 Marking of Entrance Assemblies. This requirement allows for the reuse of entrance frames when in masonry or concrete. The rationale is that a fire test does not require a test of the entrance frame and interface with the wall, but with only the door panels. An entrance frame in drywall is not acceptable unless it has a label.

The attachment to or alteration of an existing labeled entrance is permitted only if acceptable by the certifying (labeling) organization. An example is the attachment of smoke door to an existing labeled elevator entrance jamb. Unless acceptable to the certifying organization of the entrance jamb, the original label has become null and void.

The use of existing tracks, sill, and sill support and the addition of applied frames are acceptable. See also Handbook commentary on 2.11.14 through 2.11.18.

8.7.2.11 Hoistway Door Locking Devices, Access Switches, and Parking Devices

8.7.2.11.1 Interlocks. Interlocks, mechanical locks and electric contacts, parking devices, access switches, and unlocking devices should only be altered in conformance with the requirements for new equipment. In order for the devices to perform their essential functions, a brake is required with its ability to handle an overload and stop the car if the circuit is interrupted.

See also Handbook commentary on 2.12 for the device being altered.

8.7.2.11.2 Mechanical Locks and Electric Contacts. See Handbook commentary on 8.7.2.11.1.

8.7.2.11.3 Parking Devices. Where elevators are operated from within the car only, a parking device is required unless the conditions listed are satisfied. An elevator parking device is an electrical or mechanical device the function of which is to permit the opening of the hoistway door from the landing side when the car is within the landing zone. This device is not necessary on hoistway doors, which are automatically

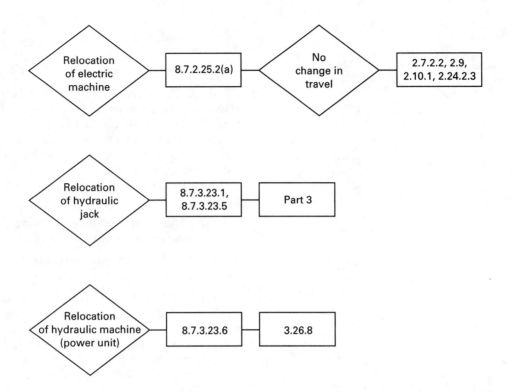

Diagram 8.7(f)   Change in Location of Driving Machine, Hydraulic Jack, or Hydraulic Machine
(Courtesy Albert Saxer)

unlocked when the car is within the landing zone. The device may also be used to close the door. An elevator parking device is used with car switch-controlled elevators to prevent inadvertent opening of the hoistway door at the usual parking floor when the elevator is located at a remote floor.

Usually, it is desired to close the hoistway door at a floor designated as the parking floor when the elevator is not in use. On car switch elevators, the hoistway doors are not normally openable from the landing side and therefore a service key is provided to allow an authorized person to unlock the door at the parking floor to allow access to the elevator.

There is a possibility that the elevator could be left unattended at a landing remote from the parking floor either with door open or closed, and therefore access at the normal parking floor would be to an open hoistway. This situation is avoided by the use of a parking device. The parking device usually consists of a mechanical lock attached to the door at the parking floor and unlocked by fixed cam on the car when the car is at the parking landing. Even though the interlock is voided by the service key, the door remains locked unless the elevator is located at the parking floor.

Usually the parking device is provided with an unlocking device similar to that which is required by this requirement and access to the hoistway when the elevator is at a remote location requires operation of the service key and the hoistway door unlocking device.

8.7.2.11.4 Access Switches and Unlocking Devices. The requirements for safe means of access to the hoistway by elevator persons are covered here.

See also Handbook commentary on 2.12.6 and 2.24.8.3 for unlocking devices and 2.12.7, 2.24.8, and 2.26.1.4 for access switches.

8.7.2.12 Power Operation. When power operation is added or altered, it is also necessary to upgrade the entrances, interlocks, and unlocking zone (door restraints), as well as the door operation. See also Handbook commentary on 8.7.2.1.1, 8.7.2.1.2, 8.7.2.1.3, and 8.7.2.1.5; and 8.10.2.3.2(a) for testing requirements.

8.7.2.13 Door Reopening Devices. Reopening devices must conform to the requirements of 2.13.4 and 2.13.5. Where firefighters' emergency operation is provided and where the device is sensitive to smoke or flame, it must be rendered inoperative on Phase I and Phase II operation. Firefighters' emergency operation must conform to the applicable requirements at the time of installation or alteration.

8.7.2.14 Car Enclosures, Car Doors and Gates, and Car Illumination. Necessary requirements are specified to ensure safe car construction. Where an alteration is only made to a component of the car (lighting device, top emergency exit, etc.), typically only the altered portion of the equipment is required to conform to the requirements for new equipment.

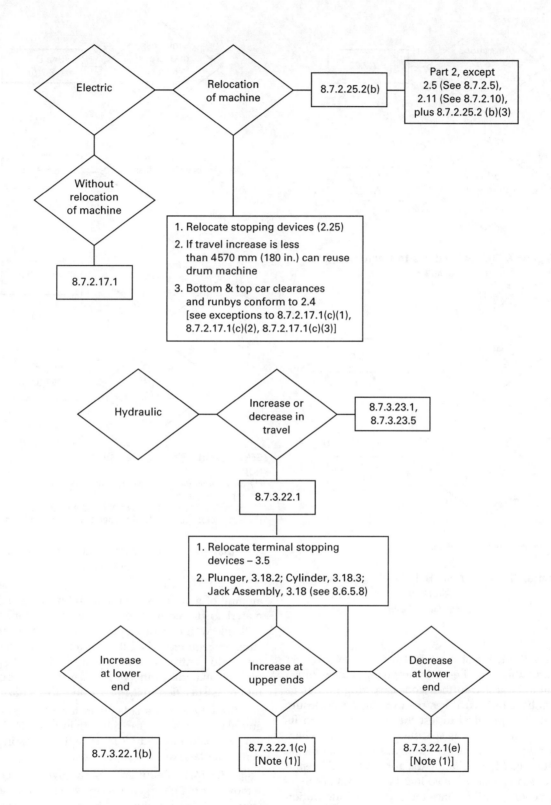

NOTE:
(1) This requirement allows reuse of Jack assemblies that were designed and installed for future travel if they conform to 3.18.

Diagram 8.7(g)   Relocation of Machine, Increase or Decrease in Travel [Adding Floor(s)]
(Courtesy Albert Saxer)

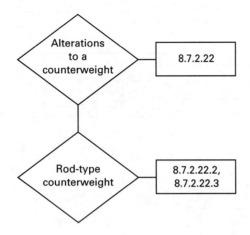

Diagram 8.7(h)   Alterations to Counterweights
(Courtesy Albert Saxer)

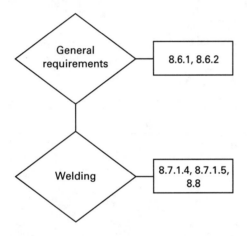

GENERAL NOTE: See 8.6.1 and 8.6.2.

Diagram 8.7(i)   Repair and Field Welding as Part of an Alteration
(Courtesy Albert Saxer)

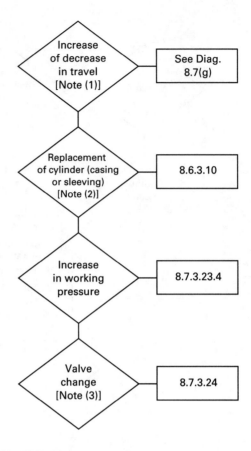

GENERAL NOTE: For tests, see 8.10.
NOTES:
(1) This requirement allows for jobs with future travel without replacing the cylinder.
(2) Spherical head and safety bulkhead are required.
(3) e.g., Motor operation to pilot operation or mechanical to electrical.

Diagram 8.7(j)   Hydraulic Elevator Alterations
(Courtesy Albert Saxer)

8.7.2.14.3(b) It is not practical to require existing car enclosures that are altered by applying a new finish material to be tested in their end-use configuration, as required by 2.14.2.1.1. When the existing car enclosure is metal, it is practical to test the new material in its end-use configuration as specified in 2.14.2.1.1. Where the existing car enclosure is retained, and it is not metal, the material added including the adhesive shall be subject to the same test specified in 2.14.2.1.1, but not in its end-use configuration. Since the end-use configuration is not tested, the flame spread for the new material, including adhesive, is more restrictive.

The limitation on the thickness of added material, when the enclosure lining is not tested in its end-use configuration, is based on similar provisions for interior finish material found in the building codes. It limits the amount of material added to a car enclosure unless the material is tested in its end-use configuration.

Textile wall covering applied to nonmetal car enclosures are required to be subject to additional test requirements, to control the potential to ignite and spread a fire if the material cannot be tested in its end-use configuration, as specified in 2.14.2.1.1 and 2.14.2.1.3.

8.7.2.14.4 In Canada where the NBCC is enforced, any alteration to the car other than those in 8.7.2.14.2 require conformance to 2.14, except for existing enclosure materials exposed to the hoistway.

8.7.2.14.5 Requirements are given to permit the removal of the emergency stop switch. The emergency stop switch is often misused by passengers or vandals.

8.7.2.15 Car Frames and Platforms

8.7.2.15.1 Alterations to Car Frames and Platforms. Any alteration to a car frame or platform requires compliance with the latest requirements. See also Handbook commentary on 2.15.1, 2.15.3, 2.15.5, 2.15.8, and 2.15.9.

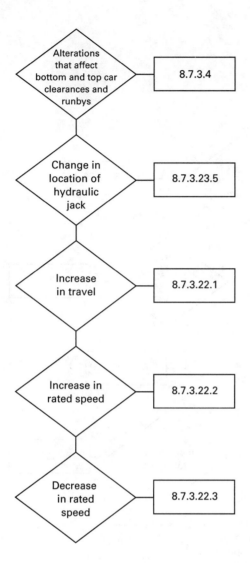

Diagram 8.7(k) Refuge Space — Hydraulic Elevator
(Courtesy Albert Saxer)

8.7.2.15.2 Increase or Decrease in Deadweight of Car. Where the load on the equipment increases or decreases more than 5%, the adequacy of the affected components must be verified. See also testing requirement in 8.10.2.3.2(d).

8.7.2.16 Capacity, Loading, and Classification

8.7.2.16.1 Change in Type of Service. The entrances (including locking devices), buffers, car frames, platforms, safeties, governors, capacity and loading, driving machines and sheaves, terminal stopping devices, operating devices, emergency operations, signaling devices, and suspension ropes must conform to present Code requirements with minor exceptions. See also the applicable portions of testing requirements in 8.10.2.3.2.

8.7.2.16.2 Change in Class of Loading. These requirements assure that the elevator is designed for the

new class of loading. See also Handbook commentary on 2.16.2 and 8.7.2.16.4.

8.7.2.16.3 Permitted Use of Freight Elevators to Carry Passengers. Requirement 2.16.4 lists the additional requirements, which must be complied with for carrying passengers. See also Handbook commentary on 2.16.4.

8.7.2.16.4 Increase in Rated Load. The adequacy of the components that are affected must be verified. See also testing requirements in 8.10.2.3.2(g).

8.7.2.17 Change in Travel or Rated Speed

8.7.2.17.1 Increase or Decrease in Travel. If the travel is changed without relocating the machine, terminal stopping devices must be relocated and proper clearances and runbys established. If the travel is increased at the top, the bottom clearance and runbys do not have to be changed. If the travel is increased at the bottom, the top clearances and runbys are not affected.

A maximum increase in travel of 4 570 mm or 180 in. is allowed for winding drum machines.

See also testing requirements in 8.10.2.3.2(k).

8.7.2.17.2 Increase in Rated Speed. When the rated speed is increased, it is necessary that those areas in which speed is a factor be upgraded. See also Handbook commentary on the various testing requirements specified and to 8.10.2.3.2(j) for tests on the equipment that is affected. Relief is given for the case where a motor is changed and the speed cannot be matched exactly.

8.7.2.17.3 Decrease in Rated Speed. Where rated speed is decreased, a change in roping may decrease runbys. Where vertical clearances and runbys are less than required by 2.4, they shall not be decreased. Governor tripping speed shall be adjusted to conform to 2.18.2.

Consideration must be given to capacity and loading and capacity and data plate revisions.

New electrical equipment and wiring must comply with current standards.

There are no tests required for a decrease in rated speed other than 8.10.2.3.2(f) for the altered governor.

8.7.2.18 Car and Counterweight Safeties. Replacement safety devices must conform to the same safety requirements that are established for new equipment. When a safety or governor is modified, the altered equipment must conform to the safety requirements that are established for new equipment. See also testing requirements in 8.10.2.3.2(e).

8.7.2.19 Speed Governors and Governor Ropes. An altered governor shall conform to the same requirements that are established for new governors except that the retention of existing smaller diameter sheaves is permissible. The governor rope must be suitable to provide the original design operation and factors of safety. See also testing requirements in 8.10.2.3.2(f).

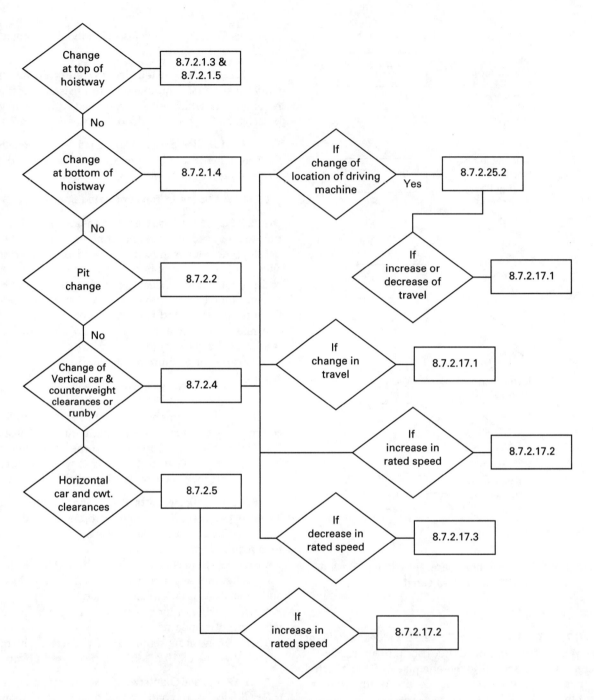

Diagram 8.7(l)   Hoistway Clearances — Electric Elevator
(Courtesy Albert Saxer)

8.7.2.20 Ascending Car Overspeed and Unintended Car Movement Protection. This is a new requirement in A17.1–2000. Any alteration or installation must conform to 2.19. See 8.10.2.2.2(cc) for test of these systems. See also Handbook commentary on 2.19.

8.7.2.21 Suspension Ropes and Their Connections. This requirement addresses alterations to and not replacement of ropes. See 8.7.2.21.2 and 8.7.2.21.3 for addition of rope equalizers or auxiliary fastening devices. See also Handbook commentary on 8.6.2.5 and 8.6.3.

8.7.2.22 Counterweights. Any alteration to a counterweight requires compliance with the requirements for new counterweights, except that rod-type counterweights may be retained. See Diagram 8.7.2.22.2, which illustrates the typical rod-type counterweight. ASME

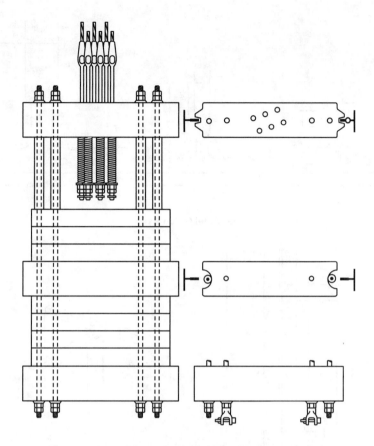

Diagram 8.7.2.22.2   Typical Rod-Type Counterweight
(Courtesy Albert Saxer)

A17.3 gives requirements for the critical areas where problems may develop. They should be closely inspected. Rod-type counterweights allow the weight section to compress upon buffer engagement. If compression is sufficient, the rope socket can be deformed resulting in a rope separation from the counterweight. See also Handbook commentary on 2.21.1.1, 2.21.1.2, 2.21.1.3, 2.21.4, and 8.7.2.3.

8.7.2.22.3   Requirements are given for installation of roller or similar guide shoes where counterweight safeties are provided.

8.7.2.23 Car and Counterweight Buffers and Bumpers. Any alteration to a buffer or bumper requires compliance with the requirements for new equipment. Because it is possible that a tag was never provided, compliance with those requirements is omitted. The adequacy of the buffer must be verified by field testing. See also testing requirement in 8.10.2.3.2(c), and Handbook commentary on 2.22.1.1, 2.22.1.2, 2.22.2, 2.22.3.1, 2.22.4, 2.22.4.1, and 2.22.4.6.

8.7.2.24 Guide Rails, Supports, and Fastenings. Changes in guide rails, supports, fastenings, or an increase in stress above 5% requires the rail system to conform to present requirements. Existing fishplates may be reused if adequate and acceptable to the authority having jurisdiction. See also testing requirements in 8.10.2.3.2(b); and Handbook commentary on 2.23.1, 2.23.2.2, 2.23.3, 2.23.4.1, 2.23.5, 2.23.5.2, 2.23.8, 2.23.9, 2.23.9.2, and 2.23.10.

Diagram 8.7.2.24 shows details for typical guide rails and older rail joints.

8.7.2.25 Driving Machines and Sheaves

8.7.2.25.1 Alterations to Driving Machines and Sheaves. Where a machine is replaced or where a component is altered, they must conform to requirements for a new machine. In addition, ropes and their connections must comply with the latest requirements. See also testing requirements in 8.10.2.3.2(h) and 8.10.2.3.2(i).

8.7.2.25.2 Change in Location of Driving Machine. If the machine is relocated and there is no change in travel, only the supports are to be verified for their adequacy. If the travel is changed and the machine is relocated, the installation is considered as new except for some modifications to the requirements affecting clearances and hoistway entrances. The safety also must be tested. See also testing requirements in 8.10.2.3.2(i).

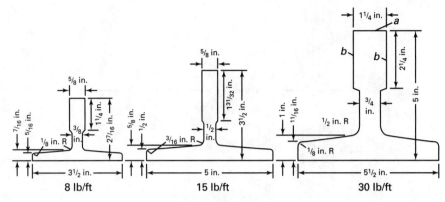

(a) Steel Guide Rail

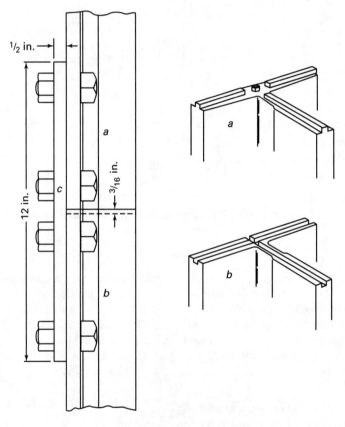

(b) Early Rail Joint

Diagram 8.7.2.24   Guide Rails
(Courtesy Albert Saxer)

8.7.2.26 Terminal Stopping Devices. Where the terminal stopping device is altered, it must comply with the latest requirements. See also Handbook commentary on 2.6, and 8.10.2.3.2(k) for testing requirements.

8.7.2.27 Operating Devices and Control Equipment

8.7.2.27.1 Addition of Top-of-Car Operating Devices. See Handbook commentary on 2.26.1.4.

8.7.2.27.2 Addition of Car-Leveling or Truck-Zoning Devices. See Handbook commentary on 2.26.1.6.

8.7.2.27.3 Change in Power Supply. It is necessary to clearly define a change in power supply and establish the equipment changes that are required to provide safe operation of the elevator. If the power at the mainline terminals of the elevator controller remains unchanged, any change of building input power will not be considered a change in power supply. Refer to the various requirements mentioned and to the Handbook commentary on them.

8.7.2.27.4 Controllers. The installation of a controller, in place of an existing controller, is considered an alteration even when a new controller is identical to the existing one, such as is required due to the original being damaged in a fire. Necessary safety requirements are specified for equipment that is controller dependent.

8.7.2.27.5 Change in Type of Motion Control. Motion control requirements are very important with respect to the safety of elevator passengers. Normally where a change is made in motion control, a complete new controller is installed. However, it is possible to change motion control by installing an auxiliary control panel (see 1.3, Definitions).

Where static-type motion control is installed, the additional requirements of 2.26.1.6.6 and 2.26.1.6.7 must be satisfied.

When this type of alteration is made, the landing openings must be properly protected. The car enclosure and car door must enclose the passengers safely. It is not necessary for the hoistway doors to be labeled. The platform dimensions must be in accordance with the requirements for a new elevator as well as terminal stopping devices, operating devices and emergency operation and signal devices.

See also testing requirements in 8.10.2.3.2(k), 8.10.2.3.2(m), and 8.10.2.3.2(t).

8.7.2.27.6 Change in Type of Operation Control. It is common in modernization work to add overlays to existing controllers. If this does not consist of a change in the operation (see 1.3, Definitions) from one of the listed types to another, this requirement would not apply.

When an alteration is made to the operation, the landing openings must be properly protected and the car and doors or gates must enclose the passengers safely. Note that it is not necessary for the hoistway doors to be labeled.

The platform dimensions must be in accordance with the requirements for a new elevator as well as the terminal stopping devices, operating devices, control equipment, and emergency operation and signaling devices.

See also the testing requirements in 8.10.2.3.2(k) and 8.10.2.3.2(m).

8.7.2.28 Emergency Operations and Signaling Devices. When an alteration is made to emergency signaling, standby power, or firefighters' service, all elevators in that group must provide the same operation as that established for new equipment. All firefighters' service keys are to be alike in a given building. See also the testing requirements in 8.10.2.3.2(l) and 8.10.2.3.2(m).

8.7.3 Alterations to Hydraulic Elevators

Refer to the requirements cited and their commentaries. See also the testing requirements in 8.10.3.3.

8.7.3.9 Machinery and Sheave Beams, Supports and Foundations. Allowance is made in order to make changes in cars, door operators, and similar equipment. In the event there are major changes, the adequacy of the building structure must be verified. See also Handbook commentary on 2.9.

8.7.3.10 Hoistway Entrances. Emergency doors are not required for hydraulic elevators without safeties and when a manually operated lowering valve is provided (see 3.19.4.4).

8.7.3.11 Hoistway Door Locking Devices. Conformance to 2.24.8.3 is not applicable because a brake is not required on hydraulic elevators.

8.7.3.12 Power Operation. See Handbook commentary on 8.7.3.10 and testing requirements in 8.10.3.3.2(a).

8.7.3.13 Car Enclosures. The exception provides for the elimination of side emergency exits on hydraulic elevators when the specified conditions are met.

8.7.3.14 Alterations to Car Frames and Platform. Any alteration to a car frame or platform requires compliance with the latest requirements. See also Handbook commentary on 3.15.

8.7.3.15 Safeties. See testing requirements in 8.10.3.3.2(e).

8.7.3.16 Governors and Governor Ropes. See testing requirements in 8.10.3.3.2(f).

8.7.3.17 Change in Type of Service. The entrances (including locking devices), buffers, rails, car frames and platforms, enclosures, car doors, safeties, capacity and loading, driving machines, valves, supply piping and fittings, tanks, terminal stopping devices, operating devices and control equipment, and ropes must conform to requirements for new elevators with a few minor exceptions. See also the testing requirements in applicable portions of 8.10.3.3.2.

Emergency doors are not required for hydraulic elevators without safeties and where a manually operated valve is provided.

8.7.3.18 Change in Class of Loading. These requirements assure that the elevator is designed for the new class of loading. See also Handbook commentary on 2.16.2 and 3.16.

8.7.3.19 Carrying of Passengers on Freight Elevators. Additional requirements, which must be complied with for carrying passengers, are found in 2.16.4.

8.7.3.20 Increase in Rated Load. See testing requirements in 8.10.3.3.2.

8.7.3.21 Increase in Deadweight of Car. See testing requirements in 8.10.3.3.2(d).

8.7.3.22 Change in Travel or Rated Speed

8.7.3.22.1 Increase or Decrease in Travel. If the travel is changed without relocating the machine, terminal stopping devices must be relocated and proper clearances and runbys established.

The plunger must have proper column strength. See also the testing requirements in 8.10.3.3.2(k) and 8.10.3.3.2(m).

8.7.3.22.2 Increase in Rated Speed. When the rated speed is increased, it is necessary that those areas where speed is a factor be upgraded. See also the testing requirements in 8.10.3.3.2(j).

8.7.3.22.3 Decrease in Rated Speed. Where the rated speed is decreased, mechanical changes may decrease runbys. Where clearances and runbys are less than required by 3.4, they shall not be decreased. Governor tripping speeds shall be adjusted to conform the requirement in 2.18.2.

Consideration must be given to capacity and loading and capacity and data plate revisions. New electrical equipment and wiring must comply with current standards. There are no tests in 8.10.3.3 for a decrease in rated speed other than 8.10.3.3.2(f) for the altered governor.

8.7.3.23 Hydraulic Equipment

8.7.3.23.4 Increase in Working Pressure. See testing requirements in 8.10.3.3.2(h).

8.7.3.23.5 Change in Location of Hydraulic Jack. See testing requirements in 8.10.3.3.2(h).

8.7.3.23.6 Change in Location of Driving Machine (Power Unit). A pressure switch in the to-from line is necessary when the jack cylinder head is above the tank to inhibit the operation of the elevator if positive pressure is lost. See also 3.26.8 and testing requirements in 8.10.3.3.2(i).

8.7.3.24 Valves, Supply Piping, and Fittings. This requirement concerns a change from a mechanically operated valve to a modern solenoid pilot operated valve. The change from a number of discrete valves to a unit (block valve) is a replacement. See also Handbook commentary on 3.19 and testing requirements in 8.10.3.3.2(o).

8.7.3.25 Suspension Ropes and Their Connections. This requirement addresses alterations to and not replacement of ropes. See also Handbook commentary on 8.6.2.5 and 8.6.3. The requirements for addition of rope equalizers or auxiliary fastening devices can be found in 8.7.2.21.2 and 8.7.2.21.3.

8.7.3.26 Counterweights. See Handbook commentary on 8.7.2.22 and 3.22.2.

8.7.3.27 Car Buffers and Bumpers. See Handbook commentary on 8.7.2.23 and 3.22.1. See also testing requirements in 8.10.3.3.2(c).

8.7.3.28 Guide Rails, Supports, and Fastenings. See Handbook commentary on 3.28. See also testing requirements in 8.10.3.3.2(b).

8.7.3.29 Tanks. See Handbook commentary on 3.24.

8.7.3.30 Terminal Stopping Devices. See testing requirements in 8.10.3.3.2(k).

8.7.3.31 Operating Devices and Control Equipment

8.7.3.31.1 Top-of-Car Operating Devices. An alteration to or addition of a top-of-car operating device must conform to the same requirements as a new elevator. See also Handbook commentary on 2.26.1.4.

8.7.3.31.2 Car-Leveling or Truck-Zoning Devices. An alteration to or addition of a car-leveling or truck-zoning device must conform to the same requirements as a new elevator. See also Handbook commentary on 3.26.3. For details on the truck-zoning device, refer to the Handbook commentary on 2.26.1.6.4, 2.26.1.6.5, and 2.26.1.6.6.

8.7.3.31.3 AntiCreep Leveling Device. An alteration to or addition of an anticreep leveling device must conform to the same requirements as a new elevator. See also Handbook commentary on 3.26.3.1.

8.7.3.31.4 Change in Power Supply. An alteration that involves a change in the power supply voltage, frequency, number of phases, direct to alternating current or vice versa or a combination of direct and alternating current must conform to: 3.26.1, various operating and control equipment; 3.26.4, electrical protective devices; 3.26.5, phase reversal and failure protection; and 3.26.6, control and operating circuits, excluding 2.26.4.4 (EMI regulations).

8.7.3.31.5 Controllers. See Handbook commentary on 8.7.2.27.4.

8.7.3.31.6 Change in Type of Motion Control. The motion control on a hydraulic elevator is the valve(s). The electrical circuits provide the input to the electrically operated pilot valves.

8.7.3.31.7 Change in Type of Operation. See Handbook commentary on 8.7.2.27.6. See also testing requirements in 8.10.2.3.2(k) and 8.10.2.3.2(m).

8.7.3.31.8 Emergency Operation and Signaling Devices. See testing requirements in 8.10.3.3.2(l).

8.7.3.31.9 Auxiliary Power Lowering Operation. The auxiliary power lowering operation will return the elevator to the lowest landing in the event of main power supply failure. If the device is altered or a new device is installed, it must meet the requirement for a new elevator. See also Handbook commentary on 3.26.10.

8.7.4 Alterations to Elevators With Other Types of Driving Machines

8.7.4.1 Rack and Pinion Elevators. The Code requirements for this equipment have not undergone substantial changes since first incorporated in the Code. The cost to comply with the present requirement, when an alteration is made, will be nominal.

8.7.4.2 Screw-Column Elevators. The Code requirements for this equipment have not undergone substantial changes since first incorporated in the Code. The cost to comply with the present requirement, when an alteration is made, will be nominal.

8.7.4.3 Hand Elevators. This technology is obsolete. Alterations are seldom made.

8.7.5 Alterations to Special Application Elevators

8.7.5.1 Inclined Elevators. The Code requirements for this equipment have not undergone substantial changes since first incorporated in the Code. The cost to comply with the present requirement, when an alteration is made, will be nominal.

8.7.5.2 Limited Use/Limited Application Elevators. At this time, there are no Code requirements for alterations to this equipment.

8.7.5.3 Private Residence Elevators. At this time, there are no Code requirements for alterations to this equipment.

8.7.5.4 Private Residence Inclined Elevators. At this time, there are no Code requirements for alterations to this equipment.

8.7.5.5 Power Sidewalk Elevators. The requirements for alterations to power sidewalk elevator are clear. Alterations are rare as the environment to which this equipment is subjected usually makes the installation of a complete new sidewalk elevator the most cost effective approach.

8.7.5.6 Rooftop Elevators. The Code requirements for this equipment have not undergone substantial changes since first incorporated in the Code. The cost to comply with the present requirement, when an alteration is made, will be nominal.

8.7.5.7 Special Purpose Personnel Elevators. The Code requirements for this equipment have not undergone substantial changes since first incorporated in the Code. The cost to comply with the present requirement, when an alteration is made, will be nominal.

8.7.5.8 Shipboard Elevators. The Code requirements for this equipment have not undergone substantial changes since first incorporated in the Code. The cost

to comply with the present requirement, when an alteration is made, will be nominal.

8.7.5.9 Mine Elevators. Mine elevators have a finite useful life. The mining operation progress usually puts them out of service in less than 10 years; therefore, alterations are rare.

8.7.6 Alterations to Escalators and Moving Walks

8.7.6.1 Escalators. Reference is made to the requirements that apply to escalators. Requirement 6.1.6 describes the operating and safety devices that are supplied for a modern escalator.

8.7.6.1.3 Protection of Floor Openings. Establishes the requirements for protection of escalator wellways against the passage of flame, heat, and smoke.

8.7.6.1.4 Protection of Trusses and Machinery Spaces Against Fire. Minimum standards of fire protection for trusses and machinery spaces of altered units are specified in 6.1.2.1.

8.7.6.1.5 Construction Requirements

8.7.6.1.5(a) Angle of Inclination. The tolerance from the designed angle of inclination for an altered escalator is limited to 1 deg to permit adjustment for field conditions.

8.7.6.1.5(b) Geometry. Any alteration of the escalator geometry must conform to current standards as established by 6.1.3.2.

8.7.6.1.5(c) Balustrades. This requires altered balustrades to conform to current standards as established by 6.1.3.3.8. Alteration to or installation of deflection devices must conform to 6.1.3.3.8.

8.7.6.1.6 Handrails. This requires altered handrails to conform to the current standard as established by 6.1.3.2.2, 6.1.3.4.1 through 6.1.3.4.4, 6.1.3.4.6, 6.1.6.3.12, and 6.1.6.4.

8.7.6.1.7 Step System. This requires conformance with current standards for the steps and step system when an alteration is made to the steps. Reduction of step width for steps having a width less than 560 mm or 22 in. is prohibited by 8.7.6.1.7(b).

8.7.6.1.8 Combplates. This requires altered combplates to conform to 6.1.6.3.13.

8.7.6.1.9 Trusses and Girders. This requires conformance with 8.7.1.4 for structural modifications of the truss or a girder. The installation of a new escalator in an existing truss is classified as a new escalator, which must comply with all the requirements in 6.1.

8.7.6.1.10 Step Wheel Tracks. This requires alterations of the escalator track to conform to the current requirements of 6.1.3.8, 6.1.3.9.4, 6.1.3.10.1, and 8.7.1.4.

8.7.6.1.11 Rated Load and Speed. Any alteration to an escalator, which increases the rated load and/or rated speed, requires conformance with all the requirements in 6.1.

8.7.6.1.12 Driving Machine, Motor, and Brake. Altered machines are required to conform to 6.1.3.9.2 (for rated load), 6.1.3.9.3 (braking requirements), 6.1.3.10.2 and 6.1.3.10.3 (factor of safety for machine parts), 6.1.3.11 (prohibition of brittle cast iron chains), 6.1.4.1 (rated speed), and 6.1.5 (requirements for driving machine, motor, and brake).

8.7.6.1.13 Operating and Safety Devices. An alteration to, or addition of, operating and/or safety devices must conform to the current requirements of 6.1.6, for that device that has been added.

8.7.6.1.14 Lighting, Access, and Electrical Work. Any alteration to or addition of lighting, access, or electrical work must conform to the current requirements established by 6.1.7, for that alteration.

8.7.6.2 Moving Walks

8.7.6.2.1 General Requirements. Reference is made to the requirements that apply to moving walks, 6.2.6.2.1 (prohibits automatic starting), 6.2.6.3.1 (emergency stop buttons), 6.2.6.3.5 (stop switch in machinery space), 6.2.6.3.6 (egress restriction), and 6.2.6.6 (tandem operation). Requirement 6.2.7.2 specifies the minimum lighting levels for safe use of moving walks.

8.7.6.2.3 Protection of Floor Openings. This establishes the requirements for protection of floor openings of moving walkways against the passage of flame, heat, and smoke.

8.7.6.2.4 Protection of Trusses and Machinery Spaces Against Fire. This establishes the requirements for fire protection when the sides or undersides of moving walk trusses or machinery spaces are altered.

8.7.6.2.5 Construction Requirements

8.7.6.2.5(a) Angle of Inclination. Increasing the angle of inclination would change the design criteria of the equipment.

8.7.6.2.5(b) Geometry. This requirement requires any alteration of the moving walk geometry to conform to current standards as established by 6.2.3.2.

8.7.6.2.5(c) Balustrades. This requires altered balustrades to conform to current standards as established by 6.2.3.3.

8.7.6.2.6 Handrails. This requires altered handrails to conform to the current standard as established by 6.2.3.2.2 and 6.2.3.4.

8.7.6.2.7 Treadway System. This requires conformance with current requirements for the treadways and treadway system if an alteration is made to the treadway. Reduction of treadway width for existing treadways less than 560 mm or 22 in. is prohibited by 6.2.3.7. See also 8.7.6.2.7(b).

8.7.6.2.8 Combplates. This requirement requires altered combplates to conform to the current standards established by 6.2.3.8.1 and 6.2.3.8.2.

8.7.6.2.9 Trusses and Girders. This requires the conformance with 6.2.3.9, 6.2.3.10.1, 6.2.3.11.1, and 8.7.1.4 for structural modifications of the truss or a girder.

8.7.6.2.10 Track System. This requires alterations of the moving walk track to conform to the current standards established by 6.2.3.9, 6.2.3.10, 6.2.3.11.1, and 8.7.1.4

8.7.6.2.11 Rated Load and Speed. Any alteration to a moving walk, which increases the rated load and/or rated speed, requires conformance with all the requirements in 6.2.

8.7.6.2.12 Driving Machine, Motor, and Brake. This requires any altered machine to conform to 6.2.3.10 (for rated load and braking requirements), 6.2.3.11 (factor of safety for machine parts), 6.2.3.12 (prohibition of brittle cast iron chains), 6.2.4 (rated speed), and 6.2.5 (requirements for driving machine, motor, and brake).

8.7.6.2.13 Operating and Safety Devices. An alteration to, or addition of, operating and/or safety devices must conform to the current requirements of 6.2.6 for that device that has been altered or added.

8.7.6.2.14 Lighting, Access, and Electrical Work. Any alteration to or addition of lighting, access, or electrical work must conform to the current requirements established by 6.2.7 for that change.

8.7.7 Alterations to Dumbwaiters and Material Lifts

8.7.7.3 Alterations to Material Lifts and Dumbwaiters with Automatic Transfer Devices. The Code requirements for this equipment have not undergone substantial changes since first incorporated in the Code. The cost to comply with the present requirement, when an alteration is made, will be nominal.

SECTION 8.8
WELDING

The requirements in 8.8 apply only when referenced by a requirement in this Code. Not all welding must comply with the requirements in 8.8. Many jurisdictions have additional requirements for licensing of welders and for all welding done at the job site. Local regulations should be reviewed.

SECTION 8.9
CODE DATA PLATE

This section applies to both new and existing installations. Equipment is required to be maintained (8.6.1.1.2) and inspected and tested (8.11.1.2) to determine that it complies with the Code under which it was installed and/or altered. This information should be readily available at the site for inspection and maintenance personnel. Therefore, it is required to be on the controller or mainline disconnect. For existing equipment, the establishment of the Code date will require some research. Several of the major manufacturers can provide the edition of the Code that the equipment was installed under. Likewise, many enforcing authorities have records that can provide this information. Where alterations have been done, both the Code under which the altered equipment was installed in addition the Code in effect at the time of the alteration along with a list of the requirements complied with must be listed on the code data plate.

Charts 8.9(a) and 8.9(b) are two "Checklists" that list the various Code requirements that are usually involved with a modernization project. In Code language, modernization is called an alteration. One checklist applies to electric elevators, the other applies to hydraulic elevators. These two types of equipment are involved in the majority of alterations. In many jurisdictions, the Code does not become effective until it is officially adopted. Therefore, two columns are provided. One lists the Code Rules from the 1996 Edition, including the 1997, 1998, 1999 and 2000 addenda; the second lists the various requirements form ASME A17.1–2000. Be sure to use those that apply to the Code in force where the equipment is located.

The intention of these "checklists" is to provide a tool for the contract engineer or contract processor to ensure compliance with Part XII or Section 8.7. The checklists will also provide the data for the person who will prepare the data tag.

SECTION 8.10
ACCEPTANCE INSPECTIONS AND TESTS

NOTE: Throughout Section 8.10, references are made to other requirements in this Code. To gain a complete understanding of a requirement, the reader should review the Handbook commentary on all referenced requirements.

All acceptance testing and inspection requirements for new and altered equipment can be found in this Section. All periodic tests can be found in Section 8.11. The purpose of the inspections and tests required by this Code is to assure that equipment conforms to the requirements in this Code. In order to fully understand and comply with these requirements, also consult: ASME A17.2.1, Inspectors' Manual for Electric Elevators; ASME A17.2.2, Inspectors' Manual for Hydraulic Elevators; and ASME A17.2.3, Inspectors' Manual for Escalators and Moving Walks. The details for an inspection and/or test will be found in those documents. Many requirements also provide specific references to items in the applicable Inspectors' Manual.

NOTE: The three separated Inspectors' Manuals were combined and published on December 31, 2001 as ASME A17.2, Guide for Inspection of Elevators, Escalators, and Moving Walks.

Another Handbook that is invaluable for making inspections and performing tests is the Elevator Industry Inspection Handbook, by Zack McCain, Jr. See Handbook commentary on Part 9 for procurement information.

8.10.1 General Requirements for Acceptance Inspections and Tests

8.10.1.1 Persons Authorized to Make Acceptance Inspections and Tests

8.10.1.1.1 The authority having jurisdiction (AHJ) is defined (see 1.3) as the organization, office, or individual responsible for enforcement of this Code. Where compliance with this Code has been mandated by legislation or regulation, the "authority having jurisdiction" is the regulatory authority.

In Canada the AHJ is often referred to as the regulatory authority which is defined (see 1.3) as the person or organization responsible for the administration and enforcement of the applicable legislation or regulation governing the design, construction, installation, operation, inspection, testing, maintenance, or alteration of equipment covered by this Code. This requirement mandates the AHJ make all acceptance test or designate the person who will make the inspections. The AHJ can stipulate that the inspections be made by a third party inspector or the installing company. Whoever makes the inspection must be QEI certified (8.10.1.1.3) except in Canada.

8.10.1.1.2 The installing contractor is required to run all the applicable test, which must be witnessed, by the AHJ or their designee. See Handbook commentary on 8.10.1.1.1 and 8.10.1.1.3.

8.10.1.1.3 ASME A17.1, Safety Code for Elevators and Escalators, covers design, construction, operation, inspection, testing, maintenance, alteration, and repair. ASME A17.2 Inspectors' Manuals supplement this Code by providing guidelines for the inspection and testing of the equipment. None of the standards, however, cover the qualifications of inspection personnel. The duties of an inspector are only covered briefly in the Introduction to the Inspectors' Manual.

The excellent safety record of elevators, escalators, and related equipment has been maintained, in part, by quality field inspections and tests. However, advancing technology and safety requirements have highlighted the need for establishing uniform criteria for the persons performing these inspections. The quality of inspections, of course, depends on the competence of the inspector. The Standard for the Qualification of Elevator Inspectors, ASME QEI-1, was published by ASME to promote uniform quality in inspections.

Chart 8.9(a) ASME A17.1 Code Data Plate Checklist
(Courtesy © 2001, National Elevator Industry Inc., Teaneck, NJ)

ELECTRIC ELEVATOR ALTERATIONS

Enter the Code under which the elevator was installed and the Code under which the elevator had been previously altered:

Alterations – General Requirements

If alteration is made to:	A17.1d–2000 Part XII, Rule	A17.1–2000 Section 8.7	Applies, check here
When any alteration is made, the entire installation must conform to ASME A17.3–1996 with Addenda	1200.1	Not applicable	A17.1d only X
Tests after alteration must conform to Part X or Section 8.10	1200.3	8.7.1.3	X
Welds made where support is involved	1200.4b & 1200.4c	8.7.1.4	
Removal of emergency stop switch — in car stop switch	1200.5b	8.7.2.14.5	
Door reopening device as part of alteration	1200.5d	8.7.2.13	
Temporary wiring		8.7.1.6	

Alteration in Hoistway or Pit

If alteration is made to:			
Hoistway enclosure, or addition of elevator to hoistway, or alteration of hoistway vent/fan	1201.1	8.7.2.1	
Hoistway electrical equipment, wiring, pipes, or ducts	1201.3	8.7.2.8	
Location or guarding of counterweight	1201.4	8.7.2.3	
Pit	1201.6	8.7.2.2	
Vertical car or counterweight clearances and runby	1201.7	8.7.2.4	
Horizontal car or counterweight clearances	1201.8	8.7.2.5	
Space below car or counterweight	1201.9	8.7.2.6	
Hoistway entrances (doors, sills, tracks, hangers, etc.)	1201.10	8.7.2.10	
Hoistway door locks or access switches	1201.11	8.7.2.11	
Power door operation (motor) (engine)	1201.12	8.7.2.12	
Rails or rail supports	1202.1	8.7.2.24	
Buffers or bumpers	1202.2	8.7.2.23	
Counterweight	1202.3	8.7.2.22	
Speed governors or governor ropes	1202.7	8.7.2.19	
Travel (rise) of elevator	1202.10	8.7.2.17.1	
Terminal stopping device	1202.11	8.7.2.26	

Alteration to the Car

If alteration is made to:	A17.1d–2000 Part XII, Rule	A17.1–2000 Section 8.7	Applies, check here
Car frame or platform	1202.4	8.7.2.15	
Change of dead weight of car	1202.4b	8.7.2.15.2	
Car enclosure (cab), car doors or gates, or car illumination	1202.5	8.7.2.14	
Car or counterweight safeties	1202.6	8.7.2.18	
Car capacity, loading or classification	1202.8	8.7.2.16	

(continued)

Chart 8.9(a)   ASME A17.1 Code Data Plate Checklist (Cont'd)

Alteration in Machine Room

If alteration is made to:			
Machine room, or means of access to machine room, or machine room doors, windows, skylights, lighting, or ventilation	1201.2	8.7.2.7	
Alterations to machine room electrical equipment, wiring pipes or ducts	1201.3	8.7.2.8	
Alterations to machinery or sheave beams, supports or foundations	1201.5	8.7.2.9	
Speed governors or governor ropes	1202.7	8.7.2.19	
Driving machine or machine sheave	1202.9	8.7.2.25	
Rated speed	1202.10	8.7.2.17.2 & 8.7.2.17.3	
Controller, power supply, top of car inspection control	1202.12	8.7.2.27	
Labeled and listed devices	1202.12	8.7.2.27	
Change in type of control (motion control)	1202.12e	8.7.2.27.5	
Change in type of operation	1202.12f	8.7.2.27.6	
Emergency operations and signaling	1202.13	8.7.2.28	

(a) The QEI standard is intended to establish uniform criteria, which will aid in the following:

(1) qualifying and training of inspection personnel for government agencies, insurance companies, and elevator companies;

(2) the accreditation of inspection agencies; and

(3) reducing the liability exposure of insurance companies, manufacturers, installers, and building owners.

It is also intended to serve as a guideline on which certification is based by detailing the expertise necessary in performing inspections.

(b) The QEI Standard covers requirements for the qualification and duties of personnel engaged in the inspection and testing of equipment within the scope of the ASME A17.1, Safety Code for Elevators and Escalators. The QEI Standard applies to persons typically employed by (but not limited to) the following:

(1) jurisdictional authorities;

(2) independent inspection agencies and elevator consultants;

(3) insurers of the equipment;

(4) manufacturers, installers, and maintainers of the equipment; and

(5) building owners and managers.

(c) The QEI Standard does not cover personnel engaged in engineering and type testing as covered in 8.3 of the ASME A17.1 Code, including inspection by laboratories in association with those tests. Under the QEI program, ASME will accredit organizations that certify inspectors who meet the requirements of ASME QEI-1.

The ASME A17.1 Code requires inspectors and inspector supervisors to be certified by an organization accredited by the ASME, except in Canada. The Code recognizes that inspections and tests should only be performed by qualified personnel. The QEI accreditation program is a recognized means of assuring that inspection personnel are qualified. For more information, or for a list of ASME accredited organizations, contact:

Secretary, QEI Committee
American Society of Mechanical Engineers
Three Park Avenue
New York, New York 10016-5990

8.10.1.3 Making Safety Devices Inoperative or Ineffective. Making safety devices inoperative or ineffective is permitted only after the installation is removed from public use. The reader is cautioned that jumpers unintentionally left on safety circuits and/or improperly used have been the cause of fatalities and serious injury. The reader should be familiar with and adhere to the jumper procedures in the latest edition of the Elevator Industry Field Employees' Safety Handbook.

8.10.2 Acceptance Inspection and Tests of Electric Elevators

An acceptance inspection and test is the initial inspection and test of new or altered equipment to check for compliance with the applicable Code requirements.

The inspector witnesses the test; he does not conduct the test. The person performing the test does not have to be QEI certified; however, the test must be witnessed by an inspector that is QEI certified, except in Canada. See Handbook commentary on 8.10.1.1.

An acceptance inspection and test is performed before the equipment is turned over for service. The purpose is to ascertain that the equipment as installed conforms

Chart 8.9(b)   ASME A17.1 Code Data Plate Checklist
(Courtesy © 2001, National Elevator Industry Inc., Teaneck, NJ)

HYDRAULIC ELEVATOR ALTERATIONS

Enter the Code under which the elevator was installed and the Code under which the elevator had been previously altered:

Alterations — General Requirements

If alteration is made to:	A17.1d–2000 Part XII, Rule	A17.1–2000 Section 8.7	Applies, check here
When any alteration is made, the entire installation must conform to ASME A17.3–1996 with Addendum	1200.1	Not applicable	A17.1d only X
Tests after alteration must conform to Part X or Section 8.10	1200.3	8.7.1.3	X
Welds made where support is involved	1200.4b & 1200.4c	8.7.1.4 & 8.7.1.5	
Removal of emergency stop switch — in car stop switch	1200.5b	8.7.3.13 & 8.7.2.14.5	
Door reopening device as part of alteration	1200.5d	8.7.2.13	
Temporary wiring		8.7.1.6	

Alteration in Hoistway or Pit

If alteration is made to:			
Hoistway enclosure, or addition of elevator to hoistway, or alteration of hoistway vent/fan	1203.1	8.7.3.1	
Hoistway electrical equipment, wiring, pipes, or ducts	1203.1c	8.7.3.8	
Location or guarding of counterweight	1203.1d	8.7.3.3	
Pit	1203.1f	8.7.3.2	
Vertical car or counterweight clearances and runby	1203.1g	8.7.3.4	
Horizontal car or counterweight clearances	1203.1h	8.7.3.5	
Space below car or counterweight	1203.1i	8.7.3.6	
Hoistway entrances (doors, sills, tracks, hangers, etc.)	1203.1j	8.7.3.10	
Hoistway door locks or access switches	1203.1k	8.7.3.11	
Power door operation (motor) (engine)	1203.1m	8.7.3.12	
Rails or rail supports	1203.2	8.7.3.28	
Buffers or bumpers	1203.2b	8.7.3.27	
Counterweight	1203.2c	8.7.3.26	
Speed governors or governor ropes	1203.2g	8.7.3.16	
Driving machine (plunger and cylinder) (jack assembly) — *single bottom cylinders should be replaced* — (Rule in 2000 Code, Section 8.6.5.8.) *see note on following page	1203.3	8.7.3.23.1, 8.7.3.23.2, 8.7.3.23.3, & 8.7.2.23.5	
Increase in working pressure (this rule appears in two places — hoistway-pit and machine room)	1203.3d	8.7.3.23.4	
Travel (rise) of elevator	1203.4a	8.7.3.22.1	

(continued)

Chart 8.9(b)   ASME A17.1 Code Data Plate Checklist (Cont'd)

Alteration to the Car

If alteration is made to:	A17.1d–2000 Part XII, Rule	A17.1–2000 Section 8.7	Applies, check here
Car frame or platform	1203.2d	8.7.3.14	
Car enclosure (cab), car doors or gates, or car illumination	1203.2e	8.7.3.13	
Car or counterweight safeties	1203.2f	8.7.3.15	
Car capacity, loading, or classification	1203.2i, 1203.2j, 1203.2k	8.7.3.17, 8.7.3.18, 8.7.3.19, & 8.7.3.20	
Change in dead weight of car	1203.2m	8.7.3.21	

Alteration in Machine Room

If alteration is made to:			
Machine room, or means of access to machine room, or machine room doors, windows, skylights, lighting, or ventilation	1203.1b	8.7.3.7	
Alterations to machine room electrical equipment, wiring pipes or ducts	1203.1c	8.7.3.8	
Alterations to machinery or sheave beams, supports, or foundations for power unit. Note: Under the 1996 Code, the plunger and cylinder are called the driving machine	1203.1e	8.7.3.9	
Speed governors or governor ropes	1203.2g	8.7.3.16	
Increase in working pressure (this rule appears in two places — hoistway-pit and machine room)	1203.3d	8.7.3.23.4	
Relocation of power unit (pump, valve, and tank)	1203.3f	8.7.3.23.6	
Rated speed	1203.4	8.7.3.22.2, 8.7.3.22.3	
Installation of valve of different type (replacement of discrete valves with a block type *is not* an alteration)	1203.5	8.7.3.24	
Tank	1203.6	8.7.3.29	
Terminal stopping device	1203.7	8.7.3.30	
Controller, power supply, top of car inspection control	1203.8	8.7.3.31	
Labeled and listed devices	1203.8	8.7.3.31	
Change in type of control (motion control)	1203.8	8.7.3.31.6	
Change in type of operation	1203.8	8.7.3.31.7	
Emergency operations and signaling	1203.8h	8.7.3.31.8	
Auxiliary power lowering unit		8.7.3.31.9	

to the applicable requirements in this edition of the Code.

8.10.2.2.2(bb) Car and Counterweight Safeties

8.10.2.2.2(bb)(1) The test on the governor overspeed switch is conducted to verify that the governor overspeed switch does trip in the up direction at a speed not greater than the governor tripping speed in the down direction as specified by 2.18.4.2.4.

8.10.2.2.2(bb)(2) Conducting the Type A safety test by manually tripping the governor at the rated speed facilitates testing without having to temporarily alter the

controls or brake system to achieve overspeeding the car up to the governor tripping speed. Since Type A safeties can only be used for rated speeds up to 150 ft/min, it is sometimes very difficult to cause them to overspeed. Also, this requires that the governors have to be arranged to be tripped by hand.

Running the safety test at rated speed verifies the operability of the safety and minimizes the potential for damage and unnecessary wear to the equipment. The governor operation is checked separately at governor tripping speed and recalibrated if necessary

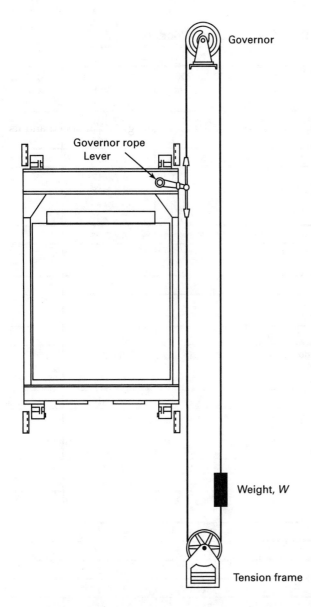

Diagram 8.10.2.2.2(bb)(2) Inertia Application for Type A Safety Device Location of Test Weight

[8.10.2.2.2(bb)(1)(b)]. There is no requirement for governor rope pull-through on Type A and Type C Safeties; therefore, there is no requirement for such a test.

The Type A safety is required to apply instantaneously and independent of governor activation. A weight is used in making this test and is positioned as shown in Diagram 8.10.2.2.2(bb)(2). The amount of weight, its location, and the actual test procedure are dependent on the governor and safety design as well as the travel of the elevator. The governor manufacturer should be consulted for the test criteria.

8.10.2.2.2(bb)(4)(b) If the switch on the car operated by the car safety mechanism is jumped out, the safety slide will be affected. If the slide is then adjusted to meet the Code requirements and a safety stop is implemented

during normal use, the slide distance could dramatically change and result in an abrupt or excessively long stopping distance.

The reason for requiring the governor overspeed switch to be inoperative during the overspeed test is to assure that the driving machine continues to be powered until the safety applies. The reason for requiring the car safety switch to be temporarily adjusted to open as close as possible to the position at which the safety mechanism is in the fully applied position is to assure that the brake does not apply any stopping force to the car, but rather that the safety brings the car to a stop.

On Type B wedge clamp safeties, the drum has to unwind considerably before the wedges are fully engaged. If the switch tripped immediately, the driving machine brake would retard the car, limited by the available traction, before the safeties were fully applied. Keeping the driving machine running during the test does not affect the sliding distances; however, more slack rope develops and the counterweight fallback is more severe. The fallback can release the safeties. When the driving machine continues to run after the test, traction generally brakes. Shutting the driving machine off during the test reduces the fallback and minimizes the potential for equipment damage. The requirement assures that the safety provides the retardation and stopping unassisted by the brake.

8.10.2.2.2(o) The test is performed to confirm compliance with 2.16.8, which requires the car to "safely lower, stop, and hold" 125% of the rated load on passenger and freight elevators authorized to carry passengers. There is no requirement to move the car in the up direction with 125% of the rated load. The upper terminal stopping device is tested with rated load.

8.10.2.2.2(v)(3) This requirement specifies the only acceptable methods of verifying the traction limits. Other methods, such as, performing a high-speed electrical stop, are not acceptable, for the following reasons:

(a) Once traction is lost, the car will continue to move uncontrolled in the hoistway until its motion is arrested due to the laws of motion. This is a fundamentally dangerous condition, since the elevator system is decoupled from the motion control system.

(b) Once traction is lost, the drive sheave will stop, and an accelerated drive sheave wear can be induced over the region of the sheave defined by the arc of contact, which is less than the full circumference of the sheave. As a result of the test, non-uniform circumferential wear of the drive sheave will be induced depending on the distance that the car moves with slipping ropes. This has an adverse effect on the elevator performance. Non-uniform wear of the drive sheave can set up an out-of-round condition, which will induce a vibration source at the driver, and will be transmitted to the car through the hoist ropes.

(c) Depending on the load in the car, position in the

hoistway, and direction of car motion when the high-speed stop is induced, an uncontrolled high-speed car could run into the terminals, resulting in a collision:

(1) A descending car with rated load can be run uncontrolled into the pit at a speed far in excess of that permitted when reduced stroke buffers are used, resulting in significant equipment damage.

(2) An empty or lightly-loaded ascending car can be run past the top terminal landing, where it will run uncontrolled into the building overhead structure, resulting in significant equipment damage.

8.10.2.2.5(c) It is not the intent of the inspection test requirements to test and verify the buffer design. This must be witnessed by an approved testing laboratory at the time buffer-type tests are being conducted. Since oil buffers are tested to meet the requirements of the engineering tests and type tests specified in 8.3, the only purpose of the field acceptance tests is to demonstrate: the plunger-return capabilities of the buffer; and the ability to retard and stop the descending mass without damage.

This is as much of a test of the car and counterweight as it is of the buffer.

The purpose of the acceptance tests is to ensure that the buffer, whose type was originally tested to assure compliance with 8.3, will safely retard and stop the descending mass, i.e., either car or counterweight, within the limits of the applicable requirements in 2.22.4.

ASME A17.1, Inquiry 78-60 stated that a fundamental safety characteristic of traction elevators is the ability to lose traction in the event that either the car or counterweight over-travels at the lower terminal and compresses the buffer; however, this is not a necessary condition for determining compliance with the buffer test requirements.

All traction elevator systems will lose traction once the required traction relation exceeds the available traction between the rope and drive sheave groove. This condition will occur when the tension in the suspension ropes leading to the mass, which had been stopped by the buffer, is reduced sufficiently to change the ratio of rope tensions on both sides of the drive sheave. However, it is difficult to accurately predict when the loss of traction will occur, if at all during the short interval of the test, due to possible embedment of the crown wires in the surface of the drive sheave grooves. In addition to those points, variations in designs and actual condition of the ropes and drive sheave grooves pose other variables.

If it is necessary to relocate the final limit switch in order to achieve full compression of the buffer, then such relocation should be performed but the switch should remain operative as required by the Code. The relocation of the switch need only be of sufficient distance to obtain full compression and yet still be capable of removing the power automatically from the driving machine motor and brake when activated by the cam. The intent

of the buffer inspection test requirements is to ensure that the buffers function properly under normal operating conditions (rated load at rated speed) and that the plunger will return after full compression.

The buffer nameplate can be checked at the time of inspection to verify whether or not the proper certified buffer has been installed for the given car speed and its rated load.

The fact that the final limit switch may be activated just prior to full compression, and it deactivates the drive motor and its brake while buffer compression is taking place, does not significantly alter the buffer performance. The inherent electrical and mechanical inertia of the system will not permit instantaneous reaction from either the motor or the brake.

In response to the specific questions raised in connection with relocating or jumping out the final limit switch and the necessity of a buffer switch, the ASME A17 Committee concluded that:

(a) The final limit switch should be relocated, when necessary, to obtain full compression of the buffer during the acceptance buffer tests. The final limit switch must never be jumped out. The A17.1 Code correctly mandates that the final limit switch remain operative even when the switch is temporarily relocated for a buffer test.

(b) A buffer switch is not required except for buffers with gas spring return oil buffers [2.22.4.5(c)]. The acceptance test, the annual buffer inspection, and the 5 year buffer test requirements provide the necessary confirmation that the buffers function as intended and as required by the Code.

8.10.2.2.5(c)(4) The buffer and emergency terminal speed-limiting device are not tested simultaneously. Testing both devices at the same time may result in an inadequate test of the buffer as the emergency terminal speed-limiting device may slow down the car too much or may not function as required. The buffer should be tested by adjusting the speed of the car to the maximum striking speed of the buffer. Then the emergency terminal speed-limiting device should be tested to assure that it will reduce the speed of the car in the required distance for that particular installation.

8.10.2.2.5(c)(6) If full compression of the buffer can be accomplished without moving the final terminal stopping device, then it is not necessary to move it. If full compression cannot be obtained, the final terminal stopping device must be relocated in order to obtain full compression. The final stopping device is not rendered inoperative for this test since it would increase the risk of damage to the driving machine by continuing to apply power after the car comes to rest.

8.10.2.3 Inspection and Test Requirements for Altered Installations. When an alteration is made to an electric elevator, (see 8.7), an inspection and/or test to determine Code compliance may be required. All requirements for inspections and/or tests for altered electric elevators are contained in 8.10.2.3. ASME A17.1 Code requirements for inspections and tests following a repair or replacement are specified in 8.6.

8.10.3 Acceptance Inspection and Tests of Hydraulic Elevators

8.10.3.2 Inspection and Test Requirements for New Installations

8.10.3.2.3(y) Requirement 3.18.1.2.3 allows ropes to have a clear coating. A visual inspection of a rope with a clear coating may not reveal all broken strands. Therefore, the requirement for a magnetic flux test capable of finding such broken strands has been added.

The use of pistons and the necessity to have seals to prevent leaks obviously will conceal one end of the indirect coupling means (ropes). Therefore, a requirement to inspect, even if disassembly is required, has been added to the Code so that the end attached to the piston is inspected.

8.10.3.2.6 Firefighters' Emergency Operation. Determination to test all the smoke detectors or a sampling of the smoke detectors is left to the judgment of the authority having jurisdiction. See also NFPA 72 for fire alarm and smoke detector testing requirements.

8.10.3.3 Inspection and Test Requirements for Altered Installations.
When an alteration is made to a hydraulic elevator (see 8.7), an inspection and/or test to determine Code compliance may be required. All requirements for inspections and/or tests for altered hydraulic elevators are contained in 8.10.3.3. ASME A17.1 Code requirements for inspections and tests following a repair or replacement are specified in 8.6.

8.10.4 Acceptance Inspection and Tests of Escalators and Moving Walks

The referenced Item and Division are included as a convenience to the user and refer to the Inspectors' Manual for Escalators and Moving Walks, ASME A17.2.3. The Inspectors' Manual contains detailed information about the referenced item to be inspected and/or tested.

8.10.4.1 Inspection and Test Requirements for New Installations

8.10.4.1.1 External Inspection and Tests

8.10.4.1.1(n) Requirement 6.1.3.3.2 requires that balustrades be designed to resist the simultaneous application of a static lateral force of 584 N/m or 40 lbf/ft and a vertical load of 730 N/m or 50 lbf/ft, both applied to the top of the handrail stand. This is a design requirement and neither the Code nor the Inspectors' Manual requires testing to verify compliance with this requirements. Conformance with design requirements may be verified by examining the manufacturer's design criteria.

There is a concern with respect to the method by which a balustrade strength test could be conducted.

From a safety standpoint, physically balancing and supporting the specified loads could be hazardous.

8.10.4.2 Inspection and Test Requirements for Altered Installations.
When an alteration is made to an escalator or moving walk (see 8.7), an inspection and/or test to determine Code compliance may be required. All requirements for inspection and/or tests for escalators and moving walks are contained in 8.10.4.2. The ASME A17.1 Code requirements for inspections and tests following a repair or replacement are specified in 5.6.

8.10.5 Acceptance Inspection and Test of Other Equipment

None of the ASME Inspector's Manuals details any of the inspection and test procedures for the equipment addressed in 8.10.5. As most of the Code requirements for this equipment are similar to the requirements for electric elevators (Part 2) or hydraulic elevators (Part 3), the applicable Inspectors' Manual should be referenced for guidance.

SECTION 8.11
PERIODIC INSPECTIONS AND TESTS

NOTE: Throughout Section 8.11, references are made to other requirements in this Code. To gain a complete understanding of a requirement, the reader should review the Handbook commentary on all referenced requirements.

All periodic inspection and testing requirements for new and altered equipment can be found in this Section. All acceptance tests can be found in Section 8.10. The purpose of the inspections and tests required by this Code is to assure that equipment conforms to the requirements in 8.11.1.2. In order to fully understand and comply with these requirements, also consult the Inspectors' Manual for Electric Elevators, ASME A17.2.1; the Inspectors' Manual for Hydraulic Elevators, ASME A17.2.2; and the Inspectors' Manual for Escalators and Moving Walks, ASME A17.2.3. The details for an inspection and/or test will be found in those documents. Many requirements also provide specific references to items in the applicable Inspectors' Manual.

NOTE: The three separated Inspectors' Manuals were combined and published on December 31, 2001 as ASME A17.2, Guide for Inspection of Elevators, Escalators, and Moving Walks.

Another Handbook that is invaluable for making inspections and performing test is the Elevator Industry Inspection Handbook, by Zack McCain, Jr. See Handbook commentary on Part 9 for procurement information.

8.11.1 General Requirements Acceptance Inspections and Tests

8.11.1.1 Persons Authorized to Make Inspections and Tests. ASME A17.1, Safety Code for Elevators and Escalators, covers design, construction, operation, inspection, testing, maintenance, alteration, and repair. ASME A17.2 Inspectors' Manuals supplments this Code by providing guidelines for the inspection and testing of the equipment. None of the standards, however, cover the qualifications of inspection personnel. The duties of an inspector are only covered briefly in the Introduction to the Inspectors' Manual.

The excellent safety record of elevators, escalators, and related equipment has been maintained, in part, by quality field inspections and tests. However, advancing technology and safety requirements have highlighted the need for establishing uniform criteria for the persons performing these inspections. The quality of inspections, of course, depends on the competence of the inspector. The Standard for the Qualification of Elevator Inspectors, ASME QEI-1, was published by ASME to promote uniform quality in inspections.

(a) The QEI standard is intended to establish uniform criteria, which will aid in the following:

(1) qualifying and training of inspection personnel for government agencies, insurance companies, and elevator companies;

(2) the accreditation of inspection agencies; and

(3) reducing the liability exposure of insurance companies, manufacturers, installers, and building owners.

It is also intended to serve as a guideline on which certification is based by detailing the expertise necessary in performing inspections.

(b) The QEI Standard covers requirements for the qualification and duties of personnel engaged in the inspection and testing of equipment within the scope of the ASME A17.1, Safety Code for Elevators and Escalators. The QEI Standard applies to persons typically employed by (but not limited to) the following:

(1) jurisdictional authorities;

(2) independent inspection agencies and elevator consultants;

(3) insurers of the equipment;

(4) manufacturers, installers, and maintainers of the equipment; and

(5) building owners and managers.

(c) The QEI Standard does not cover personnel engaged in engineering and type testing as covered in 8.3 of the ASME A17.1 Code, including inspection by laboratories in association with those tests. Under the QEI program, ASME will accredit organizations that certify inspectors who meet the requirements of ASME QEI-1.

The ASME A17.1 Code requires inspectors and inspector supervisors to be certified by an organization accredited by the ASME, except in Canada. The Code recognizes that inspections and tests should only be performed by qualified personnel. The QEI accreditation program is a recognized means of assuring that inspection personnel are qualified. For more information, or for a list of ASME accredited organizations, contact:

Secretary, QEI Committee
American Society of Mechanical Engineers
Three Park Avenue
New York, New York 10016-5990

8.11.1.1.1 Periodic Inspections. A periodic inspection is the examination and operation of equipment at specified intervals by an inspector to check for compliance with the applicable Code requirements (8.11.1.2).

The authority having jurisdiction (AHJ) is defined (see 1.3) as the organization, office, or individual responsible for enforcement of this Code. Where compliance with this Code has been mandated by legislation or regulation, the "authority having jurisdiction" is the regulatory authority.

In Canada the AHJ is often referred to as the regulatory authority which is defined (see 1.3) as the person or organization responsible for the administration and enforcement of the applicable legislation or regulation governing the design, construction, installation, operation, inspection, testing, maintenance, or alteration of equipment covered by this Code. This requirement mandates that the AHJ make all periodic inspections or designate the person who will make the inspections. The AHJ can stipulate that the inspections be made by a third party inspector or the installing company. Whoever makes the inspection must be QEI certified (8.11.1.1.3), except in Canada.

8.11.1.1.2 Periodic Tests. A periodic test is defined as a group of tests performed at common time intervals required by the authority having jurisdiction (AHJ).

The inspector does not conduct the test; they only witness the test. The person performing the test does not have to be QEI certified; however, the test must be witnessed by an inspector that is QEI certified. See Handbook commentary on 8.11.1.1.

The periodic inspections and tests are performed by persons qualified to perform such services. However, no guidance is given to determine the qualifications, as qualifications of personnel are not within the Scope of A17.1.

A record of the periodic test should be maintained by the authority having jurisdiction. Sample test records are shown in Charts 8.11.1.1.2(a), (b), and (c).

Chart 8.11.1.1.2(a) Sample Periodic Test Record for Traction Elevators

Type of Test	Car and Counterweight — Safety	Car and Counterweight — Governor	Car and Counterweight — Safety	Car and Counterweight — Governor	Brake	Emergency Terminal Speed-Limiting Device	Standby (Emergency) Power Operation	Firefighters' Service	Power Door Operation	Leveling Zone and Leveling Speed	Inner Landing Zone (Static)	Oil Buffer	Final and Normal Terminal Stopping Devices	Door Closing Force	Broken Rope, Tape or Chain Switch
Catagory	One	One	Five	Five	Five	Five	One	One	Five	Five	Five	One / Five	One	One	One
ASME A17.1 Requirement	8.11.2.2.2	8.11.2.2.3	8.11.2.3.1	8.11.2.3.2	8.11.2.3.4	8.11.2.3.5	8.11.2.2.7	8.11.2.2.6	8.11.2.3.7	8.11.2.3.8	8.11.2.3.9	8.11.2.2.1 / 8.11.2.3.3	8.11.2.2.5	8.11.2.2.8	8.11.2.2.8
Elevator Number															

DATE OF TEST

Chart 8.11.1.1.2(b)   Sample Periodic Test Record for Direct-Plunger Hydraulic Elelvators
(Without Safeties, Governors, and Oil Buffers)

Type of Test	Hydraulic Cylinder Leak Test	Hydraulic Relief Valve	Flexible Hose and Fittings	Firefighters' Service Operation	Pressure Switch	Emergency Terminal Speed Limiting Device	Power Operation of Door System	Standby (Emergency) Power Operation	Normal Terminal Stopping Device
Catagory	one	one	one	one	one	one	one	one	one
ASME A17.1 Requirement	8.11.3.2.2	8.11.3.2.1	8.11.3.2.4	8.11.3.2.3(e)	8.11.3.2.5	8.11.3.2.3(h)	8.11.3.2.3(g)	8.11.3.2.3(f)	8.11.3.2.3(a)
Elevator Number	DATE OF TEST								

8.11.1.2 Applicability of Inspection and Test Requirements. Periodic and routine inspections and tests are performed on existing equipment. An existing installation is one that has been completed or is under construction prior to the effective date of this Code.

The periodic inspection and test is performed to check for compliance with the Code in effect at the time of installation and/or alteration. This requirement for maintenance (8.6.1.1.2) and periodic inspections and test (8.11.1.2) are intended to assure the level of safety is equal to what was required when the equipment was installed and/or altered. This requirement does not refer to the ASME A17.1 Code in effect at the time of installation and/or alteration. It refers to the Code the authority having jurisdiction was enforcing at the time. That may be a version of ASME A17.1 with or without modification or a locally developed elevator code. If no code was in effect at the time of installation and/or alteration the ASME A17.1 Code in effect (see Preface to ASME A17.1 for the effective date of editions, supplements and addenda) could be used for guidance.

For example, if at the time of installation the equipment was required to conform to A17.1–1955, then that is the Code with which the equipment must comply. If at the time of installation the equipment was required to conform with A17.1–1931 and in 1978 the brake on an existing driving machine was replaced, then the equipment is checked for conformance with A17.1–1931 and A17.1–1978, Rules 208.8 and 210.8. The reason for checking for conformance with A17.1–1978, Rules 208.8 and 210.8, is that at the time the new brake was installed, it was considered by the Code to be a major alteration (see A17.1–1978, Rule 1200.2).

If the equipment is required to conform to the requirements of ASME A17.3 Safety Code for Existing Elevators and Escalators, compliance with that document must also be confirmed during routine and periodic inspection. For example, an elevator installed under the A17.1–1955 edition of the Code it would have to comply with the requirements of that Code and any additional or more stringent requirements in the A17.3 Code, such as firefighters' service.

8.11.1.3 Periodic Inspection and Test Frequencies. Inspection and testing frequencies are no longer established by ASME A17.1. However, the AHJ is required by 8.11.1.2 to establish an inspection and testing frequency. See Chart 8.11.1.3 for recommended periodic inspection and testing frequencies.

Any equipment covered by this Code that is not placed out of service must conform to all of the requirements of 8.11 for periodic inspections and tests. If an installation is out of service during a period when a routine and/or periodic inspection and test are required, then such inspection and test must be performed before the installation is returned to service. An example is an elevator out of service during the period a periodic inspection category one and category five tests were due. Before that elevator can be returned to service, the periodic inspection and category one and five tests must be performed.

8.11.1.5 Making Safety Devices Inoperative or Ineffective. Making safety devices inoperative or ineffective is permitted only after the installation is removed from public use. The reader is cautioned that jumpers unintentionally left on safety circuits and/or improperly used have been the cause of fatalities and serious injury. The reader should be familiar with and adhere to the jumper procedures in the latest edition of the Elevator Industry Field Employees' Safety Handbook.

8.11.2 Periodic Inspection and Tests of Electric Elevators

The referenced Item and Division are included as a convenience to the user and refer to the Inspectors' Manual for Electric Elevators ASME A17.2.1. The Inspectors' Manual contains detailed information about the referenced Item to be inspected and/or tested.

Chart 8.11.1.1.2(c) Sample Periodic Test Record for Escalators

Type of Test	Speed Governor (1955)	Broken Drive Chain Device (1955)	Broken Step Chain Device (1955)	Machine Space Stop Switches (1978)	Skirt Obstruction Device (1970*)	Missing Step Device (1989)	Brake Torque Test (1983)	Step Level Device (1990)	Reversal Stop Switch (1974)	Step Upthrust Device (1980)	Stop Handrail Device (1988)	Step Chain and Truss Inspect	Disconnect Motor Device (1983)	Performance Index (2000)	Clearance Step/Skirt (2000)
Catagory	One	One	One	One	Periodic	One	One	One	One	One	One	One	One	One	One
ASME A17.1 Requirement	8.11.4.2.5	8.11.4.2.6	8.11.4.2.8	8.11.4.2.2	8.11.4.1(h)	8.11.4.2.10	8.11.4.2.4	8.11.4.2.11	8.11.4.2.7	8.11.4.2.9	8.11.4.2.13	8.11.4.2.18	8.11.4.2.16	8.11.4.2.19	8.11.4.2.20
Elevator Number															

DATE OF TEST

Chart 8.11.1.3 Recommended Inspection and Test Intervals in "Months"

| Reference Section | Equipment Type | Periodic Inspections | | Periodic Tests | | | | | | | |
| | | Requirement | Interval | Category One | | Category Three | | Category Five | | | |
				Requirement	Interval	Requirement	Interval	Requirement	Interval		
8.11.2	Electric elevators	8.11.2.1	6	8.11.2.2	12	N/A	N/A	8.11.2.3	60		
8.11.3	Hydraulic elevators	8.11.3.1	6	8.11.3.2	12	8.11.3.3	36	8.11.3.4	60		
8.11.4	Escalators and moving walks	8.11.4.1	6	8.11.4.2	12	N/A	N/A	N/A	N/A		
8.11.5.1	Sidewalk elevators	8.11.2.1, 8.11.3.1	6	8.11.2.2, 8.11.3.2	12	8.11.3.3	36	8.11.2.3, 8.11.3.4	60		
8.11.5.2	Private residence elevators	8.11.2.1, 8.11.3.1	12	8.11.2.2, 8.11.3.2	12	8.11.3.3	36	8.11.2.3, 8.11.3.4	60		
8.11.5.3	Hand elevators	8.11.2.1	6	8.11.2.2	12	N/A	N/A	8.11.2.3, 8.11.3.4	60		
8.11.5.4	Dumbwaiters	8.11.2.1, 8.11.3.1	12	8.11.2.2, 8.11.3.2	12	8.11.3.3	36	8.11.2.3, 8.11.3.4	60		
8.11.5.5	Material lifts and dumbwaiters with automatic transfer devices	8.11.2.1, 8.11.3.1	12	8.11.2.2, 8.11.3.2	12	8.11.3.3	36	8.11.2.3, 8.11.3.4	60		
8.11.5.6	Special purpose personnel elevators	8.11.2.1, 8.11.3.1	6	8.11.2.2, 8.11.3.2	12	8.11.3.3	36	8.11.2.3, 8.11.3.4	60		
8.11.5.7	Inclined elevators	8.11.2.1, 8.11.3.1	6	8.11.2.2, 8.11.3.2	12	8.11.3.3	36	8.11.2.3, 8.11.3.4	60		
8.11.5.8	Shipboard elevators	8.11.2.1, 8.11.3.1	6	8.11.2.2, 8.11.3.2	12	8.11.3.3	36	8.11.2.3, 8.11.3.4	60		
8.11.5.9	Screw-column elevators	8.11.2.1, 8.11.3.1	6	8.11.2.2, 8.11.3.2	12	8.11.3.3	36	8.11.2.3, 8.11.3.4	60		
8.11.5.10	Rooftop elevators	8.11.2.1, 8.11.3.1	6	8.11.2.2, 8.11.3.2	12	8.11.3.3	36	8.11.2.3, 8.11.3.4	60		
8.11.5.12	Limited-use/limited-application elevators	8.11.2.1, 8.11.3.1	6	8.11.2.2, 8.11.3.2	12	8.11.3.3	36	8.11.2.3, 8.11.3.4	60		
8.11.5.13	Elevators used for construction	8.11.2.1, 8.11.3.1	3	8.11.2.2, 8.11.3.2	12	8.11.3.3	36	8.11.2.3, 8.11.3.4	60		

GENERAL NOTE: The intervals specified in this Table are recommended for periodic tests and inspections. Factors such as the environment, frequency and type of usage, quality of maintenance, etc., related to the equipment should be taken into account by the authority having jurisdiction prior to establishing the inspection and test intervals.

8.11.2.3 Periodic Test Requirements — Category Five

8.11.2.3.1 Car and Counterweight Safeties. The running of the category five full load safety test must be viewed in the context of a safety system whose operability and compliance with the applicable requirements were verified at the original acceptance test, followed by the category one tests whose sole purpose is to ensure the operability of the safety system at the slowest operating speed.

Conducting the category five safety test by manually tripping the governor at the rated speed facilitates testing without having to temporarily alter the controls or brake system to achieve overspeeding the car up to the governor tripping speed.

Running the category five safety test at rated speed verifies the operability of the safety and minimizes the potential for damage and unnecessary wear to the equipment while at the same time ensuring that the safety will develop average retardations consistent with the average retardations, which would have been generated at a full load, governor overspeed safety test. This latter point is verified from the comparison of the actual stopping distance of the values allowed at the rated speed, which are specified in the A17.2.1, Table 2.29.2(b), (c), or (d).

The governor operation is checked separately at governor tripping speed and recalibrated if necessary.

With each successive full-speed safety test, some wear occurs on the governor during the stopping slide. Naturally, the higher the car speed, the greater the slide and wear. In addition to the normal test, the safety may be applied a number of other times. Also, a new rope is manufactured to nominal diameter or on the positive side of the tolerance. As the rope stretches and crown progresses, the diameter is reduced. Eventually the combination of governor wear and rope diameter reduction may result in insufficient gripping force to set the safety.

The only way to ensure a safe margin for positive safety application is to test the governor and releasing carrier pull-through or pullout force on Type B Safeties. There is no requirement for governor rope pull-through on Type A and Type C Safeties; therefore, there is no requirement for such a test.

8.11.2.3.3 Oil Buffers

8.11.2.3.3(a) and (b) The buffer and emergency terminal speed-limiting device are not tested simultaneously. Testing both devices at the same time may result in an inadequate test of the buffer as the emergency terminal speed-limiting device may slow down the car too much or may not function as required. The buffer should be tested by adjusting the speed of the car to the maximum striking speed of the buffer. Then the emergency terminal speed-limiting device should be tested to assure that it will reduce the speed of the car in the required distance for that particular installation.

8.11.2.3.3(d) If full compression of the buffer can be accomplished without moving the final terminal stopping device, then it is not necessary to move it. If full compression cannot be obtained, the final terminal stopping device must be relocated in order to obtain full compression.

The final stopping device is not rendered inoperative for this test since it would increase the risk of damage to the driving machine by continuing to apply power after the car comes to rest.

8.11.2.3.10 This requirement specifies the only acceptable methods of verifying the traction limits. Other methods, such as, performing a high-speed electrical stop, are not acceptable, for the following reasons:

(a) Once traction is lost, the car will continue to move uncontrolled in the hoistway until its motion is arrested due to the laws of motion. This is a fundamentally dangerous condition, since the elevator system is decoupled from the motion control system.

(b) Once traction is lost, the drive sheave will stop, and an accelerated drive sheave wear can be induced over the region of the sheave defined by the arc of contact, which is less than the full circumference of the sheave. As a result of the test, non-uniform circumferential wear of the drive sheave will be induced depending on the distance that the car moves with slipping ropes. This has an adverse effect on the elevator performance. Non-uniform wear of the drive sheave can set up an out-of-round condition, which will induce a vibration source at the driver, and will be transmitted to the car through the hoist ropes.

(c) Depending on the load in the car, position in the hoistway, and direction of car motion when the high-speed stop is induced, an uncontrolled high-speed car could run into the terminals, resulting in a collision.

(1) A descending car with rated load can be run uncontrolled into the pit at a speed far in excess of that permitted when reduced stroke buffers are used, resulting in significant equipment damage.

(2) An empty or lightly-loaded ascending car can be run past the top terminal landing, where it will run uncontrolled into the building overhead structure, resulting in significant equipment damage.

8.11.3 Periodic Inspection and Tests of Hydraulic Elevators

The referenced Item and Division are included as a convenience to the user and refer to the Inspectors' Manual for Electric Elevators, ASME A17.2.1. The Inspectors' Manual contains detailed information about the referenced Item to be inspected and/or tested.

8.11.3.2 Periodic Test Requirements — Category One

8.11.3.2.1 Relief Valve Setting. The Code design requirement in ASME A17.1b–1995 and earlier editions for the relief valve is for initial opening at a pressure not greater than 125% of the working pressure and relieve the full capacity of the pump at not greater than 20% above the pressure that the valve opened. Therefore, if the valve opened at 125% of working pressure, the pressure could rise to 150% of working pressure while relieving the full capacity of the pump. Since it is not practical to determine the pressure that the valve first opened, the field test is specified to verify that the pressure does not exceed the highest pressure allowed by the Code.

This method of testing the relief valve also verifies the integrity of the entire system and does not require the closing of any manual valve. This is especially important where a manual valve is installed in the pit. A manual shutoff valve is not required in the pit but one is often installed to facilitate jack servicing. Pit valves should never be operated unless blocking is placed under the car to prevent it from falling if the valve fails.

8.11.3.2.2 Cylinders. The 1971 through 1978 editions of A17.1 required that a rated load be placed on the car for a minimum of 2 h. The 1978 and later editions of A17.1 required this test be performed with no load for 15 min. The reason for this change is based on the fact that in hydraulic elevator design, the working or relief pressure is approximately twice the empty car pressure and since the volume of fluid lost through a fixed opening is a function of the square root of pressure, then it follows that loss of fluid at rated load would be less than $1\frac{1}{2}$ times what it would be with an empty car and loss of fluid could be detected at this reduced pressure and time.

8.11.3.2.4 Flexible Hose and Fitting Assemblies. This requirement was revised in 1985 to increase the test frequency to yearly, now category one, and have it performed at a practical and reasonable pressure, without having to remove the hose for testing.

8.11.3.2.6 Firefighters' Service. Determination to test all the smoke detectors or a sampling of the smoke detectors is left to the judgment of the authority having jurisdiction.

8.11.3.4 Periodic Test Requirements — Category Five. ASME A17.1 Code allows multiple means of initiating the safety mechanism. All provided means to be tested for proper operation.

8.11.3.4.2 A visual inspection of a rope with a clear coating may not reveal all broken strands. Therefore, the requirement for a magnetic flux test capable of finding such broken strands has been added. The use of pistons and the necessity to have seals to prevent leaks obviously will conceal one end of the indirect coupling means (ropes). Therefore, a requirement to inspect, even if disassembly is required, has been added to the Code so that the end attached to the piston is inspected.

8.11.4 Inspection and Tests of Escalators and Moving Walks

The referenced Item and Division are included as a convenience to the user and refer to the Inspectors' Manual for Escalators and Moving Walks, ASME A17.2.3. The Inspectors' Manual contains detailed information about the referenced Item to be inspected and/or tested.

8.11.4.1 Periodic Inspection and Test Requirements
8.11.4.1(o) Requirement 6.1.3.3.2 requires that balustrades to be designed to resist the simultaneous application of a static lateral force of 584 N/m or 40 lbf/ft and a vertical load of 730 N/m or 50 lbf/ft, both applied to the top of the handrail stand. This is a design requirement and neither the Code nor the Inspectors' Manual requires testing to verify compliance with this requirements. Conformance with design requirements may be verified by examining the manufacturer's design criteria.

There is a concern with respect to the method by which a balustrade strength test could be conducted. From a safety standpoint, physically balancing and supporting the specified loads could be hazardous.

8.11.5 Acceptance Inspection and Test of Other Equipment

None of the ASME Inspector's Manuals details any of the inspection and test procedures for the equipment addressed in 8.11.5. As most of the Code requirements for this equipment are similar to the requirements for electric elevators (Part 2) or hydraulic elevators (Part 3), the applicable Inspectors' Manual should be referenced for guidance.

8.11.5.2 Private Residence Elevators and Lifts. Private residence elevators and lifts are required to be inspected and tested at the time of installation (8.10.5.2). This requirement recommends that the private residence elevators be inspected and tested at the same intervals that apply to electric and hydraulic elevators. AHJ's generally do not have the authority to enter a private residence after construction is completed. Thus, the responsibility to see that the equipment is tested and inspected is placed in the hands of the owner of the equipment.

8.11.5.13 Elevators Used for Construction

8.11.5.13.2 Periodic Test Requirements — Category One. A more stringent inspection and test program is recommended (see Diagram 8.11.1.3) for elevators used for construction due to the nature of the environment in which this equipment operates. If an elevator used for construction is moved to serve additional levels, has "jumped," or moved to another location, it should be treated as a new installation.

PART 9
REFERENCE CODES, STANDARDS, AND SPECIFICATIONS

LOCATING CODES AND STANDARDS

In-Print Codes and Standards

Locating codes and standards can be frustrating and time consuming. However, in recent years it has become increasingly easier to locate and purchase documents using the World Wide Web. Web addresses are given for all the sources of information in Part 9 of this Handbook. Most codes and standards are now available electronically (CD ROM or on the Web) or in a printed format. Purchasers should inquire as to the formats available when ordering.

Mandatory codes and standards can usually be obtained from the authority promulgating it or they can provide you with procurement information. A telephone directory including hot links to state and provincial web sites for elevator code authorities in the 50 States, and the Canadian Provinces is available from:

> Edward A. Donoghue Associates, Inc.
> Code and Safety Consultants
> http://www.eadai.com

Consensus codes are generally written as mandatory while standards are voluntary. Neither is enforceable unless adopted by the authority having jurisdiction. In the vertical transportation industry, the most widely used codes and standards documents are the Safety Codes for Elevators and Escalators, ASME A17.1; the Inspectors' Manual for Electric Elevators, ASME A17.2.1; Inspectors' Manual for Hydraulic Elevators ASME A17.2.2; Inspectors' Manual for Escalators and Moving Walks ASME A17.2.3; the Safety Code for Existing Elevators and Escalators, ASME A17.3; Standard for Elevator and Escalator Electrical Equipment, CAN/CSA-B44.1/ ASME A17.5-M; Guide for Emergency Evacuation of Passengers from Stalled Elevator Cars, ASME A17.4, and Safety Standard for Platform Lifts and Stairway Chairlifts, ASME A18.1. Additionally, there is the Handbook-A17.1 Safety Code for Elevators and Escalators by Edward A. Donoghue, CPCA, which clarifies and explains code requirements. All of these documents can be obtained from:

The American Society of Mechanical Engineers (ASME)
Order Department
22 Law Drive/Box 2300
Fairfield, NJ 07007-2300
800-THE-ASME
[in NJ (201) 882-1167]
http://www.asme.org

ASME documents are also available from Elevator World, whose address is listed later in this commentary.

ASME A17.1, Safety Code for Elevators and Escalators, contains many references to other nationally recognized codes and standards. To help the users of the ASME A17.1 Code, procurement information for those referenced codes and standards, is contained in Part 9 of the ASME A17.1 Code, and A17.1 Handbook. These documents identify 98% of the codes and standards, which will normally be encountered by the vertical transportation industry. Additional codes and standards reference material and ordering information can be found in this Part of the Handbook.

How to locate the remaining 2%? Among the best-known organizations is the American National Standards Institute (ANSI). Documents identified as an American National Standard or with the ANSI initials or logo in its heading.

Request for electronic delivery of standards — either individual sale via ANSI's Electronic Standards Store (www.ansi.org) or site license access via networking agreements should be directed to:

American National Standards Institute (ANSI)
25 West 43rd Street
New York, NY 10036
(212) 642-4900
http://www.ansi.org

Request for printed copies of standards — should be directed to:

Global Engineering Documents
15 Inverness Way East
Englewood, CO 80112-2776
(800) 854-7179
http://global.ihs.com

Other major standard writing organizations, in addition to those already mentioned, are:

American Society for Testing and Materials
  (ASTM)
100 Barr Harbor Drive
West Conshohocken, PA 19428-2959
(610) 832-9585
http://www.astm.org

National Fire Protection Association (NFPA)
One Batterymarch Park
Quincy, MA 02269-9101
(617) 770-3000 in Mass.
(800) 344-3555
http://www.nfpa.org

Underwriters Laboratories, Inc. (UL)
333 Pfingsten Road
Northbrook, IL 60062-2096
(847) 272-8800
http://www.ul.com

UL documents are available for purchase from:

comm 2000
1418 Brook Drive
Downers Grove, IL 60515
(888) 853-3503
http://www.comm-2000.com

Starting in 1992 the United States building transportation industry has been concerned with the American With Disabilities Act (ADA) and the regulations known as ADAAG, available from:

United States Architectural and
  Transportation Barriers Compliance
  Board (ATBCB)
1331 F Street, NW, Suite 1000
Washington, DC 20004-1111
(202) 272-5434
http://www.access-board.gov

ADAAG is not enforced at the local level. Local building codes usually require vertical transportation to comply with the American National Standard for Accessible and Usable Buildings and Facilities, ICC/ANSI A117.1-1998 previously CABO/ANSI A117.1–1992. These documents are available from the model building code organizations. Procurement information is listed elsewhere in this Part of the Handbook. If Federal funding is associated with the building or a tenant in the building, the Uniform Federal Accessible Standard (UFAS) may be applicable. UFAS is available from the ATBCB whose address is listed above.

A comprehensive analysis of the accessibility regulations and where they apply is contained in "ADA and Building Transportation, A Handbook on Accessibility Regulations for Elevators, Wheelchair Lifts and Escalators" by Edward A. Donoghue, CPCA. This handbook is available from:

Elevator World, Inc.
P. O. Box 6507
Mobile, AL 36660
(205) 479-4514
http://www.elevator-world.com

Elevator World also publishes the NEII Building Transportation Standards and Guidelines, NEII-1-2000. It is an invaluable reference for dimensional, performance, electrical, application, design, and evaluation of building transportation systems. NEII-1-2000 contains the industry standards and guidelines needed by architects, consultants, manufacturers, contractors, developers, building owners/operators and code officials.

Most elevator installations must also comply with the building code. Building codes in the United States have generally been based on one of the following three model building codes:

National Building Code (NBC)
Building Officials and Code Administrators
  International (BOCA)
4051 West Flossmoor Road
Country Club Hills, IL 60477
(312) 799-2300
http://www.bocai.org

Standard Building Code (SBC)
Southern Building Code Congress International
  (SBCCI)
5200 Montclair Road
Birmingham, AL 35213
(205) 591-1853
http://www.sbcci.org

Uniform Building Code (UBC)
International Conference of Building Officials
  (ICBO)
5360 South Workman Mill Road
Whittier, CA 90601
(213) 699-0541
http://www.icbo.org

In 2000, the three model building codes published the International Building Code (IBC), as a replacement for the NBC, SBC, and UBC. For further information contact one of the three model code organizations listed above or:

International Code Council (ICC)
5203 Leesburg Pike
Suite 600
Falls Church, VA 22041
(703) 931-4533
http://www.intlcode.org

The National Institute of Standards and Technology (NIST) formally the National Bureau of Standards (NBS) has a National Center for Standards and Certification Information (NCSCI). NIST-NCSCI has a collection of over 30,000 codes and standards consisting of voluntary standards, state purchasing standards, U. S. government and foreign standards. NIST-NCSCI can be contacted at:

> National Institute of Standards and
> Technology (NIST)
> National Center for Standards and
> Certification Information
> TRF Building, Room A163
> Gaithersburg, MD
> (301) 975-4040
> http://www.nist.gov

A NIST-NCSCI search will result in a list of all applicable standards, the name and address of the issuing organization, the standard number, title, price if known, and the year of publication of the latest revision. There is no charge for a NIST-NCSCI search. They will not supply you with a copy of the code or standard.

Documents of interest to manufactures on the North American continent are the Canadian Safety Code for Elevators, Escalators, Dumbwaiters, Moving Walks and Freight Platform Lifts CAN/CSA-B44-M; Elevating Devices for the Handicapped, CAN/CSA-B355-M; Elevating Devices for the Handicapped in Private Residences, CAN/CSA-B613-M; and Canadian Electrical Code, CSA C22.1. These documents as well as CAN/CSA-B44.1/ASME A17.5-M, can be purchased from:

> Canadian Standards Association (CSA)
> 178 Rexdale Boulevard
> Rexdale (Toronto), ON, Canada M9W 1R3
> (416) 747-4044
> http://www.csa.ca

CSA documents are also available from Elevator World, whose address was listed previously.

Another document is the National Building Code of Canada that can be purchased from:

> National Research Council Canada (NRCC)
> Institute for Research in Construction
> Ottawa, ON, Canada K1A 0R6
> (613) 993-2463
> http://www.irc.nrc.ca

As the world becomes smaller and smaller, international, standards will play an important role. ISO 9000, as an example, is a standard series on quality system registration and related issues that has become very popular. The American National Standards Institute (ANSI) is the United States agent for all standards issued by the International Organization for Standardization

(ISO) and International Electrotechnical Commission (IEC). Global Engineering Documents is the United State source for ISO and IEC documents and is the best source in the United States for standards published by foreign governments and private entities. Readers of Elevator World in other countries should be able to purchase ISO and IEC Standards from their National Standards Organizations. If not, they should contact ISO and IEC in Geneva, Switzerland.

> International Organization for Standards (ISO)
> Case postale 56
> CH-1211 Geneva 20
> Switzerland
> http://www.iso.ch/en/
> ISOOnline.openerpage

The Western European Common Market Standards for elevators and escalators (CEN) can be purchased in the United States from ANSI, Global Engineering Documents, or:

> AFNOR
> Tour Europe-Cedex 7
> 92080 Paris La Defense
> France
> http://www.afnor.fr

The ISO, IEC, CEN, and CAN standards are available in French and English. Be sure to specify at time of ordering.

COMMERCIAL SOURCES

Have the need for a prior edition of the Safety Code for Elevators and Escalators or other out of print documents? Typically, the original publisher does not stock out of print documents. An extensive collection of both current and out of print codes and standards is maintained and available for purchase from:

> Global Engineering Documents
> 15 Inverness Way East
> Englewood, CO 80112-2776
> (800) 624-3974
> http://global.ihs.com

and:

> Document Center, Inc.
> 1504 Industrial Way, Unit 9
> Belmont, CA 94002
> (415) 591-7600
> http://www.document-center.com

These firms also act as the sales agent for many of the organizations mentioned in the Handbook commentary.

NATIONAL STANDARDS SYSTEMS NETWORK

The National Standards Systems Network (NSSN) is a World Wide Web-based system designed to provide users with a wide range of standards information from major standards developers, including developers accredited by the American National Standards Institute, other U.S. private sector standards organizations, government agencies, including the U. S. Department of Defense, and international standards organizations. At its core is an integrated catalog database pointing to over 100,000 (and growing) standards currently in use. Users can still search for standards, and then place an order with the organizations that provide the standards. Users are able to either view or download a document, as well as order it in any other available formats.

National Standards System Network
http://www.nssn.org

NSSN Basic

Launched on February 25, 1997, NSSN-Basic is a no-charge service available via the World Wide Web. Its primary function is to provide a basic index for commercial and government standards and generate as much traffic as possible worldwide to serve as an introduction to the other NSSN services. Users can search by document number, or keywords from the title and scope. Information presented in the search results is limited to document number, title, organization, and ordering information. Orders are directed to the developing organization or its designated supplier and to ANSI for acceptance and fulfillment.

NSSN Enhanced

The Enhanced service is a fee-based service, which will go beyond NSSN Basic by offering a greater depth and breadth of information. In addition to the data available through NSSN Basic, NSSN Enhanced includes scope, referenced standards, and equivalency information. There will also be information about American National Standards and ISO standards that are under development. Development information will only be available via NSSN Enhanced. NSSN Enhanced also provides an alert service that will allow users to define interest profiles that will be used to search against new items entering the database. The alert service tracks development projects and provides information about status updates as they are entered into the NSSN database.

To fully comply with the requirements of the Safety Code for Elevators and Escalators, other Codes and Standards must also be complied with. A listing of those referenced Codes and Standards can be found in Table 9.1. When a referenced Code or Standard varies in part or in whole from a requirement found in ASME A17.1,

it is intended that the requirement in A17.1 governs. Local jurisdictions may have additional requirements; however, this Part is not intended to cover these additional requirements. When two or more codes or standards have requirements on the same subject, most jurisdictional authorities have a policy of enforcing the most stringent requirement.

The following is a list of additional publications, which will assist the reader in understanding the requirements found in the ASME A17.1 Code and some of the codes and standards referenced in this Handbook. Most of the listed publishers have catalogs, which list additional resource material.

ADAAG Manual
Publisher: United States Architectural and
  Transportation Barriers Compliance Board
131 F Street, NW, Suite 1000
Washington, DC 20004-1111
http://www.access-board.gov

ADA and Building Transportation, 2nd Edition
By: Edward A. Donoghue, CPCA
Publisher: Elevator World, Inc.
P.O. Box 6507
Mobile, Alabama 36660
http://www.elevator-world.com

The ADA Answer Book
Publisher: Building Owners and Managers
  Association (BOMA) International
1201 New York Avenue, NW, Suite 300
Washington, DC 20005
http://www.boma.org

*ADA Compliance Guide Book. A Checklist for
Your Building*
Publisher: Building Owners and Managers
  Association (BOMA) International
1201 New York Avenue, NW, Suite 300
Washington, DC 20005
http://www.boma.org

ADA Regulations
Publisher: Office of the Americans with
  Disabilities Act Civil Rights Division
United States Department of Justice
P.O. Box 66118
Washington, DC 20035-6118
http://www.usdoj.gov/crt/ada/adahom1.htm

Americans with Disabilities Act Accessibilities
  Guidelines (ADAAG)
Publisher: United States Architectural and
  Transportation Barriers Compliance Board
131 F Street, NW, Suite 1000
Washington, DC 20004-1111
http://www.access-board.gov

*Americans with Disabilities Act Title II Technical
Assistance Manual*

Publisher: Office of the Americans with
 Disabilities Act Civil Right Division
United States Department of Justice
P.O. Box 66118
Washington, DC 20035-6118
http://www.usdoj.gov/crt/ada/adahom1.htm

BOCA National Building Code Commentary, The
Publisher: Building Officials and Code
 Administrators International
4051 West Flossmoor Road
Country Club Hills, IL 60478
http://www.bocai.org

Building Material Directory
Underwriters Laboratories, Inc.
333 Pfingston Road
Northbrook, IL 60062
http://www.ul.com

*Building Transportation Standards and Guidelines,
NEII-1-2000*
Publisher: Elevator World, Inc.
P.O. Box 6507
Mobile, AL 36660
http://www.elevator-world.com

CABO/ANSI A117.1–1992 Commentary
Publisher: International Code Council
5203 Leesburg Pike, Suite 708
Falls Church, VA 22041
http://www.intlcode.org

Electric Elevators
By: Fred Hymans
Available from: Elevator World, Inc.
P.O. Box 6507
Mobile, AL 36660
http://www.elevator-world.com

*Elevator Industry Field Employees' Safety
 Handbook*
Publisher: Elevator World, Inc.
P.O. Box 6507
Mobile, AL 36660
http://www.safety.elevator-world.com

Elevator Industry Inspection Handbook
By: Zack McCain, Jr.
Publisher: Elevator World, Inc.
P.O. Box 6507
Mobile, AL 36660
http://www.elevator-world.com

Elevator Maintenance Manual
By: Zack McCain, Jr.
Publisher: Elevator World, Inc.
P.O. Box 6507
Mobile, AL 36660
http://www.elevator-world.com

Elevators
By: Frederick A. Annett

Available from: Elevator World, Inc.
P.O. Box 6507
Mobile, AL 36660
http://www.elevator-world.com

*Elevator World Educational Package and Reference
Library*
Publisher: Elevator World, Inc.
P.O. Box 6507
Mobile, AL 36660
http://www.elevator-world.com

Guide to Elevatoring, The
Publisher: Elevator World, Inc.
P.O. Box 6507
Mobile, AL 36660
http://www.elevator-world.com

FEM Terminology
Available from: Elevator World, Inc.
P.O. Box 6507
Mobile, AL 36660
http://www.elevator-world.com

Field Inspection Handbook
Publisher: McGraw Hill Book Company
New York, NY
http://www.mcgraw-hill.com

Fire Protection Handbook
Publisher: National Fire Protection Association
One Batterymarch Park
Quincy, MA 02269
http://www.nfpa.org

*IEEE Recommended Practice for Electric Power
 Systems in Commercial Buildings*
Publisher: IEEE
445 Hoes Lane
P.O. Box 1331
Piscataway, NJ 08855-1331
http://www.ieee.org

Life Safety Code® Handbook
Publisher: National Fire Protection Association
One Batterymarch Park
Quincy, Massachusetts 02269
http://www.nfpa.org

Lift Modernization Design Guide
By: Roger E. Hawkins
Publisher: Elevator World, Inc.
P.O. Box 6506
Mobile, AL 36606
http://www.elevator-world.com

Material Handling Handbook
Publisher: John Wiley and Sons
605 Third Avenue
New York, NY 10158
http://www.john-wiley.com

National Electrical Code® Handbook
Publisher: National Fire Protection Association

One Batterymarch Park
Quincy, MA 02269
http://www.nfpa.org

National Fire Alarm Code Handbook®
Publisher: National Fire Protection Association
One Batterymarch Park
Quincy, MA 02269
http://www.nfpa.org

Pedestrian Planning and Design
By: John Fruin
Available from: Elevator World, Inc.
P.O. Box 6507
Mobile, AL 36660
http://www.elevator-world.com

UFAS Retrofit Manual
Publisher: United States Architectural and
Transportation Barriers Compliance Board
131 F Street, NW, Suite 1000
Washington, DC 20004-1111
http://www.access-board.gov

Uniform Federal Accessibility Standards
Publisher: Superintendent of Documents
United States Government Printing Office
Washington, DC 20402
http://www.gpo.gov

USG Cavity Shaft Wall Systems
Publisher: United States Gypsum Company
101 South Waker
Chicago, IL 60606-4385
http://www.usg.com

Vertical Transportation Handbook, The, 3rd Edition
By: George R. Strakosch
Publisher: John Wiley and Sons, Inc.
605 Third Avenue
New York, NY 10158
http://www.john-wiley.com

Wire Rope Users Manual
Publisher: American Iron and Steel Institute
1000 16th Street, NW
Washington, DC 20036
http://www.steel.org

APPENDICES

APPENDIX A
CONTROL SYSTEM

Figure A1 is included in the Handbook commentary on 2.26 as Diagram 2.26(a).

APPENDIX B
DOOR LANDING AND UNLOCKING ZONES

Figure B1 is included in the Handbook commentary on 2.12.5 as Diagram 2.12.5.

APPENDIX C
LOCATION OF TOP EMERGENCY EXIT

Figure C1 is included in the Handbook commentary on 2.14.1.5.1 as Diagram 2.14.1.5.1(b)(2).

APPENDIX D
RATED LOAD AND CAPACITY PLATES FOR PASSENGER ELEVATORS

This Appendix is included in the Handbook commentary on 2.16.1.

APPENDIX E
CSA B44 ELEVATOR REQUIREMENTS FOR PERSONS WITH PHYSICAL DISABILITIES

The requirements in CA/CSA-B44-2000 are essentially the same as those in ICC/ANSI A117.1-1998. The handbook ADA and Building Transportation cover the requirements in ICC/ANSI A117.1. See Handbook commentary on Part 9 for procurement information.

APPENDIX F
ASCENDING CAR OVERSPEED AND UNINTENDED CAR MOVEMENT PROTECTION

Table F1 is included in the Handbook commentary on 2.19 as Chart 2.19.

Figure F1 is included in the Handbook commentary on 2.19.1 as Diagram 2.19.1.

Figure F2 is included in the Handbook commentary on 2.19.2 as Diagram 2.19.2.

APPENDIX G
TOP-OF-CAR CLEARANCE

Figure G1 is included in the Handbook commentary on 3.4.5 as Diagram 3.4.5.

Figure G2 is included in the Handbook commentary on 3.4.4 as Diagram 3.4.4.

APPENDIX H
PRIVATE RESIDENCE ELEVATOR GUARDING

Figure H1 is included in the Handbook commentary on 5.3.1.6.2 as Diagram 5.3.1.6.2.

APPENDIX I
ESCALATOR AND MOVING WALK DIAGRAMS

Figure I1 is included in the Handbook commentary on 6.1.3 as Diagram 6.1.3(a).

Figure I2 is included in the Handbook commentary on 6.1.3.2 as Diagram 6.1.3.2.

Figure I3 is included in the Handbook commentary on 6.1.3 as Diagram 6.1.3(b).

Figure I4 is included in the Handbook commentary on 6.1.3.3.6 as Diagram 6.1.3.3.6.

Figure I5 is included in the Handbook commentary on 6.1.3.3.9 as Diagram 6.1.3.3.9.

Figure I6 is included in the Handbook commentary on 6.1.3.3.10 as Diagram 6.1.3.3.10(b).

Figure I7 is included in the Handbook commentary on 6.1.3.5.2 as Diagram 6.1.3.5.2.

Figure I8 is included in the Handbook commentary on 6.1.3.5.3 as Diagram 6.1.3.5.3.

Figure I9 is included in the Handbook commentary on 6.2.3 as Diagram 6.2.3.

Figure I10 is included in the Handbook commentary on 6.2.3.5.1 as Diagram 6.2.3.5.1.

Figure I11 is included in the Handbook commentary on 6.1.5.3.1(c) as Diagram 6.1.5.3.1(c).

APPENDIX J
CSA B44 MAINTENANCE REQUIREMENTS AND INTERVALS FOR ELEVATORS, DUMBWAITERS, ESCALATORS, AND MOVING WALKS

The Appendix applies in Canadian jurisdictions only. As the intervals were arbitrarily selected are not based on sound engineering principals that take into account, age, condition design, use, environment, etc. that were not acceptable to the ASME A17 Committee.

APPENDIX K
BEVELING AND CLEARANCE REQUIREMENTS

Figure K1 is included in the Handbook commentary on 7.4.7.4 as Diagram 7.4.7.4.

APPENDIX L
INDEX OF ALTERATION REQUIREMENTS FOR ELECTRIC ELEVATORS, HYDRAULIC ELEVATORS, ESCALATORS, AND MOVING WALKS

The reader should also consult the modernization guidelines in NEII-1-2000. The pre-modernization checklists will be especially helpful in assuring compliance with 8.7. See Handbook commentary on Part 9 for procurement information.

APPENDIX M
INERTIA APPLICATION FOR TYPE A SAFETY DEVICE LOCATION OF TEST WEIGHT

Figure M1 is included in the Handbook commentary on 8.10.2.2.2(bb)(2) as Diagram 8.10.2.2.2(bb)(2).

APPENDIX N
RECOMMENDED INSPECTION AND TEST INTERVALS IN "MONTHS"

Table N1 is included in the Handbook commentary on 8.11.1.3 as Chart 8.11.1.3.

APPENDIX O
ELEVATOR CORRIDOR CALL STATION PICTOGRAPH

Figure O1 is included in the Handbook commentary on 2.27.3 as Diagram 2.27.3(a).

INDEX

References given are to the Part, Section, Requirement, Table,
Figure, and Appendix designations.

THE AUTHOR
EDWARD A. DONOGHUE

Edward A. Donoghue, CPCA, is president of Edward A. Donoghue Associates, Inc., Code and Safety Consultants, Salem, New York. From 1976 to 1989, he was Manager of Codes and Safety for the National Elevator Industry, Inc. (NEII). He is chairman of the ASME Qualification of Elevator Inspectors Committee and in 1986-87 chaired the Safety Division of ASME. He was the 1992 recipient of the ASME Safety Codes and Standards Medal.

He is a member of ASME A17 Safety Code for Elevators and Escalators Standards, Hoistway, Emergency Operations, Construction Elevator, Existing Installation, Maintenance, Dumbwaiter, Inspectors' Manual, International Standards, Sidewalk Elevator, Limited Use/Limited Application Elevator, and Evacuation Guide Committees and chairs the Editorial and Code Coordination Committees. Mr. Donoghue was a licensed master electrician and building contractor. He is the author of the "Handbook A17.1 Safety Code for Elevators and Escalators" and "ADA and Building Transportation."

He has been certified as a Professional Code Administrator by the National Academy of Code Administration and as an Elevator Inspector by NAESA, BOCA, and SBCCI. Mr. Donoghue is a member of The American Society of Mechanical Engineers (ASME) International, Association of Elevator Engineers (IAEE), American Society for Testing and Materials (ASTM), Building Officials and Code Administrators International (BOCA), Southern Building Code Congress International (SBCCI), International Conference of Building Officials (ICBO), National Association of Elevator Contractors (NAEC), National Association of Elevator Safety Authorities (NAESA), National Fire Protection Association (NFPA), and the National Conference of States on Building Codes and Standards (NCSBCS).